Arithmetic Operations:

$$ab + ac = a(b+c)$$

$$\frac{a}{b} + \frac{c}{d} = \frac{ad+bc}{bd}$$

$$\frac{a+b}{c} = \frac{a}{c} + \frac{b}{c}$$

$$\frac{\left(\frac{a}{b}\right)}{\left(\frac{c}{d}\right)} = \frac{ad}{bc}$$

$$a\left(\frac{b}{c}\right) = \frac{ab}{c}$$

$$\frac{a-b}{c-d} = \frac{b-a}{d-c}$$

$$\frac{ab+ac}{a} = b+c, \ a \neq 0$$

$$\frac{\left(\frac{a}{b}\right)}{c} = \frac{a}{bc}$$

$$\frac{a}{\left(\frac{b}{c}\right)} = \frac{ac}{b}$$

Exponents and Radicals:

$$a^0 = 1, \ a \neq 0$$

$$\frac{a^x}{a^y} = a^{x-y}$$

$$\left(\frac{a}{b}\right)^x = \frac{a^x}{b^x}$$

$$\sqrt[n]{a^m} = a^{m/n} = (\sqrt[n]{a})^m$$

$$a^{-x} = \frac{1}{a^x}$$

$$(a^x)^y = a^{xy}$$

$$\sqrt{a} = a^{1/2}$$

$$\sqrt[n]{ab} = \sqrt[n]{a}\sqrt[n]{b}$$

$$a^x a^y = a^{x+y}$$

$$(ab)^x = a^x b^x$$

$$\sqrt[n]{a} = a^{1/n}$$

$$\sqrt[n]{\left(\frac{a}{b}\right)} = \frac{\sqrt[n]{a}}{\sqrt[n]{b}}$$

Algebraic Errors to Avoid:

$\dfrac{a}{x+b} \neq \dfrac{a}{x} + \dfrac{a}{b}$ (To see this error, let a = b = x = 1.)

$\sqrt{x^2+a^2} \neq x + a$ (To see this error, let x = 3 and a = 4.)

$a - b(x-1) \neq a - bx - b$ (Remember to distribute negative signs. The equation should be $a - b(x-1) = a - bx + b$.)

$\dfrac{\left(\frac{x}{a}\right)}{b} \neq \dfrac{bx}{a}$ (To divide fractions, invert and multiply. The equation should be

$$\frac{\frac{x}{a}}{b} = \frac{\frac{x}{a}}{\frac{b}{1}} = \left(\frac{x}{a}\right)\left(\frac{1}{b}\right) = \frac{x}{ab}.)$$

$\sqrt{-x^2+a^2} \neq -\sqrt{x^2-a^2}$ (We can't factor a negative sign outside of the square root.)

$\dfrac{a+bx}{\cancel{a}} \neq 1 + bx$ (This is one of many examples of incorrect cancellation. The equation should be $\dfrac{a+bx}{a} = \dfrac{a}{a} + \dfrac{bx}{a} = 1 + \dfrac{bx}{a}$.)

$\dfrac{1}{x^{1/2}-x^{1/3}} \neq x^{-1/2} - x^{-1/3}$ (This error is a sophisticated version of the first error.)

$(x^2)^3 \neq x^5$ (The equation should be $(x^2)^3 = x^2 x^2 x^2 = x^6$.)

Conversion Table:

1 centimeter = 0.394 inches	1 joule = 0.738 foot-pounds	1 mile = 1.609 kilometers
1 meter = 39.370 inches	1 gram = 0.035 ounces	1 gallon = 3.785 liters
= 3.281 feet	1 kilogram = 2.205 pounds	1 pound = 4.448 newtons
1 kilometer = 0.621 miles	1 inch = 2.540 centimeters	1 foot-lb = 1.356 joules
1 liter = 0.264 gallons	1 foot = 30.480 centimeters	1 ounce = 28.350 grams
1 newton = 0.225 pounds	= 0.305 meters	1 pound = 0.454 kilograms

D0026275

Algebra and Trigonometry

Algebra and Trigonometry

SECOND EDITION

Roland E. Larson
Robert P. Hostetler
The Pennsylvania State University
The Behrend College

With the assistance of

David E. Heyd
The Pennsylvania State University
The Behrend College

D.C. Heath and Company
Lexington, Massachusetts   Toronto

Senior Acquisitions Editor: Mary Lu Walsh
Developmental and Senior Production Editor: Cathy Cantin
Senior Designer: Sally Steele
Editorial Assistant: Carolyn Johnson
Production Manager: Mike O'Dea
Composition: Jonathan Peck Typographers
Technical Art: Folium
Cover: Martucci Studio

Copyright © 1989 by D. C. Heath and Company.
Previous edition copyright © 1985.

All rights reserved. No part of this publication may be reproduced or transmitted in any form or by any means, electronic or mechanical, including photocopy, recording, or any information storage or retrieval system, without permission in writing from the publisher.

Published simultaneously in Canada.

Printed in the United States of America.

International Standard Book Number: 0-669-16269-8

Library of Congress Catalog Card Number: 88-817331

3   4   5   6   7   8   9   10

Preface

Success in college level mathematics courses begins with a good understanding of algebra and trigonometry. The goal of this text is to help students develop this understanding. Although we review some of the basic concepts in algebra, we assume that most students in this course will have completed two years of high school algebra.

What's New in the Second Edition

Many users of the first edition of the text have given us suggestions for improving the text. We appreciate this type of input very much and have incorporated most of the suggestions into the Second Edition. *Every* section in the text has been revised and many sections were completely rewritten. Most sections in the Second Edition have more exercises than were in the First Edition. The major changes are as follows.

In Chapter 1, we combined the first three sections of the First Edition, postponed the discussion of intervals on the real line to Section 2.7, and postponed the discussion of complex numbers until Section 2.5. In Chapter 2, we expanded the section on solving inequalities to two sections (Sections 2.7 and 2.8). Chapter 3 contains earlier coverage of lines in the plane (Section 3.3). The introduction of functions in Section 3.4 was completely rewritten. The discussion of transformations of graphs of functions has been improved (Section 3.5). We expanded the coverage of composition and inverse functions to two sections (Sections 3.6 and 3.7). In Chapter 4, we added a discussion of Descartes's Rule of Signs to Section 4.4. Chapter 6 was extensively reorganized and rewritten. More emphasis is now given to solving exponential and logarithmic equations (Section 6.4), and we included a new section on applications of exponential and logarithmic equations (Section 6.5). The first three sections of Chapter 7 were extensively rewritten to improve the flow from right-triangle trigonometry to trigonometric functions of real numbers. In Chapter 8, we now have earlier coverage of solving trigonometric equations

(Section 8.3), and the coverage of multiple-angle formulas and product-sum formulas has been condensed to one section (Section 8.5). In Chapter 10, we expanded the material on systems of inequalities and linear programming to two sections (Sections 10.4 and 10.5). The first three sections of Chapter 11 were extensively revised, and Section 11.6 now contains Cramer's Rule as well as other applications of matrices and determinants.

Features of the Second Edition

The Second Edition contains many features that we have found help students improve their skills and acquire an understanding of the material.

Graphics　Skill in visualizing a problem is a critical part of a student's ability to solve the problem. This text contains over 1225 figures. Of these, approximately 375 are in the exercise sets and approximately 400 are in the odd-numbered answers in the back of the text. The art package for the Second Edition is completely new. Every graph in the text was computer generated for the greatest possible accuracy.

Applications　Throughout the Second Edition we have included many applied problems that give students insight about the usefulness of algebra and trigonometry in a wide variety of fields including business, economics, biology, engineering, chemistry, and physics.

Examples　The Second Edition contains 760 examples, each carefully chosen to illustrate a particular concept or problem-solving technique. Each example is titled for quick reference, and many examples include color side comments to justify or explain the steps in the solution.

Exercises　Over 6375 exercises are included in the Second Edition. These are designed to build competence, skill, and understanding. Each exercise set is graded in difficulty to allow students to gain confidence as they progress. To help students develop skills in analytic geometry, we stress a graphical approach in many sections and have included numerous graphs in the exercises.

Warm up Exercises　[New in the Second Edition] We have found that students in algebra and trigonometry can benefit greatly from reinforcement of previously learned concepts. Most sections in the text contain a set of ten Warm up exercises that give students practice in the "old skills" that are necessary to master the "new skills" presented in the section. *All* of the Warm up exercises are answered in the back of the text.

Calculators　Hints and instructions for working with calculators occur in many places in the Second Edition. Because calculators have become commonplace, we no longer identify exercises that require decimal approximations.

Algebra of Calculus   Special emphasis has been given to algebraic skills that are needed in calculus. In addition to the material in Section 1.7, many other examples in the Second Edition discuss algebraic or trigonometric techniques that are used in calculus. These examples are clearly identified.

Remarks   In the Second Edition we include special instructional notes to students called *Remarks*. These appear after definitions, theorems, or examples and are designed to give additional insight, help avoid common errors, and describe generalizations.

Supplements

• For students, the *Study and Solutions Guide* by Dianna L. Zook contains detailed solutions of approximately 40% of the odd-numbered exercises in the text. Each of these is indicated in the text by a box surrounding the exercise number. This guide also contains summaries of important concepts for each section and self-tests for each chapter.

• For students, *Computer Activities for Precalculus* is an IBM-PC®* package that offers activities based on programs that enhance the learning of such topics as linear functions, quadratic functions, exponential and logarithmic functions, and trigonometric functions. Exploratory practice and directed tutorials are included as well as a function grapher.

• For instructors, the *Complete Solutions Guide* contains solutions for all of the exercises in the text.

• For instructors, the *Instructor's Guide* by Meredythe M. Burrows contains sample tests for each chapter in the text and suggestions for classroom instruction.

• For instructors, we have prepared test-generating software called *HeathTest* to accompany the text. This software will run on an IBM-PC® that has at least 256K of memory. (It will also run on many IBM compatibles.) To print the tests, the software requires an IBM graphics-compatible dot matrix printer. An Apple II®† version of the software is also available.

• For instructors, we have prepared a package containing 25 two-color transparencies of figures from the text.

*IBM is a registered trademark of International Business Machines Corp.
†Apple is a registered trademark of Apple Computer, Inc.

Acknowledgements

We would like to thank the many people who have helped us at various stages of preparing the First and Second Editions of this text. Their encouragement, criticisms, and suggestions have been invaluable to us.

Special thanks goes to the reviewers of the First and Second Editions.

Hollie Baker, Norfolk State University
Derek I. Bloomfield, Orange County Community College
Ben P. Bockstage, Broward Community College
Daniel D. Bonar, Denison University
Richard Cutts, University of Wisconsin—Stout
John E. Bruha, University of Northern Iowa
H. Eugene Hall, DeKalb Community College
Randal Hoppens, Blinn College
E. John Hornsby, Jr., University of New Orleans
William B. Jones, University of Colorado
Jimmie D. Lawson, Louisiana State University
Peter J. Livorsi, Oakton Community College
Wade T. Macey, Appalachian State University
Jerome L. Paul, University of Cincinnati
Marilyn Schiermeier, North Carolina State University
George W. Schultz, St. Petersburg Junior College
Edith Silver, Mercer County Community College
Shirley C. Sorensen, University of Maryland
Charles Stone, DeKalb Community College
Bruce Williamson, University of Wisconsin—River Falls

The mathematicians listed below responded to a survey conducted by D. C. Heath and Company that helped us outline our revision.

Holli Adams, Portland Community College
Marion Baumler, Niagara County Community College
Diane Blansett, Delta State University
Derek I. Bloomfield, Orange County Community College
Daniel D. Bonar, Denison University
John E. Bruha, University of Northern Iowa
William L. Campbell, University of Wisconsin—Platteville
John Caraluzzo, Orange County Community College
William E. Chatfield, University of Wisconsin—Platteville
Robert P. Finley, Mississippi State University
August J. Garver, University of Missouri—Rolla
Sue Goodman, University of North Carolina

Louis Hoelzle, Bucks County Community College
Randal Hoppens, Blinn College
Moana Karsteter, Florida State University
Robert C. Limburg, St. Louis Community College at Florissant Valley
Peter J. Livorsi, Oakton Community College
John Locker, University of North Alabama
Wade T. Macey, Appalachian State University
J. Kent Minichiello, Howard University
Terry Mullen, Carroll College
Richard Nation, Palomar College
William Paul, Appalachian State University
Richard A. Quint, Ventura College
Charles T. Scarborough, Mississippi State University
Shannon Schumann, University of Wyoming
Arthur E. Schwartz, Mercer County Community College
Joseph Sharp, West Georgia College
Burla J. Sims, University of Arkansas at Little Rock
James R. Smith, Appalachian State University
B. Louise Whisler, San Bernardino Valley College
Bruce Williamson, University of Wisconsin—River Falls

We would also like to thank all of the people at D. C. Heath and Company who worked with us in the development of the Second Edition, especially Mary Lu Walsh, Senior Mathematics Acquisitions Editor; Cathy Cantin, Developmental Editor and Senior Production Editor; Sally Steele, Designer; Carolyn Johnson, Editorial Assistant; and Mike O'Dea, Production Manager.

Several other people also worked on this project. David E. Heyd assisted us with the text, Dianna Zook wrote the *Study and Solutions Guide*, Helen Medley proofread the manuscript and worked the exercises, and Meredythe Burrows wrote the *Instructor's Guide*. Timothy R. Larson prepared the art and worked the exercises. Linda L. Matta proofread the galleys and typed the *Instructor's Guide*. Linda M. Bollinger proofread the galleys and typed the text manuscript, the *Study and Solutions Guide*, and the *Complete Solutions Guide*. Randall Hammond and Lisa Bickel worked the exercises.

We are grateful to our wives, Deanna Gilbert Larson and Eloise Hostetler, for their love, patience, and understanding.

If you have suggestions for improving the text, please feel free to write to us. Over the past several years we have received many useful comments from both instructors and students and we value this very much.

Roland E. Larson
Robert P. Hostetler

The Larson and Hostetler Precalculus Series

To accommodate the different methods of teaching college algebra, trigonometry, and analytic geometry, we have prepared four volumes. Each has its own supplement package. These four titles are described below.

College Algebra, Second Edition

This text is designed for a one-term course covering standard topics such as algebraic functions and their graphs, exponential and logarithmic functions, systems of equations, matrices, determinants, sequences, series, and probability.

Trigonometry, Second Edition

This text is used in a one-term course covering the trigonometric functions and their graphs, exponential and logarithmic functions, and analytic geometry (including polar coordinates and parametric equations).

Algebra and Trigonometry, Second Edition

This book combines the content of the two texts mentioned above (with the exception of polar coordinates and parametric equations). It is comprehensive enough for a two-term course, or, with careful selection, may be used in a one-term course.

Precalculus, Second Edition

With this book, students cover the algebraic, exponential and logarithmic, and trigonometric functions and their graphs, as well as analytic geometry in preparation for a course in calculus. This may be used in a one- or two-term course.

Contents

Introduction to Calculators

This text includes several examples and exercises that use a scientific calculator. As we encounter each new calculator application, we will give instructions for using a calculator efficiently. These instructions are somewhat general and may not agree precisely with the steps required by your calculator.

For use with this text, we recommend a calculator with the following features.

1. At least 8-digit display
2. Four arithmetic operations: $\boxed{+}$, $\boxed{-}$, $\boxed{\times}$, $\boxed{\div}$
3. Change sign key: $\boxed{+/-}$
4. Memory key and Recall key: $\boxed{\text{STO}}$, $\boxed{\text{RCL}}$
5. Parentheses: $\boxed{(}$, $\boxed{)}$
6. Exponential key: $\boxed{y^x}$
7. Natural logarithmic key: $\boxed{\ln x}$
8. Pi and Degree-Radian conversion: $\boxed{\pi}$, $\boxed{\text{DRG}}$
9. Inverse, reciprocal, square root: $\boxed{\text{INV}}$, $\boxed{1/x}$, $\boxed{\sqrt{x}}$
10. Trigonometric functions: $\boxed{\sin}$, $\boxed{\cos}$, $\boxed{\tan}$

One of the basic differences in calculators is their order of operations. Some calculators use an order of operations called RPN (for Reverse Polish Notation). In this text, however, all calculator steps will be given using *algebraic logic*. For example, the calculation

$$4.69[5 + 2(6.87 - 3.042)]$$

can be performed with the following steps.

$$4.69 \; \boxed{\times} \; \boxed{(} \; 5 \; \boxed{+} \; 2 \; \boxed{\times} \; \boxed{(} \; 6.87 \; \boxed{-} \; 3.042 \; \boxed{)} \; \boxed{)} \; \boxed{=}$$

This yields the value 59.35664. Without parentheses, we would work from the inside out with the sequence

6.87 $\boxed{-}$ 3.042 $\boxed{=}$ $\boxed{\times}$ 2 $\boxed{+}$ 5 $\boxed{=}$ $\boxed{\times}$ 4.69 $\boxed{=}$

to obtain the same result.

Rounding Numbers

For all their usefulness, calculators do have a problem representing numbers because they are limited to a finite number of digits. For instance, what does your calculator display when you compute 2 ÷ 3? Some calculators simply truncate (drop) the digits that exceed their display range and display .66666666. Others will round the number and display .66666667. Although the second display is more accurate, *both* of these decimal representations of 2/3 contain a rounding error.

When rounding decimals, we use the following rules.

1. Determine the number of digits of accuracy you wish to keep. The digit in the last position you keep is called the **rounding digit,** and the digit in the first position you discard is called the **decision digit.**
2. If the decision digit is 5 or greater, round up by adding 1 to the rounding digit.
3. If the decision digit is 4 or less, round down by leaving the rounding digit unchanged.

Here are some examples. Note that we round down in the first example because the decision digit is 4 or less, and we round up in the other two examples because the decision digit is 5 or greater.

Number	*Rounded to three decimal places*	
(a) $\sqrt{2} = 1.4142136\ldots$	1.414	*Round down*
(b) $\pi = 3.1415927\ldots$	3.142	*Round up*
(c) $\dfrac{7}{9} = 0.77777777\ldots$	0.778	*Round up*

One of the best ways to minimize error due to rounding is to leave numbers in your calculator until your calculations are complete. If you want to save a number for future use, store it in your calculator's memory.

C H A P T E R   1

Review of Fundamental Concepts of Algebra

1.1 The Real Number System

We begin our study of algebra with a look at the **real number system.** Real numbers are used in everyday life to describe quantities like age, miles per gallon, container size, population, and so on. To represent real numbers we use symbols such as

$$9, \quad -5, \quad 0, \quad \frac{4}{3}, \quad 0.6666\ldots, \quad 28.21, \quad \sqrt{2}, \quad \pi, \quad \text{and} \quad \sqrt[3]{-32}.$$

The set of real numbers contains some important subsets with which you need to be familiar. For instance, the numbers

$$\ldots, \quad -3, \quad -2, \quad -1, \quad 0, \quad 1, \quad 2, \quad 3, \quad \ldots$$

are called **integers.** A real number is called **rational** if it can be written as the ratio p/q of two integers, where $q \neq 0$. For instance, the numbers

$$\frac{1}{3} = 0.3333\ldots, \quad \frac{1}{8} = 0.125, \quad \text{and} \quad \frac{125}{111} = 1.126126\ldots$$

are rational. The decimal representation of a rational number either repeats (as in $3.1454545\ldots$) or terminates (as in $1/2 = 0.5$). Real numbers that

1

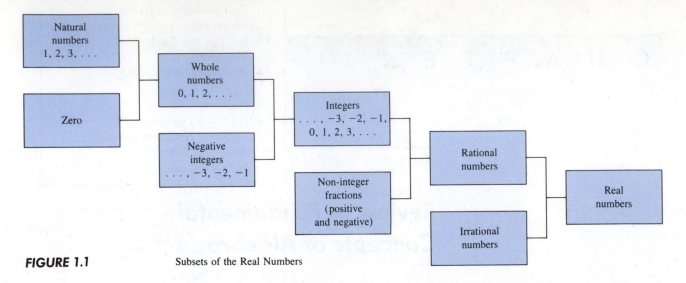

FIGURE 1.1   Subsets of the Real Numbers

cannot be written as the ratio of two integers are called **irrational.** For instance, the numbers

$$\sqrt{2} \approx 1.4142136 \quad \text{and} \quad \pi \approx 3.1415927$$

are irrational. (The symbol $\approx$ means "approximately equal to.") Several subsets of real numbers are shown in Figure 1.1.

Arithmetic Operations

There are four arithmetic operations with real numbers: **addition, multiplication, subtraction,** and **division,** denoted by the symbols $+$, $\times$ (or $\cdot$), $-$, and $\div$. Of these, addition and multiplication are considered to be the two primary arithmetic operations. We summarize their properties as follows.

Properties of Addition and Multiplication

Let a, b, and c be real numbers. Then the following properties are true.

1. Closure: $a + b$ is a real number.   *Addition*
                  $a \cdot b$ is a real number.   *Multiplication*

2. Commutative: $a + b = b + a$ *Addition*
                  $a \cdot b = b \cdot a$ *Multiplication*

3. Associative: $(a + b) + c = a + (b + c)$ *Addition*
                  $(a \cdot b) \cdot c = a \cdot (b \cdot c)$ *Multiplication*

4. Identity: $a + 0 = a = 0 + a$ *0 is the additive identity.*
                  $a \cdot 1 = a = 1 \cdot a$ *1 is the multiplicative identity.*

5. Inverse: $a + (-a) = 0 = (-a) + a$ *$-a$ is the additive inverse of a.*
                  $a\left(\dfrac{1}{a}\right) = 1 = \left(\dfrac{1}{a}\right)a, a \neq 0$ *$\dfrac{1}{a}$ is the multiplicative inverse of a.*

6. Distributive: $a(b + c) = ab + ac$ *Left Distributive Property*
                  $(a + b)c = ac + bc$ *Right Distributive Property*

Remark: Multiplication is implied when no symbol is used between two letters or groups of letters. For instance, we can write ab instead of $a \cdot b$, and $a(b + c)$ instead of $a \cdot (b + c)$.

EXAMPLE 1 *Properties of Addition and Multiplication*

(a) The statement

$$(3 + 6) + 8 = 3 + (6 + 8)$$

is justified by the Associative Property of Addition. Roughly speaking, this property tells us that parentheses are not needed when writing the sum of several real numbers. In other words, we can write this sum as $3 + 6 + 8$ without ambiguity because we obtain the same sum whether we first add 3 and 6, or 6 and 8.

(b) The statement

$$2(5 + 3) = 2 \cdot 5 + 2 \cdot 3$$

is justified by the Distributive Property, or more formally by the *left* distributive property of *multiplication over addition*.

Subtraction and division are defined as the inverse operations of addition and multiplication, respectively.

Subtraction: $a - b = a + (-b)$

Division: If $b \neq 0$, then $a \div b = a\left(\dfrac{1}{b}\right) = \dfrac{a}{b}$.

In these definitions, $-b$ is called the **negative** (or additive inverse) of b, and $1/b$ is called the **reciprocal** (or multiplicative inverse) of b. In place of $a \div b$, we often use the fraction symbol a/b. In this fractional form, a is called the **numerator** of the fraction and b is called the **denominator.**

Remark: Be sure you see the difference between the *negative of a number* and a *negative number*. If b is already negative, then its additive inverse, $-b$, is positive. For instance, if $b = -5$, then $-b = -(-5) = 5$.

The following three lists summarize the basic properties of negation, zero, and fractions. When you encounter such lists, we suggest that you not only *memorize* each property, but also try to gain an *intuitive feeling* for the validity of each.

Properties of Negation

Let a and b be real numbers. Then the following properties are true.

Properties	Examples
1. $(-1)a = -a$	1. $(-1)7 = -7$
2. $-(-a) = a$	2. $-(-6) = 6$
3. $(-a)b = -(ab) = a(-b)$	3. $(-5)3 = -(5 \cdot 3) = 5(-3)$
4. $(-a)(-b) = ab$	4. $(-2)(-6) = 12$
5. $-(a + b) = (-a) + (-b)$	5. $-(3 + 8) = (-3) + (-8)$

Properties of Zero

Let a and b be real numbers. Then the following properties are true.

1. $a + 0 = a$

   $a - 0 = a$

2. $a \cdot 0 = 0$

3. $\dfrac{0}{a} = 0, \quad a \neq 0$

4. $\dfrac{a}{0}$ is undefined.

5. If $ab = 0$, then $a = 0$ or $b = 0$.   *Factorization Principle*

Remark: The "or" in the Factorization Principle includes the possibility that both factors may be zero. This is called an **inclusive or,** and it is the way the word "or" is always used in mathematics.

Properties of Fractions

Let a, b, c, and d be real numbers with $b \neq 0$ and $d \neq 0$. Then the following properties are true.

1. **Equivalent Fractions:** $\dfrac{a}{b} = \dfrac{c}{d}$ if and only if $ad = bc$.

2. **Rules of Signs:** $-\dfrac{a}{b} = \dfrac{-a}{b} = \dfrac{a}{-b}$ and $\dfrac{-a}{-b} = \dfrac{a}{b}$

3. **Generate Equivalent Fractions:** $\dfrac{a}{b} = \dfrac{ac}{bc}, \quad c \neq 0$

4. **Add or Subtract with Like Denominators:** $\dfrac{a}{b} \pm \dfrac{c}{b} = \dfrac{a \pm c}{b}$

5. **Add or Subtract with Unlike Denominators:** $\dfrac{a}{b} \pm \dfrac{c}{d} = \dfrac{ad \pm bc}{bd}$

6. **Multiply Fractions:** $\dfrac{a}{b} \cdot \dfrac{c}{d} = \dfrac{ac}{bd}$

7. **Divide Fractions:** $\dfrac{a}{b} \div \dfrac{c}{d} = \dfrac{a}{b} \cdot \dfrac{d}{c} = \dfrac{ad}{bc}, \quad c \neq 0$

Remark: In Property 1 (equivalent fractions) the phrase "if and only if" implies two statements: If $a/b = c/d$, then $ad = bc$, and conversely, if $ad = bc$ (with $b \neq 0$ and $d \neq 0$), then $a/b = c/d$.

EXAMPLE 2 Properties of Zero and of Fractions

(a) $x - \dfrac{0}{5} = x - 0 = x$ *Properties 3 and 1 of zero*

(b) $\dfrac{x}{5} = \dfrac{3 \cdot x}{3 \cdot 5} = \dfrac{3x}{15}$ *Generate equivalent fractions*

(c) $\dfrac{x}{3} + \dfrac{2x}{5} = \dfrac{5 \cdot x + 3 \cdot 2x}{15}$ *Add fractions with unlike denominators*

(d) $\dfrac{7}{x} \div \dfrac{3}{2} = \dfrac{7}{x} \cdot \dfrac{2}{3}$ *Divide fractions*

If a, b, and c are integers such that $ab = c$, then a and b are called **factors** or **divisors** of c. For example, 2 and 3 are factors of 6 because $2 \cdot 3 = 6$. A **prime number** is a positive integer that has exactly two factors: itself and 1. For example, 2, 3, 5, 7, and 11 are prime numbers, while 1, 4, 6, 8, 9, and 10 are not. The **Fundamental Theorem of Arithmetic** states that every positive integer greater than 1 can be written as the product of prime numbers in precisely one way (disregarding order). The numbers 4, 6, 8, 9, and 10 are called **composite** because they can be written as the product of two or more prime numbers. The number 1 is neither prime nor composite.

When adding or subtracting fractions with unlike denominators, we have two options. For instance, to add 3/10 and 1/12, we could use Property 5 of fractions to obtain

$$\frac{3}{10} + \frac{1}{12} = \frac{3 \cdot 12 + 10 \cdot 1}{10 \cdot 12} = \frac{36 + 10}{120} = \frac{46}{120} = \frac{23}{60}$$

or we could use Property 4 of fractions by rewriting both fractions with like denominators. We call this the **lowest common denominator** (LCD) **method.** For instance, the lowest common denominator of 3/10 and 1/12 is 60, so we write

$$\frac{3}{10} + \frac{1}{12} = \frac{3 \cdot 6}{10 \cdot 6} + \frac{1 \cdot 5}{12 \cdot 5} = \frac{18}{60} + \frac{5}{60} = \frac{23}{60}.$$

For adding or subtracting *two* fractions, Property 5 is often more convenient. For *three or more* fractions, the LCD method is usually preferred.

EXAMPLE 3 *The LCD Method of Adding or Subtracting Fractions*

Evaluate

$$\frac{2}{15} - \frac{5}{9} + \frac{4}{5}.$$

SOLUTION

By prime factoring the denominators ($15 = 3 \cdot 5$, $9 = 3 \cdot 3$, and $5 = 5$) we see that the LCD is $3 \cdot 3 \cdot 5 = 45$. It follows that

$$\frac{2}{15} - \frac{5}{9} + \frac{4}{5} = \frac{2(3)}{45} - \frac{5(5)}{45} + \frac{4(9)}{45} = \frac{6 - 25 + 36}{45} = \frac{17}{45}.$$

An **equation** is a statement of equality between two expressions. Thus, the statement

$$a + b = c + d$$

means that the expressions $a + b$ and $c + d$ represent the same number. For instance, since $1 + 4$ and $3 + 2$ both represent the number 5 we can write $1 + 4 = 3 + 2$. Three important properties of equality are as follows.

Properties of Equality

Let a, b, and c be real numbers. Then the following properties are true.

1. **Reflexive:** $a = a$
2. **Symmetric:** If $a = b$, then $b = a$.
3. **Transitive:** If $a = b$ and $b = c$, then $a = c$.

In algebra, we often rewrite or evaluate expressions by making substitutions that are permitted under the following **Substitution Principle:** *If $a = b$, then a can be replaced by b in any expression involving a.* Two important consequences of the Substitution Principle are the following rules.

1. If $a = b$, then $a + c = b + c$.
2. If $a = b$, then $ac = bc$.

The first rule allows us to add the same number to both sides of an equation. The second allows us to multiply both sides of an equation by the same number. The converses of these two rules are called the **Cancellation Laws** for addition and multiplication.

1. If $a + c = b + c$, then $a = b$.
2. If $ac = bc$ and $c \neq 0$, then $a = b$.

The Real Number Line

The model we use to represent the real number system is called the **real number line.** It consists of a horizontal line with an arbitrary point (the **origin**) labeled 0. Numbers to the right of the origin are positive, and numbers to the left of the origin are negative, as shown in Figure 1.2. We use the term **nonnegative** to describe a number that is either positive or zero.

The Real Number Line

FIGURE 1.2

Each point on the real number line corresponds to one and only one real number and *each real number corresponds to one and only one point on the real number line.* This type of relationship is called a **one-to-one correspondence,** as shown in Figure 1.3.

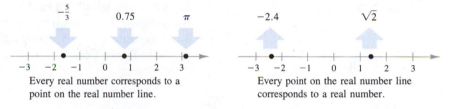

Every real number corresponds to a point on the real number line.

Every point on the real number line corresponds to a real number.

One-to-One Correspondence

FIGURE 1.3

The number associated with a point on the real number line is called the **coordinate** of the point. For example, in Figure 1.3, $-5/3$ is the coordinate of the left-most point and $\sqrt{2}$ is the coordinate of the right-most point.

Ordering the Real Numbers

One important property of real numbers is that they are **ordered.**

Definition of Order on the Real Number Line

If a and b are real numbers, then a is **less than** b if $b - a$ is positive. We denote this order by the **inequality**

$a < b$.

The symbol $a \leq b$ means that a is **less than or equal to** b.

$a < b$ if and only if a lies to the left of b.

FIGURE 1.4

Geometrically, this definition implies that $a < b$ if and only if a lies to the *left* of b on the real number line, as shown in Figure 1.4. For example, $1 < 2$ because 1 lies to the left of 2 on the real number line.

Inequalities are useful in denoting subsets of the real numbers, as shown in Examples 4 and 5.

EXAMPLE 4 *Interpreting Inequalities*

(a) The inequality $x \leq 2$ denotes all real numbers less than or equal to 2, as shown in Figure 1.5(a).
(b) The inequality $-2 \leq x < 3$ means that $x \geq -2$ *and* $x < 3$. This "double" inequality denotes all real numbers between -2 and 3, including -2 but *not* including 3, as shown in Figure 1.5(b).
(c) The inequality $x > -5$ denotes all real numbers greater than -5, as shown in Figure 1.5(c).

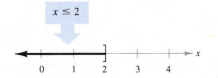
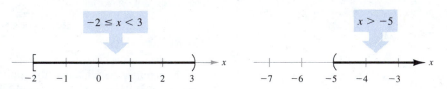

FIGURE 1.5

EXAMPLE 5 *Using Inequalities to Represent Sets of Real Numbers*

Use inequality notation to describe each of the following.

(a) c is nonnegative.
(b) b is at most 5.
(c) d is negative and greater than -3.
(d) x is positive but not more than 6.

SOLUTION

(a) "c is nonnegative" means that c is greater than or equal to zero, and we write $c \geq 0$.

(b) "b is at most 5" can be written as $b \leq 5$.

(c) "d is negative" can be written as $d < 0$, and "d is greater than -3" can be written as $-3 < d$. Combining these two inequalities produces $-3 < d < 0$.

(d) "x is positive" can be written as $0 < x$, and "x is not more than 6" can be written as $x \leq 6$. Combining these two inequalities produces $0 < x \leq 6$.

The following property of real numbers is called the **Law of Trichotomy.** It tells us that for any two real numbers a and b, *precisely* one of three orders is possible.

$$a = b, \quad a < b, \quad \text{or} \quad a > b \qquad \text{\textit{Law of Trichotomy}}$$

Absolute Value and Distance

By the **absolute value** of a real number, we mean its *magnitude* (its value disregarding its sign).

Definition of Absolute Value

If a is a real number, then the **absolute value** of a is given by

$$|a| = \begin{cases} a, & \text{if } a \geq 0 \\ -a, & \text{if } a < 0. \end{cases}$$

Be sure you see that the absolute value of a number can never be negative. For instance, if $a = -5$, then

$$|a| = |-5| = -a = -(-5) = 5$$

because $-5 < 0$. Similarly, $|0| = 0$ because $0 \geq 0$, and

$$|2 - \pi| = -(2 - \pi) = \pi - 2$$

because $(2 - \pi) < 0$.

EXAMPLE 6 *Evaluating the Absolute Value of a Number*

Evaluate the fraction

$$\frac{|x|}{x}$$

for (a) $x > 0$ and (b) $x < 0$.

SOLUTION

(a) If $x > 0$, then $|x| = x$ and we have

$$\frac{|x|}{x} = \frac{x}{x} = 1.$$

For instance, $|4|/4 = 1$.

(b) If $x < 0$, then $|x| = -x$ and we have

$$\frac{|x|}{x} = \frac{-x}{x} = -1.$$

For instance, $|-4|/(-4) = 4/(-4) = -1$.

Note that when $x = 0$, the expression $|x|/x$ is undefined.

The following list gives four useful properties of absolute value. When you see a list like this, try to formulate verbal descriptions of the properties. For instance, the third property tells us that the absolute value of a product of two numbers is equal to the product of the absolute values of the two numbers.

Properties of Absolute Value

Let a and b be real numbers. Then the following properties are true.

1. $|a| \geq 0$
2. $|-a| = |a|$
3. $|ab| = |a||b|$
4. $\left|\dfrac{a}{b}\right| = \dfrac{|a|}{|b|}, \qquad b \neq 0$

The Real Number System

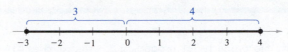

The distance between −3 and 4 is 7.

FIGURE 1.6

Absolute value can be used to define the distance between two numbers on the real number line. To see how this is done, consider the numbers −3 and 4, shown in Figure 1.6. To find the distance between these two points, we subtract *either* number from the other and then take the absolute value of the difference. For instance,

$$\text{distance} = |-3 - 4| = |-7| = 7$$

or equivalently,

$$\text{distance} = |4 - (-3)| = |7| = 7.$$

Distance Between Two Points on the Real Line

Let a and b be real numbers. The **distance between a and b** is

$$d(a, b) = |b - a| = |a - b|.$$

EXAMPLE 7 *Distance and Absolute Value*

(a) The distance between $\sqrt{7}$ and 4 is given by

$$d(\sqrt{7}, 4) = |4 - \sqrt{7}| = 4 - \sqrt{7}.$$

(b) The statement "the distance between c and −2 is at least 7" can be written

$$d(c, -2) = |c + 2| \geq 7.$$

(c) The statement "the distance between x and 2.3 is less than 1" can be written

$$d(x, 2.3) = |x - 2.3| < 1.$$

(d) The distance between −4 and the origin is given by

$$d(-4, 0) = |-4 - 0| = |-4| = 4.$$

EXERCISES 1.1

In Exercises 1–16, identify the property (or properties) illustrated in the given equation.

1. $3 + 4 = 4 + 3$
2. $x + 9 = 9 + x$
3. $-15 + 15 = 0$
4. $(x + 3) - (x + 3) = 0$
5. $\dfrac{1}{(h + 6)}(h + 6) = 1, \ h \neq -6$
6. $2\left(\tfrac{1}{2}\right) = 1$

*7. $2(x + 3) = 2x + 6$

8. $(5 + 11) \cdot 6 = 5 \cdot 6 + 11 \cdot 6$
9. $h + 0 = h$
10. $(z - 2) + 0 = z - 2$
11. $57 \cdot 1 = 57$
12. $1 \cdot (1 + x) = 1 + x$
13. $6 + (7 + 8) = (6 + 7) + 8$
14. $x + (y + 10) = (x + y) + 10$
15. $x(3y) = (x \cdot 3)y = (3x)y$
16. $\tfrac{1}{7}(7 \cdot 12) = \left(\tfrac{1}{7} \cdot 7\right)12 = 1 \cdot 12 = 12$

In Exercises 17–20, use the properties of zero to evaluate, if possible, the given expression. If the expression is undefined, state why.

17. $\dfrac{81 - (90 - 9)}{5}$
18. $10(23 - 30 + 7)$
19. $\dfrac{8}{-9 + (6 + 3)}$
20. $15 - \dfrac{3 - 3}{5}$

In Exercises 21–40, perform the indicated operation(s). (For answers that are fractions, write the result in lowest terms. For instance, write 2/4 as 1/2.)

21. $(8 - 17) + 3$
22. $3 - (-6)$
23. $10 - 6 - 2$
24. $-3(5 - 2)$
25. $(4 - 7)(-2)$
26. $(-5)(-8)$
27. $2\left(\dfrac{77}{-11}\right)$
28. $\dfrac{27 - 35}{4}$
29. $8(-6)(-2)$
30. $\tfrac{3}{16} + \tfrac{5}{16}$
31. $\tfrac{6}{7} - \tfrac{4}{7}$
32. $\tfrac{10}{11} + \tfrac{6}{33} - \tfrac{13}{66}$
33. $\tfrac{2}{5} \times \tfrac{7}{8}$
34. $4 \times \tfrac{3}{4}$
35. $\tfrac{4}{5} \times \tfrac{1}{2} \times \tfrac{3}{4}$
36. $\tfrac{2}{3} \times \tfrac{5}{8} \times \tfrac{3}{4}$
37. $\tfrac{2}{3} \div 8$
38. $\tfrac{11}{16} \div \tfrac{3}{4}$
39. $12 \div \tfrac{1}{4}$
40. $\left(\tfrac{3}{5} \div 3\right) - \left(6 \times \tfrac{4}{8}\right)$

*A boxed number indicates that a detailed solution can be found in the *Student Solution Guide*.

In Exercises 41–46, plot the two real numbers on the real number line and place the appropriate inequality sign (< or >) between them.

41. $\tfrac{3}{2}, 7$
42. $-3.5, 1$
43. $-4, -8$
44. $1, -\tfrac{16}{3}$
45. $\tfrac{5}{6}, \tfrac{2}{3}$
46. $-\tfrac{8}{7}, -\tfrac{3}{7}$

In Exercises 47–52, use inequality notation to denote the given expression.

47. x is negative.
48. y is greater than 5 and less than or equal to 12.
49. Burt's age, A, is at least 30.
50. The yield, Y, is no more than 45 bushels per acre.
51. The annual rate of inflation, R, is expected to be at least 3.5%, but no more than 6%.
52. The price, P, of unleaded gasoline is not expected to go above \$1.25 per gallon during the coming year.

In Exercises 53–56, write the given number without absolute value signs.

53. (a) $|-10|$ (b) $|3 - \pi|$
54. (a) $|0|$ (b) $|4 - \pi|$
55. (a) $\dfrac{-5}{|-5|}$ (b) $-3 - |-3|$
56. (a) $-3|-3|$ (b) $|-1| - |-2|$

In Exercises 57–64, find the distance between a and b.

57. $a = -1$, $b = 3$ (on number line from -1 to 3)

58. $a = -4$, $b = -\tfrac{3}{2}$ (on number line from -4 to -1)

59. $a = -\tfrac{5}{2}$, $b = 0$ (on number line from -3 to 0)

60. $a = \tfrac{1}{4}$, $b = \tfrac{11}{4}$ (on number line from 0 to 3)

61. $a = 126, \ b = 75$
62. $a = -126, \ b = -75$
63. $a = 9.34, \ b = -5.65$
64. $a = \tfrac{16}{5}, \ b = \tfrac{112}{75}$

In Exercises 65–70, use absolute value notation to describe the given expression.

65. The distance between x and 5 is no more than 3.

66. The distance between x and -10 is at least 6.

67. The distance between z and 3/2 is greater than 1.

68. The distance between z and 0 is less than 8.

69. y is at least six units from 0.

70. y is at most two units from a.

71. Let m and n be any two integers. Then $2m$ and $2n$ are even integers and $(2m + 1)$ and $(2n + 1)$ are odd integers.
   (a) Prove that the sum of two even integers is even.
   (b) Prove that the sum of two odd integers is even.
   (c) Prove that the product of an even integer and *any* integer is even.

72. Use the properties of real numbers to prove that if $a = -a$, then $a = 0$.

73. One worker can assemble a component in seven days and a second worker can do the same task in five days. If they work together, what fraction of the component can they assemble in two days?

74. One foot of copper wire weighs 1 ounce (1/16 pound). What is the weight of 5/8 mile of this wire? [*Note:* 1 mile = 5280 feet.]

75. A rational number expressed in decimal form is either a terminating decimal or a nonterminating decimal with a repeating pattern. Use your calculator to find the decimal form of each of the following rational numbers. If it is a nonterminating decimal, give the repeating pattern.

(a) $\frac{5}{8}$ (b) $\frac{1}{3}$ (c) $\frac{41}{333}$ (d) $\frac{6}{11}$ (e) $\frac{85}{750}$

[*Note:* For some rational numbers you need to have more decimal places than shown on a typical calculator before you can observe the repeating pattern. For example, try to find the decimal representation of 4/23.]

76. Use a calculator to complete the following table and observe that $5/n$ increases without bound as we let n approach zero.

n	10	1	0.5	0.01	0.0001	0.000001
$5/n$						

77. (a) Use a calculator to order the following numbers from smallest to largest.

$\frac{7071}{5000}$, $\frac{584}{413}$, $\sqrt{2}$, $\frac{47}{33}$, $\frac{127}{90}$

(b) Which of the rational numbers in part (a) is closest to $\sqrt{2}$?

78. Use a calculator to order the following numbers from smallest to largest.

$\frac{26}{15}$, $\sqrt{3}$, $1.73\overline{20}$, $\frac{381}{220}$, $\sqrt{10} - \sqrt{2}$

1.2 Integer Exponents

In arithmetic, multiplication by a positive integer can be described as repeated addition; that is, $3 \cdot x = x + x + x$. In this section, we look at repeated *multiplications* that can be written in **exponential form.** Here are some examples.

Repeated multiplication	*Exponential form*
$7 \cdot 7$	7^2
$a \cdot a \cdot a \cdot a \cdot a$	a^5
$(-4)(-4)(-4)$	$(-4)^3$
$(2x)(2x)(2x)(2x)$	$(2x)^4$

Exponential Notation

Let n be a positive integer. Then

$$a^n = \underbrace{a \cdot a \cdot a \cdots a}_{n \text{ factors}}$$

where n is called the **exponent** and a is called the **base.** We read a^n as "a to the nth **power**" or simply "a to the n."

Remark: It is important to recognize the difference between exponential forms like $(-2)^4$ and -2^4. In $(-2)^4$, the parentheses indicate that the exponent applies to the negative sign as well as to the 2, but in $-2^4 = -(2^4)$, the exponent applies only to the 2. Similarly, in $(5x)^3$, the parentheses indicate that the exponent applies to the 5 as well as to the x, whereas in $5x^3 = 5(x^3)$, the exponent applies only to the x.

When multiplying exponential expressions with the same base, we *add* exponents.

$$a^m \cdot a^n = a^{m+n} \qquad \text{\textit{Add exponents when multiplying}}$$

For instance, to multiply 2^2 and 2^5, we can write

$$2^2 \cdot 2^5 = \underbrace{(2 \cdot 2)}_{\substack{\text{two} \\ \text{factors}}} \cdot \underbrace{(2 \cdot 2 \cdot 2 \cdot 2 \cdot 2)}_{\substack{\text{five} \\ \text{factors}}} = \underbrace{2 \cdot 2 \cdot 2 \cdot 2 \cdot 2 \cdot 2 \cdot 2}_{\substack{\text{seven} \\ \text{factors}}} = 2^{2+5} = 2^7.$$

On the other hand, when dividing exponential expressions, we *subtract* exponents. That is,

$$\frac{a^m}{a^n} = a^{m-n}, \qquad a \neq 0. \qquad \text{\textit{Subtract exponents when dividing}}$$

For instance, dividing 2^5 by 2^2 produces

$$\frac{2^5}{2^2} = \frac{2 \cdot 2 \cdot 2 \cdot 2 \cdot 2}{2 \cdot 2} = 2 \cdot 2 \cdot 2 = 2^{5-2} = 2^3.$$

There are two special cases involving division of exponential expressions. First, if $m = n$, then

$$\frac{a^n}{a^n} = a^{n-n} = a^0 = 1, \qquad a \neq 0$$

and we say that *any nonzero number raised to the zero power is 1.* Second, if the power of the denominator is larger than the power of the numerator, then we have

$$\frac{2^2}{2^5} = \frac{2 \cdot 2}{2 \cdot 2 \cdot 2 \cdot 2 \cdot 2} = \frac{1}{2^3} = 2^{2-5} = 2^{-3}.$$

Thus, if n is a positive integer, we can write

$$\frac{1}{a^n} = a^{-n}, \qquad a \neq 0.$$

These and other properties of exponents are summarized in the following list.

Properties of Exponents

Let a, b, x, and y be real numbers, and let m and n be integers. (Assume all denominators and bases are nonzero.)

Property	*Example*
1. $a^m a^n = a^{m+n}$	1. $3^2 \cdot 3^4 = 3^{2+4} = 3^6$
2. $\dfrac{a^m}{a^n} = a^{m-n}$	2. $\dfrac{x^7}{x^4} = x^{7-4} = x^3 \cdot$
3. $\dfrac{1}{a^n} = a^{-n}$	3. $\dfrac{1}{y^4} = y^{-4}$
4. $a^0 = 1, \qquad a \neq 0$	4. $(x^2 + 1)^0 = 1$
5. $(ab)^m = a^m b^m$	5. $(5x)^4 = 5^4 x^4$
6. $(a^m)^n = a^{mn}$	6. $(y^3)^{-4} = y^{3(-4)} = y^{-12} = \dfrac{1}{y^{12}}$
7. $\left(\dfrac{a}{b}\right)^m = \dfrac{a^m}{b^m}$	7. $\left(\dfrac{2}{x}\right)^3 = \dfrac{2^3}{x^3}$
8. $|a^2| = |a|^2 = a^2$	8. $|(-2)^2| = |-2|^2 = (-2)^2 = 4$

The preceding rules hold for *all* integers m and n, not just positive ones. For instance, by Property 2, we have

$$\frac{3^4}{3^{-5}} = 3^{4-(-5)} = 3^{4+5} = 3^9.$$

EXAMPLE 1   *Using Properties of Exponents*

(a) $(-3ab^4)(4ab^{-3}) = -12(a)(a)(b^4)(b^{-3}) = -12a^2 b$

(b) $(2xy^2 z^3)^3 = 2^3(x)^3(y^2)^3(z^3)^3 = 8x^3 y^6 z^9$

(c) $3a(-4a^2 b)^0 = 3a(1) = 3a, \qquad a \neq 0, b \neq 0$

(d) $\left(\dfrac{5x^3}{y}\right)^2 = \dfrac{5^2(x^3)^2}{y^2} = \dfrac{25x^6}{y^2}$

The next example shows how expressions involving negative exponents can be converted to ones involving positive exponents.

EXAMPLE 2 Converting to Positive Exponents

(a) Because $1/a^n = a^{-n}$, we have

$$x^{-1} = \frac{1}{x}.$$

(b) $\dfrac{1}{3x^{-2}} = \dfrac{1(x^2)}{3} = \dfrac{x^2}{3}$

(c) $\dfrac{12a^3b^{-4}}{4a^{-2}b} = \dfrac{12a^3 \cdot a^2}{4b \cdot b^4} = \dfrac{3a^5}{b^5}$

EXAMPLE 3 Quotients Raised to Negative Powers

Rewrite the expression using positive exponents and simplify.

$$\left(\frac{3x^2}{y}\right)^{-2}$$

SOLUTION

$$\left(\frac{3x^2}{y}\right)^{-2} = \frac{3^{-2}(x^2)^{-2}}{y^{-2}} = \frac{3^{-2}x^{-4}}{y^{-2}} = \frac{y^2}{3^2x^4} = \frac{y^2}{9x^4}$$

Scientific Notation

Exponents provide an efficient way of writing and computing with the very large (or very small) numbers used in science. For instance, a drop of water contains more than 33 billion billion molecules. That is 33 followed by 18 zeros:

33,000,000,000,000,000,000.

It is convenient to write such numbers in **scientific notation.** This notation has the form $c \times 10^n$, where $1 \le c < 10$ and n is an integer. Thus, the number of molecules in a drop of water can be written in scientific notation as

$3.3 \times 10{,}000{,}000{,}000{,}000{,}000{,}000 = 3.3 \times 10^{19}.$

The *positive* exponent 19 indicates that the number is *large* and that the decimal point in 3.3 should be moved 19 places to the *right*. A *negative* exponent in scientific notation indicates that the number is *small* (less than

1) and the decimal point should be moved to the *left*. For instance, the mass (in grams) of one electron is approximately

$$9.0 \times 10^{-28} = 0.0000000000000000000000000009 \,.$$

28 decimal places

EXAMPLE 4 *Scientific Notation*

(a) $1.345 \times 10^2 = 134.5$

two places

(b) $0.0000782 = 7.82 \times 10^{-5}$

five places

(c) $9.36 \times 10^{-6} = 0.00000936$

six places

(d) $836,100,000.0 = 8.361 \times 10^8$

eight places

Most scientific calculators automatically switch to scientific notation when they are showing large (or small) numbers that exceed the display range. Try multiplying

$$98,900,000 \times 500.$$

If your calculator follows standard conventions, its display should be

$$\boxed{\;\textbf{4.945}\quad\textbf{10}\;}.$$

This means that $c = 4.945$ and the exponent is $n = 10$.

To *enter* numbers in scientific notation, your calculator should have an exponential entry key labeled $\boxed{\text{EE}}$ or $\boxed{\text{EXP}}$. If you were to perform the preceding multiplication using scientific notation, you could begin by writing

$$98,900,000 \times 500 = (9.89 \times 10^7)(5.0 \times 10^2)$$

and then entering

9.89 $\boxed{\text{EE}}$ 7 $\boxed{\times}$ 5 $\boxed{\text{EE}}$ 2 $\boxed{=}$.

Remark: A number such as 10^{15} should be entered as 1 $\boxed{\text{EE}}$ 15, and *not* just $\boxed{\text{EE}}$ 15. The latter sequence would be interpreted as 0×10^{15}, which is zero.

Review of Fundamental Concepts of Algebra

Exponents and Calculators

Scientific calculators are capable of evaluating exponential expressions using the $\boxed{y^x}$ key. To use this key, remember that y is the *base* and x is the *exponent*. Thus, to calculate 3^6, we can enter

3 $\boxed{y^x}$ 6 $\boxed{=}$.   *Display:* 729

Similarly, to calculate

$$\left(1 + \frac{0.09}{12}\right)^{12}$$

we enter

$\underbrace{.09 \boxed{\div} 12 \boxed{+} 1 \boxed{=}}_{\text{base}}$ $\boxed{y^x}$ $\underset{\text{exponent}}{12}$ $\boxed{=}$.   *Display:* 1.0938069

Negative exponents are entered into a calculator by pressing the change-sign key $\boxed{+/-}$ immediately after entering the exponent. For instance, to enter the number 27^{-6} we use the following sequence.

27 $\boxed{y^x}$ 6 $\boxed{+/-}$ $\boxed{=}$ *Display:* 2.5812 −09

WARM UP

Perform the indicated operations and simplify.

1. $\left(\frac{2}{3}\right)\left(\frac{3}{2}\right)$ 2. $\left(\frac{1}{4}\right)(5)(4)$
3. $3\left(\frac{2}{7}\right) + 11\left(\frac{2}{7}\right)$ 4. $\frac{1}{2} \div 2$
5. $\frac{1}{3} + \frac{1}{2} - \frac{5}{6}$ 6. $\frac{1}{3} \div \frac{1}{3}$
7. $\frac{1}{7} + \frac{1}{3} - \frac{1}{21}$ 8. $11\left(\frac{1}{4}\right) + \frac{5}{4}$
9. $\frac{1}{12} - \frac{1}{3} + \frac{1}{8}$ 10. $\left(\frac{1}{2} - \frac{1}{3}\right) \div \frac{1}{6}$

EXERCISES 1.2

In Exercises 1–16, evaluate the given expression.

1. $2^2 \cdot 2^4$ 2. $3 \cdot 3^3$ 9. $\left(-\frac{3}{5}\right)^3\left(\frac{5}{3}\right)^2$ 10. $4\left(\frac{1}{2}\right)^5$

3. $\frac{5^5}{5^2}$ 4. $\frac{2^6}{2^3}$ 11. $6^5 \cdot 6^{-3}$ 12. $6 \cdot 2^{-3} \cdot 3^{-1}$

5. $(3^3)^2$ 6. $(2^4)^2$ 13. $\frac{4 \cdot 3^{-2}}{2^{-2} \cdot 3^{-1}}$ 14. $\left(\frac{3}{4}\right)^2\left(\frac{3}{5}\right)^2$

7. $(2^3 \cdot 3^2)^2$ 8. $(-5 \cdot 4^2)^3$ 15. 3^0 16. $(-2)^0$

In Exercises 17–22, evaluate the given expression for the indicated value of x.

Expression	Value of x		Expression	Value of x
17. $-3x^3$	2	18.	$\dfrac{x^2}{2}$	6
19. $4x^{-3}$	2	20.	$7x^{-2}$	4
21. $6x^0 - (6x)^0$	10	22.	$5(-x)^3$	3

In Exercises 23–52, simplify the given expression.

23. $(3x)^2$

24. $(-5z)^3$

25. $5x^4(x^2)$

26. $(8x^4)(2x^3)$

27. $10(x^2)^2$

28. $(4x^3)^2$

29. $6y^2(2y^4)^2$

30. $(-z)^3(3z^4)$

31. $\dfrac{3x^5}{x^3}$

32. $\dfrac{25y^8}{10y^4}$

33. $\dfrac{7x^2}{x^3}$

34. $\dfrac{r^4}{r^6}$

35. $\dfrac{12(x + y)^3}{9(x + y)}$

36. $\left(\dfrac{4}{y}\right)^3\left(\dfrac{3}{y}\right)^4$

37. $(x + 5)^0$, $x \neq -5$

38. $(2x^5)^0$, $x \neq 0$

39. $(2x^2)^{-2}$

40. $(z + 2)^{-3}(z + 2)^{-1}$

41. $(-2x^2)^3(4x^3)^{-1}$

42. $(4y^{-2})(8y^4)$

43. $\left(\dfrac{x}{10}\right)^{-1}$

44. $\left(\dfrac{4}{z}\right)^{-2}$

45. $\left(\dfrac{3z^2}{x}\right)^{-2}$

46. $\left(\dfrac{x^{-3}y^4}{5}\right)^{-3}$

47. $(4a^{-2}b^3)^{-3}$

48. $(5x^2y^4z^6)^3(5x^2y^4z^6)^{-3}$

49. $(2x^2 + y^2)^4(2x^2 + y^2)^{-4}$

50. $[(x^2y^{-2})^{-1}]^{-1}$

51. $\left(\dfrac{a^{-2}}{b^{-2}}\right)\left(\dfrac{b}{a}\right)^3$

52. $(-3x^2)(4x^{-3})(\tfrac{1}{6}x)$

In Exercises 53–60, write the number in scientific notation.

53. 93,000,000

54. 900,000,000

55. 0.00000435

56. 0.000087

57. 0.004392

58. 0.0875

59. 1,637,000,000

60. 67.8

In Exercises 61–68, write the number in decimal form.

61. 1.91×10^6

62. 2.345×10^{11}

63. 6.21×10^0

64. 9.4675×10^4

65. 8.52×10^{-3}

66. 7.021×10^{-5}

67. 3.798×10^{-8}

68. 1.0909×10^{-4}

In Exercises 69–74, use a calculator to evaluate the given expressions. Round your answers to three decimal places.

69. (a) $2400(1 + 0.06)^{20}$ (b) $750\left(1 + \dfrac{0.11}{365}\right)^{800}$

70. (a) $\dfrac{(2.414 \times 10^4)^6}{(1.68 \times 10^5)^5}$ (b) $(9.3 \times 10^6)^3(6.1 \times 10^{-4})^4$

71. (a) $\dfrac{3000}{[1 + (0.05/4)]^4}$ (b) $\dfrac{(3.28 \times 10^{-6})^{10}}{(5.34 \times 10^{-3})^{25}}$

72. (a) $\dfrac{4 - 1.25^6}{[1 + (0.05/4)]^4}$ (b) $5000\left(1 + \dfrac{0.095}{365}\right)^{3650}$

73. (a) $(0.000345)(8,980,000,000)$

   (b) $\dfrac{67,000,000 + 93,000,000}{0.0052}$

74. (a) $\dfrac{848,000,000}{1,624,000}$ (b) $\dfrac{0.0000928 - 0.0000021}{0.0061}$

75. The speed of light is 11,160,000 miles per minute. The distance from the sun to the earth is 93,000,000 miles. Find the time it takes for light to travel from the sun to the earth.

76. The *per capita public debt* is defined as the gross debt divided by the population. Find the per capita debt of the United States in 1985 if the gross debt was 1823.1 billion dollars and the population was 239 million.

77. The amount A after t years in a savings account earning an annual interest rate of r compounded n times per year is

$$A = P\left(1 + \frac{r}{n}\right)^{nt}$$

where P is the (original) principal. Complete the following table for $500 deposited in an account earning 12% compounded daily. [Note that $r = 0.12$ implies an interest rate of 12%.]

t	5	10	20	30	40	50
A						

[*Hint:* If you have a programmable calculator, try using the programming feature to complete the tables in Exercises 77–79.]

78. Repeat Exercise 77 for an account that earns 10% compounded monthly.

79. Assume $3000 is deposited in an account at 8% for two years. Use the formula given in Exercise 77 to find the balance if the interest is compounded (a) annually, (b) quarterly, (c) monthly, and (d) daily.

1.3 Radicals and Rational Exponents

To denote the nth root of a real number, we use the **radical** form $\sqrt[n]{\ }$, except for $n = 2$, which is simply written as $\sqrt{\ }$.

Definition of Principal nth Root of a Real Number

If a and b are nonnegative real numbers and n is a positive integer, or if a and b are negative real numbers and n is an *odd* positive integer, then the **principal nth root of b** is defined by the statement

$$\sqrt[n]{b} = a \quad \text{if and only if} \quad a^n = b.$$

The integer n is called the **index** of the radical, and b is called the **radicand.** For $n = 2$, $\sqrt{b}$ is called the **square root** of b and for $n = 3$, $\sqrt[3]{b}$ is called the **cube root** of b.

We make three important observations about this definition.

1. *Even roots of negative numbers are not real.* For instance, $\sqrt{-9}$ is not real because there is no real number a such that $a^2 = -9$. (In Section 2.5 we will show how *square roots of negative numbers* fit into the set of complex numbers.)

2. *Even roots of positive numbers exist in pairs.* For instance, both 5 and -5 are roots of 25 because $(5)^2 = 25$ and $(-5)^2 = 25$. In general, if n is even and b positive, then there are two real nth roots of b.

$$\sqrt[n]{b} \qquad\qquad \textit{Principal nth root of b}$$
$$-\sqrt[n]{b} \qquad\qquad \textit{Negative nth root of b}$$

It is *incorrect* to write $\sqrt{81} = \pm 9$ because $\sqrt{81}$ denotes the principal (positive) square root of 81. That is,

$$\sqrt{81} \qquad \text{DOES NOT EQUAL} \qquad \pm 9.$$

3. *Odd roots of negative numbers are negative.* For instance, $\sqrt[3]{-64} = -4$ because $(-4)^3 = (-4)(-4)(-4) = -64$.

EXAMPLE 1 Roots of Real Numbers

(a) $\sqrt{121} = 11$ because $11^2 = 121$.

(b) $\sqrt[3]{\dfrac{125}{64}} = \dfrac{5}{4}$ because $\left(\dfrac{5}{4}\right)^3 = \dfrac{5^3}{4^3} = \dfrac{125}{64}$.

(c) $\sqrt[5]{-32} = -2$ because $(-2)^5 = -32$.

To simplify or evaluate expressions involving radicals, we use the following properties.

Properties of Radicals

Let m and n be positive integers and let a and b be real numbers such that the indicated roots are real. Then the following properties are true.

Property	*Example*
1. $\sqrt[n]{a^m} = (\sqrt[n]{a})^m$	1. $\sqrt[3]{8^2} = (\sqrt[3]{8})^2 = (2)^2 = 4$
2. $\sqrt[n]{a} \cdot \sqrt[n]{b} = \sqrt[n]{ab}$	2. $\sqrt{5} \cdot \sqrt{7} = \sqrt{5 \cdot 7} = \sqrt{35}$
3. $\dfrac{\sqrt[n]{a}}{\sqrt[n]{b}} = \sqrt[n]{\dfrac{a}{b}}, \quad b \neq 0$	3. $\dfrac{\sqrt[4]{27}}{\sqrt[4]{9}} = \sqrt[4]{\dfrac{27}{9}} = \sqrt[4]{3}$
4. $\sqrt[m]{\sqrt[n]{a}} = \sqrt[mn]{a}$	4. $\sqrt[3]{\sqrt{10}} = \sqrt[6]{10}$
5. $(\sqrt[n]{a})^n = a$	5. $(\sqrt[6]{15})^6 = 15$
6. For n even, $\sqrt[n]{a^n} = |a|$.	6. $\sqrt{(-12)^2} = |-12| = 12$
For n odd, $\sqrt[n]{a^n} = a$.	$\sqrt[3]{(-12)^3} = -12$

Remark: A common special case of Property 6 is $\sqrt{a^2} = |a|$.

Simplifying Radicals

An expression involving radicals is in **simplest form** when the following conditions are satisfied.

1. All possible factors have been removed from the radical sign.
2. All fractions have radical-free denominators (accomplished by a process called *rationalizing the denominator*).
3. The index for the radical has been reduced as far as possible.

To simplify radicals, we factor the radicand into factors whose powers are multiples of the index. For instance,

$$\sqrt{8} = \sqrt{2^3} = \sqrt{2^2}\sqrt{2} = 2\sqrt{2}$$

and

$$\sqrt[3]{x^7} = \sqrt[3]{x^6 \cdot x} = \sqrt[3]{(x^2)^3}\sqrt[3]{x} = x^2\sqrt[3]{x}.$$

EXAMPLE 2 Simplifying Even Roots

Simplify the following radicals.

(a) $\sqrt{75x^3}$

(b) $\sqrt[4]{(5x^2y^{-3})^4}$

SOLUTION

(a) $\sqrt{75x^3} = \sqrt{25(x^2)(3x)}$ *Find largest square factors*

$\qquad\qquad = \sqrt{25}\sqrt{x^2}\sqrt{3x}$

$\qquad\qquad = 5|x|\sqrt{3x}$ *Find roots of perfect squares*

Note that $\sqrt{x^2} = |x|$, not simply x.

(b) $\sqrt[4]{(5x^2y^{-3})^4} = |5x^2y^{-3}| = \left|\dfrac{5x^2}{y^3}\right|$

EXAMPLE 3 Simplifying Odd Roots

Simplify the following radicals.

(a) $\sqrt[3]{24a^4c^8}$

(b) $\sqrt[3]{-40x^6y^4}$

SOLUTION

(a) $\sqrt[3]{24a^4c^8} = \sqrt[3]{(8)(a^3)(c^6)(3ac^2)}$ *Find largest cube factors*

$\qquad\qquad = \sqrt[3]{8}\sqrt[3]{a^3}\sqrt[3]{c^6}\sqrt[3]{3ac^2}$

$\qquad\qquad = 2ac^2\sqrt[3]{3ac^2}$ *Find roots of perfect cubes*

(b) $\sqrt[3]{-40x^6y^4} = \sqrt[3]{(-8)(x^2)^3(y^3)(5y)}$

$\qquad\qquad = \sqrt[3]{-8}\sqrt[3]{(x^2)^3}\sqrt[3]{y^3}\sqrt[3]{5y}$

$\qquad\qquad = -2x^2y\sqrt[3]{5y}$

Rationalizing Denominators and Numerators

Another simplification technique is *rationalizing* the numerator or denominator by removing radicals from one or the other. In algebra we usually emphasize rationalizing the denominator. However, in calculus it is occasionally helpful to rationalize the numerator. In both instances, we make use of the form $a + b\sqrt{m}$ and its **conjugate** $a - b\sqrt{m}$. The product of this conjugate pair has no radical. For instance,

$$(2 + \sqrt{3})(2 - \sqrt{3}) = 2^2 - (\sqrt{3})^2 = 4 - 3 = 1.$$

Therefore, to rationalize a denominator of the form $a - b\sqrt{m}$ (or $a + b\sqrt{m}$), we multiply both numerator and denominator by the conjugate factor $a + b\sqrt{m}$ (or $a - b\sqrt{m}$). If $a = 0$, then the rationalizing factor for $\sqrt{m}$ is itself, $\sqrt{m}$.

EXAMPLE 4   Rationalizing Single-Term Denominators

Eliminate the radicals in the denominator.

(a) $\dfrac{5}{2\sqrt{3}}$

(b) $\dfrac{2}{\sqrt[3]{5}}$

SOLUTION

(a) To rationalize the denominator, we multiply *both* the numerator and the denominator by $\sqrt{3}$ to obtain

$$\frac{5}{2\sqrt{3}} = \frac{5}{2\sqrt{3}} \cdot \frac{\sqrt{3}}{\sqrt{3}} = \frac{5\sqrt{3}}{2(3)} = \frac{5\sqrt{3}}{6}.$$

(b) To rationalize the denominator in this case, we multiply both the numerator and the denominator by $\sqrt[3]{5^2}$. Note how this eliminates the radical from the denominator.

$$\frac{2}{\sqrt[3]{5}} = \frac{2}{\sqrt[3]{5}} \cdot \frac{\sqrt[3]{5^2}}{\sqrt[3]{5^2}} = \frac{2\sqrt[3]{(5)^2}}{\sqrt[3]{5^3}} = \frac{2\sqrt[3]{25}}{5}$$

EXAMPLE 5   Rationalizing a Denominator with Two Terms

Eliminate the radical in the denominator.

$$\frac{2}{3 + \sqrt{7}}$$

SOLUTION

$$\frac{2}{3 + \sqrt{7}} = \frac{2}{3 + \sqrt{7}} \cdot \frac{3 - \sqrt{7}}{3 - \sqrt{7}} \qquad \textit{Multiply numerator and denominator by conjugate}$$

$$= \frac{2(3 - \sqrt{7})}{(3)^2 - (\sqrt{7})^2}$$

$$= \frac{2(3 - \sqrt{7})}{9 - 7}$$

$$= \frac{2(3 - \sqrt{7})}{2}$$

$$= 3 - \sqrt{7} \qquad \textit{Simplify}$$

EXAMPLE 6 Rationalizing the Numerator

Eliminate the radicals in the numerator.

$$\frac{\sqrt{5} - \sqrt{7}}{2}$$

SOLUTION

$$\frac{\sqrt{5} - \sqrt{7}}{2} = \frac{\sqrt{5} - \sqrt{7}}{2} \cdot \frac{\sqrt{5} + \sqrt{7}}{\sqrt{5} + \sqrt{7}} \qquad \textit{Multiply numerator and denominator by conjugate}$$

$$= \frac{5 - 7}{2(\sqrt{5} + \sqrt{7})}$$

$$= \frac{-2}{2(\sqrt{5} + \sqrt{7})}$$

$$= \frac{-1}{\sqrt{5} + \sqrt{7}} \qquad \textit{Simplify}$$

In Example 6, do not confuse the expression $\sqrt{5} + \sqrt{7}$ with the expression $\sqrt{5 + 7}$. That is,

$$\sqrt{x + y} \qquad \text{DOES NOT EQUAL} \qquad \sqrt{x} + \sqrt{y}.$$

For example, $\sqrt{16 + 9} = \sqrt{25} = 5$, whereas $\sqrt{16} + \sqrt{9} = 4 + 3 = 7$. Also, watch out for the following version of this common error:

$$\sqrt{x^2 + y^2} \qquad \text{DOES NOT EQUAL} \qquad x + y.$$

Rational Exponents

Up to this point, our work with exponents has been restricted to *integer* exponents. In the following definition, we show how radicals are used to define *rational* exponents.

Definition of Rational Exponent

Let m be an integer, n a natural number, and b a real number such that $\sqrt[n]{b}$ is a real number. Then

$$b^{m/n} = (\sqrt[n]{b})^m = \sqrt[n]{b^m}.$$

Remark: An important special case is $b^{1/n} = \sqrt[n]{b}$, which denotes the principal nth root of b. For instance, $4^{1/2} = 2$ and $8^{1/3} = 2$.

In the definition of rational exponents, note that the denominator of the exponent is the *index* for the corresponding radical form and the numerator is the *power* of the radical (or radicand).

$$b^{m/n} = (\sqrt[n]{b})^m = \sqrt[n]{b^m}$$

When working with rational exponents, the standard properties of integer exponents (listed in the previous section) still apply. For instance,

$$2^{1/2}2^{1/3} = 2^{(1/2)+(1/3)} = 2^{5/6}.$$

EXAMPLE 7 *Changing from Radical to Exponential Form*

Write each of the following in exponential form.

(a) $\sqrt{3}$ (b) $\sqrt{(3xy)^5}$ (c) $2x\sqrt[4]{x^3}$

SOLUTION

(a) The square root of 3 can be written as 3 to the 1/2 power. That is,
$$\sqrt{3} = 3^{1/2}.$$

(b) In this case the base is $3xy$, the power is 5, and the index is 2. Thus,
$$\sqrt{(3xy)^5} = \sqrt[2]{(3xy)^5} = (3xy)^{5/2}.$$

(c) $2x\sqrt[4]{x^3} = (2x)(x^{3/4}) = 2x^{1+(3/4)} = 2x^{7/4}$

EXAMPLE 8 *Changing from Exponential to Radical Form*

Write each of the following in radical form.

(a) $(x^2 + y^2)^{3/2}$ (b) $2y^{3/4}z^{1/4}$ (c) $a^{-3/2}$

SOLUTION

(a) In this case the index is 2 and the power is 3, so we write
$$(x^2 + y^2)^{3/2} = (\sqrt{x^2 + y^2})^3 = \sqrt{(x^2 + y^2)^3}.$$

(b) Note that the factor 2 is not part of the base, so we write
$$2y^{3/4}z^{1/4} = 2(y^3z)^{1/4} = 2\sqrt[4]{y^3z}.$$

(c) First we convert to positive exponents, then to radical form, obtaining
$$a^{-3/2} = \frac{1}{a^{3/2}} = \frac{1}{\sqrt{a^3}}.$$

Remark: Rational exponents can be tricky, and you must remember that the expression $b^{m/n}$ is not defined unless $\sqrt[n]{b}$ is a real number. This restriction produces some unusual-looking results. For instance, the number $(-8)^{2/6}$ is not defined because $\sqrt[6]{-8}$ is not a real number. And yet, $(-8)^{1/3}$ is defined because $\sqrt[3]{-8} = -2$.

Rational exponents are particularly useful for evaluating roots of numbers on a calculator, for reducing the index of a radical, and for simplifying (and factoring) algebraic expressions encountered in calculus. Examples 9 through 11 demonstrate these uses.

EXAMPLE 9   *Simplifying with Rational Exponents*

Simplify the following expressions.

(a) $(27)^{2/6}$ (b) $(-32)^{-4/5}$ (c) $(-5x^{5/3})(3x^{-3/4})$

SOLUTION

(a) $(27)^{2/6} = (27)^{1/3} = \sqrt[3]{27} = 3$

(b) $(-32)^{-4/5} = (\sqrt[5]{-32})^{-4} = (-2)^{-4} = \dfrac{1}{(-2)^4} = \dfrac{1}{16}$

(c) $(-5x^{5/3})(3x^{-3/4}) = -15x^{(5/3)-(3/4)} = -15x^{11/12}$

EXAMPLE 10   *Reducing the Index of a Radical*

Simplify the following radicals, reducing the index when possible.

(a) $\sqrt[6]{a^4}$ (b) $\sqrt[3]{\sqrt{125}}$

SOLUTION

(a) $\sqrt[6]{a^4} = a^{4/6}$ *Rewrite with rational exponents*

$\qquad\quad = a^{2/3}$ *Reduce exponent*

$\qquad\quad = \sqrt[3]{a^2}$ *Rewrite in radical form*

(b) $\sqrt[3]{\sqrt{125}} = \sqrt[6]{125} = \sqrt[6]{(5)^3} = 5^{3/6} = 5^{1/2} = \sqrt{5}$

Radicals and Rational Exponents

EXAMPLE 11   *Simplifying Algebraic Expressions*

Simplify the following expressions.

(a) $(2x - 1)^{4/3}(2x - 1)^{-1/3}, \quad x \neq \dfrac{1}{2}$

(b) $\dfrac{x - 1}{(x - 1)^{-1/2}} \cdot \dfrac{\sqrt{x - 1}}{\sqrt{x - 1}}, \quad x \neq 1$

SOLUTION

(a) $(2x - 1)^{4/3}(2x - 1)^{-1/3} = (2x - 1)^{(4/3)-(1/3)}$
$$= (2x - 1)^1$$
$$= 2x - 1$$

(b) $\dfrac{x - 1}{(x - 1)^{-1/2}} \cdot \dfrac{\sqrt{x - 1}}{\sqrt{x - 1}} = \dfrac{(x - 1)^{3/2}}{(x - 1)^0} = (x - 1)^{3/2}$

Radical expressions can be combined (added or subtracted) if they are **similar**—that is, if they have the same index and radicand. For instance, $2\sqrt{3x}$, $-\sqrt{3x}$, and $\sqrt{3x}/2$ are similar but $\sqrt[3]{3x}$ and $2\sqrt{3x}$ are not similar. To determine whether two radicals are similar, you should first simplify each radical.

EXAMPLE 12   *Combining Radicals*

Simplify and combine similar radicals.

(a) $2\sqrt{48} - 3\sqrt{27}$ (b) $\sqrt[3]{16x} - \sqrt[3]{54x^4}$

SOLUTION

(a) $2\sqrt{48} - 3\sqrt{27} = 2\sqrt{16 \cdot 3} - 3\sqrt{9 \cdot 3}$ *Find square factors*
$$= 8\sqrt{3} - 9\sqrt{3} \qquad\qquad \textit{Find square roots}$$
$$= (8 - 9)\sqrt{3} \qquad\qquad \textit{Combine like terms}$$
$$= -\sqrt{3}$$

(b) $\sqrt[3]{16x} - \sqrt[3]{54x^4} = \sqrt[3]{8 \cdot 2x} - \sqrt[3]{27 \cdot x^3 \cdot 2x}$
$$= 2\sqrt[3]{2x} - 3x\sqrt[3]{2x}$$
$$= (2 - 3x)\sqrt[3]{2x}$$

Radicals and Calculators

There are two methods of evaluating radicals on most calculators. For square roots, you use the *square root key* $\boxed{\sqrt{}}$. For other roots, you should first convert the radical to exponential form and then use the *exponential key* $\boxed{y^x}$.

EXAMPLE 13 *Evaluating Radicals with a Calculator*

Use a calculator to evaluate $\sqrt[3]{56}$.

SOLUTION

First we write $\sqrt[3]{56}$ in exponential form:

$$\sqrt[3]{56} = 56^{1/3}.$$

Now there are several options.

Calculator Steps

(i) 1 $\boxed{\div}$ 3 $\boxed{=}$ $\boxed{\text{STO}}$ 56 $\boxed{y^x}$ $\boxed{\text{RCL}}$ $\boxed{=}$ *Use memory key*

(ii) 56 $\boxed{y^x}$ $\boxed{(}$ 1 $\boxed{\div}$ 3 $\boxed{)}$ $\boxed{=}$ *Use parentheses*

(iii) 56 $\boxed{y^x}$ 3 $\boxed{1/x}$ $\boxed{=}$ *Use reciprocal key*

For each of these three keystroke sequences, the answer is

$$\sqrt[3]{56} \approx 3.8258624.$$

EXAMPLE 14 *Evaluating Radicals with a Calculator*

Use a calculator to evaluate each of the following.

(a) $\sqrt[3]{-4}$ (b) $(1.2)^{-1/6}$

SOLUTION

(a) Since

$$\sqrt[3]{-4} = \sqrt[3]{(-1)(4)} = \sqrt[3]{-1} \cdot \sqrt[3]{4} = -\sqrt[3]{4}$$

we can attach the negative sign of the radicand at the end of the keystroke sequence as follows.

4 $\boxed{y^x}$ $\boxed{(}$ 1 $\boxed{\div}$ 3 $\boxed{)}$ $\boxed{=}$ $\boxed{+/-}$ *Display:* -1.5874011

(b) 1.2 $\boxed{y^x}$ $\boxed{(}$ 1 $\boxed{\div}$ 6 $\boxed{+/-}$ $\boxed{)}$ $\boxed{=}$ *Display:* 0.97007012

WARM UP

Simplify the expressions.

1. $\left(\frac{1}{3}\right)\left(\frac{2}{3}\right)^2$

2. $3(-4)^2$

3. $(-2x)^3$

4. $(-2x^3)(-3x^4)$

5. $(7x^5)(4x)$

6. $(5x^4)(25x^2)^{-1}, \quad x \neq 0$

7. $\dfrac{12z^6}{4z^2}, \quad z \neq 0$

8. $\left(\dfrac{2x}{5}\right)^2\left(\dfrac{2x}{5}\right)^{-4}, \quad x \neq 0$

9. $\left(\dfrac{3y^2}{x}\right)^0, \quad x \neq 0, y \neq 0$

10. $[(x+2)^2(x+2)^3]^2$

EXERCISES 1.3

In Exercises 1–12, fill in the missing description.

Radical Form	Rational Exponent Form
1. $\sqrt{9} = 3$	_____
2. $\sqrt[3]{64} = 4$	_____
3. _____	$32^{1/5} = 2$
4. _____	$-(144^{1/2}) = -12$
5. _____	$196^{1/2} = 14$
6. $\sqrt[3]{614.125} = 8.5$	_____
7. $\sqrt[3]{-216} = -6$	_____
8. _____	$(-243)^{1/5} = -3$
9. _____	$27^{2/3} = 9$
10. $(\sqrt[4]{81})^3 = 27$	_____
11. $\sqrt[4]{81^3} = 27$	_____
12. _____	$16^{5/4} = 32$

In Exercises 13–24, evaluate each expression without using a calculator.

13. $\sqrt{36}$

14. $\sqrt[3]{\frac{27}{8}}$

15. $-\sqrt[3]{-27}$

16. $\sqrt[3]{0}$

17. $\dfrac{4}{\sqrt{64}}$

18. $\dfrac{\sqrt[4]{81}}{3}$

19. $16^{3/2}$

20. $(\sqrt[3]{-125})^3$

21. $\sqrt[4]{562^4}$

22. $32^{-3/5}$

23. $\left(\frac{9}{4}\right)^{-1/2}$

24. $\left(-\frac{1}{64}\right)^{-1/3}$

In Exercises 25–30, simplify by removing all possible factors from the radical.

25. $\sqrt{8}$

26. $\sqrt[3]{\frac{16}{27}}$

27. $\sqrt[3]{16x^5}$

28. $\sqrt[4]{(3x^2y^3)^4}$

29. $\sqrt{75x^2y^{-4}}$

30. $\sqrt[5]{96x^5y^3}$

In Exercises 31–38, rewrite each expression by rationalizing the denominator. Simplify your answer.

31. $\dfrac{1}{\sqrt{3}}$

32. $\dfrac{5}{\sqrt{10}}$

33. $\dfrac{8}{\sqrt[3]{2}}$

34. $\dfrac{5}{\sqrt[3]{(5x)^2}}$

35. $\dfrac{2x}{5 - \sqrt{3}}$

36. $\dfrac{5}{\sqrt{14} - 2}$

37. $\dfrac{3}{\sqrt{5} + \sqrt{6}}$

38. $\dfrac{5}{2\sqrt{10} - 5}$

In Exercises 39–44, rewrite each expression by rationalizing the numerator. Simplify your answer.

39. $\dfrac{\sqrt{16}}{2}$

40. $\dfrac{\sqrt{2}}{3}$

41. $\dfrac{\sqrt{5} + \sqrt{3}}{3}$

42. $\dfrac{\sqrt{3} - \sqrt{2}}{2}$

43. $\dfrac{\sqrt{7} - 3}{4}$

44. $\dfrac{2\sqrt{3} + \sqrt{3}}{3}$

In Exercises 45–52, simplify and/or combine the given radicals.

45. $5\sqrt{x} - 3\sqrt{x}$

46. $3\sqrt{x+1} + 10\sqrt{x+1}$

47. $2\sqrt{50} + 12\sqrt{8}$

48. $4\sqrt{27} - \sqrt{75}$

49. $2\sqrt{4y} - 2\sqrt{9y} + 10\sqrt{y}$

50. $2\sqrt{80} + \sqrt{125} - \sqrt{500}$

51. $\sqrt{5x^2y}\sqrt{3y}$

52. $\sqrt[3]{\dfrac{4z^2}{y^5}}\sqrt[3]{\dfrac{2z}{y}}$

In Exercises 53–60, write the given expression as a single radical.

53. $\sqrt{50}\sqrt[3]{2}$

54. $\sqrt{40}\sqrt[3]{10}$

55. $\sqrt{x}\sqrt[3]{x}$

56. $\dfrac{\sqrt{x}}{\sqrt[3]{x}}$

57. $\sqrt{\sqrt{32}}$

58. $\sqrt{\sqrt[3]{10a^7b}}$

59. $\sqrt{\sqrt[4]{2x}}$

60. $\sqrt{\sqrt{x+1}}$

In Exercises 61–64, use fractional exponents to reduce the index of the radical.

61. $\sqrt[4]{3^2}$

62. $\sqrt[6]{x^3}$

63. $\sqrt[6]{(x+1)^4}$

64. $\sqrt[8]{(3x^2)^2}$

In Exercises 65–70, use a calculator to approximate the given number. Round your answer to four decimal places.

65. $\sqrt{57}$

66. $\sqrt[3]{45^2}$

67. $\sqrt[6]{125}$

68. $(15.25)^{-1.4}$

69. $\sqrt{75 + 3\sqrt{8}}$

70. $(2.65 \times 10^{-4})^{1/3}$

In Exercises 71–76, fill in the blank with $<$, $>$, or $=$ by finding the decimal approximation of each expression.

71. $\sqrt{5} + \sqrt{3}$ ▭ $\sqrt{5+3}$

72. $\sqrt{5} - \sqrt{3}$ ▭ $\sqrt{5-3}$

73. 5 ▭ $\sqrt{3^2 + 2^2}$

74. $\sqrt{3} \cdot \sqrt[4]{3}$ ▭ $\sqrt[8]{3}$

75. 5 ▭ $\sqrt{3^2 + 4^2}$

76. $\sqrt{\dfrac{3}{11}}$ ▭ $\dfrac{\sqrt{3}}{\sqrt{11}}$

77. To find uniform depreciation by the declining balance method, we use the formula

$$R = N\left[1 - \left(\frac{S}{C}\right)^{1/N}\right]$$

where R is the rate factor each year, N is the useful life of the item, C is the original cost, and S is the salvage value. Calculate R (to two decimal places) for each of the following.

(a) $N = 8$, $C = \$10,400$, $S = \$1500$

(b) $N = 4$, $C = \$11,200$, $S = \$3200$

78. A funnel is filled with water to a height of h. The time t it takes for the funnel to empty is given by

$$t = 0.03[12^{5/2} - (12 - h)^{5/2}], \qquad 0 \le h \le 12.$$

Find t (to two decimal places) for $h = 7$ centimeters.

1.4 *Polynomials and Special Products*

In this section we will be working with *algebraic* expressions and statements instead of *numerical* ones. An **algebraic expression** is a collection of letters called **variables** and real numbers organized in some manner by using addition, subtraction, multiplication, division, and radicals. Some examples are

$$3x - 7, \quad \frac{5xy^2 - 2x^2}{x + 2}, \quad \sqrt{x^2 - y^2}, \quad \text{and} \quad 5x^3 - 7x^2 + x - 11.$$

A **term** of an algebraic expression is any product of a real number and one or more variables raised to powers. An **algebraic sum** is an algebraic expression in which the terms are connected by the operation of *addition*. For instance, the terms of $5x^3 - 7y^2$ are $5x^3$ and $-7y^2$, because $5x^3 - 7y^2 = 5x^3 + (-7y^2)$.

Polynomials and Special Products

To **evaluate** an algebraic expression, a specific number is assigned to each variable in the expression. When evaluating expressions, we need to be aware of what numbers are *permissible* to assign to the variables. Unless otherwise stated, permissible numbers are those that yield *real* valued answers.

The set of permissible values that can be assigned to the variable in an algebraic expression is called the **domain** of the variable. As a general rule, we *exclude* from the domain numbers that yield zero denominators and numbers that yield even roots of negative numbers. For instance, the domain of x in the expression $\sqrt{x}$ is all $x \geq 0$. Similarly, the domain of x in the expression $3/(x - 2)$ is all $x \neq 2$.

Two expressions are called algebraically **equivalent** if they yield the same numbers for all x-values for which they are defined. For instance, the three expressions

$$3x + 5(x - 2), \qquad 3x + 5x + 5(-2), \qquad \text{and} \qquad 8x - 10$$

are equivalent. However, the expressions

$$\frac{x(x + 2)}{x} \qquad \text{and} \qquad x + 2$$

are not equivalent because the first expression is not defined when $x = 0$, whereas the second expression is defined when $x = 0$.

Polynomials

One of the simplest and most common kinds of algebraic expressions is the **polynomial.** Some examples are

$$2x + 5, \quad 3x^4 - 7x^2 + 2x + 4, \quad \text{and} \quad 5x^2y^2 - xy + 3.$$

The first two are *polynomials in x* and the third one is a *polynomial in x and y.* The terms of a polynomial in x have the form ax^k, where a is called the **coefficient** and k the **degree** of the term. Since a polynomial is an algebraic sum, the coefficients take on the sign between the terms. For instance,

$$2x^3 - 5x^2 + 1 = 2x^3 + (-5)x^2 + (0)x + 1$$

has coefficients 2, -5, 0, and 1.

Definition of a Polynomial in x

Let $a_0, a_1, a_2, \ldots, a_n$ be *real numbers* and let n be a *nonnegative integer*. A **polynomial in x** is an expression of the form

$$a_n x^n + a_{n-1} x^{n-1} + \cdots + a_1 x + a_0$$

where $a_n \neq 0$. The polynomial is of **degree** n. The number a_n is called the **leading coefficient.**

Remark: Polynomials with one, two, or three terms are called **monomials, binomials,** or **trinomials,** respectively.

Note in the preceding definition that the polynomial is written with descending powers of x. This is referred to as **standard form.** Some examples of polynomials rewritten in standard form are as follows.

Polynomial	Standard Form	Degree
$4x^2 - 5x^7 - 2 + 3x$	$-5x^7 + 4x^2 + 3x - 2$	7
$4 - 9x^2$	$-9x^2 + 4$	2
8	$8 \quad (8 = 8x^0)$	0

A polynomial that has all zero coefficients is called the **zero polynomial,** denoted by 0, and we do not assign a degree to this particular polynomial.

EXAMPLE 1 *Identifying a Polynomial and Its Degree*

Determine which of the following expressions is a polynomial. If it is a polynomial, give its degree.

(a) $-2 + 3x + x^2 - 2x^3$ 　　　(b) $\sqrt{x^2 - 3x}$ 　　　(c) $x^2 + 5x^{-1}$

SOLUTION

(a) In standard form, we see that $-2x^3 + x^2 + 3x - 2$ is a polynomial of degree 3.
(b) $\sqrt{x^2 - 3x}$ is not a polynomial because the radical sign indicates a non-integer power of x.
(c) $x^2 + 5x^{-1}$ is not a polynomial because of the negative exponent.

For polynomials in more than one variable, the degree of a *term* is the sum of the powers of the variables in the term. The degree of the *polynomial* is the highest degree of all its terms. For instance, the polynomial $5x^3y - x^2y^2 + 2xy - 5$ has two terms of degree 4, one term of degree 2, and one term of degree 0. The degree of the polynomial is 4.

Operations with Polynomials

We can **add** and **subtract** polynomials in much the same way that we add and subtract real numbers. We simply add or subtract the *like terms* (terms having the same variables to the same powers) by adding their coefficients. For instance, $-3xy^2$ and $5xy^2$ are like terms and their sum is given by

$$-3xy^2 + 5xy^2 = (-3 + 5)xy^2 = 2xy^2.$$

EXAMPLE 2 *Sums and Differences of Polynomials*

(a) Add $5x^3 - 7x^2 - 3$ and $x^3 + 2x^2 - x + 8$.

(b) Subtract $3x^4 - 4x^2 + 3x$ from $7x^4 - x^2 - 4x + 2$.

SOLUTION

(a) $(5x^3 - 7x^2 - 3) + (x^3 + 2x^2 - x + 8)$

$\quad = (5x^3 + x^3) + (2x^2 - 7x^2) - x + (8 - 3)$ *Group like terms*

$\quad = 6x^3 - 5x^2 - x + 5$ *Combine like terms*

(b) $(7x^4 - x^2 - 4x + 2) - (3x^4 - 4x^2 + 3x)$

$\quad = 7x^4 - x^2 - 4x + 2 - 3x^4 + 4x^2 - 3x$

$\quad = (7x^4 - 3x^4) + (4x^2 - x^2) + (-3x - 4x) + 2$ *Group like terms*

$\quad = 4x^4 + 3x^2 - 7x + 2$ *Combine like terms*

Remark: A common mistake is to fail to change the sign of *each* term inside parentheses preceded by a minus sign. For instance, note that

$$-(3x^4 - 4x^2 + 3x) = -3x^4 + 4x^2 - 3x$$

and

$$-(3x^4 - 4x^2 + 3x) \neq -3x^4 - 4x^2 + 3x. \qquad \textit{Common mistake}$$

To find the **product** of two polynomials, the left and right distributive properties are useful. For example, if we treat $(5x + 7)$ as a single quantity, we can multiply $(3x - 2)$ by $(5x + 7)$ as follows.

$$(3x - 2)(5x + 7) = 3x(5x + 7) - 2(5x + 7)$$
$$= (3x)(5x) + (3x)(7) - 2(5x) - 2(7)$$
$$= 15x^2 + 21x - 10x - 14$$

Product of **First terms**	Product of **Outer terms**	Product of **Inner terms**	Product of **Last terms**

$$= 15x^2 + 11x - 14$$

With practice, you should be able to multiply two binomials without writing this many steps.

When multiplying two polynomials, be sure to multiply *each* term of one polynomial by *each* term of the other. The following vertical arrangement is convenient.

Review of Fundamental Concepts of Algebra

$$
\begin{array}{r}
2x^2 - 3x + 5 \\
x - 4 \\
\hline
2x^3 - 3x^2 + 5x \\
- 8x^2 + 12x - 20 \\
\hline
2x^3 - 11x^2 + 17x - 20
\end{array}
$$

$\Longleftarrow \quad x(2x^2 - 3x + 5)$

$\Longleftarrow \quad -4(2x^2 - 3x + 5)$

This vertical method is further demonstrated in Example 3.

EXAMPLE 3 *Using a Vertical Arrangement to Multiply Polynomials*

Multiply $(x^2 - 2x + 2)$ by $(x^2 + 2x + 2)$.

SOLUTION

$$
\begin{array}{r}
x^2 - 2x + 2 \\
x^2 + 2x + 2 \\
\hline
x^4 - 2x^3 + 2x^2 \\
2x^3 - 4x^2 + 4x \\
2x^2 - 4x + 4 \\
\hline
x^4 + 0x^3 + 0x^2 - 0x + 4 = x^4 + 4
\end{array}
$$

$\Longleftarrow \quad x^2(x^2 - 2x + 2)$

$\Longleftarrow \quad 2x(x^2 - 2x + 2)$

$\Longleftarrow \quad 2(x^2 - 2x + 2)$

Special Products

Next we list some binomial products that should be memorized so that the factoring process in the next section will go smoothly.

Special Binomial Products

Product	Example
$(u + v)(u - v) = u^2 - v^2$	$(x + 4)(x - 4) = x^2 - 4^2$ $= x^2 - 16$
$(u + v)^2 = u^2 + 2uv + v^2$	$(x + 3)^2 = x^2 + 2(x)(3) + 3^2$ $= x^2 + 6x + 9$
$(u - v)^2 = u^2 - 2uv + v^2$	$(3x - 2)^2 = (3x)^2 - 2(3x)(2) + 2^2$ $= 9x^2 - 12x + 4$
$(u + v)^3 = u^3 + 3u^2v + 3uv^2 + v^3$	$(x + 2)^3 = x^3 + 3x^2(2) + 3x(2^2) + 2^3$ $= x^3 + 6x^2 + 12x + 8$
$(u - v)^3 = u^3 - 3u^2v + 3uv^2 - v^3$	$(x - 1)^3 = x^3 - 3x^2(1) + 3x(1^2) - 1^3$ $= x^3 - 3x^2 + 3x - 1$

EXAMPLE 4   *The Product of Conjugates*

Find the product of $(5x + 9)$ and $(5x - 9)$.

SOLUTION

The product of a sum and a difference of the *same* two terms has no middle term and it takes the form $(u + v)(u - v) = u^2 - v^2$. Thus, we have

$$(5x + 9)(5x - 9) = (5x)^2 - 9^2 = 25x^2 - 81.$$

EXAMPLE 5   *Squares and Cubes of Binomials*

Find the following products.

(a) $(6x - 5)^2$

(b) $(3x + 2)^3$

SOLUTION

(a) The square of a binomial has the form

$$(u - v)^2 = u^2 - 2uv + v^2.$$

Thus, we have

$$(6x - 5)^2 = (6x)^2 - 2(6x)(5) + 5^2$$
$$= 36x^2 - 60x + 25.$$

(b) The cube of a binomial has the form

$$(u + v)^3 = u^3 + 3u^2v + 3uv^2 + v^3.$$

Note the *decrease* of powers for u and the *increase* of powers for v. Following this pattern, we have

$$(3x + 2)^3 = (3x)^3 + 3(3x)^2(2) + 3(3x)(2)^2 + 2^3$$
$$= 27x^3 + 54x^2 + 36x + 8.$$

Occasionally the formulas for special products can be extended to cover products of two trinomials, as demonstrated in the following example.

EXAMPLE 6 *The Product of Two Trinomials*

Find the product of $(x + y - 2)$ and $(x + y + 2)$.

SOLUTION

By grouping $x + y$ in parentheses, we have

$$(x + y - 2)(x + y + 2) = [(x + y) - 2][(x + y) + 2]$$
$$= (x + y)^2 - 4$$
$$= x^2 + 2xy + y^2 - 4.$$

WARM UP

Perform the indicated operations.

1. $(7x^2)(6x)$

2. $(10z^3)(-2z^{-1})$

3. $(-3x^2)^3$

4. $-3(x^2)^3$

5. $\dfrac{27z^5}{12z^2}$

6. $\sqrt{24} \cdot \sqrt{2}$

7. $\left(\dfrac{2x}{3}\right)^{-2}$

8. $16^{3/4}$

9. $\dfrac{4}{\sqrt{8}}$

10. $\sqrt[3]{-27x^3}$

EXERCISES 1.4

In Exercises 1–6, find the degree and leading coefficient of the given polynomial.

1. $2x^2 - x + 1$

2. $-3x^4 + 2x^2 - 5$

3. $x^5 - 1$

4. 3

5. $4x^5 + 6x^4 - 3x^3 + 10x^2 - x - 1$

6. $2x$

In Exercises 7–20, perform the indicated operations and/or simplify.

7. $(6x + 5) - (8x + 15)$

8. $(2x^2 + 1) - (x^2 - 2x + 1)$

9. $-(x^3 - 2) + (4x^3 - 2x)$

10. $-(5x^2 - 1) - (-3x^2 + 5)$

11. $(15x^2 - 6) - (-8x^3 - 14x^2 - 17)$

12. $(15x^4 - 18x - 19) - (13x^4 - 5x + 15)$

13. $5z - [3z - (10z + 8)]$

14. $(y^3 + 1) - [(y^2 + 1) + (3y - 7)]$

15. $3x(x^2 - 2x + 1)$

16. $y^2(4y^2 + 2y - 3)$

17. $-5z(3z - 1)$

18. $-4x(3 - 6x^3)$

19. $(-2x)(-3x)(5x + 2)$

20. $(1 - x^3)(4x)$

In Exercises 21–62, find the given product.

21. $(x + 3)(x + 4)$

22. $(x - 5)(x + 10)$

23. $(3x - 5)(2x + 1)$ **24.** $(7x - 2)(4x - 3)$
25. $(x + 6)^2$ **26.** $(3x - 2)^2$
27. $(2x - 5y)^2$ **28.** $(5 - 8x)^2$
29. $[(x - 3) + y]^2$ **30.** $[(x + 1) - y]^2$
31. $(x + 10)(x - 10)$ **32.** $(2x + 3)(2x - 3)$
33. $(x + 2y)(x - 2y)$ **34.** $(2x + 3y)(2x - 3y)$
35. $(m - 3 + n)(m - 3 - n)$
36. $(x + y + 1)(x + y - 1)$
37. $(2r^2 - 5)(2r^2 + 5)$ **38.** $(3a^3 - 4b^2)(3a^3 + 4b^2)$
39. $(x + 1)^3$ **40.** $(x - 2)^3$
41. $(2x - y)^3$ **42.** $(3x + 2y)^3$
43. $(\sqrt{x} + \sqrt{y})(\sqrt{x} - \sqrt{y})$
44. $(5 + 2\sqrt{s})(5 - 2\sqrt{s})$
45. $(4r^3 - 3s^2)^2$ **46.** $(8y + \sqrt{z})^2$
47. $(x^3 - 2x + 1)(x - 5)$ **48.** $(x + 1)(x^2 - 1)$
49. $(x^2 + 9)(x^2 - x - 4)$ **50.** $(x - 2)(x^2 + 2x + 4)$
51. $(x + 3)(x^2 - 3x + 9)$ **52.** $(2x^2 + 3)(4x^4 - 6x^2 + 9)$
53. $(x^2 + 1)(x + 1)(x - 1)$ **54.** $(x^2 + x - 2)(x^2 - x + 2)$
55. $(x^2 + 2x + 5)(2x^2 - x - 1)$
56. $(x^3 + x + 3)(x^2 + 5x - 4)$
57. $(x + y)(x^2 + y^2)$ **58.** $(x - y + 1)(x + y - 1)$

59. $5z(x + 1) - 3z(2x - 4)$
60. $(2x - y)(x + 3y) + 3(2x - y)$
61. $(x + \sqrt{5})(x - \sqrt{5})(x + 4)$
62. $\sqrt{x}(3\sqrt{x} - 4)$

63. The probability of three successes and two failures in a certain experiment is given by

$$10p^3(1 - p)^2.$$

Find this product.

64. After three years an investment of \$500 compounded annually at an interest rate r will yield an amount

$$500(1 + r)^3.$$

Find this product.

65. The moment of inertia of a cylinder of radius r, length l, and mass m is given by

$$\frac{m}{12}(3r^2 + l^2).$$

Find this product.

66. Determine the degree of the product of two polynomials of degrees m and n.

1.5 Factoring

In Section 1.4 we showed how to multiply polynomials to get new polynomials. In this section, we show how to find factors whose product yields a given polynomial. The process of writing a polynomial as a product is called **factoring.** It is an important tool for solving equations and for reducing fractional expressions.

Unless noted otherwise, we will limit our discussion of factoring to algebraic expressions whose factors have integer coefficients. If a polynomial cannot be factored using integer coefficients, then it is said to be **prime** or **irreducible over the integers.** For instance, the polynomial $x^2 - 3$ is irreducible over the integers. [Over the *real numbers*, this polynomial can be factored as $x^2 - 3 = (x + \sqrt{3})(x - \sqrt{3})$.]

Polynomials with Common Factors

We start with polynomials that can be written as the product of a monomial and another polynomial. The technique used here is the distributive property, $a(b + c) = ab + ac$, in the *reverse* direction.

$$ab + ac = a(b + c) \qquad \text{\textit{a is a common factor}}$$

For instance, the polynomial $27x^4 - 18x^3 + 9x^2$ has $9x^2$ as a factor of each of its terms. Hence, we have

$$27x^4 - 18x^3 + 9x^2 = (9x^2)(3x^2) - (9x^2)(2x) + (9x^2)(1)$$
$$= 9x^2(3x^2 - 2x + 1).$$

EXAMPLE 1 *Removing Common Factors*

Factor each of the following.

(a) $6y^3z - 4yz^2$
(b) $(x + 2)(a + b) + (x + 2)(a - b)$

SOLUTION

(a) $2yz$ is common to both terms, so we write

$$6y^3z - 4yz^2 = (2yz)(3y^2) - (2yz)(2z)$$
$$= 2yz(3y^2 - 2z).$$

(b) In this case, the binomial factor $(x + 2)$ is common to both terms and we write

$$(x + 2)(a + b) + (x + 2)(a - b)$$
$$= (x + 2)[(a + b) + (a - b)] \qquad \text{\textit{Common factor}}$$
$$= (x + 2)[a + b + a - b] \qquad \text{\textit{Remove parentheses}}$$
$$= (x + 2)(2a). \qquad \text{\textit{Combine terms}}$$

It can be difficult to factor a polynomial of degree greater than 2. However, occasionally it is possible using the following special product formulas.

Factoring Special Polynomial Forms

Factored Form	Example
$u^2 - v^2 = (u + v)(u - v)$	$9x^2 - 4 = (3x)^2 - 2^2$
	$\qquad\qquad = (3x + 2)(3x - 2)$
$u^2 + 2uv + v^2 = (u + v)^2$	$x^2 + 6x + 9 = x^2 + 2(x)(3) + 3^2$
	$\qquad\qquad = (x + 3)^2$
$u^2 - 2uv + v^2 = (u - v)^2$	$x^2 - 6x + 9 = x^2 - 2(x)(3) + 3^2$
	$\qquad\qquad = (x - 3)^2$
$u^3 + v^3 = (u + v)(u^2 - uv + v^2)$	$x^3 + 8 = x^3 + 2^3$
	$\qquad\qquad = (x + 2)(x^2 - 2x + 4)$
$u^3 - v^3 = (u - v)(u^2 + uv + v^2)$	$27x^3 - 1 = (3x)^3 - 1^3$
	$\qquad\qquad = (3x - 1)(9x^2 + 3x + 1)$

Factoring

Difference of Two Squares

One of the easiest special polynomial forms to recognize and to factor is the difference of two squares. Think of this form as

$$u^2 \ominus v^2 = (u + v)(u - v).$$

difference   opposite signs

To recognize perfect square terms, look for coefficients that are squares of integers and variables raised to *even* powers.

EXAMPLE 2 Factoring a Polynomial

Factor $3 - 12x^2$.

SOLUTION

First we remove the common factor, 3. We then use the difference of two squares formula with $u = 1$ and $v = 2x$.

$$
\begin{aligned}
3 - 12x^2 &= 3(1 - 4x^2) && \textit{Common factor} \\
&= 3[1^2 - (2x)^2] && \textit{Difference of squares} \\
&= 3(1 + 2x)(1 - 2x)
\end{aligned}
$$

Remark: In Example 2, note that the first step in factoring a polynomial is to check for common factors. Once the common factor is removed, it is often possible to recognize patterns that were not obvious at first glance.

EXAMPLE 3 Factoring the Difference of Two Squares

Factor the following.

(a) $(x + 2)^2 - y^2$ (b) $16x^4 - y^4$

SOLUTION

(a) Using the difference of two squares formula with $u = (x + 2)$ and $v = y$, we have

$$
\begin{aligned}
(x + 2)^2 - y^2 &= [(x + 2) + y][(x + 2) - y] \\
&= (x + y + 2)(x - y + 2).
\end{aligned}
$$

(b) Applying the difference of two squares formula twice, we have

$$
\begin{aligned}
16x^4 - y^4 &= (4x^2)^2 - (y^2)^2 \\
&= (4x^2 + y^2)(4x^2 - y^2) && \textit{First application} \\
&= (4x^2 + y^2)[(2x)^2 - y^2] \\
&= (4x^2 + y^2)(2x + y)(2x - y). && \textit{Second application}
\end{aligned}
$$

Perfect Square Trinomials

A perfect square trinomial is the square of a binomial, and it has the following form.

$$u^2 + 2uv + v^2 = (u + v)^2 \quad \text{or} \quad u^2 - 2uv + v^2 = (u - v)^2$$

same sign   same sign

Note that the first and last terms are squares and the middle term is twice the product of u and v.

EXAMPLE 4 Factoring Perfect Square Trinomials

Factor the following.

(a) $16x^2 + 8x + 1$ (b) $9x^2 - 6xy + y^2$

SOLUTION

(a) Using the formula $u^2 + 2uv + v^2 = (u + v)^2$, we obtain

$$16x^2 + 8x + 1 = (4x)^2 + 2(4x)(1) + 1^2$$
$$= (4x + 1)^2$$

(b) $9x^2 - 6xy + y^2 = (3x)^2 - 2(3x)(y) + y^2 = (3x - y)^2$

Sum or Difference of Cubes

The next two formulas show that sums and differences of cubes factor easily. Pay special attention to the signs of the terms.

like signs   like signs

$$u^3 + v^3 = (u + v)(u^2 - uv + v^2) \qquad u^3 - v^3 = (u - v)(u^2 + uv + v^2)$$

unlike signs   unlike signs

EXAMPLE 5 Factoring the Sum or Difference of Cubes

Factor the following.

(a) $y^3 - 27x^3$ (b) $(x - 2)^3 + 8$

SOLUTION

(a) Consider $u = y$ and $v = 3x$. Then

$$y^3 - 27x^3 = y^3 - (3x)^3$$
$$= (y - 3x)[y^2 + y(3x) + (3x)^2]$$
$$= (y - 3x)(y^2 + 3xy + 9x^2).$$

(b) Consider $u = (x - 2)$ and $v = 2$. Then

$$(x - 2)^3 + 8 = (x - 2)^3 + 2^3$$
$$= [(x - 2) + 2][(x - 2)^2 - (x - 2)(2) + 2^2]$$
$$= (x)(x^2 - 4x + 4 - 2x + 4 + 4)$$
$$= x(x^2 - 6x + 12).$$

Trinomials with Binomial Factors

Some trinomials that are not perfect squares can be factored into the product of two binomials according to the following pattern.

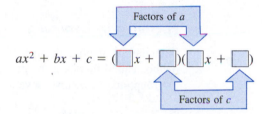

The goal is to find a combination of factors of a and factors of c so that the outer and inner products yield the middle term bx.

EXAMPLE 6   Factoring a Trinomial

Factor the trinomial $x^2 - 7x + 12$.

SOLUTION

From $ax^2 + bx + c = x^2 - 7x + 12$, we see that $a = 1$, $b = -7$, and $c = 12$. Since b is negative and c is positive, both factors of 12 must be negative. That is, $12 = (-2)(-6)$, $12 = (-1)(-12)$, or $12 = (-3)(-4)$. Therefore, the possible factorizations of $x^2 - 7x + 12$ are

$$(x - 2)(x - 6), \quad (x - 1)(x - 12), \quad \text{and} \quad (x - 3)(x - 4).$$

Testing the middle term, we find the correct factorization to be

$$x^2 - 7x + 12 = (x - 3)(x - 4).$$

EXAMPLE 7 *Factoring a Trinomial*

Factor the trinomial $2x^2 + x - 15$.

SOLUTION

For this trinomial, we have $ax^2 + bx + c = 2x^2 + x - 15$. Hence, $a = 2$ and $c = -15$, which means that the factors of -15 must have unlike signs. The eight possible factorizations are

$$(2x - 1)(x + 15) \qquad (2x + 1)(x - 15)$$
$$(2x - 3)(x + 5) \qquad (2x + 3)(x - 5)$$
$$(2x - 5)(x + 3) \qquad (2x + 5)(x - 3)$$
$$(2x - 15)(x + 1) \qquad (2x + 15)(x - 1).$$

Testing the middle term, we find the correct factorization to be

$$2x^2 + x - 15 = (2x - 5)(x + 3).$$

Factoring by Grouping

Sometimes polynomials with more than three terms can be factored by a method called **factoring by grouping.** It is not always obvious which terms to group, and sometimes several different groupings will work.

EXAMPLE 8 *Factoring by Grouping*

Factor $x^3 - 2x^2 - 3x + 6$.

SOLUTION

By grouping this polynomial as $(x^3 - 2x^2) - (3x - 6)$, we obtain the following factorization.

$$
\begin{aligned}
x^3 - 2x^2 - 3x + 6 &= (x^3 - 2x^2) - (3x - 6) &&\textit{Group terms} \\
&= x^2(x - 2) - 3(x - 2) &&\textit{Factor groups} \\
&= (x - 2)(x^2 - 3) &&\textit{Common factor}
\end{aligned}
$$

A general guideline for factoring polynomials is to consider, in order, the following factorizations.

1. Factor out any common factors.
2. Factor according to one of the special polynomial forms.

3. Factor as $ax^2 + bx + c = (mx + r)(nx + s)$.

4. Factor by grouping.

We conclude this section with an example of factoring a trinomial $ax^2 + bx + c$ by grouping. The goal is to factor the product ac into two factors whose sum is b. For instance, in $x^2 - 7x + 12 = (x - 3)(x - 4)$, the product ac is

$$(1)(12) = ac = 12 = (-3)(-4)$$

where $(-3) + (-4) = -7 = b$.

EXAMPLE 9   *Factoring a Trinomial by Grouping*

Factor $2x^2 + x - 6$.

SOLUTION

In the form $ax^2 + bx + c$, we have $ac = 2(-6) = -12$ and $b = 1$. Since $-12 = 4(-3)$, where $4 + (-3) = 1$, we *rewrite the middle term as* $x = 4x - 3x$, and obtain

$$2x^2 + x - 6 = 2x^2 + 4x - 3x - 6.$$

Using factoring by grouping, we obtain

$$
\begin{aligned}
2x^2 + x - 6 &= (2x^2 + 4x) - (3x + 6) \\
&= 2x(x + 2) - 3(x + 2) \\
&= (x + 2)(2x - 3).
\end{aligned}
$$

WARM UP

Find the products.

1. $3x(5x - 2)$
2. $-2y(y + 1)$
3. $(2x + 3)^2$
4. $(3x - 8)^2$
5. $(2x - 3)(x + 8)$
6. $(4 - 5z)(1 + z)$
7. $(2y + 1)(2y - 1)$
8. $(x + a)(x - a)$
9. $(x + 4)^3$
10. $(2x - 3)^3$

EXERCISES 1.5

In Exercises 1–6, remove the common factor.

1. $3x + 6$

2. $5y - 30$

3. $2x^3 - 6x$

4. $4x^3 - 6x^2 + 12x$

5. $(x - 1)^2 + 6(x - 1)$

6. $(3x - 5)(x + 2) + (3x - 5)(1 - x)$

In Exercises 7–12, factor the given difference of two squares.

7. $x^2 - 36$

8. $x^2 - \frac{1}{4}$

9. $16y^2 - 9$

10. $49 - 9y^2$

11. $(x - 1)^2 - 4$

12. $25 - (z + 5)^2$

In Exercises 13–18, factor the given perfect square trinomial.

13. $x^2 - 4x + 4$

14. $x^2 + 10x + 25$

15. $4t^2 + 4t + 1$

16. $9x^2 - 12x + 4$

17. $25y^2 - 10y + 1$

18. $z^2 + z + \frac{1}{4}$

In Exercises 19–32, factor the given trinomial.

19. $x^2 + x - 2$

20. $x^2 + 5x + 6$

21. $s^2 - 5s + 6$

22. $t^2 - t - 6$

23. $y^2 + y - 20$

24. $z^2 - 5z - 24$

25. $x^2 - 30x + 200$

26. $x^2 - 13x + 42$

27. $3x^2 - 5x + 2$

28. $2x^2 - x - 1$

29. $9z^2 - 3z - 2$

30. $12x^2 + 7x + 1$

31. $5x^2 + 26x + 5$

32. $5u^2 + 13u - 6$

In Exercises 33–38, factor the given sum or difference of cubes.

33. $x^3 - 8$

34. $x^3 - 27$

35. $y^3 + 64$

36. $z^3 + 125$

37. $8t^3 - 1$

38. $27x^3 + 8$

In Exercises 39–44, factor by grouping.

39. $x^3 - x^2 + 2x - 2$

40. $x^3 + 5x^2 - 5x - 25$

41. $2x^3 - x^2 - 6x + 3$

42. $5x^3 - 10x^2 + 3x - 6$

43. $6 + 2x - 3x^3 - x^4$

44. $x^5 + 2x^3 + x^2 + 2$

In Exercises 45–70, completely factor the given expression.

45. $x^3 - 4x^2$

46. $6x^2 - 54$

47. $x^2 - 2x + 1$

48. $16 + 6x - x^2$

49. $1 - 4x + 4x^2$

50. $9x^2 - 6x + 1$

51. $-2x^2 - 4x + 2x^3$

52. $2ay^3 - 7ay^2 - 15ay$

53. $9x^2 + 10x + 1$

54. $13x + 6 + 5x^2$

55. $4x(2x - 1) + (2x - 1)^2$

56. $5(3 - 4x)^2 - 8(3 - 4x)(5x - 1)$

57. $2(x + 1)(x - 3)^2 - 3(x + 1)^2(x - 3)$

58. $7(3x + 2)^2(1 - x)^2 + (3x + 2)(1 - x)^3$

59. $3x^3 + x^2 + 15x + 5$

60. $5 - x + 5x^2 - x^3$

61. $x^4 - 4x^3 + x^2 - 4x$

62. $3u - 2v + 6 - uv$

63. $25 - (z + 5)^2$

64. $(t - 1)^2 - 49$

65. $(x^2 + 1)^2 - 4x^2$

66. $(x^2 + 8)^2 - 36x^2$

67. $2t^3 - 16$

68. $(a + 1)^3 + 1$

69. $x^6 - 64$

70. $x^6 + 64$

In Exercises 71 and 72, factor the given expression by using the formula

$$(x + a)^3 = x^3 + 3x^2a + 3xa^2 + a^3.$$

71. $x^3 + 6x^2 + 12x + 8$

72. $27y^3 + 27y^2 + 9y + 1$

In Exercises 73 and 74, factor the given expression by using the formula

$$(x + a)^5 = x^5 + 5x^4a + 10x^3a^2 + 10x^2a^3 + 5xa^4 + a^5.$$

73. $x^5 - 10x^4 + 40x^3 - 80x^2 + 80x - 32$

74. $32x^5 + 80x^4 + 80x^3 + 40x^2 + 10x + 1$

1.6 *Fractional Expressions*

Quotients of algebraic expressions are called **fractional expressions.** Moreover, quotients of two *polynomials* such as

$$\frac{3x - 7}{5x^2 - 2x + 1}, \quad \frac{4}{9xy^2}, \quad \text{and} \quad \frac{x^3 - 4x}{x^2 + x - 2}$$

Fractional Expressions

are called **rational expressions.** The domain of a rational expression is the set of all real numbers for which the denominator is not zero.

The rules for operating with fractional expressions are like those for numerical fractions (see the properties of fractions in Section 1.1). Recall that a fraction is reduced to lowest terms if its numerator and denominator have no factors in common aside from ± 1. To reduce fractions, we apply the following **Cancellation Law.**

$$\frac{a \cdot b}{c \cdot b} = \frac{a}{c}, \qquad b \neq 0 \text{ and } c \neq 0$$

The key to success in simplifying rational expressions lies in your ability to *factor* polynomials. For example,

$$\frac{18xy^5}{45x^2y^2} = \frac{(9xy^2)(2y^3)}{(9xy^2)(5x)} = \frac{2y^3}{5x}.$$

EXAMPLE 1 *Reducing to Lowest Terms*

Reduce the following fraction to lowest terms.

$$\frac{x^2 + 4x - 12}{3x - 6}$$

SOLUTION

Factoring both numerator and denominator and then reducing, we have

$$\frac{x^2 + 4x - 12}{3x - 6} = \frac{(x + 6)(x - 2)}{3(x - 2)} \qquad \textit{Factor completely}$$

$$= \frac{x + 6}{3}. \qquad \textit{Reduce}$$

Remark: In Example 1, do not make the common mistake of trying to further reduce by canceling *terms*.

$$\frac{x + 6}{3} \qquad \text{DOES NOT EQUAL} \qquad \frac{x + 6^2}{3} \text{ or } x + 2.$$

Remember that to reduce fractions, we cancel *factors*, not terms.

When simplifying rational expressions, be sure to factor each polynomial completely before concluding that the numerator and denominator have no factors in common. Moreover, changing the sign of a factor may allow further reduction, as seen in part (b) of the next example.

Review of Fundamental Concepts of Algebra

EXAMPLE 2 *Simplifying Rational Expressions*

Reduce to lowest terms.

(a) $\dfrac{x^3 - 4x}{x^2 + x - 2}$

(b) $\dfrac{12 + x - x^2}{2x^2 - 9x + 4}$

SOLUTION

(a) $\dfrac{x^3 - 4x}{x^2 + x - 2} = \dfrac{x(x^2 - 4)}{(x + 2)(x - 1)}$

$= \dfrac{x(x + 2)(x - 2)}{(x + 2)(x - 1)}$ *Factor completely*

$= \dfrac{x(x - 2)}{x - 1}$ *Cancel common factors*

(b) $\dfrac{12 + x - x^2}{2x^2 - 9x + 4} = \dfrac{(4 - x)(3 + x)}{(2x - 1)(x - 4)}$ *Factor completely*

$= \dfrac{-(x - 4)(3 + x)}{(2x - 1)(x - 4)}$ $(4 - x) = -1(x - 4)$

$= -\dfrac{3 + x}{2x - 1}$ *Reduce*

To multiply and divide rational expressions, we again use the properties of fractions along with the simplification skills just demonstrated. Recall that to divide fractions we invert the divisor and multiply.

EXAMPLE 3 *Multiplying Rational Expressions*

Multiply the following rational expressions and simplify.

$$\dfrac{2x^2 + x - 6}{x^2 + 4x - 5} \cdot \dfrac{x^3 - 3x^2 + 2x}{4x^2 - 6x}$$

SOLUTION

$$\dfrac{2x^2 + x - 6}{x^2 + 4x - 5} \cdot \dfrac{x^3 - 3x^2 + 2x}{4x^2 - 6x}$$

$= \dfrac{(2x - 3)(x + 2)}{(x + 5)(x - 1)} \cdot \dfrac{x(x - 2)(x - 1)}{2x(2x - 3)}$ *Factor and reduce*

$= \dfrac{(x + 2)(x - 2)}{2(x + 5)}$

Fractional Expressions

EXAMPLE 4 *Dividing Rational Expressions*

Perform the indicated division and simplify.

$$\frac{x^3 - 8}{x^2 - 4} \div \frac{x^2 + 2x + 4}{x^3 + 8}$$

SOLUTION

After factoring numerators and denominators, we invert the second fraction and multiply the result by the first fraction.

$$\frac{x^3 - 8}{x^2 - 4} \div \frac{x^2 + 2x + 4}{x^3 + 8} = \frac{(x - 2)(x^2 + 2x + 4)}{(x + 2)(x - 2)} \cdot \frac{(x + 2)(x^2 - 2x + 4)}{x^2 + 2x + 4}$$

$$= x^2 - 2x + 4$$

Combining Rational Expressions

To add or subtract rational expressions, we use the familiar LCD (lowest common denominator) method or the basic definition

$$\frac{a}{b} \pm \frac{c}{d} = \frac{ad \pm bc}{bd}, \quad b \neq 0 \text{ and } d \neq 0.$$

This definition provides an efficient way of adding or subtracting two fractions that have no common factors in their denominators.

EXAMPLE 5 *Subtracting Rational Expressions*

Perform the indicated subtraction and simplify.

$$\frac{x}{x - 3} - \frac{2}{3x + 4}$$

SOLUTION

For these two fractions, we use the basic definition to obtain

$$\frac{x}{x - 3} - \frac{2}{3x + 4} = \frac{x(3x + 4) - 2(x - 3)}{(x - 3)(3x + 4)}$$

$$= \frac{3x^2 + 4x - 2x + 6}{(x - 3)(3x + 4)} \quad \text{\textit{Remove parentheses}}$$

$$= \frac{3x^2 + 2x + 6}{(x - 3)(3x + 4)}. \quad \text{\textit{Combine like terms}}$$

For three or more fractions, or for fractions with a repeated factor in the denominator, the LCD method works well. Recall that the LCD of several fractions consists of the product of all prime factors in the denominators, with each factor given the highest power of its occurrence in any denominator.

EXAMPLE 6 *The LCD Method for Combining Rational Expressions*

Perform the given operations and simplify.

$$\frac{3}{x-1} - \frac{2}{x} + \frac{x+3}{x^2-1}$$

SOLUTION

Using the factored denominators $(x-1)$, x, and $(x+1)(x-1)$, we see that the LCD is $x(x+1)(x-1)$. Therefore,

$$\frac{3}{x-1} - \frac{2}{x} + \frac{x+3}{(x+1)(x-1)}$$

$$= \frac{3(x)(x+1)}{x(x+1)(x-1)} - \frac{2(x+1)(x-1)}{x(x+1)(x-1)} + \frac{(x+3)(x)}{x(x+1)(x-1)}$$

$$= \frac{3(x)(x+1) - 2(x+1)(x-1) + (x+3)(x)}{x(x+1)(x-1)}$$

$$= \frac{3x^2 + 3x - 2x^2 + 2 + x^2 + 3x}{x(x+1)(x-1)}$$

$$= \frac{2x^2 + 6x + 2}{x(x+1)(x-1)}$$

$$= \frac{2(x^2 + 3x + 1)}{x(x+1)(x-1)}.$$

EXAMPLE 7 *The LCD Method for Combining Rational Expressions*

Perform the indicated operation and simplify.

$$\frac{x-2}{x^2+6x+9} - \frac{x+2}{2x^2-18}$$

SOLUTION

In this case, the factored denominators

$$(x + 3)^2$$

and

$$2(x + 3)(x - 3)$$

yield an LCD of $2(x + 3)^2(x - 3)$. Therefore,

$$\frac{x - 2}{(x + 3)^2} - \frac{x + 2}{2(x + 3)(x - 3)}$$

$$= \frac{(x - 2)(2)(x - 3)}{2(x + 3)^2(x - 3)} - \frac{(x + 2)(x + 3)}{2(x + 3)^2(x - 3)}$$

$$= \frac{(x - 2)(2)(x - 3) - (x + 2)(x + 3)}{2(x + 3)^2(x - 3)}$$

$$= \frac{2(x^2 - 5x + 6) - (x^2 + 5x + 6)}{2(x + 3)^2(x - 3)}$$

$$= \frac{x^2 - 15x + 6}{2(x + 3)^2(x - 3)}.$$

Remark: Sometimes the numerator of the answer has a factor in common with the denominator. In such cases the answer should be reduced.

Compound Fractions

So far in this section we have limited our operations to simple fractional expressions. Fractional expressions with separate fractions in the numerator, denominator, or both, are called **compound fractions.** Some examples are

$$\frac{\dfrac{2}{x} - 3}{1 - \dfrac{1}{x - 1}}, \qquad \frac{\dfrac{1}{x + h} - \dfrac{1}{x}}{h}, \qquad \text{and} \qquad \frac{\sqrt{4 - x^2} + \dfrac{x^2}{\sqrt{4 - x^2}}}{4 - x^2}$$

A compound fraction can be simplified by first combining both its numerator and its denominator into single fractions, then inverting the denominator and multiplying.

Review of Fundamental Concepts of Algebra

EXAMPLE 8 *Simplifying a Compound Fraction*

Simplify the compound fraction

$$\frac{\dfrac{2}{x} - 3}{1 - \dfrac{1}{x - 1}}$$

SOLUTION

Combining numerator and denominator fractions, we have

$$\frac{\dfrac{2}{x} - 3}{1 - \dfrac{1}{x - 1}} = \frac{\dfrac{2 - 3(x)}{x}}{\dfrac{1(x - 1) - 1}{x - 1}} \qquad \textit{Combine fractions}$$

$$= \frac{\dfrac{2 - 3x}{x}}{\dfrac{x - 2}{x - 1}} \qquad \textit{Simplify}$$

$$= \frac{2 - 3x}{x} \cdot \frac{x - 1}{x - 2} \qquad \textit{Invert denominator}$$

$$= \frac{(2 - 3x)(x - 1)}{x(x - 2)}.$$

Another way to simplify a compound fraction is to multiply each term in its numerator and denominator by the LCD of all fractions in both its numerator and denominator. Each product is then reduced to obtain a single fraction.

EXAMPLE 9 *Simplifying a Compound Fraction by Multiplying by the LCD*

Use the LCD to simplify the compound fraction

$$\frac{\dfrac{1}{x^2} - \dfrac{1}{y^2}}{\dfrac{1}{x} + \dfrac{1}{y}}.$$

SOLUTION

For the four fractions in the numerator and denominator, the LCD is x^2y^2. Multiplying each term of the numerator and denominator by this LCD yields

$$\frac{\left(\dfrac{1}{x^2} - \dfrac{1}{y^2}\right)x^2y^2}{\left(\dfrac{1}{x} + \dfrac{1}{y}\right)x^2y^2} = \frac{\left(\dfrac{1}{x^2}\right)x^2y^2 - \left(\dfrac{1}{y^2}\right)x^2y^2}{\left(\dfrac{1}{x}\right)x^2y^2 + \left(\dfrac{1}{y}\right)x^2y^2}$$

$$= \frac{y^2 - x^2}{xy^2 + x^2y}$$

$$= \frac{(y - x)(y + x)}{xy(y + x)}$$

$$= \frac{y - x}{xy}.$$

The next three examples illustrate some methods for simplifying expressions involving radicals and negative exponents. (These types of expressions occur frequently in calculus.)

EXAMPLE 10 *Simplifying Expressions with Negative Exponents*

Simplify the expression

$$x(1 - 2x)^{-3/2} + (1 - 2x)^{-1/2}.$$

SOLUTION

By rewriting the given expression with positive exponents, we obtain

$$\frac{x}{(1 - 2x)^{3/2}} + \frac{1}{(1 - 2x)^{1/2}}$$

which could then be combined by the LCD method. However, by first removing the common factor with the *smaller exponent*, the process is simpler.

$$x(1 - 2x)^{-3/2} + (1 - 2x)^{-1/2} = (1 - 2x)^{-3/2}[x + (1 - 2x)^{(-1/2)-(-3/2)}]$$

$$= (1 - 2x)^{-3/2}[x + (1 - 2x)^1]$$

$$= \frac{1 - x}{(1 - 2x)^{3/2}}$$

Note that when factoring, we subtract exponents.

Review of Fundamental Concepts of Algebra

EXAMPLE 11 *Simplifying Compound Fractions*

Simplify

$$\frac{\sqrt{4 - x^2} + \dfrac{x^2}{\sqrt{4 - x^2}}}{4 - x^2}.$$

SOLUTION

$$\frac{\sqrt{4 - x^2} + \dfrac{x^2}{\sqrt{4 - x^2}}}{4 - x^2} = \frac{\sqrt{4 - x^2} + \dfrac{x^2}{\sqrt{4 - x^2}}}{4 - x^2} \cdot \frac{\sqrt{4 - x^2}}{\sqrt{4 - x^2}}$$

$$= \frac{(4 - x^2) + x^2}{(4 - x^2)^{3/2}}$$

$$= \frac{4}{\sqrt{(4 - x^2)^3}}$$

EXAMPLE 12 *Simplifying Fractional Expressions*

Simplify

$$\frac{2x(6x - 4)^{-2/3} - \dfrac{1}{3}(6x - 4)^{1/3}}{x^2}.$$

SOLUTION

Note in this case that

$$(6x - 4)^{-2/3} = \frac{1}{(6x - 4)^{2/3}}.$$

Hence, the LCD for the given numerator is $3(6x - 4)^{2/3}$, and we obtain

$$\frac{2x(6x - 4)^{-2/3} - \dfrac{1}{3}(6x - 4)^{1/3}}{x^2}$$

$$= \frac{[2x(6x - 4)^{-2/3} - \dfrac{1}{3}(6x - 4)^{1/3}]}{x^2} \cdot \frac{(3)(6x - 4)^{2/3}}{(3)(6x - 4)^{2/3}}$$

$$= \frac{6x(1) - (6x - 4)}{3x^2(6x - 4)^{2/3}}$$

$$= \frac{4}{3x^2\sqrt[3]{(6x - 4)^2}}.$$

WARM UP

Completely factor the given polynomials.

1. $5x^2 - 15x^3$
2. $16x^2 - 9$
3. $9x^2 - 6x + 1$
4. $9 + 12y + 4y^2$
5. $z^2 + 4z + 3$
6. $x^2 - 15x + 50$
7. $3 + 8x - 3x^2$
8. $3x^2 - 46x + 15$
9. $s^3 + s^2 - 4s - 4$
10. $y^3 + 64$

EXERCISES 1.6

In Exercises 1–6, find the missing numerator so that the two fractions are equivalent.

1. $\dfrac{5}{2x} = \dfrac{()}{6x^2}$

2. $\dfrac{3}{4} = \dfrac{()}{4(x + 1)}$

3. $\dfrac{x + 1}{x} = \dfrac{()}{x(x - 2)}$

4. $\dfrac{3y - 4}{y + 1} = \dfrac{()}{y^2 - 1}$

5. $\dfrac{3x}{x - 3} = \dfrac{()}{x^2 - x - 6}$

6. $\dfrac{1 - z}{z^2} = \dfrac{()}{z^3 + z^2}$

In Exercises 7–20, reduce the given fraction to lowest terms.

7. $\dfrac{15x^2}{10x}$

8. $\dfrac{18y^2}{60y^5}$

9. $\dfrac{3xy}{xy + x}$

10. $\dfrac{9x^2 + 9xy}{xy + y^2}$

11. $\dfrac{x - 5}{10 - 2x}$

12. $\dfrac{x^2 - 25}{5 - x}$

13. $\dfrac{x^3 + 5x^2 + 6x}{x^2 - 4}$

14. $\dfrac{x^2 + 8x - 20}{x^2 + 11x + 10}$

15. $\dfrac{y^2 - 7y + 12}{y^2 + 3y - 18}$

16. $\dfrac{3 - x}{x^2 - 5x + 6}$

17. $\dfrac{2 - x + 2x^2 - x^3}{x - 2}$

18. $\dfrac{x^2 - 9}{x^3 + x^2 - 9x - 9}$

19. $\dfrac{z^3 - 8}{z^2 + 2z + 4}$

20. $\dfrac{y^3 - 2y^2 - 3y}{y^3 + 1}$

In Exercises 21–46, perform the indicated operations and simplify.

21. $\dfrac{5}{x - 1} \cdot \dfrac{x - 1}{25(x - 2)}$

22. $\dfrac{(x + 5)(x - 3)}{x + 2} \cdot \dfrac{1}{(x + 5)(x + 2)}$

23. $\dfrac{(x - 9)(x + 7)}{x + 1} \cdot \dfrac{x}{9 - x}$

24. $\dfrac{x + 13}{x^3(3 - x)} \cdot \dfrac{x(x - 3)}{5}$

25. $\dfrac{r}{r - 1} \cdot \dfrac{r^2 - 1}{r^2}$

26. $\dfrac{4y - 16}{5y + 15} \cdot \dfrac{2y + 6}{4 - y}$

27. $\dfrac{t^2 - t - 6}{t^2 + 6t + 9} \cdot \dfrac{t + 3}{t^2 - 4}$

28. $\dfrac{y^3 - 8}{2y^3} \cdot \dfrac{4y}{y^2 - 5y + 6}$

29. $\dfrac{x^2 + xy - 2y^2}{x^3 + x^2y} \cdot \dfrac{x}{x^2 + 3xy + 2y^2}$

30. $\dfrac{x^3 - y^3}{x + y} \cdot \dfrac{x^2 + y^2}{x^2 - y^2}$

31. $\dfrac{3(x + y)}{4} \div \dfrac{x + y}{2}$

32. $\dfrac{x + 2}{5(x - 3)} \div \dfrac{x - 2}{5(x - 3)}$

33. $\dfrac{\dfrac{(xy)^2}{(x + y)^2}}{\dfrac{xy}{(x + y)^3}}$

34. $\dfrac{\dfrac{x^2 - y^2}{xy}}{\dfrac{(x - y)^2}{xy}}$

35. $\dfrac{5}{x - 1} + \dfrac{x}{x - 1}$

36. $\dfrac{2x - 1}{x + 3} + \dfrac{1 - x}{x + 3}$

37. $6 - \dfrac{5}{x + 3}$

38. $\dfrac{3}{x - 1} - 5$

39. $\dfrac{3}{x - 2} + \dfrac{5}{2 - x}$

40. $\dfrac{2x}{x - 5} - \dfrac{5}{5 - x}$

41. $\dfrac{2}{x^2 - 4} - \dfrac{1}{x^2 - 3x + 2}$

42. $\dfrac{x}{x^2 + x - 2} - \dfrac{1}{x + 2}$

43. $\dfrac{1}{x^2 - x - 2} - \dfrac{x}{x^2 - 5x + 6}$

44. $\dfrac{x-1}{x^2+5x+4} + \dfrac{2}{x^2-x-2} + \dfrac{10}{x^2+2x-8}$

45. $-\dfrac{1}{x} + \dfrac{2}{x^2+1}$

46. $\dfrac{2}{x+1} + \dfrac{1-x}{x^2-2x+3}$

In Exercises 47–54, simplify the given compound fraction.

47. $\dfrac{\dfrac{x}{y}-1}{x-y}$

48. $\dfrac{x-y}{\dfrac{x}{y}-\dfrac{y}{x}}$

49. $\dfrac{\left(\dfrac{x+3}{x-3}\right)^2}{\dfrac{1}{x+3}+\dfrac{1}{x-3}}$

50. $\dfrac{\dfrac{1}{x}-\dfrac{1}{x+1}}{\dfrac{1}{x+1}}$

51. $\dfrac{\dfrac{5}{y}-\dfrac{6}{2y+1}}{\dfrac{5}{y}+4}$

52. $\dfrac{\dfrac{5}{x^2-x-2}-\dfrac{6}{x^2+x-6}}{\dfrac{4}{x^2+4x+3}}$

53. $\dfrac{\dfrac{1}{(x+h)^2}-\dfrac{1}{x^2}}{h}$

54. $\dfrac{\dfrac{x+h}{x+h+1}-\dfrac{x}{x+1}}{h}$

In Exercises 55–58, rationalize the denominator of the given fractional expression.

55. $\dfrac{3}{\sqrt{x}+1}$

56. $\dfrac{x}{\sqrt{x}-1}$

57. $\dfrac{2x}{\sqrt{x^2-1}}$

58. $\dfrac{100}{3+\sqrt{x}}$

In Exercises 59 and 60, rationalize the numerator of the given fractional expression.

59. $\dfrac{\sqrt{x+2}-\sqrt{x}}{2}$

60. $\dfrac{\sqrt{z-3}-\sqrt{z}}{3}$

In Exercises 61–68, simplify the given compound fraction.

61. $\dfrac{\sqrt{x}-\dfrac{1}{2\sqrt{x}}}{\sqrt{x}}$

62. $\dfrac{\sqrt{1-x^2}+\dfrac{x^2}{\sqrt{1-x^2}}}{1-x^2}$

63. $\dfrac{\dfrac{t^2}{\sqrt{t^2+1}}-\sqrt{t^2+1}}{t^2}$

64. $\dfrac{(2x+1)^{1/3}-\dfrac{4x}{3(2x+1)^{2/3}}}{(2x+1)^{2/3}}$

65. $\dfrac{x(x+1)^{-3/4}-(x+1)^{1/4}}{x^2}$

66. $\dfrac{3x^{1/3}-x^{-2/3}}{3x^{-2/3}}$

67. $\dfrac{-x^3(1-x^2)^{-1/2}-2x(1-x^2)^{1/2}}{x^4}$

68. $\dfrac{x^2(1+x^2)^{1/2}-x^2(1+x^2)^{-1/2}}{1+x^2}$

69. When two resistors are connected in parallel, the total resistance is given by

$$\dfrac{1}{\dfrac{1}{R_1}+\dfrac{1}{R_2}}$$

Simplify this compound fraction.

70. The approximate annual percentage rate of a monthly installment loan is given by

$$\dfrac{\dfrac{24F}{N}}{P+\dfrac{A}{12}}$$

where F is the finance charge, N is the number of monthly payments, P is the amount financed, and A is the total payment. Simplify this compound fraction.

1.7 Algebraic Errors and Some Algebra of Calculus

Algebraic Errors to Avoid

Before we wrap up our review of the fundamental concepts of algebra, let us look at some common algebraic errors. Many of these errors are made because they seem to be the *easiest* things to do.

Errors Involving Parentheses

Potential Error			Correct Form	Comment
$a - (x - b)$	DOES NOT EQUAL	$a - x - b$	$a - (x - b) = a - x + b$	Change all signs when distributing negative through parentheses.
$(a + b)^2$	DOES NOT EQUAL	$a^2 + b^2$	$(a + b)^2 = a^2 + 2ab + b^2$	Remember the middle term when squaring binomials.
$\left(\dfrac{1}{2}a\right)\left(\dfrac{1}{2}b\right)$	DOES NOT EQUAL	$\dfrac{1}{2}(ab)$	$\left(\dfrac{1}{2}a\right)\left(\dfrac{1}{2}b\right) = \dfrac{1}{4}(ab) = \dfrac{ab}{4}$	$\dfrac{1}{2}$ occurs twice as a factor.
$(3x + 6)^2$	DOES NOT EQUAL	$3(x + 2)^2$	$(3x + 6)^2 = [3(x + 2)]^2$ $= 3^2(x + 2)^2$	When factoring, apply exponents to all factors.

Errors Involving Fractions

Potential Error			Correct Form	Comment
$\dfrac{a}{x + b}$	DOES NOT EQUAL	$\dfrac{a}{x} + \dfrac{a}{b}$	Leave as $\dfrac{a}{x + b}$	Do not add denominators when adding fractions.
$\dfrac{\left(\dfrac{x}{a}\right)}{b}$	DOES NOT EQUAL	$\dfrac{bx}{a}$	$\dfrac{\left(\dfrac{x}{a}\right)}{b} = \left(\dfrac{x}{a}\right)\left(\dfrac{1}{b}\right) = \dfrac{x}{ab}$	Multiply by the reciprocal when dividing fractions.
$\dfrac{1}{a} + \dfrac{1}{b}$	DOES NOT EQUAL	$\dfrac{1}{a + b}$	$\dfrac{1}{a} + \dfrac{1}{b} = \dfrac{a + b}{ab}$	Use the definition for adding fractions.
$\dfrac{1}{3x}$	DOES NOT EQUAL	$\dfrac{1}{3}x$	$\dfrac{1}{3x} = \dfrac{1}{3} \cdot \dfrac{1}{x}$	Use the definition for multiplying fractions.
$(1/3)x$	DOES NOT EQUAL	$\dfrac{1}{3x}$	$(1/3)x = \dfrac{1}{3} \cdot x = \dfrac{x}{3}$	Be careful when using a slash to denote division.
$(1/x) + 2$	DOES NOT EQUAL	$\dfrac{1}{(x + 2)}$	$(1/x) + 2 = \dfrac{1}{x} + 2 = \dfrac{1 + 2x}{x}$	Be careful when using a slash to denote division.

Errors Involving Exponents and Radicals

Potential Error			Correct Form	Comment
$(x^2)^3$	DOES NOT EQUAL	x^5	$(x^2)^3 = x^{2 \cdot 3} = x^6$	Multiply exponents when raising an exponential form to a power.
$x^2 \cdot x^3$	DOES NOT EQUAL	x^6	$x^2 \cdot x^3 = x^{2+3} = x^5$	Add exponents when multiplying exponential forms with like bases.
$2x^3$	DOES NOT EQUAL	$(2x)^3$	$2x^3 = 2(x^3)$	Exponents have priority over coefficients.
$\dfrac{1}{x^{1/2} - x^{1/3}}$	DOES NOT EQUAL	$x^{-1/2} - x^{-1/3}$	Leave as $\dfrac{1}{x^{1/2} - x^{1/3}}$.	Do not move term-by-term from denominator to numerator.
$\sqrt{5x}$	DOES NOT EQUAL	$5\sqrt{x}$	$\sqrt{5x} = \sqrt{5}\sqrt{x}$	Radicals apply to every factor inside the radical.
$\sqrt{x^2 + a^2}$	DOES NOT EQUAL	$x + a$	Leave as $\sqrt{x^2 + a^2}$.	Do not apply radicals term-by-term.
$\sqrt{-x^2 + a^2}$	DOES NOT EQUAL	$-\sqrt{x^2 - a^2}$	Leave as $\sqrt{-x^2 + a^2}$ or write as $\sqrt{a^2 - x^2}$.	Do not factor negatives out of square roots.

Errors Involving Cancellation

Potential Error			Correct Form	Comment
$\dfrac{a + bx}{a}$	DOES NOT EQUAL	$1 + bx$	$\dfrac{a + bx}{a} = \dfrac{a}{a} + \dfrac{bx}{a} = 1 + \dfrac{b}{a}x$	Cancel common factors, *not* common terms.
$\dfrac{a + ax}{a}$	DOES NOT EQUAL	$a + x$	$\dfrac{a + ax}{a} = \dfrac{a(1 + x)}{a} = 1 + x$	Factor *before* canceling.
$1 + \dfrac{x}{2x}$	DOES NOT EQUAL	$1 + \dfrac{1}{x}$	$1 + \dfrac{x}{2x} = 1 + \dfrac{1}{2} = \dfrac{3}{2}$	Cancel common factors.

Some Algebra of Calculus

In calculus it is often necessary to take a simplified algebraic expression and "unsimplify" it. See the following list, taken from a standard calculus text.

Unusual Factoring

Expression	Useful Calculus Form of Expression	Comment
$\dfrac{5x^4}{8}$	$\dfrac{5}{8}x^4$	Write with fractional coefficient.
$\dfrac{x^2 + 3x}{-6}$	$-\dfrac{1}{6}(x^2 + 3x)$	Write with fractional coefficient.
$2x^2 - x - 3$	$2\left(x^2 - \dfrac{x}{2} - \dfrac{3}{2}\right)$	Factor out the leading coefficient.
$\dfrac{x}{2}(x + 1)^{-1/2} + (x + 1)^{1/2}$	$\dfrac{(x + 1)^{-1/2}}{2}[x + 2(x + 1)]$	Remove the factor with the negative exponent.

Inserting Factors or Terms

Expression	Useful Calculus Form of Expression	Comment
$(2x - 1)^3$	$\dfrac{1}{2}(2x - 1)^3(2)$	Multiply and divide by 2.
$7x^2(4x^3 - 5)^{1/2}$	$\dfrac{7}{12}(4x^3 - 5)^{1/2}(12x^2)$	Multiply and divide by 12.
$\dfrac{4x^2}{9} - 4y^2 = 1$	$\dfrac{x^2}{9/4} - \dfrac{y^2}{1/4} = 1$	Write with fractional denominators.
$\dfrac{x}{x + 1}$	$\dfrac{x + 1 - 1}{x + 1} = 1 - \dfrac{1}{x + 1}$	Add and subtract the same term.

Writing a Fraction as a Sum

Expression	Useful Calculus Form of Expression	Comment
$\dfrac{x + 2x^2 + 1}{\sqrt{x}}$	$x^{1/2} + 2x^{3/2} + x^{-1/2}$	Divide each term by $x^{1/2}$.
$\dfrac{1 + x}{x^2 + 1}$	$\dfrac{1}{x^2 + 1} + \dfrac{x}{x^2 + 1}$	Rewrite the fraction as the sum of fractions.
$\dfrac{2x}{x^2 + 2x + 1}$	$\dfrac{2x + 2 - 2}{x^2 + 2x + 1}$	Add and subtract a term to the numerator.
	$= \dfrac{2x + 2}{x^2 + 2x + 1} - \dfrac{2}{(x + 1)^2}$	Rewrite the fraction as the difference of fractions.
$\dfrac{x^2 - 2}{x + 1}$	$x - 1 - \dfrac{1}{x + 1}$	Use *long division*. (See Section 4.3.)
$\dfrac{x + 7}{x^2 - x - 6}$	$\dfrac{2}{x - 3} - \dfrac{1}{x + 2}$	Use the method of *partial fractions*. (See Section 5.2.)

Rewriting with Negative Exponents

Expression	Useful Calculus Form of Expression	Comment
$\dfrac{9}{5x^3}$	$\dfrac{9}{5}x^{-3}$	Move the factor to the numerator and change the sign of the exponent.
$\dfrac{7}{\sqrt{2x-3}}$	$7(2x-3)^{-1/2}$	Move the factor to the numerator and change the sign of the exponent.

The next several examples fill in some details for many of the steps given in the preceding tables.

EXAMPLE 1 Rewriting Fractions

Explain the following.

$$\frac{4x^2}{9} - 4y^2 = \frac{x^2}{9/4} - \frac{y^2}{1/4}$$

SOLUTION

To write the expression on the left side of the equation in the form given on the right, we multiply the numerator and denominator of both terms by $1/4$.

$$\frac{4x^2}{9} - 4y^2 = \frac{4x^2}{9}\left(\frac{1/4}{1/4}\right) - 4y^2\left(\frac{1/4}{1/4}\right)$$

$$= \frac{x^2}{9/4} - \frac{y^2}{1/4}$$

EXAMPLE 2 Rewriting with Negative Exponents

Rewrite the expression so that each denominator is free of the variable x.

$$\frac{2}{5x^3} - \frac{1}{\sqrt{x}} + \frac{3}{16x^2}$$

Algebraic Errors and Some Algebra of Calculus

SOLUTION

$$\frac{2}{5x^3} - \frac{1}{\sqrt{x}} + \frac{3}{16x^2} = \frac{2}{5x^3} - \frac{1}{x^{1/2}} + \frac{3}{(4x)^2}$$

$$= \frac{2}{5}x^{-3} - x^{-1/2} + 3(4x)^{-2}$$

EXAMPLE 3   Factors Involving Negative Exponents

Factor $x(x + 1)^{-1/2} + (x + 1)^{1/2}$.

SOLUTION

When multiplying like factors, we add exponents. When we factor we are undoing multiplication, and so we *subtract* exponents. Hence we have

$$x(x + 1)^{-1/2} + (x + 1)^{1/2} = (x + 1)^{-1/2}[x(x + 1)^0 + (x + 1)^1]$$
$$= (x + 1)^{-1/2}[x + (x + 1)]$$
$$= (x + 1)^{-1/2}(2x + 1).$$

EXAMPLE 4   Writing a Fraction as a Sum of Terms

Rewrite the fraction as the sum of three terms.

$$\frac{x + 2x^2 + 1}{\sqrt{x}}$$

SOLUTION

Applying the definition of addition of fractions with like denominators

$$\frac{a}{c} + \frac{b}{c} = \frac{a + b}{c}$$

in reverse, we have

$$\frac{x + 2x^2 + 1}{\sqrt{x}} = \frac{x}{x^{1/2}} + \frac{2x^2}{x^{1/2}} + \frac{1}{x^{1/2}}$$

$$= x^{1/2} + 2x^{3/2} + x^{-1/2}$$

Review of Fundamental Concepts of Algebra

EXAMPLE 5 *Writing a Fraction as a Sum of Terms*

Rewrite the fraction as the sum of two fractions.

$$\frac{2x}{x^2 + 2x + 1}$$

SOLUTION

By adding and subtracting 2 in the numerator, we have

$$\frac{2x}{x^2 + 2x + 1} = \frac{2x + 2 - 2}{x^2 + 2x + 1}$$

$$= \frac{2x + 2}{x^2 + 2x + 1} - \frac{2}{x^2 + 2x + 1}$$

$$= \frac{2x + 2}{(x + 1)^2} - \frac{2}{(x + 1)^2}$$

$$= \frac{2(x + 1)}{(x + 1)^2} - \frac{2}{(x + 1)^2}$$

$$= \frac{2}{x + 1} - \frac{2}{(x + 1)^2}.$$

EXERCISES 1.7

In Exercises 1–24, find and correct any errors.

1. $2x - (3y + 4) = 2x - 3y + 4$

2. $\dfrac{4}{16x - (2x + 1)} = \dfrac{4}{14x + 1}$

3. $5z + 3(x - 2) = 5z + 3x - 2$

4. $x(yz) = (xy)(xz)$

5. $-\dfrac{x - 3}{x - 1} = \dfrac{3 - x}{1 - x}$

6. $\dfrac{x - 1}{(5 - x)(-x)} = \dfrac{1 - x}{x(5 - x)}$

7. $a\left(\dfrac{x}{y}\right) = \dfrac{ax}{ay}$

8. $(5z)(6z) = 30z$

9. $(4x)^2 = 4x^2$

10. $\left(\dfrac{x}{y}\right)^3 = \dfrac{x^3}{y}$

11. $\sqrt{x + 9} = \sqrt{x} + 3$

12. $\sqrt{25 - x^2} = 5 - x$

13. $\dfrac{6x + y}{6x - y} = \dfrac{x + y}{x - y}$

14. $\dfrac{2x^2 + 1}{5x} = \dfrac{2x + 1}{5}$

15. $\dfrac{1}{x + y^{-1}} = \dfrac{y}{x + 1}$

16. $\dfrac{1}{a^{-1} + b^{-1}} = \left(\dfrac{1}{a + b}\right)^{-1}$

17. $x(2x - 1)^2 = (2x^2 - x)^2$

18. $x(x + 5)^{1/2} = (x^2 + 5x)^{1/2}$

19. $\sqrt[3]{x^3 + 7x^2} = x^2\sqrt[3]{x + 7}$

20. $(3x^2 - 6x)^3 = 3x(x - 2)^3$

21. $\dfrac{3}{x} + \dfrac{4}{y} = \dfrac{7}{x + y}$

22. $\dfrac{7 + 5(x + 3)}{x + 3} = 12$

23. $\dfrac{1}{2y} = (1/2)y$

24. $\dfrac{2x + 3x^2}{4x} = \dfrac{2 + 3x^2}{4}$

In Exercises 25–52, insert the required factor in the parentheses.

25. $\dfrac{3x + 2}{5} = \dfrac{1}{5}(\rule{1.5cm}{0.4pt})$

26. $\dfrac{7x^2}{10} = \dfrac{7}{10}(\rule{1.5cm}{0.4pt})$

27. $\frac{2}{3}x^2 + \frac{1}{3}x + 5 = \frac{1}{3}(\rule{1.5cm}{0.4pt})$

28. $\frac{3}{4}x + \frac{1}{2} = \frac{1}{4}(\rule{1.5cm}{0.4pt})$

29. $\frac{1}{3}x^3 + 5 = (\rule{1.5cm}{0.4pt})(x^3 + 15)$

30. $\frac{5}{2}z^2 - \frac{1}{4}z + 2 = (\rule{1.5cm}{0.4pt})(10z^2 - z + 8)$

31. $x(2x^2 + 15) = (\rule{1.5cm}{0.4pt})(2x^2 + 15)(2x)$

32. $x^2(x^3 - 1)^4 = ($ ___ $)(x^3 - 1)^4(3x^2)$

33. $x(1 - 2x^2)^3 = ($ ___ $)(1 - 2x^2)^3(-4x)$

34. $5x\sqrt[3]{1 + x^2} = ($ ___ $)\sqrt[3]{1 + x^2}(2x)$

35. $\dfrac{1}{\sqrt{x}(1 + \sqrt{x})^2} = ($ ___ $)\dfrac{1}{(1 + \sqrt{x})^2}\left(\dfrac{1}{2\sqrt{x}}\right)$

36. $\dfrac{4x + 6}{(x^2 + 3x + 7)^3} = ($ ___ $)\dfrac{1}{(x^2 + 3x + 7)^3}(2x + 3)$

37. $\dfrac{x + 1}{(x^2 + 2x - 3)^2} = ($ ___ $)\dfrac{1}{(x^2 + 2x - 3)^2}(2x + 2)$

38. $\dfrac{1}{(x - 1)\sqrt{(x - 1)^4 - 4}} = \dfrac{($ ___ $)}{(x - 1)^2\sqrt{(x - 1)^4 - 4}}$

39. $\dfrac{3}{x} + \dfrac{5}{2x^2} - \dfrac{3}{2}x = ($ ___ $)(6x + 5 - 3x^3)$

40. $\dfrac{(x - 1)^2}{169} + (y + 5)^2 = \dfrac{(x - 1)^3}{169($ ___ $)} + (y + 5)^2$

41. $\dfrac{9x^2}{25} + \dfrac{16y^2}{49} = \dfrac{x^2}{($ ___ $)} + \dfrac{y^2}{($ ___ $)}$

42. $\dfrac{3x^2}{4} - \dfrac{9y^2}{16} = \dfrac{x^2}{($ ___ $)} - \dfrac{y^2}{($ ___ $)}$

43. $\dfrac{x^2}{1/12} - \dfrac{y^2}{2/3} = \dfrac{12x^2}{($ ___ $)} - \dfrac{3y^2}{($ ___ $)}$

44. $\dfrac{x^2}{4/9} + \dfrac{y^2}{7/8} = \dfrac{9x^2}{($ ___ $)} + \dfrac{8y^2}{($ ___ $)}$

45. $\sqrt{x} + (\sqrt{x})^3 = \sqrt{x}($ ___ $)$

46. $x^{1/3} - 5x^{4/3} = x^{1/3}($ ___ $)$

47. $3(2x + 1)x^{1/2} + 4x^{3/2} = x^{1/2}($ ___ $)$

48. $(1 - 3x)^{4/3} - 4x(1 - 3x)^{1/3} = (1 - 3x)^{1/3}($ ___ $)$

49. $\dfrac{x^2}{\sqrt{x^2 + 1}} - \sqrt{x^2 + 1} = \dfrac{1}{\sqrt{x^2 + 1}}($ ___ $)$

50. $\dfrac{1}{2\sqrt{x}} + 5x^{3/2} - 10x^{5/2} = \dfrac{1}{2\sqrt{x}}($ ___ $)$

51. $\dfrac{1}{10}(2x + 1)^{5/2} - \dfrac{1}{6}(2x + 1)^{3/2} = \dfrac{(2x + 1)^{3/2}}{15}($ ___ $)$

52. $\dfrac{3}{7}(t + 1)^{7/3} - \dfrac{3}{4}(t + 1)^{4/3} = \dfrac{3(t + 1)^{4/3}}{28}($ ___ $)$

In Exercises 53–62, write the given fraction as a sum of two or more terms. (See Examples 4 and 5.)

53. $\dfrac{16 - 5x - x^2}{x}$

54. $\dfrac{x^3 - 5x^2 + 4}{x^2}$

55. $\dfrac{4x^3 - 7x^2 + 1}{x^{1/3}}$

56. $\dfrac{2x^5 - 3x^3 + 5x - 1}{x^{3/2}}$

57. $\dfrac{3 - 5x^2 - x^4}{\sqrt{x}}$

58. $\dfrac{x^3 - 5x^4}{3x^2}$

59. $\dfrac{x^2 + 4x + 8}{x^4 + 1}$

60. $\dfrac{3x^2 - 5}{x^3 + 1}$

61. $\dfrac{4x - 11}{x + 3}$

62. $\dfrac{8x^3 + 3}{2x + 5}$

CHAPTER 1 REVIEW EXERCISES

In Exercises 1–20, describe the error and then make the necessary correction.

1. $\dfrac{7}{16} + \dfrac{3}{16} = \dfrac{10}{32}$

2. $\dfrac{15}{32} - \dfrac{21}{32} = \dfrac{-6}{0}$

3. $10(4 \cdot 7) = 40 \cdot 70$

4. $\left(\dfrac{1}{3}x\right)\left(\dfrac{1}{3}y\right) = \dfrac{1}{3}xy$

5. $4\left(\dfrac{3}{7}\right) = \dfrac{12}{28}$

6. $\dfrac{2}{9} \times \dfrac{4}{9} = \dfrac{8}{9}$

7. $\dfrac{15}{16} \div \dfrac{2}{3} = \dfrac{5}{8}$

8. $\dfrac{4}{3} \div 4 = 4\left(\dfrac{3}{4}\right)$

9. $\dfrac{x - 1}{1 - x} = 1$

10. $\dfrac{-3}{-4} = -\dfrac{3}{4}$

11. $2[5 - (3 - 2)] = 2[5 - 3 - 2]$

12. $-3(-x + y) = 3x + 3y$

13. $(2x)^4 = 2x^4$

14. $\left(\dfrac{y}{8}\right)^2 = \dfrac{y^2}{8}$

15. $(5 + 8)^2 = 5^2 + 8^2$

16. $(-x)^6 = -x^6$

17. $(3^4)^4 = 3^8$

18. $\sqrt{10x} = 10\sqrt{x}$

19. $\sqrt{3^2 + 4^2} = 3 + 4$

20. $\sqrt{7x}\sqrt[3]{2} = \sqrt{14x}$

In Exercises 21–36, perform the indicated operations.

21. $-10(7 - 5)$

22. $-10(7)(-5)$

23. $-|16 - 5|$

24. $|5 - 16|$

25. $|-3| + 4(-2) - 6$

26. $(16 - 8) \div 4$

27. $\sqrt{5} \cdot \sqrt{125}$

28. $\dfrac{\sqrt{72}}{\sqrt{2}}$

29. $6[4 - 2(6 + 8)]$

30. $-4[16 - 3(7 - 10)]$

31. $\left(\dfrac{3^2}{5^2}\right)^{-3}$

32. $6^{-4}(-3)^5$

33. $2(-5)^2$

34. $\left(\dfrac{25}{16}\right)^{-1/2}$

35. $(3 \times 10^4)^2$

36. $\dfrac{1}{(4 \times 10^{-2})^3}$

Review of Fundamental Concepts of Algebra

In Exercises 37–40, use absolute value notation to describe the given statement.

37. The distance between x and 7 is at least 4.

38. The distance between x and 25 is no more than 10.

39. The distance between y and -30 is less than 5.

40. The distance between y and $1/2$ is more than 2.

In Exercises 41–70, perform the indicated operations and/or simplify.

41. $\sqrt{50} - \sqrt{18}$

42. $\sqrt[3]{9} + 10\sqrt[3]{9}$

43. $\sqrt[3]{x}(3 + 4\sqrt[3]{x^2})$

44. $(3\sqrt{5} + 2)(3\sqrt{5} - 2)$

45. $\dfrac{\sqrt{48}}{\sqrt{6}}$

46. $\dfrac{1}{\sqrt{x} - 1}$

47. $(x^2 - 2x + 1)(x^3 - 1)$

48. $(x^3 - 3x)(2x^2 + 3x + 5)$

49. $\left(x^2 - \dfrac{1}{x}\right)(x^2 + 1)$

50. $(t^5 - 3t)\left(\dfrac{1}{t^2} + t\right)$

51. $(y^2 - y)(y^2 + 1)(y^2 + y + 1)$

52. $(3z^3 + 4z)(z - 5)(z + 1)$

53. $\dfrac{x}{x^4 - 1} \cdot \dfrac{x - 1}{x^3}$

54. $\dfrac{4x^2 - 1}{(2x)(x^2 + 2x + 1)} \cdot \dfrac{x + 1}{4x^2 + 4x + 1}$

55. $\dfrac{x^2 - 4}{x^4 - 2x^2 - 8} \cdot \dfrac{x^2 + 2}{x^2}$

56. $\dfrac{x^2 - 1}{x^3 + x} \cdot \dfrac{x^4 - 1}{(x + 1)^2}$

57. $\dfrac{1}{x - 1} - \dfrac{1}{x + 2}$

58. $\dfrac{2}{x} - \dfrac{3}{x - 1} + \dfrac{4}{x + 1}$

59. $x - 1 + \dfrac{1}{x + 2} + \dfrac{1}{x - 1}$

60. $2x + \dfrac{3}{2(x - 4)} - \dfrac{1}{2(x + 2)}$

61. $x + 3 + \dfrac{6}{x - 1} + \dfrac{4}{(x - 1)^2}$

62. $\dfrac{1}{x - 1} + \dfrac{1 - x}{x^2 + x + 1}$

63. $\dfrac{1}{x} - \dfrac{x - 1}{x^2 + 1}$

64. $\dfrac{1}{6(x - 2)} - \dfrac{1}{6(x + 2)} + \dfrac{1}{3(x^2 + 2)}$

65. $\dfrac{1}{x - 2} + \dfrac{1}{(x - 2)^2} + \dfrac{1}{x + 2}$

66. $\dfrac{1}{L}\left(\dfrac{1}{y} - \dfrac{1}{L - y}\right)$, where L is a constant

67. $\dfrac{x^3}{x^3 - 1} \div \dfrac{x^2}{x^2 - 1}$

68. $\dfrac{4x - 6}{(x - 1)^2} \div \dfrac{2x^2 - 3x}{x^2 + 2x - 3}$

69. $\dfrac{\dfrac{x^2(5x - 6)}{2x + 3}}{\dfrac{5x}{2x + 3}}$

70. $\dfrac{\dfrac{x^3 + y^3}{x^2 + y^2}}{x^2 - xy + y^2}$

In Exercises 71–76, simplify the given compound fraction.

71. $\dfrac{\dfrac{1}{x} - \dfrac{1}{y}}{x^2 - y^2}$

72. $\dfrac{\dfrac{1}{x} - \dfrac{1}{y}}{\dfrac{1}{x} + \dfrac{1}{y}}$

73. $\dfrac{\dfrac{3a}{a^2} - 1}{\dfrac{a}{x} - 1}$

74. $\dfrac{\dfrac{1}{2x - 3} - \dfrac{1}{2x + 3}}{\dfrac{1}{2x} - \dfrac{1}{2x + 3}}$

75. $\dfrac{\dfrac{3}{2(x - 4)} - \dfrac{1}{2x}}{x^2 - 3x - 10}$

76. $\dfrac{1 - \dfrac{1}{1 + (1/a)}}{\dfrac{1}{a + 1} - \dfrac{1}{a - 1}}$

In Exercises 77–90, insert the missing factor(s).

77. $x^3 - x = x(\underline{})(\underline{})$

78. $3x^2 + 14x + 8 = (x + 4)(\underline{})$

79. $2x^2 + 21x + 10 = (x + 10)(\underline{})$

80. $x(x - 3) + 4(x - 3) = (x - 3)(\underline{})$

81. $x^3 - 1 = (x - 1)(\underline{})$

82. $x^6 - y^6 = (x - y)(x + y)(\underline{})(\underline{})$

83. $x^4 - 2x^2 + 1 = (x + 1)^2(\underline{})^2$

84. $z^6 + 2z^3 + 1 = (z + 1)^2(\underline{})^2$

85. $\frac{3}{4}x^2 - \frac{5}{6}x + 4 = \frac{1}{12}(\underline{})$

86. $\frac{2}{3}x^4 - \frac{3}{8}x^3 + \frac{5}{6}x^2 = \frac{x^2}{24}(\underline{})$

87. $x^3 - x^2 + 2x - 2 = (x - 1)(\underline{})$

88. $6x^3 + x^2 - 24x - 4 = (x - 2)(\underline{})(\underline{})$

89. $\dfrac{t}{\sqrt{t + 1}} - \sqrt{t + 1} = \dfrac{1}{\sqrt{t + 1}}(\underline{})$

90. $2x(x^2 - 3)^{1/3} - 5(x^2 - 3)^{4/3} = (x^2 - 3)^{1/3}(\underline{})$

91. Calculate 15^4 in two ways. First, use the exponential key $\boxed{y^x}$. Second, enter 15 and press the square key $\boxed{x^2}$ twice. Why do these two methods give the same result?

92. Calculate $\sqrt[5]{107}\sqrt[5]{1145}$ in two ways. First, use the keystroke sequence

107 $\boxed{y^x}$ .2 $\boxed{\times}$ 1145 $\boxed{y^x}$ .2 $\boxed{=}$.

Second, use the sequence

107 $\boxed{\times}$ 1145 $\boxed{=}$ $\boxed{y^x}$ .2 $\boxed{=}$.

Why do these two methods give the same result?

93. Enter any number between 0 and 1 in a calculator and press the square key $\boxed{x^2}$ repeatedly. What number does the calculator display seem to be approaching?

94. Enter any positive number other than 1 in a calculator. What number is approached as the square root key $\boxed{\sqrt{x}}$ is pressed repeatedly?

95. Use a calculator to complete the following table.

n	1	10	10^2	10^4	10^6	10^{10}
$\dfrac{5}{\sqrt{n}}$						

What number is $5/\sqrt{n}$ approaching as n increases without bound?

CHAPTER 2

Algebraic Equations and Inequalities

2.1 Linear Equations

In Chapter 1 we reviewed the fundamentals of algebra. We now want to *use* these fundamentals to solve problems that can be expressed in the form of equations or inequalities. Such problems are common in science, business, industry, and government.

Equations and Solutions of Equations

An **equation** is a statement that two algebraic expressions are equal. Some examples of equations in x are

$$3x - 5 = 7, \qquad x^2 - x - 6 = 0, \qquad \text{and} \qquad \sqrt{2x} = 4.$$

To **solve** an equation in x means that we find all values of x for which the equation is true. Such values are called **solutions.** For instance, $x = 4$ is a solution of the equation $3x - 5 = 7$, because $3(4) - 5 = 7$ is a true statement.

The solutions of an equation depend upon the kinds of numbers being considered. For instance, in the set of rational numbers the equation $x^2 = 10$ has no solution because there is no rational number whose square is 10.

However, in the set of real numbers this equation has the two solutions $\sqrt{10}$ and $-\sqrt{10}$ because $(\sqrt{10})^2 = 10$ and $(-\sqrt{10})^2 = 10$.

An equation that is true for *every* real number in the domain of the variable is called an **identity.** Two examples of identities are

$$x^2 - 9 = (x + 3)(x - 3) \qquad \text{and} \qquad \frac{x}{3x^2} = \frac{1}{3x}, \qquad x \neq 0.$$

The first equation is an identity because it is a true statement for any real value of x. The second equation is an identity because it is true for any nonzero real value of x.

An equation that is true for just *some* (or even none) of the real numbers in the domain of the variable is called a **conditional equation.** For example, the equation $x^2 - 9 = 0$ is conditional because $x = 3$ and $x = -3$ are the only values in the domain that satisfy the equation.

Linear Equations in One Variable

One of the most common types of conditional equations is a **linear equation.**

Definition of Linear Equation

A **linear equation** in one variable x is an equation that can be written in the standard form

$$ax + b = 0$$

where a and b are real numbers with $a \neq 0$.

Linear equations have exactly one solution. To see this, consider the following steps. (Remember that $a \neq 0$.)

$ax + b = 0$	*Given*
$ax = -b$	*Subtract b from both sides*
$x = -\dfrac{b}{a}$	*Divide both sides by a*

Thus, the equation $ax + b = 0$ has exactly one solution, $x = -b/a$.

To solve a conditional equation we generate a sequence of **equivalent** (and usually simpler) equations, each having the same solution(s) as the original equation. The operations that yield equivalent equations come from the Substitution Principle and the simplification techniques studied in Chapter 1.

Generating Equivalent Equations

A given equation is transformed into an *equivalent equation* by one or more of the following steps.

	Given Equation	*Equivalent Equation*
1. Remove symbols of grouping, combine like terms, or reduce fractions on both sides of the equation.	$2x - x = 4$	$x = 4$
2. Add or subtract the same quantity to both sides of the equation.	$x + 1 = 6$	$x = 5$
3. Multiply or divide both sides of the equation by the same nonzero quantity.	$2x = 6$	$x = 3$
4. Interchange the two sides of the equation.	$2 = x$	$x = 2$

EXAMPLE 1 *Solving a Linear Equation*

Solve $3x - 6 = 0$ for x.

SOLUTION

$$3x - 6 = 0 \qquad \textit{Given}$$
$$3x = 6 \qquad \textit{Add 6 to both sides}$$
$$x = 2 \qquad \textit{Divide both sides by 3}$$

When you are solving an equation, it is a good idea to **check each solution** in the *original* equation. For instance, in Example 1, we can check that $x = 2$ is a solution by substituting in the original equation $3x - 6 = 0$ to obtain

$$3(2) - 6 = 6 - 6 = 0. \qquad \textit{Check}$$

EXAMPLE 2 *Solving a Linear Equation*

Solve the equation $6(x - 1) + 4 = 3(7x + 1)$ for x.

SOLUTION

$$6(x - 1) + 4 = 3(7x + 1) \qquad \text{\textit{Given}}$$
$$6x - 6 + 4 = 21x + 3 \qquad \text{\textit{Remove parentheses}}$$
$$6x - 2 = 21x + 3 \qquad \text{\textit{Simplify}}$$
$$-15x = 5 \qquad \text{\textit{Add 2 and subtract 21x}}$$
$$x = -\frac{1}{3} \qquad \text{\textit{Divide by} -15}$$

We can check this solution by substituting in the original equation as follows.

$$6\left(-\frac{1}{3} - 1\right) + 4 \overset{?}{=} 3\left[7\left(-\frac{1}{3}\right) + 1\right]$$

$$6\left(-\frac{4}{3}\right) + 4 \overset{?}{=} 3\left[-\frac{7}{3} + 1\right]$$

$$-\frac{24}{3} + 4 \overset{?}{=} 3\left[-\frac{4}{3}\right]$$

$$-8 + 4 \overset{?}{=} -4$$

$$-4 = -4 \qquad \text{\textit{Solution checks}}$$

Students sometimes tell us that a solution looks easy when we work it out in class, but that they don't see where to begin when trying it alone. Keep in mind that no one—not even great mathematicians—can expect to look at every mathematical problem and immediately know where to begin. Many problems involve some trial and error before a solution is found. To make algebra work for you, you must put in a great deal of time, you must expect to try solution methods that end up not working, and you must learn from both your successes and your failures.

Equations Involving Fractional Expressions

To solve an equation involving fractional expressions, we find the lowest common denominator of all terms in the equation and multiply every term by this LCD. This procedure clears the equation of fractions, as shown in the following two examples.

Algebraic Equations and Inequalities

EXAMPLE 3 *Solving an Equation Involving Fractional Expressions*

Solve the following equation for x.

$$\frac{x}{3} + \frac{3x}{4} = 2$$

SOLUTION

The lowest common denominator is 12, and we have

$$\frac{x}{3} + \frac{3x}{4} = 2 \qquad \textit{Given}$$

$$(12)\frac{x}{3} + (12)\frac{3x}{4} = (12)2 \qquad \textit{Multiply by the LCD}$$

$$4x + 9x = 24 \qquad \textit{Reduce and multiply}$$

$$13x = 24 \qquad \textit{Combine like terms}$$

$$x = \frac{24}{13}. \qquad \textit{Divide by 13}$$

We leave it to you to check that the answer $x = 24/13$ satisfies the original equation.

When multiplying or dividing by a *variable* quantity it is possible to introduce an **extraneous** solution—one that does not satisfy the original equation. These "solutions" may arise when the variable quantity is zero, thus yielding an equation not equivalent to the original. The next example demonstrates the importance of a check when you have multiplied or divided by a variable.

EXAMPLE 4 *An Equation with an Extraneous Solution*

Solve the following equation for x.

$$\frac{1}{x - 2} = \frac{3}{x + 2} - \frac{6x}{x^2 - 4}$$

Linear Equations

SOLUTION

In this case, the LCD is $x^2 - 4$ or $(x + 2)(x - 2)$. Multiplying every term by this LCD and reducing produces

$$\frac{1}{x - 2}(x + 2)(x - 2) = \frac{3}{x + 2}(x + 2)(x - 2) - \frac{6x}{x^2 - 4}(x + 2)(x - 2)$$

$$x + 2 = 3(x - 2) - 6x, \; x \neq \pm 2$$
$$x + 2 = 3x - 6 - 6x$$
$$4x = -8$$
$$x = -2.$$

Since $x = -2$ yields a denominator of zero in the original equation, it is an extraneous solution. Hence, the given equation has *no solution*.

An equation with a *single fraction* on each side can be cleared of denominators by **cross multiplying,** which is equivalent to multiplying by the lowest common denominator and then reducing. For instance, in the following equation

$$\frac{2}{x - 3} = \frac{3}{x + 1}$$

the lowest common denominator is $(x - 3)(x + 1)$. Multiplying both sides of the equation by this LCD produces

$$\frac{2}{x - 3}(x - 3)(x + 1) = \frac{3}{x + 1}(x - 3)(x + 1)$$

$$2(x + 1) = 3(x - 3), \; x \neq -1, 3.$$

By comparing this equation to the original, you will see that the original numerators and denominators have been "cross multiplied." That is, the left numerator was multiplied by the right denominator and the right numerator was multiplied by the left denominator. This procedure is further demonstrated in Example 5.

Algebraic Equations and Inequalities

EXAMPLE 5 *Cross Multiplying to Solve an Equation*

Solve for y in the equation

$$\frac{3y - 2}{2y + 1} = \frac{6y - 9}{4y + 3}.$$

SOLUTION

Since this equation has a single fraction on each side, we can cross multiply to obtain the following.

$$\frac{3y - 2}{2y + 1} = \frac{6y - 9}{4y + 3} \qquad \qquad \textit{Given}$$

$$(3y - 2)(4y + 3) = (6y - 9)(2y + 1), \; y \neq -\frac{1}{2}, \; -\frac{3}{4} \qquad \textit{Cross multiply}$$

$$12y^2 + y - 6 = 12y^2 - 12y - 9 \qquad \qquad \textit{Combine like terms}$$

$$13y = -3 \qquad \qquad \textit{Divide by 13}$$

$$y = -\frac{3}{13}$$

We leave it to you to check that $y = -3/13$ is a solution of the original equation.

Up to this point, we have carefully chosen our examples so that the calculations are simple. This is rather artificial, since real-world problems frequently involve numbers that are not simple integers or fractions. In such cases a calculator is useful.

EXAMPLE 6 *Using a Calculator to Solve an Equation*

Solve for x in the equation

$$\frac{1}{9.38} - \frac{3}{x} = \frac{5}{0.3714}.$$

SOLUTION

Round-off error will be minimized if we solve for x before doing any calculations. In this case the LCD is $(9.38)(0.3714)(x)$ and we have

Exercises

$$\frac{1}{9.38} - \frac{3}{x} = \frac{5}{0.3714}$$

$$(9.38)(0.3714)(x)\left(\frac{1}{9.38} - \frac{3}{x}\right) = (9.38)(0.3714)(x)\left(\frac{5}{0.3714}\right)$$

$$0.3714x - 3(9.38)(0.3714) = (9.38)(5)(x), \; x \neq 0$$

$$0.3714x - 5(9.38)x = 3(9.38)(0.3714)$$

$$[0.3714 - 5(9.38)]x = 3(9.38)(0.3714)$$

$$x = \frac{3(9.38)(0.3714)}{0.3714 - 5(9.38)}$$

$$x \approx -0.225. \qquad \textit{Round to three places}$$

Remark: Because of round-off error, a check of a decimal solution may not yield exactly the same values for both sides of the original equation. The difference, however, will usually be quite small.

WARM UP

Perform the indicated operations and simplify your answers.

1. $(2x - 4) - (5x + 6)$

2. $(3x - 5) + (2x - 7)$

3. $2(x + 1) - (x + 2)$

4. $-3(2x - 4) + 7(x + 2)$

5. $\frac{x}{3} + \frac{x}{5}$

6. $x - \frac{x}{4}$

7. $\frac{1}{x + 1} - \frac{1}{x}$

8. $\frac{2}{x} + \frac{3}{x}$

9. $\frac{4}{x} + \frac{3}{x - 2}$

10. $\frac{1}{x + 1} - \frac{1}{x - 1}$

EXERCISES 2.1

In Exercises 1–8, determine whether the given value of x is a solution of the equation.

1. $5x - 3 = 3x + 5$
   (a) $x = 0$ (b) $x = -5$
   (c) $x = 4$ (d) $x = 10$

2. $7 - 3x = 5x - 17$
   (a) $x = -3$ (b) $x = 0$
   (c) $x = 8$ (d) $x = 3$

* 3. $3x^2 + 2x - 5 = 2x^2 - 2$
   (a) $x = -3$ (b) $x = 1$
   (c) $x = 4$ (d) $x = -5$

4. $5x^3 + 2x - 3 = 4x^3 + 2x - 11$
   (a) $x = 2$ (b) $x = -2$
   (c) $x = 0$ (d) $x = 10$

*A boxed number indicates that a detailed solution can be found in the *Student Solution Guide*.

.

5. $\dfrac{5}{2x} - \dfrac{4}{x} = 3$

    (a) $x = -\frac{1}{2}$ (b) $x = 4$

    (c) $x = 0$ (d) $x = \frac{1}{4}$

6. $3 + \dfrac{1}{x + 2} = 4$

    (a) $x = -1$ (b) $x = -2$

    (c) $x = 0$ (d) $x = 5$

7. $(x + 5)(x - 3) = 20$

    (a) $x = 3$ (b) $x = -5$

    (c) $x = 5$ (d) $x = -7$

8. $\sqrt[3]{x - 8} = 3$

    (a) $x = 2$ (b) $x = -2$

    (c) $x = 35$ (d) $x = 8$

In Exercises 9–48, solve the given equation (if possible) and check your answer.

9. $x + 10 = 15$ **10.** $7 - x = 18$

11. $7 - 2x = 15$ **12.** $7x + 2 = 16$

13. $8x - 5 = 3x + 10$ **14.** $7x + 3 = 3x - 13$

15. $2(x + 5) - 7 = 3(x - 2)$

16. $2(13t - 15) + 3(t - 19) = 0$

17. $6[x - (2x + 3)] = 8 - 5x$

18. $8(x + 2) - 3(2x + 1) = 2(x + 5)$

19. $\dfrac{5x}{4} + \dfrac{1}{2} = x - \dfrac{1}{2}$ **20.** $\dfrac{x}{5} - \dfrac{x}{2} = 3$

21. $\frac{3}{2}(z + 5) - \frac{1}{4}(z + 24) = 0$

22. $\dfrac{3x}{2} + \dfrac{1}{4}(x - 2) = 10$

23. $0.25x + 0.75(10 - x) = 3$

24. $0.60x + 0.40(100 - x) = 50$

25. $x + 8 = 2(x - 2) - x$

26. $3(x + 3) = 5(1 - x) - 1$

27. $\dfrac{100 - 4u}{3} = \dfrac{5u + 6}{4} + 6$

28. $\dfrac{17 + y}{y} + \dfrac{32 + y}{y} = 100$

29. $\dfrac{5x - 4}{5x + 4} = \dfrac{2}{3}$ **30.** $\dfrac{10x + 3}{5x + 6} = \dfrac{1}{2}$

31. $10 - \dfrac{13}{x} = 4 + \dfrac{5}{x}$ **32.** $\dfrac{15}{x} - 4 = \dfrac{6}{x} + 3$

33. $\dfrac{1}{x - 3} + \dfrac{1}{x + 3} = \dfrac{10}{x^2 - 9}$

34. $\dfrac{1}{x - 2} + \dfrac{3}{x + 3} = \dfrac{4}{x^2 + x - 6}$

35. $\dfrac{x}{x + 4} + \dfrac{4}{x + 4} + 2 = 0$

36. $\dfrac{2}{(x - 4)(x - 2)} = \dfrac{1}{x - 4} + \dfrac{2}{x - 2}$

37. $\dfrac{7}{2x + 1} - \dfrac{8x}{2x - 1} = -4$

38. $\dfrac{4}{u - 1} + \dfrac{6}{3u + 1} = \dfrac{15}{3u + 1}$

39. $\dfrac{3}{x(x - 3)} + \dfrac{4}{x} = \dfrac{1}{x - 3}$

40. $3 = 2 + \dfrac{2}{z + 2}$

41. $(x + 2)^2 + 5 = (x + 3)^2$

42. $(x + 1)^2 + 2(x - 2) = (x + 1)(x - 2)$

43. $(x + 2)^2 - x^2 = 4(x + 1)$

44. $4(x + 1) - 3x = x + 5$

45. $(2x + 1)^2 = 4(x^2 + x + 1)$

46. $6x + ax = 2x + 5$

47. $4 - 2(x - 2b) = ax + 3$

48. $5 + ax = 12 - bx$

In Exercises 49–56, determine whether the equation is conditional or an identity.

49. $3x - 10 = 4x$

50. $x^2 + 2(3x - 2) = x^2 + 8(x + 2) - 2x - 20$

51. $x^2 - 8x + 5 = (x - 4)^2 - 11$

52. $4\left[\left(x + \frac{1}{2}\right)^2 - 6\right] = 4x^2 - 4x - 23$

53. $\dfrac{x}{3} + \dfrac{x}{4} = 1$ **54.** $\dfrac{5}{x} + \dfrac{3}{x} = 24$

55. $3 + \dfrac{1}{x + 1} = \dfrac{4x}{x + 1}$

56. $2x(x^2 - 7x + 12) = 2x(x^2 - 4x) - 6(x^2 - 4x)$

In Exercises 57–62, use your calculator to solve the given equation for x and round your answer to three decimal places.

57. $0.275x + 0.725(500 - x) = 300$

58. $2.763 - 4.5(2.1x - 5.1432) = 6.32x + 5$

59. $\dfrac{x}{0.6321} + \dfrac{x}{0.0692} = 1000$

60. $\dfrac{2}{7.398} - \dfrac{4.405}{x} = \dfrac{1}{x}$

61. $(x + 5.62)^2 + 10.83 = (x + 7)^2$

62. $\dfrac{x}{2.625} + \dfrac{x}{4.875} = 1$

In Exercises 63–68, evaluate each expression in two ways. (a) Calculate entirely on your calculator by storing intermediate results, and then round the answer to two decimal places. (b) Round both the numerator and the denominator to two decimal places before dividing, and then round the final answer to two decimal places. Does the second method introduce an additional round-off error?

63. $\dfrac{1 + 0.73205}{1 - 0.73205}$

64. $\dfrac{1 + 0.86603}{1 - 0.86603}$

65. $\dfrac{2 - 1.63254}{(2.58)(0.135)}$

66. $\dfrac{2 + 0.57735}{1 + 0.57735}$

67. $\dfrac{3.33 + \dfrac{1.98}{0.74}}{4 + \dfrac{6.25}{3.15}}$

68. $\dfrac{1.73205 - 1.19195}{3 - (1.73205)(1.19195)}$

2.2 Applications

Real-world applications of algebra are usually given as verbal statements (commonly referred to as **word, story,** or **applied** problems). A truly challenging part of algebra is to learn to *translate* these oral or written descriptions into mathematical statements. The goal of this section is to provide suggestions that will increase your skill in the translation process.

Formulas

A **formula** can be thought of as an algorithm (or recipe) for performing a calculation. For instance, to find the perimeter P of a rectangle of length $L = 15$ inches and width $W = 7$ inches, we can use the formula $P = 2L + 2W$ to obtain

$$P = 2(15) + 2(7) = 30 + 14 = 44 \text{ inches.}$$

The same formula can be used to find the width W, if the perimeter and length are known. To do this, we solve the given formula for W in terms of P and L.

$P = 2L + 2W$	*Given formula*
$P - 2L = 2W$	*Subtract 2L from both sides*
$\dfrac{P - 2L}{2} = W$	*Divide both sides by 2*

Examples 1 and 2 give some additional examples of solving formulas for one variable in terms of other variables.

EXAMPLE 1 *Rewriting a Formula*

(a) Solve for r in $I = Prt$.
(b) Solve for P in $A = P + Prt$.

SOLUTION

(a) This formula gives the interest I corresponding to an initial deposit of P dollars, an annual percentage rate of r, and a time of t years. To solve for r, we have

$$I = Prt \qquad \text{\textit{Given}}$$
$$I = (Pt)r \qquad \text{\textit{Group terms}}$$
$$\frac{I}{Pt} = r. \qquad \text{\textit{Divide by Pt}}$$

(b) This formula gives the balance A of the account described in part (a).

$$A = P + Prt \qquad \text{\textit{Given}}$$
$$A = P(1 + rt) \qquad \text{\textit{Factor}}$$
$$\frac{A}{1 + rt} = P. \qquad \text{\textit{Divide by 1 + rt}}$$

EXAMPLE 2 *Rewriting a Formula*

In business, the break-even point occurs when the total revenue is equal to the total cost ($R = C$). The total revenue is given by $R = px$, and the total cost is given by $C = cx + k$, where x represents the number of units, p the (retail) price per unit, c the manufacturer's cost per unit, and k the fixed cost. Solve for x in the break-even equation

$$px = cx + k.$$

SOLUTION

$$px = cx + k \qquad \text{\textit{Given}}$$
$$px - cx = k \qquad \text{\textit{Collect like terms}}$$
$$x(p - c) = k \qquad \text{\textit{Factor}}$$
$$x = \frac{k}{p - c} \qquad \text{\textit{Divide by p - c}}$$

Applications

The following list contains widely used formulas. Knowing formulas like these will help you translate problems involving area, volume, interest, and distance into mathematical equations.

Common Formulas

	Square	Rectangle	Circle	Triangle
Area:	$A = s^2$	$A = LW$	$A = \pi r^2$	$A = \dfrac{1}{2}bh$
Perimeter:	$P = 4s$	$P = 2L + 2W$	$C = 2\pi r$	

	Cube	Rectangular Solid	Cylinder	Sphere
Volume:	$V = s^3$	$V = LWH$	$V = \pi r^2 h$	$V = \dfrac{4}{3}\pi r^3$

Temperature	Simple Interest	Balance (Savings)	Distance
$F = \dfrac{9}{5}C + 32$	$I = Prt$	$A = P(1 + r)^t$	$d = rt$

EXAMPLE 3 Using a Formula

A cylindrical can has a volume of 300 cubic centimeters (cm³) and a radius of 3 centimeters (cm), as shown in Figure 2.1. Find the height of the can.

SOLUTION

The formula for the *volume of a cylinder* is $V = \pi r^2 h$. Because we are asked to find the height of the can, we solve for h to obtain

$$h = \frac{V}{\pi r^2}.$$

Then, using $V = 300$ cm³ and $r = 3$ cm, we find the height to be

$$h = \frac{300 \text{ cm}^3}{\pi(3 \text{ cm})^2}$$

$$= \frac{300 \text{ cm}^3}{9\pi \text{ cm}^2}$$

$$\approx 10.61 \text{ cm}. \qquad \textit{Round to two decimal places}$$

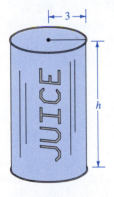

FIGURE 2.1

EXAMPLE 4 Using a Formula

A deposit of $1250 is put in a savings account yielding an annual percentage rate of 12%, compounded once a year. What is the balance in the account at the end of three years?

SOLUTION

Using the formula $A = P(1 + r)^t$ with $P = 1250$, $r = 12\% = 0.12$, and $t = 3$, we find the balance to be

$$A = P(1 + r)^t$$
$$= 1250(1 + 0.12)^3$$
$$\approx 1250(1.40493)$$
$$\approx \$1756.16.$$

Use $\boxed{y^x}$ *key*
Round to nearest cent

Solving Word Problems

There is no single rule to show how to convert word problems into mathematical statements. Most word problems are set up and solved in stages, so do not expect quick, one-step solutions. Here are some general guidelines that we have found to be helpful in solving word problems.

Guidelines for Solving Word Problems

1. Read (and reread) the problem carefully.
2. Write an informal equation or model that represents the given problem. Draw a picture, when appropriate.
3. Label the known and unknown quantities.
4. Rewrite the informal model as a formal algebraic equation.
5. Solve the equation and check the result.

Remark: Step 2 of these guidelines is the key to solving word problems. Unfortunately, it is this step that is most frequently skipped. The tendency is to try to move too quickly to the *formal* algebraic equation (Step 4).

To build an algebraic equation, it is important that you recognize the *vocabulary* of algebraic operations. We have compiled some key words and phrases that should help you identify these operations.

Creating Mathematical Models

Key Words or Phrases	Algebraic Operation	Example
equals equal to is are was will be represents	$=$	The area of a square *is* 12 square feet. *Model:* (area) = 12
sum plus greater increased by added to exceeds total of	$+$	If Eric's salary of \$22,400 is *increased by* 9%, what is his new salary? *Model:* (salary) + 9%(salary) = (new salary)
difference minus less decreased by subtracted from reduced by the remainder	$-$	All dresses in a store are *reduced by* 20%. Find the original price of a dress selling for \$36.76. *Model:* $\left(\begin{array}{c}\text{original}\\\text{price}\end{array}\right) - 20\%\left(\begin{array}{c}\text{original}\\\text{price}\end{array}\right) = \$36.76$
product multiplied by twice times percent of	$\times$	Five hundred dollars represents 25 *percent of* Lori's monthly salary. *Model:* \$500 = 25% $\times$ (monthly salary)
quotient divided by ratio per	$\div$	One number is 16 more than twice the other, and their *ratio* is 6. *Model:* $\left(\begin{array}{c}\text{larger}\\\text{number}\end{array}\right) \div \left(\begin{array}{c}\text{smaller}\\\text{number}\end{array}\right) = 6$

EXAMPLE 5 *Solving a Word Problem*

One number is one-third of another number. The difference of the two numbers is 28. Find the two numbers.

SOLUTION

$$Model: \quad \left(\begin{matrix} \text{larger} \\ \text{number} \end{matrix}\right) - \left(\begin{matrix} \text{smaller} \\ \text{number} \end{matrix}\right) = 28$$

Labels: $x = $ larger number, $\frac{1}{3}x = $ smaller number

Equation: $x - \dfrac{x}{3} = 28$

Solution:	
$3x - x = 28(3)$	*Multiply by LCD*
$2x = 84$	
$x = 42$	*Larger number*
$\dfrac{x}{3} = 14$	*Smaller number*

EXAMPLE 6 *Solving a Word Problem*

A rectangle is twice as long as it is wide, and its perimeter is 132 inches. Find the dimensions of the rectangle.

SOLUTION

In this case, a picture (see Figure 2.2) is appropriate.

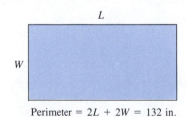

Perimeter = $2L + 2W = 132$ in.

FIGURE 2.2

Model: 2(length) + 2(width) = 132 inches

Labels: $W = $ width, $L = $ length $= 2W$

Equation: $2(2W) + 2W = 132$

Solution:	
$4W + 2W = 132$	
$6W = 132$	
$W = 22$	*Width*
$L = 2W = 44$	*Length*

Several important categories of applied problems involve percentages, discounts, distance, simple interest, coins, and mixtures. You will notice in each of the following examples that the key to the solution is understanding the basic model (informal equation) that ties the algebraic operations together.

Applications

EXAMPLE 7 *Percentages*

(a) Eighty-five dollars is what percent of $250?
(b) Two hundred is 110% of what number?

SOLUTION

(a)   *Model:* $85 = (\text{percentage}) \times (250)$

   Label: $p = \text{percentage}$

   Equation: $85 = (p)(250)$

   Solution: $p = \dfrac{85}{250} = 0.34 = 34\%$

Note that the actual calculation involves the *decimal form* of the percentage p.

(b)   *Model:* $200 = (110\%) \times (\text{number})$

   Labels: $x = \text{number}, \ 110\% = 1.1$

   Equation: $200 = 1.1(x)$

   Solution: $x = \dfrac{200}{1.1} \approx 181.82$ *Round to two decimal places*

EXAMPLE 8 *A Discount Problem*

During a fire sale, all the items are reduced by 45%. What was the original price of a clock radio now priced at $41.25?

SOLUTION

   Model: $(\text{original price}) - (\text{discount}) = \text{sale price}$

   Labels: $x = \text{original price}, \ 0.45x = \text{discount}$

   Equation: $x - 0.45x = 41.25$

   Solution: $0.55x = 41.25$

   $x = \dfrac{41.25}{0.55} = \$75$ *Original price*

Algebraic Equations and Inequalities

In the following distance problems, note the use of the basic formula $d = rt$. That is, (distance) = (rate) × (time). Equivalent forms are $r = d/t$ and $t = d/r$.

EXAMPLE 9 *Distance Problems*

(a) Chad and Pam are running in a 10-kilometer race. Chad runs at 12 kilometers per hour and Pam at 10 kilometers per hour. In how many minutes will they be 2/3 kilometer apart?

(b) Greg traveled 20 miles downstream by boat in the same time it took him to travel 12 miles upstream. If the speed of the boat in still water is 10 miles per hour, what was the speed of the stream?

SOLUTION

(a) *Model:* (Chad's distance) − (Pam's distance) = $\dfrac{2}{3}$

  Labels: Chad: d_1 = (rate)(time) = $(12)(t)$
  Pam: d_2 = (rate)(time) = $(10)(t)$

  Equation: $12t - 10t = \dfrac{2}{3}$

  Solution: $2t = \dfrac{2}{3}$

  $t = \dfrac{1}{3}$ hour = 20 minutes

(b) *Model:* time downstream = time upstream

  $$\frac{\text{distance downstream}}{\text{rate downstream}} = \frac{\text{distance upstream}}{\text{rate upstream}}$$

  Labels: x = rate of stream
  $10 + x$ = rate of boat downstream
  $10 - x$ = rate of boat upstream
  20 = distance downstream
  12 = distance upstream

  Equation: $\dfrac{20}{10 + x} = \dfrac{12}{10 - x}$

  Solution: $20(10 - x) = 12(10 + x), x \neq \pm 10$
  $200 - 20x = 120 + 12x$
  $80 = 32x$
  $x = 2.5$ mph   *Rate of stream*

EXAMPLE 10   A Coin Problem

Kim has saved $21.20 in dimes and quarters. If there are 119 coins in all, how many of each coin has she saved?

SOLUTION

$$Model: \quad \left(\begin{matrix} \text{dime} \\ \text{value} \end{matrix}\right)(\text{number}) + \left(\begin{matrix} \text{quarter} \\ \text{value} \end{matrix}\right)(\text{number}) = \text{total value}$$

Labels: dime value = 0.10
quarter value = 0.25
number of dimes $= x$
number of quarters $= 119 - x$

Equation: $0.10(x) + 0.25(119 - x) = 21.20$

Solution: $0.1x - 0.25x + 29.75 = 21.20$
$-0.15x = -8.55$
$x = 57$ dimes
$119 - x = 62$ quarters

EXAMPLE 11   A Simple Interest Problem

Eric received $10,000 from a trust fund and invested it in two ways. Some is invested at $9\frac{1}{2}\%$ and the rest at 11%. How much is invested at each rate if he receives $1038.50 simple interest per year?

SOLUTION

Simple interest problems are based on the formula $I = rP$, where I is the interest, r is the annual percentage rate, and P is the principal.

Model: $(r_1)(P_1) + (r_2)(P_2) = I$

Labels: $r_1 = 9\frac{1}{2}\% = 0.095,$ $\qquad P_1 = x$
$r_2 = 11\% = 0.11,$ $\qquad P_2 = 10,000 - x$
$I = \text{interest} = 1038.50$

Equation: $0.095(x) + 0.11(10,000 - x) = 1038.50$

Solution: $0.095x - 0.11x + 1100 = 1038.50$
$-0.015x = -61.50$
$x = \$4100$ at $9\frac{1}{2}\%$
$10,000 - x = \$5900$ at 11%

EXAMPLE 12 *A Mixture Problem*

A pharmacist needs to strengthen a 15% alcohol solution so that it contains 32% alcohol. How much pure alcohol should be added to 200 milliliters of the 15% solution? (See Figure 2.3)

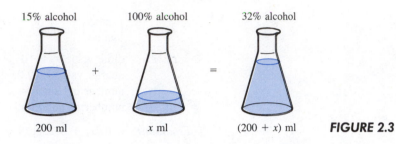

15% alcohol   100% alcohol   32% alcohol

200 ml x ml $(200 + x)$ ml   **FIGURE 2.3**

SOLUTION

Model: $\quad p_1\left(\dfrac{\text{original}}{\text{solution}}\right) + p_2\left(\dfrac{\text{pure}}{\text{alcohol}}\right) = p_3\left(\dfrac{\text{final}}{\text{solution}}\right)$

Labels: original solution: $p_1 = 0.15$, amount $= 200$
        pure alcohol: $p_2 = 1.00$, amount $= x$
        final solution: $p_3 = 0.32$, amount $= 200 + x$

Equation: $\quad 0.15(200) + 1.00(x) = 0.32(200 + x)$

Solution:
$$30 + x = 64 + 0.32x$$
$$0.68x = 34$$
$$x = 50 \text{ ml} \qquad \textit{Pure alcohol}$$

EXAMPLE 13 *A Mixture Problem*

A department store has $30,000 of inventory in 12-inch and 19-inch color TVs. The profit on a 12-inch set is 22%, while the profit on a 19-inch set is 40%. If the profit on the entire stock is 35%, how much was invested in each type of TV?

SOLUTION

Model: $\quad p_1\left(\dfrac{\text{inventory}}{\text{value}}\right) + p_2\left(\dfrac{\text{inventory}}{\text{value}}\right) = p_3\left(\dfrac{\text{total}}{\text{inventory}}\right)$

Labels: 12-inch TV: $p_1 = 0.22$, value $= x$
        19-inch TV: $p_2 = 0.40$, value $= 30{,}000 - x$
   total inventory: $p_3 = 0.35$, value $= 30{,}000$

Equation: $0.22(x) + 0.40(30{,}000 - x) = 0.35(30{,}000)$

Solution: $0.22x - 0.4x + 12{,}000 = 10{,}500$

$-0.18x = -1500$

$x = \$8333.33$ *12-inch models*

$30{,}000 - x = \$21{,}666.67$ *19-inch models*

WARM UP

Solve the given equations (if possible) and check your answers.

1. $3x - 42 = 0$

2. $64 - 16x = 0$

3. $2 - 3x = 14 + x$

4. $7 + 5x = 7x - 1$

5. $5[1 + 2(x + 3)] = 6 - 3(x - 1)$

6. $2 - 5(x - 1) = 2[x + 10(x - 1)]$

7. $\dfrac{x}{3} + \dfrac{x}{2} = \dfrac{1}{3}$

8. $\dfrac{2}{x} + \dfrac{2}{5} = 1$

9. $1 - \dfrac{2}{z} = \dfrac{z}{z + 3}$

10. $\dfrac{x}{x + 1} - \dfrac{1}{2} = \dfrac{4}{3}$

EXERCISES 2.2

In Exercises 1–24, solve for the indicated variable.

1. Area of a Triangle
Solve for h: $A = \frac{1}{2}bh$

2. Perimeter of a Rectangle
Solve for L: $P = 2L + 2W$

3. Volume of a Rectangular Solid
Solve for L: $V = LWH$

4. Volume of a Right Circular Cylinder
Solve for h: $V = \pi r^2 h$

5. Markup
Solve for C: $S = C + RC$

6. Discount
Solve for L: $S = L - RL$

7. Investment at Simple Interest
Solve for R: $A = P + PRT$

8. Investment at Compound Interest
Solve for P: $A = P\left(1 + \dfrac{R}{n}\right)^{nT}$

9. Area of a Trapezoid
Solve for b: $A = \frac{1}{2}(a + b)h$

10. Area of a Sector of a Circle
Solve for θ: $A = \dfrac{\pi r^2 \theta}{360}$

11. Volume of a Spherical Segment
Solve for r: $V = \frac{1}{3}\pi h^2(3r - h)$

12. Volume of an Oblate Spheroid
Solve for b: $V = \frac{4}{3}\pi a^2 b$

13. Thermal Expansion
Solve for α: $L = L_0[1 + \alpha(\Delta t)]$

14. Freely Falling Body
Solve for a: $h = v_0 t + \frac{1}{2}at^2$

15. Newton's Law of Universal Gravitation
Solve for m_2: $F = \alpha\dfrac{m_1 m_2}{r^2}$

16. Heat Flow
Solve for t_1: $H = \dfrac{KA(t_2 - t_1)}{L}$

17. Lensmaker's Equation

Solve for R_1: $\dfrac{1}{f} = (n - 1)\left(\dfrac{1}{R_1} - \dfrac{1}{R_2}\right)$

18. Thermometer Construction

Solve for B_γ: $L = \dfrac{\nu_0}{A_0}(B_m - B_\gamma)t$

19. Arithmetic Progression
Solve for n: $L = a + (n - 1)d$

20. Arithmetic Progression

Solve for a: $S = \dfrac{n}{2}[2a + (n - 1)d]$

21. Geometric Progression

Solve for r: $S = \dfrac{rL - a}{r - 1}$

22. Prismoidal Formula
Solve for S_1: $V = \frac{1}{6}H(S_0 + 4S_1 + S_2)$

23. Capacitance in Series Circuits

Solve for C_1: $C = \dfrac{1}{\dfrac{1}{C_1} + \dfrac{1}{C_2}}$

24. Inductance in Parallel Circuits

Solve for L_3: $L = \dfrac{1}{\dfrac{1}{L_1} + \dfrac{1}{L_2} + \dfrac{1}{L_3}}$

In Exercises 25–30, write an algebraic expression for the given verbal expression.

25. The sum of two consecutive natural numbers.

26. The product of two natural numbers whose sum is 25.

27. The distance traveled in t hours by a car traveling at 50 miles per hour.

28. The amount of acid in x gallons of a 20% acid solution.

29. The perimeter of a rectangle whose width is x and whose length is twice the width.

30. The total cost of producing x units for which the fixed costs are $1200 and the cost per unit is $25.

31. The sum of two consecutive natural numbers is 525. Find the two numbers.

32. Find three consecutive natural numbers whose sum is 804.

33. One number is five times another number. The difference between the two numbers is 148. Find the numbers.

34. One number is one-fifth of another number. The difference between the two numbers is 76. Find the numbers.

35. Find two consecutive integers whose product is 5 less than the square of the smaller number.

36. Find two consecutive natural numbers such that the difference of their reciprocals is one-fourth the reciprocal of the smaller number.

37. Jean was 30 years old when her daughter Ruth was born.
(a) How old will Ruth be when her age is one-third that of Jean's age?
(b) How old will Ruth be when Ruth's and Jean's combined ages total 100?

38. Tom's weekly paycheck is $45 more than Bill's. Their two paychecks total $627. Find the amount of each paycheck.

39. A rectangle is 1.5 times as long as it is wide, and its perimeter is 75 inches (see figure). Find the dimensions of the rectangle.

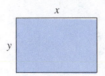

FIGURE FOR 39

40. A picture frame has a total perimeter of 3 feet (see figure). The width of the frame is 0.62 times its height. Find the dimensions of the frame.

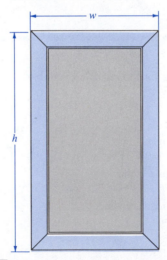

FIGURE FOR 40

41. To get an A in a course a student must have an average of at least 90 on four tests that have 100 points each. A student's scores on the first three tests were 87, 92, and 84. What must the student score on the fourth test to get an A for the course?

42. Repeat Exercise 41, assuming that the fourth test score counts twice as much as each of the other scores.

43. What is 30% of 45? **44.** What is 175% of 360?

45. What is 0.045% of 2,650,000?

46. 432 is what percent of 1600?

47. 459 is what percent of 340?

48. 12 is $\frac{1}{2}$% of what number?

49. 70 is 40% of what number?

50. $825 is 250% of what number?

51. A family has annual loan payments equaling 58.6% of their annual income. During the year, their loan payments total $13,077.75. What is their income?

52. The price of a swimming pool has been discounted 16.5%. The sale price is $849. Find the original list price of the pool.

53. Two cars start at a given point and travel in the same direction at average speeds of 40 miles per hour and 55 miles per hour. How much time must elapse before the two cars are 5 miles apart?

54. Students are traveling in two cars to a football game 135 miles away. The first car leaves on time and travels at an average speed of 45 miles per hour. The second car starts $\frac{1}{2}$ hour later and travels at an average speed of 55 miles per hour. How long will it take the second carload of students to catch up to the first car? Will the second car catch up to the first car before the first car arrives at the game?

55. Two families meet at a park for a picnic. At the end of the day one family travels east at an average speed of 42 miles per hour and the other travels west at an average speed of 50 miles per hour. Both families have approximately 160 miles to travel.
 (a) Find the time it takes each family to get home.
 (b) Find the time that will have elapsed when they are 100 miles apart.
 (c) Find the distance the eastbound family has to travel after the westbound family has arrived home.

56. The driver of a large truck traveled at an average speed of 55 miles per hour on a 200-mile trip to pick up a load of freight. On the return trip (with the truck fully loaded), the average speed was 40 miles per hour. Find the average speed for the round trip.

57. An executive flew in the corporate jet to a meeting in a city 1500 miles away (see figure). After traveling the same amount of time on the return flight, the pilot mentioned that they still had 300 miles to go. If the air speed of the plane was 600 miles per hour, how fast was the wind blowing? (Assume that the wind direction was parallel to the flight path and constant all day.)

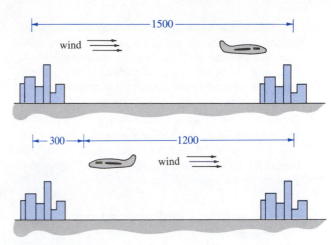

FIGURE FOR 57

58. Light travels at the speed of 3.0×10^8 meters per second. Find the time in minutes required for light to travel from the sun to the earth (a distance of 1.5×10^{11} meters).

59. Radio waves travel at the same speed as light, 3.0×10^8 meters per second. Find the time required for a radio wave to travel from mission control in Houston to NASA astronauts on the surface of the moon 3.86×10^8 meters away.

60. Find the distance to a star that is 50 light years (distance traveled by light in one year) away. Use the fact that light travels at 186,000 miles per second.

61. John has $10.75 in quarters and half dollars. He has 35 coins in all. How many of each coin does John have?

62. Nancy has $175 in $5 and $10 bills. She has a total of 23 bills. How many of each denomination does she have?

63. Denise invests $12,000 in two funds paying $10\frac{1}{2}$% and 13% simple interest. The total annual interest is $1447.50. How much is invested in each fund?

64. Robert invests $25,000 in two funds paying 11% and $12\frac{1}{2}$% simple interest. The total annual interest is $2975.00. How much is invested in each fund?

65. Jack invested $12,000 in a fund paying $9\frac{1}{2}$% simple interest and $8000 in a fund where the interest rate is variable. At the end of the year Jack received notification that his total interest for both funds was $2054.40. Find the equivalent simple interest rate on the variable rate fund.

66. Mary has $10,000 on deposit earning simple interest with the interest rate linked to the *prime rate*. Because of a drop in the prime rate, the rate on Mary's investment dropped by $1\frac{1}{2}$% for the last quarter of the year. Her annual earnings on the fund are $1112.50. Find the interest rate for the first three quarters of the year and the last quarter.

67. A company has fixed costs of $10,000 per month and variable costs of $8.50 per unit manufactured. The company has $85,000 available to cover the monthly costs. How many units can they manufacture? (*Fixed costs* are those that occur regardless of the level of production. *Variable costs* depend on the level of production.)

68. Repeat Exercise 67 for a variable cost of $9.30 per unit.

69. The diameter of the cylindrical portion of a propane gas tank is 4 feet (see figure). The total volume of the tank (including the hemispherical ends) is $152\pi/3$ cubic feet. Find the total length of the tank.

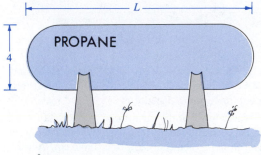

FIGURE FOR 69

70. A trough is 12 feet long, 3 feet deep, and 3 feet wide (see figure). Find the depth of the water when the trough contains 70 gallons. (1 gallon ≈ 0.13368 cubic feet)

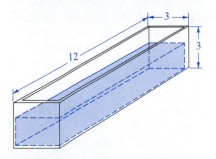

FIGURE FOR 70

71. Using the values from the following table, determine the amounts of Solutions 1 and 2, respectively, needed to obtain the desired amount and concentration of the final mixture.

	Concentration			Amount of Final Solution
	Solution 1	Solution 2	Final Solution	
(a)	10%	30%	25%	100 gal
(b)	25%	50%	30%	5 L
(c)	15%	45%	30%	10 qt
(d)	70%	90%	75%	25 gal

72. A 55-gallon barrel contains a mixture with a concentration of 40%. How much of this mixture must be withdrawn and replaced by 100% concentrate to bring the mixture up to 75% concentration?

73. A farmer mixed gasoline and oil to have 2 gallons of mixture for his two-cycle chain saw engine. This mixture was 32 parts gasoline and 1 part two-cycle oil. How much gasoline must be added to bring the mixture to 40 parts gasoline and 1 part oil?

74. A grocer mixes two kinds of nuts, costing $2.49 per pound and $3.89 per pound, respectively, to make 100 pounds of a mixture costing $3.19 per pound. How much of each kind of nut was put into the mixture?

2.3 *Quadratic Equations*

So far in this chapter we have been working primarily with linear equations. We now concentrate on **quadratic equations.**

Definition of a Quadratic Equation

A **quadratic equation** in x is an equation that can be written in the standard form

$$ax^2 + bx + c = 0$$

where a, b, and c are real numbers with $a \neq 0$.

We will discuss three methods for solving quadratic equations: factoring, completing the square, and the quadratic formula. Discussion of the quadratic formula is postponed until the next section.

Solving Quadratic Equations by Factoring

Solution by factoring is based upon the Factorization Principle given in Section 1.1.

If $ab = 0$, then $a = 0$ or $b = 0$.

To use this principle, we rewrite the standard form of a quadratic equation as the product of two linear factors. Then we find the solutions of the quadratic equation by setting each linear factor equal to zero. For instance, the solutions of the equation $x^2 - 3x - 10 = 0$ are found as follows.

$$
\begin{aligned}
x^2 - 3x - 10 &= 0 && \text{\textit{Standard form}} \\
(x - 5)(x + 2) &= 0 && \text{\textit{Factored form}} \\
x - 5 = 0 \quad &\text{or} \quad x + 2 = 0 && \text{\textit{Set each factor equal to zero}} \\
x = 5 \quad &\text{or} \quad x = -2 && \text{\textit{Solutions}}
\end{aligned}
$$

Be sure you see that the Factorization Principle works *only* for equations written in standard form (in which the right side of the equation is zero). Therefore, all terms must be collected on one side *before* factoring. For instance, in the equation

$$(x - 5)(x + 2) = 8$$

it is *incorrect* to set each factor equal to 8. Can you solve this equation correctly?

Algebraic Equations and Inequalities

EXAMPLE 1 *Solving Quadratic Equations by Factoring*

Solve the following quadratic equations.

(a) $2x^2 + 9x + 7 = 3$

(b) $6x^2 - 3x = 0$

SOLUTION

(a)

$2x^2 + 9x + 7 = 3$	*Given*
$2x^2 + 9x + 4 = 0$	*Standard form*
$(2x + 1)(x + 4) = 0$	*Factored form*
$2x + 1 = 0, \qquad x + 4 = 0$	*Set factors to zero*
$x = -\dfrac{1}{2} \qquad x = -4$	*Solutions*

(b)

$6x^2 - 3x = 0$	*Standard form*
$3x(2x - 1) = 0$	*Factored form*
$3x = 0, \qquad 2x - 1 = 0$	*Set factors to zero*
$x = 0 \qquad\quad x = \dfrac{1}{2}$	*Solutions*

Try to develop the habit of checking your solutions in the original equation. For instance, in Example 1(a), we can check the two solutions as follows.

$$2\left(-\frac{1}{2}\right)^2 + 9\left(-\frac{1}{2}\right) + 7 = \frac{1}{2} - \frac{9}{2} + 7 = 3 \qquad x = -\frac{1}{2} \text{ checks}$$

$$2(-4)^2 + 9(-4) + 7 = 32 - 36 + 7 = 3 \qquad x = -4 \text{ checks}$$

If the two factors of a quadratic expression are identical, then the corresponding solution is called a **double** or **repeated** solution. This occurs in the next example.

EXAMPLE 2 *A Quadratic Equation with a Repeated Solution*

Solve the equation $9x^2 - 6x + 1 = 0$.

SOLUTION

$9x^2 - 6x + 1 = 0$	*Standard form*
$(3x - 1)^2 = 0$	*Factored form*
$3x - 1 = 0, \qquad 3x - 1 = 0$	*Set factors to zero*
$x = \dfrac{1}{3} \qquad\quad x = \dfrac{1}{3}$	*Repeated solution*

The special quadratic form $x^2 = d$, where $d \geq 0$, could be solved by rewriting as $x^2 - d = 0$ and then factoring to obtain

$$(x + \sqrt{d})(x - \sqrt{d}) = 0 \qquad \Longrightarrow \qquad x = \pm\sqrt{d}.$$

However, an easier way to find these two solutions is to simply take the square root of both sides of the equation $x^2 = d$. (Don't forget the "$\pm$" symbol.)

EXAMPLE 3 *Solving the Quadratic Form $x^2 = d$*

Solve the following quadratic equations.

(a) $4x^2 = 12$

(b) $(x - 3)^2 = 7$

SOLUTION

(a) $4x^2 = 12$ *Given*

  $x^2 = 3$ *Divide both sides by 4*

  $x = \pm\sqrt{3}$ *Take square root of both sides*

Note that $x^2 - 3 = 0$ factors as $(x + \sqrt{3})(x - \sqrt{3}) = 0$, which gives the same two solutions.

(b) In this case, an extra step is needed after taking square roots.

  $(x - 3)^2 = 7$ *Given equation*

  $x - 3 = \pm\sqrt{7}$ *Take square root of both sides*

  $x = 3 \pm \sqrt{7}$ *Add 3 to both sides*

Thus, $x = 3 + \sqrt{7}$ or $x = 3 - \sqrt{7}$.

Solving Quadratic Equations by Completing the Square

The equation in Example 3(b) was given in the form $(x - 3)^2 = 7$ so that we could find the solution by taking the square root of both sides. Suppose, however, that the equation $(x - 3)^2 = 7$ had been given in the standard form

  $x^2 - 6x + 2 = 0.$ *Standard form*

This equation is equivalent to the original and thus has the same two solutions, $x = 3 \pm \sqrt{7}$. However, the left side of the equation is not factorable, and we cannot find its solutions unless we can *reverse* the steps shown above. A procedure for doing this incorporates the following process.

Completing the Square

To **complete the square** for the expression

$$x^2 + bx$$

we add $(b/2)^2$, which is the square of half the coefficient of x. Consequently,

$$x^2 + bx + \left(\frac{b}{2}\right)^2 = \left(x + \frac{b}{2}\right)^2.$$

When solving quadratic equations by completing the square, we must add $(b/2)^2$ to *both sides* in order to maintain equality. This is demonstrated in Example 4.

EXAMPLE 4 Completing the Square: Leading Coefficient Is 1

Solve the equation $x^2 - 6x + 2 = 0$ by completing the square. Compare the solutions to those obtained in Example 3(b).

SOLUTION

$$
\begin{array}{ll}
x^2 - 6x + 2 = 0 & \textit{Given} \\
x^2 - 6x = -2 & \textit{Subtract 2 from both sides} \\
x^2 - 6x + 3^2 = -2 + 3^2 & \textit{Add } 3^2 \textit{ to both sides} \\
\qquad \text{(half)}^2 & \\
x^2 - 6x + 9 = 7 & \textit{Simplify} \\
(x - 3)^2 = 7 & \textit{Perfect square trinomial} \\
x - 3 = \pm\sqrt{7} & \textit{Take square root of both sides} \\
x = 3 \pm\sqrt{7} & \textit{Solutions}
\end{array}
$$

If the leading coefficient of a quadratic is not *1*, we must divide both sides of the equation by this coefficient *before* completing the square, as shown in the following example.

EXAMPLE 5 Completing the Square: Leading Coefficient Is Not 1

Solve $3x^2 - 4x - 5 = 0$.

SOLUTION

In this case the quadratic expression is not factorable and the leading coefficient is not 1. Hence, we need to divide both sides by 3.

$$3x^2 - 4x - 5 = 0 \qquad \text{\textit{Given}}$$

$$3x^2 - 4x = 5 \qquad \text{\textit{Add 5 to both sides}}$$

$$x^2 - \frac{4}{3}x = \frac{5}{3} \qquad \text{\textit{Divide both sides by 3}}$$

$$x^2 - \frac{4}{3}x + \left(\frac{2}{3}\right)^2 = \frac{5}{3} + \left(\frac{2}{3}\right)^2 \qquad \text{\textit{Add (2/3)}}^2 \text{ \textit{to both sides}}$$

$$\underbrace{\phantom{x^2 - \frac{4}{3}x}}_{(\text{half})^2}$$

$$\left(x - \frac{2}{3}\right)^2 = \frac{19}{9}$$

$$x - \frac{2}{3} = \pm\frac{\sqrt{19}}{3} \qquad \text{\textit{Take square root of both sides}}$$

$$x = \frac{2}{3} \pm \frac{\sqrt{19}}{3} \qquad \text{\textit{Solutions}}$$

Completing the square has many uses. In the next section we will see how it is used to develop a general formula for solving quadratic equations. In Section 5.4 we will use it to write equations of conics. In the remainder of this section we use it to rewrite algebraic expressions in forms that simplify calculus operations.

When rewriting algebraic expressions by completing the square, we must add *and subtract* the quantity $(b/2)^2$, as demonstrated in the next example.

EXAMPLE 6 *Completing the Square within an Algebraic Expression*

For the following expression, rewrite the denominator as the sum or difference of two squares.

$$\frac{1}{x^2 - 2x - 3}$$

SOLUTION

For this expression, we complete the square of the denominator as follows.

$$\begin{aligned}
x^2 - 2x - 3 &= x^2 - 2x + 1^2 - 3 - 1^2 \qquad &\text{\textit{Add and subtract 1}}^2 \\
&= (x^2 - 2x + 1) - 4 \qquad &\text{\textit{Group terms}} \\
&= (x - 1)^2 - 2^2 \qquad &\text{\textit{Difference of two squares}}
\end{aligned}$$

Therefore, the original expression can be written

$$\frac{1}{x^2 - 2x - 3} = \frac{1}{(x - 1)^2 - 2^2}.$$

Algebraic Equations and Inequalities

Remark: Although the quadratic $x^2 - 2x - 3$ in Example 6 is factorable, some operations in calculus are simpler with the completed square form than with the factored form.

EXAMPLE 7 *Completing the Square within an Algebraic Expression*

Rewrite the expression inside the radical as the sum or difference of two squares.

$$\frac{1}{\sqrt{3x - x^2}}$$

SOLUTION

The negative coefficient for x^2 requires an additional step to complete the square.

$$3x - x^2 = -(x^2 - 3x) \qquad \textit{Factor out} -1$$

$$= -\left[x^2 - 3x + \left(\frac{3}{2}\right)^2 - \left(\frac{3}{2}\right)^2 \right] \qquad \textit{Add and subtract } (3/2)^2$$

$$= -\left[\left(x - \frac{3}{2}\right)^2 - \left(\frac{3}{2}\right)^2 \right] \qquad \textit{Difference of two squares}$$

$$= \left(\frac{3}{2}\right)^2 - \left(x - \frac{3}{2}\right)^2 \qquad \textit{Simplify}$$

Therefore, the original expression can be written

$$\frac{1}{\sqrt{3x - x^2}} = \frac{1}{\sqrt{(3/2)^2 - [x - (3/2)]^2}}.$$

WARM UP

Simplify the given expressions.

1. $\sqrt{\frac{7}{50}}$

2. $\sqrt{32}$

3. $\sqrt{7^2 + 3 \cdot 7^2}$

4. $\sqrt{\frac{1}{4} + \frac{3}{8}}$

Factor the algebraic expressions.

5. $3x^2 + 7x$

6. $4x^2 - 25$

7. $16 - (x - 11)^2$

8. $x^2 + 7x - 18$

9. $10x^2 + 13x - 3$

10. $6x^2 - 73x + 12$

EXERCISES 2.3

In Exercises 1–10, write the equation in standard quadratic form and identify the constants a, b, and c.

1. $2x^2 = 3 - 5x$ **2.** $4x^2 - 2x = 9$

3. $x^2 = 25x$ **4.** $10x^2 = 90$

5. $(x - 3)^2 = 2$ **6.** $12 - 3(x + 7)^2 = 0$

7. $x(x + 2) = 3x^2 + 1$ **8.** $x^2 + 1 = \dfrac{x - 3}{2}$

9. $\dfrac{3x^2 - 10}{5} = 12x$ **10.** $x(x + 5) = 2(x + 5)$

In Exercises 11–20, solve the quadratic equation by factoring.

11. $6x^2 + 3x = 0$ **12.** $9x^2 - 1 = 0$

13. $x^2 - 2x - 8 = 0$ **14.** $x^2 - 10x + 9 = 0$

15. $x^2 + 10x + 25 = 0$ **16.** $16x^2 + 56x + 49 = 0$

17. $3 + 5x - 2x^2 = 0$ **18.** $2x^2 = 19x + 33$

19. $x^2 + 2ax + a^2 = 0$ **20.** $(x + a)^2 - b^2 = 0$

In Exercises 21–30, solve the equation by taking the square root of both sides.

21. $x^2 = 16$ **22.** $x^2 = 144$

23. $x^2 = 7$ **24.** $x^2 = 27$

25. $3x^2 = 36$ **26.** $9x^2 = 25$

27. $(x - 12)^2 = 18$ **28.** $(x + 13)^2 = 21$

29. $(x - 7)^2 = (x + 3)^2$ **30.** $(x + 5)^2 = (x + 4)^2$

In Exercises 31–40, solve the quadratic equation by completing the square.

31. $x^2 - 2x = 0$ **32.** $x^2 + 4x = 0$

33. $x^2 + 4x - 32 = 0$ **34.** $x^2 - 2x - 3 = 0$

35. $x^2 + 6x + 2 = 0$ **36.** $x^2 + 8x + 14 = 0$

37. $9x^2 - 18x + 3 = 0$ **38.** $9x^2 - 12x - 14 = 0$

39. $8 + 4x - x^2 = 0$ **40.** $4x^2 - 4x - 99 = 0$

In Exercises 41–64, solve the quadratic equation.

41. $x^2 = 64$ **42.** $7x^2 = 32$

43. $x^2 - 2x - 1 = 0$ **44.** $x^2 - 6x + 4 = 0$

45. $16x^2 - 9 = 0$ **46.** $11x^2 + 33x = 0$

47. $4x^2 - 12x + 9 = 0$ **48.** $x^2 - 14x + 49 = 0$

49. $(x + 3)^2 = 81$ **50.** $(x - 5)^2 = 8$

51. $4x = 4x^2 - 3$ **52.** $80 + 6x = 9x^2$

53. $50 + 5x = 3x^2$ **54.** $144 - 73x + 4x^2 = 0$

55. $12x = x^2 + 27$ **56.** $26x = 8x^2 + 15$

57. $x^2 - x - \frac{11}{4} = 0$ **58.** $x^2 + 3x - \frac{3}{4} = 0$

59. $50x^2 - 60x - 7 = 0$ **60.** $9x^2 + 12x + 3 = 0$

61. $(x + 3)^2 - 4 = 0$ **62.** $a^2x^2 - b^2 = 0$

63. $(x + 1)^2 = x^2$ **64.** $(x + 1)^2 = 4x^2$

In Exercises 65–70, complete the square for the quadratic portion of the algebraic expression.

65. $\dfrac{1}{x^2 + 2x + 5}$ **66.** $\dfrac{1}{x^2 - 4x + 13}$

67. $\dfrac{1}{x^2 - 4x - 12}$ **68.** $\dfrac{4}{4x^2 + 4x - 3}$

69. $\dfrac{1}{\sqrt{6x - x^2}}$ **70.** $\dfrac{1}{\sqrt{16 - 6x - x^2}}$

2.4   The Quadratic Formula and Applications

In Section 2.3 we introduced the technique of completing the square to solve irreducible quadratic equations. In that setting, we were required to complete the square for each equation separately. In this section we will use the procedure for completing the square *once* in a general setting to obtain the **quadratic formula,** a short cut for solving quadratic equations.

Algebraic Equations and Inequalities

Often in mathematics you are taught the long way of doing a problem first. Then, the longer technique is used to develop shorter techniques. The long way stresses understanding and the short way stresses efficiency.

Derivation of the Quadratic Formula

$$ax^2 + bx + c = 0 \qquad \textit{Given, } a \neq 0$$

$$ax^2 + bx = -c \qquad \textit{Subtract c from both sides}$$

$$x^2 + \frac{b}{a}x = -\frac{c}{a} \qquad \textit{Divide both sides by a}$$

$$x^2 + \frac{b}{a}x + \left(\frac{b}{2a}\right)^2 = -\frac{c}{a} + \left(\frac{b}{2a}\right)^2 \qquad \textit{Complete the square}$$

$$\underbrace{\qquad}_{\text{(half)}^2}$$

$$\left(x + \frac{b}{2a}\right)^2 = \frac{b^2 - 4ac}{4a^2} \qquad \textit{Simplify}$$

$$x + \frac{b}{2a} = \pm\sqrt{\frac{b^2 - 4ac}{4a^2}} \qquad \textit{Take square root of both sides}$$

$$x = -\frac{b}{2a} \pm \frac{\sqrt{b^2 - 4ac}}{2|a|} \qquad \textit{Solutions}$$

Since $\pm 2|a|$ represents the same numbers as $\pm 2a$, we can omit the absolute value sign, as shown in the following statement. You should memorize this formula.

The Quadratic Formula

The solutions of a quadratic equation in the standard form

$$ax^2 + bx + c = 0, \qquad a \neq 0$$

are given by the **quadratic formula**

$$x = \frac{-b \pm \sqrt{b^2 - 4ac}}{2a}.$$

The quantity under the radical sign, $b^2 - 4ac$, is called the **discriminant** of the quadratic expression $ax^2 + bx + c$. It is used to determine the nature of the solutions of a quadratic equation.

Solutions of a Quadratic Equation

The quadratic equation $ax^2 + bx + c = 0$, $a \neq 0$, has the following solutions.

1. If $b^2 - 4ac > 0$, then there are *two distinct real solutions*

$$x = \frac{-b \pm \sqrt{b^2 - 4ac}}{2a}.$$

2. If $b^2 - 4ac = 0$, then there is *one repeated real solution*

$$x = -\frac{b}{2a}.$$

3. If $b^2 - 4ac < 0$, then there are *no real solutions*.

Remark: If the discriminant of a quadratic equation is negative, as in case 3 above, then its square root is imaginary (not a real number) and the quadratic formula yields two complex solutions. We will study this case in Section 2.5, after we have introduced complex numbers.

EXAMPLE 1   *Using the Discriminant*

Use the discriminant to determine the number of real solutions of each of the following quadratic equations.

(a) $4x^2 - 20x + 25 = 0$

(b) $13x^2 + 7x + 1 = 0$

(c) $5x^2 = 8x$

SOLUTION

(a) Since $a = 4$, $b = -20$, and $c = 25$, the discriminant is

$$b^2 - 4ac = 400 - 4(4)(25) = 400 - 400 = 0.$$

Therefore, there is *one* repeated real solution.

(b) In this case, $a = 13$, $b = 7$, and $c = 1$. Thus, the discriminant is

$$b^2 - 4ac = 49 - 4(13)(1) = 49 - 52 = -3 < 0$$

and we conclude that there are *no* real solutions.

(c) In standard form the equation is

$$5x^2 - 8x = 0$$

with $a = 5$, $b = -8$, and $c = 0$. Thus, the discriminant is

$$b^2 - 4ac = 64 - 4(5)(0) = 64 > 0$$

and we conclude that there are *two* real solutions.

Algebraic Equations and Inequalities

EXAMPLE 2 *Using the Quadratic Formula: Two Distinct Solutions*

Use the quadratic formula to solve $x^2 + 3x = 9$.

SOLUTION

In standard form, the equation is

$$x^2 + 3x - 9 = 0$$

where $a = 1$, $b = 3$, and $c = -9$. By the quadratic formula, we have

$$
\begin{aligned}
x &= \frac{-b \pm \sqrt{b^2 - 4ac}}{2a} \\
&= \frac{-3 \pm \sqrt{(3)^2 - 4(1)(-9)}}{2(1)} \\
&= \frac{-3 \pm \sqrt{45}}{2} \\
&= \frac{-3 \pm 3\sqrt{5}}{2}.
\end{aligned}
$$

Thus, the solutions are

$$x = \frac{-3 + 3\sqrt{5}}{2} \quad \text{and} \quad x = \frac{-3 - 3\sqrt{5}}{2}.$$

EXAMPLE 3 *Using the Quadratic Formula: One Repeated Solution*

Use the quadratic formula to solve $8x^2 - 24x + 18 = 0$.

SOLUTION

To simplify the calculations, we begin by dividing both sides of the equation by 2.

$$
\begin{aligned}
8x^2 - 24x + 18 &= 0 && \textit{Given} \\
4x^2 - 12x + 9 &= 0 && \textit{Divide both sides by 2}
\end{aligned}
$$

Thus, $a = 4$, $b = -12$, and $c = 9$, and by the quadratic formula we obtain

$$
\begin{aligned}
x &= \frac{-b \pm \sqrt{b^2 - 4ac}}{2a} \\
&= \frac{-(-12) \pm \sqrt{(-12)^2 - 4(4)(9)}}{2(4)} \\
&= \frac{12 \pm \sqrt{0}}{8} = \frac{3}{2}.
\end{aligned}
$$

Note that there is only one (repeated) solution. This occurs because the discriminant is zero.

The discriminant in Example 3 is a perfect square (zero in this case), and we could have *factored* the quadratic as

$$4x^2 - 12x + 9 = (2x - 3)^2 = 0$$

to conclude that the solution is $x = 3/2$. Since factoring is easier than applying the quadratic formula, try factoring first. If, however, factors cannot be readily found, then use the quadratic formula. For instance, try solving the quadratic equation $x^2 - x - 12 = 0$ in two ways—by factoring and by the quadratic formula—to see that you get the same solutions either way.

When using a calculator to evaluate the quadratic formula, you should get in the habit of using the memory key. This will minimize round-off error.

EXAMPLE 4   *Using a Calculator to Evaluate the Quadratic Formula*

Solve the quadratic equation $16.3x^2 - 197.6x + 7.042 = 0$.

SOLUTION

In this case, $a = 16.3$, $b = -197.6$, $c = 7.042$, and we have

$$x = \frac{-b \pm \sqrt{b^2 - 4ac}}{2a}$$

$$= \frac{-(-197.6) \pm \sqrt{(-197.6)^2 - 4(16.3)(7.042)}}{2(16.3)}.$$

To evaluate these solutions, we begin by calculating the square root of the discriminant as follows.

Calculator Steps	*Display*
197.6 $\boxed{+/-}$ $\boxed{x^2}$ $\boxed{-}$ 4 $\boxed{\times}$ 16.3 $\boxed{\times}$ 7.042 $\boxed{=}$ $\boxed{\sqrt{}}$	196.43478

Storing this result and, using the recall key, we find the two solutions.

$$x \approx \frac{197.6 + 196.43478}{2(16.3)} \approx 12.087 \qquad \textit{Add stored value}$$

$$x \approx \frac{197.6 - 196.43478}{2(16.3)} \approx 0.036 \qquad \textit{Subtract stored value}$$

Algebraic Equations and Inequalities

Applications Involving Quadratic Equations

Keep the guidelines for solving word problems (Section 2.2) in mind as you study the next few examples and work the exercises for this section.

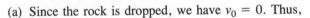

EXAMPLE 5 *Falling Objects*

The height s (in feet) of a falling object is given by the **position equation**

$$s = -16t^2 + v_0 t + s_0$$

where t is the time (in seconds), s_0 is the initial height (when $t = 0$), and v_0 is the initial velocity. A rock is at the top of a 300-foot river gorge. How long will it take the rock to hit the water below if (a) it is dropped, and (b) it is thrown vertically downward at 50 feet per second? (See Figure 2.4.)

SOLUTION

In both cases, the initial height is $s_0 = 300$. Moreover, since the water level is represented by $s = 0$, we can find the time that the rock hits the water by solving for t in the equation $-16t^2 + v_0 t + 300 = 0$.

(a) Since the rock is dropped, we have $v_0 = 0$. Thus,

$$-16t^2 + 300 = 0$$
$$16t^2 = 300$$
$$t^2 = \frac{300}{16} = \frac{75}{4}$$
$$t = \pm\frac{\sqrt{75}}{2}.$$

Finally, choosing the positive time value, we have

$$t = \frac{\sqrt{75}}{2} \approx 4.33 \text{ seconds.}$$

(b) In this case, the rock is thrown downward at 50 feet per second and the initial velocity is $v_0 = -50$. (The negative velocity denotes the downward direction.) Thus, to find the time it hits the water, we solve $-16t^2 - 50t + 300 = 0$, obtaining

$$t = \frac{50 \pm \sqrt{(-50)^2 - 4(-16)(300)}}{-32} = \frac{50 \pm \sqrt{21700}}{-32}.$$

Again, choosing the positive time value, we have

$$t \approx 3.04 \text{ seconds.}$$

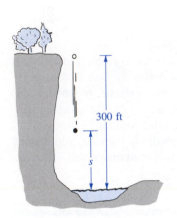

s is the height at time *t*.

FIGURE 2.4

Remark: When solving applied problems, try to get in the habit of asking yourself if your answer seems reasonable. For instance, in Example 5 it is reasonable that the rock that was thrown downward would hit the water sooner than the one that was dropped.

EXAMPLE 6 An Application Involving Area

A picture is 3 inches longer than it is wide and has an area of 120 square inches. It is to be enclosed in a frame that has a uniform width of 2 inches, as shown in Figure 2.5. What are the dimensions of the frame?

SOLUTION

Model: area of picture = (width)(length) = 120

Labels: W = picture width
L = picture length
$W + 4$ = frame width
$L + 4$ = frame length

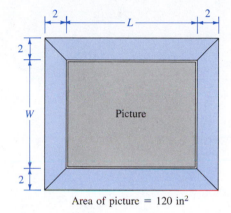

Area of picture = 120 in²

FIGURE 2.5

Since the picture is 3 inches longer than it is wide, it follows that $L = W + 3$.

Equation: $W(W + 3) = 120$
$$W^2 + 3W - 120 = 0$$

The quadratic formula produces

$$W = \frac{-3 \pm \sqrt{9 + 480}}{2}$$

$$= \frac{-3 \pm \sqrt{489}}{2}.$$

Choosing the positive root, we have

$$W = \frac{-3 + \sqrt{489}}{2} \approx 9.56 \text{ inches}$$

$$L = W + 3 \approx 12.56 \text{ inches.}$$

Finally, the frame has the following dimensions.

$W + 4 \approx 13.56$ inches   *Width of frame*
$L + 4 = W + 7 \approx 16.56$ inches   *Length of frame*

EXAMPLE 7 *An Application Involving the Pythagorean Theorem*

An L-shaped sidewalk from building A to building B on a college campus is 400 meters long, as shown in Figure 2.6. By cutting diagonally across the grass, students shorten the walking distance to 300 meters. What are the lengths of the two legs of the existing sidewalk?

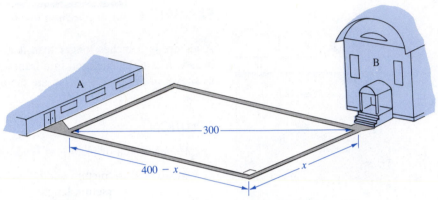

FIGURE 2.6

SOLUTION

Model: $a^2 + b^2 = c^2$ *Pythagorean Theorem*

Labels: a = length of one leg = x
b = length of other leg = $400 - x$
c = length of diagonal = 300

Equation:
$$x^2 + (400 - x)^2 = (300)^2$$
$$2x^2 - 800x + 160{,}000 = 90{,}000$$
$$2x^2 - 800x + 70{,}000 = 0$$
$$x^2 - 400x + 35{,}000 = 0$$

By the quadratic formula, we have

$$x = \frac{400 \pm \sqrt{(-400)^2 - 4(1)(35{,}000)}}{2(1)}$$

$$= \frac{400 \pm \sqrt{20{,}000}}{2}$$

$$= \frac{400 \pm 100\sqrt{2}}{2}$$

$$= 200 \pm 50\sqrt{2}.$$

Quadratic Formula and Applications

Both solutions are positive, and it does not matter which one we choose. If we let

$$x = 200 + 50\sqrt{2} \approx 270.7 \text{ meters}$$

then the length of the other leg is

$$400 - x \approx 400 - 270.7 \approx 129.3 \text{ meters.}$$

Try choosing the other value of x to see that the same two lengths result.

EXAMPLE 8   *Reduced Rates*

A ski club chartered a bus for a ski trip at a cost of $480. In an attempt to lower the bus fare per skier, the club invited nonmembers to go along. After five nonmembers joined the trip, the fare per skier decreased by $4.80. How many club members are going on the bus?

SOLUTION

Model: (cost per skier)(number of skiers) = 480

Labels: x = number of ski club members
$x + 5$ = number of skiers

Original cost per skier = $\dfrac{480}{x}$

New cost per skier = $\dfrac{480}{x} - 4.80$

Equation:

$$\left(\frac{480}{x} - 4.80\right)(x + 5) = 480$$

$$\left(\frac{480 - 4.8x}{x}\right)(x + 5) = 480$$

$$(480 - 4.8x)(x + 5) = 480x, \; x \neq 0$$

$$480x - 4.8x^2 - 24x + 2400 = 480x$$

$$-4.8x^2 - 24x + 2400 = 0$$

$$x^2 + 5x - 500 = 0$$

$$(x + 25)(x - 20) = 0$$

$$x = 20 \text{ or } -25$$

Choosing the positive value of x, we have

$$x = 20 \text{ ski club members.}$$

Algebraic Equations and Inequalities

WARM UP

Simplify the given expressions.

1. $\sqrt{9 - 4(3)(-12)}$

2. $\sqrt{36 - 4(2)(3)}$

3. $\sqrt{12^2 - 4(3)(4)}$

4. $\sqrt{15^2 + 4(9)(12)}$

Solve the given quadratic equations by factoring.

5. $x^2 - x - 2 = 0$

6. $2x^2 + 3x - 9 = 0$

7. $x^2 - 4x = 5$

8. $2x^2 + 13x = 7$

9. $x^2 = 5x - 6$

10. $x(x - 3) = 4$

EXERCISES 2.4

In Exercises 1–8, use the discriminant to determine the number of real solutions of the quadratic equation.

1. $4x^2 - 4x + 1 = 0$

2. $2x^2 - x - 1 = 0$

3. $3x^2 + 4x + 1 = 0$

4. $x^2 + 2x + 4 = 0$

5. $2x^2 - 5x + 5 = 0$

6. $3x^2 - 6x + 3 = 0$

7. $\frac{1}{5}x^2 + \frac{6}{5}x - 8 = 0$

8. $\frac{1}{3}x^2 - 5x + 25 = 0$

In Exercises 9–34, use the quadratic formula to solve the equation.

9. $2x^2 + x - 1 = 0$

10. $2x^2 - x - 1 = 0$

11. $16x^2 + 8x - 3 = 0$

12. $25x^2 - 20x + 3 = 0$

13. $2 + 2x - x^2 = 0$

14. $x^2 - 10x + 22 = 0$

15. $x^2 + 14x + 44 = 0$

16. $6x = 4 - x^2$

17. $x^2 + 8x - 4 = 0$

18. $4x^2 - 4x - 4 = 0$

19. $12x - 9x^2 = -3$

20. $16x^2 + 22 = 40x$

21. $36x^2 + 24x - 7 = 0$

22. $3x + x^2 - 1 = 0$

23. $4x^2 + 4x = 7$

24. $16x^2 - 40x + 5 = 0$

25. $28x - 49x^2 = 4$

26. $9x^2 + 24x + 16 = 0$

27. $25h^2 + 80h + 61 = 0$

28. $8t = 5 + 2t^2$

29. $(y - 5)^2 = 2y$

30. $(z + 6)^2 = -2z$

31. $\frac{1}{x} - \frac{1}{x + 1} = 3$

32. $\frac{x}{x^2 - 4} + \frac{1}{x + 2} = 3$

33. $\frac{20 - x}{x} = x$

34. $\frac{4}{x} - \frac{5}{3} = \frac{x}{6}$

In Exercises 35–40, use a calculator to solve the given equation. Round your answers to three decimal places.

35. $5.1x^2 - 1.7x - 3.2 = 0$

36. $10.4x^2 + 8.6x + 1.2 = 0$

37. $7.06x^2 - 4.85x + 0.50 = 0$

38. $-0.005x^2 + 0.101x - 0.193 = 0$

39. $422x^2 - 506x - 347 = 0$

40. $2x^2 - 2.50x - 0.42 = 0$

41. Find two numbers whose sum is 100 and whose product is 2500.

42. Find two consecutive positive integers whose product is 72.

43. Find two consecutive positive integers such that the sum of their squares is 113.

44. Find two consecutive even integers whose product is 440.

In Exercises 45–48, use the cost equation to find the number of units x that a manufacturer can produce for the given cost C. (Round your answer to the nearest positive integer.)

45. $C = 0.125x^2 + 20x + 5000$ $\qquad C = \$14{,}000$

46. $C = 0.5x^2 + 15x + 5000$ $\qquad C = \$11{,}500$

47. $C = 800 + 0.04x + 0.0002x^2$ $\qquad C = \$1680$

48. $C = 800 - 10x + \frac{x^2}{4}$ $\qquad C = \$896$

In Exercises 49–52, find the time when the object hits the ground under the given conditions. (Use the position equation given in Example 5.)

49. The object is dropped from a balloon at a height of 1600 feet.

50. The object is thrown vertically upward from ground level with an initial velocity of 60 feet per second.

51. The object is dropped at a height of 64 feet from a balloon rising vertically at the rate of 16 feet per second. (The balloon's velocity becomes the object's initial velocity.)

52. The object is dropped from the top of the Washington Monument (a height of approximately 550 feet).

53. A rancher has 200 feet of fencing to enclose two adjacent rectangular corrals (see figure). Find the dimensions such that the enclosed area will be 1400 square feet.

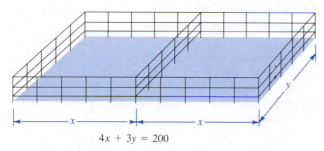

$$4x + 3y = 200$$

FIGURE FOR 53

54. A rectangular classroom seats 72 students. If the seats were rearranged with three more seats in each row, the classroom would have two fewer rows. Find the original number of seats in each row.

55. Two brothers must mow a rectangular lawn 100 feet by 200 feet. Each wants to mow no more than half of the lawn. The first starts by mowing around the outside of the lawn. How wide a strip must he mow on each of the four sides? Approximately how many times must he go around the lawn if the mower has a 24-inch cut?

56. Repeat Exercise 55 assuming that the first brother agrees to mow three-fourths of the lawn.

57. An open box is to be made from a square piece of material by cutting 2-inch squares from each corner and turning up the sides (see figure). The volume of the finished box is to be 200 cubic inches. Find the size of the original piece of material.

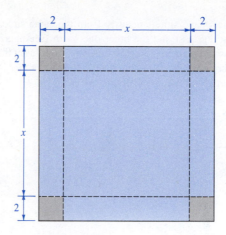

FIGURE FOR 57

58. Repeat Exercise 57, assuming that a 4-inch square is cut from each corner of the material.

59. An open box with a square base is to be constructed from 108 square inches of material (see figure). What should be the dimensions of the base if the height of the box is to be 3 inches? [*Hint:* The surface area is given by $S = x^2 + 4xh$.]

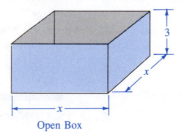

Open Box   **FIGURE FOR 59**

60. A small commuter airline flies to three cities whose locations form the vertices of a right triangle (see figure). The total flight distance (from City A to City B to City C and back to City A) is 1400 miles. It is 600 miles between the two cities that are farthest apart. Find the other two distances between cities.

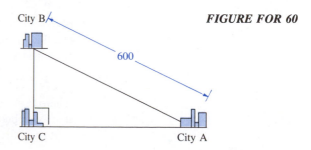

FIGURE FOR 60

61. Two planes leave simultaneously from the same airport, one flying due east and the other due south (see figure). The eastbound plane is flying 50 miles per hour faster than the southbound plane. After three hours the planes are 2440 miles apart. Find the speed of each plane.

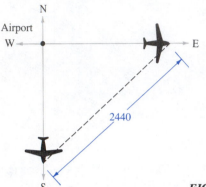

FIGURE FOR 61

62. A family drove 1080 miles to their vacation lodge. Because of increased traffic density, their average speed on the return trip was decreased by 6 miles per hour and the trip took $2\frac{1}{2}$ hours longer. Determine their average speed on the way to the lodge.

63. A windlass is used to tow a boat to the dock. The rope is attached to the boat at a point 15 feet below the level of the windlass (see figure). Find the distance from the boat to the dock when there is 75 feet of rope out.

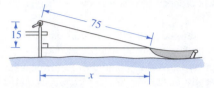

FIGURE FOR 63

64. A college charters a bus for $1700 to take a group of students to the World's Fair. When six more students join the trip, the cost per student drops by $7.50. How many students were in the original group?

65. A business is now selling 2000 units per month at $10.00 per unit. It can sell 250 more units per month for each $0.25 reduction in price. What price per unit will yield a monthly revenue of $36,000?

66. A lump sum of $10,000 is invested for two years at $r\%$ compounded annually. At the end of the two-year period the investment has increased to $12,321. Find the annual percentage rate r.

67. Repeat Exercise 66, assuming that the investment increased to $11,990.25 at the end of the two-year period.

2.5 Complex Numbers

In Section 2.4 we noted that if the discriminant $b^2 - 4ac$ of a quadratic is less than zero, then there are no real solutions to the equation $ax^2 + bx + c = 0$. Moreover, by the square root method (see Example 3 of Section 2.3), we see that the set of real numbers is not large enough to accommodate the solutions to an equation as simple as

$$x^2 = -1$$

because $x = \sqrt{-1}$ is not a real number. To overcome this deficiency, mathematicians created an expanded system of numbers using the **imaginary unit** *i*, defined as

$$i = \sqrt{-1}$$

where $i^2 = -1$. By adding real numbers to real multiples of this imaginary unit, we obtain the set of **complex numbers.** Each complex number can be written in the **standard form,** $a + bi$.

Definition of a Complex Number

For real numbers a and b, the number

$a + bi$

is called a **complex number.** If $a = 0$ and $b \neq 0$, then the complex number bi is called a **pure imaginary number.**

The set of real numbers is a subset of the set of complex numbers because every real number a can be written as a complex number using $b = 0$. That is, for every real number a, we can write $a = a + 0i$.

Two complex numbers $a + bi$ and $c + di$, written in standard form, are **equal** to each other

$$a + bi = c + di$$

Equality of two complex numbers

if and only if $a = c$ and $b = d$.

Operations with Complex Numbers

To add (or subtract) two complex numbers, we add (or subtract) the real and imaginary parts of the numbers separately.

Addition and Subtraction of Complex Numbers

If $a + bi$ and $c + di$ are two complex numbers written in standard form, then their sum and difference are defined as follows.

$$\text{Sum:} \quad (a + bi) + (c + di) = (a + c) + (b + d)i$$
$$\text{Difference:} \quad (a + bi) - (c + di) = (a - c) + (b - d)i$$

The **additive identity** in the complex number system is zero (the same as in the real number system). Furthermore, the **additive inverse** of the complex number $a + bi$ is

$$-(a + bi) = -a - bi.$$

Additive inverse

Thus, we have

$$(a + bi) + (-a - bi) = 0 + 0i = 0.$$

EXAMPLE 1 *Adding and Subtracting Complex Numbers*

Write the following sums and differences in standard form.

(a) $(3 - i) + (2 + 3i)$ (b) $2i + (-4 - 2i)$

(c) $3 - (-2 + 3i) + (-5 + i)$

SOLUTION

(a)
$$\begin{aligned}
(3 - i) + (2 + 3i) &= 3 - i + 2 + 3i \\
&= 3 + 2 - i + 3i \\
&= (3 + 2) + (-1 + 3)i \\
&= 5 + 2i
\end{aligned}$$

(b)
$$\begin{aligned}
2i + (-4 - 2i) &= 2i - 4 - 2i \\
&= -4 + 2i - 2i \\
&= -4
\end{aligned}$$

(c)
$$\begin{aligned}
3 - (-2 + 3i) + (-5 + i) &= 3 + 2 - 3i - 5 + i \\
&= 3 + 2 - 5 - 3i + i \\
&= 0 - 2i \\
&= -2i
\end{aligned}$$

Remark: Note in Example 1(b) that the sum of two complex numbers can be a real number.

Many of the properties of real numbers are valid for complex numbers as well. Here are some.

    Associative Property of Addition and Multiplication
    Commutative Property of Addition and Multiplication
    Distributive Property of Multiplication over Addition

Notice how these properties are used when two complex numbers are multiplied.

$$\begin{aligned}
(a + bi)(c + di) &= a(c + di) + bi(c + di) && \text{\textit{Distributive Law}} \\
&= ac + (ad)i + (bc)i + (bd)i^2 && \text{\textit{Distributive Law}} \\
&= ac + (ad)i + (bc)i + (bd)(-1) && \text{\textit{Definition of i}} \\
&= ac - bd + (ad)i + (bc)i && \text{\textit{Commutative Law}} \\
&= (ac - bd) + (ad + bc)i && \text{\textit{Associative Law}}
\end{aligned}$$

Rather than trying to memorize this multiplication rule, you may just want to remember how the distributive property is used to derive it. The procedure is similar to multiplying two polynomials and combining like terms.

EXAMPLE 2 *Multiplying Complex Numbers*

Find the following products.

(a) $(i)(-3i)$
(b) $(2 - i)(4 + 3i)$
(c) $(3 + 2i)(3 - 2i)$

SOLUTION

(a) $(i)(-3i) = -3i^2 = -3(-1) = 3$

(b) $(2 - i)(4 + 3i) = 8 + 6i - 4i - 3i^2$ *Binomial product*
$$= 8 + 6i - 4i - 3(-1)$$ *$i^2 = -1$*
$$= 8 + 3 + 6i - 4i$$ *Collect terms*
$$= 11 + 2i$$ *Combine like terms*

(c) $(3 + 2i)(3 - 2i) = 9 - 6i + 6i - 4i^2$
$$= 9 - 4(-1) = 9 + 4$$
$$= 13$$

Complex Conjugates

We see from Example 2(c) that the product of complex numbers can be a real number. This occurs with pairs of complex numbers of the forms $a + bi$ and $a - bi$, called **complex conjugates.** In general, we have the following.

$$(a + bi)(a - bi) = a^2 - abi + abi - b^2i^2$$
$$= a^2 - b^2(-1)$$
$$= a^2 + b^2$$

Complex conjugates can be used to divide two complex numbers. That is, to find the quotient

$$\frac{a + bi}{c + di}, \qquad c \text{ and } d \text{ not both zero}$$

we multiply the numerator and denominator by the conjugate of the denominator to obtain

$$\frac{a + bi}{c + di} = \frac{a + bi}{c + di}\left(\frac{c - di}{c - di}\right) = \frac{(ac + bd) + (bc - ad)i}{c^2 + d^2}.$$

This procedure is demonstrated in Example 3.

Algebraic Equations and Inequalities

EXAMPLE 3 *Dividing Complex Numbers*

Write the following in standard form.

(a) $\dfrac{1}{1 + i}$

(b) $\dfrac{2 + 3i}{4 - 2i}$

SOLUTION

(a) $\dfrac{1}{1 + i} = \dfrac{1}{1 + i}\left(\dfrac{1 - i}{1 - i}\right) = \dfrac{1 - i}{1^2 - i^2}$

$$= \dfrac{1 - i}{1 - (-1)} = \dfrac{1 - i}{2} = \dfrac{1}{2} - \dfrac{1}{2}i$$

(b) $\dfrac{2 + 3i}{4 - 2i} = \dfrac{2 + 3i}{4 - 2i}\left(\dfrac{4 + 2i}{4 + 2i}\right) = \dfrac{8 + 4i + 12i + 6i^2}{16 - 4i^2}$

$$= \dfrac{8 - 6 + 16i}{16 + 4} = \dfrac{1}{20}(2 + 16i)$$

$$= \dfrac{1}{10} + \dfrac{4}{5}i$$

When using the quadratic formula or the square root method to solve quadratic equations, we often obtain a result like $\sqrt{-3}$, which we know is not a real number. By factoring out $i = \sqrt{-1}$, we can write this number in the standard form, as follows.

$$\sqrt{-3} = \sqrt{3(-1)} = \sqrt{3}\sqrt{-1} = \sqrt{3}\,i$$

We call $\sqrt{3}\,i$ the principal square root of -3.

Principal Square Root of a Negative Number

For $a > 0$, the **principal square root** of $-a$ is defined as

$$\sqrt{-a} = \sqrt{a}\,i.$$

In this definition we are using the rule $\sqrt{ab} = \sqrt{a}\sqrt{b}$, for $a > 0$ and $b < 0$. This rule is not valid if *both* a and b are negative. For example,

$$\sqrt{-5}\sqrt{-5} = (\sqrt{5}\,i)(\sqrt{5}\,i) = 5i^2 = -5$$

whereas

$$\sqrt{(-5)(-5)} = \sqrt{25} = 5.$$

Consequently, $\sqrt{(-5)(-5)} \neq \sqrt{-5}\sqrt{-5}$.

Complex Numbers

Remark: When working with square roots of negative numbers, be sure to convert to standard form *before* multiplying.

EXAMPLE 4   Writing Complex Numbers in Standard Form

(a) $\sqrt{-3}\sqrt{-12} = \sqrt{3}\,i\sqrt{12}\,i = \sqrt{36}\,i^2 = 6(-1) = -6$

(b) $\sqrt{-48} - \sqrt{-27} = \sqrt{48}\,i - \sqrt{27}\,i = 4\sqrt{3}\,i - 3\sqrt{3}\,i = \sqrt{3}\,i$

(c) $(-1 + \sqrt{-3})^2 = (-1 + \sqrt{3}\,i)^2$

$$= (-1)^2 - 2\sqrt{3}\,i + (\sqrt{3})^2(i^2)$$

$$= 1 - 2\sqrt{3}\,i + 3(-1)$$

$$= -2 - 2\sqrt{3}\,i$$

Example 5 shows how the principal square root of a negative number is used to represent complex solutions of a quadratic equation.

EXAMPLE 5   Complex Solutions of a Quadratic Equation

Solve the equation $3x^2 - 2x + 5 = 0$.

SOLUTION

By the quadratic formula, we have

$$x = \frac{-(-2) \pm \sqrt{(-2)^2 - 4(3)(5)}}{2(3)}$$

$$= \frac{2 \pm \sqrt{-56}}{6}$$

$$= \frac{2 \pm 2\sqrt{14}\,i}{6}$$

$$= \frac{1}{3} \pm \frac{\sqrt{14}}{3}\,i.$$

When working with polynomials of degree higher than 2, we occasionally need to raise i to the third or higher power. The pattern looks like this:

$$i^1 = i$$
$$i^2 = -1$$
$$i^3 = i^2 \cdot i = -i$$
$$i^4 = i^2 \cdot i^2 = (-1)(-1) = 1$$
$$i^5 = i^4 \cdot i = i.$$

Algebraic Equations and Inequalities

Since the pattern begins to repeat after the fourth power, we can compute the value of i^n for any natural number n. We simply factor out the multiples of 4 in the exponent and compute the remaining portion. For example,

$$i^{38} = i^{36} \cdot i^2$$
$$= (i^4)^9 \cdot i^2$$
$$= (1)^9(-1) = -1.$$

EXAMPLE 6 Finding Powers of i

Find the following powers of i.

(a) i^{10}
(b) i^{73}

SOLUTION

(a) Since 8 is the largest multiple of 4 that is less than 10, we write

$$i^{10} = i^8 \cdot i^2 = (i^4)^2 \cdot i^2 = (1)^2 \cdot (-1) = -1.$$

(b) Since 72 is the largest multiple of 4 that is less than 73, we write

$$i^{73} = i^{72} \cdot i^1 = (i^4)^{18} \cdot i = (1)i = i.$$

WARM UP

Simplify the given expressions.

1. $\sqrt{12}$
2. $\sqrt{500}$
3. $\sqrt{20} - \sqrt{5}$
4. $\sqrt{27} - \sqrt{243}$
5. $\sqrt{24}\sqrt{6}$
6. $2\sqrt{18}\sqrt{32}$
7. $\dfrac{1}{\sqrt{3}}$
8. $\dfrac{2}{\sqrt{2}}$

Solve the given quadratic equations.

9. $x^2 + x - 1 = 0$
10. $x^2 + 2x - 1 = 0$

EXERCISES 2.5

1. Write out the first 16 positive integer powers of i (that is, i, i^2, i^3, ..., i^{16}), and express each as i, $-i$, 1, or -1.

2. Express each of the following powers of i as i, $-i$, 1, or -1.

   (a) i^{40} (b) i^{25}

   (c) i^{50} (d) i^{67}

In Exercises 3–6, find real numbers a and b so that the equation is true.

3. $a + bi = -10 + 6i$ **4.** $a + bi = 13 + 4i$

5. $(a - 1) + (b + 3)i = 5 + 8i$

6. $(a + 6) + 2bi = 6 - 5i$

In Exercises 7–18, write the complex number in standard form and find its complex conjugate.

7. $4 + \sqrt{-9}$ **8.** $3 + \sqrt{-16}$

9. $2 - \sqrt{-27}$ **10.** $1 + \sqrt{-8}$

11. $\sqrt{-75}$ **12.** 45

13. $-6i + i^2$ **14.** $4i^2 - 2i^3$

15. $-5i^5$ **16.** $(-i)^3$

17. 8 **18.** $(\sqrt{-4})^2 - 5$

In Exercises 19–56, perform the indicated operation and write the result in standard form.

19. $(5 + i) + (6 - 2i)$

20. $(13 - 2i) + (-5 + 6i)$

21. $(8 - i) - (4 - i)$

22. $(3 + 2i) - (6 + 13i)$

23. $(-2 + \sqrt{-8}) + (5 - \sqrt{-50})$

24. $(8 + \sqrt{-18}) - (4 + 3\sqrt{2}\, i)$

25. $-(\frac{3}{2} + \frac{5}{2}i) + (\frac{5}{3} + \frac{11}{3}i)$

26. $(1.6 + 3.2i) + (-5.8 + 4.3i)$

27. $\sqrt{-6} \cdot \sqrt{-2}$

28. $\sqrt{-5} \cdot \sqrt{-10}$

29. $(\sqrt{-10})^2$

30. $(\sqrt{-75})^3$

31. $(1 + i)(3 - 2i)$

32. $(6 - 2i)(2 - 3i)$

33. $(4 + 5i)(4 - 5i)$

34. $(6 + 7i)(6 - 7i)$

35. $6i(5 - 2i)$

36. $-8i(9 + 4i)$

37. $-8(5 - 2i)$

38. $(\sqrt{5} - \sqrt{3}\, i)(\sqrt{5} + \sqrt{3}\, i)$

39. $(\sqrt{14} + \sqrt{10}\, i)(\sqrt{14} - \sqrt{10}\, i)$

40. $(3 + \sqrt{-5})(7 - \sqrt{-10})$

41. $(4 + 5i)^2$

42. $(2 - 3i)^3$

43. $\dfrac{4}{4 - 5i}$

44. $\dfrac{3}{1 - i}$

45. $\dfrac{2 + i}{2 - i}$ **46.** $\dfrac{8 - 7i}{1 - 2i}$

47. $\dfrac{6 - 7i}{i}$ **48.** $\dfrac{8 + 20i}{2i}$

49. $\dfrac{1}{(2i)^3}$ **50.** $\dfrac{1}{(4 - 5i)^2}$

51. $\dfrac{5}{(1 + i)^3}$ **52.** $\dfrac{(2 - 3i)(5i)}{2 + 3i}$

53. $\dfrac{(21 - 7i)(4 + 3i)}{2 - 5i}$ **54.** $\dfrac{1}{i^3}$

55. $(2 + 3i)^2 + (2 - 3i)^2$ **56.** $(1 - 2i)^2 - (1 + 2i)^2$

In Exercises 57–64, use the quadratic formula to solve the quadratic equation.

57. $x^2 - 2x + 2 = 0$ **58.** $x^2 + 6x + 10 = 0$

59. $4x^2 + 16x + 17 = 0$ **60.** $9x^2 - 6x + 37 = 0$

61. $4x^2 + 16x + 15 = 0$ **62.** $9x^2 - 6x - 35 = 0$

63. $16t^2 - 4t + 3 = 0$ **64.** $5s^2 + 6s + 3 = 0$

65. Prove that the sum of a complex number and its conjugate is a real number.

66. Prove that the difference of a complex number and its conjugate is an imaginary number.

67. Prove that the product of a complex number and its conjugate is a real number.

68. Prove that the conjugate of the product of two complex numbers is the product of their conjugates.

69. Prove that the conjugate of the sum of two complex numbers is the sum of their conjugates.

2.6 *Other Types of Equations*

In this section we extend our techniques for solving equations to include special types of nonlinear and nonquadratic equations.

We have only two basic methods for solving nonlinear equations—*factoring* and the *quadratic formula*. So the main goal of this section is to learn to *rewrite* nonlinear equations in factorable or quadratic form. In this section, we will list only real solutions. Complex solutions will be discussed in detail in Section 4.5.

Because factoring polynomials is so crucial in this section, you may want to review the special forms listed in Section 1.5.

EXAMPLE 1 *Solving a Polynomial Equation by Factoring*

Solve $3x^4 = 48x^2$.

SOLUTION

$$3x^4 - 48x^2 = 0 \qquad \text{\textit{Collect terms to left side}}$$
$$(3x^2)(x^2 - 16) = 0 \qquad \text{\textit{Common monomial factor}}$$
$$3(x^2)(x + 4)(x - 4) = 0 \qquad \text{\textit{Factor difference of squares}}$$

Now, by setting each *variable* factor equal to zero, we obtain the following solutions.

$$x^2 = 0, \qquad x + 4 = 0, \qquad x - 4 = 0$$
$$x = 0 \qquad\qquad x = -4 \qquad\qquad x = 4.$$

Thus, the solutions are $x = 0, -4$, and 4.

Remark: A mistake often made when factoring equations like that given in Example 1 is dividing both sides of the equation by the variable factor x^2 before attempting to solve the equation. This loses the solution, $x = 0$. Be sure to factor completely and *then* set each variable factor equal to zero.

EXAMPLE 2 *Solving a Polynomial Equation by Factoring*

Solve $x^3 - 3x^2 - 3x + 9 = 0$.

Other Types of Equations

SOLUTION

We use factoring by grouping to solve this equation.

$$x^3 - 3x^2 - 3x + 9 = 0 \qquad \text{\textit{Given}}$$
$$x^2(x - 3) - 3(x - 3) = 0 \qquad \text{\textit{Group terms}}$$
$$(x - 3)(x^2 - 3) = 0 \qquad \text{\textit{Factor out } (x - 3)}$$
$$x - 3 = 0, \qquad x^2 - 3 = 0 \qquad \text{\textit{Set factors to zero}}$$
$$x = 3 \qquad \qquad x = \pm\sqrt{3} \qquad \text{\textit{Solve for } x}$$

The solutions are $x = 3$, $\sqrt{3}$, and $-\sqrt{3}$.

The equation $x^4 - 3x^2 + 2 = 0$ is said to be of *quadratic type* because the left side can be written in the form

$$(x^2)^2 - 3(x^2) + 2 = u^2 - 3u + 2$$

where $u = x^2$. In general, an equation is of **quadratic type** if it can be written in the form

$$au^2 + bu + c = 0$$

where $a \neq 0$ and u is some variable expression. Once an equation is written in this form, we can solve for the variable u by factoring or by the quadratic formula.

EXAMPLE 3 *Solving an Equation of Quadratic Type*

Solve $x^4 - 3x^2 + 2 = 0$.

SOLUTION

This equation is of quadratic type with $u = x^2$. Thus, we factor the left side of the equation as the product of two second-degree polynomials.

$$x^4 - 3x^2 + 2 = 0 \qquad \text{\textit{Given}}$$
$$(x^2 - 1)(x^2 - 2) = 0 \qquad \text{\textit{Factor polynomial}}$$
$$x^2 - 1 = 0, \qquad x^2 - 2 = 0 \qquad \text{\textit{Set factors to zero}}$$
$$x^2 = 1 \qquad \qquad x^2 = 2$$
$$x = \pm 1 \qquad \qquad x = \pm\sqrt{2} \qquad \text{\textit{Solve for } x}$$

The solutions are $x = -1$, 1, $-\sqrt{2}$, and $\sqrt{2}$.

EXAMPLE 4 *An Equation Involving Rational Exponents*

Solve $2x^{2/3} + 3x^{1/3} - 9 = 0$.

SOLUTION

$$2x^{2/3} + 3x^{1/3} - 9 = 0$$
$$2(x^{1/3})^2 + 3(x^{1/3}) - 9 = 0$$
$$(2x^{1/3} - 3)(x^{1/3} + 3) = 0$$

Setting these factors equal to zero produces the following.

$$2x^{1/3} - 3 = 0, \qquad\qquad x^{1/3} + 3 = 0$$

$$x^{1/3} = \frac{3}{2} \qquad\qquad x^{1/3} = -3$$

$$x = \left(\frac{3}{2}\right)^3 = \frac{27}{8} \qquad\qquad x = (-3)^3 = -27$$

A check will show that both solutions are valid.

The remaining equations in this section require manipulations that can introduce extraneous roots. Operations like squaring both sides of an equation, raising both sides to a rational power, or multiplying both sides by a variable quantity all have this potential danger. When you use any of these operations, you must check each "solution" in the original equation.

EXAMPLE 5 *An Equation Involving a Radical*

Solve $\sqrt{2x + 7} - x = 2$.

SOLUTION

It is best to isolate the radical before squaring both sides.

$\sqrt{2x + 7} - x = 2$	*Given*
$\sqrt{2x + 7} = x + 2$	*Isolate radical*
$2x + 7 = x^2 + 4x + 4$	*Square both sides*
$0 = x^2 + 2x - 3$	*Standard form*
$0 = (x + 3)(x - 1)$	*Factor*
$x + 3 = 0, \qquad x - 1 = 0$	*Set factors to zero*
$x = -3 \qquad\quad x = 1$	*Possible solutions*

Other Types of Equations

SOLUTION CHECK

$$\sqrt{2(-3) + 7} + 3 = \sqrt{1} + 3 \neq 2 \qquad \text{\textit{-3 is not a solution}}$$

$$\sqrt{2(1) + 7} - 1 = \sqrt{9} - 1 = 2 \qquad \text{\textit{1 is a solution}}$$

Thus, the only solution is $x = 1$.

For equations with two or more radicals, it may be necessary to repeat the "isolate a radical and square both sides" routine demonstrated in the preceding example.

EXAMPLE 6   *An Equation Involving Two Radicals*

Solve $\sqrt{2x + 6} - \sqrt{x + 4} = 1$.

SOLUTION

$\sqrt{2x + 6} - \sqrt{x + 4} = 1$	*Given*
$\sqrt{2x + 6} = 1 + \sqrt{x + 4}$	*Isolate radical*
$2x + 6 = 1 + 2\sqrt{x + 4} + (x + 4)$	*Square both sides*
$x + 1 = 2\sqrt{x + 4}$	*Isolate radical*
$x^2 + 2x + 1 = 4(x + 4)$	*Square both sides*
$x^2 - 2x - 15 = 0$	*Standard form*
$(x - 5)(x + 3) = 0$	*Factor*
$x - 5 = 0, \qquad x + 3 = 0$	*Set factors to zero*
$x = 5 \qquad\qquad x = -3$	*Possible solutions*

SOLUTION CHECK

$$\sqrt{2(5) + 6} - \sqrt{5 + 4} = 4 - 3 = 1 \qquad \text{\textit{5 is a solution}}$$

$$\sqrt{2(-3) + 6} - \sqrt{-3 + 4} = 0 - 1 \neq 1 \qquad \text{\textit{-3 is not a solution}}$$

Thus, the only solution is $x = 5$.

Algebraic Equations and Inequalities

EXAMPLE 7 An Equation Involving a Rational Exponent

Solve $(x^2 - x - 4)^{3/4} = 8$.

SOLUTION

Because the left side of the equation is raised to the 3/4 power, we begin by raising both sides of the equation to the *reciprocal* power, 4/3. Note that $8^{4/3} = (\sqrt[3]{8})^4 = 2^4 = 16$.

$$(x^2 - x - 4)^{3/4} = 8 \qquad \text{\textit{Given}}$$
$$(x^2 - x - 4) = (8)^{4/3} \qquad \text{\textit{Raise both sides to 4/3 power}}$$
$$x^2 - x - 4 = 16 \qquad \text{\textit{Simplify}}$$
$$x^2 - x - 20 = 0 \qquad \text{\textit{Standard form}}$$
$$(x - 5)(x + 4) = 0 \qquad \text{\textit{Factor}}$$
$$x - 5 = 0, \qquad x + 4 = 0 \qquad \text{\textit{Set factors to zero}}$$
$$x = 5 \qquad\qquad x = -4 \qquad \text{\textit{Solutions}}$$

A check will show that both solutions are valid.

EXAMPLE 8 An Equation Involving Fractions

Solve the equation

$$\frac{2}{x} = \frac{3}{x - 2} - 1.$$

SOLUTION

We can get rid of fractions by multiplying through by the LCD, $x(x - 2)$.

$$x(x - 2)\frac{2}{x} = x(x - 2)\frac{3}{x - 2} - x(x - 2)(1)$$
$$2(x - 2) = 3x - x(x - 2), \; x \neq 0, 2$$
$$2x - 4 = -x^2 + 5x$$
$$x^2 - 3x - 4 = 0$$
$$(x - 4)(x + 1) = 0$$
$$x - 4 = 0, \qquad x + 1 = 0$$
$$x = 4 \qquad\qquad x = -1$$

A check will show that both of these solutions are valid.

Exercises

EXAMPLE 9   *An Equation Involving Absolute Value*

Find all real solutions of $|x^2 - 3x| = -4x + 6$.

SOLUTION

Since the variable expression inside the absolute value signs can be positive or negative, we must solve two equations.

$$
\begin{array}{ll}
x^2 - 3x = -4x + 6 & -(x^2 - 3x) = -4x + 6 \\
x^2 + x - 6 = 0 & x^2 - 7x + 6 = 0 \\
(x + 3)(x - 2) = 0 & (x - 1)(x - 6) = 0 \\
x + 3 = 0, \quad x - 2 = 0 & x - 1 = 0, \quad x - 6 = 0 \\
x = -3 \qquad x = 2 & x = 1 \qquad x = 6
\end{array}
$$

SOLUTION CHECK

$$|(-3)^2 - 3(-3)| = -4(-3) + 6 \qquad \text{\textit{−3 is a solution}}$$

$$|2^2 - 3(2)| \ne -4(2) + 6 \qquad \text{\textit{2 is not a solution}}$$

$$|1^2 - 3(1)| = -4(1) + 6 \qquad \text{\textit{1 is a solution}}$$

$$|6^2 - 3(6)| \ne -4(6) + 6 \qquad \text{\textit{6 is not a solution}}$$

Thus, $x = -3$ and $x = 1$ are the only real solutions.

WARM UP

Find the real solution(s) of each of the given equations.

1. $x^2 - 22x + 121 = 0$

2. $x(x - 20) + 3(x - 20) = 0$

3. $(x + 20)^2 = 625$

4. $5x^2 + x = 0$

5. $3x^2 + 4x - 4 = 0$

6. $12x^2 + 8x - 55 = 0$

7. $x^2 + 4x - 5 = 0$

8. $4x^2 + 4x - 15 = 0$

9. $\dfrac{4}{x} - \dfrac{2}{3} = \dfrac{5}{x} - 2$

10. $\dfrac{1}{x - 2} + 10 = \dfrac{6}{x - 2}$

EXERCISES 2.6

In Exercises 1–66, find all solutions of the given equation. Check your answers in the original equation.

1. $4x^4 - 18x^2 = 0$

2. $20x^3 - 125x = 0$

3. $x^3 - 2x^2 - 3x = 0$

4. $2x^4 - 15x^3 + 18x^2 = 0$

5. $x^4 - 81 = 0$

6. $x^6 - 64 = 0$

7. $5x^3 + 30x^2 + 45x = 0$

8. $9x^4 - 24x^3 + 16x^2 = 0$

9. $x^3 - 3x^2 - x + 3 = 0$

10. $x^3 + 2x^2 + 3x + 6 = 0$

11. $x^4 - x^3 + x - 1 = 0$

12. $x^4 + 2x^3 - 8x - 16 = 0$

13. $x^4 - 10x^2 + 9 = 0$

14. $x^4 - 29x^2 + 100 = 0$

15. $x^4 + 5x^2 - 36 = 0$

16. $x^4 - 4x^2 + 3 = 0$

17. $4x^4 - 65x^2 + 16 = 0$

18. $36t^4 + 29t^2 - 7 = 0$

19. $x^6 + 7x^3 - 8 = 0$

20. $x^6 + 3x^3 + 2 = 0$

21. $\dfrac{1}{t^2} + \dfrac{8}{t} + 15 = 0$

22. $6\left(\dfrac{s}{s+1}\right)^2 + 5\left(\dfrac{s}{s+1}\right) - 6 = 0$

23. $2x + 9\sqrt{x} - 5 = 0$

24. $6x - 7\sqrt{x} - 3 = 0$

25. $5 - 3x^{1/3} - 2x^{2/3} = 0$

26. $9t^{2/3} + 24t^{1/3} + 16 = 0$

27. $\sqrt{2x} - 10 = 0$

28. $4\sqrt{x} - 3 = 0$

29. $\sqrt{x - 10} - 4 = 0$

30. $\sqrt{5 - x} - 3 = 0$

31. $\sqrt[3]{2x + 5} + 3 = 0$

32. $\sqrt[3]{3x + 1} - 5 = 0$

33. $x = \sqrt{11x - 30}$

34. $2x - \sqrt{15 - 4x} = 0$

35. $-\sqrt{26 - 11x} + 4 = x$

36. $x + \sqrt{31 - 9x} = 5$

37. $\sqrt{x + 1} - 3x = 1$

38. $\sqrt{x + 5} = \sqrt{x - 5}$

39. $\sqrt{x} + \sqrt{x - 20} = 10$

40. $\sqrt{x} - \sqrt{x - 5} = 1$

41. $\sqrt{x + 5} + \sqrt{x - 5} = 10$

42. $2\sqrt{x + 1} - \sqrt{2x + 3} = 1$

43. $(x - 5)^{2/3} = 16$

44. $(x + 3)^{4/3} = 16$

45. $(x + 3)^{3/2} = 8$

46. $(x^2 + 2)^{2/3} = 9$

47. $(x^2 - 5)^{2/3} = 16$

48. $(x^2 - x - 22)^{4/3} = 16$

49. $3x(x - 1)^{1/2} + 2(x - 1)^{3/2} = 0$

50. $4x^2(x - 1)^{1/3} + 6x(x - 1)^{4/3} = 0$

51. $8x^2(x^2 - 1)^{1/3} + 3(x^2 - 1)^{4/3} = 0$

52. $3x(2x - 1)^{1/2} + (2x - 1)^{3/2} = 0$

53. $x = \dfrac{3}{x} + \dfrac{1}{2}$

54. $4x + 1 = \dfrac{3}{x}$

55. $\dfrac{1}{x} = \dfrac{4}{x - 1} + 1$

56. $x + \dfrac{9}{x + 1} = 5$

57. $\dfrac{4}{x + 1} - \dfrac{3}{x + 2} = 1$

58. $\dfrac{x + 1}{3} - \dfrac{x + 1}{x + 2} = 0$

59. $|x + 1| = 2$

60. $|x - 2| = 3$

61. $|2x - 1| = 5$

62. $|3x + 2| = 7$

63. $|x| = x^2 + x - 3$

64. $|x^2 + 6x| = 3x + 18$

65. $|x - 10| = x^2 - 10x$

66. $|x + 1| = x^2 - 5$

In Exercises 67–70, use a calculator to find the real solutions of the equation. Round your answers to three decimal places.

67. $3.2x^4 - 1.5x^2 - 2.1 = 0$

68. $7.08x^6 + 4.15x^3 - 9.6 = 0$

69. $1.8x - 6\sqrt{x} - 5.6 = 0$

70. $4x^{2/3} + 8x^{1/3} + 3.6 = 0$

71. An airline offers daily flights between Chicago and Denver. The total monthly cost of these flights is given by

$$C = \sqrt{0.2x + 1}$$

where C is measured in millions of dollars and x is measured in thousands of passengers. The total cost of the flights for a certain month is 2.5 million dollars. How many passengers flew that month?

72. The demand equation for a certain product is given by

$$p = 40 - \sqrt{x - 1}$$

where x is the number of units demanded per day and p is the price per unit. Find the demand if the price is set at $34.70.

73. A power station is on one side of a river that is $\frac{1}{2}$ mile wide. A factory is 6 miles downstream on the other side of the river. It costs $18 per foot to run power lines overland and $24 per foot to run them underwater. The total cost of the project is $616,877.27. Find the length x as labeled in the figure.

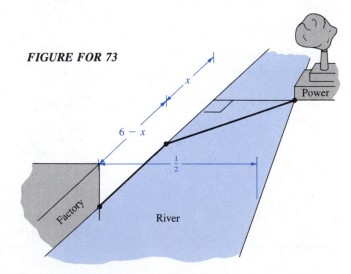

FIGURE FOR 73

74. The diagonal of a rectangle is 2 inches longer than the length which, in turn, is 4 inches longer than the width. Find the length of the diagonal.

75. Determine the length and width of a rectangle with a perimeter of 28 inches and a diagonal of 10 inches.

76. An equation describing a circuit containing inductance i and capacitance C is

$$i = \pm\sqrt{\frac{1}{LC}}\sqrt{Q^2 - q}.$$

Solve this equation for Q.

77. The surface area of a cone is given by

$$S = \pi r\sqrt{r^2 + h^2}.$$

Solve this equation for h.

2.7 Linear Inequalities

Simple inequalities were introduced in Section 1.1 in the context of *order* on the real line. There, we used inequality signs to compare two numbers and to denote subsets of real numbers. For instance, the simple inequality

$$x \geq 3$$

denotes all numbers x greater than or equal to 3.

In this section we expand our work with inequalities to include more involved statements such as

$$5x - 7 < 3x + 9 \qquad \text{or} \qquad -3 \leq 6x - 1 < 3.$$

As with an equation, we *solve* an inequality by finding the set of all real numbers for which the inequality is true. In most cases, these solution sets consist of intervals on the real line and we call them **solution intervals** of the inequality. Solution intervals can be denoted by the following **interval notation.**

Interval Notation for Subsets of Real Numbers (Bounded)

The following intervals on the real number line are called **bounded intervals.**

Notation	Interval Type	Inequality	Graph
$[a, b]$	closed	$a \leq x \leq b$	
(a, b)	open	$a < x < b$	
$[a, b)$	half-open	$a \leq x < b$	
$(a, b]$	half-open	$a < x \leq b$	

Interval Notation for Subsets of Real Numbers (Unbounded)

The following intervals on the real number line are called **unbounded intervals.**

Notation	Interval Type	Inequality	Graph
$[a, \infty)$	half-open	$x \geq a$	
(a, ∞)	open	$x > a$	
$(-\infty, b]$	half-open	$x \leq b$	
$(-\infty, b)$	open	$x < b$	
$(-\infty, \infty)$	entire real line		

Remark: The symbols ∞ (**positive infinity**) and $-\infty$ (**negative infinity**) are not real numbers. They are simply convenient symbols used to denote the *unboundedness* of an interval such as $(1, \infty)$. Note also that an **open interval** (a, b) excludes its endpoints a and b, whereas a **closed interval** includes them. **Half-open intervals** include just one of the endpoints.

EXAMPLE 1 *Intervals and Inequalities*

Write an inequality to represent each of the following intervals and state whether the interval is bounded or unbounded.

(a) $(-3, 5]$ 　　　　　　(b) $(-3, \infty)$ 　　　　　　(c) $[0, 2]$

SOLUTION

(a) $(-3, 5]$ corresponds to $-3 < x \leq 5$. 　　　*Bounded*
(b) $(-3, \infty)$ corresponds to $-3 < x$. 　　　*Unbounded*
(c) $[0, 2]$ corresponds to $0 \leq x \leq 2$. 　　　*Bounded*

Solving Inequalities

A **linear inequality** in x, like a linear equation, is an inequality in which the variable x occurs only to the first power. For instance, $2x - 6 \leq 4x - 7$ is a linear inequality, whereas $x^2 + 4x - 6 < 0$ is not a linear inequality. The procedures used to solve linear inequalities are much like those for solving linear equations. There is, however, an important difference. That is, when both sides of an inequality are multiplied (or divided) by a negative number, we must *reverse* the direction of the inequality. Here is an example.

$$-2 < 5 \qquad \text{\textit{Given inequality}}$$
$$(-3)(-2) > (-3)(5) \qquad \text{\textit{Multiply both sides by} } -3 \text{ \textit{and reverse inequality}}$$
$$6 > -15 \qquad \text{\textit{New inequality}}$$

Properties of Inequalities

Let a, b, c, and d be real numbers. Then the following properties are true.

Property	Example
1. **Transitive** $a < b$ and $b < c \implies a < c$	Since $-2 < 5$ and $5 < 7$, then $-2 < 7$.
2. **Addition of Inequalities** $a < b$ and $c < d \implies a + c < b + d$	Since $2 < 4$ and $3 < 5$, then $2 + 3 < 4 + 5$.
3. **Addition of a Constant** $a < b \implies a + c < b + c$	Since $-3 < 7$, then $-3 + 2 < 7 + 2$.
4. **Multiplying by a Constant** (i) For $c > 0$, $a < b \implies ac < bc$	Since $5 > 0$ and $3 < 9$, then $3(5) < 9(5)$.
(ii) For $c < 0$, $a < b \implies ac > bc$	Since $-5 < 0$ and $3 < 9$, then $3(-5) > 9(-5)$.

Remark: Each of the above properties is true if the symbol $<$ is replaced by $\leq$.

EXAMPLE 2   Solving a Linear Inequality

Solve the inequality $5x - 7 > 3x + 9$.

SOLUTION

$$5x - 7 > 3x + 9 \qquad \textit{Given}$$
$$5x > 3x + 16 \qquad \textit{Add 7 to both sides}$$
$$5x - 3x > 16 \qquad \textit{Subtract 3x from both sides}$$
$$2x > 16 \qquad \textit{Combine terms}$$
$$x > 8 \qquad \textit{Divide both sides by 2}$$

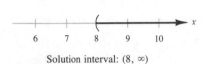

Solution interval: $(8, \infty)$

FIGURE 2.7

Thus, the solution interval is $(8, \infty)$, as shown in Figure 2.7. ▬▬

Remark: The five inequalities forming the solution steps of Example 2 are all **equivalent** in the sense that each has the same solution set.

Checking the solution set of an inequality is not as simple as checking the solutions of an equation. (There are usually too many x-values to substitute back into the original inequality.) We can, however, get an indication of the validity of a solution set by substituting a few convenient values of x. For instance, in Example 2 we found the solution of $5x - 7 > 3x + 9$ to be $x > 8$. Try checking that $x = 9$ satisfies the original inequality, whereas $x = 7$ does not.

EXAMPLE 3   Solving a Linear Inequality

Solve the inequality

$$1 - \frac{3x}{2} \geq x - 4.$$

SOLUTION

$$1 - \frac{3x}{2} \geq x - 4 \qquad \textit{Given}$$
$$2 - 3x \geq 2x - 8 \qquad \textit{Multiply both sides by LCD}$$
$$-3x \geq 2x - 10 \qquad \textit{Subtract 2 from both sides}$$
$$-5x \geq -10 \qquad \textit{Subtract 2x from both sides}$$
$$x \leq 2 \qquad \textit{Divide both sides by} -5 \textit{ and reverse inequality}$$

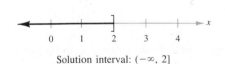

Solution interval: $(-\infty, 2]$

FIGURE 2.8

Thus, the solution interval is $(-\infty, 2]$, as shown in Figure 2.8. ▬▬

Sometimes it is convenient to write two inequalities as a "double" inequality. For instance, we can write the two inequalities $-3 \le 6x - 1$ and $6x - 1 < 3$ more simply as $-3 \le 6x - 1 < 3$. This form allows us to solve the two given inequalities together, as demonstrated in Example 4.

EXAMPLE 4 *Solving a Double Inequality*

Solve the inequality $-3 \le 6x - 1 < 3$.

SOLUTION

$$-3 \le 6x - 1 < 3 \qquad \qquad \textit{Given}$$
$$-2 \le 6x < 4 \qquad \qquad \textit{Add 1}$$
$$-\frac{1}{3} \le x < \frac{2}{3} \qquad \qquad \textit{Divide by 6 and reduce}$$

Thus, the solution interval is $[-1/3, 2/3)$, as shown in Figure 2.9.

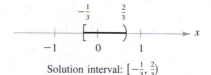

Solution interval: $\left[-\frac{1}{3}, \frac{2}{3}\right)$

FIGURE 2.9

The double inequality in Example 4 could have been solved in two parts, as

$$-3 \le 6x - 1 \qquad \textit{and} \qquad 6x - 1 < 3$$
$$-2 \le 6x \qquad \qquad \qquad 6x < 4$$
$$-\frac{1}{3} \le x \qquad \qquad \qquad x < \frac{2}{3}$$

and the intervals put together as $-1/3 \le x < 2/3$.

When combining two inequalities to form a double inequality, be sure that the inequalities satisfy the Transitive Law for inequalities. For instance, it is *incorrect* to combine the inequalities $3 < x$ and $x \le -1$ as

$$3 < x \le -1. \qquad \qquad \textit{Incorrect}$$

This "inequality" is obviously wrong because 3 is not less than -1.

EXAMPLE 5   An Application of Inequalities

A subcompact car can be rented from Company A for $180 per week with no extra charge for mileage. A similar car can be rented from Company B for $100 per week, plus 20 cents for each mile driven. How many miles must a person drive in a week to make the rental fee for Company A less than that for Company B?

SOLUTION

$$\text{Model:} \quad \text{(weekly cost from B)} > \text{(weekly cost from A)}$$

$$\text{Label:} \quad m = \text{number of miles driven in one week}$$

$$\text{Inequality:} \quad 100 + 0.20m > 180$$

$$\text{Solution:} \quad 0.20m > 80$$
$$m > 400 \text{ miles}$$

Thus, the car from Company A is cheaper if the person plans to drive more than 400 miles in a week.

Inequalities Involving Absolute Values

Many important uses of inequalities involve absolute values. In Section 1.1 we used absolute value to denote the distance between two points on the real number line. That is,

$$d(a, b) = |a - b|$$
$$= \text{distance between points } a \text{ and } b.$$

This means that

$$d(x, 0) = |x|$$
$$= \text{distance between } x \text{ and } 0.$$

As a consequence, we give the following interpretation of two basic types of inequalities involving absolute values.

Two Basic Types of Inequalities Involving Absolute Value

Let a be a positive real number.

$|x| < a$ if and only if $-a < x < a$.

$|x| > a$ if and only if $x < -a \text{ or } x > a$.

Inequality	Interpretation	Graph
$|x| < a$	All numbers x whose distance from 0 is *less* than a	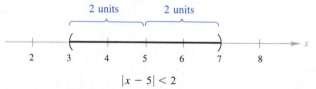
$|x| > a$	All numbers x whose distance from 0 is *greater* than a	

Remark: Note that $-a < x < a$ *means that* $-a < x$ *and* $x < a$.

EXAMPLE 6 An Inequality Involving Absolute Value

Solve the inequality $|x - 5| < 2$.

SOLUTION

The solution set of this inequality is the set of all numbers x whose distance from 5 is less than 2 units, as shown in Figure 2.10.

FIGURE 2.10

$$|x - 5| < 2 \qquad \text{\textit{Given}}$$
$$-2 < x - 5 < 2 \qquad \text{\textit{Interpret absolute value}}$$
$$3 < x < 7 \qquad \text{\textit{Add 5}}$$

Thus, the solution interval is $(3, 7)$.

EXAMPLE 7 An Inequality Involving Absolute Value

Solve the inequality $|x + 3| \geq 7$.

SOLUTION

The solution set of this inequality is the set of all numbers x that are at least 7 units away from -3, as shown in Figure 2.11.

$|x + 3| \geq 7$		*Given*
$x + 3 \leq -7$ *or* $x + 3 \geq 7$		*Interpret absolute value*
$x \leq -10$ $x \geq 4$		*Solve separately*

Therefore, the inequality is satisfied if x lies in the interval $(-\infty, -10]$ *or* in the interval $[4, \infty)$.

$|x + 3| \geq 7$

FIGURE 2.11

EXAMPLE 8 An Inequality Involving Absolute Value

Solve the inequality

$$\left| 2 - \frac{x}{3} \right| < 0.01.$$

SOLUTION

$\left\| 2 - \frac{x}{3} \right\| < 0.01$	*Given*
$-0.01 < 2 - \frac{x}{3} < 0.01$	*Interpret absolute value*
$-2.01 < \frac{-x}{3} < -1.99$	*Subtract 2*
$6.03 > x > 5.97$	*Multiply by -3 and reverse both inequalities*

Therefore, the solution interval is $(5.97, 6.03)$.

WARM UP

Which of the two given numbers is larger?

1. $-\frac{1}{2}, -7$

2. $-\frac{1}{3}, -\frac{1}{6}$

3. $-\pi, -3$

4. $-6, -\frac{13}{2}$

Use inequality notation to denote each of the given statements.

5. x is nonnegative.

6. z is strictly between -3 and 10.

7. P is no more than 2.

8. W is at least 200.

Evaluate each expression for the given values of x.

9. $|x - 10|, x = 12, x = 3$

10. $|2x - 3|, x = \frac{3}{2}, x = 1$

EXERCISES 2.7

In Exercises 1–4, determine whether or not the given value of x satisfies the inequality.

1. $5x - 12 > 0$

   (a) $x = 3$ (b) $x = -3$

   (c) $x = \frac{5}{2}$ (d) $x = \frac{3}{2}$

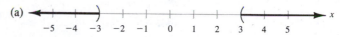

(a)

2. $x + 1 < \dfrac{2x}{3}$

   (a) $x = 0$ (b) $x = 4$

   (c) $x = -4$ (d) $x = -3$

(b)

3. $0 < \dfrac{x - 2}{4} < 2$

   (a) $x = 4$ (b) $x = 10$

   (c) $x = 0$ (d) $x = \frac{7}{2}$

(c)

4. $-1 < \dfrac{3 - x}{2} \le 1$

   (a) $x = 0$ (b) $x = \sqrt{5}$

   (c) $x = 1$ (d) $x = 5$

(d)

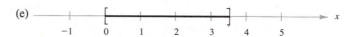

(e)

In Exercises 5–12, match the given inequality with its graph. [The graphs are labeled (a)–(h), at right.]

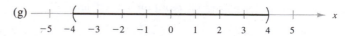

(f)

5. $x < 4$ **6.** $x \ge 6$

7. $-2 < x \le 5$ **8.** $0 \le x \le \frac{7}{2}$

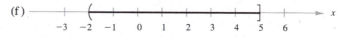

(g)

9. $|x| < 4$ **10.** $|x| > 3$

11. $|x - 5| > 2$ **12.** $|x + 6| < 3$

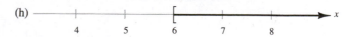

(h)

Algebraic Equations and Inequalities

In Exercises 13–46, solve the inequality and sketch the solution on the real number line.

13. $4x < 12$

14. $2x > 3$

15. $-10x < 40$

16. $-6x > 15$

17. $x - 5 \geq 7$

18. $x + 7 \leq 12$

19. $4(x + 1) < 2x + 3$

20. $2x + 7 < 3$

21. $2x - 1 \geq 0$

22. $3x + 1 \geq 2$

23. $4 - 2x < 3$

24. $6x - 4 \leq 2$

25. $1 < 2x + 3 < 9$

26. $-8 \leq 1 - 3(x - 2) < 13$

27. $-4 < \dfrac{2x - 3}{3} < 4$

28. $0 \leq \dfrac{x + 3}{2} < 5$

29. $\dfrac{3}{4} > x + 1 > \dfrac{1}{4}$

30. $-1 < -\dfrac{x}{3} < 1$

31. $|x| < 5$

32. $|2x| < 6$

33. $\left|\dfrac{x}{2}\right| > 3$

34. $|5x| > 10$

35. $|x - 20| \leq 4$

36. $|x - 7| < 6$

37. $|x - 20| \geq 4$

38. $|x + 14| + 3 > 17$

39. $\left|\dfrac{x - 3}{2}\right| \geq 5$

40. $|1 - 2x| < 5$

41. $|9 - 2x| - 2 < -1$

42. $\left|1 - \dfrac{2x}{3}\right| < 1$

43. $2|x + 10| \geq 9$

44. $3|4 - 5x| \leq 9$

45. $|x - 5| < 0$

46. $|x - 5| \geq 0$

In Exercises 47–52, find the interval(s) on the real number line for which the radicand is nonnegative (greater than or equal to zero).

47. $\sqrt{x - 5}$

48. $\sqrt{x - 10}$

49. $\sqrt{x + 3}$

50. $\sqrt[4]{6x + 15}$

51. $\sqrt[4]{7 - 2x}$

52. $\sqrt{3 - x}$

In Exercises 53–60, use absolute value notation to define each interval (or pair of intervals) on the real line.

53.

54.

55.

56.

57. All real numbers within 10 units of 12.

58. All real numbers at least 5 units from 8.

59. All real numbers whose distances from -3 are more than 5.

60. All real numbers whose distances from -6 are no more than 7.

61. P dollars is invested at a simple interest rate of r. The balance in the account after t years is given by

$A = P + Prt.$

In order for an investment of $1000 to grow to *more than* $1250 in two years, what must the interest rate be?

62. A doughnut shop at a shopping mall sells a dozen doughnuts for $2.95. Beyond the fixed costs (for rent, utilities, and insurance) of $150 per day it costs $1.45 for enough materials (flour, sugar, and so on) and labor to produce a dozen doughnuts. If the daily profit varies between $50 and $200, between what levels (in dozens) do the daily sales vary?

63. The revenue for selling x units of a product is

$R = 115.95x.$

The cost of producing x units is

$C = 95x + 750.$

In order to obtain a profit, the revenue must be greater than the cost. For what values of x will this product return a profit?

64. A utility company has a fleet of vans. The annual operating cost per van is

$$C = 0.32m + 2300$$

where m is the number of miles traveled by a van in a year. What number of miles will yield an annual operating cost that is less than $10,000?

65. The heights, h, of two-thirds of the members of a certain population satisfy the inequality

$$\left| \frac{h - 68.5}{2.7} \right| \leq 1$$

where h is measured in inches. Determine the interval on the real line in which these heights lie.

66. A certain electronic device is to be operated in an environment with relative humidity h in the interval defined by

$$|h - 50| \leq 30.$$

What are the minimum and maximum relative humidities for the operation of this device?

67. Given two real numbers a and b, such that $a > b > 0$, prove that

$$\frac{1}{a} < \frac{1}{b}.$$

2.8 *Other Types of Inequalities*

To solve a polynomial inequality like

$$x^2 - x + 6 < 0$$

we use the principle that a polynomial can change signs only at its zeros (the values that make the polynomial zero). Between two consecutive zeros a polynomial must be entirely positive or entirely negative. This means that when the real zeros of a polynomial are put in order, they divide the real line into intervals in which the polynomial has no sign changes. We call these zeros the **critical numbers** and the resulting intervals the **test intervals** for the inequality. For example, the polynomial

$$x^2 - x - 6 = (x + 2)(x - 3)$$

has two zeros, $x = -2$ and $x = 3$, and these zeros divide the real line into three test intervals:

$$(-\infty, -2), \quad (-2, 3), \quad \text{and} \quad (3, \infty).$$

To solve the original inequality, we need only test one value from each of these test intervals, as demonstrated in Example 1.

Algebraic Equations and Inequalities

EXAMPLE 1 Solving a Quadratic Inequality

Solve the inequality $x^2 < x + 6$.

SOLUTION

$$x^2 < x + 6 \qquad \textit{Given}$$
$$x^2 - x - 6 < 0 \qquad \textit{Standard form}$$
$$(x - 3)(x + 2) < 0 \qquad \textit{Factor}$$

Critical numbers: $x = -2, x = 3$

Test intervals: $(-\infty, -2), (-2, 3), (3, \infty)$

Test: Is $(x - 3)(x + 2) < 0$?

To test an interval, we first choose a convenient number in the interval and then compute the sign of $x^2 - x - 6$. The results are shown in Figure 2.12.

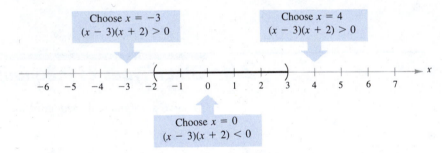

Choose $x = -3$
$(x - 3)(x + 2) > 0$

Choose $x = 4$
$(x - 3)(x + 2) > 0$

Choose $x = 0$
$(x - 3)(x + 2) < 0$

FIGURE 2.12

Since the inequality $(x - 3)(x + 2) < 0$ is satisfied only by the middle test interval, we conclude that the solution set of the inequality is the interval $(-2, 3)$.

In Example 1, note that the first step in solving a polynomial inequality is to write it in standard form (with the polynomial on the left and zero on the right).

EXAMPLE 2 *Solving a Cubic Inequality*

Solve the inequality $2x^3 + 5x^2 \geq 12x$.

SOLUTION

$$2x^3 + 5x^2 \geq 12x \qquad \text{\textit{Given}}$$
$$2x^3 + 5x^2 - 12x \geq 0 \qquad \text{\textit{Standard form}}$$
$$x(2x - 3)(x + 4) \geq 0 \qquad \text{\textit{Factor}}$$

Critical numbers: $x = -4, \; x = 0, \; x = \frac{3}{2}$

Test intervals: $(-\infty, -4), \; (-4, 0), \; (0, \frac{3}{2}), \; (\frac{3}{2}, \infty)$

Test: Is $x(2x - 3)(x + 4) \geq 0$?   (See Figure 2.13.)

Solution intervals: $[-4, 0], \; [\frac{3}{2}, \infty)$

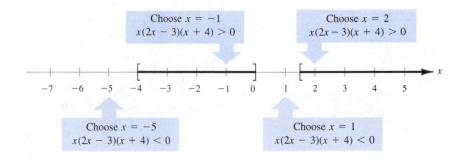

FIGURE 2.13

Remark: In Example 2, note that the polynomial $2x^3 + 5x^2 - 12x$ is zero at the critical numbers, $-4, 0,$ and $3/2$. Since we are asked to find the values of x for which $2x^3 + 5x^2 - 12x$ is greater than *or equal* to zero, we include $x = -4, 0,$ and $3/2$ in the solution intervals.

Example 3 shows how the quadratic formula can be used to solve a nonfactorable inequality.

EXAMPLE 3 *Using the Quadratic Formula*

Solve the inequality $x^2 + 4x + 1 \leq 0$.

SOLUTION

By the quadratic formula, we find the zeros of $x^2 + 4x + 1$ to be

$$x = \frac{-4 \pm \sqrt{16 - 4}}{2} = \frac{-4 \pm 2\sqrt{3}}{2} = -2 \pm \sqrt{3}.$$

Critical numbers: $x = -2 - \sqrt{3}, x = -2 + \sqrt{3}$

Test intervals: $(-\infty, -2 - \sqrt{3}), (-2 - \sqrt{3}, -2 + \sqrt{3}),$
$(-2 + \sqrt{3}, \infty)$

Test: Is $x^2 + 4x + 1 \leq 0$? (See Figure 2.14.)

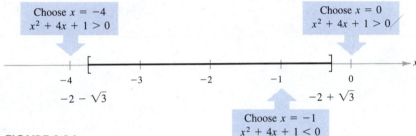

Choose $x = -4$
$x^2 + 4x + 1 > 0$

Choose $x = 0$
$x^2 + 4x + 1 > 0$

$-2 - \sqrt{3}$

$-2 + \sqrt{3}$

Choose $x = -1$
$x^2 + 4x + 1 < 0$

FIGURE 2.14

Since we are asked to find the *x*-values for which $x^2 + 4x + 1$ is less than *or equal* to zero, we conclude that the solution set of the original inequality is the *closed* interval

$$[-2 - \sqrt{3}, -2 + \sqrt{3}].$$

The concept of critical numbers can be extended to inequalities involving fractional expressions. Specifically, an expression that is the ratio of two polynomials can change signs only at its *zeros* (the values that make the numerator zero) or *undefined* values (the values that make the denominator zero). For example, the critical numbers of the inequality

$$\frac{x - 1}{(x - 2)(x + 3)} < 0$$

are $x = 1$, $x = 2$, and $x = -3$ because these are the only values at which the fractional expression is either zero or undefined. In Example 4 note that we begin by writing the inequality in *standard form*.

Other Types of Inequalities

EXAMPLE 4 *An Inequality Involving Fractions*

Solve the inequality

$$\frac{2x - 7}{x - 5} \le 3.$$

SOLUTION

$$\frac{2x - 7}{x - 5} \le 3 \qquad\qquad\qquad \textit{Given}$$

$$\frac{2x - 7}{x - 5} - 3 \le 0 \qquad\qquad \textit{Standard form}$$

$$\frac{2x - 7 - 3x + 15}{x - 5} \le 0 \qquad \textit{Combine terms}$$

$$\frac{-x + 8}{x - 5} \le 0 \qquad\qquad \textit{Simplify}$$

Critical numbers: $x = 5, x = 8$

Test intervals: $(-\infty, 5), (5, 8), (8, \infty)$

Test: Is $\dfrac{-x + 8}{x - 5} \le 0$?   (See Figure 2.15.)

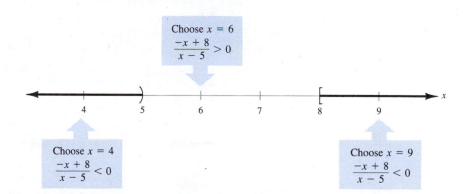

FIGURE 2.15

Solution intervals: $(-\infty, 5), [8, \infty)$

Note that we include the critical number 8 in the solution intervals, but we do not include $x = 5$ since it yields a zero denominator in the original inequality.

Algebraic Equations and Inequalities

EXAMPLE 5 *Another Inequality Involving Fractions*

Solve the inequality

$$\frac{x}{x-2} > \frac{1}{x+3}.$$

SOLUTION

$$\frac{x}{x-2} > \frac{1}{x+3} \qquad \textit{Given}$$

$$\frac{x}{x-2} - \frac{1}{x+3} > 0 \qquad \textit{Standard form}$$

$$\frac{x^2 + 3x - x + 2}{(x-2)(x+3)} > 0 \qquad \textit{Combine fractions}$$

$$\frac{x^2 + 2x + 2}{(x-2)(x+3)} > 0 \qquad \textit{Simplify}$$

The discriminant $b^2 - 4ac = 4 - 8 = -4 < 0$ indicates that the numerator has no zeros. Therefore, $x = 2$ and $x = -3$ are the only critical numbers.

Critical numbers: $x = -3, x = 2$

Test intervals: $(-\infty, -3), (-3, 2), (2, \infty)$ (See Figure 2.16.)

Solution intervals: $(-\infty, -3), (2, \infty)$

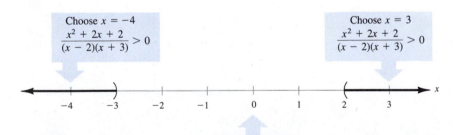

FIGURE 2.16

Applications of Inequalities

In Examples 6 and 7, we give two applications of nonlinear inequalities.

EXAMPLE 6 *Finding the Domain of a Variable*

Find the domain of x in the expression $\sqrt{64 - 4x^2}$.

SOLUTION

Since $\sqrt{64 - 4x^2}$ has real values if and only if the radicand $64 - 4x^2$ is nonnegative, we solve the inequality $64 - 4x^2 \geq 0$.

$$64 - 4x^2 \geq 0 \qquad \text{\textit{Standard form}}$$
$$16 - x^2 \geq 0 \qquad \text{\textit{Divide both sides by 4}}$$
$$(4 - x)(4 + x) \geq 0 \qquad \text{\textit{Factor}}$$

Critical numbers: $\quad x = -4, x = 4$

Test intervals: $\quad (-\infty, -4), (-4, 4), (4, \infty)$

A test shows that $64 - 4x^2 \geq 0$ in the *closed* interval $[-4, 4]$. Thus, the domain of x is the interval $[-4, 4]$.

EXAMPLE 7 *The Height of a Projectile*

A projectile is fired straight upward from ground level with an initial velocity of 384 feet per second. During what time period will its height exceed 2000 feet? (See Figure 2.17.)

SOLUTION

Recall from Section 2.4 that the position of an object moving in a vertical path is given by $s = -16t^2 + v_0 t + s_0$, where s is the height in feet and t is the time in seconds. In this case, $s_0 = 0$ and $v_0 = 384$. Thus, we need to solve the inequality $-16t^2 + 384t > 2000$.

$$-16t^2 + 384t > 2000 \qquad \text{\textit{Given}}$$
$$t^2 - 24t < -125 \qquad \text{\textit{Divide by -16 and reverse inequality}}$$
$$t^2 - 24t + 125 < 0 \qquad \text{\textit{Standard form}}$$

By the quadratic formula the critical numbers are

$$t = \frac{24 \pm \sqrt{76}}{2} = \frac{24 \pm 2\sqrt{19}}{2} = 12 \pm \sqrt{19} \approx 7.64 \text{ or } 16.36.$$

A test will verify that the height of the projectile will exceed 2000 feet during the time interval

$$7.64 \text{ sec} < t < 16.36 \text{ sec}.$$

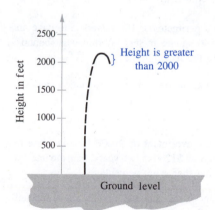

} Height is greater than 2000

Ground level

FIGURE 2.17

WARM UP

Solve the inequalities.

1. $-\dfrac{y}{3} > 2$

2. $-6z < 27$

3. $-3 \le 2x + 3 < 5$

4. $-3x + 5 \ge 20$

5. $10 > 4 - 3(x + 1)$

6. $3 < 1 + 2(x - 4) < 7$

7. $2|x| \le 7$

8. $|x - 3| > 1$

9. $|x + 4| > 2$

10. $|2 - x| \le 4$

EXERCISES 2.8

In Exercises 1–28, solve the inequality and graph the solution on the real number line.

1. $x^2 \le 9$

2. $x^2 < 5$

3. $x^2 > 4$

4. $(x - 3)^2 \ge 1$

5. $(x + 2)^2 < 25$

6. $(x + 6)^2 \le 8$

7. $x^2 + 4x + 4 \ge 9$

8. $x^2 - 6x + 9 < 16$

9. $x^2 + x < 6$

10. $x^2 + 2x > 3$

11. $3(x - 1)(x + 1) > 0$

12. $6(x + 2)(x - 1) < 0$

13. $x^2 + 2x - 3 < 0$

14. $x^2 - 4x - 1 > 0$

15. $4x^3 - 6x^2 < 0$

16. $4x^3 - 12x^2 > 0$

17. $x^3 - 4x \ge 0$

18. $2x^3 - x^4 \le 0$

19. $\dfrac{1}{x} > x$

20. $\dfrac{1}{x} < 4$

21. $\dfrac{x + 6}{x + 1} < 2$

22. $\dfrac{x + 12}{x + 2} \ge 3$

23. $\dfrac{3x - 5}{x - 5} > 4$

24. $\dfrac{5 + 7x}{1 + 2x} < 4$

25. $\dfrac{4}{x + 5} > \dfrac{1}{2x + 3}$

26. $\dfrac{5}{x - 6} > \dfrac{3}{x + 2}$

27. $\dfrac{1}{x - 3} \le \dfrac{9}{4x + 3}$

28. $\dfrac{1}{x} \ge \dfrac{1}{x + 3}$

In Exercises 29–36, find the domain of x in the given expression.

29. $\sqrt[4]{4 - x^2}$

30. $\sqrt{x^2 - 4}$

31. $\sqrt{x^2 - 7x + 12}$

32. $\sqrt{144 - 9x^2}$

33. $\sqrt{12 - x - x^2}$

34. $\sqrt{x^2 + 4}$

35. $\sqrt{x^2 - 3x + 3}$

36. $\sqrt[4]{-x^2 + 2x - 2}$

37. A projectile is fired straight upward from ground level with an initial velocity of 160 feet per second.
    (a) At what instant will it be back at ground level?
    (b) During what time period will its height exceed 384 feet?

38. Repeat Exercise 37 for an initial velocity of 128 feet per second, and in part (b) find the time period in which the height of the projectile is less than 128 feet.

39. A rectangle with a perimeter of 100 meters is to have an area of at least 500 square meters. Within what bounds must the length of the rectangle lie?

40. P dollars invested at interest rate r compounded annually increases to an amount

    $$A = P(1 + r)^2$$

    in two years. If an investment of \$1000 is to increase to an amount greater than \$1200 in two years, then the interest rate must be greater than what percentage?

41. When two resistors of resistance R_1 and R_2 are connected in parallel, the total resistance R satisfies the equation

    $$\frac{1}{R} = \frac{1}{R_1} + \frac{1}{R_2}.$$

    Find R_1 for a parallel circuit in which $R_2 = 2$ ohms and R must be at least 1 ohm.

42. The sum of a real number and six times its reciprocal must be at least 6. For what real numbers will this condition be met?

CHAPTER 2 REVIEW EXERCISES

In Exercises 1–34, solve the given equation.

1. $3x - 2(x + 5) = 10$ **2.** $4x + 2(7 - x) = 5$

3. $5x^4 - 12x^3 = 0$ **4.** $4x^3 - 6x = 0$

5. $6x = 3x^2$ **6.** $2x = \dfrac{2}{x^3}$

7. $3\left(1 - \dfrac{1}{5t}\right) = 0$ **8.** $3x^2 + 1 = 0$

9. $2 - x^{-2} = 0$ **10.** $2 + 8x^{-2} = 0$

11. $\dfrac{8}{x^3} = 1$ **12.** $-t^2 + 6t = 4$

13. $4t^3 - 12t^2 + 8t = 0$ **14.** $12t^3 - 84t^2 + 120t = 0$

15. $(x - 1)(2x - 3) + (x^2 - 3x + 2) = 0$

16. $\dfrac{(t^2 + 2t + 2) - (t + 1)(2t + 2)}{(t^2 + 2t + 2)^2} = 0$

17. $\dfrac{(x - 1)(2x) - x^2}{(x - 1)^2} = 0$ **18.** $\dfrac{1}{(t + 1)^2} = 1$

19. $\dfrac{4}{(x - 4)^2} = 1$ **20.** $\dfrac{1}{x - 2} = 3$

21. $|x - 5| = 10$ **22.** $|2x + 3| = 7$

23. $|x^2 - 3| = 2x$ **24.** $|x^2 - 6| = x$

25. $\sqrt{x + 4} = 3$ **26.** $\sqrt{x - 2} - 8 = 0$

27. $(x - 1)^{2/3} - 25 = 0$ **28.** $(x + 2)^{3/4} = 27$

29. $(x + 4)^{1/2} + 5x(x + 4)^{3/2} = 0$

30. $8x^2(x^2 - 4)^{1/3} + (x^2 - 4)^{4/3} = 0$

31. $2(x - 2)\sqrt{x} + \dfrac{(x - 2)^2}{2\sqrt{x}} = 0$

32. $3\sqrt{3x} + \dfrac{3(x + 2)}{2\sqrt{3x}} = 0$

33. $\sqrt{2x + 3} + \sqrt{x - 2} = 2$

34. $5\sqrt{x} - \sqrt{x - 1} = 6$

In Exercises 35–46, solve the given inequality.

35. $\frac{1}{2}(3 - x) > \frac{1}{3}(2 - 3x)$

36. $\dfrac{x}{5} - 6 \leq -\dfrac{x}{2} + 6$

37. $x^2 - 4 \leq 0$

38. $x^2 - 2x \geq 3$

39. $\dfrac{x - 5}{3 - x} < 0$

40. $\dfrac{2}{x + 1} \leq \dfrac{3}{x - 1}$

41. $|x - 2| < 1$

42. $|x| \leq 4$

43. $\left|x - \frac{3}{2}\right| \geq \frac{3}{2}$

44. $|x + 3| > 4$

45. $x^3 - x^2 < 2x$

46. $x^3 + 2x^2 + x > 0$

In Exercises 47–52, solve the given equation for the indicated variable.

47. Solve for r: $V = \frac{1}{3}\pi r^2 h$

48. Solve for r: $A = 2\pi r^2 + 2\pi rh$

49. Solve for t: $S = V_0 t - 16t^2$

50. Solve for X: $Z = \sqrt{R^2 - X^2}$

51. Solve for p: $L = \dfrac{k}{3\pi r^2 p}$

52. Solve for v: $E = 2kw\left(\dfrac{v}{2}\right)^2$

In Exercises 53–64, perform the indicated operations and write the result in standard form.

53. $(7 + 5i) + (-4 + 2i)$

54. $-(6 - 2i) + (-8 + 3i)$

55. $\left(\dfrac{\sqrt{2}}{2} - \dfrac{\sqrt{2}}{2}i\right) - \left(\dfrac{\sqrt{2}}{2} + \dfrac{\sqrt{2}}{2}i\right)$

56. $(13 - 8i) - 5i$

57. $5i(13 - 8i)$

58. $(1 + 6i)(5 - 2i)$

59. $(10 - 8i)(2 - 3i)$

60. $i(6 + i)(3 - 2i)$

61. $\dfrac{6 + i}{i}$

62. $\dfrac{3 + 2i}{5 + i}$

63. $\dfrac{4}{-3i}$

64. $\dfrac{1}{(2 + i)^4}$

65. The distance from a spacecraft to the horizon is 1000 miles. Find x, the altitude of the craft (see figure). Assume that the radius of the earth is 4000 miles.

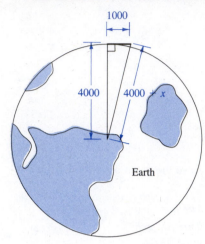

1000

4000   4000 + x

Earth

FIGURE FOR 65

66. Each week a salesperson must make a 180-mile trip to pick up supplies. If she were to increase her average speed by 5 miles per hour, the trip would take 24 minutes less than usual. Find her usual average speed.

67. A group of farmers agree to share equally in the cost of a $48,000 piece of machinery. If they could find two more farmers to join the group, each person's share of the cost would decrease by $4000. How many farmers are presently in the group?

68. Find a positive real number such that the sum of the number and its reciprocal is 26/5.

69. The period of a pendulum is given by

$$T = 2\pi\sqrt{\frac{L}{32}}$$

where T is time in seconds and L is the length (of the pendulum) in feet. If the period is to be at least 2 seconds, determine the minimum length of the pendulum.

CHAPTER 3

Functions and Graphs

3.1 The Cartesian Plane

Just as we can represent real numbers by points on the real line, we can represent ordered pairs of real numbers by points in a plane. This plane is called the **rectangular coordinate system,** or the **Cartesian plane,** after the French mathematician René Descartes (1596–1650).

The Cartesian plane is formed by two real lines intersecting at right angles, as shown in Figure 3.1. The horizontal real line is usually called the *x*-axis, and the vertical real line is usually called the *y*-axis. The point of intersection of these two axes is called the **origin,** and the axes divide the plane into four parts called **quadrants.**

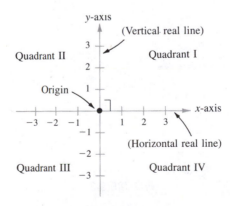

FIGURE 3.1　　　　　The Cartesian Plane

139

Functions and Graphs

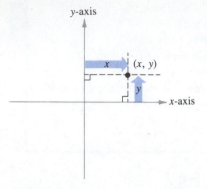

FIGURE 3.2

Each point in the plane corresponds to an **ordered pair** (x, y) of real numbers x and y, called **coordinates** of the point. The **x-coordinate** (or **abscissa**) represents the directed distance from the y-axis to the point, and the **y-coordinate** (or **ordinate**) represents the directed distance from the x-axis to the point, as shown in Figure 3.2.

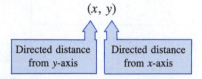

Remark: It is customary to use the notation (x, y) to denote both a point in the plane and an open interval on the real line. The nature of a specific problem will show which of the two we are talking about.

EXAMPLE 1 *Plotting Points in the Cartesian Plane*

Locate the points $(-1, 2)$, $(3, 4)$, $(0, 0)$, $(3, 0)$, and $(-2, -3)$ in the Cartesian plane.

SOLUTION

To plot the point $(-1, 2)$ we envision a vertical line through -1 on the x-axis and a horizontal line through 2 on the y-axis. The intersection of these two lines is the point $(-1, 2)$, as shown in Figure 3.3. The other four points can be plotted in a similar way.

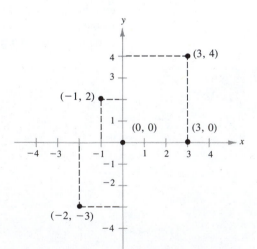

FIGURE 3.3

The Cartesian Plane

The value of the rectangular coordinate system is that it allows us to visualize relationships between the variables x and y. Today, Descartes's ideas are in common use in virtually every scientific and business-related field.

EXAMPLE 2 An Application of the Rectangular Coordinate System

The prime interest rate in the United States from 1976 to 1986 is given in Table 3.1. Plot these points on a rectangular coordinate system.

TABLE 3.1

Year	1976	1977	1978	1979	1980	1981
Rate	6.84	6.83	9.06	12.67	15.26	18.87

Year	1982	1983	1984	1985	1986
Rate	14.86	10.79	11.51	9.93	8.33

SOLUTION

The points are shown in Figure 3.4. The break in the x-axis indicates that we have omitted the numbers between 0 and 1976.

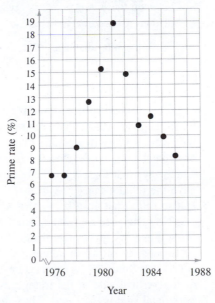

FIGURE 3.4   Prime Interest Rate

Functions and Graphs

The Distance Between Two Points in the Plane

We know from Section 1.1, that the distance d between two points a and b on a number line is simply

$$d = |a - b|.$$

This same rule is used to find the distance between two points that lie on the same *vertical* or *horizontal* line in the plane.

EXAMPLE 3 *Finding Horizontal and Vertical Distances*

(a) Find the distance between the points $(1, -1)$ and $(1, 4)$.
(b) Find the distance between the points $(-3, -1)$ and $(1, -1)$.

SOLUTION

(a) Because the x-coordinates are equal, we visualize a vertical line through the points $(1, -1)$ and $(1, 4)$, as shown in Figure 3.5. The distance between these two points is given by the absolute value of the difference of their y-coordinates. That is,

$$\text{vertical distance} = |4 - (-1)| = 5. \qquad \textit{Subtract y-coordinates}$$

(b) Because the y-coordinates are equal, we visualize a horizontal line through the points $(-3, -1)$ and $(1, -1)$, as shown in Figure 3.5. The distance between these two points is given by the absolute value of the difference of their x-coordinates. That is,

$$\text{horizontal distance} = |1 - (-3)| = 4 \qquad \textit{Subtract x-coordinates}$$

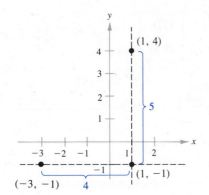

FIGURE 3.5

The technique used in Example 3 can be used to develop a general formula for finding the distance between two points in the plane. This general formula will work for any two points, even if they do not lie on the same vertical or horizontal line. To develop the formula, we use the Pythagorean Theorem, which says that for a right triangle with hypotenuse c and sides a and b, we have the relationship $a^2 + b^2 = c^2$, as shown in Figure 3.6. (The converse is also true. That is, if $a^2 + b^2 = c^2$, then the triangle is a right triangle.)

Now, suppose we want to determine the distance d between the two points (x_1, y_1) and (x_2, y_2) in the plane. With these two points, a right triangle can be formed, as shown in Figure 3.7. Note that the third vertex of the triangle is (x_1, y_2). Since (x_1, y_1) and (x_1, y_2) lie on the same vertical line, the length of the vertical side of the triangle is $|y_2 - y_1|$. Similarly, the length of the horizontal side is $|x_2 - x_1|$. Thus, by the Pythagorean Theorem, we have

$$d^2 = |x_2 - x_1|^2 + |y_2 - y_1|^2.$$

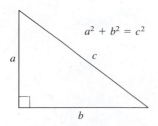

Pythagorean Theorem

FIGURE 3.6

The Cartesian Plane

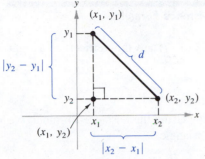

Distance Between Two Points

FIGURE 3.7

Since the distance d must be positive, we choose the positive square root and write

$$d = \sqrt{|x_2 - x_1|^2 + |y_2 - y_1|^2}.$$

Finally, replacing $|x_2 - x_1|^2$ and $|y_2 - y_1|^2$ by the equivalent expressions $(x_2 - x_1)^2$ and $(y_2 - y_1)^2$ gives us the following formula for the distance between any two points in a coordinate plane.

The Distance Formula

The distance d between the points (x_1, y_1) and (x_2, y_2) in the coordinate plane is

$$d = \sqrt{(x_2 - x_1)^2 + (y_2 - y_1)^2}.$$

Remark: Note that for the special case in which the two points lie on the same vertical or horizontal line, the Distance Formula still works. For instance, applying the Distance Formula to the points $(1, -1)$ and $(1, 4)$, we obtain

$$d = \sqrt{(1 - 1)^2 + [4 - (-1)]^2} = \sqrt{5^2} = 5$$

which is the same result we obtained in Example 3.

EXAMPLE 4   *Finding the Distance Between Two Points*

Find the distance between the points $(-2, 1)$ and $(3, 4)$.

SOLUTION

Letting $(x_1, y_1) = (-2, 1)$ and $(x_2, y_2) = (3, 4)$, we apply the Distance Formula to obtain

$$\begin{aligned}
d &= \sqrt{[3 - (-2)]^2 + (4 - 1)^2} \\
&= \sqrt{5^2 + 3^2} \\
&= \sqrt{25 + 9} \\
&= \sqrt{34} \approx 5.83.
\end{aligned}$$

See Figure 3.8.

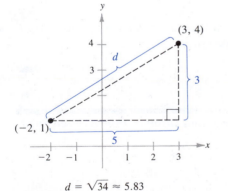

$d = \sqrt{34} \approx 5.83$

FIGURE 3.8

In Example 4, the figure provided was not essential to the solution of the problem. *Nevertheless,* we recommend that you include graphs with your problem solutions.

Functions and Graphs

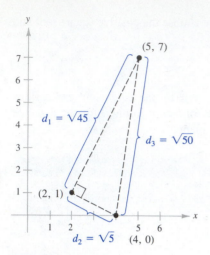

FIGURE 3.9

EXAMPLE 5   An Application of the Distance Formula

Show that the points (2, 1), (4, 0), and (5, 7) are the vertices of a right triangle.

SOLUTION

The three points are plotted in Figure 3.9. Using the Distance Formula, we find the lengths of the three sides of the triangle.

$$d_1 = \sqrt{(5 - 2)^2 + (7 - 1)^2} = \sqrt{9 + 36} = \sqrt{45}$$
$$d_2 = \sqrt{(4 - 2)^2 + (0 - 1)^2} = \sqrt{4 + 1} = \sqrt{5}$$
$$d_3 = \sqrt{(5 - 4)^2 + (7 - 0)^2} = \sqrt{1 + 49} = \sqrt{50}$$

Since $d_1^2 + d_2^2 = 45 + 5 = 50 = d_3^2$, we can conclude from the Pythagorean Theorem that the triangle is a right triangle.

EXAMPLE 6   Testing for Collinearity

Determine whether the points (−2, 3), (1, −2), and (4, −5) are collinear (lie on the same line) or form the vertices of a triangle.

SOLUTION

By the Distance Formula, we have

$$d_1 = \sqrt{(1 + 2)^2 + (-2 - 3)^2} = \sqrt{34} \approx 5.83$$
$$d_2 = \sqrt{(4 - 1)^2 + (-5 + 2)^2} = \sqrt{18} \approx 4.24$$
$$d_3 = \sqrt{(4 + 2)^2 + (-5 - 3)^2} = \sqrt{100} = 10.$$

Since no two distances add up to the third, the points are not collinear. Thus, they form a triangle, as shown in Figure 3.10.

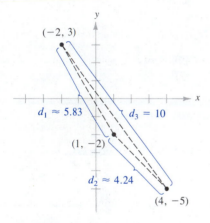

FIGURE 3.10

EXAMPLE 7   Finding Points at a Specified Distance from a Given Point

Find x so that the distance between $(x, 3)$ and $(2, -1)$ is 5.

SOLUTION

As we begin this problem we do not know how many values of x satisfy the given requirements. Even so, we can use the Distance Formula to find the distance to be

The Cartesian Plane

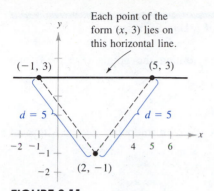

Each point of the form $(x, 3)$ lies on this horizontal line.

$(-1, 3)$ $(5, 3)$

$d = 5$ $d = 5$

$(2, -1)$

FIGURE 3.11

$$\sqrt{(x - 2)^2 + (3 + 1)^2} = 5.$$

Squaring both sides of this equation, we obtain

$$(x^2 - 4x + 4) + 16 = 25$$
$$x^2 - 4x - 5 = 0$$
$$(x - 5)(x + 1) = 0$$
$$x = 5 \text{ or } -1.$$

We see that there are two solutions, and we conclude that both of the points $(5, 3)$ and $(-1, 3)$ lie five units from the point $(2, -1)$, as shown in Figure 3.11.

The Midpoint Formula

Next we introduce a formula for finding the midpoint of a line segment joining two points. The coordinates of the midpoint are simply the average values of the coordinates of the two endpoints.

The Midpoint Formula

The **midpoint** of the line segment joining the points (x_1, y_1) and (x_2, y_2) is

$$\left(\frac{x_1 + x_2}{2}, \frac{y_1 + y_2}{2}\right).$$

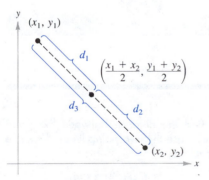

(x_1, y_1)

d_1

$\left(\dfrac{x_1 + x_2}{2}, \dfrac{y_1 + y_2}{2}\right)$

d_3 d_2

(x_2, y_2)

Midpoint Formula

FIGURE 3.12

PROOF

Using Figure 3.12, we need to show that

$$d_1 = d_2 \quad \text{and} \quad d_1 + d_2 = d_3.$$

By the Distance Formula, we obtain

$$d_1 = \sqrt{\left(\frac{x_1 + x_2}{2} - x_1\right)^2 + \left(\frac{y_1 + y_2}{2} - y_1\right)^2} = \frac{1}{2}\sqrt{(x_2 - x_1)^2 + (y_2 - y_1)^2}$$

$$d_2 = \sqrt{\left(x_2 - \frac{x_1 + x_2}{2}\right)^2 + \left(y_2 - \frac{y_1 + y_2}{2}\right)^2} = \frac{1}{2}\sqrt{(x_2 - x_1)^2 + (y_2 - y_1)^2}$$

$$d_3 = \sqrt{(x_2 - x_1)^2 + (y_2 - y_1)^2}.$$

Thus, it follows that $d_1 = d_2$ and $d_1 + d_2 = d_3$.

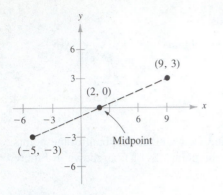

FIGURE 3.13

EXAMPLE 8 *Finding the Midpoint of a Line Segment*

Find the midpoint of the line segment joining the points $(-5, -3)$ and $(9, 3)$.

SOLUTION

Figure 3.13 shows the two given points and their midpoint. By the Midpoint Formula, we have

$$\text{midpoint} = \left(\frac{-5 + 9}{2}, \frac{-3 + 3}{2} \right)$$
$$= (2, 0).$$

WARM UP

Simplify the given expressions.

1. $\sqrt{(2-6)^2 + [1-(-2)]^2}$

2. $\sqrt{(1-4)^2 + (-2-1)^2}$

3. $\dfrac{4 + (-2)}{2}$

4. $\dfrac{-1 + (-3)}{2}$

5. $\sqrt{18} + \sqrt{45}$

6. $\sqrt{12} + \sqrt{44}$

Solve for x or y.

7. $\sqrt{(4-x)^2 + (5-2)^2} = \sqrt{58}$

8. $\sqrt{(8-6)^2 + (y-5)^2} = 2\sqrt{5}$

9. $\dfrac{x + 3}{2} = 7$

10. $\dfrac{-2 + y}{2} = 1$

EXERCISES 3.1

In Exercises 1–4, sketch the polygon with the indicated vertices.

1. Triangle: $(-1, 1)$, $(2, -1)$, $(3, 4)$

2. Triangle: $(0, 3)$, $(-1, -2)$, $(4, 8)$

3. Square: $(2, 4)$, $(5, 1)$, $(2, -2)$, $(-1, 1)$

4. Parallelogram: $(5, 2)$, $(7, 0)$, $(1, -2)$, $(-1, 0)$

In Exercises 5–8, find the distance between the given points. [*Note:* In each case the two points lie on the same horizontal or vertical line.]

5. $(6, -3)$, $(6, 5)$

6. $(1, 4)$, $(8, 4)$

7. $(-3, -1)$, $(2, -1)$

8. $(-3, -4)$, $(-3, 6)$

In Exercises 9–12, (a) find the length of the two sides of the right triangle and use the Pythagorean Theorem to find the length of the hypotenuse, and (b) use the Distance Formula to find the length of the hypotenuse of the triangle.

9.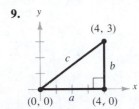

10.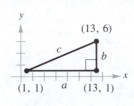

11.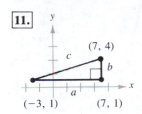

12.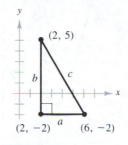

In Exercises 13–24, (a) plot the points, (b) find the distance between the points, and (c) find the midpoint of the line segment joining the points.

13. $(1, 1)$, $(9, 7)$

14. $(1, 12)$, $(6, 0)$

15. $(-4, 10)$, $(4, -5)$

16. $(-7, -4)$, $(2, 8)$

17. $(-1, 2)$, $(5, 4)$

18. $(2, 10)$, $(10, 2)$

19. $\left(\frac{1}{2}, 1\right)$, $\left(-\frac{5}{2}, \frac{4}{3}\right)$

20. $\left(-\frac{1}{3}, -\frac{1}{3}\right)$, $\left(-\frac{1}{6}, -\frac{1}{2}\right)$

21. $(6.2, 5.4)$, $(-3.7, 1.8)$

22. $(-16.8, 12.3)$, $(5.6, 4.9)$

23. $(-36, -18)$, $(48, -72)$

24. $(1.451, 3.051)$, $(5.906, 11.360)$

In Exercises 25–28, show that the given points form the vertices of the indicated polygon. (A rhombus is a parallelogram whose sides are all of the same length.)

25. Right triangle: $(4, 0)$, $(2, 1)$, $(-1, -5)$

26. Isosceles triangle: $(1, -3)$, $(3, 2)$, $(-2, 4)$

27. Rhombus: $(0, 0)$, $(1, 2)$, $(2, 1)$, $(3, 3)$

28. Parallelogram: $(0, 1)$, $(3, 7)$, $(4, 4)$, $(1, -2)$

In Exercises 29 and 30, find x so that the distance between the points is 13.

29. $(1, 2)$, $(x, -10)$

30. $(-8, 0)$, $(x, 5)$

In Exercises 31 and 32, find y so that the distance between the points is 17.

31. $(0, 0)$, $(8, y)$

32. $(-8, 4)$, $(7, y)$

In Exercises 33 and 34, find a relationship between x and y so that (x, y) is equidistant from the two given points.

33. $(4, -1)$, $(-2, 3)$

34. $\left(3, \frac{5}{2}\right)$, $(-7, -1)$

In Exercises 35–42, determine the quadrant(s) in which (x, y) is located so that the given conditions are satisfied.

35. $x > 0$ and $y < 0$

36. $x < 0$ and $y < 0$

37. $x > 0$ and $y > 0$

38. $x < 0$ and $y > 0$

39. $x = -4$ and $y > 0$

40. $x > 2$ and $y = 3$

41. $y < -5$

42. $x > 4$

43. A line segment has (x_1, y_1) as one endpoint and (x_m, y_m) as its midpoint. Find the other endpoint (x_2, y_2) of the line segment in terms of x_1, y_1, x_m, and y_m.

44. Use the result of Exercise 43 to find the endpoint of a line segment if the other endpoint and midpoint are, respectively,
(a) $(-3, 2)$ and $(1, 7)$ (b) $(-5, 11)$ and $(2, 4)$.

45. Use the Midpoint Formula twice to find the three points that divide the line segment joining (x_1, y_1) and (x_2, y_2) into four parts.

46. Use the result of Exercise 45 to find the points that divide the line segment joining the given points into four equal parts.
(a) $(1, -2)$, $(4, -1)$ (b) $(-2, -3)$, $(0, 0)$

In Exercises 47 and 48, use the Midpoint Formula to estimate the sales of a company for 1983, given the sales in 1980 and 1986. Assume the sales followed a linear pattern.

47.

Year	1980	1986
Sales	$520,000	$740,000

48.

Year	1980	1986
Sales	$4,200,000	$5,650,000

Functions and Graphs

49. Find the price of corn for the following dates (see figure).
  (a) May 19   (b) July 7
  (c) August 11   (d) September 15

50. Find the *decrease* in corn prices from May 19 to September 15.

51. Find the trade deficit for the following months (see figure).
  (a) February   (b) December

52. Find the percentage *increase* in the trade deficit from October to November.

Price of Corn

FIGURE FOR 49 and 50

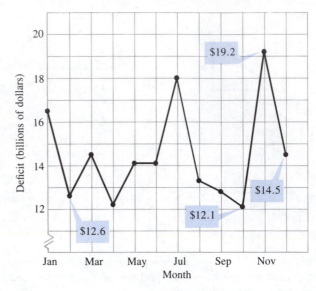

U.S. Trade Deficit

FIGURE FOR 51 and 52

3.2 Graphs of Equations

News magazines frequently show graphs comparing the rate of inflation, the gross national product, wholesale prices, or the unemployment rate to the time of year. Industrial firms and businesses use graphs to report their monthly production and sales statistics. Such graphs provide a simple geometric picture of the way one quantity changes with respect to another.

Frequently, the relationship between two quantities is expressed in the form of an equation. In this section, we introduce the basic procedure for determining the geometric picture associated with an algebraic equation.

For an equation in variables x and y, a point (a, b) is a **solution point** if the substitution of $x = a$ and $y = b$ satisfies the equation. Of course, most equations have many solution points. For example, the equation

$$3x + y = 5$$

has solution points $(0, 5)$, $(1, 2)$, $(2, -1)$, $(3, -4)$, and so on. The set of all solution points of a given equation is called the **graph** of the equation.

The Point-Plotting Method of Graphing

To sketch a graph of an equation by point-plotting, use the following steps.

1. If possible, rewrite the equation by isolating one of the variables.
2. Make up a table of several solution points.
3. Plot these points in the coordinate plane.
4. Connect the points with a smooth curve.

EXAMPLE 1   Sketching the Graph of an Equation

Sketch a graph of the equation $3x + y = 5$.

SOLUTION

In this case we isolate variable y, to get

$$y = 5 - 3x.$$

Using negative, zero, and positive values for x, we obtain the following table of values (solution points).

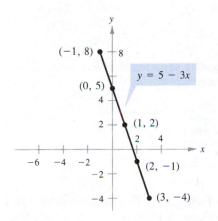

FIGURE 3.14

x	-1	0	1	2	3
$y = 5 - 3x$	8	5	2	-1	-4

Next, we plot these points and connect them as shown in Figure 3.14. It appears that the graph is a straight line. (We will study lines extensively in Section 3.3.)

Step 4 of the point-plotting method can be difficult. For instance, how would you connect the four points in Figure 3.15? Without further information about the equation, any one of the three graphs in Figure 3.16 would be reasonable.

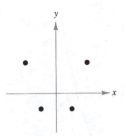

FIGURE 3.15

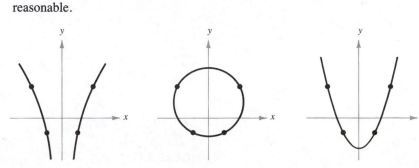

FIGURE 3.16

With too few solution points, we could grossly misrepresent the graph of a given equation. Just how many points should be plotted? For a straight-line graph, two points are sufficient. For more complicated graphs, we need many more points. More sophisticated techniques will be discussed in later sections, but for now plot enough points so as to reveal the essential behavior of the graph. A programmable calculator is useful for determining the many solution points needed for an accurate graph.

EXAMPLE 2 *Sketching the Graph of an Equation*

Sketch the graph of the equation $y = x^2 - 2$.

SOLUTION

First, we make a table of values by choosing several convenient values of x and calculating the corresponding values of y.

x	-2	-1	0	1	2	3
$y = x^2 - 2$	2	-1	-2	-1	2	7

Next, we plot these points, as shown in Figure 3.17. Finally, we connect the points by a smooth curve, as shown in Figure 3.18.

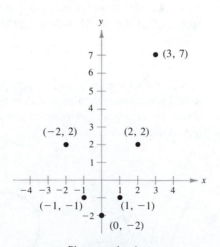

Plot several points.

FIGURE 3.17

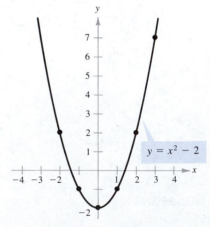

Connect points with a smooth curve.

FIGURE 3.18

Intercepts of a Graph

When choosing points to plot, we suggest you start with those that are easiest to calculate. Two such points have zero as either their x- or y-coordinate. These points are called **intercepts,** because they are points at which the graph intersects the x- or y-axis.

Definition of Intercepts

The point $(a, 0)$ is called an **x-intercept** of the graph of an equation if it is a solution point of the equation. To find the x-intercepts, let y be zero and solve the equation for x.

The point $(0, b)$ is called a **y-intercept** of the graph of an equation if it is a solution point of the equation. To find the y-intercepts, let x be zero and solve the equation for y.

Remark: Some texts denote the x-intercept as the x-coordinate of the point $(a, 0)$ rather than the point itself. Unless it is necessary to make a distinction, we will use "intercept" to mean either the point or the coordinate.

Of course, it is possible that a particular graph will have no intercepts or several intercepts. For instance, consider the three graphs in Figure 3.19.

FIGURE 3.19

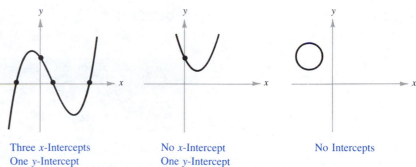

Three x-Intercepts   No x-Intercept   No Intercepts
One y-Intercept   One y-Intercept

EXAMPLE 3   *Finding x- and y-Intercepts*

Find the x- and y-intercepts for the graph of $y^2 - 3 = x$.

SOLUTION

Let $y = 0$. Then $-3 = x$.

> *x-intercept:* $(-3, 0)$

Let $x = 0$. Then $y^2 - 3 = 0$ has solutions $y = \pm\sqrt{3}$.

> *y-intercepts:* $(0, \sqrt{3})$, $(0, -\sqrt{3})$

See Figure 3.20.

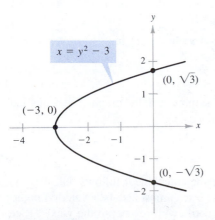

$x = y^2 - 3$

$(0, \sqrt{3})$

$(-3, 0)$

$(0, -\sqrt{3})$

FIGURE 3.20

Functions and Graphs

Symmetry

The graph shown in Figure 3.20 is said to be "symmetric" with respect to the *x*-axis. This means that if the Cartesian plane were folded along the *x*-axis, the portion of the graph above the *x*-axis would coincide with the portion below the *x*-axis. Symmetry with respect to the *y*-axis can be described in a similar manner.

Knowing the symmetry of a graph *before* attempting to sketch it is beneficial, because we then need only half as many solution points as we would otherwise. We define the three basic types of symmetry as follows. See Figure 3.21.

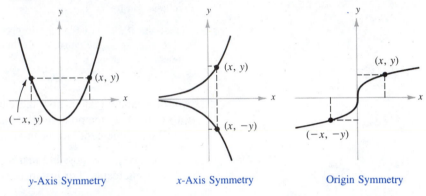

y-Axis Symmetry   x-Axis Symmetry   Origin Symmetry

FIGURE 3.21

Definition of Symmetry with Respect to the Coordinate Axes and the Origin

A graph is said to be **symmetric with respect to the y-axis** if, whenever (x, y) is on the graph, $(-x, y)$ is also on the graph.

A graph is said to be **symmetric with respect to the x-axis** if, whenever (x, y) is on the graph, $(x, -y)$ is also on the graph.

A graph is said to be **symmetric with respect to the origin** if, whenever (x, y) is on the graph, $(-x, -y)$ is also on the graph.

Suppose we apply this definition of symmetry to the graph of the equation $y = x^2 - 2$. Replacing *x* with $-x$ produces

$$y = (-x)^2 - 2$$
$$y = x^2 - 2.$$

Since this substitution does not change the equation, it follows that if (x, y) is a solution point of the equation, then $(-x, y)$ must also be a solution point. Thus, the graph of $y = x^2 - 2$ is symmetric with respect to the *y*-axis. (See Figure 3.22.)

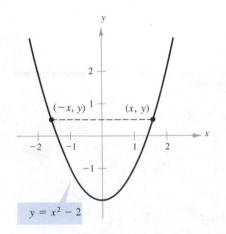

$y = x^2 - 2$

y-Axis Symmetry

FIGURE 3.22

A similar test can be made for symmetry with respect to the x-axis or to the origin. These three tests are summarized as follows.

Tests for Symmetry

1. The graph of an equation is symmetric with respect to the *y-axis*, if replacing x with $-x$ yields an equivalent equation.
2. The graph of an equation is symmetric with respect to the *x-axis*, if replacing y with $-y$ yields an equivalent equation.
3. The graph of an equation is symmetric with respect to the *origin*, if replacing x with $-x$ *and* y with $-y$ yields an equivalent equation.

EXAMPLE 4   *Using Intercepts and Symmetry as Sketching Aids*

Use intercepts and symmetry to sketch the graph of $x - y^2 = 1$.

SOLUTION

Intercepts: Letting $x = 0$, we see that $-y^2 = 1$ or $y^2 = -1$ has no real solutions. Hence, there are no y-intercepts. Letting $y = 0$, we obtain $x = 1$. Thus, the x-intercept is $(1, 0)$.

Symmetry: Replacing y with $-y$ yields

$$x - (-y)^2 = 1$$
$$x - y^2 = 1$$

which means that the graph is symmetric with respect to the x-axis.

Using symmetry we need only find solution points above the x-axis and then reflect them to obtain the desired graph (see Figure 3.23).

y	0	1	2
$x = y^2 + 1$	1	2	5

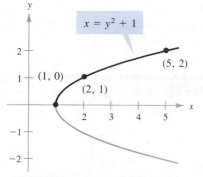

First plot the points above the x-axis, then use symmetry to complete the graph.

FIGURE 3.23

EXAMPLE 5 *Using Intercepts and Symmetry as Sketching Aids*

Use intercepts and symmetry to sketch the graph of $y = x^3 - 4x$.

SOLUTION

Intercepts: Letting $y = 0$, we have

$$0 = x^3 - 4x = x(x^2 - 4).$$

Setting both factors equal to zero and solving for x, we obtain $x = 0$ and $x = \pm 2$. Thus, the x-intercepts are $(0, 0)$, $(2, 0)$, and $(-2, 0)$. Letting $x = 0$, we obtain $y = 0$, which tells us that the y-intercept is $(0, 0)$.

Symmetry: Replacing x with $-x$ and y with $-y$ yields

$$-y = (-x)^3 - 4(-x)$$
$$-y = -x^3 + 4x$$
$$y = x^3 - 4x \qquad\qquad \textit{Multiply by } -1$$

which is the original equation. Thus, the graph of $y = x^3 - 4x$ is symmetric with respect to the origin.

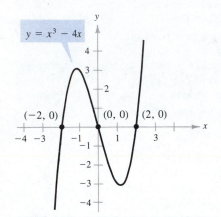

FIGURE 3.24

Using the intercepts, symmetry, and the following table of values, we obtain the graph shown in Figure 3.24.

x	0	1	2	3
$y = x^3 - 4x$	0	-3	0	15

Not all equations have graphs that are symmetric with respect to one of the axes or the origin. For instance, the graph of the equation shown in Example 6 has none of these three types of symmetry.

EXAMPLE 6 *Sketching the Graph of an Equation*

Sketch the graph of $y = |4x - x^2|$.

SOLUTION

Intercepts: Letting $x = 0$ yields $y = 0$, which means that $(0, 0)$ is a y-intercept. Letting $y = 0$ yields $x = 0$ and $x = 4$, which means that $(0, 0)$ and $(4, 0)$ are x-intercepts.

Graphs of Equations

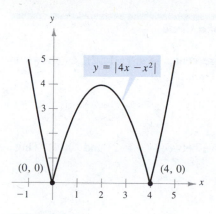

FIGURE 3.25

Symmetry: This equation fails all three tests for symmetry and consequently its graph is not symmetric with respect to either axis or to the origin.

The absolute value sign indicates that y is always nonnegative, and we obtain the following table of values.

x		-1	0	1	2	3	4	5
$y = \lvert 4x - x^2 \rvert$		5	0	3	4	3	0	5

The graph is shown in Figure 3.25.

Circles

So far in this section we have studied the point-plotting method and two additional concepts (intercepts and symmetry) that can be used to streamline the graphing procedure.

Another graphing aid is *equation recognition*, the ability to recognize the general shape of a graph simply by looking at its equation. A circle is one type of graph whose equation is easily recognized.

Figure 3.26 shows a circle of radius r with center at (h, k). The point (x, y) is on this circle if and only if its distance from the center (h, k) is r. This means that a **circle** consists of the set of all points (x, y) that are at a given positive distance r from a fixed point (h, k). Expressing this relationship by means of the Distance Formula, we have

$$\sqrt{(x - h)^2 + (y - k)^2} = r.$$

By squaring both sides of this equation, we obtain the following definition.

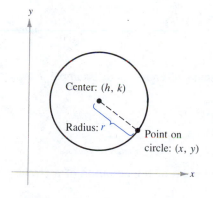

FIGURE 3.26

Standard Form of the Equation of a Circle

The **standard form of the equation of a circle** with radius r and center at (h, k) is

$$(x - h)^2 + (y - k)^2 = r^2.$$

Remark: The standard form of the equation of a circle whose *center is the origin* is simply

$$x^2 + y^2 = r^2.$$

Functions and Graphs

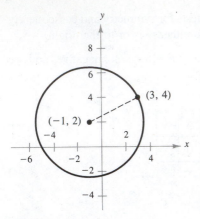

FIGURE 3.27

EXAMPLE 7   *Finding an Equation of a Circle*

The point (3, 4) lies on a circle whose center is at $(-1, 2)$, as shown in Figure 3.27. Find an equation for the circle.

SOLUTION

The radius r of the circle is the distance between $(-1, 2)$ and (3, 4). Thus, we have

$$r = \sqrt{[3 - (-1)]^2 + (4 - 2)^2}$$
$$= \sqrt{16 + 4}$$
$$= \sqrt{20}.$$

We conclude that the standard equation for this circle is

$$[x - (-1)]^2 + (y - 2)^2 = (\sqrt{20})^2$$
$$(x + 1)^2 + (y - 2)^2 = 20.$$

If we remove parentheses in the standard equation in Example 7, we obtain

$$(x + 1)^2 + (y - 2)^2 = 20 \qquad \textit{Standard form}$$
$$x^2 + 2x + 1 + y^2 - 4y + 4 = 20$$
$$x^2 + y^2 + 2x - 4y - 15 = 0 \qquad \textit{General form}$$

where the last equation is in the **general form of the equation of a circle:**

$$Ax^2 + Ay^2 + Dx + Ey + F = 0, \qquad A \neq 0.$$

The general form of the equation of a circle is less useful than the standard form. For instance, it is not immediately apparent from the general equation of the circle in Example 7 that the center is at $(-1, 2)$ and the radius is $\sqrt{20}$. To graph the equation of a circle, it is best to write the equation in standard form. We do this by **completing the square** (Section 2.3), as demonstrated in the following example.

EXAMPLE 8   *Sketching a Circle*

Sketch the circle whose general equation is

$$4x^2 + 4y^2 + 20x - 16y + 37 = 0.$$

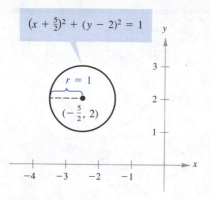

FIGURE 3.28

SOLUTION

To complete the square we first divide by 4 so that the coefficients of x^2 and y^2 are both 1.

$$4x^2 + 4y^2 + 20x - 16y + 37 = 0 \qquad \textit{General form}$$

$$x^2 + y^2 + 5x - 4y + \frac{37}{4} = 0 \qquad \textit{Divide by 4}$$

$$(x^2 + 5x + \quad) + (y^2 - 4y + \quad) = -\frac{37}{4} \qquad \textit{Group terms}$$

$$\left(x^2 + 5x + \frac{25}{4}\right) + (y^2 - 4y + 4) = -\frac{37}{4} + \frac{25}{4} + 4 \qquad \textit{Complete square}$$

$$\underbrace{\qquad}_{(\text{half})^2} \qquad \underbrace{\qquad}_{(\text{half})^2}$$

$$\left(x + \frac{5}{2}\right)^2 + (y - 2)^2 = 1 \qquad \textit{Standard form}$$

Therefore, the circle is centered at $(-5/2, 2)$, and its radius is 1, as shown in Figure 3.28.

Remark: The general equation $Ax^2 + Ay^2 + Dx + Ey + F = 0$ may not always represent a circle. Such an equation will have no solution points if the procedure of completing the square yields the *impossible* result

$$(x - h)^2 + (y - k)^2 = \text{negative number}.$$

Moreover, the general equation $Ax^2 + Ay^2 + Dx + Ey + F = 0$ will have exactly one solution point if the procedure of completing the square yields the result $(x - h)^2 + (y - k)^2 = 0$. The point (h, k) is the only solution point for this equation.

WARM UP

Solve for y in terms of x.

1. $3x - 5y = 2$

2. $x^2 - 4x + 2y - 5 = 0$

Solve for x.

3. $x^2 - 4x + 4 = 0$

4. $(x - 1)(x + 5) = 0$

5. $x^3 - 9x = 0$

6. $x^4 - 8x^2 + 16 = 0$

Simplify the equations.

7. $-y = (-x)^3 + 4(-x)$

8. $(-x)^2 + (-y)^2 = 4$

9. $y = 4(-x)^2 + 8$

10. $(-y)^2 = 3(-x)^2 + 4$

EXERCISES 3.2

In Exercises 1–6, determine whether the indicated points lie on the graph of the given equation.

1. $y = \sqrt{x} + 4$
  (a) $(0, 2)$ (b) $(5, 3)$

2. $y = x^2 - 3x + 2$
  (a) $(2, 0)$ (b) $(-2, 8)$

3. $2x - y - 3 = 0$
  (a) $(1, 2)$ (b) $(1, -1)$

4. $x^2 + y^2 = 20$
  (a) $(3, -2)$ (b) $(-4, 2)$

5. $x^2y - x^2 + 4y = 0$
  (a) $\left(1, \frac{1}{5}\right)$ (b) $\left(2, \frac{1}{2}\right)$

6. $y = \dfrac{1}{x^2 + 1}$
  (a) $(0, 0)$ (b) $(3, 0.1)$

In Exercises 7–10, find the constant C such that the given ordered pair is a solution point of the equation.

7. $y = x^2 + C$, $(2, 6)$ **8.** $y = Cx^3$, $(-4, 8)$
9. $y = C\sqrt{x + 1}$, $(3, 8)$
10. $x + C(y + 2) = 0$, $(4, 3)$

In Exercises 11–18, find the x- and y-intercepts of the graph of the given equation.

11. $y = x - 5$ **12.** $y = (x - 1)(x - 3)$
13. $y = x^2 + x - 2$ **14.** $y = 4 - x^2$
15. $y = x\sqrt{x + 2}$ **16.** $xy = 4$
17. $xy - 2y - x + 1 = 0$ **18.** $x^2y - x^2 + 4y = 0$

In Exercises 19–26, check for symmetry with respect to both axes and the origin.

19. $x^2 - y = 0$ **20.** $xy^2 + 10 = 0$
21. $x - y^2 = 0$ **22.** $y = \sqrt{9 - x^2}$
23. $y = x^3$ **24.** $xy = 4$
25. $y = \dfrac{x}{x^2 + 1}$ **26.** $y = x^4 - x^2 + 3$

In Exercises 27–32, match the given equation with its graph. [The graphs are labeled (a)–(f).]

27. $y = 4 - x$ **28.** $y = x^2 + 2x$
29. $y = \sqrt{4 - x^2}$ **30.** $y = \sqrt{x}$
31. $y = x^3 - x$ **32.** $y = |x| - 2$

(a)

(b)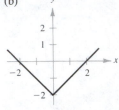

(c)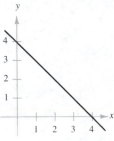

(d)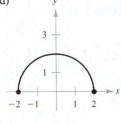

(e)

(f)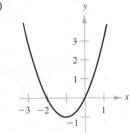

In Exercises 33–52, sketch the graph of the given equation. Identify any intercepts and test for symmetry.

33. $y = -3x + 2$ **34.** $y = 2x - 3$
35. $y = 1 - x^2$ **36.** $y = x^2 - 1$
37. $y = x^2 - 4x + 3$ **38.** $y = -x^2 - 4x$
39. $y = x^3 + 2$ **40.** $y = x^3 - 1$
41. $y = x(x - 2)^2$ **42.** $y = \dfrac{4}{x^2 + 1}$
43. $y = \sqrt{x - 3}$ **44.** $y = \sqrt{1 - x}$
45. $y = \sqrt[3]{x}$ **46.** $y = \sqrt[3]{x + 1}$
47. $y = |x - 2|$ **48.** $y = 4 - |x|$
49. $x = y^2 - 1$ **50.** $x = y^2 - 4$
51. $x^2 + y^2 = 4$ **52.** $x^2 + y^2 = 16$

In Exercises 53–60, find the standard form of the equation of the specified circle.

53. Center: $(0, 0)$; radius: 3

54. Center: $(0, 0)$; radius: 5

55. Center: $(2, -1)$; radius: 4

56. Center: $(0, \frac{1}{3})$; radius: $\frac{1}{3}$

57. Center: $(-1, 2)$; solution point: $(0, 0)$

58. Center: $(3, -2)$; solution point: $(-1, 1)$

59. Endpoints of a diameter: $(0, 0)$, $(6, 8)$

60. Endpoints of a diameter: $(-4, -1)$, $(4, 1)$

In Exercises 61–68, write the given equation of the circle in standard form and sketch its graph.

61. $x^2 + y^2 - 2x + 6y + 6 = 0$

62. $x^2 + y^2 - 2x + 6y - 15 = 0$

63. $x^2 + y^2 - 2x + 6y + 10 = 0$

64. $3x^2 + 3y^2 - 6y - 1 = 0$

65. $2x^2 + 2y^2 - 2x - 2y - 3 = 0$

66. $4x^2 + 4y^2 - 4x + 2y - 1 = 0$

67. $16x^2 + 16y^2 + 16x + 40y - 7 = 0$

68. $x^2 + y^2 - 4x + 2y + 3 = 0$

In Exercises 69–71, (a) sketch a graph to compare the given data and the model for that data, and (b) use the model to predict y for the year 1990.

69. The following table gives the consumer price index (CPI) for selected years from 1970 to 1985. In the base year of 1967, CPI = 100.

Year	1970	1972	1974	1976	1978	1980
CPI	116.3	125.3	147.7	170.5	195.3	247.0

Year	1981	1982	1983	1984	1985	1990
CPI	272.3	288.6	297.4	311.1	322.2	?

A mathematical model for the CPI during this period is

$$y = 15.21t + 247.45$$

where y represents the CPI and t represents time in years with $t = 0$ corresponding to 1980.

70. The following table gives the life expectancy of a child (at birth) for selected years from 1920 to 1980.

Year	1920	1930	1940	1950
Life expectancy	54.1	59.7	62.9	68.2

Year	1960	1970	1980	1990
Life expectancy	69.7	70.8	73.7	?

A mathematical model for the life expectancy during this period is

$$y = 0.31t + 65.59$$

where y represents the life expectancy and t represents time in years with $t = 0$ corresponding to 1950.

71. The following table gives the per capita public debt for the United States for selected years from 1950 to 1985.

Year	1950	1960	1970
Per capita debt	$1688.30	$1572.31	$1807.09

Year	1980	1985	1990
Per capita debt	$3969.55	$7614.15	?

A mathematical model for the per capita debt during this period is

$$y = 9.84t^2 - 200.2t + 1970.4$$

where y represents the per capita debt and t is time in years with $t = 0$ corresponding to 1950.

3.3 Lines in the Plane

In this section, we study lines and their equations. Throughout this text, we follow the convention of using the term **line** to mean a *straight* line.

The Slope of a Line

The **slope** of a nonvertical line represents the number of units a line rises or falls vertically for each unit of horizontal change from left to right.

For instance, consider the two points (x_1, y_1) and (x_2, y_2) on the line shown in Figure 3.29. As we move from left to right along this line, a change of $(y_2 - y_1)$ units in the vertical direction corresponds to a change of $(x_2 - x_1)$ units in the horizontal direction. That is,

$$y_2 - y_1 = \text{the change in } y$$

and

$$x_2 - x_1 = \text{the change in } x.$$

The slope of the line is given by the ratio of these two changes.

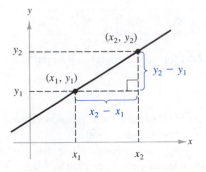

FIGURE 3.29

Definition of the Slope of a Line

The **slope** m of the nonvertical line passing through the points (x_1, y_1) and (x_2, y_2) is

$$m = \frac{y_2 - y_1}{x_2 - x_1}$$

where $x_1 \neq x_2$.

When this formula is used, the *order of subtraction* is important. Given two points on a line, we are free to label either one of them as (x_1, y_1), and the other as (x_2, y_2). However, once this is done, we must form the numerator and denominator using the same order of subtraction.

$$\underbrace{m = \frac{y_2 - y_1}{x_2 - x_1}}_{\text{Correct}} \qquad \underbrace{m = \frac{y_1 - y_2}{x_1 - x_2}}_{\text{Correct}} \qquad \underbrace{m = \frac{y_2 - y_1}{x_1 - x_2}}_{\text{Incorrect}}$$

EXAMPLE 1 *Finding the Slope of a Line Passing Through Two Points*

Find the slopes of the lines passing through the following pairs of points.

(a) $(-2, 0)$ and $(3, 1)$ (b) $(-1, 2)$ and $(2, 2)$ (c) $(0, 4)$ and $(1, -1)$

SOLUTION

(a) The slope of the line through $(-2, 0)$ and $(3, 1)$ is

$$m = \frac{y_2 - y_1}{x_2 - x_1} \quad \longleftarrow \text{Difference in } y\text{-values}$$
$$ \quad \longleftarrow \text{Difference in } x\text{-values}$$

$$= \frac{1 - 0}{3 - (-2)}$$

$$= \frac{1}{3 + 2}$$

$$= \frac{1}{5}.$$

(b) The slope of the line through $(-1, 2)$ and $(2, 2)$ is

$$m = \frac{2 - 2}{2 - (-1)} = \frac{0}{3} = 0.$$

(c) The slope of the line through $(0, 4)$ and $(1, -1)$ is

$$m = \frac{-1 - 4}{1 - 0} = \frac{-5}{1} = -5.$$

The graphs of the three lines are shown in Figure 3.30.

(a)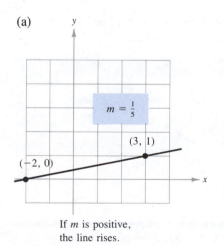

$m = \frac{1}{5}$

$(3, 1)$

$(-2, 0)$

If m is positive,
the line rises.

(b)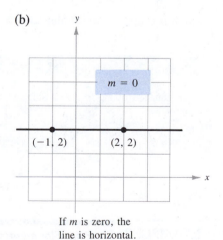

$m = 0$

$(-1, 2)$ $(2, 2)$

If m is zero, the
line is horizontal.

(c)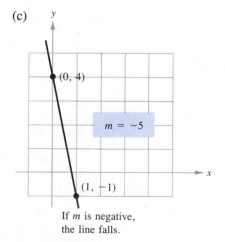

$(0, 4)$

$m = -5$

$(1, -1)$

If m is negative,
the line falls.

FIGURE 3.30

Functions and Graphs

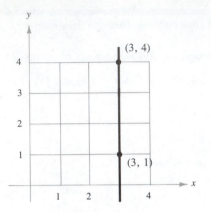

If the line is vertical,
the slope is undefined.

FIGURE 3.31

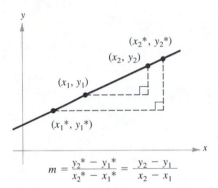

$$m = \frac{y_2{}^* - y_1{}^*}{x_2{}^* - x_1{}^*} = \frac{y_2 - y_1}{x_2 - x_1}$$

Any two points on a line can be
used to determine the slope of the line.

FIGURE 3.32

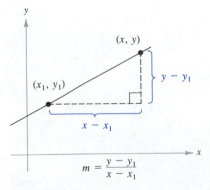

$$m = \frac{y - y_1}{x - x_1}$$

FIGURE 3.33

In Example 1, note that as we move *from left to right:*

1. A line with positive slope ($m > 0$) *rises.*
2. A line with negative slope ($m < 0$) *falls.*
3. A line with zero slope ($m = 0$) is *horizontal.*

Note that the definition of slope does not apply to vertical lines. For instance, consider the points $(3, 4)$ and $(3, 1)$ on the vertical line shown in Figure 3.31. In attempting to find the slope of this line, we obtain

$$m = \frac{4 - 1}{3 - 3}. \qquad \textit{Undefined division by zero}$$

Since division by zero is not defined, we do not define the slope of a vertical line.

Any two points on a line can be used to calculate its slope. This can be verified from the similar triangles shown in Figure 3.32. Recall that the ratios of corresponding sides of similar triangles are equal.

Equations of Lines

If we know the slope of a line and one point on the line, we can determine the equation of the line. For instance, in Figure 3.33, let (x_1, y_1) be a point on the line whose slope is m. If (x, y) is any *other* point on the line, then

$$\frac{y - y_1}{x - x_1} = m.$$

This equation, involving the two variables x and y, can be rewritten in the form

$$y - y_1 = m(x - x_1)$$

which is called the **point-slope form** of the equation of a line.

Point-Slope Form of the Equation of a Line

An equation of the line with slope m and passing through the point (x_1, y_1) is given by

$$y - y_1 = m(x - x_1).$$

EXAMPLE 2 *The Point-Slope Form of the Equation of a Line*

Find an equation of the line with slope 3 and passing through the point $(1, -2)$.

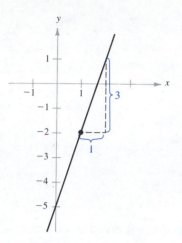

FIGURE 3.34

SOLUTION

Using the point-slope form with $(x_1, y_1) = (1, -2)$ and $m = 3$, we have

$$y - y_1 = m(x - x_1) \qquad \textit{Point-slope form}$$
$$y - (-2) = 3(x - 1)$$
$$y + 2 = 3x - 3$$
$$y = 3x - 5.$$

See Figure 3.34. ▬▬▬

The point-slope form can be used to find the equation of a line passing through two points (x_1, y_1) and (x_2, y_2). First, we find the slope of the line

$$m = \frac{y_2 - y_1}{x_2 - x_1}.$$

Then, we use the point-slope form to obtain the equation

$$y - y_1 = \frac{y_2 - y_1}{x_2 - x_1}(x - x_1).$$

This is sometimes called the **two-point form** of the equation of a line.

EXAMPLE 3   A Linear Model for Sales Prediction

The total United States sales (including inventories) during the first two quarters of 1984 were 1596 and 1649 billion dollars, respectively. (a) Write a linear equation giving the total sales y in terms of the quarter x. (b) Use the equation to predict the total sales during the fourth quarter of 1984.

SOLUTION

(a) In Figure 3.35 we let $(1, 1596)$ and $(2, 1649)$ be two points on the line representing the total United States sales. The slope of the line passing through these two points is

$$m = \frac{1649 - 1596}{2 - 1} = 53.$$

By the point-slope form, the equation of this line is

$$y - y_1 = m(x - x_1)$$
$$y - 1596 = 53(x - 1)$$
$$y = 53x - 53 + 1596 = 53x + 1543.$$

(b) Using the equation from part (a), we estimate the fourth quarter ($x = 4$) sales to be

$$y = (53)(4) + 1543 = 1755 \text{ billion dollars.} \qquad ▬▬▬$$

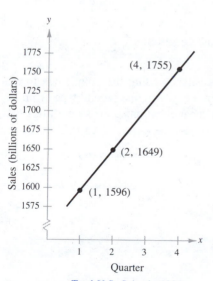

Total U.S. Sales in 1984

FIGURE 3.35

Functions and Graphs

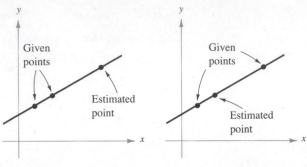

FIGURE 3.36   Linear Extrapolation   Linear Interpolation

The prediction method illustrated in Example 3 is called **linear extrapolation.** Note in Figure 3.36 that the estimated point does not lie between the given points. When the estimated point lies *between* two given points, we call the procedure **linear interpolation.**

Sketching Graphs of Lines

Many problems in coordinate (or analytic) geometry can be classified in two basic categories.

1. Given a graph (or parts of it), find its equation.
2. Given an equation, find its graph.

For lines, the first problem is solved easily by using the point-slope form. However, this formula is not convenient for solving the second type of problem. The form better suited to graphing linear equations is the **slope-intercept form** of the equation of a line. To derive the slope-intercept form, we write

$$y - y_1 = m(x - x_1) \qquad \text{\textit{Point-slope form}}$$
$$y = mx - mx_1 + y_1$$
$$y = mx + b \qquad \text{\textit{Slope-intercept form}}$$

where $b = y_1 - mx_1$.

The Slope-Intercept Form of the Equation of a Line

The graph of the equation

$$y = mx + b$$

is a line with *slope m* and *y-intercept* $(0, b)$.

EXAMPLE 4 Graphing Linear Equations

Sketch the graphs of the following linear equations.

(a) $y = 2x + 1$

(b) $y = 2$

(c) $x + y = 2$

SOLUTION

(a) Since $b = 1$, the y-intercept is $(0, 1)$. Moreover, since the slope is $m = 2$, this line *rises* two units for each unit the line moves to the right, as shown in Figure 3.37(a).

(b) By writing the equation $y = 2$ in the form $y = (0)x + 2$, we see that the y-intercept is $(0, 2)$ and the slope is zero. A zero slope implies that the line is horizontal, as shown in Figure 3.37(b).

(c) By writing the equation $x + y = 2$ in slope-intercept form, $y = -x + 2$, we see that the y-intercept is $(0, 2)$. Moreover, since the slope is $m = -1$, this line *falls* one unit for each unit the line moves to the right, as shown in Figure 3.37(c).

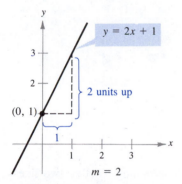

(a) When m is positive, the line rises.

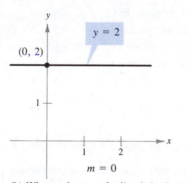

(b) When m is zero, the line is horizontal.

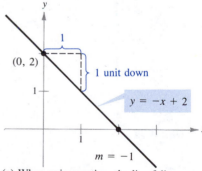

(c) When m is negative, the line falls.

FIGURE 3.37

From the slope-intercept form of the equation of a line, we see that a horizontal line ($m = 0$) has an equation of the form

$$y = (0)x + b \qquad \text{or} \qquad y = b. \qquad \textit{Horizontal line}$$

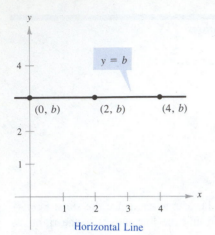

Horizontal Line

FIGURE 3.38

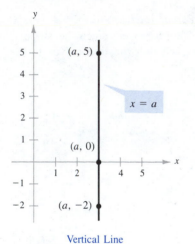

Vertical Line

FIGURE 3.39

This is consistent with the fact that each point on a horizontal line through $(0, b)$ has a y-coordinate of b, as shown in Figure 3.38.

Similarly, each point on a vertical line through $(a, 0)$ has an x-coordinate of a, as shown in Figure 3.39. Hence a vertical line has an equation of the form

$$x = a. \qquad \textcolor{blue}{\textit{Vertical line}}$$

This equation cannot be written in the slope-intercept form, because the slope of a vertical line is undefined. However, *every* line has an equation that can be written in the **general form**

$$Ax + By + C = 0 \qquad \textcolor{blue}{\textit{General form}}$$

where A and B are not *both* zero. If $A = 0$ (and $B \neq 0$), the equation can be reduced to the form $y = b$, which represents a horizontal line. If $B = 0$ (and $A \neq 0$), the general equation can be reduced to the form $x = a$, which represents a vertical line.

For convenience, we list the five most common forms of equations of lines.

Summary of Equations of Lines

1. General form: $Ax + By + C = 0$
2. Vertical line: $x = a$
3. Horizontal line: $y = b$
4. Slope-intercept form: $y = mx + b$
5. Point-slope form: $y - y_1 = m(x - x_1)$

Parallel and Perpendicular Lines

The slope of a line is a convenient tool for determining whether two lines are parallel or perpendicular.

Parallel Lines

Two distinct nonvertical lines are **parallel** if and only if their slopes are equal.

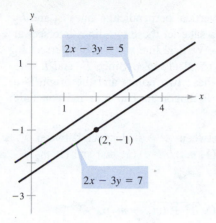

FIGURE 3.40

EXAMPLE 5 *Equations of Parallel Lines*

Find an equation of the line that passes through the point $(2, -1)$ and is parallel to the line $2x - 3y = 5$, as shown in Figure 3.40.

SOLUTION

Writing the given equation in slope-intercept form, we have

$$2x - 3y = 5 \qquad \textit{Given equation}$$
$$3y = 2x - 5$$
$$y = \frac{2}{3}x - \frac{5}{3}. \qquad \textit{Slope-intercept form}$$

Therefore, the given line has a slope of $m = 2/3$. Since any line parallel to the given line must also have a slope of $2/3$, the required line through $(2, -1)$ has the equation

$$y - (-1) = \frac{2}{3}(x - 2)$$
$$3(y + 1) = 2(x - 2)$$
$$3y + 3 = 2x - 4$$
$$-2x + 3y = -7$$
$$2x - 3y = 7.$$

Note the similarity to the original equation $2x - 3y = 5$.

Perpendicular Lines

Two nonvertical lines are **perpendicular** if and only if their slopes are related by the equation

$$m_1 = -\frac{1}{m_2}.$$

PROOF

Recall that the phrase "if and only if" is a way of stating two rules in one. One rule says, "If two nonvertical lines are perpendicular, then their slopes must be negative reciprocals." The other rule is the converse, which says, "If two lines have slopes that are negative reciprocals, they must be perpendicular." We will prove the first of these two rules and leave the proof of the converse as an exercise (see Exercise 74).

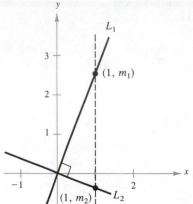

The slopes of perpendicular lines are negative reciprocals of each other.

FIGURE 3.41

Assume that we are given two nonvertical perpendicular lines L_1 and L_2 with slopes m_1 and m_2. For simplicity's sake let these two lines intersect at the origin, as shown in Figure 3.41. The vertical line $x = 1$ will intersect L_1 and L_2 at the respective points $(1, m_1)$ and $(1, m_2)$. Since L_1 and L_2 are perpendicular, the triangle formed by these two points and the origin is a right triangle. Thus, we can apply the Pythagorean Theorem and conclude that

$$\left(\begin{array}{c} \text{distance between} \\ (0, 0) \text{ and } (1, m_1) \end{array} \right)^2 + \left(\begin{array}{c} \text{distance between} \\ (0, 0) \text{ and } (1, m_2) \end{array} \right)^2 = \left(\begin{array}{c} \text{distance between} \\ (1, m_1) \text{ and } (1, m_2) \end{array} \right)^2.$$

Using the Distance Formula, we have

$$(\sqrt{1 + m_1^2})^2 + (\sqrt{1 + m_2^2})^2 = (\sqrt{0^2 + (m_1 - m_2)^2})^2$$
$$1 + m_1^2 + 1 + m_2^2 = (m_1 - m_2)^2$$
$$2 + m_1^2 + m_2^2 = m_1^2 - 2m_1m_2 + m_2^2$$
$$2 = -2m_1m_2$$
$$-1 = m_1m_2$$
$$-\frac{1}{m_2} = m_1.$$

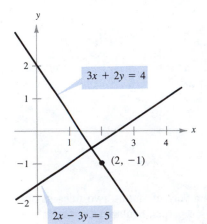

FIGURE 3.42

EXAMPLE 6   *Equations of Perpendicular Lines*

Find an equation of the line that passes through the point $(2, -1)$ and is perpendicular to the line $2x - 3y = 5$.

SOLUTION

From Example 5, the given line has slope $2/3$. Hence, any line perpendicular to this line must have a slope of $-3/2$. Therefore, the required line through $(2, -1)$ has the equation

$$y - (-1) = -\frac{3}{2}(x - 2)$$
$$2(y + 1) = -3(x - 2)$$
$$3x + 2y = 4.$$

The graphs of both lines are shown in Figure 3.42.

WARM UP

Simplify the expressions.

1. $\dfrac{4 - (-5)}{-3 - (-1)}$

2. $\dfrac{-5 - 8}{0 - (-3)}$

3. Find $-1/m$ for $m = 4/5$.

4. Find $-1/m$ for $m = -2$.

Solve for y in terms of x.

5. $2x - 3y = 5$

6. $4x + 2y = 0$

7. $y - (-4) = 3[x - (-1)]$

8. $y - 7 = \frac{2}{3}(x - 3)$

9. $y - (-1) = \dfrac{3 - (-1)}{2 - 4}(x - 4)$

10. $y - 5 = \dfrac{3 - 5}{0 - 2}(x - 2)$

EXERCISES 3.3

In Exercises 1–6, estimate the slope of the given line from its graph.

1.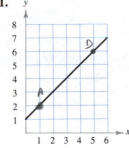

2.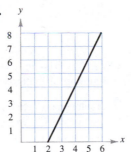

5.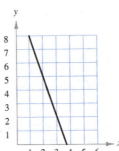

6.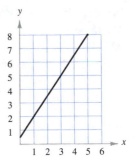

3.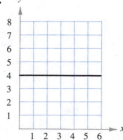

4.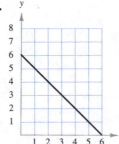

In Exercises 7 and 8, sketch the graph of the lines through the given point with the indicated slope. Make the sketches on the same set of coordinate axes.

Point	Slopes
7. $(2, 3)$	(a) 0 (b) 1 (c) 2 (d) -3
8. $(-4, 1)$	(a) 3 (b) -3 (c) $\frac{1}{2}$ (d) undefined

In Exercises 9–14, plot the points and find the slope of the line passing through each pair of points.

9. $(-3, -2)$, $(1, 6)$ **10.** $(2, 4)$, $(4, -4)$
11. $(-6, -1)$, $(-6, 4)$ **12.** $(0, -10)$, $(-4, 0)$
13. $(1, 2)$, $(-2, -2)$ **14.** $\left(\frac{7}{8}, \frac{3}{4}\right)$, $\left(\frac{5}{4}, -\frac{1}{4}\right)$

In Exercises 15–18, determine if the lines L_1 and L_2 passing through the given pairs of points are parallel, perpendicular, or neither.

15. L_1: $(0, -1)$, $(5, 9)$ **16.** L_1: $(-2, -1)$, $(1, 5)$
L_2: $(0, 3)$, $(4, 1)$ L_2: $(1, 3)$, $(5, -5)$
17. L_1: $(3, 6)$, $(-6, 0)$ **18.** L_1: $(4, 8)$, $(-4, 2)$
L_2: $(0, -1)$, $\left(5, \frac{7}{3}\right)$ L_2: $(3, -5)$, $\left(-1, \frac{1}{3}\right)$

In Exercises 19–24, use the given point on the line and the slope of the line to find three additional points that the line passes through. (The solution is not unique.)

Point	Slope
19. $(2, 1)$	$m = 0$
20. $(-3, 4)$	m undefined
21. $(5, -6)$	$m = 1$
22. $(10, -6)$	$m = -1$
23. $(-8, 1)$	m undefined
24. $(-3, -1)$	$m = 0$

In Exercises 25–30, find the slope and y-intercept (if possible) of the line specified by the given equation.

25. $5x - y + 3 = 0$ **26.** $2x + 3y - 9 = 0$
27. $5x - 2 = 0$ **28.** $3y + 5 = 0$
29. $7x + 6y - 30 = 0$ **30.** $x - y - 10 = 0$

In Exercises 31–38, find an equation for the line passing through the given points and sketch a graph of the line.

31. $(5, -1)$, $(-5, 5)$ **32.** $(4, 3)$, $(-4, -4)$
33. $\left(2, \frac{1}{2}\right)$, $\left(\frac{1}{2}, \frac{5}{4}\right)$ **34.** $(-1, 4)$, $(6, 4)$
35. $(-8, 1)$, $(-8, 7)$ **36.** $(1, 1)$, $\left(6, -\frac{2}{3}\right)$
37. $(1, 0.6)$, $(-2, -0.6)$ **38.** $(-8, 0.6)$, $(2, -2.4)$

In Exercises 39–48, find an equation of the line that passes through the given point and has the indicated slope. Sketch the graph of the line.

Point	Slope
39. $(0, -2)$	$m = 3$
40. $(0, 10)$	$m = -1$
41. $(-3, 6)$	$m = -2$
42. $(0, 0)$	$m = 4$
43. $(4, 0)$	$m = -\frac{1}{3}$
44. $(-2, -5)$	$m = \frac{3}{4}$
45. $(6, -1)$	m is undefined
46. $(-10, 4)$	$m = 0$
47. $\left(4, \frac{5}{2}\right)$	$m = \frac{4}{3}$
48. $\left(-\frac{1}{2}, \frac{3}{2}\right)$	$m = -3$

49. Prove that the line with intercepts $(a, 0)$ and $(0, b)$ has the following equation:

$$\frac{x}{a} + \frac{y}{b} = 1, \quad a \neq 0, \quad b \neq 0.$$

50. Prove that the slope of the line given by $ax + by + c = 0$ is $-a/b$, provided $b \neq 0$.

In Exercises 51–56, use the result of Exercise 49 to write an equation of the indicated line.

51. x-intercept: $(2, 0)$
y-intercept: $(0, 3)$
52. x-intercept: $(-3, 0)$
y-intercept: $(0, 4)$
53. x-intercept: $\left(-\frac{1}{6}, 0\right)$
y-intercept: $\left(0, -\frac{2}{3}\right)$
54. x-intercept: $\left(-\frac{2}{3}, 0\right)$
y-intercept: $(0, -2)$
55. point on line: $(1, 2)$
x-intercept: $(a, 0)$
y-intercept: $(0, a)$
$(a \neq 0)$
56. point on line: $(-3, 4)$
x-intercept: $(a, 0)$
y-intercept: $(0, a)$
$(a \neq 0)$

In Exercises 57–62, write an equation of the line through the given point (a) parallel to the given line and (b) perpendicular to the given line.

	Point	Line
57.	$(2, 1)$	$4x - 2y = 3$
58.	$(-3, 2)$	$x + y = 7$
59.	$(-6, 4)$	$3x + 4y = 7$
60.	$(\frac{7}{8}, \frac{3}{4})$	$5x + 3y = 0$
61.	$(-1, 0)$	$y = -3$
62.	$(2, 5)$	$x = 4$

63. Find the equation of the line giving the relationship between the temperature in degrees Celsius, C, and degrees Fahrenheit, F. Remember that water freezes at 0° Celsius (32° Fahrenheit) and boils at 100° Celsius (212° Fahrenheit).

64. Use the result of Exercise 63 to complete the following table.

C		$-10°$	$10°$			$177°$
F	$0°$			$68°$	$90°$	

65. An individual buys a $1000 corporate bond paying $9\frac{1}{2}\%$ simple interest. Write a linear equation giving the total interest I earned if the bond is held for t years.

66. A small business purchases a piece of equipment for $875. After five years the equipment will be outdated and have no value. Write a linear equation giving the value V of the equipment during the five years it will be used.

67. A store is offering a 15% discount on all items in its inventory. Write a linear equation giving the sale price S for an item with a list price L.

68. A manufacturer pays its assembly line workers $11.50 per hour. In addition, workers receive a piecework rate of $0.75 per unit produced. Write a linear equation for the hourly wages W in terms of the number of units produced per hour x.

69. A salesperson receives a monthly salary of $2500 plus a commission of 7% of his sales. Write a linear equation for the salesperson's monthly wage W in terms of his monthly sales S.

70. A sales representative uses her own car as she travels for her company. The cost to the company is $95 per day for lodging and meals plus $0.22 per mile driven. Write a linear equation giving the daily cost C to the company in terms of x, the number of miles driven.

71. A contractor purchases a piece of equipment for $36,500. The equipment requires an average expenditure of $5.25 per hour for fuel and maintenance, and the operator is paid $11.50 per hour.
   (a) Write a linear equation giving the total cost C of operating this equipment for t hours. (Include the purchase cost for the equipment.)
   (b) If customers are charged $27 per hour of machine use, write an equation for the revenue R derived from t hours of use.
   (c) Use the formula for profit ($P = R - C$) to write an equation for the profit derived from t hours of use.
   (d) (*Break-Even Point*) Use the result of part (c) to find the number of hours this equipment must be used to yield a profit of 0 dollars.

72. A real estate office handles an apartment complex with 50 units. When the rent per unit is $380 per month, all 50 units are occupied. However, when the rent is $425 per month, the average number of occupied units drops to 47. Assume that the relationship between the monthly rent p and the demand x is linear.
   (a) Write the equation of the line giving the demand x in terms of the rent p.
   (b) (*Linear Extrapolation*) Use this equation to predict the number of units occupied if the rent is raised to $455.
   (c) (*Linear Interpolation*) Predict the number of units occupied if the rent is lowered to $395.

73. Prove that if two distinct lines have equal slopes, they must be parallel.

74. Prove that if two nonvertical lines have slopes that are negative reciprocals, they must be perpendicular.

3.4 Functions

Many everyday phenomena involve two quantities that are related to each other by some rule of correspondence. Here are some examples.

1. The simple interest I earned on \$1000 for one year is related to the annual percentage rate r by the formula $I = 1000r$.
2. The distance d traveled on a bicycle in two hours is related to the speed s of the bicycle by the formula $d = 2s$.
3. The area A of a circle is related to its radius r by the formula $A = \pi r^2$.

Not all correspondences between two quantities have simple mathematical formulas. For instance, we commonly match up quantities such as NFL starting quarterbacks with touchdown passes, days of the year with the Dow-Jones Industrial Average, and hours of the day with temperature. In each of these cases, however, there is some rule of correspondence that matches each item from one set with exactly one item from a different set. Such a rule of correspondence is called a **function.**

Definition of a Function

A **function** f from a set A to a set B is a rule of correspondence that assigns to each element x in the set A exactly one element y in the set B.

Remark: The set A is called the **domain** (or set of inputs) of the function f, and the set B contains the **range** (or set of outputs).

To get a better idea of this definition, look at the function illustrated in Figure 3.43. This function can be represented more efficiently by the following set of ordered pairs:

$$\{(1, 9°), \quad (2, 13°), \quad (3, 15°), \quad (4, 15°), \quad (5, 12°), \quad (6, 4°)\}$$

where the first coordinate is the input and the second is the output. From this list and Figure 3.43, we note the following characteristics of a function.

1. Each element in A must be matched with an element in B.
2. Some elements in B may not be matched with any element in A.
3. Two or more elements of A may be matched with the same element of B.

The converse of the third statement is not true. That is, an element of A (the domain) cannot be matched with two different elements of B. This is illustrated in Example 1.

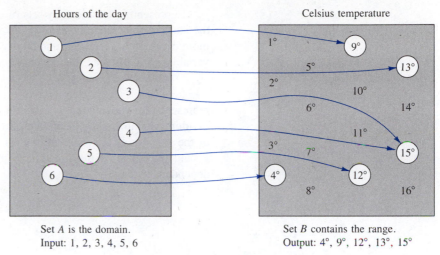

Hours of the day   Celsius temperature

Set A is the domain.   Set B contains the range.
Input: 1, 2, 3, 4, 5, 6   Output: 4°, 9°, 12°, 13°, 15°

FIGURE 3.43   Function from Set A to Set B

EXAMPLE 1 *Testing for Functions*

Let $A = \{a, b, c\}$ and $B = \{1, 2, 3, 4, 5\}$. Determine which of the following sets of ordered pairs or figures represents a function from set A to set B.

(a) $\{(a, 2), (b, 3), (c, 4)\}$ (b) $\{(a, 4), (b, 5)\}$

(c)   (d)

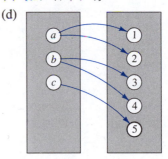

SOLUTION

(a) This collection of ordered pairs *does* represent a function from A to B. Each element of A is matched with exactly one element of B.

(b) This collection of ordered pairs *does not* represent a function from A to B. Not all elements of A are matched with an element of B.

(c) This figure *does* represent a function from A to B. It does not matter that each element of A is matched with the same element in B.

(d) This figure *does not* represent a function from A to B. The element a in A is matched with *two* elements, 1 and 2, of B. This is also true of the element b.

Representing functions by sets of ordered pairs is a common practice in the study of *discrete mathematics*. In algebra or in calculus, however, it is more common to represent functions by equations or formulas involving two variables. For instance, the equation

$$y = x^2$$

represents the variable y as a function of the variable x. We call x the **independent variable** and y the **dependent variable.** In this context, the domain of the function is the set of all values taken on by the independent variable x, and the range of the function is the set of all values taken on by the dependent variable y.

EXAMPLE 2 *Testing for Functions Represented by Equations*

Determine which of the following equations represent y as a function of x.

(a) $x^2 + y = 1$

(b) $-x + y^2 = 1$

SOLUTION

In each case, to determine whether y is a function of x, it is helpful to solve for y in terms of x.

(a) Solving for y, we obtain

$$x^2 + y = 1 \qquad \text{Given equation}$$
$$y = 1 - x^2. \qquad \text{Solve for } y$$

To each value of x there corresponds just one value for y; hence y *is* a function of x.

(b) In this case, solving for y yields

$$-x + y^2 = 1 \qquad \text{Given equation}$$
$$y^2 = 1 + x$$
$$y = \pm\sqrt{1 + x}. \qquad \text{Solve for } y$$

The $\pm$ indicates that to a given value of x there correspond *two* values for y. Hence y *is not* a function of x.

Function Notation

When using an equation to represent a function it is convenient to name the function so that it can be easily referenced. For example, we know that the equation $y = 1 - x^2$, Example 2(a), describes y as a function of x. Suppose we give this function the name "f." Then we can use the following **function notation.**

Input	Output	Equation
x	$f(x)$	$f(x) = 1 - x^2$

The symbol $f(x)$ is read as the **value of f at x** or simply "f of x." This corresponds to the y-value for a given x. Thus, we can (and often do) write $y = f(x)$.

Keep in mind that f is the *name* of the function, while $f(x)$ is the *value* of the function at x. For instance, the function given by

$$f(x) = 3 - 2x$$

has *function values* denoted by $f(-1)$, $f(0)$, $f(2)$, and so on. To find these values, we substitute the specified input values into the given equation, as follows.

For $x = -1$, $f(-1) = 3 - 2(-1) = 3 + 2 = 5.$
For $x = 0$, $f(0) = 3 - 2(0) = 3 + 0 = 3.$
For $x = 2$, $f(2) = 3 - 2(2) = 3 - 4 = -1.$

Although we generally use f as a convenient function name and x as the independent variable, we can use other letters. For instance, $f(x) = x^2 - 4x + 7$, $f(t) = t^2 - 4t + 7$, and $g(s) = s^2 - 4s + 7$ all define the same function. In fact, the role of the independent variable in a function is simply that of a "placeholder." Consequently, the above function can be properly described by the form

$$f(\ ) = (\ )^2 - 4(\ ) + 7$$

where parentheses are used in place of a letter. Therefore, to evaluate $f(-2)$, we simply place -2 in each set of parentheses:

$$\begin{aligned} f(-2) &= (-2)^2 - 4(-2) + 7 \\ &= 4 + 8 + 7 \\ &= 19. \end{aligned}$$

Similarly, the value of $f(3x)$ is

$$\begin{aligned} f(3x) &= (3x)^2 - 4(3x) + 7 \\ &= 9x^2 - 12x + 7. \end{aligned}$$

EXAMPLE 3 *Evaluating Functions*

Let $g(x) = 4x - x^2$ and find the following.

(a) $g(2)$ (b) $g(x + 2)$

SOLUTION

(a) Replacing x with 2, we obtain

$$g(2) = 4(2) - (2)^2 = 8 - 4 = 4.$$

(b) Replacing x with $x + 2$, we obtain

$$\begin{aligned} g(x + 2) &= 4(x + 2) - (x + 2)^2 \\ &= 4x + 8 - (x^2 + 4x + 4) \\ &= -x^2 + 4. \end{aligned}$$

Remark: Example 3 shows that $g(x + 2) \neq g(x) + g(2)$ because $-x^2 + 4 \neq 4x - x^2 + 4$. In general, $g(u + v)$ is not equal to $g(u) + g(v)$.

Sometimes a function is defined using more than one equation. An illustration is given in Example 4.

EXAMPLE 4 *A Function Defined by Two Equations*

Evaluate the function given by

$$f(x) = \begin{cases} x^2 + 1, & x < 0 \\ x - 1, & x \geq 0 \end{cases}$$

at $x = -1, 0$, and 1.

SOLUTION

Since $x = -1 < 0$, we use $f(x) = x^2 + 1$ to obtain

$$f(-1) = (-1)^2 + 1 = 2.$$

For $x = 0$, we use $f(x) = x - 1$ to obtain

$$f(0) = (0) - 1 = -1.$$

For $x = 1$, we use $f(x) = x - 1$ to obtain

$$f(1) = (1) - 1 = 0.$$

Finding the Domain of a Function

The domain of a function may be explicitly described along with the function, or it may be *implied* by the expression used to define the function. The **implied domain** is the set of all real numbers for which the expression is defined. For instance, the function given by

$$f(x) = \frac{1}{x^2 - 4}$$

has an implied domain which consists of all real x other than $x = \pm 2$. These two values are excluded from the domain because division by zero is undefined. Another common type of implied domain is that used to avoid even roots of negative numbers. For example, the function given by

$$f(x) = \sqrt{x}$$

is defined only for $x \geq 0$. Hence, its implied domain is the interval $[0, \infty)$.

The *ranges* of such functions are more difficult to find, and can best be obtained from their graphs, which we will study in Section 3.5.

EXAMPLE 5 Finding the Domain of a Function

Find the domain of each of the functions.

(a) f: $\{(-3, 0), (-1, 4), (0, 2), (2, 2), (4, -1)\}$

(b) Volume of a sphere: $V = \frac{4}{3}\pi r^3$

(c) $g(x) = \dfrac{1}{x + 5}$ (d) $h(x) = \sqrt{4 - x^2}$

SOLUTION

(a) The domain of f consists of all first coordinates in the set of ordered pairs, and is therefore the set

domain = $\{-3, -1, 0, 2, 4\}$.

(b) For the volume of a sphere we must choose nonnegative values for the radius (independent variable) r. Thus, the domain is the set of all real numbers r such that $r \geq 0$.

(c) Excluding x-values that yield zero in the denominator, the domain of g is the set of all real numbers $x \neq -5$.

(d) We choose x-values for which $4 - x^2 \geq 0$. Using the methods of Section 2.8, we conclude that $-2 \leq x \leq 2$. Thus, the domain is the interval $[-2, 2]$.

Remark: In Example 5(b), note that the domain of a function may be implied by the physical context. For instance, from the equation $V = \frac{4}{3}\pi r^3$, we would have no reason to restrict r to nonnegative values, but the physical context tells us that a sphere cannot have a negative radius.

Applications

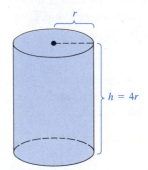

FIGURE 3.44

EXAMPLE 6   The Dimensions of a Container

The marketing manager of a soft drink company wants to design a can with a height that is four times the radius of the can.

(a) Express the volume of the can as a function of the radius r.
(b) Express the volume of the can as a function of the height h.

SOLUTION

The volume of a right circular cylinder is given by the formula

$$V = \pi(\text{radius})^2(\text{height}).$$

(a) If the radius is r, then the height is $4r$, as shown in Figure 3.44. The volume as a function of r is given by

$$V = \pi r^2(4r) = 4\pi r^3.$$

(b) If the height is h and $h = 4r$, then $r = h/4$. The volume as a function of h is

$$V = \pi\left(\frac{h}{4}\right)^2 h = \frac{\pi h^3}{16}.$$

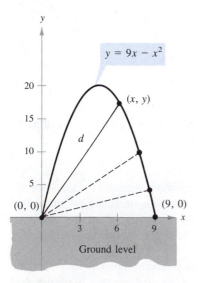

FIGURE 3.45

EXAMPLE 7   The Path of a Projectile

A projectile moves in a parabolic path given by $y = 9x - x^2$, as shown in Figure 3.45.

(a) For any point (x, y) on the path, express the distance d from (x, y) to the launch point $(0, 0)$ as a function of x.
(b) Find the domain of d.

SOLUTION

(a) By the Distance Formula, we know that

$$d = \sqrt{(x_2 - x_1)^2 + (y_2 - y_1)^2} = \sqrt{(x - 0)^2 + (y - 0)^2}.$$

Since $y = 9x - x^2$, it follows that as a function of x, d is given by

$$d = \sqrt{x^2 + (9x - x^2)^2} = \sqrt{x^4 - 18x^3 + 82x^2}.$$

(b) The domain of d is restricted to the x-values between the launch point $(0, 0)$ and the point of impact (the other x-intercept). Solving

$$0 = 9x - x^2 = x(9 - x)$$

we find the other x-intercept to be $(9, 0)$. Consequently, $0 \leq x \leq 9$, and the domain of d is the interval $[0, 9]$.

One of the basic definitions in calculus employs the ratio

$$\frac{f(x + h) - f(x)}{h}, \qquad h \neq 0$$

called a **difference quotient.** The next example shows how such an expression can be evaluated.

EXAMPLE 8   *Evaluating a Difference Quotient*

For the function given by $f(x) = x^2 - 4x + 7$, find

$$\frac{f(x + h) - f(x)}{h}.$$

SOLUTION

$$\frac{f(x + h) - f(x)}{h} = \frac{[(x + h)^2 - 4(x + h) + 7] - [x^2 - 4x + 7]}{h}$$

$$= \frac{x^2 + 2xh + h^2 - 4x - 4h + 7 - x^2 + 4x - 7}{h}$$

$$= \frac{2xh + h^2 - 4h}{h}$$

$$= \frac{h(2x + h - 4)}{h}$$

$$= 2x + h - 4$$

Summary of Function Terminology

Function

A relationship between two variables such that to each value of the independent variable there corresponds exactly one value of the dependent variable.

Function Notation $y = f(x)$

f is the **name** of the function.
y is the **dependent variable.**
x is the **independent variable.**
$f(x)$ is the **value of the function at x.**

Domain

The set of all values (inputs) of the independent variable for which the function is defined. If x is in the domain of f, we say that f is **defined** at x. If x is not in the domain of f, we say that f is **undefined** at x.

Range

The set of all values (outputs) assumed by the dependent variable (that is, the set of all function values).

Implied Domain

If f is defined by an algebraic expression and the domain is not specified, then the implied domain consists of all real numbers for which the expression is defined.

WARM UP

Simplify the expression.

1. $2(-3)^3 + 4(-3) - 7$

2. $4(-1)^2 - 5(-1) + 4$

3. $(x + 1)^2 + 3(x + 1) - 4 - (x^2 + 3x - 4)$

4. $(x - 2)^2 - 4(x - 2) - (x^2 - 4)$

Solve for y in terms of x.

5. $2x + 5y - 7 = 0$

6. $y^2 = x^2$

Solve the inequality.

7. $x^2 - 4 \geq 0$

8. $9 - x^2 \geq 0$

9. $x^2 + 2x + 1 \geq 0$

10. $x^2 - 3x + 2 \geq 0$

EXERCISES 3.4

In Exercises 1–4, fill in the blanks using the specified function and the given value of the independent variable. (The symbol Δx represents a single variable and is read "delta x." This symbol is commonly used in calculus to denote a small change in x.)

1. $f(x) = 6 - 4x$
   (a) $f(3) = 6 - 4($ ⬚ $)$
   (b) $f(-7) = 6 - 4($ ⬚ $)$
   (c) $f(t) = 6 - 4($ ⬚ $)$
   (d) $f(c + 1) = 6 - 4($ ⬚ $)$

2. $g(x) = x^2 - 2x$
   (a) $g(2) = ($ ⬚ $)^2 - 2($ ⬚ $)$
   (b) $g(-3) = ($ ⬚ $)^2 - 2($ ⬚ $)$
   (c) $g(t + 1) = ($ ⬚ $)^2 - 2($ ⬚ $)$
   (d) $g(x + \Delta x) = ($ ⬚ $)^2 - 2($ ⬚ $)$

3. $f(s) = \dfrac{1}{s + 1}$
   (a) $f(4) = \dfrac{1}{(\ ⬚\ ) + 1}$
   (b) $f(0) = \dfrac{1}{(\ ⬚\ ) + 1}$
   (c) $f(4x) = \dfrac{1}{(\ ⬚\ ) + 1}$
   (d) $f(x + h) = \dfrac{1}{(\ ⬚\ ) + 1}$

4. $f(t) = \sqrt{25 - t^2}$
   (a) $f(3) = \sqrt{25 - (\ ⬚\ )^2}$
   (b) $f(5) = \sqrt{25 - (\ ⬚\ )^2}$
   (c) $f(x + 5) = \sqrt{25 - (\ ⬚\ )^2}$
   (d) $f(2 + h) = \sqrt{25 - (\ ⬚\ )^2}$

In Exercises 5–16, evaluate the function at the specified value of the independent variable and simplify the results.

5. $f(x) = 2x - 3$
   (a) $f(1)$
   (b) $f(-3)$
   (c) $f(x - 1)$
   (d) $f(\frac{1}{4})$

6. $g(y) = 7 - 3y$
   (a) $g(0)$
   (b) $g(\frac{7}{3})$
   (c) $g(s)$
   (d) $g(s + 2)$

7. $h(t) = t^2 - 2t$
   (a) $h(2)$
   (b) $h(-1)$
   (c) $h(x + 2)$
   (d) $h(1.5)$

8. $V(r) = \frac{4}{3}\pi r^3$
   (a) $V(3)$
   (b) $V(0)$
   (c) $V(\frac{3}{2})$
   (d) $V(2r)$

9. $f(y) = 3 - \sqrt{y}$
   (a) $f(4)$
   (b) $f(100)$
   (c) $f(4x^2)$
   (d) $f(0.25)$

10. $f(x) = \sqrt{x + 8} + 2$
   (a) $f(-8)$
   (b) $f(1)$
   (c) $f(x - 8)$
   (d) $f(h + 8)$

11. $q(x) = \dfrac{1}{x^2 - 9}$
   (a) $q(4)$
   (b) $q(0)$
   (c) $q(3)$
   (d) $q(y + 3)$

12. $q(t) = \dfrac{2t^2 + 3}{t^2}$
   (a) $q(2)$
   (b) $q(0)$
   (c) $q(x)$
   (d) $q(-x)$

13. $f(x) = \dfrac{|x|}{x}$
   (a) $f(2)$
   (b) $f(-2)$
   (c) $f(x^2)$
   (d) $f(x - 1)$

14. $f(x) = |x| + 4$
   (a) $f(2)$
   (b) $f(-2)$
   (c) $f(x^2)$
   (d) $f(x + \Delta x) - f(x)$

15. $f(x) = \begin{cases} 2x + 1, & x < 0 \\ 2x + 2, & x \ge 0 \end{cases}$
   (a) $f(-1)$
   (b) $f(0)$
   (c) $f(1)$
   (d) $f(2)$

16. $f(x) = \begin{cases} x^2 + 2, & x \le 1 \\ 2x^2 + 2, & x > 1 \end{cases}$
   (a) $f(-2)$
   (b) $f(0)$
   (c) $f(1)$
   (d) $f(2)$

In Exercises 17–22, find the indicated difference quotient and simplify your answer.

17. $f(x) = x^2 - x + 1$

$\dfrac{f(2 + h) - f(2)}{h}$

18. $f(x) = 5x - x^2$

$\dfrac{f(5 + h) - f(5)}{h}$

19. $f(x) = x^3$

$\dfrac{f(x + \Delta x) - f(x)}{\Delta x}$

20. $f(x) = 2x$

$\dfrac{f(x + \Delta x) - f(x)}{\Delta x}$

21. $g(x) = 3x - 1$

$\dfrac{g(x) - g(3)}{x - 3}$

22. $f(t) = \dfrac{1}{t}$

$\dfrac{f(t) - f(1)}{t - 1}$

In Exercises 23–28, find all real values of x such that $f(x) = 0$.

23. $f(x) = 15 - 3x$

24. $f(x) = \dfrac{3x - 4}{5}$

25. $f(x) = x^2 - 9$

26. $f(x) = x^3 - x$

27. $f(x) = \dfrac{3}{x - 1} + \dfrac{4}{x - 2}$

28. $f(x) = a + \dfrac{b}{x}$

In Exercises 29–38, find the domain of the function.

29. $f(x) = 5x^2 + 2x - 1$

30. $g(x) = 1 - 2x^2$

31. $h(t) = \dfrac{4}{t}$

32. $s(y) = \dfrac{3y}{y + 5}$

33. $g(y) = \sqrt{y - 10}$

34. $f(t) = \sqrt[3]{t + 4}$

35. $f(x) = \sqrt[4]{1 - x^2}$

36. $h(x) = \dfrac{10}{x^2 - 2x}$

37. $g(x) = \dfrac{1}{x} - \dfrac{3}{x + 2}$

38. $f(s) = \dfrac{\sqrt{s - 1}}{s - 4}$

In Exercises 39–48, identify the equations that determine y as a function of x.

39. $x^2 + y^2 = 4$

40. $x = y^2$

41. $x^2 + y = 4$

42. $x + y^2 = 4$

43. $2x + 3y = 4$

44. $x^2 + y^2 - 2x - 4y + 1 = 0$

45. $y^2 = x^2 - 1$

46. $y = \sqrt{x + 5}$

47. $x^2y - x^2 + 4y = 0$

48. $xy - y - x - 2 = 0$

In Exercises 49 and 50, determine which of the sets of ordered pairs represents a function from A to B. Give reasons for your answers.

49. $A = \{0, 1, 2, 3\}$ and $B = \{-2, -1, 0, 1, 2\}$
   (a) $\{(0, 1), (1, -2), (2, 0), (3, 2)\}$
   (b) $\{(0, -1), (2, 2), (1, -2), (3, 0), (1, 1)\}$
   (c) $\{(0, 0), (1, 0), (2, 0), (3, 0)\}$
   (d) $\{(0, 2), (3, 0), (1, 1)\}$

50. $A = \{\alpha, \beta, \gamma\}$ and $B = \{w, x, y, z\}$
   (a) $\{(\alpha, x), (\gamma, y), (\gamma, z), (\beta, z)\}$
   (b) $\{(\alpha, x), (\beta, y), (\gamma, z)\}$
   (c) $\{(x, \alpha), (w, \alpha), (y, \gamma), (z, \beta)\}$
   (d) $\{(\gamma, w), (\beta, w), (\alpha, z)\}$

In Exercises 51–54, assume that the domain of f is the set $A = \{-2, -1, 0, 1, 2\}$. Determine the set of ordered pairs representing the function f.

51. $f(x) = x^2$

52. $f(x) = \dfrac{2x}{x^2 + 1}$

53. $f(x) = \sqrt{x + 2}$

54. $f(x) = |x + 1|$

In Exercises 55–58, find the value(s) of x for which $f(x) = g(x)$.

55. $f(x) = x^2$, $g(x) = x + 2$

56. $f(x) = x^2 + 2x + 1$, $g(x) = 3x + 3$

57. $f(x) = \sqrt{3x} + 1$, $g(x) = x + 1$

58. $f(x) = x^4 - 2x^2$, $g(x) = 2x^2$

59. Express the volume V of a cube as a function of the length e of one of its edges.

60. Express the circumference C of a circle as a function of its (a) radius r, and (b) diameter d.

61. Express the area A of a circle as a function of its circumference C.

62. Express the area A of an equilateral triangle as a function of the length s of one of its sides.

63. An open box is to be made from a square piece of material, 12 inches on a side, by cutting equal squares from each corner and turning up the sides (see figure). Write the volume V of the box as a function of x.

64. A rectangular package to be sent by a postal service can have a maximum combined length and girth (perimeter of a cross section) of 108 inches (see figure). Write the volume of the package as a function of x.

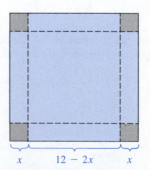

FIGURE FOR 63

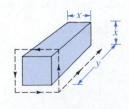

FIGURE FOR 64

65. A right triangle is formed in the first quadrant by the *x*- and *y*-axes and a line through the point (1, 2) (see figure). Write the area of the triangle as a function of *x*, and determine the domain of the function.

66. A rectangle is bounded by the *x*-axis and the semicircle $y = \sqrt{25 - x^2}$ (see figure). Write the area of the rectangle as a function of *x*, and determine the domain of the function.

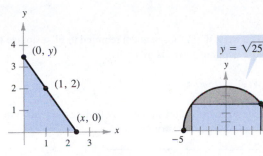

FIGURE FOR 65   **FIGURE FOR 66**

67. A balloon carrying a transmitter ascends vertically from a point 2000 feet from the receiving station (see figure). Let *d* be the distance between the balloon and the receiving station. Express the height of the balloon as a function of *d*.

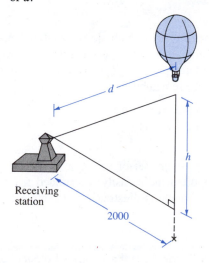

FIGURE FOR 67

68. A mechanical arm is used to fill containers as they move on a conveyor belt (see figure). Express the length *L* of the arm as a function of *x*, as labeled in the figure.

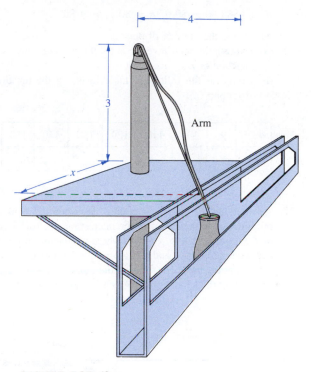

FIGURE FOR 68

69. A company produces a product for which the variable cost is $12.30 per unit and the fixed costs are $98,000. The product sells for $17.98. Let *x* be the number of units produced.
  (a) Write the total cost *C* as a function of the number of units produced.
  (b) Write the revenue *R* as a function of the number of units produced.
  (c) Write the profit *P* as a function of the number of units produced. (*Note: P = R − C.*)

70. The inventor of a new game believes that the variable cost for producing the game is $0.95 per unit and the fixed costs are $6000. The inventor sells each game for $1.69. Let *x* be the number of games sold.
  (a) Write the total cost *C* as a function of the number of games sold.
  (b) Write the average cost per unit $\bar{C} = C/x$ as a function of *x*.

71. For groups of 80 or more people, a charter bus company determines the rate per person according to the following formula:

    Rate = $8.00 − $0.05(n − 80), n ≥ 80

    where n is the number of persons.
    (a) Express the total revenue R for the bus company as a function of n.
    (b) Complete the following table by using the function from part (a).

n	90	100	110	120	130	140	150
R(n)							

72. Assume that the amount of money deposited in a bank is proportional to the square of the interest rate the bank pays on the money. That is, $d = kr^2$, where d is the total deposit, r is the interest rate, and k is the proportionality constant. Assuming the bank can reinvest the money for a return of 18%, write the bank's profit P as a function of the interest rate r.

73. The force F (in tons) of water pressure against the face of a dam is a function of the depth y of the water given by

    $$F(y) = 149.76\sqrt{10}y^{5/2}.$$

    Complete the following table by finding the force for various water levels.

y	5	10	20	30	40
F(y)					

74. The work W (in foot-pounds) required to fill a storage tank to a depth of y feet is given by

    $$W(y) = 25\pi\left(4y^2 - \frac{y^3}{3}\right).$$

    Complete the following table by finding the work required to fill the tank to various levels.

y	1	2	4	6	8
W(y)					

3.5 Graphs of Functions

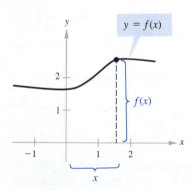

FIGURE 3.46

In Section 3.4 we discussed functions from an algebraic (or analytic) point of view. Here we look at functions from a geometric perspective. The **graph of a function** f is the collection of ordered pairs (x, f(x)) such that x is in the domain of f. As you study this section, remember that

$$x = \text{the directed distance from the } y\text{-axis}$$
$$f(x) = \text{the directed distance from the } x\text{-axis}$$

as shown in Figure 3.46.

We noted in Section 3.4 that the *range* (set of values assumed by the dependent variable) of a function is often more easily determined from its graph than from its equation. This technique is illustrated in Example 1.

EXAMPLE 1 *Finding Domain and Range from the Graph of a Function*

Use the graph of the function f, shown in Figure 3.47, to find (a) the domain of f, (b) f(−1) and f(2), and (c) the range of f.

Graphs of Functions

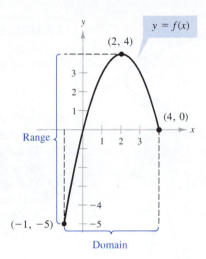

FIGURE 3.47

SOLUTION

(a) Because the graph does not extend beyond $x = -1$ (on the left) and $x = 4$ (on the right), the domain of f is all x in the interval $[-1, 4]$.

(b) Since $(-1, -5)$ is a point on the graph, it follows that

$$f(-1) = -5.$$

Similarly, since $(2, 4)$ is a point on the graph, it follows that

$$f(2) = 4.$$

(c) Because the graph does not extend below $f(-1) = -5$ nor above $f(2) = 4$, the range of f is the interval $[-5, 4]$.

Remark: Note in Figure 3.47 that the solid dots representing the points $(-1, -5)$ and $(4, 0)$ indicate that the graph terminates at these points.

By the definition of a function, at most one y-value corresponds to a given x-value. It follows, then, that a vertical line can intersect the graph of a function at most once. This observation provides us with a convenient visual test for functions.

Vertical Line Test for Functions

A set of points in a coordinate plane is the graph of y as a function of x if and only if no vertical line intersects the graph at more than one point.

Functions and Graphs

(a)

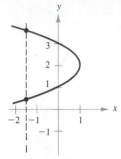

EXAMPLE 2 *Vertical Line Test for Functions*

Which of the graphs in Figure 3.48 represents y as a function of x?

SOLUTION

(a) This *is not* a graph of y as a function of x because we can find a vertical line that intersects the graph twice.
(b) This *is* a graph of y as a function of x because every vertical line intersects the graph at most once.
(c) This *is* a graph of y as a function of x. (Note that if a vertical line does not intersect the graph, it simply means that the function is undefined for this particular value of x.)

(b)

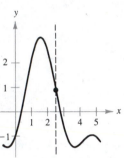

(c)

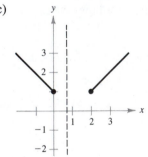

FIGURE 3.48

The more we know about the graph of a function, the more we know about the function itself. Consider the graph shown in Figure 3.49. As we move *from left to right*, this graph falls for values of x between -2 and 0, is constant from $x = 0$ to $x = 2$, and then rises for values of x between 2 and 4. From these observations, we say that the function is:

decreasing on the interval $(-2, 0)$,

constant on the interval $(0, 2)$, and

increasing on the interval $(2, 4)$.

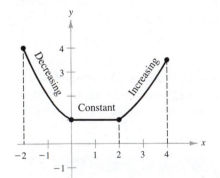

FIGURE 3.49

Definition of Increasing, Decreasing, and Constant Functions

A function f is **increasing** on an interval if, for any x_1 and x_2 in the interval,

$$x_1 < x_2 \quad \text{implies} \quad f(x_1) < f(x_2).$$

A function f is **decreasing** on an interval if, for any x_1 and x_2 in the interval,

$$x_1 < x_2 \quad \text{implies} \quad f(x_1) > f(x_2).$$

A function f is **constant** on an interval if, for every x_1 and x_2 in the interval,

$$f(x_1) = f(x_2).$$

Graphs of Functions

EXAMPLE 3 *Increasing and Decreasing Functions*

In Figure 3.50 determine the open intervals on which each function is increasing, decreasing, or constant.

SOLUTION

(a) Although it might appear that there is an interval about zero over which this function is constant, we see that if $x_1 < x_2$, then $f(x_1) = x_1^3 < x_2^3 = f(x_2)$, and we conclude that the function is increasing for all x.

(b) This function is

increasing on the interval $(-\infty, -1)$,
decreasing on the interval $(-1, 1)$, and
increasing on the interval $(1, \infty)$.

(c) This function is

increasing on the interval $(-\infty, 0)$,
constant on the interval $(0, 2)$, and
decreasing on the interval $(2, \infty)$.

(a)

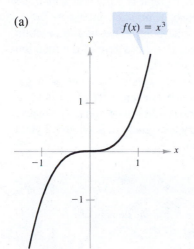

$f(x) = x^3$

(b)

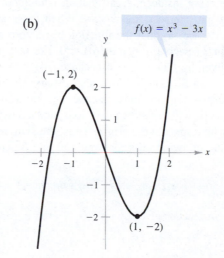

$f(x) = x^3 - 3x$

(c)

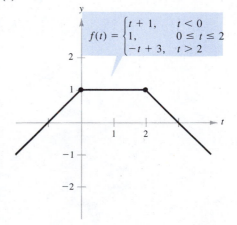

$$f(t) = \begin{cases} t + 1, & t < 0 \\ 1, & 0 \le t \le 2 \\ -t + 3, & t > 2 \end{cases}$$

FIGURE 3.50

Remark: The points at which a function changes its increasing, decreasing, or constant behavior are especially important in producing an accurate graph of the function. These points often identify *maximum* or *minimum* values of the function. Techniques for finding the exact location of these special points are developed in calculus.

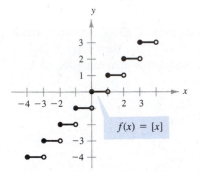

Greatest Integer Function

FIGURE 3.51

EXAMPLE 4   *The Greatest Integer Function*

The **greatest integer function** is denoted by $[x]$ and is defined by

$$f(x) = [x] = \text{the greatest integer less than or equal to } x.$$

The graph of this function is shown in Figure 3.51. Note that the graph of f jumps vertically one unit at each integer and is constant (a horizontal line segment) between each pair of consecutive integers. Some values of this function are

$$[-1] = -1, \quad [-0.5] = -1, \quad [1] = 1, \quad \text{and} \quad [1.5] = 1.$$

The range of the greatest integer function is the set of all integers.

Aids for Sketching Graphs of Functions

Intercepts, symmetry, and equation recognition, which we used for sketching the graph of an equation, can also be used to sketch the graph of a function.

For instance, recall that an x-intercept of a graph is a point at which the graph intersects the x-axis. When the graph of a function intersects the x-axis, we say that the function has a zero at that point. That is, the function f has a **zero** at $x = a$ if $f(a) = 0$. For instance, $x = 4$ is a zero of the function given by $f(x) = x - 4$ because $f(4) = 4 - 4 = 0$.

In Section 3.2, we also discussed different types of symmetry. In the terminology of functions, we say that a function is **even** if its graph is symmetric with respect to the y-axis, and a function is **odd** if its graph is symmetric with respect to the origin. Thus, the symmetry tests given in Section 3.2 yield the following tests for even and odd functions.

Test for Even and Odd Functions

A function given by $y = f(x)$ is **even** if, for each x in the domain of f,

$$f(-x) = f(x).$$

A function given by $y = f(x)$ is **odd** if, for each x in the domain of f,

$$f(-x) = -f(x).$$

EXAMPLE 5 **Even and Odd Functions**

Determine whether the following functions are even, odd, or neither.

(a) $g(x) = x^3 - x$ (b) $h(x) = x^2 + 1$ (c) $f(x) = x^3 - 1$

SOLUTION

(a) This function is odd because

$$g(-x) = (-x)^3 - (-x) \qquad \textit{Substitute } -x \textit{ for } x$$
$$= -x^3 + x$$
$$= -(x^3 - x)$$
$$= -g(x).$$

(b) This function is even because

$$h(-x) = (-x)^2 + 1 = x^2 + 1 = h(x).$$

(c) By substituting $-x$ for x, we have

$$f(-x) = (-x)^3 - 1 = -x^3 - 1.$$

Because $f(x) = x^3 - 1$ and $-f(x) = -x^3 + 1$, we conclude that $f(-x) \neq f(x)$ and $f(-x) \neq -f(x)$. Hence, the function is neither even nor odd.

The graphs of these three functions are shown in Figure 3.52.

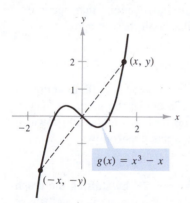

(a) Odd function
     (symmetric to origin)

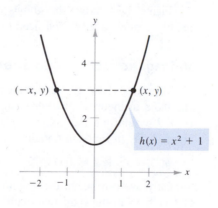

(b) Even function
     (symmetric to y-axis)

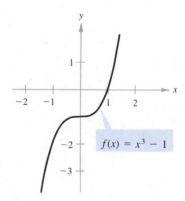

(c) Neither even nor odd

FIGURE 3.52

Functions and Graphs

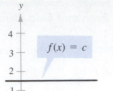
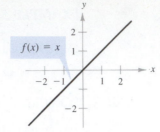
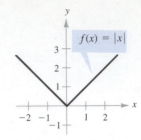

(a)   Constant function   (b)   Identity function   (c)   Absolute value function

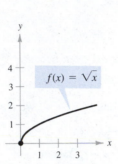
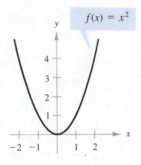
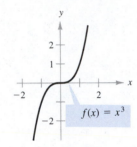

(d)   Square root function   (e)   Squaring function   (f)   Cubing function

FIGURE 3.53

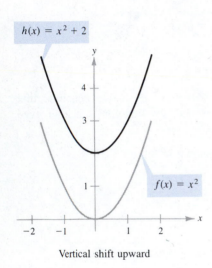

Vertical shift upward

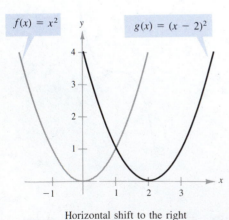

Horizontal shift to the right

FIGURE 3.54

Figure 3.53 shows the graphs of some basic functions that occur frequently in applications. You need to be familiar with these graphs.

Shifting, Reflecting, and Stretching Graphs

Many functions have graphs that are simple transformations of familiar graphs like those in Figure 3.53. For example, we can obtain the graph of $h(x) = x^2 + 2$ by shifting the graph of $f(x) = x^2$ upward two units, as shown in Figure 3.54. In function notation, h and f are related as follows.

$$h(x) = x^2 + 2 = f(x) + 2 \qquad \textit{Upward shift of 2}$$

Similarly, we can obtain the graph of $g(x) = (x - 2)^2$ by shifting the graph of $f(x) = x^2$ to the right two units. In this case, the functions g and f have the following relationship.

$$g(x) = (x - 2)^2 = f(x - 2) \qquad \textit{Right shift of 2}$$

We summarize these two types of **horizontal** and **vertical shifts** of the graphs of functions as follows.

Vertical and Horizontal Shifts

Let c be a positive real number. **Vertical** and **horizontal shifts** in the graph of $y = f(x)$ are represented as follows.

Vertical shift c units **upward:**	$h(x) = f(x) + c$
Vertical shift c units **downward:**	$h(x) = f(x) - c$
Horizontal shift c units to the **right:**	$h(x) = f(x - c)$
Horizontal shift c units to the **left:**	$h(x) = f(x + c)$

Some graphs can be obtained from a *combination* of vertical and horizontal shifts. This is demonstrated in part (c) of the next example.

EXAMPLE 6   *Shifts in the Graph of a Function*

Use the graph of $f(x) = x^3$ to sketch the graph of each of the following functions.

(a) $g(x) = x^3 + 1$ (b) $h(x) = (x - 1)^3$

(c) $k(x) = (x + 2)^3 + 1$

SOLUTION

Relative to the graph of $f(x) = x^3$, the graph of $g(x) = x^3 + 1$ is an upward shift of one unit, the graph of $h(x) = (x - 1)^3$ is a right shift of one unit, and the graph of $k(x) = (x + 2)^3 + 1$ involves a left shift of two units and an upward shift of one unit. The graphs are shown in Figure 3.55.

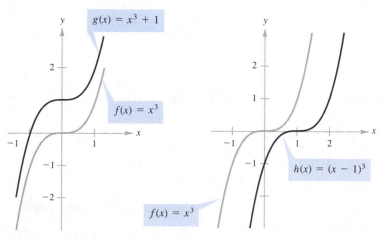

(a)   Vertical shift: 1 up

(b)   Horizontal shift: 1 right

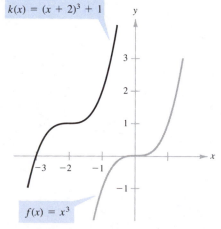

(c)   Horizontal shift: 2 left
Vertical shift: 1 up

FIGURE 3.55

Functions and Graphs

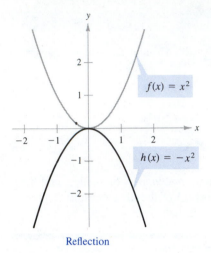

$f(x) = x^2$

$h(x) = -x^2$

Reflection

FIGURE 3.56

The second common type of transformation is called a **reflection.** For instance, if we assume that the *x*-axis represents a mirror, then the graph of $h(x) = -x^2$ is the mirror image (or reflection) of the graph of $f(x) = x^2$, as shown in Figure 3.56.

Reflections in the Coordinate Axes

Reflections, in the coordinate axes, of the graph of $y = f(x)$ are represented as follows.

Reflection in the *x*-axis:	$h(x) = -f(x)$
Reflection in the *y*-axis:	$h(x) = f(-x)$

EXAMPLE 7 *Reflections and Shifts*

Sketch a graph of each of the following functions.

(a) $g(x) = -\sqrt{x}$ (b) $h(x) = \sqrt{-x}$ (c) $k(x) = -\sqrt{x + 2}$

SOLUTION

(a) Relative to the graph of $f(x) = \sqrt{x}$, the graph of *g* is a reflection in the *x*-axis because

$$g(x) = -\sqrt{x} = -f(x).$$

(b) The graph of *h* is a reflection of the graph of $f(x) = \sqrt{x}$ in the *y*-axis because

$$h(x) = \sqrt{-x} = f(-x).$$

(c) From the equation

$$k(x) = -\sqrt{x + 2} = -f(x + 2)$$

we conclude that the graph of *k* is, first, a left shift of two units, followed by a reflection in the *x*-axis.

The graphs of all three functions are shown in Figure 3.57.

Graphs of Functions

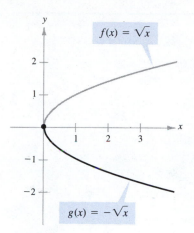

(a)   Reflection in *x*-axis

(b)   Reflection in *y*-axis

(c)   Shift and reflection

FIGURE 3.57

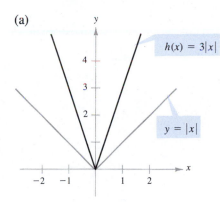

(a)

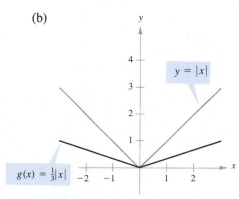

(b)

FIGURE 3.58

Horizontal shifts, vertical shifts, and reflections are called **rigid** transformations because the basic shape of the graph is unchanged. These transformations change only the *position* of the graph in the *xy*-plane. **Nonrigid** transformations are those which cause a *distortion*—a change in the shape of the original graph. For instance, a nonrigid transformation of the graph of $y = f(x)$ is represented by $y = cf(x)$, where the transformation is a **vertical stretch** if $c > 1$ and a **vertical shrink** if $0 < c < 1$.

EXAMPLE 8   *Nonrigid Transformations*

Sketch a graph of each of the following.

(a) $h(x) = 3|x|$

(b) $g(x) = \dfrac{1}{3}|x|$

SOLUTION

(a) Relative to the graph of $f(x) = |x|$, the graph of

$$h(x) = 3|x| = 3f(x)$$

is a vertical stretch (multiply each *y*-value by 3) of the graph of *f*.

(b) Similarly, the equation

$$g(x) = \frac{1}{3}|x| = \frac{1}{3}f(x)$$

indicates that the graph of *h* is a vertical shrink of the graph of *f*.

The graphs of both functions are shown in Figure 3.58.

WARM UP

1. Find $f(2)$ for $f(x) = -x^3 + 5x$.

2. Find $f(6)$ for $f(x) = x^2 - 6x$.

3. Find $f(-x)$ for $f(x) = \dfrac{3}{x}$.

4. Find $f(-x)$ for $f(x) = x^2 + 3$.

Solve for x.

5. $x^3 - 16x = 0$

6. $2x^2 - 3x + 1 = 0$

Find the domain of the function.

7. $g(x) = \dfrac{4}{x - 4}$

8. $f(x) = \dfrac{2x}{x^2 - 9x + 20}$

9. $h(t) = \sqrt[4]{5 - 3t}$

10. $f(t) = t^3 + 3t - 5$

EXERCISES 3.5

In Exercises 1–6, determine the domain and range of the given function.

1. $f(x) = \sqrt{x - 1}$

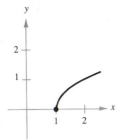

2. $f(x) = 4 - x^2$

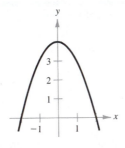

5. $f(x) = \sqrt{25 - x^2}$

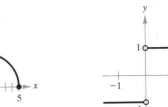

6. $f(x) = |x|/x$

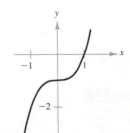

In Exercises 7–12, use the vertical line test to determine if y is a function of x.

3. $f(x) = \sqrt{x^2 - 4}$

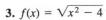

4. $f(x) = |x - 2|$

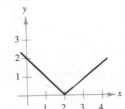

7. $y = x^2$

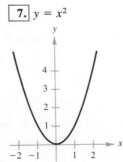

8. $y = x^3 - 1$

9. $x - y^2 = 0$

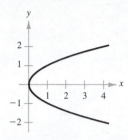

10. $x^2 + y^2 = 9$

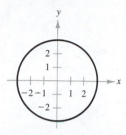

17. $f(x) = 3x^4 - 6x^2$

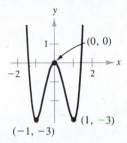

(0, 0)

$(-1, -3)$ $(1, -3)$

18. $f(x) = x^{2/3}$

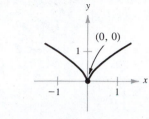

(0, 0)

11. $x^2 = xy - 1$

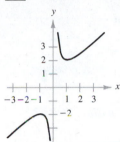

12. $x = |y|$

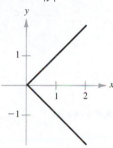

19. $f(x) = x\sqrt{x + 3}$

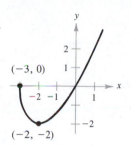

$(-3, 0)$

$(-2, -2)$

20. $f(x) = |x + 1| + |x - 1|$

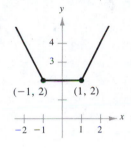

$(-1, 2)$ $(1, 2)$

In Exercises 13–20, (a) determine the intervals over which the function is increasing, decreasing, or constant, and (b) determine if the function is even, odd, or neither.

13. $f(x) = 2x$

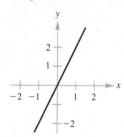

14. $f(x) = x^2 - 2x$

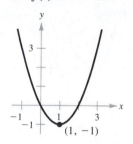

$(1, -1)$

In Exercises 21–26, determine whether the given function is even, odd, or neither.

21. $f(x) = x^6 - 2x^2 + 3$

22. $h(x) = x^3 - 5$

23. $g(x) = x^3 - 5x$

24. $f(x) = x\sqrt{1 - x^2}$

25. $f(t) = t^2 + 2t - 3$

26. $g(s) = 4s^{2/3}$

15. $f(x) = x^3 - 3x^2$

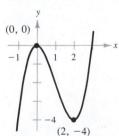

(0, 0)

-4

$(2, -4)$

16. $f(x) = \sqrt{x^2 - 4}$

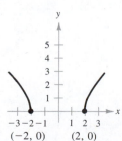

$(-2, 0)$ $(2, 0)$

In Exercises 27–38, sketch the graph of the given function and determine whether the function is even, odd, or neither.

27. $f(x) = 3$

28. $g(x) = x$

29. $f(x) = 5 - 3x$

30. $h(x) = x^2 - 4$

31. $g(s) = \dfrac{s^3}{4}$

32. $f(t) = -t^4$

33. $f(x) = \sqrt{1 - x}$

34. $f(x) = x^{3/2}$

35. $g(t) = \sqrt[3]{t} - 1$

36. $f(x) = |x + 2|$

37. $f(x) = \begin{cases} x + 3, & x \le 0 \\ 3, & 0 < x \le 2 \\ 2x - 1, & x > 2 \end{cases}$

38. $f(x) = \begin{cases} 2x + 1, & x \le -1 \\ x^2 - 2, & x > -1 \end{cases}$

▸In Exercises 39–48, sketch the graph of the function and determine the interval(s) (if any) on the real axis for which $f(x) \ge 0$.

39. $f(x) = 4 - x$

40. $f(x) = 4x + 2$

41. $f(x) = x^2 - 9$

42. $f(x) = x^2 - 4x$

43. $f(x) = 1 - x^4$

44. $f(x) = \sqrt{x} + 2$

45. $f(x) = x^2 + 1$

46. $f(x) = -(1 + |x|)$

47. $f(x) = -5$

48. $f(x) = \frac{1}{2}(2 + |x|)$

49. Sketch (on the same set of coordinate axes) a graph of f for $c = -2, 0,$ and 2.
(a) $f(x) = \frac{1}{2}x + c$
(b) $f(x) = \frac{1}{2}(x - c)$
(c) $f(x) = \frac{1}{2}(cx)$

50. Sketch (on the same set of coordinate axes) a graph of f for $c = -2, 0,$ and 2.
(a) $f(x) = x^3 + c$
(b) $f(x) = (x - c)^3$
(c) $f(x) = (x - 2)^3 + c$

51. Use the graph of $f(x) = \sqrt{x}$ (see figure) to sketch the graph of each of the following.
(a) $y = \sqrt{x} + 2$
(b) $y = -\sqrt{x}$
(c) $y = \sqrt{x - 2}$
(d) $y = \sqrt{x + 3}$
(e) $y = 2 - \sqrt{x - 4}$
(f) $y = \sqrt{2x}$

52. Use the graph of $f(x) = \sqrt[3]{x}$ (see figure) to sketch the graph of each of the following.
(a) $y = \sqrt[3]{x} - 1$ (b) $y = \sqrt[3]{x + 1}$
(c) $y = \sqrt[3]{x - 1}$ (d) $y = -\sqrt[3]{x - 2}$
(e) $y = \sqrt[3]{x + 1} - 1$ (f) $y = \frac{1}{2}\sqrt[3]{x}$

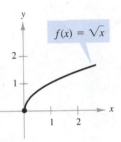

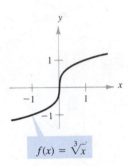

FIGURE FOR 51   **FIGURE FOR 52**

53. Use the graph of $f(x) = x^2$ [see Figure 3.53(e)] to write formulas for the functions whose graphs are shown in parts (a) and (b).

(a)
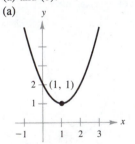

(b)
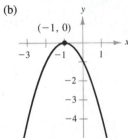

54. Use the graph of $f(x) = x^3$ [see Figure 3.53(f)] to write formulas for the functions whose graphs are shown in parts (a) and (b).

(a)
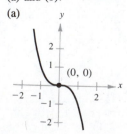

(b)
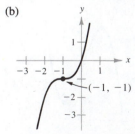

In Exercises 55–58, write the height h of the given rectangle as a function of x.

55.

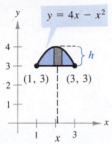

$y = 4x - x^2$

$(1, 3)$ $(3, 3)$

56.

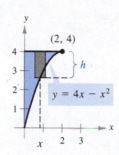

$(2, 4)$

$y = 4x - x^2$

57.

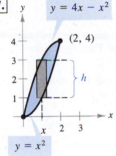

$y = 4x - x^2$

$(2, 4)$

$y = x^2$

58.

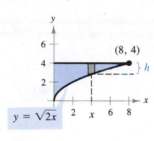

$(8, 4)$

$y = \sqrt{2x}$

In Exercises 59 and 60, write the length L of the given rectangle as a function of y.

59.

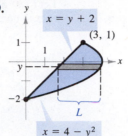

$x = y + 2$

$(3, 1)$

L

$x = 4 - y^2$

60.

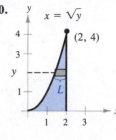

$x = \sqrt{y}$

$(2, 4)$

L

61. Prove that a function of the form

$$f(x) = a_{2n+1}x^{2n+1} + a_{2n-1}x^{2n-1} + \cdots + a_3x^3 + a_1x$$

is odd.

62. Prove that a function of the form

$$f(x) = a_{2n}x^{2n} + a_{2n-2}x^{2n-2} + \cdots + a_2x^2 + a_0$$

is even.

3.6 Combinations of Functions

Just as two real numbers can be combined by the operations of addition, subtraction, multiplication, and division to form other real numbers, two functions can be combined to create new functions. For example, if

$$f(x) = 2x - 3 \quad \text{and} \quad g(x) = x^2 - 1,$$

we can form the sum, difference, product, and quotient of f and g as follows.

$$f(x) + g(x) = (2x - 3) + (x^2 - 1) = x^2 + 2x - 4 \qquad \textit{Sum}$$

$$f(x) - g(x) = (2x - 3) - (x^2 - 1) = -x^2 + 2x - 2 \qquad \textit{Difference}$$

$$f(x)g(x) = (2x - 3)(x^2 - 1) = 2x^3 - 3x^2 - 2x + 3 \qquad \textit{Product}$$

$$\frac{f(x)}{g(x)} = \frac{2x - 3}{x^2 - 1}, \qquad x \neq \pm 1 \qquad \textit{Quotient}$$

The domain of an arithmetic combination of functions f and g consists of all real numbers *common* to the domains of f and g. In the case of $f(x)/g(x)$, there is the further restriction that $g(x) \neq 0$.

Functions and Graphs

Definition of Sum, Difference, Product, and Quotient of Functions

Let f and g be two functions with overlapping domains. Then for all x common to both domains the sum, difference, product, and quotient of f and g are defined as follows.

1. *Sum:* $(f + g)(x) = f(x) + g(x)$
2. *Difference:* $(f - g)(x) = f(x) - g(x)$
3. *Product:* $(fg)(x) = f(x) \cdot g(x)$
4. *Quotient:* $\left(\dfrac{f}{g}\right)(x) = \dfrac{f(x)}{g(x)}, \qquad g(x) \neq 0$

EXAMPLE 1   *Evaluating the Sum and Difference of Two Functions*

Given $f(x) = 2x + 1$ and $g(x) = x^2 + 2x - 1$, find the following.

(a) $(f + g)(2)$ (b) $(f - g)(2)$

SOLUTION

(a) Since

$$(f + g)(x) = f(x) + g(x)$$
$$= (2x + 1) + (x^2 + 2x - 1)$$
$$= x^2 + 4x$$

it follows that

$$(f + g)(2) = 2^2 + 4(2) = 12.$$

(b) Since

$$(f - g)(x) = f(x) - g(x)$$
$$= (2x + 1) - (x^2 + 2x - 1)$$
$$= -x^2 + 2$$

it follows that

$$(f - g)(2) = -(2)^2 + 2 = -2.$$

In Example 1, both f and g have domains that consist of all real numbers. Thus, the domain of their sum and difference is also the set of all real numbers. Remember that any restrictions on the domains of f or g must be taken into account when forming the sum, difference, product, or quotient of f and g. For instance, the domain of $f(x) = 1/x$ is all $x \neq 0$, and the domain of $g(x) = \sqrt{x}$ is $[0, \infty)$. This implies that the domain of $f + g$ is $(0, \infty)$.

EXAMPLE 2 *The Quotient of Two Functions*

Find $(f/g)(x)$ and $(g/f)(x)$ for the functions

$$f(x) = \sqrt{x} \qquad \text{and} \qquad g(x) = \sqrt{4 - x^2}.$$

Then find the domains of f/g and g/f.

SOLUTION

The quotient of f and g is given by

$$\left(\frac{f}{g}\right)(x) = \frac{f(x)}{g(x)} = \frac{\sqrt{x}}{\sqrt{4 - x^2}}$$

and the quotient of g and f is given by

$$\left(\frac{g}{f}\right)(x) = \frac{g(x)}{f(x)} = \frac{\sqrt{4 - x^2}}{\sqrt{x}}.$$

The domain of f is $[0, \infty)$ and the domain of g is $[-2, 2]$. The intersection of these two domains is $[0, 2]$. Thus, we have the following domains for f/g and g/f.

Domain of $\dfrac{f}{g}$: $[0, 2)$

Domain of $\dfrac{g}{f}$: $(0, 2]$

Can you see why these two domains differ slightly?

Composition of Functions

Another way of combining two functions is to form the **composition** of one with the other. For instance, if $f(x) = x^2$ and $g(x) = x + 1$, then the composition of f with g is given by

$$f(g(x)) = f(x + 1) = (x + 1)^2.$$

We denote this **composite function** as $f \circ g$.

Definition of Composition of Two Functions

The **composition of the functions** f and g is given by

$(f \circ g)(x) = (f(g(x)).$

The domain of $f \circ g$ is the set of all x in the domain of g such that $g(x)$ is in the domain of f. (See Figure 3.59.)

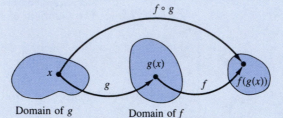

Domain of g Domain of f **FIGURE 3.59**

EXAMPLE 3 *Forming the Composition of f with g*

Find $(f \circ g)(x)$ for

$f(x) = \sqrt{x}, \qquad x \geq 0$

and

$g(x) = x - 1, \qquad x \geq 1.$

If possible, find $(f \circ g)(2)$ and $(f \circ g)(0)$.

SOLUTION

To find $(f \circ g)(x)$, we write

$$(f \circ g)(x) = f(g(x)) \qquad \text{\textit{Definition of } } f \circ g$$
$$= f(x - 1) \qquad \text{\textit{Definition of } } g(x)$$
$$= \sqrt{x - 1}. \qquad \text{\textit{Definition of } } f(x)$$

The domain of $f \circ g$ is $[1, \infty)$. Thus,

$$(f \circ g)(2) = \sqrt{2 - 1} = 1$$

is defined, but $(f \circ g)(0)$ is not defined because 0 is not in the domain of $f \circ g$.

The composition of f with g is generally *not* the same as the composition of g with f. This is illustrated in Example 4.

EXAMPLE 4 Composition of Functions

Given $f(x) = x + 2$ and $g(x) = 4 - x^2$, find the following.

(a) $(f \circ g)(x)$ (b) $(g \circ f)(x)$

SOLUTION

(a) The composition of f with g is given by

$$
\begin{aligned}
(f \circ g)(x) &= f(g(x)) & & \text{\textit{Definition of } } f \circ g \\
&= f(4 - x^2) & & \text{\textit{Definition of } } g(x) \\
&= (4 - x^2) + 2 & & \text{\textit{Definition of } } f(x) \\
&= -x^2 + 6. & & \text{\textit{Simplify}}
\end{aligned}
$$

(b) The composition of g with f is given by

$$
\begin{aligned}
(g \circ f)(x) &= g(f(x)) & & \text{\textit{Definition of } } g \circ f \\
&= g(x + 2) & & \text{\textit{Definition of } } f(x) \\
&= 4 - (x + 2)^2 & & \text{\textit{Definition of } } g(x) \\
&= 4 - (x^2 + 4x + 4) & & \text{\textit{Expand}} \\
&= -x^2 - 4x. & & \text{\textit{Simplify}}
\end{aligned}
$$

Note that $(f \circ g)(x) \neq (g \circ f)(x)$.

EXAMPLE 5 A Case in Which $f \circ g = g \circ f$

Given $f(x) = 2x + 3$ and $g(x) = \frac{1}{2}(x - 3)$, find the following.

(a) $(f \circ g)(x)$ (b) $(g \circ f)(x)$

SOLUTION

(a) $(f \circ g)(x) = f(g(x)) = f\left(\frac{1}{2}(x - 3)\right)$

$$= 2\left[\frac{1}{2}(x - 3)\right] + 3 = x - 3 + 3 = x$$

(b) $(g \circ f)(x) = g(f(x)) = g(2x + 3)$

$$= \frac{1}{2}[(2x + 3) - 3] = \frac{1}{2}(2x) = x$$

Remark: In Example 5, note that the two composite functions $f \circ g$ and $g \circ f$ happen to be equal, and both represent the identity function. That is, $(f \circ g)(x) = (g \circ f)(x) = x$. We will return to this special case in the next section.

In Examples 4 and 5 we formed the composite of two given functions. In calculus, it is also important to be able to identify two functions that *make up a given* composite function. For instance, the function h given by $h(x) = (3x - 5)^3$ is the composite of f with g, where $f(x) = x^3$ and $g(x) = 3x - 5$. That is,

$$h(x) = (3x - 5)^3 = [g(x)]^3 = f(g(x)).$$

Basically, to "decompose" a composite function we look for an "inner" and an "outer" function. In the function h above, $g(x) = 3x - 5$ is the inner function and $f(x) = x^3$ is the outer function.

EXAMPLE 6 *Identifying a Composite Function*

Express the function

$$h(x) = \frac{1}{(x - 2)^2}$$

as a composition of two functions f and g.

SOLUTION

One way to write h as a composite of two functions is to take the inner function to be

$$g(x) = x - 2$$

and the outer function to be

$$f(x) = \frac{1}{x^2} = x^{-2}.$$

Then we can write

$$h(x) = \frac{1}{(x - 2)^2} = (x - 2)^{-2} = f(x - 2) = f(g(x)).$$

EXAMPLE 7 *An Application of a Composite Function*

The number of bacteria in a refrigerated food is given by

$$N(T) = 20T^2 - 80T + 500, \qquad 2 \le T \le 14$$

where T is the temperature of the food. When the food is removed from refrigeration, the temperature is given by

$$T(t) = 4t + 2, \qquad 0 \le t \le 3$$

where t is the time in hours. Find the following.

Combinations of Functions

(a) The composite function $N(T(t))$.

(b) The number of bacteria in the food when $t = 2$.

(c) The time when the bacteria count reaches 2000.

SOLUTION

(a)
$$
\begin{aligned}
N(T(t)) &= 20(4t + 2)^2 - 80(4t + 2) + 500 \\
&= 20(16t^2 + 16t + 4) - 320t - 160 + 500 \\
&= 320t^2 + 320t + 80 - 320t - 160 + 500 \\
&= 320t^2 + 420
\end{aligned}
$$

(b) When $t = 2$,

$$
N = 320(2)^2 + 420 = 1280 + 420 = 1700.
$$

(c) The bacteria count reaches $N = 2000$ when

$$
320t^2 + 420 = 2000
$$
$$
320t^2 = 1580
$$
$$
t^2 = \frac{1580}{320} = \frac{79}{16}
$$
$$
t = \frac{\sqrt{79}}{4} \approx 2.222 \text{ hours.}
$$

WARM UP

Perform the indicated operations and simplify the results.

1. $\dfrac{1}{x} + \dfrac{1}{1 - x}$

2. $\dfrac{2}{x + 3} - \dfrac{2}{x - 3}$

3. $\dfrac{3}{x - 2} - \dfrac{2}{x(x - 2)}$

4. $\dfrac{x}{x - 5} + \dfrac{1}{3}$

5. $(x - 1)\left(\dfrac{1}{\sqrt{x^2 - 1}}\right)$

6. $\left(\dfrac{x}{x^2 - 4}\right)\left(\dfrac{x^2 - x - 2}{x^2}\right)$

7. $(x^2 - 4) \div \left(\dfrac{x + 2}{5}\right)$

8. $\left(\dfrac{x}{x^2 + 3x - 10}\right) \div \left(\dfrac{x^2 + 3x}{x^2 + 6x + 5}\right)$

9. $\dfrac{(1/x) + 5}{3 - (1/x)}$

10. $\dfrac{(x/4) - (4/x)}{x - 4}$

EXERCISES 3.6

In Exercises 1–8, find (a) $(f + g)(x)$, (b) $(f - g)(x)$, (c) $(fg)(x)$, and (d) $(f/g)(x)$. What is the domain of f/g?

1. $f(x) = x + 1,$ $g(x) = x - 1$

2. $f(x) = 2x - 5,$ $g(x) = 1 - x$

3. $f(x) = x^2,$ $g(x) = 1 - x$

4. $f(x) = 2x - 5,$ $g(x) = 5$

5. $f(x) = x^2 + 5,$ $g(x) = \sqrt{1 - x}$

6. $f(x) = \sqrt{x^2 - 4},$ $g(x) = \dfrac{x^2}{x^2 + 1}$

7. $f(x) = \dfrac{1}{x},$ $g(x) = \dfrac{1}{x^2}$

8. $f(x) = \dfrac{x}{x + 1},$ $g(x) = x^3$

In Exercises 9–20, evaluate the indicated function for $f(x) = x^2 + 1$ and $g(x) = x - 4$.

9. $(f + g)(3)$ **10.** $(f - g)(-2)$

11. $(f - g)(2t)$ **12.** $(f + g)(t - 1)$

13. $(fg)(4)$ **14.** $(fg)(-6)$

15. $\left(\dfrac{f}{g}\right)(5)$ **16.** $\left(\dfrac{f}{g}\right)(0)$

17. $(f - g)(0)$ **18.** $(f + g)(1)$

19. $\left(\dfrac{f}{g}\right)(-1) - g(3)$ **20.** $(2f)(5)$

In Exercises 21–24, find (a) $f \circ g$, (b) $g \circ f$, and (c) $f \circ f$.

21. $f(x) = x^2,$ $g(x) = x - 1$

22. $f(x) = 3x,$ $g(x) = 2x + 1$

23. $f(x) = 3x + 5,$ $g(x) = 5 - x$

24. $f(x) = x^3,$ $g(x) = \dfrac{1}{x}$

In Exercises 25–32, find (a) $f \circ g$ and (b) $g \circ f$.

25. $f(x) = \sqrt{x + 4},$ $g(x) = x^2$

26. $f(x) = \sqrt[3]{x - 1},$ $g(x) = x^3 + 1$

27. $f(x) = \frac{1}{3}x - 3,$ $g(x) = 3x + 1$

28. $f(x) = x^4,$ $g(x) = x^4$

29. $f(x) = \sqrt{x},$ $g(x) = \sqrt{x}$

30. $f(x) = 2x - 3,$ $g(x) = 2x - 3$

31. $f(x) = |x|,$ $g(x) = x + 6$

32. $f(x) = x^{2/3},$ $g(x) = x^6$

In Exercises 33–36, use the graphs of f and g (see figure) to evaluate the indicated functions.

33. (a) $(f + g)(3)$ (b) $\left(\dfrac{f}{g}\right)(2)$

34. (a) $(f - g)(1)$ (b) $(fg)(4)$

35. (a) $(f \circ g)(2)$ (b) $(g \circ f)(2)$

36. (a) $(f \circ g)(1)$ (b) $(g \circ f)(3)$

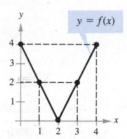 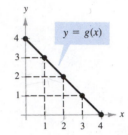

FIGURE FOR 33–36

In Exercises 37–44, find two functions f and g such that $(f \circ g)(x) = h(x)$. (There are many correct answers to these exercises.)

37. $h(x) = (2x + 1)^2$ **38.** $h(x) = (1 - x)^3$

39. $h(x) = \sqrt[3]{x^2 - 4}$ **40.** $h(x) = \sqrt{9 - x}$

41. $h(x) = \dfrac{1}{x + 2}$ **42.** $h(x) = \dfrac{4}{(5x + 2)^2}$

43. $h(x) = (x + 4)^2 + 2(x + 4)$

44. $h(x) = (x + 3)^{3/2}$

In Exercises 45–48, determine the domain of (a) f, (b) g, and (c) $f \circ g$.

45. $f(x) = \sqrt{x},$ $g(x) = x^2 + 1$

46. $f(x) = \dfrac{1}{x},$ $g(x) = x + 3$

47. $f(x) = \dfrac{3}{x^2 - 1},$ $g(x) = x + 1$

48. $f(x) = 2x + 3,$ $g(x) = \dfrac{x}{2}$

49. A pebble is dropped into a calm pond, causing ripples in the form of concentric circles. The radius (in feet) of the outer ripple is given by $r(t) = 0.6t$, where t is time in seconds after the pebble strikes the water. The area of the circle is given by the function $A(r) = \pi r^2$. Find and interpret $(A \circ r)(t)$.

50. The weekly cost of producing x units in a manufacturing process is given by the function $C(x) = 60x + 750$. The number of units produced in t hours is given by $x(t) = 50t$. Find and interpret $(C \circ x)(t)$.

51. An air traffic controller spots two planes at the same altitude flying toward each other (see figure). One plane is 150 miles from point P and is moving at 450 miles per hour. The other plane is 200 miles from point P and is moving at 450 miles per hour. Write the distance s between the planes as a function of time t.

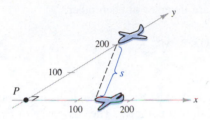

FIGURE FOR 51

52. Prove that the product of two even functions is an even function.

53. Prove that the product of two odd functions is an even function.

54. Prove that the product of an odd function and an even function is odd.

55. Given a function f, prove that $g(x)$ is even and $h(x)$ is odd where

$$g(x) = \tfrac{1}{2}[f(x) + f(-x)]$$

and

$$h(x) = \tfrac{1}{2}[f(x) - f(-x)].$$

56. Use the result of Exercise 55 to prove that any function can be written as a sum of even and odd functions. [*Hint:* Add the two equations in Exercise 55.]

57. Use the result of Exercise 55 to write each of the following functions as a sum of even and odd functions.

(a) $f(x) = x^2 - 2x + 1$ (b) $f(x) = \dfrac{1}{x + 1}$

3.7 Inverse Functions

In Example 5 of the previous section, we noted that for

$$f(x) = 2x + 3 \quad \text{and} \quad g(x) = \frac{1}{2}(x - 3)$$

both composite functions $f \circ g$ and $g \circ f$ equal the identity function. That is,

$$f(g(x)) = x = g(f(x)).$$

Note that the functions f and g have the effect of "undoing" each other, as shown in Figure 3.60. We call such functions **inverses** of each other.

FIGURE 3.60

Definition of Inverse Functions

Two functions f and g are **inverses** of each other if

$$f(g(x)) = x \quad \text{for every } x \text{ in the domain of } g$$

and

$$g(f(x)) = x \quad \text{for every } x \text{ in the domain of } f.$$

We denote g by f^{-1} (read "f inverse"). Thus, $f(f^{-1}(x)) = x$ and $f^{-1}(f(x)) = x$.

Functions and Graphs

Remark: For inverse functions f and g, the range of g must be *equal to* the domain of f, and vice versa.

Don't be confused by the use of -1 to denote the inverse function f^{-1}. Whenever we write $f^{-1}(x)$, we will *always* be referring to the inverse of the function f and *not* to the reciprocal of f.

EXAMPLE 1 *Verifying Inverse Functions*

Show that the following functions are inverses of each other.

$$f(x) = 2x^3 - 1 \qquad \text{and} \qquad g(x) = \sqrt[3]{\frac{x+1}{2}}$$

SOLUTION

First note that both composite functions exist because the domain and range of both f and g consist of the set of all real numbers. The composite of f with g is given by

$$f(g(x)) = f\left(\sqrt[3]{\frac{x+1}{2}}\right) = 2\left(\sqrt[3]{\frac{x+1}{2}}\right)^3 - 1$$
$$= 2\left(\frac{x+1}{2}\right) - 1$$
$$= x + 1 - 1 = x.$$

The composite of g with f is given by

$$g(f(x)) = g(2x^3 - 1) = \sqrt[3]{\frac{(2x^3 - 1) + 1}{2}}$$
$$= \sqrt[3]{\frac{2x^3}{2}}$$
$$= \sqrt[3]{x^3} = x.$$

Since $f(g(x)) = x = g(f(x))$, f and g are inverses of each other.

EXAMPLE 2 *Testing for Inverse Functions*

Which of the functions

$$g(x) = \frac{x-2}{5} \qquad \text{and} \qquad h(x) = \frac{5}{x} + 2$$

is the inverse of the function

$$f(x) = \frac{5}{x-2}?$$

SOLUTION

The composition of f with g is

$$f(g(x)) = f\left(\frac{x-2}{5}\right) = \frac{5}{[(x-2)/5] - 2}$$

$$= \frac{25}{(x-2) - 10}$$

$$= \frac{25}{x - 12}$$

$$\neq x.$$

The composition of f with h is

$$f(h(x)) = f\left(\frac{5}{x} + 2\right) = \frac{5}{(5/x) + 2 - 2}$$

$$= \frac{5}{5/x}$$

$$= x.$$

Thus, it appears that f and h are inverses of each other, which is confirmed by the fact that

$$h(f(x)) = h\left(\frac{5}{x-2}\right) = \frac{5}{5/(x-2)} + 2$$

$$= x - 2 + 2$$

$$= x.$$

Check to see that the domain of f is the same as the range of h and vice versa.

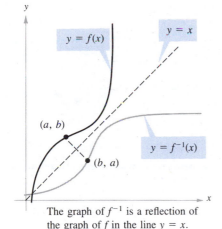

The graph of f^{-1} is a reflection of the graph of f in the line $y = x$.

FIGURE 3.61

The graphs of f and f^{-1} are related to each other in the following way. If the point (a, b) lies on the graph of f, then the point (b, a) lies on the graph of f^{-1} and vice versa. This means that the graph of f^{-1} is a reflection of the graph of f in the line $y = x$, as shown in Figure 3.61.

The Existence of an Inverse Function

A function need not have an inverse. For instance, the function $f(x) = x^2$ has no inverse [assuming a domain of $(-\infty, \infty)$]. In order for a function to have an inverse, it is necessary that the function be **one-to-one,** which means that no two elements in the domain of f correspond to the same element in the range of f.

Definition of One-to-One Function

A function f is **one-to-one** if, for a and b in its domain,

$$f(a) = f(b) \quad \text{implies that} \quad a = b.$$

The function $f(x) = x + 1$ *is* one-to-one because

$$a + 1 = b + 1$$

implies that a and b must be equal. However, the function $f(x) = x^2$ is *not* one-to-one because

$$a^2 = b^2$$

does not imply that $a = b$. For instance, $(-1)^2 = 1^2$ and yet $-1 \neq 1$.

The following theorem tells us that a function has an inverse if and only if the function is one-to-one.

Existence of an Inverse Function

A function f has an inverse function f^{-1} if and only if f is one-to-one.

EXAMPLE 3 Testing for One-to-One Functions

Which of the following functions are one-to-one and therefore have an inverse?

(a) $f(x) = x^3 + 1$ (b) $g(x) = x^2 - x$

(c) $h(x) = \sqrt{x}$

SOLUTION

(a) Let a and b be real numbers with $f(a) = f(b)$. Then we have

$$
\begin{aligned}
a^3 + 1 &= b^3 + 1 & &\text{\textit{Set }} f(a) = f(b) \\
a^3 &= b^3 \\
a &= b.
\end{aligned}
$$

Therefore, $f(a) = f(b)$ implies that $a = b$, and we conclude that $f(x) = x^3 + 1$ *is* one-to-one, and hence has an inverse.

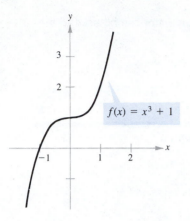

(a) *f* is one-to-one.

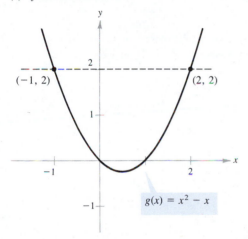

(b) *g* is not one-to-one.

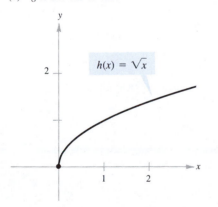

(c) *h* is one-to-one.

FIGURE 3.62

(b) Since

$$g(-1) = (-1)^2 - (-1) = 2$$

and

$$g(2) = 2^2 - 2 = 2$$

we have two distinct inputs matched with the same output. Therefore, *g* is *not* a one-to-one function, and hence has no inverse.

(c) Let *a* and *b* be nonnegative real numbers with $h(a) = h(b)$. Then we have

$$\sqrt{a} = \sqrt{b} \qquad \text{\textit{Set} } h(a) = h(b)$$
$$a = b.$$

Therefore, $h(a) = h(b)$ implies that $a = b$, and we conclude that $h(x) = \sqrt{x}$ *is* one-to-one, and hence has an inverse.

From its graph, it is easy to tell whether a function of *x* is one-to-one. We simply check to see that every *horizontal* line intersects the graph of the function at most once. For instance, Figure 3.62 shows the graphs of the three functions given in Example 3. Note that for the graph of $g(x) = x^2 - x$ it is possible to find a horizontal line that intersects the graph twice.

Two special types of functions that do pass the **horizontal line test** are those that are increasing or decreasing on their entire domains.

1. If *f* is *increasing* on its entire domain, then *f* is one-to-one.
2. If *f* is *decreasing* on its entire domain, then *f* is one-to-one.

Finding the Inverse of a Function

Now that we have a test to determine whether a function has an inverse, we need a procedure for *finding* the inverse.

Finding the Inverse of a Function

To find the inverse of *f*, use the following steps.

1. Test to see that *f* is one-to-one.
2. Write the function in the form $y = f(x)$ and solve for *x* in terms of *y* to obtain $x = f^{-1}(y)$.
3. Check to see that the domain of *f* is the range of f^{-1}, and that the domain of f^{-1} is the range of *f*.

In the following example, remember that any letter can be used to represent the independent variable. For instance,

$$f^{-1}(x) = x^2 + 4 \qquad \text{and} \qquad f^{-1}(y) = y^2 + 4$$

represent the same function.

EXAMPLE 4 Finding the Inverse of a Function

Find the inverse (if it exists) of

$$f(x) = \frac{5 - 3x}{2}.$$

SOLUTION

From Figure 3.63 we see that f is one-to-one, and therefore has an inverse. To begin, we write the function in the form $y = f(x)$. Then we solve for x in terms of y as follows.

$$y = \frac{5 - 3x}{2} \qquad\qquad \textit{Write in form } y = f(x)$$

$$2y = 5 - 3x$$

$$3x = 5 - 2y$$

$$x = \frac{5 - 2y}{3} \qquad\qquad \textit{Solve for } x$$

Therefore, the inverse of f is

$$f^{-1}(y) = \frac{5 - 2y}{3}.$$

Or, using x as the independent variable, f^{-1} can be written as

$$f^{-1}(x) = \frac{5 - 2x}{3}.$$

The domain and range of both f and f^{-1} consist of all real numbers.

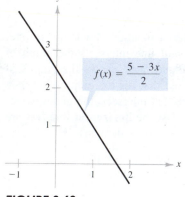

$$f(x) = \frac{5 - 3x}{2}$$

FIGURE 3.63

EXAMPLE 5 Finding the Inverse of a Function

Find the inverse of the function

$$f(x) = \sqrt{2x - 3}$$

and sketch the graphs of f and f^{-1}.

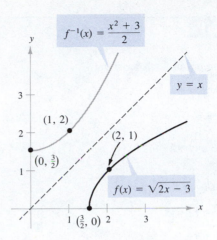

FIGURE 3.64

SOLUTION

The graph of f is shown in Figure 3.64. To find the inverse of f, we proceed as follows.

$$y = \sqrt{2x - 3}$$
$$y^2 = 2x - 3$$
$$y^2 + 3 = 2x$$
$$x = \frac{y^2 + 3}{2} \qquad \text{Solve for } x$$
$$f^{-1}(y) = \frac{y^2 + 3}{2} \qquad y \geq 0$$

Or, using x as the independent variable, f^{-1} can be written as

$$f^{-1}(x) = \frac{x^2 + 3}{2}. \qquad x \geq 0$$

The graph of f^{-1} is the reflection of the graph of f in the line $y = x$, as shown in Figure 3.64. Note that the domain of f is the interval $[3/2, \infty)$ and the range of f is the interval $[0, \infty)$. Moreover, the domain of f^{-1} is the interval $[0, \infty)$ and the range of f^{-1} is the interval $[3/2, \infty)$.

The problem of finding the inverse of a function can be difficult (or even impossible) for two reasons. First, given $y = f(x)$, it may be algebraically difficult to solve for x in terms of y. Second, if f is not one-to-one, then f^{-1} does not exist.

In later chapters we will study two important classes of inverse functions: logarithmic functions and inverse trigonometric functions.

WARM UP

Find the domain of each function.

1. $f(x) = \sqrt[3]{x + 1}$

2. $f(x) = \sqrt{x + 1}$

3. $g(x) = \dfrac{2}{x^2 - 2x}$

4. $h(x) = \dfrac{x}{3x + 5}$

Simplify the expressions.

5. $2\left(\dfrac{x + 5}{2}\right) - 5$

6. $7 - 10\left(\dfrac{7 - x}{10}\right)$

7. $\sqrt[3]{2\left(\dfrac{x^3}{2} - 2\right) + 4}$

8. $(\sqrt[5]{x + 2})^5 - 2$

Solve for x in terms of y.

9. $y = \dfrac{2x - 6}{3}$

10. $y = \sqrt[3]{2x - 4}$

EXERCISES 3.7

In Exercises 1–10, (a) show that f and g are inverse functions by showing that $f(g(x)) = x$ and $g(f(x)) = x$, and (b) graph f and g on the same set of coordinate axes.

1. $f(x) = 2x,$ $g(x) = \dfrac{x}{2}$

2. $f(x) = x - 5,$ $g(x) = x + 5$

3. $f(x) = 5x + 1,$ $g(x) = \dfrac{x - 1}{5}$

4. $f(x) = 3 - 4x,$ $g(x) = \dfrac{3 - x}{4}$

5. $f(x) = x^3,$ $g(x) = \sqrt[3]{x}$

6. $f(x) = \dfrac{1}{x},$ $g(x) = \dfrac{1}{x}$

7. $f(x) = \sqrt{x - 4},$ $g(x) = x^2 + 4, x \geq 0$

8. $f(x) = 9 - x^2, x \geq 0,$ $g(x) = \sqrt{9 - x}, x \leq 9$

9. $f(x) = 1 - x^3,$ $g(x) = \sqrt[3]{1 - x}$

10. $f(x) = \dfrac{1}{1 + x}, x \geq 0,$ $g(x) = \dfrac{1 - x}{x}, 0 < x \leq 1$

In Exercises 11–20, determine whether the function is one-to-one.

11.

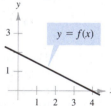

12.

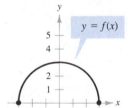

13.

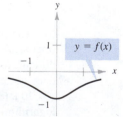

14.

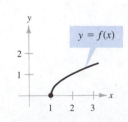

15. $g(x) = \dfrac{4 - x}{6}$ **16.** $f(x) = 10$

17. $h(x) = |x + 4|$ **18.** $g(x) = (x + 5)^3$

19. $f(x) = -\sqrt{16 - x^2}$ **20.** $f(x) = (x + 2)^2$

In Exercises 21–30, find the inverse of the one-to-one function f. Then graph both f and f^{-1} on the same coordinate plane.

21. $f(x) = 2x - 3$ **22.** $f(x) = 3x$

23. $f(x) = x^5$ **24.** $f(x) = x^3 + 1$

25. $f(x) = \sqrt{x}$

26. $f(x) = x^2, \quad x \geq 0$

27. $f(x) = \sqrt{4 - x^2}, \quad 0 \leq x \leq 2$

28. $f(x) = \dfrac{4}{x}$

29. $f(x) = \sqrt[3]{x - 1}$

30. $f(x) = x^{3/5}$

In Exercises 31–46, determine whether the given function is one-to-one. If it is, find its inverse.

31. $f(x) = x^4$ **32.** $f(x) = \dfrac{1}{x^2}$

33. $g(x) = \dfrac{x}{8}$ **34.** $f(x) = 3x + 5$

35. $p(x) = -4$ **36.** $f(x) = \dfrac{3x + 4}{5}$

37. $f(x) = (x + 3)^2, \quad x \geq -3$

38. $q(x) = (x - 5)^2$

39. $h(x) = \dfrac{1}{x}$

40. $f(x) = |x - 2|, \quad x \leq 2$

41. $f(x) = \sqrt{2x + 3}$

42. $f(x) = \sqrt{x - 2}$

43. $g(x) = x^2 - x^4$

44. $f(x) = \dfrac{x^2}{x^2 + 1}$

45. $f(x) = 25 - x^2, \; x \leq 0$

46. $f(x) = ax + b, \quad a \neq 0$

In Exercises 47 and 48, use the graph of the function f to complete the table and to sketch the graph of f^{-1}.

47.

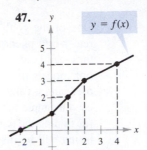

x	0	1	2	3	4
$f^{-1}(x)$					

48.

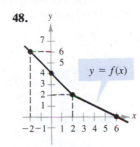

x	0	2	4	6
$f^{-1}(x)$				

In Exercises 49–52, use the functions

$$f(x) = \tfrac{1}{8}x - 3 \quad \text{and} \quad g(x) = x^3$$

to find the indicated value.

49. $(f^{-1} \circ g^{-1})(1)$

50. $(g^{-1} \circ f^{-1})(-3)$

51. $(f^{-1} \circ f^{-1})(6)$

52. $(g^{-1} \circ g^{-1})(-4)$

In Exercises 53–56, use the functions

$$f(x) = x + 4 \quad \text{and} \quad g(x) = 2x - 5$$

to find the indicated functions.

53. $g^{-1} \circ f^{-1}$

54. $f^{-1} \circ g^{-1}$

55. $(f \circ g)^{-1}$

56. $(g \circ f)^{-1}$

57. Prove that if f is a one-to-one odd function, then f^{-1} is an odd function.

58. Prove that if f and g are one-to-one functions, then $(f \circ g)^{-1}(x) = (g^{-1} \circ f^{-1})(x)$.

3.8 *Variation and Mathematical Models*

One of the goals of applied mathematics is to find equations that describe real-world phenomena. We call such equations **mathematical models,** and they can be developed in two basic ways: experimentally and theoretically. For models that are developed experimentally, we usually have a collection of *x*- and *y*-values and then try to *fit a curve* (or line) to the points (x, y). We used a simple form of curve fitting in Section 3.3, when we studied a method for finding the equation of a line through two points. Other examples of curve fitting will be discussed in Chapter 10.

In Sections 2.2 and 2.4 we studied ways to set up linear and quadratic models for a variety of applications. In this section we will study models related to the concept of **variation.**

Direct Variation

The following statements are equivalent.

1. *y* **varies directly** as *x*.
2. *y* is **directly proportional** to *x*.
3. $y = kx$ for some constant *k*.

We call *k* the **constant of variation** or the **constant of proportionality.**

In the mathematical model for direct variation, *y* is a *linear* function of *x*. That is,

$$y = kx.$$

To set up a mathematical model, we are usually given specific values of *x* and *y* which then enables us to find the value of the constant *k*. This procedure is demonstrated in Example 1.

EXAMPLE 1 Direct Variation

Hooke's Law for a spring states that the distance a spring is stretched (or compressed) varies directly as the force on the spring. A force of 20 pounds stretches the spring 4 inches.

(a) Write an equation relating the distance stretched to the force applied.
(b) How far will a force of 30 pounds stretch the spring?

SOLUTION

(a) We begin by letting

d = distance spring is stretched (in inches)
F = force (in pounds).

Since distance varies directly as force, we have the model

$d = kF$.

To find the value of the constant k, we use the fact that $d = 4$ when $F = 20$. Substituting these values into the given model produces

$$\begin{array}{cc} d & F \\ \downarrow & \downarrow \\ 4 = & k(20) \end{array}$$

which implies that $k = 4/20 = 1/5$. Thus, the equation relating distance and force is

$$d = \frac{1}{5}F.$$

(b) When $F = 30$, the distance is

$$d = \frac{1}{5}F = \frac{1}{5}(30) = 6 \text{ inches}.$$

See Figure 3.65.

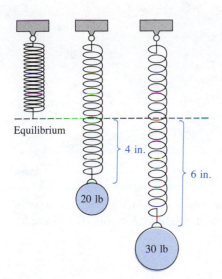

Equilibrium

4 in.

6 in.

20 lb

30 lb

FIGURE 3.65

A second type of direct variation relates one variable to a *power* of another variable.

Direct Variation as nth Power

The following statements are equivalent.

1. y **varies directly as the *n*th power** of x.
2. y is **directly proportional to the *n*th power** of x.
3. $y = kx^n$ for some constant k.

EXAMPLE 2 *Direct Variation as a Power*

The distance a ball rolls down an inclined plane is directly proportional to the square of the time it rolls. During the first second the ball rolls 8 feet.

(a) Write an equation relating the distance traveled to the time.
(b) How far will the ball roll during the first 3 seconds?

Functions and Graphs

SOLUTION

(a) Letting d be the distance (in feet) the ball rolls and t be the time (in seconds), we obtain the model

$$d = kt^2.$$

Now, since $d = 8$ when $t = 1$, we can see that $k = 8$. Thus, the equation relating distance to time is

$$d = 8t^2.$$

(b) When $t = 3$, the distance traveled is

$$d = 8(3^2) = 8(9) = 72 \text{ feet.}$$

See Figure 3.66.

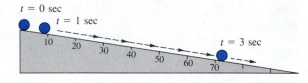

FIGURE 3.66

In Examples 1 and 2 the direct variations were such that an *increase* in one variable corresponded to an *increase* in the other variable. For example, in the model

$$d = \frac{1}{5}F, \qquad F > 0$$

an increase in F results in an increase in d. You should not, however, assume that this always occurs with direct variation. For example, in the model

$$y = -3x$$

an increase in x results in a *decrease* in y, and yet we say that y varies directly as x.

A third type of variation is called **inverse variation.**

Inverse Variation

The following statements are equivalent.

1. y **varies inversely** as x.   2. y is **inversely proportional** to x.

3. $y = \dfrac{k}{x}$ for some constant k.

Remark: If x and y are related by an equation of the form

$$y = \frac{k}{x^n}$$

then we say that y varies inversely as the nth power of x (or y is inversely proportional to the nth power of x).

EXAMPLE 3 *Inverse Variation*

A gas law states that the volume of an enclosed gas varies directly as the temperature *and* inversely as the pressure. The pressure of a gas is 0.75 kilograms per square centimeter when the temperature is 294° K and the volume is 8000 cubic centimeters.

(a) Write an equation relating the pressure, temperature, and volume of this gas.
(b) Find the pressure when the temperature is 300° K and the volume is 7000 cubic centimeters.

SOLUTION

(a) Let

$$V = \text{volume (in cubic centimeters)}$$
$$P = \text{pressure (in kilograms per square centimeter)}$$
$$T = \text{temperature (in degrees Kelvin)}.$$

Since V varies directly as T *and* inversely as P, we have the model

$$V = \frac{kT}{P}.$$

Note that the same constant of proportionality can be used for the direct variation of T and the inverse variation of P. Now, since $P = 0.75$ when $T = 294$ and $V = 8000$, we have

$$8000 = \frac{k(294)}{0.75}$$

$$\frac{8000(0.75)}{294} = k$$

$$k = \frac{6000}{294} = \frac{1000}{49}.$$

Thus, the equation relating pressure, temperature, and volume is

$$V = \frac{1000}{49}\left(\frac{T}{P}\right).$$

(b) When $T = 300$ and $V = 7000$, the pressure is

$$P = \frac{1000}{49}\left(\frac{300}{7000}\right) = \frac{300}{343} \approx 0.87 \text{ kilogram per square centimeter.}$$

Remark: In Example 3, note that when a direct and inverse variation occur in the same statement, we couple them with the word "and."

To describe two different *direct* variations in the same statement, we use the word **jointly.**

Joint Variation

The following statements are equivalent.

1. z **varies jointly** as x and y.
2. z is **jointly proportional** to x and y.
3. $z = kxy$ for some constant k.

Remark: If x, y, and z are related by an equation of the form

$$z = kx^n y^m$$

then we say z varies jointly as the nth power of x and the mth power of y.

EXAMPLE 4 *Joint Variation*

The *simple* interest for a certain savings account is jointly proportional to the time and the principal. After one quarter (three months) the interest on a principal of $5000 is $106.25.

(a) Write an equation relating the interest, principal, and time.
(b) Find the interest after three quarters.

SOLUTION

(a) Let I = interest (in dollars), P = principal (in dollars), and t = time (in years). Since I is jointly proportional to P and t, we have the model

$$I = kPt.$$

For $I = 106.25$, $P = 5000$, and $t = 1/4$, we have

$$106.25 = k(5000)\left(\frac{1}{4}\right) \quad \text{which implies that} \quad k = \frac{4(106.25)}{5000} = 0.085.$$

Thus, the equation relating interest, principal, and time is

$$I = 0.085Pt$$

which is the familiar equation for simple interest where the constant of proportionality, 0.085, represents an annual percentage rate of 8.5%.

(b) When P = $5000 and $t = 3/4$, the interest is

$$I = (0.085)(5000)\left(\frac{3}{4}\right) = \$318.75.$$

WARM UP

Solve for k.

1. $15 = k(45)$

2. $9 = k(4^2)$

3. $20 = \dfrac{k(15)}{32}$

4. $30 = \dfrac{k(0.2)}{0.5}$

5. $110 = k(27)(0.4)$

6. $210 = k(4^2)(16)$

Find the indicated value.

7. Let $d = 2.7r$. Find d when $r = 10$.

8. Let $s = 3tp^3$. Find s when $t = 2$ and $p = 1/3$.

9. Let $R = 4t/h$. Find R when $t = 7$ and $h = 13$.

10. Let $M = 14rst$. Find M when $r = 0.01$, $s = 150$, and $t = 7.5$.

EXERCISES 3.8

In Exercises 1–14, find a mathematical model for the verbal statement.

1. A varies directly as the square of r.

2. V varies directly as the cube of e.

3. y varies inversely as the square of x.

4. h varies inversely as the square root of s.

5. z is proportional to the cube root of u.

6. x is inversely proportional to $t + 1$.

7. z varies jointly as u and v.

8. V varies jointly as l, w, and h.

9. F varies directly as g and inversely as the square of r.

10. z is jointly proportional to the square of x and the cube of y.

11. *(Boyle's Law)* For a constant temperature, the pressure P of a gas is inversely proportional to the volume V of the gas.

12. *(Newton's Law of Cooling)* The rate of change R of the temperature of an object is proportional to the difference between the temperature T of the object and the temperature T_e of the environment in which the object is placed.

13. *(Newton's Law of Universal Gravitation)* The gravitational attraction F between two objects of masses m_1 and m_2 is proportional to the product of the masses and inversely proportional to the square of the distance r between the objects.

14. *(Logistics Growth)* The rate of growth R of a population is jointly proportional to the size S of the population and the difference between S and the maximum size L that the environment can support.

In Exercises 15–30, find a mathematical model representing the given statement. (In each case determine the constant of proportionality.)

15. y varies directly as x. ($y = 25$ when $x = 10$.)

16. y is directly proportional to x. ($y = 8$ when $x = 24$.)

17. A varies directly as the square of r. ($A = 9\pi$ when $r = 3$.)

18. s is directly proportional to the square of t. ($s = 64$ when $t = 2$.)

19. y varies inversely as x. ($y = 3$ when $x = 25$.)

20. y is inversely proportional to x. ($y = 7$ when $x = 4$.)

21. h is inversely proportional to the third power of t. ($h = 3/16$ when $t = 4$.)

22. R varies inversely as the square of s. ($R = 80$ when $s = 1/5$.)

23. z varies jointly as x and y. ($z = 64$ when $x = 4$ and $y = 8$.)

24. z is jointly proportional to x and y. ($z = 32$ when $x = 10$ and $y = 16$.)

25. F is jointly proportional to r and the third power of s. ($F = 4158$ when $r = 11$ and $s = 3$.)

26. P varies directly as x and inversely as the square of y. ($P = 28/3$ when $x = 42$ and $y = 9$.)

27. z varies directly as the square of x and inversely as y. ($z = 6$ when $x = 6$ and $y = 4$.)

28. v varies jointly as p and q and inversely as the square of s. ($v = 1.5$ when $p = 4.1$, $q = 6.3$, and $s = 1.2$.)

29. S varies directly as L and inversely as $L - S$. ($S = 4$ when $L = 6$.)

30. P is jointly proportional to S and $L - S$. ($P = 10$ when $S = 4$ and $L = 6$.)

In Exercises 31–34, use Hooke's Law as stated in Example 1 of this section.

31. A force of 50 pounds stretches a spring 5 inches.
(a) How far will a force of 20 pounds stretch the spring?
(b) What force is required to stretch the spring 1.5 inches?

32. Repeat Exercise 31 assuming that a force of 50 pounds stretches the spring 3 inches.

33. The coiled spring of a toy supports the weight of a child. The spring compresses a distance of 1.9 inches under the weight of a 25-pound child. The toy will not work properly if its spring is compressed more than 3 inches. What is the weight of the heaviest child who should be allowed to use the toy?

34. An overhead garage door has two springs, one on each side of the door. A force of 15 pounds is required to stretch each spring 1 foot. Because of a pulley system, the springs stretch only one-half the distance the door travels. The door moves a total of 8 feet and the springs are at their natural length when the door is open. Find the combined lifting force applied to the door by the springs.

In Exercises 35 and 36, use the fact that the diameter of a particle moved by a stream varies approximately as the square of the velocity of the stream.

35. A stream with a velocity of 1/4 mile per hour can move coarse sand particles of about 0.02 inch diameter. What must the velocity be to carry particles with a diameter of 0.12 inch?

36. A stream of velocity v can move particles of diameter d or less. By what factor does d increase when the velocity is doubled?

37. Neglecting air resistance, the distance s that an object falls varies directly as the square of the time t it has been falling. An object falls a distance of 144 feet in 3 seconds. How far will it fall in 5 seconds?

38. A company has found that the demand for its product varies inversely as the price of the product. When the price is $3.75, the demand is 500 units. Approximate the demand for a price of $4.25

39. The illumination from a light source varies inversely as the square of the distance from the light source. When the distance from a light source is doubled, how does the illumination change?

40. The resistance of a wire carrying electrical current is directly proportional to its length and inversely proportional to its cross-sectional area. If #28 copper wire (which has a diameter of 0.0126 inch) has a resistance of 66.17 ohms per thousand feet, what length of #28 copper wire will produce a resistance of 33.5 ohms?

41. A 14-foot piece of copper wire produces a resistance of 0.05 ohms. Use Exercise 40 to find the diameter of the wire.

42. The load that can be safely supported by a horizontal beam varies jointly as the width of the beam and the square of its depth and inversely as the length of the beam. Determine what happens to the safe load under the following conditions.
(a) The width and length of the beam are doubled.
(b) The width and depth of the beam are doubled.
(c) All three of the dimensions are doubled.
(d) The depth of the beam is halved.

43. The velocity v of a fluid flowing in a conduit is inversely proportional to the cross-sectional area of the conduit. (Assume the volume of the flow per unit of time is held constant.) Determine the change of velocity of water flowing from a hose when a person places a finger over the end of the hose to decrease its cross-sectional area by 25%.

44. Use the fluid velocity equation of Exercise 43 to determine the effect on the velocity of a stream when it is dredged to increase its cross-sectional area by one-third.

CHAPTER 3 REVIEW EXERCISES

In Exercises 1–6, find (a) the distance between the two points, (b) the coordinates of the midpoint of the line segment between the two points, (c) an equation of the line through the two points, and (d) an equation of the circle whose diameter is the line segment between the two points.

1. $(0, 0)$, $(0, 10)$
2. $(-1, 4)$, $(2, 0)$
3. $(2, 1)$, $(14, 6)$
4. $(-2, 2)$, $(3, -10)$
5. $(-1, 0)$, $(6, 2)$
6. $(1, 6)$, $(4, 2)$

In Exercises 7–10, find t so that the three points are collinear.

7. $(-2, 5)$, $(0, t)$, $(1, 1)$
8. $(-6, 1)$, $(1, t)$, $(10, 5)$
9. $(1, -4)$, $(t, 3)$, $(5, 10)$
10. $(-3, 3)$, $(t, -1)$, $(8, 6)$

In Exercises 11–14, show that the given points form the vertices of the indicated polygon.

11. Parallelogram: $(1, 1)$, $(8, 2)$, $(9, 5)$, $(2, 4)$
12. Isosceles triangle: $(4, 5)$, $(1, 0)$, $(-1, 2)$
13. Right triangle: $(-1, -1)$, $(10, 7)$, $(2, 18)$
14. Square: $(-4, 0)$, $(1, -3)$, $(4, 2)$, $(-1, 5)$

In Exercises 15–24, find the intercepts of the given graph and check for symmetry with respect to each of the coordinate axes and the origin.

15. $2y^2 = x^3$

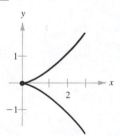

16. $x^2 + (y + 2)^2 = 4$

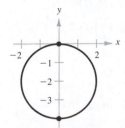

17. $y = \dfrac{x^4}{4} - 2x^2$

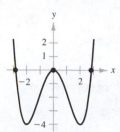

18. $y = \dfrac{x^3}{4} - 3x$

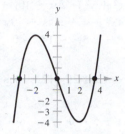

19. $y = x\sqrt{4 - x^2}$

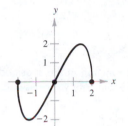

20. $y = x\sqrt{x + 3}$

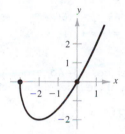

21. $y = x^3 - 3x^2$

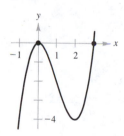

22. $x^3 + y^3 - 3xy = 0$

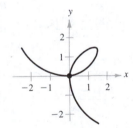

23. $y^2 = \dfrac{x^3}{4 - x}$

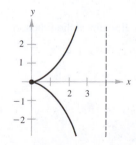

24. $x^{2/3} + y^{2/3} = 4^{2/3}$

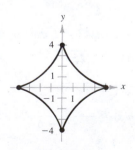

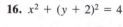

Functions and Graphs

In Exercises 25–28, determine the center and radius of the circle. Then, sketch its graph.

25. $x^2 + y^2 - 12x - 8y + 43 = 0$
26. $x^2 + y^2 - 20x - 10y + 100 = 0$
27. $4x^2 + 4y^2 - 4x - 40y + 92 = 0$
28. $5x^2 + 5y^2 - 14y = 0$

In Exercises 29–42, sketch a graph of the equation.

29. $y - 2x - 3 = 0$ **30.** $3x + 2y + 6 = 0$
31. $x - 5 = 0$ **32.** $y = 8 - |x|$
33. $y = \sqrt{5 - x}$ **34.** $y = \sqrt{x + 2}$
35. $y + 2x^2 = 0$ **36.** $y = x^2 - 4x$
37. $y = \sqrt{25 - x^2}$ **38.** $x^2 + y^2 = 10$
39. $x^2 + y^2 = 16$ **40.** $y = -(x - 4)^2$
41. $y = \frac{1}{4}(x + 1)^3$ **42.** $y = 4 - (x - 4)^2$

In Exercises 43–46, evaluate the function at the specified values of the independent variable. Simplify your answers.

43. $f(x) = x^2 + 1$
  (a) $f(2)$ (b) $f(-4)$
  (c) $f(t^2)$ (d) $-f(x)$

44. $g(x) = x^{4/3}$
  (a) $g(8)$ (b) $g(t + 1)$
  (c) $\dfrac{g(8) - g(1)}{8 - 1}$ (d) $g(-x)$

45. $h(x) = 6 - 5x^2$
  (a) $h(2)$ (b) $h(x + 3)$
  (c) $\dfrac{h(4) - h(2)}{4 - 2}$ (d) $\dfrac{h(x + \Delta x) - h(x)}{\Delta x}$

46. $f(t) = \sqrt[4]{t}$
  (a) $f(16)$ (b) $f(t + 5)$
  (c) $\dfrac{f(16) - f(0)}{16}$ (d) $f(t + \Delta t)$

In Exercises 47–52, determine the domain of the function.

47. $f(x) = \sqrt{25 - x^2}$ **48.** $f(x) = 3x + 4$
49. $g(s) = \dfrac{5}{3s - 9}$ **50.** $f(x) = \sqrt{x^2 + 8x}$
51. $h(x) = \dfrac{x}{x^2 - x - 6}$ **52.** $h(t) = |t + 1|$

In Exercises 53–58, (a) find f^{-1}, (b) sketch the graphs of f and f^{-1} on the same coordinate plane, and (c) verify that $f^{-1}(f(x)) = x = f(f^{-1}(x))$.

53. $f(x) = \frac{1}{2}x - 3$
54. $f(x) = 5x - 7$
55. $f(x) = \sqrt{x + 1}$
56. $f(x) = x^3 + 2$
57. $f(x) = x^2 - 5, \quad x \geq 0$
58. $f(x) = \sqrt[3]{x + 1}$

In Exercises 59–62, restrict the domain of the function f to an interval where the function is increasing and determine f^{-1} over that interval.

59. $f(x) = 2(x - 4)^2$
60. $f(x) = |x - 2|$
61. $f(x) = \sqrt{x^2 - 4}$
62. $f(x) = x^{4/3}$

In Exercises 63–70, let
$$f(x) = 3 - 2x, \quad g(x) = \sqrt{x}, \quad \text{and} \quad h(x) = 3x^2 + 2$$
and find the indicated value.

63. $(f - g)(4)$
64. $(f + h)(5)$
65. $(fh)(1)$
66. $\left(\dfrac{g}{h}\right)(1)$
67. $(h \circ g)(7)$
68. $(g \circ f)(-2)$
69. $g^{-1}(3)$
70. $(h \circ f^{-1})(1)$

In Exercises 71–74, find a mathematical model representing the given statement. (In each case determine the constant of proportionality.)

71. F is jointly proportional to x and the square root of y. ($F = 6$ when $x = 9$ and $y = 4$.)

72. R varies inversely as the cube of x. ($R = 128$ when $x = 2$.)

73. z varies directly as the square of x and inversely as y. ($z = 16$ when $x = 5$ and $y = 2$.)

74. w varies jointly as x and y and inversely as the cube of z. ($w = 44/9$ when $x = 12$, $y = 11$, and $z = 6$.)

75. The velocity of a ball thrown vertically upward from ground level is given by

$$v(t) = -32t + 48$$

where t is time in seconds and v is velocity in feet per second.
(a) Find the velocity when $t = 1$.
(b) Find the time when the ball reaches its maximum height. [*Hint:* Find the time when $v(t) = 0$.]
(c) Find the velocity when $t = 2$.

76. A company produces a product for which the variable cost is $5.35 per unit and the fixed costs are $16,000. The company sells the product for $8.20, and can sell all that it produces.
(a) Find the total cost as a function of x, the number of units produced.
(b) Find the profit as a function of x.

77. A wire 24 inches long is to be cut into four pieces to form a rectangle whose shortest side has a length of x. Express the area A of the rectangle as a function of x. Determine the domain of the function and sketch its graph over that domain.

78. The power P produced by a wind turbine is proportional to the cube of the wind speed S. A wind speed of 27 miles per hour produces a power output of 750 kilowatts. Find the output for a wind speed of 40 miles per hour.

CHAPTER 4

Polynomial Functions: Graphs and Zeros

4.1 Quadratic Functions

Throughout this chapter we will study polynomial functions, the most widely used functions in algebra.

Definition of Polynomial Function

Let n be a nonnegative integer and let $a_n, a_{n-1}, \ldots, a_2, a_1, a_0$ be real numbers with $a_n \neq 0$. The function given by

$$f(x) = a_n x^n + a_{n-1} x^{n-1} + \cdots + a_2 x^2 + a_1 x + a_0$$

is called a **polynomial function of x with degree n.**

The polynomial function $f(x) = a$, $a \neq 0$, has degree 0 and is called a **constant function.** The polynomial function $f(x) = ax + b$, $a \neq 0$, has degree 1 and is called a **linear function.** In the previous chapter, we saw that the graph of a constant function is a horizontal line and the graph of the linear function $f(x) = ax + b$ is a line whose slope is a and whose y-intercept is $(0, b)$.

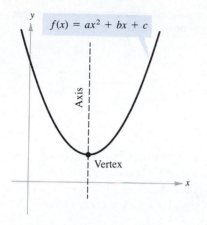

$f(x) = ax^2 + bx + c$

Axis

Vertex

$a > 0$: Parabola opens upward

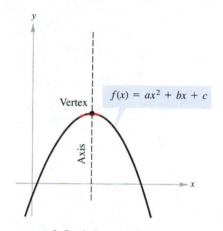

Vertex

$f(x) = ax^2 + bx + c$

Axis

$a < 0$: Parabola opens downward

FIGURE 4.1

In this section we look at second-degree polynomial functions, called **quadratic functions.**

Definition of Quadratic Function

Let a, b, and c be real numbers with $a \neq 0$. The function of x given by

$$f(x) = ax^2 + bx + c$$

is called a **quadratic function.** The graph of a quadratic function is called a **parabola.**

All parabolas are symmetric with respect to a line called the **axis of symmetry,** or simply the **axis** of the parabola. The point where the axis intersects the parabola is called the **vertex** of the parabola, as shown in Figure 4.1. If $a > 0$, then the graph of $f(x) = ax^2 + bx + c$ is a parabola that opens upward, and if $a < 0$, the graph is a parabola that opens downward.

The simplest type of quadratic function is $f(x) = ax^2$. Its graph is a parabola whose vertex is $(0, 0)$. If $a > 0$, then the vertex is the *minimum* point on the graph, and if $a < 0$, then the vertex is the *maximum* point on the graph, as shown in Figure 4.2.

When sketching the graph of $f(x) = ax^2$, it is helpful to use the graph of $y = x^2$ as a reference. (You may want to review the material presented in Section 3.5.)

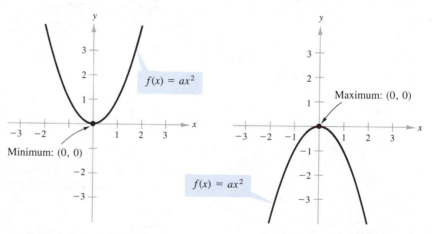

$f(x) = ax^2$

Minimum: $(0, 0)$

$a > 0$: Parabola opens upward

Maximum: $(0, 0)$

$f(x) = ax^2$

$a < 0$: Parabola opens downward

FIGURE 4.2

EXAMPLE 1 *Sketching the Graphs of Simple Quadratic Functions*

Sketch the graphs of the following.

(a) $f(x) = \dfrac{1}{3}x^2$ 　　　　　　　　　　　　　(b) $g(x) = 2x^2$

SOLUTION

(a) Compared with $y = x^2$, each output of f "shrinks" by a factor of $1/3$. The result is a parabola that opens upward and is broader than the parabola represented by $y = x^2$, as shown in Figure 4.3.

(b) In this case, each output of g "stretches" by a factor of 2, creating the more narrow parabola shown in Figure 4.4.

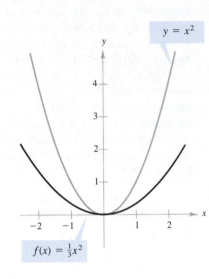

FIGURE 4.3

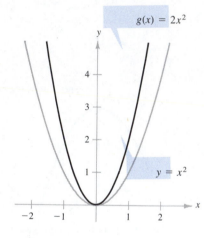

FIGURE 4.4

Remark: In Example 1 note that the coefficient a determines how widely the parabola given by $f(x) = ax^2$ opens. If $|a|$ is small, the parabola opens more widely than if $|a|$ is large.

Recall from Section 3.5 that the graphs of $y = f(x \pm c)$, $y = f(x) \pm c$, and $y = -f(x)$ are rigid transformations of the graph of $y = f(x)$.

$y = f(x \pm c)$	*Horizontal shift*
$y = f(x) \pm c$	*Vertical shift*
$y = -f(x)$	*Reflection*

Note how we use these transformations in Example 2.

EXAMPLE 2   *Sketching a Parabola*

Sketch the graphs of the following.

(a) $f(x) = -x^2 + 1$

(b) $g(x) = (x + 2)^2 - 3$

SOLUTION

(a) Compared to the graph of $y = x^2$, the negative coefficient in $f(x) = -x^2 + 1$ turns the graph *downward* and the positive constant term shifts the vertex *up one unit*. The graph of f is shown in Figure 4.5. Note that the axis of the parabola is the y-axis and the vertex is $(0, 1)$.

(b) Compared to the graph of $y = x^2$, the graph of $g(x) = (x + 2)^2 - 3$ represents a horizontal shift of two units *to the left* and a vertical shift of three units *down*, as shown in Figure 4.6. Note that the axis of the parabola is the vertical line $x = -2$, and the vertex is $(-2, -3)$.

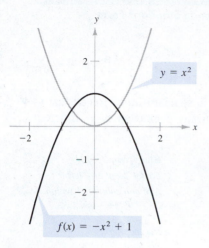

FIGURE 4.5

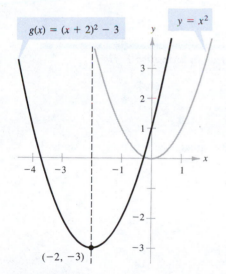

FIGURE 4.6

The equation in Example 2(b) is written in the **standard form** $f(x) = a(x - h)^2 + k$. This form is especially convenient for sketching a parabola because it identifies the vertex of the parabola.

Standard Form of a Quadratic Function

The quadratic function

$$f(x) = a(x - h)^2 + k, \qquad a \neq 0$$

is said to be in **standard form.** The graph of f is a parabola whose axis is the vertical line $x = h$ and whose vertex is the point (h, k). If $a > 0$, the parabola opens upward and if $a < 0$, the parabola opens downward.

To write a quadratic function in standard form we use the process of *completing the square*, discussed in Section 2.3.

EXAMPLE 3 Writing a Quadratic Function in Standard Form

Sketch the graph of $f(x) = 2x^2 + 8x + 7$.

SOLUTION

We begin by writing the quadratic function in standard form. Recall that to complete the square we must first factor out any coefficient of x^2 different from 1.

$$
\begin{aligned}
f(x) &= 2x^2 + 8x + 7 & &\textit{Given form} \\
&= 2(x^2 + 4x) + 7 & &\textit{Factor 2 out of x terms} \\
&= 2(x^2 + 4x + 4 - 4) + 7 & &\textit{Add and subtract 4 within} \\
& \qquad\quad \underset{2^2}{\underbrace{}} & &\textit{parentheses} \\
&= 2(x^2 + 4x + 4) - 2(4) + 7 & &\textit{Regroup terms} \\
&= 2(x^2 + 4x + 4) - 8 + 7 & &\textit{Simplify} \\
&= 2(x + 2)^2 - 1 & &\textit{Standard form}
\end{aligned}
$$

From the standard form, we see that the graph of f is a parabola that opens upward with vertex $(-2, -1)$. This corresponds to a left shift of two units and a downward shift of one unit relative to the graph of $y = 2x^2$, as shown in Figure 4.7.

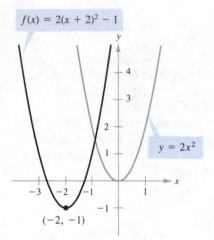

$f(x) = 2(x + 2)^2 - 1$

$y = 2x^2$

$(-2, -1)$

FIGURE 4.7

EXAMPLE 4 Writing a Quadratic Function in Standard Form

Sketch a graph of $f(x) = -x^2 + 6x - 8$.

SOLUTION

As in Example 3, we begin by writing the quadratic function in standard form.

$$
\begin{aligned}
f(x) &= -x^2 + 6x - 8 \\
&= -(x^2 - 6x) - 8 \\
&= -(x^2 - 6x + 9 - 9) - 8 \\
& \qquad\quad \underset{3^2}{\underbrace{}} \\
&= -(x^2 - 6x + 9) - (-9) - 8 \\
&= -(x^2 - 6x + 9) + 9 - 8 \\
&= -(x - 3)^2 + 1
\end{aligned}
$$

Thus, the graph of f is a parabola that opens downward with vertex at $(3, 1)$, as shown in Figure 4.8.

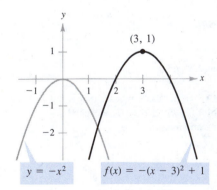

$(3, 1)$

$y = -x^2$

$f(x) = -(x - 3)^2 + 1$

FIGURE 4.8

Quadratic Functions

In the next example, we reverse the problem. That is, instead of sketching the graph of a given quadratic function, we show how to find a quadratic function that represents a given parabola.

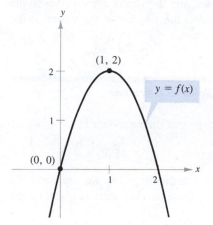

FIGURE 4.9

EXAMPLE 5 *Finding the Equation of a Parabola*

Find an equation for the parabola whose vertex is (1, 2) and passes through the point (0, 0), as shown in Figure 4.9.

SOLUTION

Since the parabola has a vertex at $(h, k) = (1, 2)$, the equation must have the form

$$f(x) = a(x - 1)^2 + 2. \qquad \textit{Standard form}$$

Because the parabola passes through the point (0, 0), it follows that $f(0) = 0$. Thus, we obtain

$$0 = a(0 - 1)^2 + 2 \implies a = -2$$

which implies that the equation is

$$f(x) = -2(x - 1)^2 + 2 = -2x^2 + 4x.$$

To find the x-intercepts of the graph of $f(x) = ax^2 + bx + c$ we must solve the equation $ax^2 + bx + c = 0$. For example, from the equation

$$-2x^2 + 4x = -2x(x - 2) = 0$$

we see that the parabola shown in Figure 4.9 has x-intercepts at (0, 0) and (2, 0). If $ax^2 + bx + c$ does not factor, you can use the Quadratic Formula to find the x-intercepts. Remember, however, that a parabola may have no x-intercepts.

Applications of Quadratic Functions

Many applications involve finding the maximum or minimum value of the quadratic function $f(x) = ax^2 + bx + c$, $a \neq 0$. By completing the square we can write $f(x)$ in the form

$$f(x) = a\left(x + \frac{b}{2a}\right)^2 + \left(c - \frac{b^2}{4a}\right).$$

Polynomial Functions: Graphs and Zeros

Thus, $f(-b/2a)$ is either a maximum or minimum value of the function. Specifically, if $a > 0$, then the graph of f is a parabola opening upward and the vertex corresponds to the *minimum* value $f(-b/2a)$. If $a < 0$, then the graph of f is a parabola opening downward and the vertex corresponds to the *maximum* value $f(-b/2a)$. (Remember that the y-value of the vertex is the minimum or maximum value of the function.)

EXAMPLE 6 *Finding the Maximum Value of a Product*

The sum of two numbers is 18. How large can their product be?

SOLUTION

Since it is given that the sum of the two numbers is 18, we represent the numbers as x and $18 - x$. The product of these two numbers is given by the quadratic function

$$f(x) = x(18 - x) = 18x - x^2.$$

In standard form, this equation is

$$f(x) = -(x - 9)^2 + 81$$

and we see that the graph of f is a parabola opening downward with vertex (highest point) at the point $(9, 81)$, as shown in Figure 4.10. Therefore, we conclude that the maximum value of the product of the two numbers is 81.

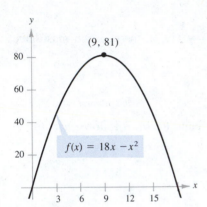

FIGURE 4.10

EXAMPLE 7 *Finding the Maximum Height of an Object*

An object is propelled straight upward from a height of 6 feet with an initial velocity of 32 feet per second. The height at any time t is given by

$$s(t) = -16t^2 + 32t + 6$$

where $s(t)$ is measured in feet and t in seconds. Find the maximum height attained by the object before it begins falling to the ground.

SOLUTION

The maximum height of the object occurs at the vertex of the graph of s, as shown in Figure 4.11. Writing $s(t)$ in standard form, we obtain

$$\begin{aligned}
s(t) &= -16t^2 + 32t + 6 \\
&= -16(t^2 - 2t) + 6 \\
&= -16(t^2 - 2t + 1 - 1) + 6 \\
&= -16(t - 1)^2 + 22.
\end{aligned}$$

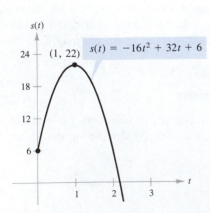

FIGURE 4.11

Thus, the vertex occurs at (1, 22), and we conclude that the maximum height of the object is 22 feet, and this occurs when $t = 1$ second.

WARM UP

Solve the quadratic equations by factoring.

1. $2x^2 + 11x - 6 = 0$

2. $5x^2 - 12x - 9 = 0$

3. $3 + x - 2x^2 = 0$

4. $x^2 + 20x + 100 = 0$

Solve the quadratic equations by completing the square.

5. $x^2 - 6x + 4 = 0$

6. $x^2 + 4x + 1 = 0$

7. $2x^2 - 16x + 25 = 0$

8. $3x^2 + 30x + 74 = 0$

Use the Quadratic Formula to solve the quadratic equations.

9. $x^2 + 3x + 3 = 0$

10. $x^2 + 3x - 3 = 0$

EXERCISES 4.1

In Exercises 1–6, match the given quadratic function with the correct graph. [The graphs are labeled (a)–(f).]

1. $f(x) = (x - 3)^2$

2. $f(x) = (x + 5)^2$

3. $f(x) = x^2 - 4$

4. $f(x) = 5 - x^2$

5. $f(x) = 4 - (x - 1)^2$

6. $f(x) = (x + 2)^2 - 2$

(c)
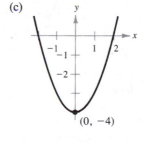

(d)
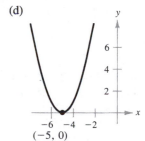

(a)
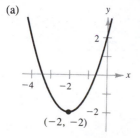

(b)
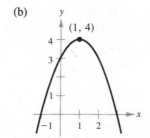

(e)
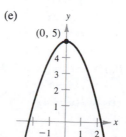

(f)
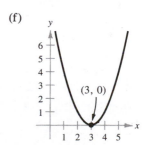

In Exercises 7–12, find an equation for the given parabola.

7.

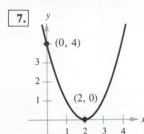

8.

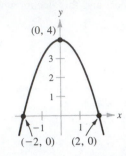

9.

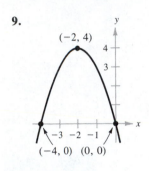

10.

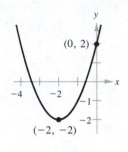

11.

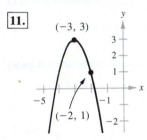

12.

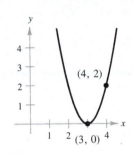

In Exercises 13–30, sketch the graph of the given quadratic function. Identify the vertex and intercepts.

13. $f(x) = x^2 - 5$

14. $f(x) = \frac{1}{2}x^2 - 4$

15. $f(x) = 16 - x^2$

16. $h(x) = 25 - x^2$

17. $f(x) = (x + 5)^2 - 6$

18. $f(x) = (x - 6)^2 + 3$

19. $h(x) = x^2 - 8x + 16$

20. $g(x) = x^2 + 2x + 1$

21. $f(x) = -(x^2 + 2x - 3)$

22. $g(x) = x^2 + 8x + 11$

23. $f(x) = x^2 - x + \frac{5}{4}$

24. $f(x) = x^2 + 3x + \frac{1}{4}$

25. $f(x) = -x^2 + 2x + 5$

26. $f(x) = -x^2 - 4x + 1$

27. $h(x) = 4x^2 - 4x + 21$

28. $f(x) = 2x^2 - x + 1$

29. $f(x) = 2x^2 - 16x + 31$

30. $g(x) = \frac{1}{2}(x^2 + 4x - 2)$

In Exercises 31–34, find the quadratic function that has the indicated vertex and whose graph passes through the given point.

31. Vertex: (3, 4); Point: (1, 2)

32. Vertex: (2, 3); Point: (0, 2)

33. Vertex: (5, 12); Point: (7, 15)

34. Vertex: (−2, −2); Point: (−1, 0)

In Exercises 35–40, find two quadratic functions whose graphs have the given x-intercepts. (One function has a graph that opens upward and the other has a graph that opens downward.)

35. (−1, 0), (3, 0)

36. $\left(-\frac{5}{2}, 0\right)$, (2, 0)

37. (0, 0), (10, 0)

38. (4, 0), (8, 0)

39. (−3, 0), $\left(-\frac{1}{2}, 0\right)$

40. (−5, 0), (5, 0)

In Exercises 41–44, find two positive real numbers that satisfy the given requirements.

41. The sum is 110 and the product is a maximum.

42. The sum is S and the product is a maximum.

43. The sum of the first and twice the second is 24 and the product is a maximum.

44. The sum of two numbers is 50 and the product is a maximum.

In Exercises 45 and 46, find the length and width of a rectangle of maximum area for the given perimeter.

45. Perimeter: 100 feet

46. Perimeter: P units

47. A rancher has 200 feet of fencing to enclose two adjacent rectangular corrals (see figure). What dimensions will produce a maximum enclosed area?

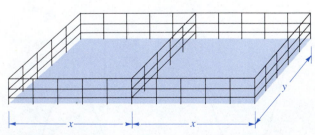

FIGURE FOR 47

48. An indoor physical fitness room consists of a rectangular region with a semicircle on each end (see figure). The perimeter of the room is to be a 200-meter running track. What dimensions will produce a maximum area of the rectangle?

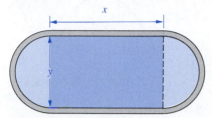

FIGURE FOR 48

49. Find the number of units x that produce a maximum revenue R for $R = 900x - 0.1x^2$.

50. A manufacturer of lighting fixtures has daily production costs of

$$C = 800 - 10x + 0.25x^2.$$

How many fixtures x should be produced each day to yield a minimum cost?

51. Let x be the amount (in hundreds of dollars) a company spends on advertising, and let P be the profit, where

$$P = 230 + 20x - 0.5x^2.$$

What expenditure for advertising gives the maximum profit?

52. The path of the diver shown in the figure is parabolic. Find the equation of the path.

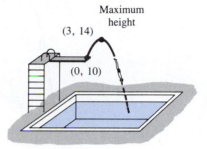

Maximum height

(3, 14)

(0, 10)

FIGURE FOR 52

53. Complete the square for the quadratic function $f(x) = ax^2 + bx + c$ $(a \neq 0)$ and show that the vertex is at

$$\left(-\frac{b}{2a}, -\frac{b^2 - 4ac}{4a} \right).$$

54. Use Exercise 53 to verify the vertices found in Exercises 19 and 20.

55. Assume that the function $f(x) = ax^2 + bx + c$ $(a \neq 0)$ has two real zeros. Show that the x-coordinate of the vertex of the graph is the average of the zeros of f. [*Hint:* Use the Quadratic Formula.]

4.2 Polynomial Functions of Higher Degree

At this point you should be able to sketch an accurate graph of polynomial functions of degrees 0, 1, and 2.

Function	Graph
$f(x) = a$	Horizontal line
$f(x) = ax + b$	Line of slope a
$f(x) = ax^2 + bx + c$	Parabola

The graphs of polynomial functions of degree greater than 2 are more difficult to sketch. However, in this section we show how to recognize some of the basic features of the graphs of polynomial functions. Using these features, together with point-plotting, intercepts, and symmetry, you should be able to make reasonably accurate sketches.

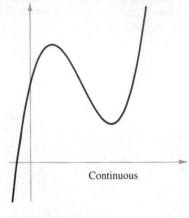

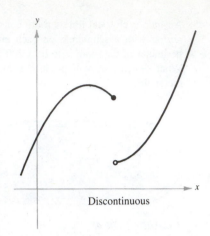

Continuous

Discontinuous

FIGURE 4.12

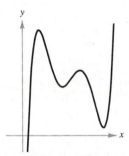

Polynomial functions have smooth, rounded graphs.

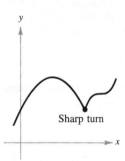

Sharp turn

The graph of a polynomial function *cannot* have a sharp pointed turn.

FIGURE 4.13

The first feature is that a polynomial function is **continuous.** Essentially, this means that the graph of a polynomial function has no breaks, as shown in Figure 4.12.

The second feature is that the graph of a polynomial function has only smooth, rounded turns, as shown in Figure 4.13.

The simplest polynomial functions to graph are monomials of the form

$$f(x) = x^n,$$

where n is an integer greater than zero. From Figure 4.14, we see that when n *is even* the graph is similar to the graph of $f(x) = x^2$ and when n *is odd* the graph is similar to the graph of $f(x) = x^3$. Moreover, the greater the value of n, the flatter the graph on the interval $[-1, 1]$.

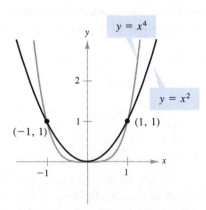

If n is even, the graph of $y = x^n$ *touches* axis at x-intercept.

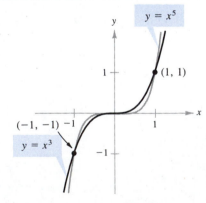

If n is odd, the graph of $y = x^n$ *crosses* axis at x-intercept.

FIGURE 4.14

EXAMPLE 1 *Sketching Transformations of Monomial Functions*

Sketch the graph of the following polynomial functions.

(a) $f(x) = -x^5$ (b) $g(x) = x^4 + 1$ (c) $h(x) = (x + 1)^4$

SOLUTION

(a) Since the degree of f is odd, the graph is similar to the graph of $y = x^3$. Moreover, the negative coefficient turns the graph upside down. Plotting the intercept $(0, 0)$ and the points $(1, -1)$ and $(-1, 1)$, we obtain the graph shown in Figure 4.15.

(b) In this case, the graph of g is an upward shift, by one unit, of the graph of $y = x^4$ (see Figure 4.14). Thus, we obtain the graph shown in Figure 4.16.

(c) The graph of h is a left shift, by one unit, of the graph of $y = x^4$ and it is shown in Figure 4.17.

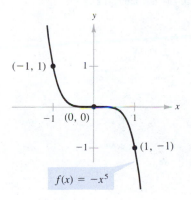

FIGURE 4.15

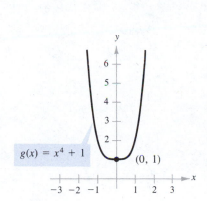

FIGURE 4.16

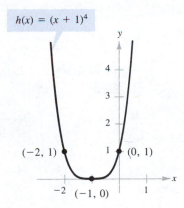

FIGURE 4.17

In Example 1, note that the three graphs eventually rise or fall without bound as x moves to the right or left. Symbolically, we write

$$f(x) \to \infty \quad \text{as} \quad x \to \infty$$

to mean that $f(x)$ increases without bound as x moves to the right without bound. (We use the infinity symbol ∞ to indicate unboundedness.) Whether the graph of a polynomial function eventually rises or falls can be determined by the function's degree (odd or even) and by its leading coefficient, as indicated in the **Leading Coefficient Test.**

Polynomial Functions: Graphs and Zeros

Leading Coefficient Test

As x moves without bound to the left or to the right, the graph of the polynomial function $f(x) = a_n x^n + \cdots + a_1 x + a_0$ eventually rises or falls in the following manner.

1. When n is *odd*:

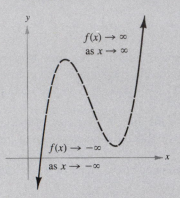

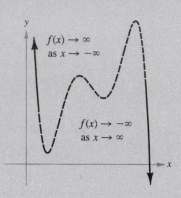

If $a_n > 0$, the graph falls to the left and rises to the right.   If $a_n < 0$, the graph rises to the left and falls to the right.

2. When n is *even*:

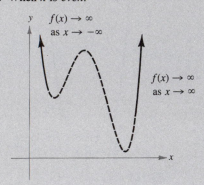

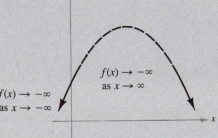

If $a_n > 0$, the graph rises to the left and right.   If $a_n < 0$, the graph falls to the left and right.

Remark: The dashed portions in the graphs for the Leading Coefficient Test indicate that the test determines *only* the right and left behavior of the graph.

EXAMPLE 2 *Applying the Leading Coefficient Test*

Use the Leading Coefficient Test to determine the right and left behavior of the graphs of the following polynomial functions.

(a) $f(x) = -x^3 + 4x$ (b) $f(x) = x^4 - 5x^2 + 4$ (c) $f(x) = x^5 - x$

SOLUTION

(a) Because the degree is odd and the leading coefficient is negative, the graph rises to the left and falls to the right, as shown in Figure 4.18.
(b) Because the degree is even and the leading coefficient is positive, the graph rises to the left and right, as shown in Figure 4.19.
(c) Because the degree is odd and the leading coefficient is positive, the graph falls to the left and rises to the right, as shown in Figure 4.20.

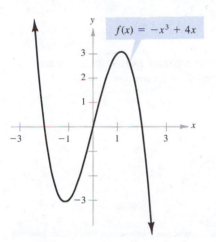

FIGURE 4.18

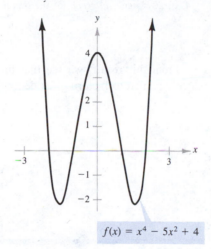

FIGURE 4.19

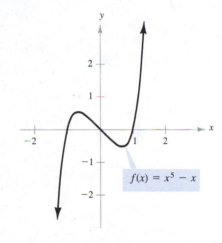

FIGURE 4.20

Zeros of Polynomial Functions

It can be shown that for a polynomial function f of degree n, the following are true.

1. The graph of f has, at most, $n - 1$ turning points. (Turning points are points at which the graph changes from increasing to decreasing or vice versa.)
2. f has, at most, n real zeros. (We will discuss this result in detail in Section 4.5 when we present the Fundamental Theorem of Algebra.)

Polynomial Functions: Graphs and Zeros

Finding the zeros of polynomial functions is one of the most important problems in algebra. There is a strong interplay between graphical and algebraic approaches to this problem. Sometimes we use information about the graph of a function to help find its zeros, and in other cases we use information about the zeros of a function to help sketch its graph.

Real Zeros of Polynomial Functions

If f is a polynomial function and a is a real number, then the following statements are equivalent.

1. $x = a$ is a *zero* of the function f.
2. $x = a$ is a *solution* of the polynomial equation $f(x) = 0$.
3. $(x - a)$ is a *factor* of the polynomial $f(x)$.
4. $(a, 0)$ is an *x-intercept* of the graph of the function f.

From this result, we see that finding zeros of polynomial functions is closely related to factoring, as demonstrated in Examples 3, 4, and 5.

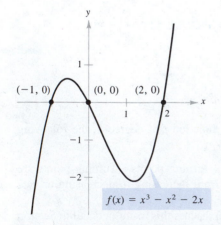

FIGURE 4.21

EXAMPLE 3 *Finding Zeros of a Third-Degree Polynomial Function*

Find all real zeros of $f(x) = x^3 - x^2 - 2x$.

SOLUTION

By factoring, we have

$$f(x) = x^3 - x^2 - 2x$$
$$= x(x^2 - x - 2)$$
$$= x(x - 2)(x + 1).$$

Thus, the real zeros are $x = 0$, $x = 2$, and $x = -1$, and the corresponding x-intercepts are $(0, 0)$, $(2, 0)$, and $(-1, 0)$. (See Figure 4.21 and note the two turning points.)

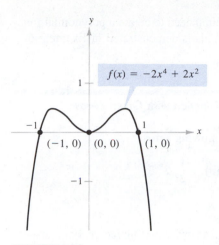

FIGURE 4.22

EXAMPLE 4 Finding Zeros of a Fourth-Degree Polynomial Function

Find all real zeros of $f(x) = -2x^4 + 2x^2$.

SOLUTION

In this case, factoring produces

$$
\begin{aligned}
f(x) &= -2x^4 + 2x^2 \\
&= -2x^2(x^2 - 1) \\
&= -2x^2(x - 1)(x + 1).
\end{aligned}
$$

Thus, the real zeros are $x = 0$, $x = 1$, and $x = -1$, and the corresponding intercepts are $(0, 0)$, $(1, 0)$, and $(-1, 0)$. (See Figure 4.22 and note the three turning points.)

In Example 4, the real zero arising from $-2x^2 = 0$ is called a **repeated zero**. In general, we say a factor $(x - a)^k$ yields a repeated zero $x = a$ of **multiplicity k**. If k is odd, then the graph *crosses* the x-axis at $x = a$. If k is even, then the graph *touches* (does not cross) the x-axis at $x = a$. Note how this occurs in Figure 4.22.

EXAMPLE 5 Finding Zeros of a Fifth-Degree Polynomial Function

Find all real zeros of $f(x) = -\frac{1}{4}x^5 + \frac{3}{4}x^3 + x$.

SOLUTION

Factoring produces

$$
\begin{aligned}
f(x) &= -\frac{1}{4}x^5 + \frac{3}{4}x^3 + x \\
&= \frac{1}{4}(-x^5 + 3x^3 + 4x) \\
&= -\frac{1}{4}x(x^4 - 3x^2 - 4) \\
&= -\frac{1}{4}x(x^2 - 4)(x^2 + 1) \\
&= -\frac{1}{4}x(x - 2)(x + 2)(x^2 + 1).
\end{aligned}
$$

Thus, the real zeros are $x = 0$, $x = 2$, and $x = -2$, and the corresponding intercepts are $(0, 0)$, $(2, 0)$, and $(-2, 0)$. Note that $x^2 + 1 = 0$ has no real solutions and so, produces no real zeros of the function. (See Figure 4.23.)

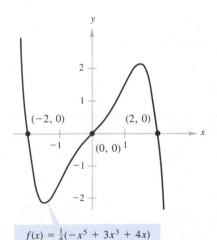

$f(x) = \frac{1}{4}(-x^5 + 3x^3 + 4x)$

FIGURE 4.23

To find mathematical models, we often need to create a polynomial function that has a particular set of zeros. This is demonstrated in Example 6.

EXAMPLE 6 Finding a Polynomial Function with Given Zeros

Find a polynomial function with the following zeros.

(a) $-2, -1, 1, 2$ 　　　　　　　　　　(b) $-\dfrac{1}{2}, 3, 3$

SOLUTION

(a) For each of the given zeros, we form a corresponding factor. For instance, the zero given by $x = -2$ corresponds to the factor $(x + 2)$. Thus, we can write the function as

$$f(x) = (x + 2)(x + 1)(x - 1)(x - 2)$$
$$= (x^2 - 4)(x^2 - 1)$$
$$= x^4 - 5x^2 + 4.$$

(b) Note that the zero $x = -1/2$ corresponds to either $\left(x + \frac{1}{2}\right)$ or $(2x + 1)$. To avoid fractions, we choose the second factor and write

$$f(x) = (2x + 1)(x - 3)^2$$
$$= (2x + 1)(x^2 - 6x + 9)$$
$$= 2x^3 - 11x^2 + 12x + 9.$$

In Example 7, we show how the Leading Coefficient Test and zeros of polynomial functions can be used as sketching aids.

EXAMPLE 7 Sketching the Graph of a Polynomial Function

Sketch the graph of $f(x) = 3x^4 - 4x^3$.

SOLUTION

Because the leading coefficient is positive and the degree is even, we know that the graph eventually rises to the right and to the left, as shown in Figure 4.24. By factoring $f(x)$,

$$f(x) = 3x^4 - 4x^3$$
$$= x^3(3x - 4)$$

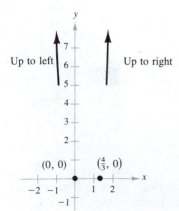

FIGURE 4.24

we see that the zeros of f are $x = 0$ and $x = 4/3$ (both of odd multiplicity). Thus, the x-intercepts occur at $(0, 0)$ and $(4/3, 0)$. Finally, we plot a few additional points as indicated in the accompanying table and obtain the graph shown in Figure 4.25.

x	-1	0.5	1	1.5
$f(x)$	7	-0.3125	-1	1.6875

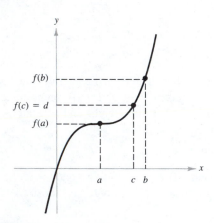

If d lies between $f(a)$ and $f(b)$, then there exists c between a and b such that $f(c) = d$.

FIGURE 4.26

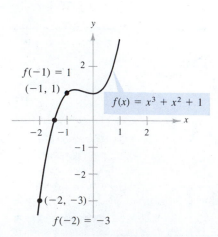

f has a zero between -2 and -1.

FIGURE 4.27

$f(x) = 3x^4 - 4x^3$

FIGURE 4.25

The Intermediate Value Theorem

The next theorem, called the **Intermediate Value Theorem,** informs us of the existence of real zeros of polynomial functions. The theorem implies that if $(a, f(a))$ and $(b, f(b))$ are two points on the graph of a polynomial such that $f(a) \neq f(b)$, then for any number d between $f(a)$ and $f(b)$ there must be a number c between a and b such that $f(c) = d$. (See Figure 4.26.)

Intermediate Value Theorem

If f is a polynomial function such that $a < b$ and $f(a) \neq f(b)$, then f takes on every value between $f(a)$ and $f(b)$ in the interval $[a, b]$.

This theorem helps us locate the real zeros of polynomial functions in the following way. If we can find a value $x = a$ where a polynomial function is positive, and another value $x = b$ where it is negative, then we can conclude that the function has at least one real zero between these two values. For example, the function $f(x) = x^3 + x^2 + 1$ is negative when $x = -2$ and positive when $x = -1$. It follows from the Intermediate Value Theorem that f must have a real zero somewhere between -2 and -1, as shown in Figure 4.27.

Polynomial Functions: Graphs and Zeros

EXAMPLE 8 *Approximating Zeros of a Polynomial Function*

Use the Intermediate Value Theorem to approximate the real zero of $f(x) = x^3 - x^2 + 1$.

SOLUTION

We begin by computing a few function values as follows.

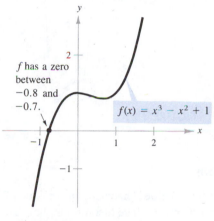

f has a zero between −0.8 and −0.7.

$f(x) = x^3 - x^2 + 1$

FIGURE 4.28

x	-2	-1	0	1
$f(x)$	-11	-1	1	1

Since $f(-1)$ is negative and $f(0)$ is positive, we conclude from the Intermediate Value Theorem that the function has a zero between -1 and 0. To pinpoint this zero more closely, we divide the interval $[-1, 0]$ into tenths and evaluate the function at each point. After doing this, we find that

$$f(-0.8) = -0.152 \quad \text{and} \quad f(-0.7) = 0.167.$$

Thus, f must have a zero between -0.8 and -0.7, as shown in Figure 4.28. By continuing the process we can approximate this zero to any desired accuracy.

WARM UP

Factor the expressions completely.

1. $12x^2 + 7x - 10$
2. $25x^3 - 60x^2 + 36x$
3. $12z^4 + 17z^3 + 5z^2$
4. $y^3 + 125$
5. $x^3 + 3x^2 - 4x - 12$
6. $x^3 + 2x^2 + 3x + 6$

Find all real solutions.

7. $5x^2 + 8 = 0$
8. $x^2 - 6x + 4 = 0$
9. $4x^2 + 4x - 11 = 0$
10. $x^4 - 18x^2 + 81 = 0$

EXERCISES 4.2

In Exercises 1–8, match the polynomial functions with the correct graph. [The graphs are labeled (a)–(h).]

1. $f(x) = -3x + 5$

2. $f(x) = x^2 - 2x$

3. $f(x) = -2x^2 - 8x - 9$

4. $f(x) = 3x^3 - 9x + 1$

5. $f(x) = -\frac{1}{3}x^3 + x - \frac{2}{3}$

6. $f(x) = -\frac{1}{4}x^4 + 2x^2$

7. $f(x) = 3x^4 + 4x^3$

8. $f(x) = x^5 - 5x^3 + 4x$

(a)
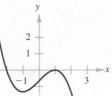

(b)
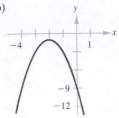

(c)
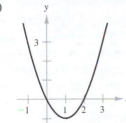

(d)
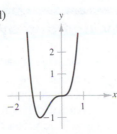

(e)
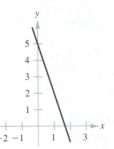

(f)
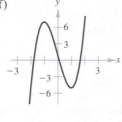

(g)
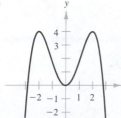

(h)
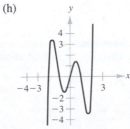

In Exercises 9–18, determine the right-hand and left-hand behavior of the graph of the polynomial function.

9. $f(x) = 2x^2 - 3x + 1$

10. $f(x) = \frac{1}{3}x^3 + 5x$

11. $g(x) = 5 - \frac{7}{2}x - 3x^2$

12. $f(x) = -2.1x^5 + 4x^3 - 2$

13. $f(x) = 2x^5 - 5x + 7.5$

14. $h(x) = 1 - x^6$

15. $f(x) = 6 - 2x + 4x^2 - 5x^3$

16. $f(x) = \dfrac{3x^4 - 2x + 5}{4}$

17. $h(t) = -\frac{2}{3}(t^2 - 5t + 3)$

18. $f(s) = -\frac{7}{8}(s^3 + 5s^2 - 7s + 1)$

In Exercises 19–34, find all the real zeros of the polynomial function.

19. $f(x) = x^2 - 25$

20. $f(x) = 49 - x^2$

21. $h(t) = t^2 - 6t + 9$

22. $f(x) = x^2 + 10x + 25$

23. $f(x) = x^2 + x - 2$

24. $f(x) = \frac{1}{2}x^2 + \frac{5}{2}x - \frac{3}{2}$

25. $f(x) = 3x^2 - 12x + 3$

26. $g(x) = 5(x^2 - 2x - 1)$

27. $f(t) = t^3 - 4t^2 + 4t$

28. $f(x) = x^4 - x^3 - 20x^2$

29. $g(t) = \frac{1}{2}t^4 - \frac{1}{2}$

30. $f(x) = x^5 + x^3 - 6x$

31. $f(x) = 2x^4 - 2x^2 - 40$

32. $g(t) = t^5 - 6t^3 + 9t$

33. $f(x) = 5x^4 + 15x^2 + 10$

34. $f(x) = x^3 - 4x^2 - 25x + 100$

In Exercises 35–44, find a polynomial function that has the given zeros.

35. 0, 10

36. 0, −3

37. 2, −6

38. −4, 5

39. 0, −2, −3

40. 0, 2, 5

41. 4, −3, 3, 0

42. −2, −1, 0, 1, 2

43. $1 + \sqrt{3}, 1 - \sqrt{3}$

44. $2, 4 + \sqrt{5}, 4 - \sqrt{5}$

45. Use the graph of $y = x^3$ to sketch the graphs of the following functions.

(a) $f(x) = (x - 2)^3$

(b) $f(x) = x^3 - 2$

(c) $f(x) = (x - 2)^3 - 2$

(d) $f(x) = -\frac{1}{2}x^3$

46. Use the graph of $y = x^4$ to sketch the graphs of the following functions.

(a) $f(x) = (x + 3)^4$

(b) $f(x) = x^4 - 3$

(c) $f(x) = 4 - x^4$

(d) $f(x) = \frac{1}{2}(x - 1)^4$

In Exercises 47–58, sketch the graph of the given function.

47. $f(x) = -\frac{3}{2}$

48. $h(x) = \frac{1}{3}x - 3$

49. $f(t) = \frac{1}{4}(t^2 - 2t + 15)$

50. $g(x) = -x^2 + 10x - 16$

51. $f(x) = x^3 - 3x^2$

52. $f(x) = 1 - x^3$

53. $f(x) = x^3 - 4x$

54. $f(x) = \frac{1}{4}x^4 - 2x^2$

55. $g(t) = -\frac{1}{4}(t - 2)^2(t + 2)^2$

56. $f(x) = x^2(x - 4)$

57. $f(x) = 1 - x^6$

58. $g(x) = 1 - (x + 1)^6$

In Exercises 59–62, follow the procedure given in Example 8 to estimate the zero of $f(x)$ in the given interval $[a, b]$. (Give your approximation to the nearest tenth.)

59. $f(x) = x^3 + x - 1$, $\quad [0, 1]$

60. $f(x) = x^5 + x + 1$, $\quad [-1, 0]$

61. $f(x) = x^4 - 10x^2 - 11$, $\quad [3, 4]$

62. $f(x) = -x^3 + 3x^2 + 9x - 2$, $\quad [4, 5]$

4.3 Polynomial Division and Synthetic Division

Up to this point in the text we have added, subtracted, and multiplied polynomials. In this section, we look at a procedure for *dividing* polynomials. This procedure has many important applications and is especially valuable in factoring and finding the zeros of polynomial functions.

To begin, suppose that we are given the graph of

$$f(x) = 6x^3 - 19x^2 + 16x - 4.$$

Notice that a zero of f occurs at $x = 2$, as shown in Figure 4.29. [Try verifying this by evaluating $f(x)$ at $x = 2$.] Since $x = 2$ is a zero of the polynomial function f, we know that $(x - 2)$ is a factor of $f(x)$. This means that there exists a second-degree polynomial $q(x)$ such that

$$f(x) = (x - 2) \cdot q(x).$$

To find $q(x)$, we use a process called **long division of polynomials.** Study the following example to see how this process works.

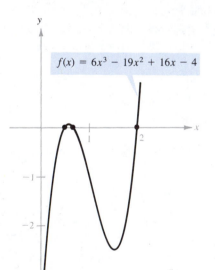

$f(x) = 6x^3 - 19x^2 + 16x - 4$

FIGURE 4.29

EXAMPLE 1 *Long Division of Polynomials*

Divide the polynomial

$$f(x) = 6x^3 - 19x^2 + 16x - 4$$

by $x - 2$, and use the result to factor $f(x)$ completely.

SOLUTION

Partial quotients

$$
\begin{array}{r}
6x^2 - 7x + 2 \\
x - 2\overline{)6x^3 - 19x^2 + 16x - 4} \\
\underline{6x^3 - 12x^2} \\
-7x^2 + 16x \\
\underline{-7x^2 + 14x} \\
2x - 4 \\
\underline{2x - 4} \\
0
\end{array}
$$

Multiply: $6x^2(x - 2)$

Subtract

Multiply: $-7x(x - 2)$

Subtract

Multiply: $2(x - 2)$

Subtract

We see that

$$6x^3 - 19x^2 + 16x - 4 = (x - 2)(6x^2 - 7x + 2)$$

and by factoring the quadratic $6x^2 - 7x + 2$, we have

$$6x^3 - 19x^2 + 16x - 4 = (x - 2)(2x - 1)(3x - 2).$$

Note that this factorization agrees with the graph of f (Figure 4.29) in that the three x-intercepts occur at

$$x = 2, \quad x = \frac{1}{2}, \quad \text{and} \quad x = \frac{2}{3}.$$

In Example 1, the polynomial $x - 2$ is a factor of the polynomial $6x^3 - 19x^2 + 16x - 4$, and the long division process produced a remainder of zero. Often, long division will produce a nonzero remainder. For instance, if we divide $x^2 + 3x + 5$ by $x + 1$, we obtain the following.

$$
\begin{array}{r}
x + 2 \quad \longleftarrow \text{Quotient} \\
\text{Divisor} \longrightarrow x + 1\overline{)x^2 + 3x + 5} \quad \longleftarrow \text{Dividend} \\
\underline{x^2 + x} \\
2x + 5 \\
\underline{2x + 2} \\
3 \quad \longleftarrow \text{Remainder}
\end{array}
$$

In fractional form, we write this result as

$$
\overbrace{\underbrace{\frac{x^2 + 3x + 5}{x + 1}}_{\text{Divisor}}}^{\text{Dividend}} = \overbrace{x + 2}^{\text{Quotient}} + \overbrace{\underbrace{\frac{3}{x + 1}}_{\text{Divisor}}}^{\text{Remainder}}
$$

This example illustrates the following well-known theorem called the **Division Algorithm.**

Polynomial Functions: Graphs and Zeros

The Division Algorithm

If $f(x)$ and $d(x)$ are polynomials such that $d(x) \neq 0$, and the degree of $d(x)$ is less than or equal to the degree of $f(x)$, then there exist unique polynomials $q(x)$ and $r(x)$ such that

$$f(x) = d(x)q(x) + r(x)$$

Dividend   Divisor   Quotient   Remainder

where $r(x) = 0$ or the degree of $r(x)$ is less than the degree of $d(x)$. If the remainder $r(x)$ is zero, then we say that $d(x)$ **divides evenly** into $f(x)$.

Remark: The Division Algorithm can also be written as

$$\frac{f(x)}{d(x)} = q(x) + \frac{r(x)}{d(x)}.$$

In the Division Algorithm the rational expression $f(x)/d(x)$ is called **improper** because the degree of $f(x)$ is greater than or equal to the degree of $d(x)$. On the other hand, the rational expression $r(x)/d(x)$ is called **proper** because the degree of $r(x)$ is less than the degree of $d(x)$.

EXAMPLE 2 Long Division of Polynomials

Divide $x^3 - 1$ by $x - 1$.

SOLUTION

Because there is no x^2-term or x-term in the dividend, we line up the subtraction by using zero coefficients (or leaving a space) for the missing terms.

$$
\begin{array}{r}
x^2 + x + 1 \\
x - 1 \overline{)x^3 + 0x^2 + 0x - 1} \\
\underline{x^3 - x^2} \\
x^2 \\
\underline{x^2 - x} \\
x - 1 \\
\underline{x - 1} \\
0
\end{array}
$$

Thus, $x - 1$ divides evenly into $x^3 - 1$ and we can write

$$\frac{x^3 - 1}{x - 1} = x^2 + x + 1.$$

EXAMPLE 3 *Long Division of Polynomials*

Divide $2x^4 + 4x^3 - 5x^2 + 3x - 2$ by $x^2 + 2x - 3$.

SOLUTION

$$
\begin{array}{r}
2x^2 \qquad\quad + 1 \\
x^2 + 2x - 3\overline{)2x^4 + 4x^3 - 5x^2 + 3x - 2} \\
\underline{2x^4 + 4x^3 - 6x^2} \\
x^2 + 3x - 2 \\
\underline{x^2 + 2x - 3} \\
x + 1
\end{array}
$$

Note that the first subtraction eliminated two terms from the dividend. When this happens, the quotient skips a term. Thus, we can write

$$
\frac{2x^4 + 4x^3 - 5x^2 + 3x - 2}{x^2 + 2x - 3} = 2x^2 + 1 + \frac{x + 1}{x^2 + 2x - 3}.
$$

Synthetic Division

There is a nice shortcut for long division by polynomials of the form $x - k$. The shortcut is called **synthetic division.** To see how synthetic division works, we take another look at Example 1.

$$
\begin{array}{r}
6x^2 - 7x + 2 \\
x - 2\overline{)6x^3 - 19x^2 + 16x - 4} \\
\underline{6x^3 - 12x^2} \\
-7x^2 + 16x \\
\underline{-7x^2 + 14x} \\
2x - 4 \\
\underline{2x - 4} \\
0
\end{array}
$$

We can retain the essential steps of this division tableau by using only the coefficients, as follows.

$$
\begin{array}{rrrr}
 & 6 & -7 & 2 \\
-2\overline{)6} & -19 & 16 & -4 \\
\underline{6} & -12 & & \\
 & -7 & 16 & \\
 & \underline{-7} & 14 & \\
 & & 2 & -4 \\
 & & \underline{2} & -4 \\
 & & & 0
\end{array}
$$

Since the coefficients shown in color are duplicates of those in the quotient or the dividend, we can omit them and condense vertically.

$$
\begin{array}{r}
6 \quad\;\; -7 \quad\;\; 2 \\
-2\overline{)6 \quad -19 \quad 16 \quad -4} \\
-12 \quad 14 \quad -4 \\
\hline
0
\end{array}
$$

Now, we move the quotient to the bottom row.

$$
\begin{array}{r}
-2\overline{)6 \quad -19 \quad 16 \quad -4} \\
-12 \quad 14 \quad -4 \\
\hline
6 \quad\;\; -7 \quad\;\; 2 \quad\;\; 0
\end{array}
$$

Finally, we change from subtraction to addition (and reduce the likelihood of errors) by changing the sign of the divisor and of row two. This produces the following synthetic division array.

$$
\begin{array}{r}
2\,\lfloor 6 \quad -19 \quad 16 \quad -4 \\
12 \quad -14 \quad\;\; 4 \\
\hline
6 \quad\;\; -7 \quad\;\; 2 \quad\;\; 0
\end{array}
$$

We summarize the pattern for synthetic division of a cubic polynomial as follows. (The pattern for higher-degree polynomials is similar.)

Synthetic Division (for a Cubic Polynomial)

To divide $ax^3 + bx^2 + cx + d$ by $x - k$, use the following pattern.

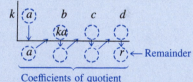

Coefficients of quotient

Vertical Pattern: Add terms.
Diagonal Pattern: Multiply by k.

Remark: Note that synthetic division works *only* for divisors of the form $x - k$. You cannot use synthetic division to divide a polynomial by a quadratic such as $x^2 - 3$.

EXAMPLE 4   *Using Synthetic Division*

Use synthetic division to divide $x^4 - 10x^2 - 2x + 4$ by $x + 3$.

SOLUTION

We can set up the array as follows. (Note that we include a zero for each missing term in the dividend.)

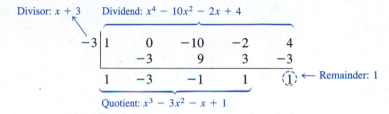

Thus, we have

$$\frac{x^4 - 10x^2 - 2x + 4}{x + 3} = x^3 - 3x^2 - x + 1 + \frac{1}{x + 3}.$$

The remainder obtained in the synthetic division process has an important interpretation, as given in the following Remainder Theorem.

The Remainder Theorem

If a polynomial $f(x)$ is divided by $x - k$, then the remainder is

$r = f(k)$.

PROOF

From the Division Algorithm, we have

$f(x) = (x - k)q(x) + r(x)$

and since either $r(x) = 0$ or the degree of $r(x)$ is less than the degree of $x - k$, we know that $r(x)$ must be a constant. That is, $r(x) = r$. Now, by evaluating $f(x)$ at $x = k$, we have

$f(k) = (k - k)q(k) + r = (0)q(k) + r = r.$

EXAMPLE 5 *Evaluating a Polynomial by the Remainder Theorem*

Use the Remainder Theorem to evaluate

$$f(x) = 3x^3 + 8x^2 + 5x - 7$$

at $x = -2$.

SOLUTION

Using synthetic division, we obtain the following.

$$
\begin{array}{r|rrrr}
-2 & 3 & 8 & 5 & -7 \\
   &   & -6 & -4 & -2 \\
\hline
   & 3 & 2 & 1 & -9
\end{array}
$$

Since the remainder is $r = -9$, we conclude that

$$f(-2) = -9.$$

This means that $(-2, -9)$ is a point on the graph of f. Try checking this by substituting $x = -2$ in the original function.

We now use the Remainder Theorem to prove the following important result.

Factor Theorem

A polynomial $f(x)$ has a factor $(x - k)$ if and only if $f(k) = 0$.

PROOF

Using the Division Algorithm with the factor $(x - k)$, we have

$$f(x) = (x - k)q(x) + r(x).$$

By the Remainder Theorem, $r(x) = r = f(k)$, and we have

$$f(x) = (x - k)q(x) + f(k)$$

where $q(x)$ is a polynomial of lesser degree than $f(x)$. If $f(k) = 0$, then

$$f(x) = (x - k)q(x)$$

and we see that $(x - k)$ is a factor of $f(x)$. Conversely, if $(x - k)$ is a factor of $f(x)$, then division of $f(x)$ by $(x - k)$ yields a remainder of 0. Hence, by the Remainder Theorem, we have $f(k) = 0$.

EXAMPLE 6   Using Synthetic Division to Find Factors of a Polynomial

Show that $(x - 2)$ and $(x + 3)$ are factors of the polynomial

$$f(x) = 2x^4 + 7x^3 - 4x^2 - 27x - 18.$$

Then find the remaining factors of $f(x)$.

SOLUTION

Using synthetic division with 2 and -3 successively, we obtain the following.

$$
\begin{array}{r|rrrrr}
2 & 2 & 7 & -4 & -27 & -18 \\
  &   & 4 & 22 & 36 & 18 \\
\hline
  & 2 & 11 & 18 & 9 & 0
\end{array}
$$

0 remainder $\Longrightarrow$ $(x - 2)$ is a factor

$$
\begin{array}{r|rrrr}
-3 & 2 & 11 & 18 & 9 \\
   &   & -6 & -15 & -9 \\
\hline
   & 2 & 5 & 3 & 0
\end{array}
$$

0 remainder $\Longrightarrow$ $(x + 3)$ is a factor

Since the resulting quadratic factors as

$$2x^2 + 5x + 3 = (2x + 3)(x + 1),$$

the complete factorization of $f(x)$ is

$$f(x) = (x - 2)(x + 3)(2x + 3)(x + 1).$$

Horner's Method

Synthetic division gives us a method for evaluating polynomials that is very useful with a calculator. Reconsider the polynomial function in Example 5,

$$f(x) = 3x^3 + 8x^2 + 5x - 7.$$

Synthetic division by $(x - k)$ yields the following.

$$
\begin{array}{r|cccc}
k & 3 & 8 & 5 & -7 \\
  &   & 3k & (3k + 8)k & [(3k + 8)k + 5]k \\
\hline
  & 3 & 3k + 8 & (3k + 8)k + 5 & [(3k + 8)k + 5]k - 7
\end{array}
$$

Hence, by the Remainder Theorem, we know that

$$f(k) = [(3k + 8)k + 5]k - 7.$$

In terms of x, we can now write

$$3x^3 + 8x^2 + 5x - 7 = [(3x + 8)x + 5]x - 7.$$

We call this **Horner's Method** of writing a polynomial. It can be applied to any polynomial by successively factoring out x from each nonconstant term, as demonstrated in the following example.

EXAMPLE 7 *Horner's Method*

Use Horner's Method to rewrite the polynomial function

$$f(x) = 5x^4 - 3x^3 + x^2 - 8x + 7.$$

SOLUTION

$$
\begin{aligned}
f(x) &= 5x^4 - 3x^3 + x^2 - 8x + 7 \\
     &= (5x^3 - 3x^2 + x - 8)x + 7 && \textit{Factor x from first four terms} \\
     &= [(5x^2 - 3x + 1)x - 8]x + 7 && \textit{Factor x from first three terms} \\
     &= ([(5x - 3)x + 1]x - 8)x + 7 && \textit{Factor x from first two terms}
\end{aligned}
$$

Notice how easily we can evaluate $f(k)$ for the polynomial function in Example 7 by entering the following calculator steps.

$$5 \; \boxed{\times} \; k \; \boxed{-} \; 3 \; \boxed{=}$$

$$\boxed{\times} \; k \; \boxed{+} \; 1 \; \boxed{=}$$

$$\boxed{\times} \; k \; \boxed{-} \; 8 \; \boxed{=}$$

$$\boxed{\times} \; k \; \boxed{+} \; 7 \; \boxed{=}$$

If k is a large number or a decimal value, we could save time by storing k and then using the $\boxed{\text{RCL}}$ key each time k is needed.

Evaluating a Polynomial with a Calculator

To evaluate the polynomial

$$f(x) = a_n x^n + a_{n-1} x^{n-1} + \cdots + a_1 x + a_0$$

with a calculator (having algebraic logic), use the following key sequence.

$$x \; \boxed{\text{STO}} \; a_n \; \boxed{\times} \; \boxed{\text{RCL}} \; \boxed{+} \; a_{n-1} \; \boxed{=}$$

$$\boxed{\times} \; \boxed{\text{RCL}} \; \boxed{+} \; a_{n-2} \; \boxed{=}$$

$$\vdots \qquad\qquad \vdots$$

$$\boxed{\times} \; \boxed{\text{RCL}} \; \boxed{+} \; a_1 \; \boxed{=}$$

$$\boxed{\times} \; \boxed{\text{RCL}} \; \boxed{+} \; a_0 \; \boxed{=}$$

If $a_i = 0$, then omit $\boxed{+}$ a_i, and if $a_i < 0$, use $\boxed{-}$ instead of $\boxed{+}$.

Remark: Note how the calculator routine *repeats* the five-stroke sequence $\boxed{\times} \; \boxed{\text{RCL}} \; \boxed{+} \; a_i \; \boxed{=}$.

WARM UP

Write the expressions in standard polynomial form.

1. $(x - 1)(x^2 + 2) + 5$

2. $(x^2 - 3)(2x + 4) + 8$

3. $(x^2 + 1)(x^2 - 2x + 3) - 10$

4. $(x + 6)(2x^3 - 3x) - 5$

Factor the polynomials.

5. $x^2 - 4x + 3$

6. $4x^3 - 10x^2 + 6x$

Find a polynomial function that has the given zeros.

7. $0, 3, 4$

8. $-6, 1$

9. $-3, 1 + \sqrt{2}, 1 - \sqrt{2}$

10. $1, -2, 2 + \sqrt{3}, 2 - \sqrt{3}$

EXERCISES 4.3

In Exercises 1–14, divide by long division.

	Dividend	Divisor
1.	$2x^2 + 10x + 12$	$x + 3$
2.	$5x^2 - 17x - 12$	$x - 4$
3.	$4x^3 - 7x^2 - 11x + 5$	$4x + 5$
4.	$6x^3 - 16x^2 + 17x - 6$	$3x - 2$
5.	$x^4 + 5x^3 + 6x^2 - x - 2$	$x + 2$
6.	$x^3 + 4x^2 - 3x - 12$	$x^2 - 3$
7.	$7x + 3$	$x + 2$
8.	$8x - 5$	$2x + 1$
9.	$6x^3 + 10x^2 + x + 8$	$2x^2 + 1$
10.	$x^3 - 9$	$x^2 + 1$
11.	$x^4 + 3x^2 + 1$	$x^2 - 2x + 3$
12.	$x^5 + 7$	$x^3 - 1$
13.	$2x^3 - 4x^2 - 15x + 5$	$(x - 1)^2$
14.	x^4	$(x - 1)^3$

	Dividend	Divisor
19.	$-x^3 + 75x - 250$	$x + 10$
20.	$3x^3 - 16x^2 - 72$	$x - 6$
21.	$5x^3 - 6x^2 + 8$	$x - 4$
22.	$5x^3 + 6x + 8$	$x + 2$
23.	$10x^4 - 50x^3 - 800$	$x - 6$
24.	$x^5 - 13x^4 - 120x + 80$	$x + 3$
25.	$x^3 + 512$	$x + 8$
26.	$5x^3$	$x + 3$
27.	$-3x^4$	$x - 2$
28.	$-3x^4$	$x + 2$
29.	$5 - 3x + 2x^2 - x^3$	$x + 1$
30.	$180x - x^4$	$x - 6$
31.	$4x^3 + 16x^2 - 23x - 15$	$x + \frac{1}{2}$
32.	$3x^3 - 4x^2 + 5$	$x - \frac{3}{2}$

In Exercises 15–32, divide by synthetic division.

	Dividend	Divisor
15.	$3x^3 - 17x^2 + 15x - 25$	$x - 5$
16.	$5x^3 + 18x^2 + 7x - 6$	$x + 3$
17.	$4x^3 - 9x + 8x^2 - 18$	$x + 2$
18.	$9x^3 - 16x - 18x^2 + 32$	$x - 2$

In Exercises 33–40, use synthetic division to show that x is a solution of the third-degree polynomial equation, and use the result to factor the polynomial completely.

33. $x^3 - 7x + 6 = 0, \quad x = 2$

34. $x^3 - 28x - 48 = 0, \quad x = -4$

35. $2x^3 - 15x^2 + 27x - 10 = 0, \quad x = \frac{1}{2}$

36. $48x^3 - 80x^2 + 41x - 6 = 0, \quad x = \frac{2}{3}$

37. $x^3 - 3x^2 + 2 = 0, \quad x = 1 + \sqrt{3}$

38. $x^3 - x^2 - 13x - 3 = 0$, $x = 2 - \sqrt{5}$

39. $x^3 + 3x^2 - 3x - 1$, $x = -2 - \sqrt{3}$

40. $x^3 + 2x^2 - 2x - 4$, $x = \sqrt{2}$

In Exercises 41–44, express the function in the form $f(x) = (x - k)q(x) + r$ for the given value of k, and demonstrate that $f(k) = r$.

41. $f(x) = x^3 - x^2 - 14x + 11$, $k = 4$

42. $f(x) = \frac{1}{3}(15x^4 + 10x^3 - 6x^2 + 17x + 14)$, $k = -\frac{2}{3}$

43. $f(x) = x^3 + 3x^2 - 2x - 14$, $k = \sqrt{2}$

44. $f(x) = 4x^3 - 6x^2 - 12x - 4$, $k = 1 - \sqrt{3}$

In Exercises 45–50, use synthetic division to find the required function values.

45. $f(x) = 4x^3 - 13x + 10$
  (a) $f(1)$ (b) $f(-2)$
  (c) $f\left(\frac{1}{2}\right)$ (d) $f(8)$

46. $g(x) = x^6 - 4x^4 + 3x^2 + 2$
  (a) $g(2)$ (b) $g(-4)$
  (c) $g(3)$ (d) $g(-1)$

47. $h(x) = 3x^3 + 5x^2 - 10x + 1$
  (a) $h(3)$ (b) $h\left(\frac{1}{3}\right)$
  (c) $h(-2)$ (d) $h(-5)$

48. $f(x) = 0.4x^4 - 1.6x^3 + 0.7x^2 - 2$
  (a) $f(1)$ (b) $f(-2)$
  (c) $f(5)$ (d) $f(-10)$

49. $f(x) = x^3 - 2x^2 - 11x + 52$
  (a) $f(5)$ (b) $f(-4)$
  (c) $f(1.2)$ (d) $f(2)$

50. $g(x) = x^3 - x^2 + 25x - 25$
  (a) $g(5)$ (b) $g\left(\frac{1}{5}\right)$
  (c) $g(-1.5)$ (d) $g(-1)$

In Exercises 51–54, use Horner's Method to find the required function values.

51. $f(x) = x^3 - 6x^2 + 12x - 8$
  (a) $f(5)$ (b) $f(-4.5)$

52. $f(x) = 3x^4 + 6x^3 - 10x^2 - 7x + 2$
  (a) $f(-4.8)$ (b) $f(0.02)$

53. $f(x) = -5x^4 + 8.5x^3 + 10x - 3$
  (a) $f(1.08)$ (b) $f(-5.4)$

54. $f(x) = -2x^5 + 4x^3 - 6x^2 + 10$
  (a) $f(4)$ (b) $f(-3.7)$

4.4 Real Zeros of Polynomial Functions

In this and the following section we will present some additional aids for finding zeros of polynomial functions.

Descartes's Rule of Signs

In Section 4.2, we noted that an *n*th degree polynomial function can have, *at most*, *n* real zeros. Of course, many *n*th degree polynomials do not have that many real zeros. For instance, $f(x) = x^2 + 1$ has no real zeros, and $f(x) = x^3 + 1$ has only one real zero. The following theorem, called **Descartes's Rule of Signs,** sheds more light on the number of real zeros a polynomial can have.

Descartes's Rule of Signs

Let $f(x) = a_n x^n + a_{n-1} x^{n-1} + \cdots + a_2 x^2 + a_1 x + a_0$ be a polynomial with real coefficients and $a_0 \neq 0$.

1. The number of *positive real zeros* of f is either equal to the number of variations in sign of $f(x)$ or is less than that number by an even integer.
2. The number of *negative real zeros* of f is either equal to the number of variations in sign of $f(-x)$ or is less than that number by an even integer.

Remark: Note that when there is only one variation in sign, Descartes's Rule of Signs guarantees the existence of exactly one positive (or negative) real zero.

By a **variation in sign,** we mean that two consecutive coefficients have opposite signs. For example, the polynomial

$$
\begin{array}{ccc}
{\scriptstyle +\,\text{to}\,-} & {\scriptstyle +\,\text{to}\,-} & \\
f(x) = 3x^3 - 5x^2 + 6x - 4 \\
{\scriptstyle -\,\text{to}\,+}
\end{array}
$$

has *three* variations in sign, whereas

$$
\begin{aligned}
f(-x) &= 3(-x)^3 - 5(-x)^2 + 6(-x) - 4 \\
&= -3x^3 - 5x^2 - 6x - 4
\end{aligned}
$$

has no variations in sign.

EXAMPLE 1 Using Descartes's Rule of Signs

Apply Descartes's Rule of Signs to

$$f(x) = 2x^4 + 7x^3 - 4x^2 - 27x - 18.$$

SOLUTION

Because $f(x)$ has only *one* variation in sign, we know that f must have *exactly one* positive real zero. Moreover, because

$$
\begin{aligned}
f(-x) &= 2(-x)^4 + 7(-x)^3 - 4(-x)^2 - 27(-x) - 18 \\
&= 2x^4 - 7x^3 - 4x^2 + 27x - 18
\end{aligned}
$$

has *three* variations in sign, f has either three negative zeros or one negative zero. This result agrees with Example 6 in Section 4.3, where the zeros were determined to be 2, -3, $-3/2$, and -1.

EXAMPLE 2 *Using Descartes's Rule of Signs*

Apply Descartes's Rule of Signs to

$$f(x) = x^3 - x + 1.$$

SOLUTION

Because $f(x)$ has two variations in sign, it follows that f can have either two or no positive real zeros. Moreover, because

$$f(-x) = -x^3 + x + 1$$

has one variation in sign, we know that f has exactly one negative real zero.

The Rational Zero Test

The second test presented in this section is called the **Rational Zero Test.** This test relates the possible rational zeros of a polynomial (having integer coefficients) to the leading coefficient and to the constant term of the polynomial.

The Rational Zero Test

If the polynomial

$$f(x) = a_n x^n + a_{n-1} x^{n-1} + \cdots + a_2 x^2 + a_1 x + a_0$$

has *integer* coefficients, then every rational zero of f has the form

$$\text{rational zero} = \frac{p}{q}$$

where p and q have no common factors other than 1, and

p = a factor of the constant term a_0
q = a factor of the leading coefficient a_n.

PROOF

We begin by assuming that p/q is a rational zero of f and that p/q is written in reduced form (that is, p and q have no common factors other than 1). Since $f(p/q) = 0$, we have

$$a_n\left(\frac{p}{q}\right)^n + a_{n-1}\left(\frac{p}{q}\right)^{n-1} + \cdots + a_2\left(\frac{p}{q}\right)^2 + a_1\left(\frac{p}{q}\right) + a_0 = 0$$

$$a_n p^n + a_{n-1} p^{n-1} q + \cdots + a_2 p^2 q^{n-2} + a_1 p q^{n-1} + a_0 q^n = 0.$$

Now we rewrite this equation in two convenient forms.

$$p(a_n p^{n-1} + a_{n-1} p^{n-2} q + \cdots + a_2 pq^{n-2} + a_1 q^{n-1}) = -a_0 q^n$$
$$q(a_{n-1} p^{n-1} + \cdots + a_2 p^2 q^{n-3} + a_1 pq^{n-2} + a_0 q^{n-1}) = -a_n p^n$$

From the first equation we see that p is a factor of $a_0 q^n$. However, since p and q have no factors in common, it follows that p must be a factor of a_0. Similarly, from the second equation we see that q is a factor of $a_n p^n$, and again, since p and q have no factors in common, it follows that q must be a factor of a_n.

To use the Rational Zero Test, we first list all rational numbers whose numerators are factors of the constant term and whose denominators are factors of the leading coefficient.

$$\text{Possible rational zeros} = \frac{\text{factors of constant term}}{\text{factors of leading coefficient}}$$

Having formed this list of *possible rational zeros*, we use a trial and error method to determine which, if any, are actual zeros of the polynomial. Note that when the leading coefficient is 1, then the possible rational zeros are simply the factors of the constant term. This case is illustrated in Examples 3 and 4.

EXAMPLE 3   Rational Zero Test with Leading Coefficient of 1

Find the rational zeros of $f(x) = x^3 + x + 1$.

SOLUTION

Note that ± 1 are the only factors of the leading coefficient.

$$\textcircled{1}x^3 + x + \textcircled{1}$$

Factors of constant term: ± 1
Factors of leading coefficient: ± 1

Hence, the possible rational zeros are simply the factors of the constant term.

Possible rational zeros: ± 1

By testing these possible zeros, we see that neither works.

$$f(1) = (1)^3 + 1 + 1 = 3$$
$$f(-1) = (-1)^3 + (-1) + 1 = -1$$

We conclude that the given polynomial has *no* rational zeros.

EXAMPLE 4 *Rational Zero Test with Leading Coefficient of 1*

Find the rational zeros of

$$f(x) = x^4 - x^3 + x^2 - 3x - 6.$$

SOLUTION

For $f(x) = x^4 - x^3 + x^2 - 3x - 6$, the leading coefficient is 1; hence, the possible rational zeros are the factors of the constant term.

Possible rational zeros: $\pm 1, \pm 2, \pm 3, \pm 6$

A test of these possible zeros would show that $x = -1$ and $x = 2$ are the only two that work. Check the others to be sure.

If the leading coefficient of a polynomial is not 1, the list of possible rational zeros could increase dramatically. In such cases the search can be shortened in several ways. A programmable calculator can be used to speed up the calculations. For some functions a rough sketch may give a good estimate of the location of the zeros. A third option is to use synthetic division. For instance, in Example 4 we can *verify* that $x = -1$ and $x = 2$ are zeros of

$$f(x) = x^4 - x^3 + x^2 - 3x - 6$$

in the following way.

$$
\begin{array}{r|rrrrr}
-1 & 1 & -1 & 1 & -3 & -6 \\
   &   & -1 & 2 & -3 & 6 \\
\hline
   & 1 & -2 & 3 & -6 & 0
\end{array}
$$

$$
\begin{array}{r|rrrr}
2 & 1 & -2 & 3 & -6 \\
  &   & 2 & 0 & 6 \\
\hline
  & 1 & 0 & 3 & 0
\end{array}
$$

Thus, we have

$$f(x) = (x + 1)(x - 2)(x^2 + 3).$$

Since the factor $(x^2 + 3)$ has no real zeros, we conclude that $x = -1$ and $x = 2$ are the *only* real zeros of $f(x)$.

Finding the first zero is often the hardest part. After that the search is simplified by using the lower-degree polynomial obtained in synthetic division. This is demonstrated in Example 5.

Real Zeros of Polynomial Functions

EXAMPLE 5 *Using the Rational Zero Test*

Find the rational zeros of $f(x) = 2x^3 + 3x^2 - 8x + 3$.

SOLUTION

Since the leading coefficient is 2 and the constant term is 3, we have

Possible rational zeros:
$$\frac{\text{factors of } 3}{\text{factors of } 2} = \frac{\pm 1, \pm 3}{\pm 1, \pm 2} = \pm 1, \pm 3, \pm \frac{1}{2}, \pm \frac{3}{2}$$

By synthetic division, we determine that $x = 1$ is a zero.

$$
\begin{array}{r|rrrr}
1 & 2 & 3 & -8 & 3 \\
  &   & 2 & 5 & -3 \\
\hline
  & 2 & 5 & -3 & 0
\end{array}
$$

Thus, $f(x)$ factors as

$$f(x) = (x - 1)(2x^2 + 5x - 3)$$
$$= (x - 1)(2x - 1)(x + 3)$$

and we conclude that the zeros of f are $x = 1$, $1/2$, and -3.

In the next example, we make a sketch to aid our search for the zeros of a polynomial function.

EXAMPLE 6 *Using the Rational Zero Test*

Find all the zeros of $f(x) = 10x^3 - 15x^2 - 16x + 12$.

SOLUTION

Since the leading coefficient is 10 and the constant term is 12, we have a long list of possible rational zeros.

Possible rational zeros:
$$\frac{\text{factors of } 12}{\text{factors of } 10} = \frac{\pm 1, \pm 2, \pm 3, \pm 4, \pm 6, \pm 12}{\pm 1, \pm 2, \pm 5, \pm 10}$$

With so many possibilities (32, in fact), it is worth our time to stop and make a rough sketch of this function. From Figure 4.30, it looks like three reasonable choices would be $x = -6/5$, $x = 1/2$, and $x = 2$. Testing these by synthetic division shows that only $x = 2$ works.

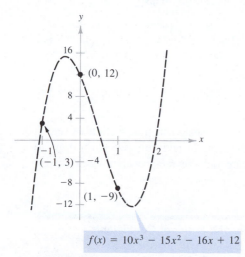

$f(x) = 10x^3 - 15x^2 - 16x + 12$

FIGURE 4.30

$$\begin{array}{r|rrrr} 2 & 10 & -15 & -16 & 12 \\ & & 20 & 10 & -12 \\ \hline & 10 & 5 & -6 & 0 \end{array}$$

Thus, we have

$$f(x) = (x - 2)(10x^2 + 5x - 6).$$

Using the Quadratic Formula, we find that the two additional zeros are irrational numbers.

$$x = \frac{-5 + \sqrt{265}}{20} \approx 0.5639 \qquad x = \frac{-5 - \sqrt{265}}{20} \approx -1.0639$$

Bounds for Real Zeros of Polynomials

The third test for zeros of a polynomial function is related to the sign pattern in the last row of the synthetic division algorithm. This test can give us an upper or a lower bound of the real zeros of f. A real number b is an **upper bound** for the real zeros of f if no zeros are greater than b. Similarly, b is a **lower bound** if no real zeros of f are less than b.

Lower and Upper Bound Rule

Let $f(x)$ be a polynomial with real coefficients and a positive leading coefficient. Suppose $f(x)$ is divided by $x - c$, using synthetic division.

1. If $c > 0$ and each number in the last row is either positive or zero, then c is an *upper bound* for the real zeros of $f(x)$.
2. If $c < 0$ and the numbers in the last row are alternately positive and negative (zero entries count as positive or negative), then c is a *lower bound* for the real zeros of $f(x)$.

In Example 7 we use all three tests presented in this section to search for the real zeros of a polynomial function. In addition, we show how to handle rational coefficients by factoring out the reciprocal of their lowest common denominator (LCD).

EXAMPLE 7 A Polynomial Function with Rational Coefficients

Find the real zeros of

$$f(x) = x^3 - \frac{2}{3}x^2 + \frac{1}{2}x - \frac{1}{3}.$$

Real Zeros of Polynomial Functions

SOLUTION

To find the rational zeros, we rewrite $f(x)$ by factoring out the fraction 1/LCD. Since the LCD of the coefficients is 6, we obtain

$$f(x) = \frac{6}{6}x^3 - \frac{4}{6}x^2 + \frac{3}{6}x - \frac{2}{6} = \frac{1}{6}(6x^3 - 4x^2 + 3x - 2).$$

Now, the zeros of f will coincide with those of

$$g(x) = 6x^3 - 4x^2 + 3x - 2$$

which has the following possible rational zeros.

Possible rational zeros:

$$\frac{\text{factors of 2}}{\text{factors of 6}} = \frac{\pm 1, \ \pm 2}{\pm 1, \ \pm 2, \ \pm 3, \ \pm 6}$$

$$= \pm 1, \ \pm\frac{1}{2}, \ \pm\frac{1}{3}, \ \pm\frac{1}{6}, \ \pm\frac{2}{3}, \ \pm 2$$

Since $f(x)$ has three variations in sign and $f(-x)$ has none, we conclude by Descartes's Rule of Signs that there are three or one positive real zeros and no negative zeros. Trying $x = 1$, we obtain the following.

$$\begin{array}{r|rrrr}
1 & 6 & -4 & 3 & -2 \\
  &   & 6 & 2 & 5 \\
\hline
  & 6 & 2 & 5 & 3
\end{array}$$

Since the last row has all positive entries, we know that $x = 1$ is an upper bound for the real zeros. Thus, we restrict our search to zeros between 0 and 1. Choosing $x = 2/3$, we obtain the following.

$$\begin{array}{r|rrrr}
\frac{2}{3} & 6 & -4 & 3 & -2 \\
  &   & 4 & 0 & 2 \\
\hline
  & 6 & 0 & 3 & 0
\end{array}$$

Thus, $f(x)$ factors as

$$f(x) = \frac{1}{6}\left(x - \frac{2}{3}\right)(6x^2 + 3)$$

$$= \frac{1}{6}\left(\frac{1}{3}\right)(3x - 2)(3)(2x^2 + 1)$$

$$= \frac{1}{6}(3x - 2)(2x^2 + 1).$$

Since $2x^2 + 1$ has no real zeros, we conclude that $x = 2/3$ is the only real zero of $f(x)$.

Polynomial Functions: Graphs and Zeros

Before concluding this section, we list two additional hints that should be helpful.

1. If the terms of $f(x)$ have a common monomial factor, it should be factored out before applying the tests in this section. For instance, by writing

$$f(x) = x^4 - 5x^3 + 3x^2 + x$$
$$= x(x^3 - 5x^2 + 3x + 1)$$

we see that $x = 0$ is a zero of f and the remaining zeros can be obtained by analyzing the cubic factor.

2. If you are able to find all but two zeros of $f(x)$, then you are home free because you can always use the Quadratic Formula on the remaining quadratic factor. For instance, if you have succeeded in writing

$$f(x) = x^4 - 5x^3 + 3x^2 + x$$
$$= x(x - 1)(x^2 - 4x - 1)$$

then you can apply the Quadratic Formula to $x^2 - 4x - 1$ to find the two remaining zeros.

WARM UP

Find a polynomial function with integer coefficients having the given zeros.

1. $-1, \frac{2}{3}, 3$

2. $-2, 0, \frac{3}{4}, 2$

Divide by synthetic division.

3. $\dfrac{x^5 - 9x^3 + 5x + 18}{x + 3}$

4. $\dfrac{3x^4 + 17x^3 + 10x^2 - 9x - 8}{x + (2/3)}$

Use the given zero to find all the real zeros of f.

5. $f(x) = 2x^3 + 11x^2 + 2x - 4, \quad x = \frac{1}{2}$

6. $f(x) = 6x^3 - 47x^2 - 124x - 60, \quad x = 10$

7. $f(x) = 4x^3 - 13x^2 - 4x + 6, \quad x = -\frac{3}{4}$

8. $f(x) = 10x^3 + 51x^2 + 48x - 28, \quad x = \frac{2}{5}$

Find all real solutions.

9. $x^4 - 3x^2 + 2 = 0$

10. $x^4 - 7x^2 + 12 = 0$

EXERCISES 4.4

In Exercises 1–10, use Descartes's Rule of Signs to determine the possible number of positive and negative zeros of the function.

1. $f(x) = x^3 + 3$
2. $g(x) = x^3 + 3x^2$
3. $h(x) = 3x^4 + 2x^2 + 1$
4. $h(x) = 2x^4 - 3x + 2$
5. $g(x) = 2x^3 - 3x^2 - 3$
6. $f(x) = 4x^3 - 3x^2 + 2x - 1$
7. $f(x) = -5x^3 + x^2 - x + 5$
8. $g(x) = 5x^5 + 10x$
9. $h(x) = 4x^2 - 8x + 3$
10. $f(x) = 3x^3 + 2x^2 + x + 3$

In Exercises 11–16, use the Rational Zero Test to list all possible rational zeros of f and verify that the zeros of f shown on the graph are contained in the list.

11. $f(x) = x^3 + x^2 - 4x - 4$

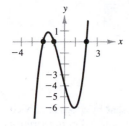

12. $f(x) = -3x^3 + 20x^2 - 36x + 16$

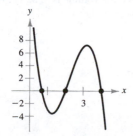

13. $f(x) = -4x^3 + 15x^2 - 8x - 3$

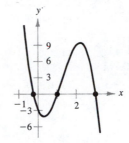

14. $f(x) = 4x^3 - 12x^2 - x + 15$

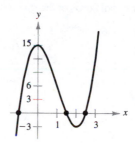

15. $f(x) = -2x^4 + 13x^3 - 21x^2 + 2x + 8$

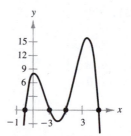

16. $f(x) = 4x^4 - 17x^2 + 4$

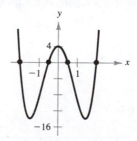

In Exercises 17–20, use synthetic division to determine if the given x-value is an upper bound of the zeros of f, a lower bound of the zeros of f, or neither.

17. $f(x) = x^4 - 4x^3 + 15$
(a) $x = 4$ (b) $x = -1$ (c) $x = 3$

18. $f(x) = 2x^3 - 3x^2 - 12x + 8$
(a) $x = 2$ (b) $x = 4$ (c) $x = -1$

19. $f(x) = x^4 - 4x^3 + 16x - 16$
(a) $x = -1$ (b) $x = -3$ (c) $x = 5$

20. $f(x) = 2x^4 - 8x + 3$
(a) $x = 1$ (b) $x = 3$ (c) $x = -4$

In Exercises 21–36, find the real zeros of the given function.

21. $f(x) = x^3 - 6x^2 + 11x - 6$

22. $f(x) = x^3 - 7x - 6$

23. $g(x) = x^3 - 4x^2 - x + 4$

24. $h(x) = x^3 - 9x^2 + 20x - 12$

25. $h(t) = t^3 + 12t^2 + 21t + 10$

26. $f(x) = x^3 + 6x^2 + 12x + 8$

27. $f(x) = x^3 - 4x^2 + 5x - 2$

28. $p(x) = x^3 - 9x^2 + 27x - 27$

29. $C(x) = 2x^3 + 3x^2 - 1$

30. $f(x) = 3x^3 - 19x^2 + 33x - 9$

31. $f(x) = 4x^3 - 3x - 1$

32. $f(z) = 12z^3 - 4z^2 - 27z + 9$

33. $f(y) = 4y^3 + 3y^2 + 8y + 6$

34. $g(x) = 3x^3 - 2x^2 + 15x - 10$

35. $f(x) = x^4 - 3x^2 + 2$

36. $P(t) = t^4 - 7t^2 + 12$

In Exercises 37–44, find all real solutions of the given polynomial equation.

37. $z^4 - z^3 - 2z - 4 = 0$

38. $x^4 - x^3 - 29x^2 - x - 30 = 0$

39. $x^4 - 13x^2 - 12x = 0$

40. $2y^4 + 7y^3 - 26y^2 + 23y - 6 = 0$

41. $2x^4 - 11x^3 - 6x^2 + 64x + 32 = 0$

42. $x^5 - x^4 - 3x^3 + 5x^2 - 2x = 0$

43. $x^5 - 7x^4 + 10x^3 + 14x^2 - 24x = 0$

44. $6x^4 - 11x^3 - 51x^2 + 99x - 27 = 0$

In Exercises 45–48, (a) list the possible rational zeros of f, (b) sketch the graph of f so that some of the possible zeros in part (a) can be disregarded, and then (c) determine all real zeros of f.

45. $f(x) = 32x^3 - 52x^2 + 17x + 3$

46. $f(x) = 6x^3 - x^2 - 13x + 8$

47. $f(x) = 4x^3 + 7x^2 - 11x - 18$

48. $f(x) = 2x^3 + 5x^2 - 21x - 10$

In Exercises 49–52, find the rational zeros of the polynomial function.

49. $P(x) = x^4 - \frac{25}{4}x^2 + 9$

50. $f(x) = x^3 - \frac{3}{2}x^2 - \frac{23}{2}x + 6$

51. $f(x) = x^3 - \frac{1}{4}x^2 - x + \frac{1}{4}$

52. $f(z) = z^3 + \frac{11}{6}z^2 - \frac{1}{2}z - \frac{1}{3}$

53. An open box is to be made from a rectangular piece of material, 9 inches by 5 inches, by cutting equal squares from each corner and turning up the sides (see figure). Find the dimensions of the box, given that the volume is to be 18 cubic inches.

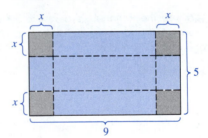

FIGURE FOR 53

54. A rectangular package to be sent by a postal service can have a maximum combined length and girth (perimeter of a cross section) of 108 inches (see figure). Find the dimensions of the package, given that the volume is to be 11,664 cubic inches. [*Hint:* $V = 108x^2 - 4x^3$.]

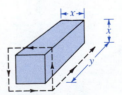

FIGURE FOR 54

4.5 *Complex Zeros and the Fundamental Theorem of Algebra*

We have been using the fact that an nth degree polynomial function can have, at most, n real zeros. In this section we improve upon that result and show that, in the complex number system, every nth degree polynomial function has *precisely n zeros.* This important result is derived from the **Fundamental Theorem of Algebra,** first proved by the famous German mathematician Carl Friedrich Gauss, 1777–1855.

The Fundamental Theorem of Algebra

If f is a polynomial function of degree $n > 0$, then f has at least one zero in the complex number system.

Using the Fundamental Theorem and the equivalence of zeros and factors, we can prove the following theorem.

Linear Factorization Theorem

If $f(x)$ is a polynomial of degree $n > 0$, then $f(x)$ has precisely n linear factors

$$f(x) = a(x - c_1)(x - c_2) \cdots (x - c_n)$$

where $c_1, c_2, \ldots, c_n$ are complex numbers and a is the leading coefficient of $f(x)$.

PROOF

Using the Fundamental Theorem, we know that f must have at least one zero c_1. Consequently, $(x - c_1)$ is a factor of $f(x)$, and we have

$$f(x) = (x - c_1)f_1(x).$$

If the degree of $f_1(x)$ is greater than zero, we again apply the Fundamental Theorem to conclude that f_1 must have a zero c_2, which implies that

$$f(x) = (x - c_1)(x - c_2)f_2(x).$$

It is clear that the degree of $f_1(x)$ is $n - 1$, that the degree of $f_2(x)$ is $n - 2$, and that we can repeatedly apply the Fundamental Theorem n times until we obtain

$$f(x) = a(x - c_1)(x - c_2) \cdots (x - c_n)$$

where a is the leading coefficient of the polynomial $f(x)$.

Polynomial Functions: Graphs and Zeros

Note that neither the Fundamental Theorem nor the Linear Factorization Theorem tells us how to find the zeros or factors of a polynomial. We call such theorems **existence theorems.** To find the zeros of a polynomial function, we still rely on the techniques developed in earlier sections.

Remember that the n zeros of a polynomial function can be real or complex, and they may be repeated. We illustrate some of these cases in the following example.

EXAMPLE 1 *Zeros of Polynomial Functions*

(a) The first-degree function

$$f(x) = x - 2$$

has exactly *one* zero: $x = 2$.

(b) Counting multiplicity, the second-degree function

$$f(x) = x^2 - 6x + 9 = (x - 3)(x - 3)$$

has exactly *two* zeros: $x = 3$ and $x = 3$.

(c) The third-degree function

$$f(x) = x^3 + 4x = x(x - 2i)(x + 2i)$$

has exactly *three* zeros: $x = 0$, $x = -2i$, and $x = 2i$.

(d) The fourth-degree function

$$f(x) = x^4 - 1 = (x - 1)(x + 1)(x - i)(x + i)$$

has exactly *four* zeros: $x = -1$, $x = 1$, $x = -i$, and $x = i$.

Example 2 shows how we can use the methods of the previous sections (Descartes's Rule of Signs, Rational Zero Test, synthetic division, and factoring) to find all the zeros of a polynomial function, including the complex zeros.

EXAMPLE 2 *Finding the Zeros of a Polynomial Function*

Write the polynomial function

$$f(x) = x^5 + x^3 + 2x^2 - 12x + 8$$

as the product of linear factors and list all of its zeros.

SOLUTION

Descartes's Rule of Signs indicates two or no positive real zeros and one negative real zero. Moreover, the possible rational zeros are ± 1, ± 2, ± 4, and ± 8. Synthetic division produces the following.

$$
\begin{array}{r|rrrrrr}
1 & 1 & 0 & 1 & 2 & -12 & 8 \\
  &   & 1 & 1 & 2 & 4 & -8 \\
\hline
  & 1 & 1 & 2 & 4 & -8 & 0
\end{array}
\quad\Longrightarrow\quad \textit{1 is a zero}
$$

$$
\begin{array}{r|rrrrr}
1 & 1 & 1 & 2 & 4 & -8 \\
  &   & 1 & 2 & 4 & 8 \\
\hline
  & 1 & 2 & 4 & 8 & 0
\end{array}
\quad\Longrightarrow\quad \textit{1 is a repeated zero}
$$

$$
\begin{array}{r|rrrr}
-2 & 1 & 2 & 4 & 8 \\
   &   & -2 & 0 & -8 \\
\hline
   & 1 & 0 & 4 & 0
\end{array}
\quad\Longrightarrow\quad \textit{-2 is a zero}
$$

Thus, we have

$$
\begin{aligned}
f(x) &= x^5 + x^3 + 2x^2 - 12x + 8 \\
     &= (x - 1)(x - 1)(x + 2)(x^2 + 4).
\end{aligned}
$$

Finally, by factoring $x^2 + 4$, we have

$$
f(x) = (x - 1)(x - 1)(x + 2)(x - 2i)(x + 2i)
$$

which gives the five zeros

$$
x = 1, \quad 1, \quad -2, \quad 2i, \quad \text{and} \quad -2i.
$$

In Example 2, note that the two complex zeros are **conjugates.** That is, they are of the form

$$
a + bi \qquad \text{and} \qquad a - bi.
$$

This is not a coincidence, as the following theorem indicates.

Complex Zeros Occur in Conjugate Pairs

If $a + bi$ ($b \neq 0$) is a zero of a polynomial function, with real coefficients, then the conjugate $a - bi$ is also a zero of the function.

Remark: Be sure you see that this result is only true if the polynomial function has *real* coefficients. For instance, the result applies to the function $f(x) = x^2 + 1$, but not to the function $g(x) = x - i$.

EXAMPLE 3 *Finding a Polynomial with Given Zeros*

Find a *fourth-degree* polynomial function, with real coefficients, that has -1, -1, and $3i$ as zeros.

SOLUTION

Since $3i$ is a zero, we know that $-3i$ is also a zero. Thus, by the Linear Factorization Theorem, $f(x)$ can be written as

$$f(x) = a(x + 1)(x + 1)(x - 3i)(x + 3i).$$

For simplicity, we let $a = 1$, and we obtain

$$f(x) = (x^2 + 2x + 1)(x^2 + 9)$$
$$= x^4 + 2x^3 + 10x^2 + 18x + 9.$$

EXAMPLE 4 *Finding a Polynomial with Given Zeros*

Find a *cubic* polynomial function f, with real coefficients, that has 2 and $1 - i$ as zeros, and such that $f(1) = 3$.

SOLUTION

Because $1 - i$ is a zero of f, so is $1 + i$. Therefore, we have

$$f(x) = a(x - 2)[x - (1 - i)][x - (1 + i)]$$
$$= a(x - 2)[x^2 - x(1 - i) - x(1 + i) + 1 - i^2]$$
$$= a(x - 2)(x^2 - 2x + 2)$$
$$= a(x^3 - 4x^2 + 6x - 4).$$

To find the value of a, we use the fact that $f(1) = 3$ and obtain $f(1) = a(1 - 4 + 6 - 4) = 3$. Thus, $a = -3$ and we conclude that

$$f(x) = -3(x^3 - 4x^2 + 6x - 4)$$
$$= -3x^3 + 12x^2 - 18x + 12.$$

The Linear Factorization Theorem tells us that we can write any nth degree polynomial as the product of n linear factors

$$f(x) = a(x - c_1)(x - c_2)(x - c_3) \cdots (x - c_n).$$

However, this result includes the possibility that some of the values of c_i are complex. The following result tells us that even if we do not want to get involved with "complex factors," we can still write $f(x)$ as the product of linear and/or quadratic factors.

Factors of a Polynomial

Every polynomial of degree $n > 0$ with real coefficients can be written as the product of linear and quadratic factors with real coefficients, where the quadratic factors have no real zeros.

PROOF

To begin, we use the Linear Factorization Theorem to conclude that $f(x)$ can be *completely* factored in the form

$$f(x) = a(x - c_1)(x - c_2)(x - c_3) \cdots (x - c_n).$$

If each c_i is real, there is nothing more to prove. If any c_i is complex ($c_i = a + bi$, $b \neq 0$), then because the coefficients of $f(x)$ are real we know that the conjugate $c_j = a - bi$ is also a zero. By multiplying the corresponding factors, we obtain

$$(x - c_i)(x - c_j) = [x - (a + bi)][x - (a - bi)]$$
$$= x^2 - 2ax + (a^2 + b^2)$$

where each coefficient is real. This completes the proof.

A quadratic factor with no real zeros is said to be **irreducible over the reals.** Be sure you see that this is not the same as being *irreducible over the rationals*. For example, the quadratic

$$x^2 + 1 = (x - i)(x + i)$$

is irreducible over the reals (and therefore over the rationals). On the other hand, the quadratic

$$x^2 - 2 = (x - \sqrt{2})(x + \sqrt{2})$$

is irreducible over the rationals, but it is *reducible* over the reals.

EXAMPLE 5 *Factoring a Polynomial*

Write the polynomial $f(x) = x^4 - x^2 - 20$

(a) as the product of factors that are irreducible over the *rationals*,
(b) as the product of linear factors and quadratic factors that are irreducible over the *reals*, and
(c) in completely factored form.

SOLUTION

(a) We begin by factoring the polynomial into the product of two quadratic polynomials.

$$x^4 - x^2 - 20 = (x^2 - 5)(x^2 + 4)$$

Both of these factors are irreducible over the rationals.

(b) By factoring over the reals, we have

$$x^4 - x^2 - 20 = (x + \sqrt{5})(x - \sqrt{5})(x^2 + 4)$$

where the quadratic factor is irreducible over the reals.

(c) In completely factored form, we have

$$x^4 - x^2 - 20 = (x + \sqrt{5})(x - \sqrt{5})(x + 2i)(x - 2i).$$

EXAMPLE 6 *Finding the Zeros of a Polynomial Function*

Find all zeros of

$$f(x) = x^4 - 3x^3 + 6x^2 + 2x - 60,$$

given that $1 + 3i$ is a zero of f.

SOLUTION

Since complex zeros occur in pairs, we know that $1 - 3i$ is also a zero of f. This means that both

$$[x - (1 + 3i)] \quad \text{and} \quad [x - (1 - 3i)]$$

are factors of $f(x)$. Multiplying these two factors produces

$$\begin{aligned}[x - (1 + 3i)][x - (1 - 3i)] &= [(x - 1) - 3i][(x - 1) + 3i] \\ &= (x - 1)^2 - 9i^2 \\ &= x^2 - 2x + 10.\end{aligned}$$

Using long division, we can divide $x^2 - 2x + 10$ into $f(x)$ to obtain the following.

$$\begin{array}{r} x^2 - x - 6 \\ x^2 - 2x + 10{\overline{\smash{\big)}\,x^4 - 3x^3 + 6x^2 + 2x - 60}} \\ \underline{x^4 - 2x^3 + 10x^2} \\ -x^3 - 4x^2 + 2x \\ \underline{-x^3 + 2x^2 - 10x} \\ -6x^2 + 12x - 60 \\ \underline{-6x^2 + 12x - 60} \end{array}$$

Exercises

Therefore, we have

$$f(x) = (x^2 - 2x + 10)(x^2 - x - 6)$$
$$= (x^2 - 2x + 10)(x - 3)(x + 2)$$

and we conclude that the zeros of f are

$$x = 1 + 3i, \quad 1 - 3i, \quad 3, \quad \text{and} \quad -2.$$

Throughout this chapter, we have basically stated the results and examples in terms of *zeros of polynomial functions*. Be sure you see that they could also have been stated in terms of *solutions of polynomial equations*. This is true because the zeros of the polynomial function

$$f(x) = a_n x^n + a_{n-1} x^{n-1} + \cdots + a_2 x^2 + a_1 x + a_0$$

are precisely the solutions of the polynomial equation

$$a_n x^n + a_{n-1} x^{n-1} + \cdots + a_2 x^2 + a_1 x + a_0 = 0.$$

WARM UP

Write each complex number in standard form and give its complex conjugate.

1. $4 - \sqrt{-29}$ **2.** $-5 - \sqrt{-144}$

3. $-1 + \sqrt{-32}$ **4.** $6 + \sqrt{-1/4}$

Perform the indicated operations and write the answers in standard form.

5. $(-3 + 6i) - (10 - 3i)$ **6.** $(12 - 4i) + 20i$

7. $(4 - 2i)(3 + 7i)$ **8.** $(2 - 5i)(2 + 5i)$

9. $\dfrac{1 + i}{1 - i}$ **10.** $(3 + 2i)^3$

EXERCISES 4.5

In Exercises 1–26, find all the zeros of the function and write the polynomial as a product of linear factors.

1. $f(x) = x^2 + 25$ **2.** $f(x) = x^2 - x + 56$

3. $h(x) = x^2 - 4x + 1$ **4.** $g(x) = x^2 + 10x + 23$

5. $f(x) = x^4 - 81$ **6.** $f(y) = y^4 - 625$

7. $f(z) = z^2 - 2z + 2$

8. $h(x) = x^3 - 3x^2 + 4x - 2$

9. $g(x) = x^3 - 6x^2 + 13x - 10$

10. $f(x) = x^3 - 2x^2 - 11x + 52$

11. $f(t) = t^3 - 3t^2 - 15t + 125$

12. $f(x) = x^3 + 11x^2 + 39x + 29$

13. $f(x) = x^3 + 24x^2 + 214x + 740$

14. $f(s) = 2s^3 - 5s^2 + 12s - 5$

15. $f(x) = 16x^3 - 20x^2 - 4x + 15$

16. $f(x) = 9x^3 - 15x^2 + 11x - 5$

17. $h(x) = x^3 - x + 6$

18. $h(x) = x^3 + 9x^2 + 27x + 35$

19. $f(x) = 5x^3 - 9x^2 + 28x + 6$

20. $g(x) = 3x^3 - 4x^2 + 8x + 8$

21. $g(x) = x^4 - 4x^3 + 8x^2 - 16x + 16$

22. $h(x) = x^4 + 6x^3 + 10x^2 + 6x + 9$

23. $f(x) = x^4 + 10x^2 + 9$

24. $f(x) = x^4 + 29x^2 + 100$

25. $f(x) = 2x^4 + 5x^3 + 4x^2 + 5x + 2$

26. $g(x) = x^5 - 8x^4 + 28x^3 - 56x^2 + 64x - 32$

In Exercises 27–36, find a polynomial with integer coefficients that has the given zeros.

27. $1, 5i, -5i$

28. $4, 3i, -3i$

29. $2, 4 + i, 4 - i$

30. $6, -5 + 2i, -5 - 2i$

31. $i, -i, 6i, -6i$

32. $2, 2, 2, 4i, -4i$

33. $-5, -5, 1 + \sqrt{3}i$

34. $\frac{2}{3}, -1, 3 + \sqrt{2}i$

35. $\frac{3}{4}, -2, -\frac{1}{2} + i$

36. $0, 0, 4, 1 + i$

In Exercises 37–40, write the polynomial (a) as the product of factors that are irreducible over the *rationals*, (b) as the product of linear and quadratic factors that are irreducible over the *reals*, and (c) in completely factored form.

37. $f(x) = x^4 + 6x^2 - 27$

38. $f(x) = x^4 - 2x^3 - 3x^2 + 12x - 18$

[*Hint:* One factor is $x^2 - 6$.]

39. $f(x) = x^4 - 4x^3 + 5x^2 - 2x - 6$

[*Hint:* One factor is $x^2 - 2x - 2$.]

40. $f(x) = x^4 - 3x^3 - x^2 - 12x - 20$

[*Hint:* One factor is $x^2 + 4$.]

In Exercises 41–50, use the given zero of f to find all the zeros of f.

41. $f(x) = 2x^3 + 3x^2 + 50x + 75, \quad r = 5i$

42. $f(x) = x^3 + x^2 + 9x + 9, \quad r = 3i$

43. $f(x) = 2x^4 - x^3 + 7x^2 - 4x - 4, \quad r = 2i$

44. $g(x) = x^3 - 7x^2 - x + 87, \quad r = 5 + 2i$

45. $g(x) = 4x^3 + 23x^2 + 34x - 10, \quad r = -3 + i$

46. $h(x) = 3x^3 - 4x^2 + 8x + 8, \quad r = 1 - \sqrt{3}i$

47. $f(x) = x^4 + 3x^3 - 5x^2 - 21x + 22, \quad r = -3 + \sqrt{2}i$

48. $f(x) = x^3 + 4x^2 + 14x + 20, \quad r = -1 - 3i$

49. $h(x) = 8x^3 - 14x^2 + 18x - 9, \quad r = (1 - \sqrt{5}i)/2$

50. $f(x) = 25x^3 - 55x^2 - 54x - 18, \quad r = (-2 + \sqrt{2}i)/5$

51. Find a quadratic function f (with integer coefficients) that has $\pm\sqrt{b}\,i$ as zeros. Assume that b is a positive integer.

4.6 Approximation Techniques for Zeros of Polynomial Functions

Throughout history mathematicians have devoted a great deal of time and effort to developing methods for finding the zeros of polynomial functions. We have not looked at all of these methods in this chapter, but we have looked at the basic ones. These include graphical methods, factorization methods, the Rational Zero Test, and Descartes's Rule of Signs.

Generally, the higher the degree of a polynomial function, the more difficult it is to find its zeros. In practical applications involving polynomials of degree 3 or greater, we often must be content with an approximation technique for finding zeros. Most approximation methods involve an **iterative process,** meaning that the method is applied repeatedly to obtain better and better approximations.

In this section we look at one of the simpler approximation methods, the **Bisection Method.** To apply this method, we must find two x-values—one at which the function is positive and one at which it is negative. By the Intermediate Value Theorem, we know that the function has at least one zero between the two x-values, as shown in Figure 4.31.

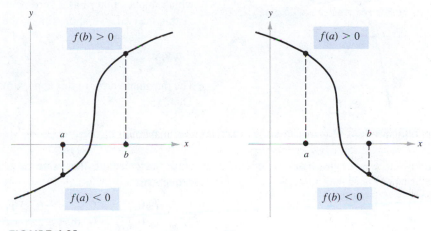

FIGURE 4.31

In Figure 4.31, we can see that a zero must occur somewhere in the interval (a, b). To apply the Bisection Method, we cut this interval in half and consider the two intervals

$$\left(a, \frac{a + b}{2}\right) \quad \text{and} \quad \left(\frac{a + b}{2}, b\right).$$

Depending upon the value of $f(x)$ at the midpoint $(a + b)/2$, we apply the Bisection Method again to one of these intervals. The process is illustrated in Example 1.

EXAMPLE 1 *Using the Bisection Method*

Use the Bisection Method to approximate the real zero of

$$f(x) = x^3 - x^2 + 1$$

to within 0.001 unit.

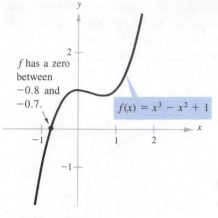

FIGURE 4.32

f has a zero between -0.8 and -0.7.

$f(x) = x^3 - x^2 + 1$

SOLUTION

We begin by making a sketch of f, as shown in Figure 4.32. In Example 8 of Section 4.2, we discovered that

$$f(-0.8) = -0.152 < 0 \quad \text{and} \quad f(-0.7) = 0.167 > 0$$

which implies that f has a zero between -0.8 and -0.7. Using the midpoint of this interval, we approximate the zero to be

$$c = \frac{-0.8 + (-0.7)}{2} = -0.75.$$

The maximum error of this approximation is one-half the length of the interval. That is,

$$\text{maximum error} \leq \frac{-0.7 - (-0.8)}{2} = 0.05.$$

Now, we calculate the value of $f(-0.75)$ and make one of the following deductions.

1. If $f(-0.75) = 0$, then -0.75 is a zero of f.
2. If $f(-0.75) > 0$, then a zero occurs between -0.8 and -0.75.
3. If $f(-0.75) < 0$, then a zero occurs between -0.75 and -0.7.

Because $f(-0.75) = 0.015625$ is positive, we choose $(-0.8, -0.75)$ as our new interval. The results of several iterations are shown in Table 4.1. After the seventh iteration, the maximum error is less than 0.001, and we approximate the zero of f to be

$$c = -0.75546875 \approx -0.755. \qquad \textit{To three-place accuracy}$$

TABLE 4.1

	a	c	b	$f(a)$	$f(c)$	$f(b)$	Maximum Error
1	-0.8	-0.75	-0.7	-0.1520	0.0156	0.1670	0.05
2	-0.8	-0.775	-0.75	-0.1520	-0.0661	0.0156	0.025
3	-0.775	-0.7625	-0.75	-0.0661	-0.0247	0.0156	0.0125
4	-0.7625	-0.7563	-0.75	-0.0247	-0.0044	0.0156	0.0063
5	-0.7563	-0.7531	-0.75	-0.0044	0.0056	0.0156	0.0031
6	-0.7563	-0.7547	-0.7531	-0.0044	0.0006	0.0056	0.0016
7	-0.7563	-0.7555	-0.7547	-0.0044	-0.0019	0.0006	0.0008

By continuing the process in Table 4.1, we could approximate the zero to *any* desired accuracy, and we say that the sequence of successively better approximations *converges* to the zero of the function. The convergence of the Bisection Method is relatively slow and several iterations are usually necessary to obtain a very fine accuracy.

In Example 1, the function had only one real zero. However, the Bisection Method can be applied just as well to functions that have several real zeros, as demonstrated in the following example.

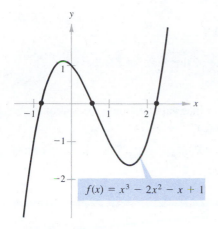

FIGURE 4.33

$f(x) = x^3 - 2x^2 - x + 1$

EXAMPLE 2   *Using the Bisection Method*

Use the Bisection Method to approximate each of the real zeros of

$$f(x) = x^3 - 2x^2 - x + 1$$

to within 0.001 unit.

SOLUTION

From Figure 4.33, we see that this function has three real zeros: one in the interval $[-1, 0]$, one in the interval $[0, 1]$, and one in the interval $[2, 3]$.

The following BASIC program can be used to approximate the first of these three zeros.

BASIC PROGRAM FOR BISECTION METHOD

```
10 REM   DEFINE THE FUNCTION
20         DEF FNY(X)=X^3-2*X^2-X+1
30 REM   A = LEFT ENDPOINT, B = RIGHT ENDPOINT
40         A = -1
50         B = 0
60         C = (A+B)/2
70 REM   ME = MAXIMUM ERROR
80         ME = (B-A)/2
90         PRINT "A = ";A,
                 "C = ";C,
                 "B = ";B,
                 "MAX ERR = ";ME
100 REM  TEST THE SIZE OF THE ERROR
110        IF ME < .001 GOTO 150
120 REM  TEST THE SIGN OF F(C)
130        IF FNY(A)*FNY(C) < 0 THEN B = C: GOTO 60
140        IF FNY(B)*FNY(C) < 0 THEN A = C: GOTO 60
150        END
```

PRINTOUT FROM BASIC PROGRAM

```
A = -1.0000 C = -0.5000 B =  0.0000 MAX ERR = 0.5000
A = -1.0000 C = -0.7500 B = -0.5000 MAX ERR = 0.2500
A = -1.0000 C = -0.8750 B = -0.7500 MAX ERR = 0.1250
A = -0.8750 C = -0.8125 B = -0.7500 MAX ERR = 0.0625
A = -0.8125 C = -0.7813 B = -0.7500 MAX ERR = 0.0313
A = -0.8125 C = -0.7969 B = -0.7813 MAX ERR = 0.0156
A = -0.8125 C = -0.8047 B = -0.7969 MAX ERR = 0.0078
A = -0.8047 C = -0.8008 B = -0.7969 MAX ERR = 0.0039
A = -0.8047 C = -0.8027 B = -0.8008 MAX ERR = 0.0020
A = -0.8027 C = -0.8018 B = -0.8008 MAX ERR = 0.0010
```

From this printout, we approximate one of the zeros of f to be

$$x \approx -0.802.$$

By rerunning the program (with different interval endpoints) we approximate the other two zeros to be $x \approx 0.556$ and $x \approx 2.247$.

EXERCISES 4.6

In Exercises 1–8, use the Bisection Method to approximate the indicated real zero(s) of the function to within 0.01 unit.

1. $f(x) = x^3 + x - 1$ **2.** $f(x) = x^5 + x - 1$ **4.** $f(x) = 4x^3 - 12x^2 + 12x - 3$

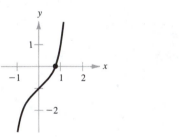

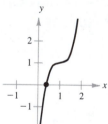

3. $f(x) = 2x^3 - 6x^2 + 6x - 1$ **5.** $f(x) = -x^3 + 3x^2 - x + 1$

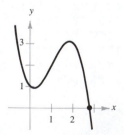

6. $f(x) = x^3 - 3x - 1$ **7.** $f(x) = x^4 - x - 3$

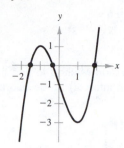

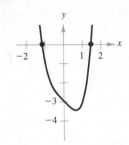

8. $f(x) = x^3 + 2x + 1$

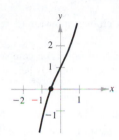

In Exercises 9–12, find the zero of the function (in the given interval) to the nearest hundredth.

9. $f(x) = 4x^3 + 14x - 8$,   [0, 1]
10. $f(x) = 4x^3 - 14x^2 - 2$,   [3, 4]
11. $f(x) = 7x^4 - 42x^3 + 43x^2 + 216x - 324$,   [1, 2]
12. $f(x) = 3x^4 - 12x^3 + 27x^2 + 4x - 4$,   [0, 1]

13. The concentration of a certain chemical in the bloodstream t hours after injection into muscle tissue is given by

$$C = \frac{3t^2 + t}{t^3 + 50}.$$

The concentration is greatest when $3t^4 + 2t^3 - 300t - 50 = 0$. Approximate this time to the nearest tenth of an hour.

14. The ordering and transportation cost C of the components used in manufacturing a certain product is given by

$$C = 100\left(\frac{200}{x^2} + \frac{x}{x + 30}\right), \qquad 1 \le x$$

where C is measured in thousands of dollars, and x is the order size in hundreds. The cost is a minimum when $3x^3 - 40x^2 - 2400x - 36000 = 0$. Approximate the optimal order size to the nearest hundred units.

CHAPTER 4 REVIEW EXERCISES

In Exercises 1–10, sketch the graph of the function.

1. $f(x) = \left(x + \frac{3}{2}\right)^2 + 1$ **2.** $f(x) = (x - 4)^2 - 4$
3. $f(x) = -(x - 1)^3$ **4.** $f(x) = (x + 1)^3$
5. $g(x) = x^4 - x^3 - 2x^2$ **6.** $h(x) = -2x^3 - x^2 + x$
7. $f(t) = t^3 - 3t$ **8.** $f(x) = -x^3 + 3x - 2$
9. $f(x) = x(x + 3)^2$ **10.** $f(t) = t^4 - 4t^2$

In Exercises 11–20, find the maximum or minimum value of the quadratic function.

11. $g(x) = x^2 - 2x$ **12.** $f(x) = x^2 + 8x + 10$
13. $f(x) = 6x - x^2$ **14.** $h(x) = 3 + 4x - x^2$
15. $f(x) = x^2 + 3x - 8$ **16.** $h(x) = 4x^2 + 4x + 13$
17. $f(t) = -2t^2 + 4t + 1$ **18.** $g(z) = -3z^2 - 12z + 4$
19. $h(x) = x^2 + 5x - 4$ **20.** $f(x) = 4x^2 + 4x + 5$

In Exercises 21–28, perform the indicated division.

21. $\dfrac{24x^2 - x - 8}{3x - 2}$ **22.** $\dfrac{4x + 7}{3x - 2}$

23. $\dfrac{x^4 + x^3 - x^2 + 2x}{x^2 + 2x}$

24. $\dfrac{5x^3 - 13x^2 - x + 2}{x^2 - 3x + 1}$

25. $\dfrac{x^4 - 3x^2 + 2}{x^2 - 1}$ **26.** $\dfrac{3x^4}{x^2 - 1}$

27. $\dfrac{x^4 - 3x^3 + 4x^2 - 6x + 3}{x^2 + 2}$

28. $\dfrac{6x^4 + 10x^3 + 13x^2 - 5x + 2}{2x^2 - 1}$

In Exercises 29–34, use synthetic division to perform the indicated division.

29. $\dfrac{0.25x^4 - 4x^3}{x - 2}$

30. $\dfrac{2x^3 + 2x^2 - x + 2}{x - (1/2)}$

31. $\dfrac{6x^4 - 4x^3 - 27x^2 + 18x}{x - (2/3)}$

32. $\dfrac{0.1x^3 + 0.3x^2 - 0.5}{x - 5}$

33. $\dfrac{2x^3 - 5x^2 + 12x - 5}{x - (1 + 2i)}$

34. $\dfrac{9x^3 - 15x^2 + 11x - 5}{x - [(1/3) + (2/3)i]}$

In Exercises 35–38, use synthetic division to determine whether the given values of x are zeros of the function.

35. $f(x) = 2x^3 + 3x^2 - 20x - 21$
    (a) $x = 4$ (b) $x = -1$
    (c) $x = -\frac{7}{2}$ (d) $x = 0$

36. $f(x) = 20x^4 + 9x^3 - 14x^2 - 3x$
    (a) $x = -1$ (b) $x = \frac{3}{4}$
    (c) $x = 0$ (d) $x = 1$

37. $f(x) = 2x^3 + 7x^2 - 18x - 30$
    (a) $x = 1$ (b) $x = \frac{5}{2}$
    (c) $x = -3 + \sqrt{3}$ (d) $x = 0$

38. $f(x) = 3x^3 - 26x^2 + 364x - 232$
    (a) $x = 4 - 10i$ (b) $x = 4$
    (c) $x = \frac{2}{3}$ (d) $x = -1$

In Exercises 39–42, use synthetic division to find the specified value of the function.

39. $g(x) = 2x^4 - 17x^3 + 58x^2 - 77x + 26$
    (a) $g(-2)$ (b) $g\left(\frac{1}{2}\right)$

40. $h(x) = 5x^5 - 2x^4 - 45x + 18$
    (a) $h(2)$ (b) $h(\sqrt{3})$

41. $f(x) = x^4 + 10x^3 - 24x^2 + 20x + 44$
    (a) $f(-3)$ (b) $f(\sqrt{2}\,i)$

42. $g(t) = 2t^5 - 5t^4 - 8t + 20$
    (a) $g(-4)$ (b) $g(\sqrt{2})$

In Exercises 43 and 44, find a fourth-degree polynomial with the given zeros.

43. $-1, -1, \frac{1}{3}, -\frac{1}{2}$ 44. $\frac{2}{3}, 4, \sqrt{3}\,i, -\sqrt{3}\,i$

In Exercises 45–50, find all zeros of the function.

45. $f(x) = 4x^3 - 11x^2 + 10x - 3$

46. $f(x) = 10x^3 + 21x^2 - x - 6$

47. $f(x) = 6x^3 - 5x^2 + 24x - 20$

48. $f(x) = x^3 - 1.3x^2 - 1.7x + 0.6$

49. $f(x) = 6x^4 - 25x^3 + 14x^2 + 27x - 18$

50. $f(x) = 5x^4 + 126x^2 + 25$

In Exercises 51–54, use the Bisection Method to find the zero of the function (in the given interval) to the nearest hundredth.

51. $f(x) = x^4 + 2x - 1$, $[0, 1]$

52. $g(x) = x^3 - 3x^2 + 3x + 2$, $[-1, 0]$

53. $h(x) = x^3 - 6x^2 + 12x - 10$, $[3, 4]$

54. $f(x) = x^5 + 2x^3 - 3x - 20$, $[1, 2]$

55. Find the number of units x that produce a maximum revenue R for $R = 900x - 0.1x^2$.

56. Let x be the amount (in hundreds of dollars) a company spends on advertising, and let P be the profit, where

$$P = 230 + 20x - \tfrac{1}{2}x^2.$$

What amount of advertising will yield a maximum profit?

57. A real estate office handles 50 apartment units. When the rent is \$360 per month, all units are occupied. However, for each \$20 increase in rent, one unit becomes vacant. Each occupied unit requires an average of \$12 per month for service and repairs. What rent should be charged to realize the most profit?

58. The sum of one positive number and twice a second positive number is 100. Find the two numbers so that their product is a maximum.

59. A rectangle is inscribed in the region bounded by the x-axis, the y-axis, and the graph of $x + 2y - 6 = 0$ (see figure). Find the coordinates (x, y) that yield a maximum area for the rectangle.

60. A spherical tank of radius 50 feet (see figure) will be two-thirds full when the depth of the fluid is $x + 50$ feet, where

$$3x^3 - 22,500x + 250,000 = 0.$$

Use the Bisection Method to approximate x to within 0.01 unit.

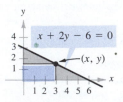

FIGURE FOR 59

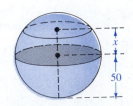

FIGURE FOR 60

CHAPTER 5

Rational Functions and Conic Sections

5.1 Rational Functions and Their Graphs

A **rational function** is one that can be written in the form

$$f(x) = \frac{p(x)}{q(x)}$$

where $p(x)$ and $q(x)$ are polynomials and $q(x)$ is not the zero polynomial. In this section we assume $p(x)$ and $q(x)$ have no common factors. Some examples of rational functions are

$$f(x) = \frac{1}{x + 2}, \quad g(x) = \frac{x - 1}{(x + 1)(x - 2)}, \quad \text{and} \quad h(x) = \frac{x}{x^2 + 1}.$$

Unlike polynomial functions, whose domains consist of all real numbers, rational functions often have restricted domains to avoid division by zero. In the examples above, the domain of f excludes $x = -2$, and the domain of g excludes $x = 2$ and $x = -1$. On the other hand, the domain of h is all real numbers because there are no real values of x that make the denominator $x^2 + 1$ equal to zero.

In general, the *domain* of a rational function of x includes all real numbers except x-values that make the denominator zero. Much of our discussion of rational functions will focus on their graphical behavior near these x-values.

279

EXAMPLE 1   *Finding the Domain of a Rational Function*

Find the domain of

$$f(x) = \frac{1}{x}$$

and assess the behavior of f near any excluded x-values.

SOLUTION

Because the denominator is zero when $x = 0$, the domain of f is all real numbers except $x = 0$. To assess the behavior of f near this excluded value, we evaluate $f(x)$ to the left and right of $x = 0$, as indicated in the following two tables.

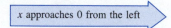

x approaches 0 from the left

x	-1	-0.5	-0.1	-0.01	-0.001	$\to 0$
$f(x)$	-1	-2	-10	-100	-1000	$\to -\infty$

x approaches 0 from the right

x	$0 \leftarrow$	0.001	0.01	0.1	0.5	1
$f(x)$	$\infty \leftarrow$	1000	100	10	2	1

Note that as x approaches 0 *from the left* $f(x)$ decreases without bound, whereas as x approaches 0 *from the right* $f(x)$ increases without bound. The graph of f is shown in Figure 5.1.

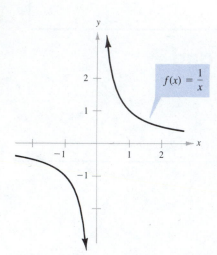

$f(x) = \dfrac{1}{x}$

FIGURE 5.1

The behavior of $f(x) = 1/x$ near $x = 0$ is denoted by

$f(x) \to -\infty$ as $x \to 0^-$ *f(x) decreases without bound as x approaches 0 from the left.*

$f(x) \to \infty$ as $x \to 0^+$ *f(x) increases without bound as x approaches 0 from the right.*

and we call the line $x = 0$ a **vertical asymptote** of the graph of f, as shown in Figure 5.2.

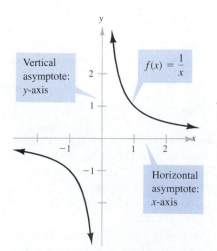

Vertical asymptote: *y*-axis

$f(x) = \dfrac{1}{x}$

Horizontal asymptote: *x*-axis

FIGURE 5.2

Rational Functions and Their Graphs

From Figure 5.2, we see that the graph of f also has a **horizontal asymptote**—the line $y = 0$. This means that the values of $f(x) = 1/x$ approach zero as x increases or decreases without bound. We denote this behavior as follows.

$f(x) \rightarrow 0$ as $x \rightarrow -\infty$ *$f(x)$ approaches 0 as x decreases without bound.*

$f(x) \rightarrow 0$ as $x \rightarrow \infty$ *$f(x)$ approaches 0 as x increases without bound.*

These two types of asymptotes are summarized next.

Definition of Vertical and Horizontal Asymptotes

1. The line $x = a$ is a **vertical asymptote** of the graph of f if

$$f(x) \rightarrow \infty \quad \text{or} \quad f(x) \rightarrow -\infty$$

as $x \rightarrow a$, either from the right or from the left.

2. The line $y = b$ is a **horizontal asymptote** of the graph of f if

$$f(x) \rightarrow b$$

as $x \rightarrow \infty$ or $x \rightarrow -\infty$.

Though the graph of a rational function will never intersect its vertical asymptote, it may intersect its horizontal asymptote. However, eventually (as $x \rightarrow \infty$ or $x \rightarrow -\infty$) the distance between a horizontal asymptote and the points on the graph must approach zero. Figure 5.3 shows the horizontal and vertical asymptotes of the graphs of three rational functions.

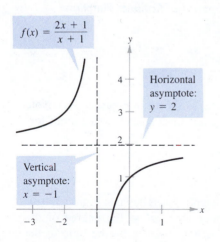

$f(x) = \dfrac{2x + 1}{x + 1}$

Horizontal asymptote: $y = 2$

Vertical asymptote: $x = -1$

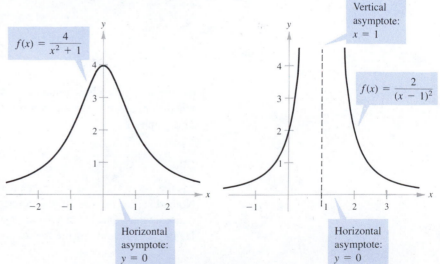

$f(x) = \dfrac{4}{x^2 + 1}$

Horizontal asymptote: $y = 0$

Vertical asymptote: $x = 1$

$f(x) = \dfrac{2}{(x - 1)^2}$

Horizontal asymptote: $y = 0$

FIGURE 5.3

Remark: The graphs of $f(x) = 1/x$ (see Figure 5.2) and $f(x) = (2x + 1)/$ $(x + 1)$ (Figure 5.3) are called **hyperbolas.** We will say more about this type of curve in Sections 5.3 and 5.4.

A rational function can have, at most, one horizontal asymptote, and it need not have any. For instance, the function

$$f(x) = \frac{x^2 + 1}{x + 1}$$

has no horizontal asymptote. To determine the horizontal asymptote (if any), we use the following theorem.

Horizontal Asymptote of a Rational Function

Let f be the rational function given by

$$f(x) = \frac{a_n x^n + a_{n-1} x^{n-1} + \cdots + a_1 x + a_0}{b_m x^m + b_{m-1} x^{m-1} + \cdots + b_1 x + b_0}.$$

1. If $n < m$, then the graph of f has the x-axis as a horizontal asymptote.
2. If $n = m$, then the graph of f has the line $y = a_n/b_m$ as a horizontal asymptote.
3. If $n > m$, then the graph of f has no horizontal asymptote.

We can apply this theorem by comparing the degrees of the numerator and denominator, as illustrated in the following example.

EXAMPLE 2 Finding Horizontal Asymptotes of Rational Functions

(a) The graph of

$$f(x) = \frac{2x}{3x^2 + 1}$$

has the x-axis as a horizontal asymptote, as shown in Figure 5.4(a). Note that the degree of the numerator is *less than* the degree of the denominator.

(b) The graph of

$$f(x) = \frac{2x^2}{3x^2 + 1}$$

has the line $y = 2/3$ as a horizontal asymptote, as shown in Figure 5.4(b). Note that the degree of the numerator is *equal to* the degree of the denominator, and the horizontal asymptote is given by the ratio of the leading coefficients of the numerator and denominator.

Rational Functions and Their Graphs

(a)

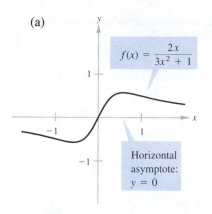

(b)

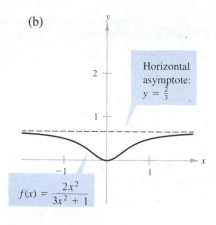

(c)

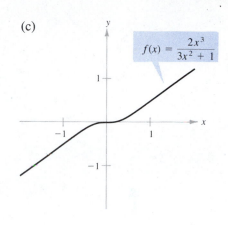

FIGURE 5.4

(c) The graph of

$$f(x) = \frac{2x^3}{3x^2 + 1}$$

has no horizontal asymptote because the degree of the numerator is *greater than* the degree of the denominator. See Figure 5.4(c).

To sketch the graph of a rational function, we suggest the following guidelines.

Guidelines for Graphing Rational Functions

Let $f(x) = p(x)/q(x)$, where $p(x)$ and $q(x)$ are polynomials with no common factors.

1. Find and plot the y-intercept (if any) by evaluating $f(0)$.
2. Find the zeros of the numerator (if any) by solving the equation $p(x) = 0$. Then plot the corresponding x-intercepts.
3. Find the zeros of the denominator (if any) by solving the equation $q(x) = 0$. Then sketch the corresponding vertical asymptotes.
4. Find and sketch the horizontal asymptote (if any) using the theorem on horizontal asymptotes.
5. Plot at least one point both *between and beyond* each x-intercept and vertical asymptote.
6. Use smooth curves to complete the graph between and beyond the vertical asymptotes.

Remark: Testing for symmetry can be useful, especially for simple rational functions. For example, the graph of $f(x) = 1/x$ is symmetrical with respect to the origin, and the graph of $g(x) = 1/x^2$ is symmetrical with respect to the y-axis.

We illustrate the use of these guidelines in the next several examples.

EXAMPLE 3 *Sketching the Graph of a Rational Function*

Sketch the graph of

$$g(x) = \frac{3}{x - 2}.$$

SOLUTION

We begin by noting that the numerator and denominator have no common factors. For this function, we have the following.

FIGURE 5.5

y-intercept: $(0, -3/2)$, from $g(0) = -3/2$.

x-intercept: None, since $3 \neq 0$.

Vertical Asymptote: $x = 2$, zero of denominator

Horizontal Asymptote: $y = 0$, degree of $p(x) <$ degree of $q(x)$

Additional Points:

x	-4	1	3	5
$g(x)$	-0.5	-3	3	1

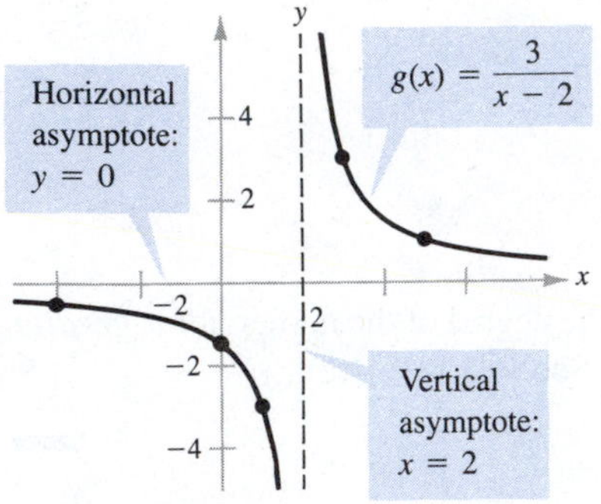

By plotting the intercepts, asymptotes, and a few additional points, we obtain the graph shown in Figure 5.5.

Remark: Note that the graph of g is a vertical stretch and a right shift of the graph of $f(x) = 1/x$ because

$$g(x) = \frac{3}{x - 2} = 3\left(\frac{1}{x - 2}\right) = 3 f(x - 2).$$

EXAMPLE 4 *Sketching the Graph of a Rational Function*

Sketch the graph of

$$f(x) = \frac{2x - 1}{x}.$$

SOLUTION

y-intercept: None, because $x = 0$ is not in the domain.

x-intercept: $(1/2, 0)$, from $2x - 1 = 0$.

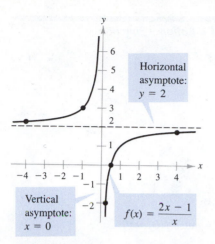

FIGURE 5.6

Vertical Asymptote: $x = 0$, zero of denominator

Horizontal Asymptote: $y = 2$, degree of $p(x)$ = degree of $q(x)$

Additional Points:

x	-4	-1	$\frac{1}{4}$	4
$f(x)$	2.25	3	-2	1.75

The graph of f is shown in Figure 5.6.

Because the zeros of the numerator and denominator give such useful information about the graph of a rational function, it is a good idea to completely factor both the numerator and the denominator as a preliminary step in the graphing process.

EXAMPLE 5   *Sketching the Graph of a Rational Function*

Sketch the graph of

$$f(x) = \frac{x}{x^2 - x - 2}.$$

SOLUTION

By factoring the denominator, we have

$$f(x) = \frac{x}{x^2 - x - 2} = \frac{x}{(x + 1)(x - 2)}.$$

y-intercept: $(0, 0)$, because $f(0) = 0$.

x-intercept: $(0, 0)$

Vertical Asymptotes: $x = -1$, $x = 2$

Horizontal Asymptote: $y = 0$

Additional Points:

x	-3	-0.5	1	3
$f(x)$	-0.3	0.4	-0.5	0.75

The graph is shown in Figure 5.7.

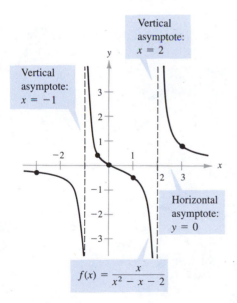

FIGURE 5.7

Rational Functions and Conic Sections

EXAMPLE 6 *Sketching the Graph of a Rational Function*

Sketch the graph of

$$f(x) = \frac{2(x^2 - 9)}{x^2 - 4}.$$

SOLUTION

In factored form, we have

$$f(x) = \frac{2(x^2 - 9)}{x^2 - 4} = \frac{2(x - 3)(x + 3)}{(x - 2)(x + 2)}.$$

> *y-intercept:* $(0, 9/2)$, because $f(0) = 9/2$.
>
> *x-intercepts:* $(-3, 0)$ and $(3, 0)$
>
> *Vertical Asymptotes:* $x = -2$, $x = 2$
>
> *Horizontal Asymptote:* $y = 2$
>
> *Symmetry:* With respect to y-axis, since $f(-x) = f(x)$.
>
> *Additional Points:*

x	0.5	2.5	6
$f(x)$	4.67	−2.44	1.69

Using the *symmetry* with respect to the *y*-axis, we sketch the graph as shown in Figure 5.8.

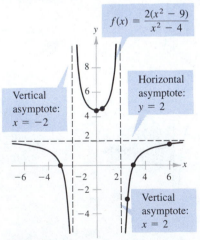

FIGURE 5.8

Rational Functions and Their Graphs

If the degree of the numerator of a rational function is exactly *one more* than the degree of its denominator, then the graph of the function has a **slant asymptote.** For example, the graphs of

$$f(x) = \frac{x^2 - x}{x + 1} \quad \text{and} \quad g(x) = \frac{-x^3 + x^2 + 4}{x^2}$$

have slant asymptotes as shown in Figure 5.9.

To find the equation of a slant asymptote, we use long division. For instance, by dividing $x + 1$ into $x^2 - x$, we have

$$f(x) = \frac{x^2 - x}{x + 1} = \underbrace{x - 2} + \frac{2}{x + 1}.$$

Slant asymptote
$(y = x - 2)$

This procedure is demonstrated in Example 7.

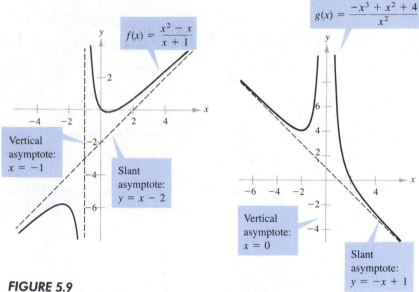

FIGURE 5.9

EXAMPLE 7 **A Rational Function with a Slant Asymptote**

Sketch the graph of

$$f(x) = \frac{x^2 - x - 2}{x - 1}.$$

SOLUTION

As a preliminary step, we write $f(x)$ in two different ways. Factoring the numerator

$$f(x) = \frac{x^2 - x - 2}{x - 1} = \frac{(x - 2)(x + 1)}{x - 1}$$

allows us to recognize the x-intercepts, and long division

$$f(x) = \frac{x^2 - x - 2}{x - 1} = x - \frac{2}{x - 1}$$

allows us to recognize that the line $y = x$ is a slant asymptote of the graph of f.

y-intercept: $(0, 2)$, because $f(0) = 2$.

x-intercepts: $(-1, 0)$ and $(2, 0)$

Vertical Asymptote: $x = 1$

Slant Asymptote: $y = x$

Additional Points:

x	-2	0.5	1.5	3
$f(x)$	-1.33	4.5	-2.5	2

FIGURE 5.10

The graph of f is shown in Figure 5.10.

Finally, note that it is possible for the graph of a rational function to cross its horizontal or its slant asymptote, as shown in Figure 5.11.

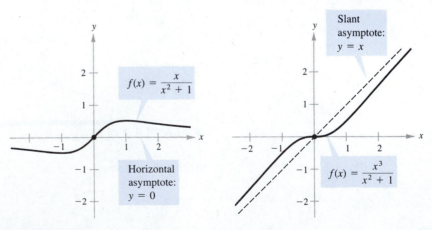

FIGURE 5.11

Exercises

WARM UP

Factor the polynomials.

1. $x^2 - 3x - 10$

2. $x^2 - 7x + 10$

3. $x^3 + 4x^2 + 3x$

4. $x^3 - 4x^2 - 2x + 8$

Sketch the graph of each equation.

5. $y = 2$

6. $x = -1$

7. $y = x - 2$

8. $y = -x + 1$

Use long division to write the improper rational expression as the sum of a polynomial and a proper rational expression.

9. $\dfrac{x^2 + 5x + 6}{x - 4}$

10. $\dfrac{x^2 + 5x + 6}{x + 4}$

EXERCISES 5.1

In Exercises 1–8, match the rational function with its graph. [The graphs are labeled (a)–(h).]

1. $f(x) = \dfrac{2}{x + 1}$

2. $f(x) = \dfrac{1}{x - 4}$

3. $f(x) = \dfrac{x + 1}{x}$

4. $f(x) = \dfrac{1 - 2x}{x}$

5. $f(x) = \dfrac{x - 2}{x - 1}$

6. $f(x) = -\dfrac{x + 2}{x + 1}$

7. $f(x) = \dfrac{x^2 + 1}{x}$

8. $f(x) = \dfrac{x^2 - 2x}{x - 1}$

(c)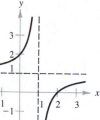

(d)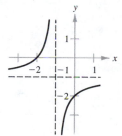

(e)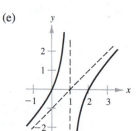

(f)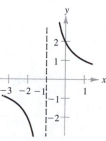

(a)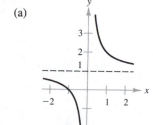

(b)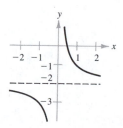

(g)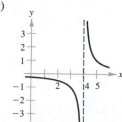

(h)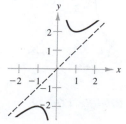

In Exercises 9–18, find the domain of the function and identify any horizontal, vertical, or slant asymptotes.

9. $f(x) = \dfrac{1}{x^2}$

10. $f(x) = \dfrac{4}{(x-2)^3}$

11. $f(x) = \dfrac{2+x}{2-x}$

12. $f(x) = \dfrac{1-5x}{1+2x}$

13. $f(x) = \dfrac{x^3}{x^2-1}$

14. $f(x) = \dfrac{2x^2}{x+1}$

15. $f(x) = \dfrac{3x^2+1}{x^2+9}$

16. $f(x) = \dfrac{3x^2+x-5}{x^2+1}$

17. $f(x) = \dfrac{5x^4}{x^2+1}$

18. $f(x) = \dfrac{x^2}{x+1}$

In Exercises 19 and 20, use the given figures to sketch the required graphs.

19. (a) $y = f(x) - 2 = \dfrac{4}{x^2} - 2$

(b) $y = f(x-2) = \dfrac{4}{(x-2)^2}$

(c) $y = -f(x) = -\dfrac{4}{x^2}$

(d) $y = \dfrac{1}{4}f(x) = \dfrac{1}{x^2}$

20. (a) $y = g(x) + 1 = \dfrac{8}{x^3} + 1$

(b) $y = g(x+2) = \dfrac{8}{(x+2)^3}$

(c) $y = -g(x) = -\dfrac{8}{x^3}$

(d) $y = \dfrac{1}{8}g(x) = \dfrac{1}{x^3}$

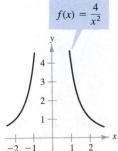

FIGURE FOR 19

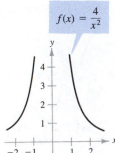

FIGURE FOR 20

In Exercises 21–44, sketch the graph of the rational function. As sketching aids, check for intercepts, symmetry, vertical asymptotes, and horizontal asymptotes.

21. $f(x) = \dfrac{1}{x+2}$

22. $f(x) = \dfrac{1}{x-3}$

23. $h(x) = \dfrac{-1}{x+2}$

24. $g(x) = \dfrac{1}{3-x}$

25. $f(x) = \dfrac{x+1}{x+2}$

26. $f(x) = \dfrac{x-2}{x-3}$

27. $f(x) = \dfrac{2+x}{1-x}$

28. $f(x) = \dfrac{3-x}{2-x}$

29. $f(t) = \dfrac{3t+1}{t}$

30. $f(t) = \dfrac{1-2t}{t}$

31. $g(x) = \dfrac{1}{x+2} + 2$

32. $h(x) = \dfrac{1}{x-3} + 1$

33. $C(x) = \dfrac{5+2x}{1+x}$

34. $P(x) = \dfrac{1-3x}{1-x}$

35. $f(x) = \dfrac{x^2}{x^2+9}$

36. $f(x) = 2 - \dfrac{3}{x^2}$

37. $h(x) = \dfrac{x^2}{x^2-9}$

38. $g(x) = \dfrac{x}{x^2-9}$

39. $g(s) = \dfrac{s}{s^2+1}$

40. $h(t) = \dfrac{4}{t^2+1}$

41. $f(x) = -\dfrac{1}{(x-2)^2}$

42. $g(x) = -\dfrac{x}{(x-2)^2}$

43. $f(x) = \dfrac{3x}{x^2-x-2}$

44. $f(x) = \dfrac{2x}{x^2+x-2}$

In Exercises 45–54, sketch the graph of the rational function. As sketching aids, check for intercepts, symmetry, vertical asymptotes, and *slant* asymptotes.

45. $f(x) = \dfrac{2x^2+1}{x}$

46. $f(x) = \dfrac{1-x^2}{x}$

47. $g(x) = \dfrac{x^2+1}{x}$

48. $h(x) = \dfrac{x^2}{x-1}$

49. $f(x) = \dfrac{x^3}{x^2-1}$

50. $g(x) = \dfrac{x^3}{2x^2-8}$

51. $f(x) = \dfrac{x^2-x+1}{x-1}$

52. $f(x) = \dfrac{2x^2-5x+5}{x-2}$

53. $f(x) = \dfrac{x^2+5x+8}{x+3}$

54. $f(x) = \dfrac{2x^2+x}{x+1}$

55. The game commission introduces 50 deer into newly acquired state game lands. The population of the herd is given by

$$N = \frac{10(5 + 3t)}{1 + 0.04t}, \qquad 0 \le t$$

where t is time in years.
(a) Find the population when t is 5, 10, and 25.
(b) What is the limiting size of the herd as time increases?

56. The cost of producing x units is $C = 0.2x^2 + 10x + 5$, and therefore the average cost per unit is

$$\overline{C} = \frac{C}{x} = \frac{0.2x^2 + 10x + 5}{x}, \qquad 0 < x.$$

Sketch the graph of the average cost function, and estimate the number of units that should be produced to minimize the average cost per unit.

57. Psychologists have developed mathematical models to predict performance as a function of the number of trials n for a certain task. Consider the learning curve given by

$$P = \frac{0.5 + 0.9(n - 1)}{1 + 0.9(n - 1)}, \qquad 0 < n$$

where P is the percentage of correct responses after n trials.
(a) Complete the following table for this model.

n	1	2	3	4	5	6	7	8	9	10
P										

(b) According to this model, what is the limiting percentage of correct responses as n increases?

58. The region shown in the figure is bounded by $y = x^2$, $y = 0$, and $x = 2$. It can be shown that the area A of this region is approximated by

$$f(n) = \frac{4}{3}\left(\frac{2n^3 + 3n^2 + n}{n^3}\right)$$

where n is a positive integer. As n increases without bound, $f(n)$ approaches the exact area of the region. Find A.

59. A right triangle is formed in the first quadrant by the x-axis, the y-axis, and a line segment through the point $(2, 3)$. (See figure.)
(a) Show that an equation of the line segment is

$$y = \frac{3(x - a)}{2 - a}, \qquad 0 \le x \le a.$$

(b) Show that the area of the triangle is

$$A = \frac{-3a^2}{2(2 - a)}.$$

(c) Sketch the graph of the area function of part (b), and from the graph estimate the value of a that yields a minimum area.

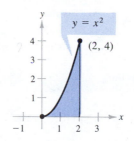

FIGURE FOR 58

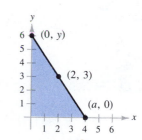

FIGURE FOR 59

5.2 Partial Fractions

In Section 1.6, we learned how to combine two or more rational expressions using a lowest common denominator (LCD). In this section, we reverse the problem (a useful procedure in calculus) and write a given rational expression as the sum of two or more simpler rational expressions. For example, the rational expression $(x + 7)/(x^2 - x - 6)$ can be written as the sum of two fractions with first-degree denominators. That is,

$$\frac{x + 7}{x^2 - x - 6} = \frac{2}{x - 3} + \frac{-1}{x + 2}.$$

Each fraction on the right side of the equation is a **partial fraction,** and together they make up the **partial fraction decomposition** of the left side.

In Chapter 4 we noted that it is theoretically possible to write any polynomial as the product of linear and irreducible quadratic factors. For instance,

$$x^5 + x^4 - x - 1 = (x - 1)(x + 1)^2(x^2 + 1)$$

where $(x - 1)$ is a linear factor, $(x + 1)$ is a repeated linear factor, and $(x^2 + 1)$ is an irreducible quadratic factor.

We can use this factorization to find the partial fraction decomposition of any rational expression having $x^5 + x^4 - x - 1$ as its denominator. Specifically, if $N(x)$ is a polynomial of degree 4 or less, then the partial fraction decomposition of $N(x)/(x^5 + x^4 - x - 1)$ has the form

$$\frac{N(x)}{x^5 + x^4 - x - 1} = \frac{N(x)}{(x - 1)(x + 1)^2(x^2 + 1)}$$

$$= \frac{A}{x - 1} + \frac{B}{x + 1} + \frac{C}{(x + 1)^2} + \frac{Dx + F}{x^2 + 1}.$$

Note that the factor $(x + 1)^2$ results in *two* fractions: one for $(x + 1)$ and one for $(x + 1)^2$. If $(x + 1)^3$ were a factor, then we would use three fractions: one for $(x + 1)$, one for $(x + 1)^2$, and one for $(x + 1)^3$. In general, the number of fractions resulting from a repeated factor is equal to the exponent of the factor.

The following guidelines summarize the steps used to find the partial fraction decomposition of a rational expression.

Decomposition of N(x)/D(x) into Partial Fractions

1. *Divide if improper:* If $N(x)/D(x)$ is an improper fraction, then divide the denominator into the numerator to obtain

$$\frac{N(x)}{D(x)} = (\text{polynomial}) + \frac{N_1(x)}{D(x)}$$

and apply Steps 2, 3, and 4 to the proper rational expression $N_1(x)/D(x)$.

2. *Factor denominator:* Completely factor the denominator into factors of the form

$$(px + q)^m \qquad \text{and} \qquad (ax^2 + bx + c)^n$$

where $(ax^2 + bx + c)$ is irreducible.

3. *Linear factors:* For *each* factor of the form $(px + q)^m$, the partial fraction decomposition must include the following sum of m fractions.

$$\frac{A_1}{(px + q)} + \frac{A_2}{(px + q)^2} + \cdots + \frac{A_m}{(px + q)^m}$$

4. *Quadratic factors:* For *each* factor of the form $(ax^2 + bx + c)^n$, the partial fraction decomposition must include the following sum of n fractions.

$$\frac{B_1x + C_1}{ax^2 + bx + c} + \frac{B_2x + C_2}{(ax^2 + bx + c)^2} + \cdots + \frac{B_nx + C_n}{(ax^2 + bx + c)^n}$$

Algebraic techniques for determining the constants in the numerators of the partial fractions are demonstrated in the examples that follow.

EXAMPLE 1   *Distinct Linear Factors*

Write the partial fraction decomposition for

$$\frac{x + 7}{x^2 - x - 6}.$$

SOLUTION

Since $x^2 - x - 6 = (x - 3)(x + 2)$, we include one partial fraction with a constant numerator for each linear factor of the denominator and write

$$\frac{x + 7}{x^2 - x - 6} = \frac{A}{x - 3} + \frac{B}{x + 2}.$$

Multiplying both sides of this equation by the lowest common denominator, $(x - 3)(x + 2)$, leads to the **basic equation**

$$x + 7 = A(x + 2) + B(x - 3). \qquad \textit{Basic equation}$$

Because this equation is true for all x, we can substitute any *convenient* values of x which will help us determine the constants A and B. Values of x that are especially convenient are ones that make the factors $(x + 2)$ and $(x - 3)$ equal to zero. For instance, if we let $x = -2$, then

$$-2 + 7 = A(0) + B(-5)$$
$$5 = -5B$$
$$-1 = B.$$

To solve for A, we let $x = 3$ and obtain

$$3 + 7 = A(5) + B(0)$$
$$10 = 5A$$
$$2 = A.$$

Therefore, the decomposition is

$$\frac{x + 7}{x^2 - x - 6} = \frac{2}{x - 3} - \frac{1}{x + 2}$$

as indicated at the beginning of this section. Try checking this result by combining the two partial fractions on the right side of the equation.

EXAMPLE 2   *Repeated Linear Factors*

Write the partial fraction decomposition for

$$\frac{5x^2 + 20x + 6}{x^3 + 2x^2 + x}.$$

SOLUTION

Since the denominator factors as

$$x^3 + 2x^2 + x = x(x^2 + 2x + 1) = x(x + 1)^2$$

we include one fraction with a constant numerator for each power of x and $(x + 1)$ and write

$$\frac{5x^2 + 20x + 6}{x(x + 1)^2} = \frac{A}{x} + \frac{B}{x + 1} + \frac{C}{(x + 1)^2}.$$

Multiplying by the LCD, $x(x + 1)^2$, leads to the basic equation

$$5x^2 + 20x + 6 = A(x + 1)^2 + Bx(x + 1) + Cx. \qquad \textit{Basic equation}$$

Letting $x = -1$ eliminates the A and B terms and yields

$$5 - 20 + 6 = 0 + 0 - C$$
$$C = 9.$$

Letting $x = 0$ eliminates the B and C terms and yields

$$6 = A(1) + 0 + 0$$
$$6 = A.$$

At this point, we have exhausted the most convenient choices for x, so to find the value of B, we use *any other value* for x along with the known values of A and C. Thus, using $x = 1$, $A = 6$, and $C = 9$, we have

$$5 + 20 + 6 = A(4) + B(2) + C$$
$$31 = 6(4) + 2B + 9$$
$$-2 = 2B$$
$$-1 = B.$$

Therefore, the partial fraction decomposition is

$$\frac{5x^2 + 20x + 6}{x(x + 1)^2} = \frac{6}{x} - \frac{1}{x + 1} + \frac{9}{(x + 1)^2}.$$

The procedure used to solve for the constants $A, B, C, \ldots$ in Examples 1 and 2 works well when the factors of the denominator are linear. However, when the denominator contains irreducible quadratic factors, we use a different procedure, which involves expanding the basic equation. We demonstrate this technique in Examples 3 and 4.

EXAMPLE 3 *Distinct Linear and Quadratic Factors*

Write the partial fraction decomposition for

$$\frac{3x^2 + 4x + 4}{x^3 + 4x}.$$

SOLUTION

Since the denominator factors as

$$x^3 + 4x = x(x^2 + 4)$$

we include one partial fraction with a constant numerator and one partial fraction with a linear numerator and write

$$\frac{3x^2 + 4x + 4}{x^3 + 4x} = \frac{A}{x} + \frac{Bx + C}{x^2 + 4}.$$

Multiplying by the LCD, $x(x^2 + 4)$, yields the basic equation

$$3x^2 + 4x + 4 = A(x^2 + 4) + (Bx + C)x. \qquad \textit{Basic equation}$$

Expanding this basic equation and collecting like terms produces

$$3x^2 + 4x + 4 = Ax^2 + 4A + Bx^2 + Cx$$
$$= (A + B)x^2 + Cx + 4A.$$

Finally, because two polynomials are equal if and only if the coefficients of like terms are equal,

$$3x^2 + 4x + 4 = (A + B)x^2 + Cx + 4A$$

we obtain the following equations.

$$3 = A + B, \quad 4 = C, \quad \text{and} \quad 4 = 4A$$

Thus, $A = 1$ and $C = 4$. Moreover, substituting $A = 1$ in the equation $3 = A + B$, we have

$$3 = 1 + B$$
$$2 = B.$$

Therefore, the partial fraction decomposition is

$$\frac{3x^2 + 4x + 4}{x^3 + 4x} = \frac{1}{x} + \frac{2x + 4}{x^2 + 4}.$$

Rational Functions and Conic Sections

EXAMPLE 4 *Repeated Quadratic Factors*

Write the partial fraction decomposition for

$$\frac{8x^3 + 13x}{(x^2 + 2)^2}.$$

SOLUTION

We include one partial fraction with a linear numerator for each power of $(x^2 + 2)$, and write

$$\frac{8x^3 + 13x}{(x^2 + 2)^2} = \frac{Ax + B}{x^2 + 2} + \frac{Cx + D}{(x^2 + 2)^2}.$$

Multiplying by the LCD, $(x^2 + 2)^2$, yields the basic equation

$$
\begin{aligned}
8x^3 + 13x &= (Ax + B)(x^2 + 2) + Cx + D \qquad \textit{Basic equation}\\
&= Ax^3 + 2Ax + Bx^2 + 2B + Cx + D\\
&= Ax^3 + Bx^2 + (2A + C)x + (2B + D).
\end{aligned}
$$

Equating coefficients of like terms,

$$8x^3 + 0x^2 + 13x + 0 = Ax^3 + Bx^2 + (2A + C)x + (2B + D)$$

produces

$$8 = A, \quad 0 = B, \quad 13 = 2A + C, \quad \text{and} \quad 0 = 2B + D.$$

Finally, using the values $A = 8$ and $B = 0$, we have the following.

$$
\begin{aligned}
13 &= 2A + C = 2(8) + C & 0 &= 2B + D = 2(0) + D\\
-3 &= C & 0 &= D
\end{aligned}
$$

Therefore, we conclude that

$$\frac{8x^3 + 13x}{(x^2 + 2)^2} = \frac{8x}{x^2 + 2} + \frac{-3x}{(x^2 + 2)^2}.$$

Remark: By equating coefficients of like terms in Examples 3 and 4, we obtained several equations involving A, B, C, and D, which we solved by a method known as *substitution*. In Chapter 10 we will discuss a more general method for solving systems of equations.

The following guidelines summarize the two techniques we have used to solve for A, B, C, . . . in the basic equation.

Partial Fractions

Guidelines for Solving the Basic Equation

Linear Factors:

1. Substitute the *zeros* of the distinct linear factors into the basic equation.
2. For repeated linear factors, use the coefficients determined in Step 1 to rewrite the basic equation. Then substitute *other* convenient values for x and solve for the remaining coefficients.

Quadratic Factors:

1. Expand the basic equation.
2. Collect terms according to powers of x.
3. Equate the coefficients of like terms to obtain equations involving A, B, C, and so on.
4. Use substitution to solve for A, B, C, ...

Keep in mind that for *improper* rational expressions like

$$\frac{N(x)}{D(x)} = \frac{2x^3 + x^2 - 7x + 7}{x^2 + x - 2}$$

you must first divide to obtain the form

$$\frac{N(x)}{D(x)} = (\text{polynomial}) + \frac{N_1(x)}{D(x)}.$$

The proper rational expression $N_1(x)/D(x)$ is then decomposed into its partial fractions by the usual methods.

WARM UP

Find the sums and simplify.

1. $\dfrac{2}{x} + \dfrac{3}{x + 1}$

2. $\dfrac{5}{x + 2} + \dfrac{3}{x}$

3. $\dfrac{7}{x - 2} - \dfrac{3}{2x - 1}$

4. $\dfrac{2}{x + 5} - \dfrac{5}{x + 12}$

5. $\dfrac{1}{x - 3} + \dfrac{3}{(x - 3)^2} - \dfrac{5}{(x - 3)^3}$

6. $\dfrac{-5}{x + 2} + \dfrac{4}{(x + 2)^2}$

7. $\dfrac{-3}{x} + \dfrac{3x - 1}{x^2 + 3}$

8. $\dfrac{5}{x + 1} - \dfrac{x - 6}{x^2 + 5}$

9. $\dfrac{3}{x^2 + 1} + \dfrac{x - 3}{(x^2 + 1)^2}$

10. $\dfrac{x}{x^2 + x + 1} - \dfrac{x - 1}{(x^2 + x + 1)^2}$

EXERCISES 5.2

In Exercises 1–36, write the partial fraction decomposition for the rational expression.

1. $\dfrac{1}{x^2 - 1}$

2. $\dfrac{1}{4x^2 - 9}$

3. $\dfrac{1}{x^2 + x}$

4. $\dfrac{3}{x^2 - 3x}$

5. $\dfrac{1}{2x^2 + x}$

6. $\dfrac{5}{x^2 + x - 6}$

7. $\dfrac{3}{x^2 + x - 2}$

8. $\dfrac{x + 1}{x^2 + 4x + 3}$

9. $\dfrac{5 - x}{2x^2 + x - 1}$

10. $\dfrac{3x^2 - 7x - 2}{x^3 - x}$

11. $\dfrac{x^2 + 12x + 12}{x^3 - 4x}$

12. $\dfrac{x + 2}{x(x - 4)}$

13. $\dfrac{4x^2 + 2x - 1}{x^2(x + 1)}$

14. $\dfrac{2x - 3}{(x - 1)^2}$

15. $\dfrac{x - 1}{x^3 + x^2}$

16. $\dfrac{4x^2 - 1}{2x(x + 1)^2}$

17. $\dfrac{3x}{(x - 3)^2}$

18. $\dfrac{6x^2 + 1}{x^2(x - 1)^3}$

19. $\dfrac{x^2 - 1}{x(x^2 + 1)}$

20. $\dfrac{x}{(x - 1)(x^2 + x + 1)}$

21. $\dfrac{x^2}{x^4 - 2x^2 - 8}$

22. $\dfrac{2x^2 + x + 8}{(x^2 + 4)^2}$

23. $\dfrac{x}{16x^4 - 1}$

24. $\dfrac{x^2 - 4x + 7}{(x + 1)(x^2 - 2x + 3)}$

25. $\dfrac{x^2 + x + 2}{(x^2 + 2)^2}$

26. $\dfrac{x^3}{(x + 2)^2(x - 2)^2}$

27. $\dfrac{x^2 + 5}{(x + 1)(x^2 - 2x + 3)}$

28. $\dfrac{x + 1}{x^3 + x}$

29. $\dfrac{2x^3 - 4x^2 - 15x + 5}{x^2 - 2x - 8}$

30. $\dfrac{x^3 - x + 3}{x^2 + x - 2}$

31. $\dfrac{x^4}{(x - 1)^3}$

32. $\dfrac{x^2 - x}{x^2 + x + 1}$

33. $\dfrac{1}{a^2 - x^2}$, a is a constant

34. $\dfrac{1}{x(x + a)}$, a is a constant

35. $\dfrac{1}{y(L - y)}$, L is a constant

36. $\dfrac{1}{(x + 1)(n - x)}$, n is a positive integer

5.3 Conic Sections

Conic sections were discovered during the classical Greek period, which lasted from 600 to 300 B.C. By the beginning of the Alexandrian period, enough was known of conics for Apollonius (262–190 B.C.) to produce an eight-volume work on the subject.

This early Greek study was largely concerned with the geometrical properties of conics. It was not until the early seventeenth century that the broad applicability of conics became apparent and played a prominent role in the early development of calculus.

A **conic section** (or simply **conic**) can be described as the intersection of a plane and a double-napped cone. Notice from Figure 5.12 that in the formation of the four basic conics, the intersecting plane does not pass through the vertex of the cone. When the plane does pass through the vertex, we call the resulting figure a **degenerate conic,** as shown in Figure 5.13.

There are several ways to approach the study of conics. We could begin by defining conics in terms of the intersections of planes and cones, as the

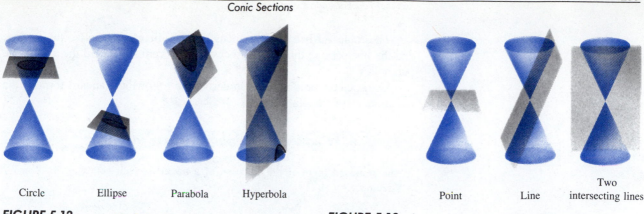

Circle   Ellipse   Parabola   Hyperbola

FIGURE 5.12   Conic Sections

Point   Line   Two intersecting lines

FIGURE 5.13   Degenerate Conics

Greeks did, or we could define them algebraically, in terms of the general second-degree equation

$$Ax^2 + Bxy + Cy^2 + Dx + Ey + F = 0.$$

However, we will use a third approach, in which each of the conics is defined as a *locus* (collection) of points satisfying a certain geometric property. For example, in Section 3.2, we saw how the definition of a circle as *the collection of all points (x, y) that are equidistant from a fixed point (h, k)* led easily to the standard equation of a circle, $(x - h)^2 + (y - k)^2 = r^2$.

We will restrict our study of conics in this section to parabolas with vertices at the origin and ellipses and hyperbolas with centers at the origin. In the following section, we will look at the more general cases.

Parabolas

In Section 4.1, we determined that the graph of the quadratic function $f(x) = ax^2 + bx + c$ is a parabola that opens upward or downward. The following definition of a parabola is more general in the sense that it is independent of the orientation of the parabola.

Definition of a Parabola

A **parabola** is the set of all points (x, y) that are equidistant from a fixed line called the **directrix** and a fixed point called the **focus** (not on the line). See Figure 5.14.

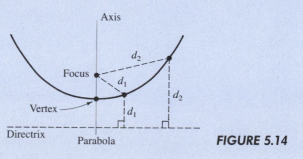

FIGURE 5.14

The midpoint between the focus and the directrix is called the **vertex,** and the line passing through the focus and the vertex is called the **axis** of the parabola.

Using this definition, we can derive the following standard form of the equation of a parabola.

Standard Equation of a Parabola (Vertex at Origin)

The **standard form of the equation of a parabola** with vertex at $(0, 0)$ and directrix $y = -p$ is

$$x^2 = 4py, \quad p \neq 0. \qquad \text{\textit{Vertical axis}}$$

For directrix $x = -p$, the equation is

$$y^2 = 4px, \quad p \neq 0. \qquad \text{\textit{Horizontal axis}}$$

The focus is on the axis p units (directed distance) from the vertex.

PROOF

Since the two cases are similar, we give a proof for the first case only. Suppose the directrix ($y = -p$) is parallel to the x-axis. In Figure 5.15 we assume $p > 0$, and since p is the directed distance from the vertex to the focus, the focus must lie above the vertex. Since the point (x, y) is equidistant from $(0, p)$ and $y = -p$, we can apply the distance formula to obtain

$$\sqrt{(x - 0)^2 + (y - p)^2} = y + p$$
$$x^2 + (y - p)^2 = (y + p)^2$$
$$x^2 + y^2 - 2py + p^2 = y^2 + 2py + p^2$$
$$x^2 = 4py.$$

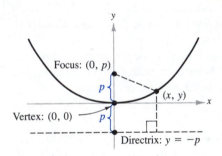

FIGURE 5.15

EXAMPLE 1 Finding the Focus of a Parabola

Find the focus of the parabola whose equation is

$$y = -2x^2.$$

SOLUTION

Since the squared term in the equation involves x, we know that the axis is vertical, and we have the standard form

$$x^2 = 4py.$$

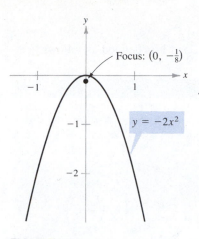

FIGURE 5.16

Writing the given equation in this form, we have

$$x^2 = -\frac{1}{2}y$$

$$x^2 = 4\left(-\frac{1}{8}\right)y.$$

Thus, we have $p = -1/8$. Since p is negative, the parabola opens downward (see Figure 5.16), and the focus of the parabola is

$$(0, p) = \left(0, -\frac{1}{8}\right).$$

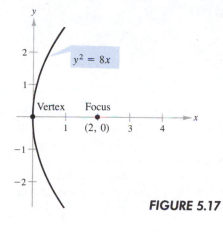

EXAMPLE 2 A Parabola with a Horizontal Axis

Write the standard form of the equation of the parabola with vertex at the origin and focus at (2, 0).

SOLUTION

The axis of the parabola is horizontal, passing through (0, 0) and (2, 0), as shown in Figure 5.17. Thus, we consider the form

$$y^2 = 4px.$$

Since $p = 2$, the equation is

$$y^2 = 4(2)x$$
$$y^2 = 8x.$$

FIGURE 5.17

Remark: Be sure you understand that the term *parabola* is a technical term used in mathematics and does not simply refer to *any* U-shaped curve.

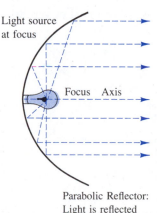

Parabolic Reflector: Light is reflected in parallel rays.

FIGURE 5.18

Parabolas occur in a wide variety of applications. For instance, a parabolic reflector can be formed by revolving a parabola about its axis. The resulting surface has the property that all incoming rays parallel to the axis are reflected through the focus of the parabola—this is the principle behind the construction of the parabolic mirrors used in reflecting telescopes. Conversely, the light rays emanating from the focus of a parabolic reflector used in a flashlight are all parallel to one another, as shown in Figure 5.18.

Ellipses

Another type of conic is called an **ellipse,** and it is defined as follows.

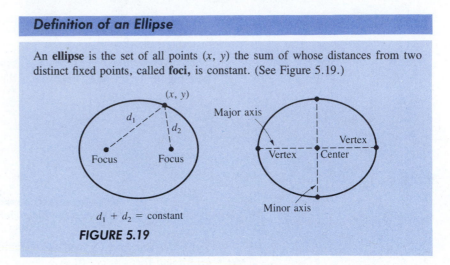

Definition of an Ellipse

An **ellipse** is the set of all points (x, y) the sum of whose distances from two distinct fixed points, called **foci,** is constant. (See Figure 5.19.)

$d_1 + d_2 =$ constant

FIGURE 5.19

The line through the foci intersects the ellipse at two points, called the **vertices.** The chord joining the vertices is called the **major axis,** and its midpoint is called the **center** of the ellipse. The chord perpendicular to the major axis at the center is called the **minor axis** of the ellipse.

You can visualize the definition of an ellipse by imagining two thumbtacks placed at the foci, as shown in Figure 5.20. If the ends of a fixed length of string are fastened to the thumbtacks and the string is drawn taut with a pencil, the path traced by the pencil will be an ellipse.

The standard form of the equation of an ellipse takes one of two forms, depending upon whether the major axis is horizontal or vertical.

FIGURE 5.20

Standard Equation of an Ellipse (Center at Origin)

The **standard form of the equation of an ellipse** with center at the origin and major and minor axes of lengths $2a$ and $2b$, respectively, is

$$\frac{x^2}{a^2} + \frac{y^2}{b^2} = 1 \qquad \text{or} \qquad \frac{x^2}{b^2} + \frac{y^2}{a^2} = 1, \ 0 < b < a.$$

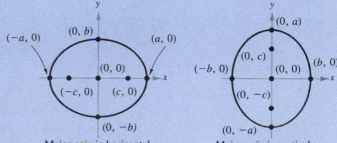

Major axis is horizontal.
Minor axis is vertical.

Major axis is vertical.
Minor axis is horizontal.

FIGURE 5.21

The vertices and foci lie on the major axis, a and c units, respectively, from the center. Moreover, a, b, and c are related by the equation $c^2 = a^2 - b^2$. (See Figure 5.21.)

EXAMPLE 3 *Finding the Standard Equation of an Ellipse*

Find the standard form of the equation of the ellipse that has a major axis of length 6 and foci at $(-2, 0)$ and $(2, 0)$, as shown in Figure 5.22.

SOLUTION

Since the foci occur at $(-2, 0)$ and $(2, 0)$, the center of the ellipse is $(0, 0)$, and the major axis is horizontal. Thus, the ellipse has an equation of the form

$$\frac{x^2}{a^2} + \frac{y^2}{b^2} = 1.$$

Since the length of the major axis is 6, we have

$$2a = 6 \qquad\qquad \textit{Length of major axis}$$

which implies that $a = 3$. Moreover, the distance from the center to either focus is $c = 2$. Finally, we have

$$b^2 = a^2 - c^2 = 3^2 - 2^2 = 9 - 4 = 5$$

which yields the equation

$$\frac{x^2}{9} + \frac{y^2}{5} = 1.$$

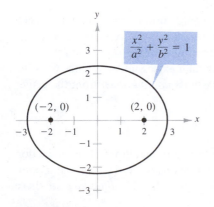

FIGURE 5.22

EXAMPLE 4 *Sketching an Ellipse*

Sketch the ellipse given by $4x^2 + y^2 = 36$, and identify the vertices.

SOLUTION

We begin by writing the equation in standard form.

$$4x^2 + y^2 = 36 \qquad \textit{Given equation}$$

$$\frac{4x^2}{36} + \frac{y^2}{36} = \frac{36}{36}$$

$$\frac{x^2}{3^2} + \frac{y^2}{6^2} = 1 \qquad \textit{Standard form}$$

Since the denominator of the y^2 term is larger than the denominator of the x^2 term, we conclude that the major axis is vertical. Moreover, since $a = 6$, the vertices are $(0, -6)$ and $(0, 6)$. Finally, since $b = 3$, the endpoints of the minor axis are $(-3, 0)$ and $(3, 0)$, as shown in Figure 5.23.

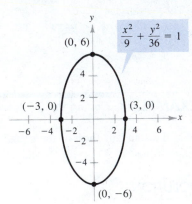

FIGURE 5.23

In Example 4 note that from the standard form of the equation we can sketch the ellipse by locating the endpoints of the two axes. Since 3^2 is the denominator of the x^2 term, we move three units to the *right and left* of the center to locate the endpoints of the horizontal axis. Similarly, since 6^2 is the denominator of the y^2 term, we move six units *up and down* from the center to locate the endpoints of the vertical axis.

Hyperbolas

The definition of a **hyperbola** is similar to that of an ellipse. The distinction is that, for an ellipse, the *sum* of the distances between the foci and a point on the ellipse is constant, while, for a hyperbola, the *difference* of these distances is constant.

Definition of a Hyperbola

A **hyperbola** is the set of all points (x, y), the difference of whose distances from two distinct fixed points, called **foci,** is constant. (See Figure 5.24.)

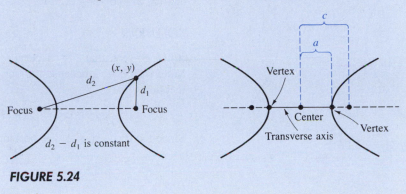

FIGURE 5.24

The graph of a hyperbola has two disconnected parts, called **branches.** The line through the two foci intersects the hyperbola at two points, called **vertices.** The line segment connecting the vertices is called the **transverse axis,** and the midpoint of the transverse axis is called the **center** of the hyperbola.

Standard Equation of a Hyperbola (Center at Origin)

The **standard form of the equation of a hyperbola** with center at $(0, 0)$ is

$$\frac{x^2}{a^2} - \frac{y^2}{b^2} = 1 \quad \text{or} \quad \frac{y^2}{a^2} - \frac{x^2}{b^2} = 1.$$

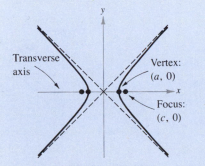
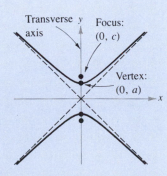

FIGURE 5.25

The vertices and foci are, respectively, a and c units from the center, and $b^2 = c^2 - a^2$. (See Figure 5.25.)

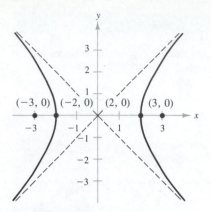

FIGURE 5.26

EXAMPLE 5 *Finding the Standard Equation of a Hyperbola*

Find the standard form of the equation of the hyperbola with foci at $(-3, 0)$ and $(3, 0)$ and vertices at $(-2, 0)$ and $(2, 0)$, as shown in Figure 5.26.

SOLUTION

It can be seen that $c = 3$ because the foci are three units from the center. Moreover, $a = 2$ because the vertices are two units from the center. Thus, it follows that

$$b^2 = c^2 - a^2 = 3^2 - 2^2 = 9 - 4 = 5.$$

Since the transverse axis is horizontal, the standard form of the equation is

$$\frac{x^2}{a^2} - \frac{y^2}{b^2} = 1.$$

Finally, substituting $a^2 = 4$ and $b^2 = 5$, we have

$$\frac{x^2}{4} - \frac{y^2}{5} = 1.$$

An important aid in sketching the graph of a hyperbola is the determination of its **asymptotes,** as shown in Figure 5.27. Each hyperbola has two asymptotes that intersect at the center of the hyperbola. Furthermore, the asymptotes pass through the corners of a rectangle of dimensions $2a$ by $2b$. The line segment of length $2b$, joining $(0, b)$ and $(0, -b)$ [or $(-b, 0)$ and $(b, 0)$], is referred to as the **conjugate axis** of the hyperbola.

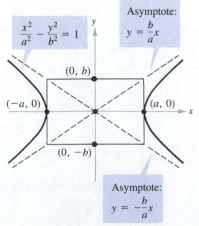

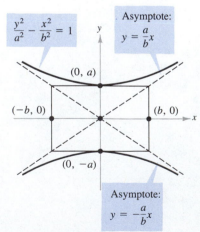

FIGURE 5.27   Transverse axis is horizontal.   Transverse axis is vertical.

Asymptotes of a Hyperbola (Center at Origin)

The asymptotes of a hyperbola with center at (0, 0) are

$$y = \frac{b}{a}x \quad \text{and} \quad y = -\frac{b}{a}x$$

Transverse axis is horizontal

or

$$y = \frac{a}{b}x \quad \text{and} \quad y = -\frac{a}{b}x.$$

Transverse axis is vertical

EXAMPLE 6   *Sketching the Graph of a Hyperbola*

Sketch the graph of the hyperbola whose equation is $4x^2 - y^2 = 16$.

SOLUTION

We begin by rewriting the equation in standard form.

$$4x^2 - y^2 = 16 \qquad \textit{Given equation}$$

$$\frac{4x^2}{16} - \frac{y^2}{16} = \frac{16}{16}$$

$$\frac{x^2}{2^2} - \frac{y^2}{4^2} = 1 \qquad \textit{Standard form}$$

Because the x^2-term is positive, we conclude that the transverse axis is horizontal and the vertices occur at $(-2, 0)$ and $(2, 0)$. Moreover, the endpoints of the conjugate axis occur at $(0, -4)$ and $(0, 4)$, and we are able to sketch the rectangle shown in Figure 5.28. Finally, by drawing the asymptotes through the corners of this rectangle, we complete the sketch shown in Figure 5.29.

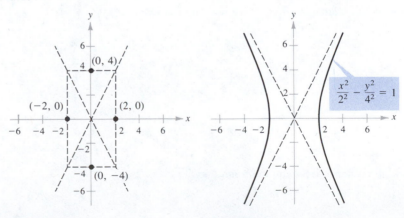

FIGURE 5.28   **FIGURE 5.29**

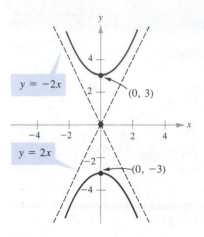

FIGURE 5.30

EXAMPLE 7 *Finding the Standard Equation of a Hyperbola*

Find the standard form of the equation of the hyperbola having vertices at $(0, -3)$ and $(0, 3)$ and with asymptotes $y = -2x$ and $y = 2x$, as shown in Figure 5.30.

SOLUTION

Since the transverse axis is vertical, we have asymptotes of the form

$$y = \frac{a}{b}x \quad \text{and} \quad y = -\frac{a}{b}x.$$

Thus, we have

$$\frac{a}{b} = 2$$

and since $a = 3$, we can determine that $b = 3/2$. Finally, we see that the hyperbola has the following equation.

$$\frac{y^2}{3^2} - \frac{x^2}{(3/2)^2} = 1$$

WARM UP

Rewrite the equations so that they have no fractions.

1. $\dfrac{x^2}{16} + \dfrac{y^2}{9} = 1$

2. $\dfrac{x^2}{32} + \dfrac{4y^2}{32} = \dfrac{32}{32}$

3. $\dfrac{x^2}{1/4} - \dfrac{y^2}{4} = 1$

4. $\dfrac{3x^2}{1/9} + \dfrac{4y^2}{9} = 1$

Solve for c. (Assume $c > 0$.)

5. $c^2 = 3^2 - 1^2$

6. $c^2 = 2^2 + 3^2$

7. $c^2 + 2^2 = 4^2$

8. $c^2 - 1^2 = 2^2$

Find the distance between the point and the origin.

9. $(0, -4)$

10. $(-2, 0)$

EXERCISES 5.3

In Exercises 1–8, match the equation with its graph. [The graphs are labeled (a)–(h).]

1. $x^2 = 4y$

2. $x^2 = -4y$

3. $y^2 = 4x$

4. $y^2 = -4x$

5. $\dfrac{x^2}{1} + \dfrac{y^2}{4} = 1$

6. $\dfrac{x^2}{4} + \dfrac{y^2}{1} = 1$

7. $\dfrac{x^2}{1} - \dfrac{y^2}{4} = 1$

8. $\dfrac{y^2}{4} - \dfrac{x^2}{1} = 1$

(a)
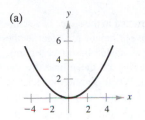

(b)
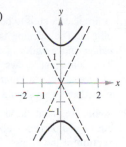

(c)
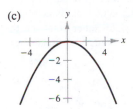

(d)
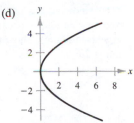

(e)
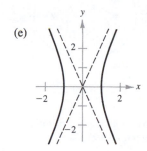

(f)
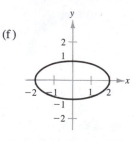

(g)
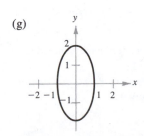

(h)
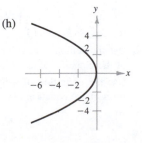

In Exercises 9–16, find the vertex and focus of the parabola and sketch its graph.

9. $y = 4x^2$

10. $y = 2x^2$

11. $y^2 = -6x$

12. $y^2 = 3x$

13. $x^2 + 8y = 0$

14. $x + y^2 = 0$

15. $y^2 - 8x = 0$

16. $x^2 + 12y = 0$

In Exercises 17–26, find an equation of the specified parabola with vertex at the origin.

17. Focus: $(0, -3/2)$

18. Focus: $(2, 0)$

19. Focus: $(-2, 0)$

20. Focus: $(0, -2)$

21. Directrix: $y = -1$

22. Directrix: $x = 3$

23. Directrix: $y = 2$

24. Directrix: $x = -2$

25. Horizontal axis and passes through the point $(4, 6)$

26. Vertical axis and passes through the point $(-2, -2)$

In Exercises 27–34, find the center and vertices of the ellipse and sketch its graph.

27. $\dfrac{x^2}{25} + \dfrac{y^2}{16} = 1$

28. $\dfrac{x^2}{144} + \dfrac{y^2}{169} = 1$

29. $\dfrac{x^2}{16} + \dfrac{y^2}{25} = 1$

30. $\dfrac{x^2}{169} + \dfrac{y^2}{144} = 1$

31. $\dfrac{x^2}{9} + \dfrac{y^2}{5} = 1$

32. $\dfrac{x^2}{28} + \dfrac{y^2}{64} = 1$

33. $5x^2 + 3y^2 = 15$

34. $x^2 + 4y^2 = 4$

In Exercises 35–42, find an equation of the specified ellipse with center at the origin.

35. Vertices: $(0, \pm 2)$; Minor axis of length 2

36. Vertices: $(\pm 2, 0)$; Minor axis of length 3

37. Vertices: $(\pm 5, 0)$; Foci: $(\pm 2, 0)$

38. Vertices: $(0, \pm 8)$; Foci: $(0, \pm 4)$

39. Foci: $(\pm 5, 0)$; Major axis of length 12

40. Foci: $(\pm 2, 0)$; Major axis of length 8

41. Vertices: $(0, \pm 5)$; Passes through the point $(4, 2)$

42. Major axis vertical;
Passes through points $(0, 4)$ and $(2, 0)$

In Exercises 43–50, find the center and vertices of the hyperbola and sketch its graph, using asymptotes as an aid.

43. $x^2 - y^2 = 1$

44. $\dfrac{x^2}{9} - \dfrac{y^2}{16} = 1$

45. $\dfrac{y^2}{1} - \dfrac{x^2}{4} = 1$

46. $\dfrac{y^2}{9} - \dfrac{x^2}{1} = 1$

47. $\dfrac{y^2}{25} - \dfrac{x^2}{144} = 1$

48. $\dfrac{x^2}{36} - \dfrac{y^2}{4} = 1$

49. $2x^2 - 3y^2 = 6$

50. $3y^2 - 5x^2 = 15$

In Exercises 51–58, find an equation of the specified hyperbola with center at the origin.

51. Vertices: $(0, \pm 2)$; Foci: $(0, \pm 4)$

52. Vertices: $(\pm 3, 0)$; Foci: $(\pm 5, 0)$

53. Vertices: $(\pm 1, 0)$; Asymptotes: $y = \pm 3x$

54. Vertices: $(0, \pm 3)$; Asymptotes: $y = \pm 3x$

55. Foci: $(0, \pm 8)$; Asymptotes: $y = \pm 4x$

56. Foci: $(\pm 10, 0)$; Asymptotes: $y = \pm\frac{3}{4}x$

57. Vertices: $(0, \pm 3)$; Passes through the point $(-2, 5)$

58. Vertices: $(\pm 2, 0)$; Passes through the point $(3, \sqrt{3})$

59. The receiver in a parabolic television dish antenna is 3 feet from the vertex and is located at the focus (see figure). Find an equation of a cross section of the reflector. (Assume that the dish is directed upward and the vertex is at the origin.)

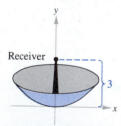

FIGURE FOR 59

60. Each cable of a suspension bridge is suspended (in the shape of a parabola) between two towers that are 400 feet apart and 50 feet above the roadway (see figure). The cables touch the roadway midway between the towers. Find an equation for the parabolic shape of each cable.

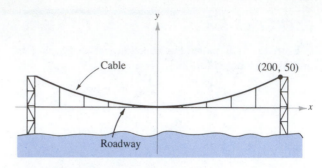

FIGURE FOR 60

61. A fireplace arch is to be constructed in the shape of a semi-ellipse. The opening is to have a height of 2 feet at the center and a width of 5 feet along the base (see figure). The contractor draws the outline of the ellipse by the method shown in Figure 5.20. Where should the tacks be placed and what should be the length of the piece of string?

62. A line segment through a focus with endpoints on the ellipse and perpendicular to the major axis is called a **latus rectum** of the ellipse. Therefore, an ellipse has two latus recta. Knowing the length of the latus recta is helpful in sketching an ellipse because it yields other points on the curve (see figure). Show that the length of each latus rectum is $2b^2/a$.

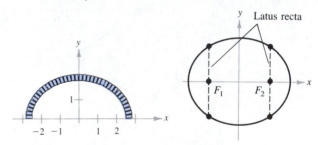

FIGURE FOR 61   **FIGURE FOR 62**

63. Sketch the graph of each ellipse, making use of the endpoints of the latus recta (see Exercise 62).

(a) $\dfrac{x^2}{4} + \dfrac{y^2}{1} = 1$ (b) $6x^2 + 4y^2 = 1$

(c) $5x^2 + 3y^2 = 15$

64. Use the definition of an ellipse to derive the standard form of the equation of an ellipse.

65. Use the definition of a hyperbola to derive the standard form of the equation of a hyperbola.

5.4 Conic Sections and Translations

In Section 5.3, we looked at conic sections whose graphs were in *standard position*. In this section, we will study the equations of conic sections that have been shifted vertically or horizontally in the plane. The following summary lists the standard forms of the equations of the four basic conics.

Standard Forms of Equations of Conics

Circle (r = radius)

$$(x - h)^2 + (y - k)^2 = r^2$$

Ellipse ($2a$ = major axis length, $2b$ = minor axis length):

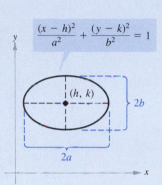

$$\frac{(x - h)^2}{a^2} + \frac{(y - k)^2}{b^2} = 1$$

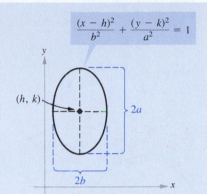

$$\frac{(x - h)^2}{b^2} + \frac{(y - k)^2}{a^2} = 1$$

FIGURE 5.31

Hyperbola ($2a$ = transverse axis length, $2b$ = conjugate axis length)

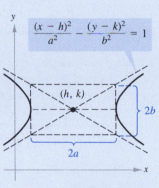

$$\frac{(x - h)^2}{a^2} - \frac{(y - k)^2}{b^2} = 1$$

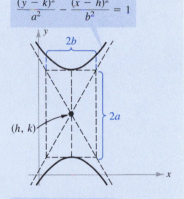

$$\frac{(y - k)^2}{a^2} - \frac{(x - h)^2}{b^2} = 1$$

FIGURE 5.32

Parabola (p = directed distance from vertex to focus)

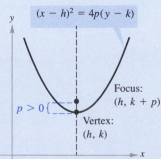

$$(x - h)^2 = 4p(y - k)$$

Focus: $(h, k + p)$

Vertex: (h, k)

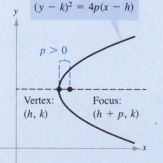

$$(y - k)^2 = 4p(x - h)$$

$p > 0$

Vertex: (h, k) Focus: $(h + p, k)$

FIGURE 5.33

To write the equation of a conic in standard form, we complete the square as demonstrated in Examples 1, 2, and 3.

EXAMPLE 1 *Finding the Standard Form of a Parabola*

Find the vertex and focus of the parabola given by

$$x^2 - 2x + 4y - 3 = 0.$$

SOLUTION

$x^2 - 2x + 4y - 3 = 0$	*Given equation*
$x^2 - 2x = -4y + 3$	*Group terms*
$x^2 - 2x + 1 = -4y + 3 + 1$	*Add 1 to both sides*
$(x - 1)^2 = -4y + 4$	*Completed square form*
$(x - 1)^2 = 4(-1)(y - 1)$	*Standard form*
$(x - h)^2 = 4p(y - k)$	

From this standard form, we see that $h = 1$, $k = 1$, and $p = -1$. Since the axis is vertical and p is negative, the parabola opens downward. The vertex and focus are

Vertex: $(h, k) = (1, 1)$

Focus: $(h, k + p) = (1, 0)$.

The graph of this parabola is shown in Figure 5.34.

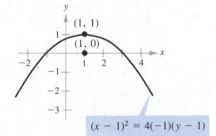

$(x - 1)^2 = 4(-1)(y - 1)$

FIGURE 5.34

Remark: Note in Example 1 that p is the *directed distance* from the vertex to the focus. Because the axis of the parabola is vertical and $p = -1$, the focus is one unit *below* the vertex, and the parabola opens downward.

EXAMPLE 2 *Sketching an Ellipse*

Sketch the graph of the ellipse whose equation is

$$x^2 + 4y^2 + 6x - 8y + 9 = 0.$$

SOLUTION

$x^2 + 4y^2 + 6x - 8y + 9 = 0$	*Given equation*
$(x^2 + 6x + \quad) + (4y^2 - 8y + \quad) = -9$	*Group terms*
$(x^2 + 6x + \quad) + 4(y^2 - 2y + \quad) = -9$	*Factor 4 out of y-terms*
$(x^2 + 6x + 9) + 4(y^2 - 2y + 1) = -9 + 9 + 4(1)$	*Add 9 and 4 to both sides*
$(x + 3)^2 + 4(y - 1)^2 = 4$	*Completed square form*
$\dfrac{(x + 3)^2}{4} + \dfrac{(y - 1)^2}{1} = 1$	*Standard form*

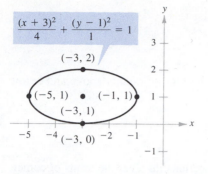

FIGURE 5.35

$$\frac{(x-h)^2}{a^2} + \frac{(y-k)^2}{b^2} = 1$$

From this standard form, we see that the center is $(h, k) = (-3, 1)$. Since the denominator of the x-term is $4 = a^2 = 2^2$, the endpoints of the major axis lie two units to the right and left of the center. Similarly, since the denominator of the y-term is $1 = b^2 = 1^2$, the endpoints of the minor axis lie one unit up and down from the center. The graph of this ellipse is shown in Figure 5.35.

EXAMPLE 3 *Sketching the Graph of a Hyperbola*

Sketch the graph of the hyperbola given by the equation

$$y^2 - 4x^2 + 4y + 24x - 41 = 0.$$

SOLUTION

$y^2 - 4x^2 + 4y + 24x - 41 = 0$	*Given equation*
$(y^2 + 4y + \quad) - (4x^2 - 24x + \quad) = 41$	*Group terms*
$(y^2 + 4y + \quad) - 4(x^2 - 6x + \quad) = 41$	*Factor 4 out of x-terms*
$(y^2 + 4y + 4) - 4(x^2 - 6x + 9) = 41 + 4 - 4(9)$	*Add 4, subtract 36*
$(y + 2)^2 - 4(x - 3)^2 = 9$	*Completed square form*
$\dfrac{(y + 2)^2}{9} - \dfrac{4(x - 3)^2}{9} = 1$	*Divide both sides by 9*
$\dfrac{(y + 2)^2}{9} - \dfrac{(x - 3)^2}{9/4} = 1$	*Change 4 to $\dfrac{1}{1/4}$*
$\dfrac{(y + 2)^2}{3^2} - \dfrac{(x - 3)^2}{(3/2)^2} = 1$	*Standard form*
$\dfrac{(y - h)^2}{a^2} - \dfrac{(x - h)^2}{b^2} = 1$	

From the standard form, we see that the transverse axis is vertical and the center lies at $(h, k) = (3, -2)$. Since the denominator of the y-term is $a^2 = 3^2$, we know that the vertices occur three units above and below the center.

 Vertices: $(3, -5)$ and $(3, 1)$

To sketch the hyperbola, we draw a rectangle whose top and bottom pass through the vertices. Since the denominator of the x-term is $b^2 = (3/2)^2$, we locate the sides of the rectangle $3/2$ units to the right and left of center, as shown in Figure 5.36. Finally, we sketch the asymptotes by drawing lines through the opposite corners of the rectangle. Using these asymptotes, we complete the graph of the hyperbola, as shown in Figure 5.36.

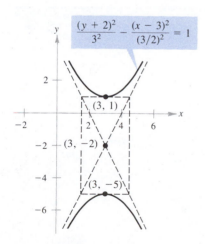

FIGURE 5.36

Rational Functions and Conic Sections

To find the foci in Example 3 we first find c.

$$c^2 = a^2 + b^2 = 9 + \frac{9}{4} = \frac{45}{4} \implies c = \frac{3\sqrt{5}}{2}$$

Because the transverse axis is vertical, the foci lie c units above and below the center.

$$\textit{Foci:} \quad \left(3, -2 + \frac{3\sqrt{5}}{2}\right) \quad \text{and} \quad \left(3, -2 - \frac{3\sqrt{5}}{2}\right)$$

An interesting application of conic sections involves the orbits of comets in our solar system. Of the 610 comets identified prior to 1970, 245 have elliptical orbits, 295 have parabolic orbits, and 70 have hyperbolic orbits. For example, Halley's Comet has an elliptical orbit, and we can predict the reappearance of this comet every 75 years. The center of the sun is a focus of each of these orbits, and each orbit has a vertex at the point where the comet is closest to the sun, as shown in Figure 5.37.

If p is the distance between the vertex and the focus, and v is the speed of the comet at the vertex, then the orbit is:

$$\text{an } \textit{ellipse} \text{ if} \quad v < \sqrt{\frac{2GM}{p}}$$

$$\text{a } \textit{parabola} \text{ if} \quad v = \sqrt{\frac{2GM}{p}}$$

$$\text{a } \textit{hyperbola} \text{ if} \quad v > \sqrt{\frac{2GM}{p}}$$

where M is the mass of the sun and $G \approx 6.67(10^{-8})$ cm^3/(gm $\cdot$ sec^2).

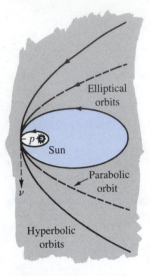

FIGURE 5.37

WARM UP

Describe the conic represented by each equation.

1. $\dfrac{x^2}{4} - \dfrac{y^2}{4} = 1$

2. $\dfrac{x^2}{9} + \dfrac{y^2}{1} = 1$

3. $2x + y^2 = 0$

4. $\dfrac{x^2}{9} - \dfrac{y^2}{4} = 1$

5. $\dfrac{x^2}{4} + \dfrac{y^2}{16} = 1$

6. $4x^2 + 4y^2 = 25$

7. $\dfrac{y^2}{4} - \dfrac{x^2}{2} = 1$

8. $x^2 - 6y = 0$

9. $3x - y^2 = 0$

10. $\dfrac{x^2}{9/4} + \dfrac{y^2}{4} = 1$

EXERCISES 5.4

In Exercises 1–12, find the vertex, focus, and directrix of the parabola, and sketch its graph.

1. $(x - 1)^2 + 8(y + 2) = 0$

2. $(x + 3) + (y - 2)^2 = 0$

3. $\left(y + \frac{1}{2}\right)^2 = 2(x - 5)$ **4.** $\left(x + \frac{1}{2}\right)^2 = 4(y - 3)$

5. $y = \frac{1}{4}(x^2 - 2x + 5)$ **6.** $y = -\frac{1}{6}(x^2 + 4x - 2)$

7. $4x - y^2 - 2y - 33 = 0$

8. $y^2 + x + y = 0$

9. $y^2 + 6y + 8x + 25 = 0$

10. $x^2 - 2x + 8y + 9 = 0$

11. $y^2 - 4y - 4x = 0$ **12.** $y^2 - 4x - 4 = 0$

In Exercises 13–20, find an equation of the specified parabola.

13. Vertex: (3, 2); Focus: (1, 2)

14. Vertex: (−1, 2); Focus: (−1, 0)

15. Vertex: (0, 4); Directrix: $y = 2$

16. Vertex: (−2, 1); Directrix: $x = 1$

17. Focus: (2, 2); Directrix: $x = -2$

18. Focus: (0, 0); Directrix: $y = 4$

19.

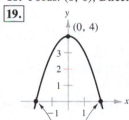

20.

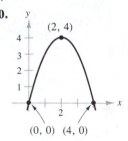

In Exercises 21–28, find the center, foci, and vertices of the ellipse, and sketch the graph.

21. $\dfrac{(x - 1)^2}{9} + \dfrac{(y - 5)^2}{25} = 1$

22. $(x + 2)^2 + 4(y + 4)^2 = 1$

23. $9x^2 + 4y^2 + 36x - 24y + 36 = 0$

24. $9x^2 + 4y^2 - 36x + 8y + 31 = 0$

25. $16x^2 + 25y^2 - 32x + 50y + 16 = 0$

26. $9x^2 + 25y^2 - 36x - 50y + 61 = 0$

27. $12x^2 + 20y^2 - 12x + 40y - 37 = 0$

28. $36x^2 + 9y^2 + 48x - 36y + 43 = 0$

In Exercises 29–36, find an equation for the specified ellipse.

29. Vertices: (0, 2), (4, 2); Minor axis of length 2

30. Foci: (0, 0), (4, 0); Major axis of length 8

31. Foci: (0, 0), (0, 8); Major axis of length 16

32. Center: (2, −1); Vertex: $\left(2, \frac{1}{2}\right)$; Minor axis of length 2

33. Vertices: (3, 1), (3, 9); Minor axis of length 6

34. Center: (3, 2); $a = 3c$; Foci: (1, 2), (5, 2)

35. Center: (0, 4); $a = 2c$; Vertices: (−4, 4), (4, 4)

36. Vertices: (5, 0), (5, 12);
Endpoints of the minor axis: (0, 6), (10, 6)

In Exercises 37–46, find the center, vertices, and foci of the hyperbola and sketch the graph, using asymptotes as an aid.

37. $\dfrac{(x - 1)^2}{4} - \dfrac{(y + 2)^2}{1} = 1$

38. $\dfrac{(x + 1)^2}{144} - \dfrac{(y - 4)^2}{25} = 1$

39. $(y + 6)^2 - (x - 2)^2 = 1$

40. $\dfrac{(y - 1)^2}{1/4} - \dfrac{(x + 3)^2}{1/9} = 1$

41. $9x^2 - y^2 - 36x - 6y + 18 = 0$

42. $x^2 - 9y^2 + 36y - 72 = 0$

43. $9y^2 - x^2 + 2x + 54y + 62 = 0$

44. $16y^2 - x^2 + 2x + 64y + 63 = 0$

45. $x^2 - 9y^2 + 2x - 54y - 80 = 0$

46. $9x^2 - y^2 + 54x + 10y + 55 = 0$

In Exercises 47–54, find an equation for the specified hyperbola.

47. Vertices: (2, 0), (6, 0); Foci: (0, 0), (8, 0)

48. Vertices: (2, 3), (2, −3); Foci: (2, 5), (2, −5)

49. Vertices: (4, 1), (4, 9); Foci: (4, 0), (4, 10)

50. Vertices: (−2, 1), (2, 1); Foci: (−3, 1), (3, 1)

51. Vertices: (2, 3), (2, −3); Passes through the point (0, 5)

52. Vertices: (−2, 1), (2, 1); Passes through the point (4, 3)

53. Vertices: (0, 2), (6, 2);
Asymptotes: $y = \frac{2}{3}x$, $y = 4 - \frac{2}{3}x$

54. Vertices: (3, 0), (3, 4);
Asymptotes: $y = \frac{2}{3}x$, $y = 4 - \frac{2}{3}x$

In Exercises 55–62, classify the graph of each equation as a circle, a parabola, an ellipse, or a hyperbola.

55. $x^2 + y^2 - 6x + 4y + 9 = 0$

56. $x^2 + 4y^2 - 6x + 16y + 21 = 0$

57. $4x^2 - y^2 - 4x - 3 = 0$

58. $y^2 - 4y - 4x = 0$

59. $4x^2 + 3y^2 + 8x - 24y + 51 = 0$

60. $4y^2 - 2x^2 - 4y - 8x - 15 = 0$

61. $25x^2 - 10x - 200y - 119 = 0$

62. $4x^2 + 4y^2 - 16y + 15 = 0$

63. An earth satellite in a 100-mile-high circular orbit around the earth has a velocity of approximately 17,500 miles per hour. If this velocity is multiplied by $\sqrt{2}$, then the satellite will have the minimum velocity necessary to escape the earth's gravity and it will follow a parabolic path with the center of the earth as the focus (see figure).

(a) Find the escape velocity of the satellite.

(b) Find an equation of its path (assume the radius of the earth is 4000 miles).

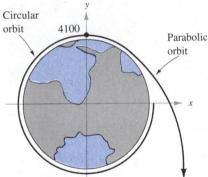

Circular orbit

4100

Parabolic orbit

FIGURE FOR 63

64. Water is flowing from a horizontal pipe 48 feet above the ground. The falling stream of water has the shape of a parabola whose vertex (0, 48) is at the end of the pipe (see figure). The stream of water strikes the ground at the point $(10\sqrt{3}, 0)$. Find the equation of the path taken by the water.

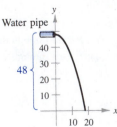

Water pipe

48

FIGURE FOR 64

In Exercises 65–69, *e* is called the **eccentricity** of the ellipse and is defined by $e = c/a$. It measures the flatness of the ellipse.

65. Find an equation of the ellipse with vertices $(\pm 5, 0)$ and eccentricity $e = 3/5$.

66. Find an equation of the ellipse with vertices $(0, \pm 8)$ and eccentricity $e = 1/2$.

67. The earth moves in an elliptical orbit with the sun at one of the foci. The length of half of the major axis is 93 million miles and the eccentricity is 0.017. Find the least and greatest distances of the earth from the sun.

68. The first artificial satellite to orbit the earth was Sputnik I (launched by Russia in 1957). Its highest point above the earth's surface was 583 miles, and its lowest point was 132 miles (see figure). Assume that the center of the earth is the focus of the elliptical orbit and the radius of the earth is 4000 miles. Find the eccentricity of the orbit.

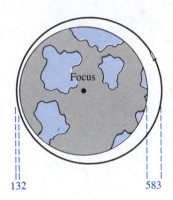

Focus

132   583   *FIGURE FOR 68*

69. Show that the equation of an ellipse can be written as

$$\frac{(x-h)^2}{a^2} + \frac{(y-k)^2}{a^2(1-e^2)} = 1.$$

Note that as *e* approaches zero the ellipse approaches a circle of radius *a*.

CHAPTER 5 REVIEW EXERCISES

In Exercises 1–20, analyze the equation and sketch its graph.

1. $y = \dfrac{2 + x}{1 - x}$

2. $y = \dfrac{x - 3}{x - 2}$

3. $y = \dfrac{x}{x^2 + 1}$

4. $y = \dfrac{2x}{x^2 + 4}$

5. $y = \dfrac{x^2}{x^2 + 1}$

6. $y = \dfrac{2x^2}{x^2 + 4}$

7. $y = \dfrac{x}{x^2 - 1}$

8. $y = \dfrac{2x}{x^2 - 4}$

9. $y = \dfrac{2x^2}{x^2 - 4}$

10. $y = \dfrac{x^2 + 1}{x + 1}$

11. $y = \dfrac{2x^3}{x^2 + 1}$

12. $xy - 4 = 0$

13. $5y^2 - 4x^2 = 20$

14. $y^2 - 12y - 8x + 20 = 0$

15. $x^2 - 6x + 2y + 9 = 0$

16. $4x^2 + y^2 - 16x + 15 = 0$

17. $x^2 + y^2 - 2x - 4y + 5 = 0$

18. $16x^2 + 16y^2 - 16x + 24y - 3 = 0$

19. $x^2 + 9y^2 + 10x - 18y + 25 = 0$

20. $x^2 - 9y^2 + 10x + 18y + 7 = 0$

In Exercises 21–28, write the partial fraction decomposition for the rational expression.

21. $\dfrac{4 - x}{x^2 + 6x + 8}$

22. $\dfrac{-x}{x^2 + 3x + 2}$

23. $\dfrac{x^2}{x^2 + 2x - 15}$

24. $\dfrac{9}{x^2 - 9}$

25. $\dfrac{x^2 + 2x}{x^3 - x^2 + x - 1}$

26. $\dfrac{4x - 2}{3(x - 1)^2}$

27. $\dfrac{3x^3 + 4x}{(x^2 + 1)^2}$

28. $\dfrac{4x^2}{(x - 1)(x^2 + 1)}$

In Exercises 29–32, find an equation of the specified parabola.

29. Vertex: (4, 2); Focus: (4, 0)

30. Vertex: (2, 0); Focus: (0, 0)

31. Vertex: (0, 2); Passes through point $(-1, 0)$; Horizontal axis

32. Vertex: (2, 2); Directrix: $y = 0$

In Exercises 33–36, find an equation of the specified ellipse.

33. Vertices: $(-3, 0)$, (7, 0); Foci: (0, 0), (4, 0)

34. Vertices: (2, 0), (2, 4); Foci: (2, 1), (2, 3)

35. Vertices: $(0, \pm6)$; Passes through the point (2, 2)

36. Vertices: (0, 1), (4, 1); Endpoints of minor axis: (2, 0), (2, 2)

In Exercises 37–40, find an equation of the specified hyperbola.

37. Vertices: $(0, \pm1)$; Foci: $(0, \pm3)$

38. Vertices: (2, 2), $(-2, 2)$; Foci: (4, 2), $(-4, 2)$

39. Foci: (0, 0), (8, 0); Asymptotes: $y = \pm2(x - 4)$

40. Foci: $(3, \pm2)$; Asymptotes: $y = \pm2(x - 3)$

41. The cost in millions of dollars for the government to seize $p\%$ of a certain illegal drug as it enters the country is given by

$$C = \dfrac{528p}{100 - p}, \qquad 0 \le p < 100.$$

(a) Find the cost of seizing 25%.

(b) Find the cost of seizing 50%.

(c) Find the cost of seizing 75%.

(d) What does the cost approach as p approaches 100?

42. A business has a cost of $C = 0.5x + 500$ for producing x units. The average cost per unit is

$$\overline{C} = \frac{C}{x} = \frac{0.5x + 500}{x}, \qquad 0 < x.$$

Determine the average cost per unit as x increases without bound. (Find the horizontal asymptote.)

43. A cross section of a large parabolic antenna is given by $y = x^2/200$, $0 \le x \le 100$ (see figure). The receiving and transmitting equipment is positioned at the focus. Find the coordinates of the focus.

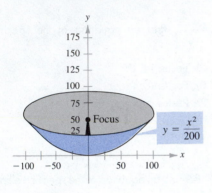

FIGURE FOR 43

44. A semi-elliptical archway is to be formed over the entrance to an estate. The arch is to set on pillars that are 10 feet apart and is to have a height (atop the pillars) of 4 feet (see figure). Where should the foci be placed in order to sketch the elliptic arch?

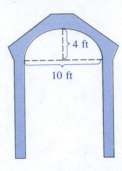

FIGURE FOR 44

CHAPTER 6

Exponential and Logarithmic Functions

6.1 Exponential Functions

Thus far in the text, we have dealt only with **algebraic functions,** which include polynomial functions and rational functions. In this chapter we study two types of nonalgebraic functions—*exponential* functions and *logarithmic* functions. These functions are called **transcendental functions.** (The trigonometric and inverse trigonometric functions to be discussed in Chapter 7 are also transcendental.)

Exponential functions are widely used in describing economic and physical phenomena such as compound interest, population growth, memory retention, and decay of radioactive material. Exponential functions involve a *constant base* and a *variable exponent* such as

$$f(x) = 2^x \qquad \text{or} \qquad g(x) = 3^{-x}.$$

The general definition of the **exponential function with base a** is as follows.

Definition of Exponential Function

The **exponential function f with base a** is denoted by

$$f(x) = a^x$$

where $a > 0$, $a \neq 1$, and x is any real number.

Remark: We exclude the base $a = 1$ because it yields $f(x) = 1^x = 1$. This is a constant function, not an exponential function.

In Sections 1.2 and 1.3 we learned to evaluate a^x for integer and rational values of x. For example, we know that

$$8^3 = 8 \cdot 8 \cdot 8 = 512 \quad \text{and} \quad 8^{2/3} = (\sqrt[3]{8})^2 = (2)^2 = 4.$$

However, to evaluate 8^x for any real x, we need to interpret forms with *irrational* exponents, such as $8^{\sqrt{2}}$ and 8^π. A technical definition of such forms is beyond the scope of this text. For our purposes it is sufficient to think of

$$a^{\sqrt{2}} \quad \text{(where } \sqrt{2} \approx 1.414214\text{)}$$

as that value which has the successively closer approximations

$$a^{1.4}, \ a^{1.41}, \ a^{1.414}, \ a^{1.4142}, \ a^{1.41421}, \ \ldots \ .$$

Consequently, we assume in this text that a^x exists for all real x and that the properties of exponents (Section 1.2) can be extended to cover exponential functions. For instance,

$$a^{-x} = \frac{1}{a^x} = \left(\frac{1}{a}\right)^x.$$

EXAMPLE 1 Graphs of $y = a^x$

On the same coordinate plane, sketch the graphs of the following functions.

(a) $f(x) = 2^x$ (b) $g(x) = 4^x$

SOLUTION

Table 6.1 lists some values for each function, and Figure 6.1 shows their graphs.

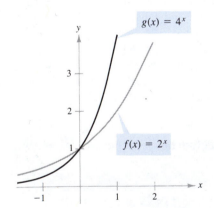

FIGURE 6.1

TABLE 6.1

x	-2	-1	0	1	2	3
(a) 2^x	$\dfrac{1}{4}$	$\dfrac{1}{2}$	1	2	4	8
(b) 4^x	$\dfrac{1}{16}$	$\dfrac{1}{4}$	1	4	16	64

EXAMPLE 2 Graphs of $y = a^{-x}$

On the same coordinate plane, sketch the graphs of the following functions.

(a) $F(x) = 2^{-x}$

(b) $G(x) = \left(\dfrac{1}{4}\right)^x$

SOLUTION

To evaluate $F(x)$, we write

$$F(x) = 2^{-x} = \frac{1}{2^x} = \left(\frac{1}{2}\right)^x.$$

Table 6.2 lists some values for each function, and Figure 6.2 shows their graphs.

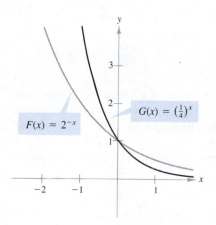

$G(x) = \left(\frac{1}{4}\right)^x$

$F(x) = 2^{-x}$

FIGURE 6.2

TABLE 6.2

x	-3	-2	-1	0	1	2
(a) 2^{-x}	8	4	2	1	$\dfrac{1}{2}$	$\dfrac{1}{4}$
(b) $\left(\dfrac{1}{4}\right)^x$	64	16	4	1	$\dfrac{1}{4}$	$\dfrac{1}{16}$

Comparing the functions in Examples 1 and 2, observe that

$$F(x) = 2^{-x} = f(-x)$$

and

$$G(x) = \left(\frac{1}{4}\right)^x = 4^{-x} = g(-x).$$

Consequently, the graph of F is a reflection (in the y-axis) of the graph of f. The graphs of G and g have the same relationship. This is verified by a comparison of the graphs in Figures 6.1 and 6.2.

Remark: Examples 1 and 2 point out that the function $f(x) = a^x$ *increases* if $a > 1$ and *decreases* if $0 < a < 1$.

Exponential and Logarithmic Functions

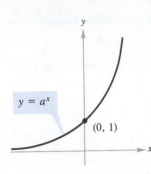

Graph of $y = a^x$
- Domain: $(-\infty, \infty)$
- Range: $(0, \infty)$
- Intercept: $(0, 1)$
- Increasing
- x-axis is a horizontal asymptote $(a^x \to 0$ as $x \to -\infty)$
- Continuous

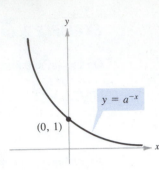

Graph of $y = a^{-x}$
- Domain: $(-\infty, \infty)$
- Range: $(0, \infty)$
- Intercept: $(0, 1)$
- Decreasing
- x-axis is a horizontal asymptote $(a^{-x} \to 0$ as $x \to \infty)$
- Continuous
- Reflection of graph of $y = a^x$ about y-axis

Characteristics of the Exponential Functions a^x and a^{-x} $(a > 1)$

FIGURE 6.3

(a)

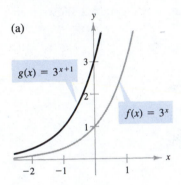

(b)

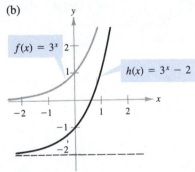

(c)

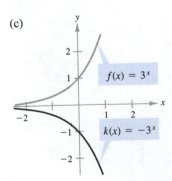

The graphs in Figures 6.1 and 6.2 are typical of the exponential functions a^x and a^{-x}. They have one y-intercept and one horizontal asymptote (the x-axis), and they are continuous. The basic characteristics of these exponential functions are summarized in Figure 6.3.

In the following example, we use the graph of a^x to sketch the graphs of functions of the form $f(x) = b \pm a^{x+c}$.

EXAMPLE 3 *Sketching Graphs of Exponential Functions*

Sketch the graph of each of the following.

(a) $g(x) = 3^{x+1}$

(b) $h(x) = 3^x - 2$

(c) $k(x) = -3^x$

SOLUTION

The graph of each of these three functions is similar to the graph of $f(x) = 3^x$, as shown in Figure 6.4.

(a) Because $g(x) = 3^{x+1} = f(x + 1)$, the graph of g can be obtained by shifting the graph of f one unit to the left.
(b) Because $h(x) = 3^x - 2 = f(x) - 2$, the graph of h can be obtained by shifting the graph of f down two units.
(c) Because $k(x) = -3^x = -f(x)$, the graph of k can be obtained by reflecting the graph of f in the x-axis.

FIGURE 6.4

To evaluate exponential functions with a calculator, you need to use the exponential key $\boxed{y^x}$, where y is the base and x is the exponent. The base is entered first, then the exponent, as shown in the following keystroke sequences.

Number	Keystrokes	Display
$(1.085)^3$	1.085 $\boxed{y^x}$ 3 $\boxed{=}$	1.277289
$12^{5/7}$	12 $\boxed{y^x}$ $\boxed{(}$ 5 $\boxed{\div}$ 7 $\boxed{)}$ $\boxed{=}$	5.899888
$2^{-\pi}$	2 $\boxed{y^x}$ π $\boxed{+/-}$ $\boxed{=}$	0.1133147

EXAMPLE 4   Sketching the Graph of an Exponential Function

Sketch a graph of $f(x) = 2^{1-x^2}$.

SOLUTION

First, note that the graph of f is symmetric with respect to the y-axis because

$$f(-x) = 2^{1-(-x)^2} = 2^{1-x^2} = f(x).$$

Using this symmetry and the table of values

x	0	1	2
$f(x)$	2	1	$\dfrac{1}{8}$

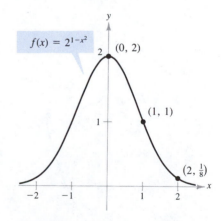

FIGURE 6.5

we obtain the graph shown in Figure 6.5. (Use your calculator to verify the shape of the graph between 0 and 1.)

EXAMPLE 5 An Application Involving Radioactive Decay

Let y represent the mass of a particular radioactive element whose half-life is 25 years. After t years, the mass (in grams) is given by

$$y = 10\left(\frac{1}{2}\right)^{t/25}.$$

(a) What is the initial mass (when $t = 0$)?
(b) How much of the initial mass is present after 80 years?

SOLUTION

(a) When $t = 0$, the mass is

$$y = 10\left(\frac{1}{2}\right)^{0} = 10(1) = 10 \text{ grams.}$$

(b) When $t = 80$, the mass is

$$y = 10\left(\frac{1}{2}\right)^{80/25}$$
$$= 10(0.5)^{3.2}$$
$$\approx 1.088 \text{ grams.}$$

The graph of this function is shown in Figure 6.6.

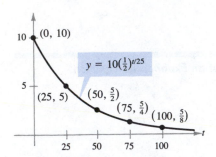

Radioactive Half-life of 25 Years

FIGURE 6.6

The Natural Base e

We used an unspecified base a to introduce exponential functions. It happens that in many applications the convenient choice for a base is the irrational number

$$e \approx 2.71828. \dots$$

called the **natural base.** It is also the base most frequently used in calculus. The function $f(x) = e^x$ is called the **natural exponential function.** Its graph is shown in Figure 6.7.

Remark: Be sure you see that for the exponential function $f(x) = e^x$, e is the constant $2.71828. \dots$, whereas x is the variable.

Some calculators have a natural exponential key $\boxed{e^x}$. On such a calculator you can evaluate e^2 by entering the sequence $2 \boxed{e^x}$. Other calculators require the two-key sequence $\boxed{\text{INV}} \boxed{\ln x}$ to evaluate exponential functions. On such a calculator you can evaluate e^2 using the following sequence.

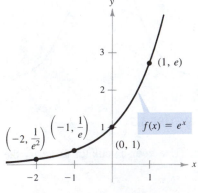

FIGURE 6.7

$2 \boxed{\text{INV}} \boxed{\ln x}$ *Display* 7.3890561

Similarly, to evaluate e^{-1}, enter the sequence

1 $\boxed{+/-}$ $\boxed{\text{INV}}$ $\boxed{\ln x}$.   *Display* 0.3678794

Example 6 gives an indication of how the irrational number e arises in applications. We will refer to this example when we develop the formula for continuous compounding of interest.

EXAMPLE 6 *Approximation of the Number e*

Evaluate the expression

$$\left(1 + \frac{1}{n}\right)^n$$

for several large values of n to see that the values approach $e \approx 2.71828$ as n increases without bound.

SOLUTION

Using the keystroke sequence

n $\boxed{1/x}$ $\boxed{+}$ 1 $\boxed{=}$ $\boxed{y^x}$ n $\boxed{=}$

we obtain the values shown in the following table.

n	10	100	1000	10,000	100,000	1,000,000
$\left(1 + \dfrac{1}{n}\right)^n$	2.59374	2.70481	2.71692	2.71815	2.71827	2.71828

From this table, it seems reasonable to conclude that

$$\left(1 + \frac{1}{n}\right)^n \to e \quad \text{as} \quad n \to \infty.$$

Compound Interest

One of the most familiar examples of exponential growth is that of an investment earning *compound interest*. Suppose a principal P is invested at an annual percentage rate r, compounded once a year. If the interest is added to the principal at the end of the year, then the balance is

$$P_1 = P + Pr = P(1 + r).$$

This pattern of multiplying the previous principal by $1 + r$ is then repeated each successive year, as shown in Table 6.3.

Exponential and Logarithmic Functions

TABLE 6.3

Time in years	Balance after each compounding
0	$P = P$
1	$P_1 = P(1 + r)$
2	$P_2 = P_1(1 + r) = P(1 + r)(1 + r) = P(1 + r)^2$
3	$P_3 = P_2(1 + r) = P(1 + r)^2(1 + r) = P(1 + r)^3$
⋮	⋮
n	$P_n = P(1 + r)^n$

To accommodate more frequent (quarterly, monthly, or daily) compounding of interest, we let n be the number of compoundings per year and t the number of years. Then the rate per compounding is r/n and the account balance after t years is

$$A = P\left(1 + \frac{r}{n}\right)^{nt}.$$

Amount with n compoundings per year

If we let the number of compoundings, n, increase without bound, we approach what is called **continuous compounding.** In the formula for n compoundings per year, let $m = n/r$. This produces

$$A = P\left(1 + \frac{r}{n}\right)^{nt} = P\left(1 + \frac{1}{m}\right)^{mrt} = P\left[\left(1 + \frac{1}{m}\right)^{m}\right]^{rt}.$$

As m increases without bound, we know from Example 6 that

$$\left(1 + \frac{1}{m}\right)^{m} \to e.$$

Hence, for continuous compounding, it follows that

$$P\left[\left(1 + \frac{1}{m}\right)^{m}\right]^{rt} \to P[e]^{rt}$$

and we write $A = Pe^{rt}$.

We summarize the two formulas for compound interest as follows.

Formulas for Compound Interest

After t years, the balance A in an account with principal P and annual percentage rate r (expressed as a decimal) is given by the following formulas.

1. For n compoundings per year: $A = P\left(1 + \frac{r}{n}\right)^{nt}$

2. For continuous compounding: $A = Pe^{rt}$

EXAMPLE 7 *Finding the Balance for Compound Interest*

A sum of $9000 is invested at an annual percentage rate of 8.5%, compounded annually. Find the balance in the account after three years.

SOLUTION

In this case, $P = 9000$, $r = 8.5\% = 0.085$, $n = 1$, and $t = 3$. Using the formula

$$A = P\left(1 + \frac{r}{n}\right)^{nt}$$

we have

$$A = 9000(1 + 0.085)^3 = 9000(1.085)^3 \approx \$11{,}495.60.$$

EXAMPLE 8 *Compounding n Times and Continuously*

A total of $12,000 is invested at an annual percentage rate of 9%. Find the balance after five years if it is compounded

(a) quarterly

(b) continuously.

SOLUTION

(a) For quarterly compoundings, we have $n = 4$. Thus, in five years at 9%, the balance is

$$A = P\left(1 + \frac{r}{n}\right)^{nt} = 12{,}000\left(1 + \frac{0.09}{4}\right)^{4(5)} = \$18{,}726.11.$$

(b) Compounded continuously, the balance is

$$A = Pe^{rt} = 12{,}000e^{0.09(5)} = \$18{,}819.75.$$

Note that continuous compounding yields

$$\$18{,}819.75 - \$18{,}726.11 = \$93.64$$

more than quarterly compounding.

Exponential and Logarithmic Functions

EXAMPLE 9 Population Growth

The number of fruit flies in an experimental population after t hours is given by

$$Q(t) = 20e^{0.03t}, \qquad t \geq 0.$$

(a) Find the initial number of fruit flies in the population.
(b) How large is the population of fruit flies after 72 hours?
(c) Sketch the graph of Q.

SOLUTION

(a) To find the initial population, we evaluate $Q(t)$ at $t = 0$.

$$Q(0) = 20e^{0.03(0)} = 20e^0 = 20(1) = 20 \text{ flies}$$

(b) After 72 hours, the population size is

$$Q(72) = 20e^{(0.03)(72)} = 20e^{2.16} \approx 173 \text{ flies}.$$

(c) To sketch the graph of Q, we evaluate $Q(t)$ for several values of t (rounded to the nearest integer), and plot the corresponding points, as shown in Figure 6.8.

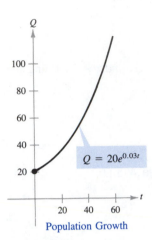

Population Growth

FIGURE 6.8

t	0	5	10	20	40	60
$20e^{0.03t}$	20	23	27	36	66	121

Remark: Many animal populations have a growth pattern described by the function $Q(t) = ce^{kt}$ where c is the original population, $Q(t)$ is the population at time t, and k is a constant determined by the rate of growth. Informally, we can write this as *(Then)* = *(Now)*(e^{kt}).

WARM UP

Use the properties of exponents to simplify the expressions.

1. $5^{2x}(5^{-x})$

2. $3^{-x}(3^{3x})$

3. $\dfrac{4^{5x}}{4^{2x}}$

4. $\dfrac{10^{2x}}{10^x}$

5. $(4^x)^2$

6. $(4^{2x})^5$

7. $\left(\dfrac{2^x}{3^x}\right)^{-1}$

8. $(4^{6x})^{1/2}$

9. $(2^{3x})^{-1/3}$

10. $(16^x)^{1/4}$

EXERCISES 6.1

In Exercises 1–14, use a calculator to evaluate the given quantity. Round your answers to three decimal places.

1. $(3.4)^{5.6}$

2. $(1.005)^{400}$

3. $1000(1.06)^{-5}$

4. $5000(2^{-1.5})$

5. $\sqrt[4]{763}$

6. $\sqrt[3]{4395}$

7. $8^{2\pi}$

8. $5^{-\pi}$

9. $100^{\sqrt{2}}$

10. $0.6^{\sqrt{3}}$

11. e^2

12. $e^{1/2}$

13. $e^{-3/4}$

14. $e^{3.2}$

In Exercises 15–22, match the exponential function with its graph. [The graphs are labeled (a)–(h).]

15. $f(x) = 3^x$

16. $f(x) = -3^x$

17. $f(x) = 3^{-x}$

18. $f(x) = -3^{-x}$

19. $f(x) = 3^x - 4$

20. $f(x) = 3^x + 1$

21. $f(x) = -3^{x-2}$

22. $f(x) = 3^{x-2}$

(a)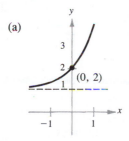

(b)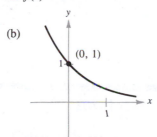

(c)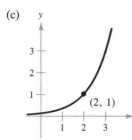

(d)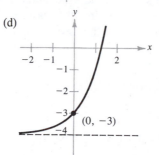

(e)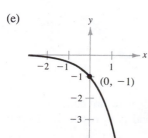

(f)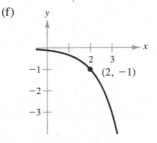

(g)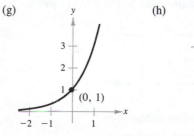

(h)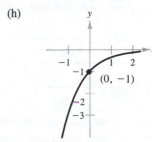

In Exercises 23–40, sketch the graph of the given exponential function.

23. $g(x) = 5^x$

24. $f(x) = \left(\frac{3}{2}\right)^x$

25. $f(x) = \left(\frac{1}{5}\right)^x = 5^{-x}$

26. $h(x) = \left(\frac{3}{2}\right)^{-x}$

27. $h(x) = 5^{x-2}$

28. $g(x) = \left(\frac{3}{2}\right)^{x+2}$

29. $g(x) = 5^{-x} - 3$

30. $f(x) = \left(\frac{3}{2}\right)^{-x} + 2$

31. $y = 2^{-x^2}$

32. $y = 3^{-x^2}$

33. $y = 3^{|x|}$

34. $y = 3^{-|x|}$

35. $s(t) = 2^{-t} + 3$

36. $s(t) = \frac{1}{4}(3^{-t})$

37. $f(x) = e^{2x}$

38. $h(x) = e^{x-2}$

39. $g(x) = 1 + e^{-x}$

40. $N(t) = 1000e^{-0.2t}$

In Exercises 41–44, complete the following table to determine the balance A for P dollars invested at rate r for t years and compounded n times per year.

n	1	2	4	12	365	Continuous compounding
A						

41. $P = \$2500$, $r = 12\%$, $t = 10$ years

42. $P = \$1000$, $r = 10\%$, $t = 10$ years

43. $P = \$2500$, $r = 12\%$, $t = 20$ years

44. $P = \$1000$, $r = 10\%$, $t = 40$ years

In Exercises 45–48, complete the following table to determine the amount of money P that should be invested at rate r to produce a final balance of $100,000 in t years.

t	1	10	20	30	40	50
P						

45. $r = 9\%$, compounded continuously

46. $r = 12\%$, compounded continuously

47. $r = 10\%$, compounded monthly

48. $r = 7\%$, compounded daily

49. The demand equation for a certain product is given by

$$p = 500 - 0.5e^{0.004x}.$$

Find the price p for a demand of (a) $x = 1000$ units and (b) $x = 1500$ units.

50. The demand equation for a certain product is given by

$$p = 5000\left(1 - \frac{4}{4 + e^{-0.002x}}\right).$$

Find the price p for a demand of (a) $x = 100$ units and (b) $x = 500$ units.

51. A certain type of bacteria increases according to the model

$$P(t) = 100e^{0.2197t}$$

where t is time in hours. Find (a) $P(0)$, (b) $P(5)$, and (c) $P(10)$.

52. The population of a town increases according to the model

$$P(t) = 2500e^{0.0293t}$$

where t is time in years, with $t = 0$ corresponding to 1980. Use the model to approximate the population in (a) 1985, (b) 1990, and (c) 2000.

53. Given the exponential function $f(x) = a^x$, show that
(a) $f(u + v) = f(u) \cdot f(v)$ (b) $f(2x) = [f(x)]^2$.

6.2 Logarithmic Functions

In Section 6.1, we saw that the exponential function a^x increases if $a > 1$ and decreases if $0 < a < 1$. In either case, a^x is one-to-one and must have an inverse function. We call this inverse function the **logarithmic function with base a.**

Definition of Logarithmic Function

For $x > 0$ and $0 < a \neq 1$,

$\quad y = \log_a x \quad$ if and only if $\quad x = a^y.$

The function given by

$\quad f(x) = \log_a x$

is called the **logarithmic function with base a.**

Remark: Note that the equations $y = \log_a x$ and $x = a^y$ are equivalent. The first equation is in logarithmic form and the second is in exponential form.

Since $\log_a x$ is the inverse function of a^x, it follows that the domain of $\log_a x$ is the range of a^x, $(0, \infty)$. In other words, $\log_a x$ is defined only if x is positive.

When evaluating logarithms, remember that

A LOGARITHM IS AN EXPONENT.

This means that $\log_a x$ is the exponent to which a must be raised to obtain x. For instance,

$$\log_2 8 = 3$$

because 2 must be raised to the third power to obtain 8.

EXAMPLE 1   *Evaluating Logarithms*

(a) $\log_2 32 = 5$ because $2^5 = 32$.

(b) $\log_3 27 = 3$ because $3^3 = 27$.

(c) $\log_4 2 = \dfrac{1}{2}$ because $4^{1/2} = \sqrt{4} = 2$.

(d) $\log_{10} \dfrac{1}{100} = -2$ because $10^{-2} = \dfrac{1}{10^2} = \dfrac{1}{100}$.

(e) $\log_3 1 = 0$ because $3^0 = 1$.

(f) $\log_2 2 = 1$ because $2^1 = 2$.

The following properties follow directly from the definition of the logarithmic function with base a.

Properties of Logarithms

1. $\log_a 1 = 0$ because $a^0 = 1$.
2. $\log_a a = 1$ because $a^1 = a$.
3. $\log_a a^x = x$ because $a^x = a^x$.

To sketch the graph of $y = \log_a x$, we can use the fact that the graphs of inverse functions are reflections of each other in the line $y = x$, as demonstrated in Example 2.

EXAMPLE 2 Graphs of Exponential and Logarithmic Functions

On the same coordinate plane, sketch the graphs of

(a) $f(x) = 2^x$ 　　　　　　　　　　(b) $g(x) = \log_2 x.$

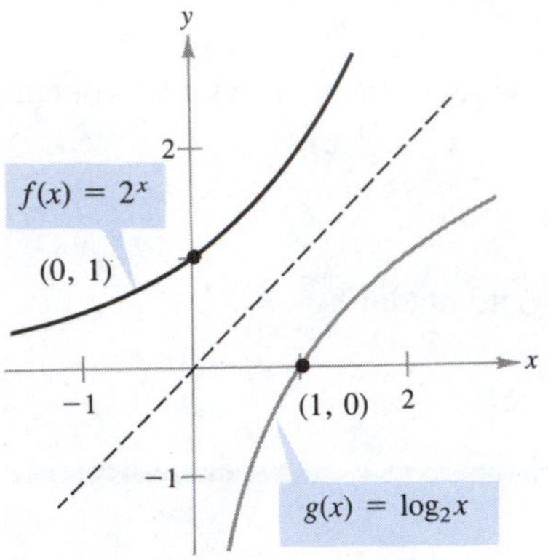

SOLUTION

(a) For $f(x) = 2^x$, we make up the following table of values.

x	-2	-1	0	1	2	3
$f(x) = 2^x$	$\dfrac{1}{4}$	$\dfrac{1}{2}$	1	2	4	8

By plotting these points and connecting them with a smooth curve, we have the graph shown in Figure 6.9.

(b) Since $g(x) = \log_2 x$ is the inverse of $f(x) = 2^x$, the graph of g is obtained by reflecting the graph of f in the line $y = x$, as shown in Figure 6.9.

Inverse Functions

FIGURE 6.9

The logarithmic function with base 10 is called the **common logarithmic function.** On most calculators, this function is denoted by $\boxed{\log}$. You can tell whether this key denotes base 10 by entering 10 $\boxed{\log}$. The display should be 1.

EXAMPLE 3 Sketching the Graph of a Logarithmic Function

Sketch the graph of the common logarithmic function $y = \log_{10} x.$

SOLUTION

We begin by making the following table of values. Note that some of the values can be obtained without a calculator, while others require a calculator. We plot the corresponding points and sketch the graph shown in Figure 6.10.

	Without Calculator				With Calculator		
x	$\dfrac{1}{100}$	$\dfrac{1}{10}$	1	10	2	5	8
$\log_{10} x$	-2	-1	0	1	0.301	0.699	0.903

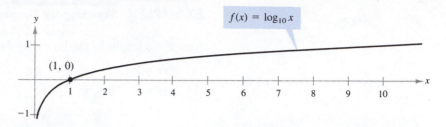

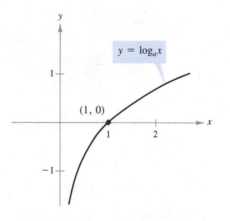

FIGURE 6.10

The nature of the graph in Figure 6.10 is typical of functions of the form $f(x) = \log_a x$, $a > 1$. They have one x-intercept and one vertical asymptote. Notice how slowly the graph rises for $x > 1$. In Figure 6.10 we would need to move out to $x = 1000$ before the graph rises to $y = 3$. We summarize the basic characteristics of logarithmic graphs in Figure 6.11.

$y = \log_a x$

$(1, 0)$

Graph of $y = \log_a x$, $a > 1$
- Domain: $(0, \infty)$
- Range: $(-\infty, \infty)$
- Intercept: $(1, 0)$
- Increasing
- y-axis is a vertical asymptote
  ($\log_a x \to -\infty$ as $x \to 0^+$)
- Continuous
- Reflection of graph of $y = a^x$
  about the line $y = x$

FIGURE 6.11

Remark: In Figure 6.11, note that the vertical asymptote occurs at $x = 0$, where $\log_a x$ is undefined.

In the following example we use the graph of $\log_a x$ to sketch the graphs of functions of the form $y = b \pm \log_a(x + c)$.

EXAMPLE 4 *Sketching the Graphs of Logarithmic Functions*

Sketch the graphs of the following functions.

(a) $g(x) = \log_4(x - 1)$

(b) $h(x) = 2 + \log_4 x$

SOLUTION

The graph of each of these functions is similar to the graph of $f(x) = \log_4 x$, as shown in Figure 6.12.

(a) Because $g(x) = \log_4(x - 1) = f(x - 1)$ the graph of g can be obtained by shifting the graph of f one unit to the right.

(b) Because $h(x) = 2 + \log_4 x = 2 + f(x)$ the graph of h can be obtained by shifting the graph of f two units up.

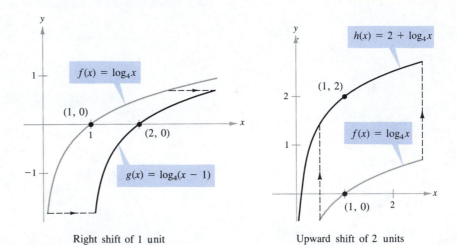

Right shift of 1 unit   Upward shift of 2 units

FIGURE 6.12

The function $f(x) = \log_a(bx + c)$ has a domain which consists of all x such that $bx + c > 0$. The vertical asymptote occurs when $bx + c = 0$, and the x-intercept occurs when $bx + c = 1$.

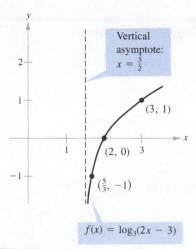

Vertical
asymptote:
$x = \frac{3}{2}$

$(3, 1)$

$(2, 0)$

$(\frac{5}{3}, -1)$

$f(x) = \log_3(2x - 3)$

FIGURE 6.13

EXAMPLE 5 *Sketching the Graph of a Logarithmic Function*

Sketch the graph of $f(x) = \log_3(2x - 3)$.

SOLUTION

The function f is defined for all x such that $2x - 3 > 0$, which implies that $x > 3/2$. Thus, the domain of f is the interval $(3/2, \infty)$. The vertical asymptote is $x = 3/2$ (the solution to $2x - 3 = 0$), and the x-intercept occurs when $x = 2$ (the solution to $2x - 3 = 1$). Using this information together with a couple of additional points, we obtain the graph shown in Figure 6.13.

The Natural Logarithmic Function

As with exponential functions, the most widely used base for logarithmic functions is the number e. We call the logarithmic function with base e the **natural logarithmic function** and denote it by the special symbol ln x, read as "el en of x."

The Natural Logarithmic Function

The function defined by

$$f(x) = \log_e x = \ln x, \qquad x > 0$$

is called the **natural logarithmic function.**

The three properties of logarithms listed at the beginning of this section are also valid for natural logarithms.

Properties of Natural Logarithms

1. $\ln 1 = 0$ because $e^0 = 1$.
2. $\ln e = 1$ because $e^1 = e$.
3. $\ln e^x = x$ because $e^x = e^x$.

Exponential and Logarithmic Functions

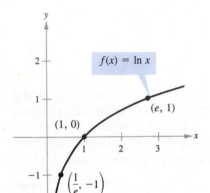

FIGURE 6.14

EXAMPLE 6 Evaluating the Natural Logarithmic Function

(a) $\ln \dfrac{1}{e} = \ln e^{-1} = -1$ *Property 3*

(b) $\ln e^2 = 2$ *Property 3*

On most calculators, the natural logarithm is denoted by $\boxed{\ln x}$.

Number	Calculator Steps	Display
(c) $\ln 2$	2 $\boxed{\ln x}$	0.6931472
(d) $\ln 0.3$	.3 $\boxed{\ln x}$	-1.2039728
(e) $\ln(-1)$	1 $\boxed{+/-}$ $\boxed{\ln x}$	

Be sure you see that $\ln(-1)$ gives an error. This occurs because the domain of $\ln x$ is the set of positive real numbers. (See Figure 6.14.) Hence, $\ln(-1)$ is undefined.

The graph of the natural logarithmic function is shown in Figure 6.14.

EXAMPLE 7 Finding the Domain of Logarithmic Functions

Find the domain of the following functions.

(a) $f(x) = \ln(x - 2)$

(b) $g(x) = \ln(2 - x)$·

(c) $h(x) = \ln x^2$

SOLUTION

(a) Because $\ln(x - 2)$ is defined only if $x - 2 > 0$, it follows that the domain of f is $(2, \infty)$.

(b) Because $\ln(2 - x)$ is defined only if $2 - x > 0$, it follows that the domain of g is $(-\infty, 2)$. The graph of g is shown in Figure 6.15.

(c) Because $\ln x^2$ is defined only if $x^2 > 0$, it follows that the domain of h is all real numbers except $x = 0$.

FIGURE 6.15

EXAMPLE 8 An Application

Students participating in a psychological experiment attended several lectures on a subject. Every month for a year after that, the students were tested to see how much of the material they remembered. The average scores for the group were given by the *human memory model*

$$f(t) = 75 - 6 \ln(t + 1), \qquad 0 \le t \le 12$$

where t is the time in months.

(a) What was the average score on the original ($t = 0$) exam?
(b) What was the average score at the end of $t = 2$ months?
(c) What was the average score at the end of $t = 6$ months?
(d) Sketch the graph of f.

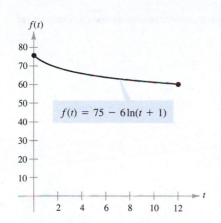

$f(t) = 75 - 6\ln(t + 1)$

FIGURE 6.16

SOLUTION

(a) The original average score was

$$f(0) = 75 - 6 \ln(0 + 1) = 75 - 6(0) = 75.$$

(b) After two months, the average score was

$$f(2) = 75 - 6 \ln 3 \approx 75 - 6(1.0986) \approx 68.4.$$

(c) After six months, the average score was

$$f(6) = 75 - 6 \ln 7 \approx 75 - 6(1.9459) \approx 63.3.$$

(d) Several points are shown in the following table, and the graph of f is shown in Figure 6.16.

t	0	1	2	6	8	12
$f(t)$	75	70.8	68.4	63.3	61.8	59.6

Change of Base

Although 10 and e are the most frequently used bases, we occasionally need to evaluate logarithms to other bases. In such cases the following *change of base formula* is useful. (This formula is derived in Example 10, Section 6.4.)

Change of Base Formula

Let a, b, and x be positive real numbers such that $a \neq 1$ and $b \neq 1$. Then $\log_a x$ is given by

$$\log_a x = \frac{\log_b x}{\log_b a}.$$

Remark: One way to look at the change of base formula is that logarithms to base a are simply *constant multiples* of logarithms to base b. The constant multiplier is $1/(\log_b a)$.

EXAMPLE 9 Changing Bases

Use *common* logarithms to evaluate the following.

(a) $\log_4 30$ (b) $\log_2 14$

SOLUTION

(a) Using the change of base formula with $a = 4$, $b = 10$, and $x = 30$, we convert to common logarithms and obtain

$$\log_4 30 = \frac{\log_{10} 30}{\log_{10} 4} \approx \frac{1.47712}{0.60206} \approx 2.4534.$$

(b) Using the change of base formula with $a = 2$, $b = 10$, and $x = 14$, we convert to common logarithms and obtain

$$\log_2 14 = \frac{\log_{10} 14}{\log_{10} 2} \approx \frac{1.14613}{0.30103} \approx 3.8074.$$

EXAMPLE 10 Changing Bases

Use *natural* logarithms to evaluate the following.

(a) $\log_4 30$ (b) $\log_2 14$

SOLUTION

(a) Using the change of base formula with $a = 4$, $b = e$, and $x = 30$, we convert to natural logarithms and obtain

$$\log_4 30 = \frac{\ln 30}{\ln 4} \approx \frac{3.40120}{1.38629} \approx 2.4534.$$

(b) Using the change of base formula with $a = 2$, $b = e$, and $x = 14$, we convert to natural logarithms and obtain

$$\log_2 14 = \frac{\ln 14}{\ln 2} \approx \frac{2.63906}{0.693147} \approx 3.8074.$$

Note that the results agree with those obtained in Example 9, using common logarithms.

WARM UP

Solve for x.

1. $2^x = 8$

2. $4^x = 1$

3. $10^x = 0.1$

4. $e^x = e$

Evaluate the given expressions. (Round your answers to three decimal places.)

5. e^2

6. e^{-1}

Describe how the graph of g is related to the graph of f.

7. $g(x) = f(x + 2)$

8. $g(x) = -f(x)$

9. $g(x) = -1 + f(x)$

10. $g(x) = f(-x)$

EXERCISES 6.2

In Exercises 1–16, evaluate the given expression without using a calculator.

1. $\log_2 16$

2. $\log_4 64$

3. $\log_5\left(\frac{1}{25}\right)$

4. $\log_2\left(\frac{1}{8}\right)$

5. $\log_{16} 4$

6. $\log_{27} 9$

7. $\log_7 1$

8. $\log_{10} 1000$

9. $\log_{10} 0.01$

10. $\log_{10} 10$

11. $\ln e^3$

12. $\ln \dfrac{1}{e}$

13. $\ln e^{-2}$

14. $\ln 1$

15. $\log_a a^2$

16. $\log_a \dfrac{1}{a}$

In Exercises 17–26, use the definition of a logarithm to write the given equation in logarithmic form. For instance, the logarithmic form of $2^3 = 8$ is $\log_2 8 = 3$.

17. $5^3 = 125$

18. $8^2 = 64$

19. $81^{1/4} = 3$

20. $9^{3/2} = 27$

21. $6^{-2} = \frac{1}{36}$

22. $10^{-3} = 0.001$

23. $e^3 = 20.0855\ldots$

24. $e^0 = 1$

25. $e^x = 4$

26. $u^v = w$

In Exercises 27–32, use a calculator to evaluate the logarithm. Round your answer to three decimal places.

27. $\log_{10} 345$ **28.** $\log_{10}\left(\frac{4}{5}\right)$

29. $\ln 18.42$ **30.** $\ln \sqrt{42}$

31. $\ln(1 + \sqrt{3})$ **32.** $\ln(\sqrt{5} - 2)$

In Exercises 33–36, demonstrate that f and g are inverses of each other by sketching their graphs on the same coordinate plane.

33. $f(x) = 3^x$, $g(x) = \log_3 x$

34. $f(x) = 5^x$, $g(x) = \log_5 x$

35. $f(x) = e^x$, $g(x) = \ln x$

36. $f(x) = 10^x$, $g(x) = \log_{10} x$

In Exercises 37–42, use the graph of $y = \ln x$ to match the given function to its graph. [The graphs are labeled (a)–(f).]

37. $f(x) = \ln x + 2$ **38.** $f(x) = -\ln x$

39. $f(x) = -\ln(x + 2)$ **40.** $f(x) = \ln(x - 1)$

41. $f(x) = \ln(1 - x)$ **42.** $f(x) = -\ln(-x)$

(a)
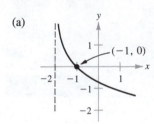

(b)
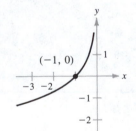

(c)
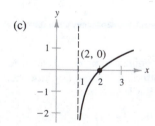

(d)
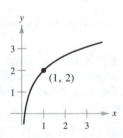

(e)
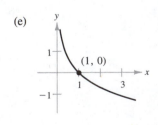

(f)
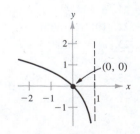

In Exercises 43–50, find the domain, vertical asymptote, and x-intercept of the logarithmic function, and sketch its graph.

43. $f(x) = \log_4 x$ **44.** $g(x) = \log_6 x$

45. $h(x) = \log_4(x - 3)$ **46.** $f(x) = -\log_6(x + 2)$

47. $f(x) = \ln(x - 2)$ **48.** $h(x) = \ln(x + 1)$

49. $g(x) = \ln(-x)$ **50.** $f(x) = \ln(3 - x)$

In Exercises 51–54, use the change of base formula to write the given logarithm as a multiple of a common logarithm. For instance, $\log_2 3 = (1/\log_{10} 2)\log_{10} 3$.

51. $\log_3 5$ **52.** $\log_4 10$

53. $\log_2 x$ **54.** $\ln 5$

In Exercises 55–58, use the change of base formula to write the given logarithm as a multiple of a natural logarithm. For instance, $\log_2 3 = (1/\ln 2)\ln 3$.

55. $\log_3 5$ **56.** $\log_4 10$

57. $\log_2 x$ **58.** $\log_{10} 5$

In Exercises 59–66, evaluate the logarithm using the change of base formula. Do the problem twice, once with common logarithms and once with natural logarithms. Round your answer to three decimal places.

59. $\log_3 7$ **60.** $\log_7 4$

61. $\log_{1/2} 4$ **62.** $\log_4(0.55)$

63. $\log_9(0.4)$ **64.** $\log_{20} 125$

65. $\log_{15} 1250$ **66.** $\log_{1/3}(0.015)$

67. Students in a mathematics class were given an exam and then retested monthly with an equivalent exam. The average score for the class was given by the human memory model

$$f(t) = 80 - 17 \log_{10}(t + 1), \qquad 0 \le t \le 12$$

where t is the time in months.
(a) What was the average score on the original exam ($t = 0$)?
(b) What was the average score after four months?
(c) What was the average score after ten months?

68. The time (in hours) necessary for a certain object to cool $10°$ is

$$t = \frac{10 \ln(1/2)}{\ln(3/4)}.$$

Find t.

69. The population of a town will double in

$$t = \frac{10 \ln 2}{\ln 67 - \ln 50}$$

years. Find t.

70. The work (in foot-pounds) done in compressing an initial volume of 9 cubic feet at a pressure of 15 pounds per square inch to a volume of 3 cubic feet is

$$W = 19,440(\ln 9 - \ln 3).$$

Find W.

71. The time in years for the world population to double if it is increasing at a continuous rate of r is given by

$$t = \frac{\ln 2}{r}.$$

Complete the following table.

r	0.005	0.010	0.015	0.020	0.025	0.030
t						

72. A principal P invested at $9\frac{1}{2}\%$ and compounded continuously, increases to an amount KP after t years, where t is given by

$$t = \frac{\ln K}{0.095}.$$

(a) Complete the following table.

K	1	2	3	4	6	8	10	12
t								

(b) Use the table in part (a) to graph this function.

73. (a) Use a calculator to complete the following table for the function

$$f(x) = \frac{\ln x}{x}.$$

x	1	5	10	10^2	10^4	10^6
$f(x)$						

(b) Use the table in part (a) to determine what $f(x)$ approaches as x increases without bound.

6.3 *Properties of Logarithms*

We know from the previous section that the logarithmic function with base a is the *inverse* of the exponential function with base a. Thus, it makes sense that the properties of exponents should have corresponding properties involving logarithms. For instance, the exponential property

$$a^0 = 1$$

has the corresponding logarithmic property

$$\log_a 1 = 0.$$

In this section we will study the logarithmic properties that correspond to the following three exponential properties.

1. $a^n a^m = a^{n+m}$

2. $\dfrac{a^n}{a^m} = a^{n-m}$

3. $(a^n)^m = a^{nm}$

Exponential and Logarithmic Functions

Properties of Logarithms

Let a be a positive real number such that $a \neq 1$, and let n be a real number. If u and v are positive real numbers, then the following properties are true.

1. $\log_a(uv) = \log_a u + \log_a v$

2. $\log_a \dfrac{u}{v} = \log_a u - \log_a v$

3. $\log_a u^n = n \log_a u$

PROOF

We give a proof of Property 1 and leave the other two proofs for you. To prove Property 1, let

$$x = \log_a u \qquad \text{and} \qquad y = \log_a v.$$

The corresponding exponential forms of these two equations are

$$a^x = u \qquad \text{and} \qquad a^y = v.$$

Multiplying u and v produces $uv = a^x a^y = a^{x+y}$. The corresponding logarithmic form of $uv = a^{x+y}$ is

$$\log_a(uv) = x + y.$$

Hence, $\log_a(uv) = \log_a u + \log_a v$.

Remark: Note that there is no general property that can be used to simplify $\log_a(u \pm v)$. Specifically,

$$\log_a(x + y) \text{ DOES NOT EQUAL } \log_a x + \log_a y.$$

The natural logarithmic versions of these three properties are as follows.

1. $\ln(uv) = \ln u + \ln v$

2. $\ln \dfrac{u}{v} = \ln u - \ln v$

3. $\ln u^n = n \ln u$

EXAMPLE 1 *Using Properties of Logarithms*

Given $\ln 2 \approx 0.693$, $\ln 3 \approx 1.099$, and $\ln 7 \approx 1.946$, use the properties of logarithms to approximate the following.

(a) $\ln 6$ \hspace{4cm} (b) $\ln \dfrac{7}{27}$

SOLUTION

(a) $\ln 6 = \ln(2 \cdot 3)$

$= \ln 2 + \ln 3$ \hspace{3cm} *Property 1*

$\approx 0.693 + 1.099$

$= 1.792$

(b) $\ln \dfrac{7}{27} = \ln 7 - \ln 27$ \hspace{2cm} *Property 2*

$= \ln 7 - \ln 3^3$

$= \ln 7 - 3 \ln 3$ \hspace{2cm} *Property 3*

$\approx 1.946 - 3(1.099)$

$= -1.351$

EXAMPLE 2 *Using Properties of Logarithms*

Use the properties of logarithms to verify that

$$-\ln \frac{1}{2} = \ln 2.$$

SOLUTION

$$-\ln \frac{1}{2} = -\ln(2^{-1}) = -(-1) \ln 2 = \ln 2$$

Try checking this result on your calculator.

The properties of logarithms are useful for rewriting logarithmic expressions in forms that simplify the operations of algebra and calculus. This is true because they convert complicated products, quotients, and exponential forms into simpler sums, differences, and products, respectively. Examples 3, 4, and 5 illustrate some cases.

EXAMPLE 3 *Rewriting the Logarithm of a Product*

Use the properties of logarithms to rewrite

$$\log_{10} 5x^2y$$

as the sum of logarithms.

SOLUTION

$$
\begin{aligned}
\log_{10} 5x^2y &= \log_{10} 5 + \log_{10} x^2y \\
&= \log_{10} 5 + \log_{10} x^2 + \log_{10} y \\
&= \log_{10} 5 + 2 \log_{10} x + \log_{10} y
\end{aligned}
$$

EXAMPLE 4 *Rewriting the Logarithm of a Quotient*

Use the properties of logarithms to rewrite

$$\ln \frac{\sqrt{3x - 5}}{7x^3}$$

as the sum and/or difference of logarithms.

SOLUTION

$$
\begin{aligned}
\ln \frac{\sqrt{3x - 5}}{7x^3} &= \ln(3x - 5)^{1/2} - \ln 7x^3 \\
&= \ln(3x - 5)^{1/2} - (\ln 7 + \ln x^3) \\
&= \frac{1}{2} \ln(3x - 5) - \ln 7 - 3 \ln x
\end{aligned}
$$

EXAMPLE 5 *Expanding a Logarithmic Expression*

Use the properties of logarithms to rewrite

$$\ln\left(\frac{5x^2}{y}\right)^3$$

as the sum and/or difference of logarithms.

SOLUTION

$$\ln\left(\frac{5x^2}{y}\right)^3 = 3 \ln \frac{5x^2}{y}$$
$$= 3(\ln 5x^2 - \ln y)$$
$$= 3(\ln 5 + \ln x^2 - \ln y)$$
$$= 3(\ln 5 + 2 \ln x - \ln y)$$
$$= 3 \ln 5 + 6 \ln x - 3 \ln y$$

In Examples 3, 4, and 5, we used the properties of logarithms to *expand* logarithmic expressions. In Examples 6 and 7, we reverse the procedure and use the properties of logarithms to *condense* logarithmic expressions.

EXAMPLE 6 *Condensing a Logarithmic Expression*

Rewrite as the logarithm of a single quantity.

$$\frac{1}{2} \log_{10} x - 3 \log_{10}(x + 1)$$

SOLUTION

$$\frac{1}{2} \log_{10} x - 3 \log_{10}(x + 1) = \log_{10} x^{1/2} - \log_{10}(x + 1)^3$$
$$= \log_{10} \frac{\sqrt{x}}{(x + 1)^3}$$

EXAMPLE 7 *Condensing a Logarithmic Expression*

Rewrite as the logarithm of a single quantity.

$$2 \ln(x + 2) - \frac{1}{3}(\ln x + \ln y)$$

SOLUTION

$$2 \ln(x + 2) - \frac{1}{3}(\ln x + \ln y) = \ln(x + 2)^2 - (\ln x^{1/3} + \ln y^{1/3})$$
$$= \ln(x + 2)^2 - \ln(xy)^{1/3}$$
$$= \ln \frac{(x + 2)^2}{\sqrt[3]{xy}}$$

Exponential and Logarithmic Functions

When applying the properties of logarithms to a logarithmic function, you should be careful to check the domain of the function. For example, the domain of $f(x) = \ln x^2$ is all real $x \neq 0$, whereas the domain of $g(x) = 2 \ln x$ is all real $x > 0$.

WARM UP

Evaluate the expressions without using a calculator.

1. $\log_7 49$

2. $\log_2\left(\frac{1}{32}\right)$

3. $\ln \dfrac{1}{e^2}$

4. $\log_{10} 0.001$

Simplify the expressions.

5. $e^2 e^3$

6. $\dfrac{e^2}{e^3}$

7. $(e^2)^3$

8. $(e^2)^0$

Rewrite the expressions in exponential form.

9. $\dfrac{1}{x^2}$

10. $\sqrt{x}$

EXERCISES 6.3

In Exercises 1–20, use the properties of logarithms to write the expression as a sum, difference, or multiple of logarithms.

1. $\log_2 5x$

2. $\log_4 10z$

3. $\log_3 \dfrac{5}{x}$

4. $\log_5 \dfrac{y}{2}$

5. $\log_8 x^4$

6. $\log_6 z^{-3}$

7. $\ln \sqrt{z}$

8. $\ln \sqrt[3]{t}$

9. $\log_2 xyz$

10. $\ln \dfrac{xy}{z}$

11. $\ln \sqrt{a-1}$

12. $\log_3\left(\dfrac{x^2-1}{x^3}\right)^3$

13. $\ln z(z-1)^2$

14. $\log_4 \sqrt{\dfrac{x^2}{y^3}}$

15. $\log_b \dfrac{x^2}{y^2 z^3}$

16. $\log_b \dfrac{\sqrt{x}\,y^4}{z^4}$

17. $\ln \sqrt[3]{x/y}$

18. $\ln \dfrac{x}{\sqrt{x^2+1}}$

19. $\log_9 \dfrac{x^4\sqrt{y}}{z^5}$

20. $\ln \sqrt{x^2(x+2)}$

In Exercises 21–40, write the expression as the logarithm of a single quantity.

21. $\ln x + \ln 2$

22. $\ln y + \ln z$

23. $\log_4 z - \log_4 y$

24. $\log_5 8 - \log_5 t$

25. $2 \log_2(x+4)$

26. $-4 \log_6 2x$

27. $\ln x - 3 \ln(x+1)$

28. $2 \ln 8 + 5 \ln z$

29. $\frac{1}{3} \log_3 5x$

30. $\frac{3}{2} \log_7(z-2)$

31. $\log_3(x-2) - \log_3(x+2)$

32. $3 \ln x + 2 \ln y - 4 \ln z$

33. $\ln x - 2[\ln(x+2) + \ln(x-2)]$

34. $4[\ln z + \ln(z+5)] - 2 \ln(z-5)$

35. $\frac{1}{3}[2 \ln(x + 3) + \ln x - \ln(x^2 - 1)]$

36. $2[\ln x - \ln(x + 1) - \ln(x - 1)]$

37. $\frac{1}{3}[\ln y + 2 \ln(y + 4)] - \ln(y - 1)$

38. $\frac{1}{2}[\ln(x + 1) + 2 \ln(x - 1)] + 3 \ln x$

39. $2 \ln 3 - \frac{1}{2} \ln(x^2 + 1)$

40. $\frac{3}{2} \log_4 5t^6 - \frac{3}{4} \log_4 t^4$

In Exercises 41–56, approximate the logarithm using the properties of logarithms, given $\log_b 2 \approx 0.3562$, $\log_b 3 \approx 0.5646$, and $\log_b 5 \approx 0.8271$.

41. $\log_b 6$

42. $\log_b 15$

43. $\log_b\left(\frac{3}{2}\right)$

44. $\log_b\left(\frac{5}{3}\right)$

45. $\log_b 25$

46. $\log_b 18$ $(18 = 2 \cdot 3^2)$

47. $\log_b \sqrt{2}$

48. $\log_b\left(\frac{9}{2}\right)$

49. $\log_b 40$

50. $\log_b \sqrt[3]{75}$

51. $\log_b\left(\frac{1}{4}\right)$

52. $\log_b(3b^2)$

53. $\log_b \sqrt{5b}$

54. $\log_b 1$

55. $\log_b \dfrac{(4.5)^3}{\sqrt{3}}$

56. $\log_b \dfrac{4}{15^2}$

In Exercises 57–62, find the exact value of the logarithm.

57. $\log_3 9$

58. $\log_6 \sqrt[3]{6}$

59. $\log_4 16^{1.2}$

60. $\log_5\left(\frac{1}{125}\right)$

61. $\ln e^{4.5}$

62. $\ln \sqrt[4]{e^3}$

In Exercises 63–70, use the properties of logarithms to simplify the given logarithmic expression.

63. $\log_4 8$

64. $\log_5\left(\frac{1}{15}\right)$

65. $\log_7 \sqrt{70}$

66. $\log_2(4^2 \cdot 3^4)$

67. $\log_5\left(\frac{1}{250}\right)$

68. $\log_{10}\left(\frac{9}{300}\right)$

69. $\ln(5e^6)$

70. $\ln \dfrac{6}{e^2}$

71. Prove that $\log_b \dfrac{u}{v} = \log_b u - \log_b v$.

72. Prove that $\log_b u^n = n \log_b u$.

73. Use a calculator to demonstrate that

$$\frac{\ln x}{\ln y} \neq \ln \frac{x}{y} = \ln x - \ln y$$

by completing the following table.

x	y	$\dfrac{\ln x}{\ln y}$	$\ln \dfrac{x}{y}$	$\ln x - \ln y$
1	2			
3	4			
10	5			
4	0.5			

6.4 Solving Exponential and Logarithmic Equations

So far in this chapter, we have focused our study on the definitions, graphs, and properties of exponential and logarithmic functions. Here we will concentrate on procedures for *solving equations* involving these exponential and logarithmic functions. As a simple example, consider the exponential equation

$$2^x = 32.$$

We can obtain the solution by rewriting the equation in the form

$$2^x = 2^5.$$

Since exponential functions are one-to-one, we can equate exponents of like bases to obtain

$$x = 5.$$

Though this method works in some cases, it does not work for an equation as simple as

$$2^x = 7.$$

Because $2^2 = 4$ and $2^3 = 8$, we see that x lies between 2 and 3. To actually solve for x, we take the logarithm (with base 2) of both sides to obtain

$$\log_2 2^x = \log_2 7$$
$$x = \log_2 7$$
$$= \frac{\ln 7}{\ln 2}$$ *Change of base*
$$\approx 2.81.$$

Guidelines for Solving Exponential and Logarithmic Equations

1. *To solve an exponential equation*, first isolate the exponential expression, then take the logarithm of both sides and solve for the variable.
2. *To solve a logarithmic equation*, rewrite the equation in exponential form and solve for the variable.

Note that these two guidelines are based on the following inverse properties of exponential and logarithmic functions.

Base a	*Base e*
1. $\log_a a^x = x$	$\ln e^x = x$
2. $a^{\log_a x} = x$	$e^{\ln x} = x$

Solving Exponential Equations

EXAMPLE 1 *Solving an Exponential Equation*

Solve for x in the equation $e^x = 72$.

SOLUTION

Taking the natural logarithm of both sides, we obtain

$$\ln e^x = \ln 72$$
$$x = \ln 72 \approx 4.277.$$

EXAMPLE 2 *Solving an Exponential Equation*

Solve for x in the equation $3^x = 0.026$.

SOLUTION

Taking the logarithm (with base 3) of both sides, we obtain

$$\log_3 3^x = \log_3 0.026$$
$$x = \log_3 0.026$$
$$= \frac{\ln 0.026}{\ln 3} \qquad \text{\textit{Change of base}}$$
$$\approx -3.322.$$

ALTERNATIVE SOLUTION

Rather than use logarithms with base 3, we could just as easily have taken the *natural logarithm* of both sides to obtain

$$\ln 3^x = \ln 0.026$$
$$x \ln 3 = \ln 0.026$$
$$x = \frac{\ln 0.026}{\ln 3} \approx -3.322.$$

Remember that the first step in solving an exponential equation is to isolate the exponential expression. This is demonstrated in Example 3.

EXAMPLE 3 *Solving an Exponential Equation*

Solve for x in the equation $4e^{2x} = 5$.

SOLUTION

We isolate the exponential by dividing both sides of the equation by 4 to obtain

$$e^{2x} = \frac{5}{4}.$$

Then, taking the natural logarithm of both sides, we have

$$\ln e^{2x} = \ln \frac{5}{4}$$
$$2x = \ln \frac{5}{4}$$
$$x = \frac{1}{2} \ln \frac{5}{4} \approx 0.112.$$

When an equation involves two or more exponential expressions, we can still use a procedure similar to that demonstrated in the first three examples. However, the algebra is a bit more complicated. Study the next two examples carefully.

EXAMPLE 4 *Solving an Exponential Equation*

Solve the equation

$$4^{x+3} = 7^x.$$

SOLUTION

Taking the natural logarithm of both sides, we obtain the following.

$$\ln 4^{x+3} = \ln 7^x \qquad \text{\textit{Take ln of both sides}}$$
$$(x + 3) \ln 4 = x \ln 7 \qquad \text{\textit{Property 3}}$$
$$x \ln 4 + 3 \ln 4 = x \ln 7 \qquad \text{\textit{Distributive Property}}$$
$$x \ln 4 - x \ln 7 = -3 \ln 4 \qquad \text{\textit{Collect like terms}}$$
$$x(\ln 4 - \ln 7) = -3 \ln 4 \qquad \text{\textit{Factor out x}}$$

$$x = \frac{-3 \ln 4}{\ln 4 - \ln 7} \qquad \text{\textit{Divide}}$$

$$x \approx 7.432$$

EXAMPLE 5 *Solving an Exponential Equation*

Solve for x in the equation

$$e^x + 2e^{-x} = 3.$$

SOLUTION

Some preliminary algebra is helpful. By multiplying both sides of the equation by e^x, we can eliminate the negative exponent and obtain an equation of quadratic type. Note how this works.

$$e^x + 2e^{-x} = 3 \qquad \text{\textit{Given}}$$
$$e^x(e^x + 2e^{-x}) = 3e^x \qquad \text{\textit{Multiply both sides by }} e^x$$
$$(e^x)^2 + 2 = 3e^x \qquad \text{\textit{Distributive Property}}$$
$$(e^x)^2 - 3e^x + 2 = 0 \qquad \text{\textit{Quadratic form}}$$
$$(e^x - 2)(e^x - 1) = 0 \qquad \text{\textit{Factor}}$$
$$e^x - 2 = 0 \qquad e^x - 1 = 0 \qquad \text{\textit{Set factors to zero}}$$
$$e^x = 2 \qquad e^x = 1$$
$$x = \ln 2 \qquad x = 0 \qquad \text{\textit{Solutions}}$$

Solving Logarithmic Equations

To solve a logarithmic equation such as

$$\ln x = 3 \qquad \textit{Logarithmic form}$$

we write the equation in exponential form as follows.

$$e^{\ln x} = e^3 \qquad \textit{Exponentiate both sides}$$
$$x = e^3 \qquad \textit{Exponential form}$$

This procedure is sometimes called *exponentiating* both sides of an equation.

EXAMPLE 6 *Solving a Logarithmic Equation*

Solve for x in the equation $2 \ln 3x = 4$.

SOLUTION

$$
\begin{aligned}
2 \ln 3x &= 4 &&\textit{Given} \\
\ln 3x &= 2 &&\textit{Divide both sides by 2} \\
3x &= e^2 &&\textit{Exponential form} \\
x &= \tfrac{1}{3}e^2 &&\textit{Divide both sides by 3}
\end{aligned}
$$

The techniques used to solve equations involving logarithmic expressions can produce extraneous solutions, as demonstrated in Example 7.

EXAMPLE 7 *Solving a Logarithmic Equation*

Solve for x in the equation $\ln(x - 2) + \ln(2x - 3) = 2 \ln x$.

SOLUTION

$$
\begin{aligned}
\ln(x - 2) + \ln(2x - 3) &= 2 \ln x &&\textit{Given} \\
\ln(x - 2)(2x - 3) &= \ln x^2 &&\textit{Properties 1 and 3} \\
\ln(2x^2 - 7x + 6) &= \ln x^2
\end{aligned}
$$

Now, because the natural logarithmic function is one-to-one, we can write

$$
\begin{aligned}
2x^2 - 7x + 6 &= x^2 \\
x^2 - 7x + 6 &= 0 &&\textit{Quadratic form} \\
(x - 6)(x - 1) &= 0 &&\textit{Factor} \\
x - 6 = 0 \quad x - 1 &= 0 &&\textit{Set factors to zero} \\
x = 6 \qquad x &= 1
\end{aligned}
$$

Finally, by checking these two "solutions" in the original equation, we find that $x = 1$ is not valid. Can you see why? Thus, the only solution is $x = 6$.

Exponential and Logarithmic Functions

EXAMPLE 8 *Solving a Logarithmic Equation*

Solve for x in the equation

$$\log_{10}(4x + 2) - \log_{10}(x - 1) = 1.$$

SOLUTION

$$\log_{10}(4x + 2) - \log_{10}(x - 1) = 1$$

$$\log_{10}\left(\frac{4x + 2}{x - 1}\right) = 1 \qquad \textit{Logarithmic form}$$

$$\frac{4x + 2}{x - 1} = 10 \qquad \textit{Exponential form}$$

$$4x + 2 = 10(x - 1)$$
$$4x - 10x = -10 - 2$$
$$-6x = -12$$
$$x = 2$$

EXAMPLE 9 *Finding the Zeros of a Logarithmic Function*

Find the zeros of the function

$$f(x) = 2 \ln(2x - 1) - \ln 9.$$

SOLUTION

We begin by setting $f(x)$ equal to zero and solving the resulting equation for x.

$$2 \ln(2x - 1) - \ln 9 = 0$$
$$2 \ln(2x - 1) = \ln 9$$
$$\ln(2x - 1) = \frac{1}{2} \ln 9$$
$$\ln(2x - 1) = \ln 9^{1/2}$$
$$\ln(2x - 1) = \ln 3$$
$$2x - 1 = 3$$
$$2x = 4$$
$$x = 2$$

Thus, f has one zero, $x = 2$. Try checking this by substituting $x = 2$ in the original function.

In the next example, we prove the *change of base formula* presented in Section 6.2.

EXAMPLE 10 The Change of Base Formula

Prove the change of base formula given in Section 6.2.

$$\log_a x = \frac{\log_b x}{\log_b a}$$

SOLUTION

We begin by letting

$$y = \log_a x$$

and writing the equivalent exponential form

$$a^y = x.$$

Now, taking the logarithm *with base b* of both sides, we have

$$\log_b a^y = \log_b x$$

$$y \log_b a = \log_b x$$

$$y = \frac{\log_b x}{\log_b a}$$

$$\log_a x = \frac{\log_b x}{\log_b a}.$$

When solving exponential or logarithmic equations, the following properties are useful.

1. $x = y$ if and only if $\log_a x = \log_a y$.
2. $x = y$ if and only if $a^x = a^y, a > 0, a \neq 1$.
3. $a = b$ if and only if $a^x = b^x, x \neq 0$.

Can you see where these properties were used in the examples in this section?

WARM UP

Solve for x.

1. $x \ln 2 = \ln 3$

2. $(x - 1) \ln 4 = 2$

3. $2xe^2 = e^3$

4. $4xe^{-1} = 8$

5. $x^2 - 4x + 5 = 0$

6. $2x^2 - 3x + 1 = 0$

Simplify the expressions.

7. $\log_{10} 100^x$

8. $\log_4 64^x$

9. $\ln e^{2x}$

10. $\ln e^{-x^2}$

EXERCISES 6.4

In Exercises 1–10, solve for x.

1. $4^x = 16$

2. $3^x = 243$

3. $7^x = \frac{1}{49}$

4. $8^x = 4$

5. $\left(\frac{3}{4}\right)^x = \frac{27}{64}$

6. $3^{x-1} = 27$

7. $\log_4 x = 3$

8. $\log_5 5x = 2$

9. $\log_{10} x = -1$

10. $\ln(2x - 1) = 0$

In Exercises 11–16, apply the inverse properties of ln x and e^x to simplify the given expression.

11. $\ln e^{x^2}$

12. $\ln e^{2x-1}$

13. $e^{\ln(5x+2)}$

14. $-1 + \ln e^{2x}$

15. $e^{\ln x^2}$

16. $-8 + e^{\ln x^3}$

In Exercises 17–48, solve the given exponential equation.

17. $10^x = 42$

18. $10^x = 570$

19. $\frac{1}{3}10^{2x} = 12$

20. $8(10^{3x}) = 12$

21. $3(10^{x-1}) = 2$

22. $2^{3x} = 565$

23. $e^x = 10$

24. $e^x = 6500$

25. $500e^{-x} = 300$

26. $1000e^{-4x} = 75$

27. $3e^{3x/2} = 40$

28. $6e^{1-x} = 25$

29. $25e^{2x+1} = 962$

30. $(1.003)^{365t} = 2$

31. $100\left(1 + \frac{0.09}{4}\right)^{4x} = 300$

32. $\frac{1250}{(1.04)^x} = 500$

33. $e^{0.09t} = 3$

34. $e^{0.125t} = 8$

35. $\left(1 + \frac{0.10}{12}\right)^{12t} = 2$

36. $\left(1 + \frac{0.065}{365}\right)^{365t} = 4$

37. $\frac{10,000}{1 + 19e^{-t/5}} = 2000$

38. $80e^{-t/2} + 20 = 70$

39. $\left(\frac{1}{1.0775}\right)^N = 0.2247$

40. $3^{2x+1} = 5^{x+2}$

41. $10^{7-x} = 5^{x+1}$

42. $4^{x^2} = 100$

43. $3(1 + e^{2x}) = 4$

44. $20(100 - e^{x/2}) = 500$

45. $\frac{400}{1 + e^{-x}} = 200$

46. $\frac{3000}{2 + e^{-2x}} = 1200$

47. $\frac{e^x + e^{-x}}{e^x - e^{-x}} = 2$

48. $\frac{e^x + e^{-x}}{2} = 2$

In Exercises 49–64, solve the given logarithmic equation.

49. $\ln x = 5$

50. $\ln 2x = -1$

51. $2 \ln x = 7$

52. $3 \ln 5x = 10$

53. $2 \ln 4x = 0$

54. $6 \ln(x + 1) = 2$

55. $\log_{10}(z - 3) = 2$

56. $\log_{10} x^2 = 20$

57. $\ln x + \ln(x - 2) = 1$

58. $\ln \sqrt{x + 2} = 1$

59. $\log_{10}(x + 4) - \log_{10} x = \log_{10}(x + 2)$

60. $\log_{10} x - \log_{10}(2x - 1) = 0$

61. $\ln x + \ln(x + 3) = 1$

62. $\log_2(x + 5) - \log_2(x - 2) = 3$

63. $\ln x^2 = (\ln x)^2$

64. $\log_4 x - \log_4(x - 1) = \frac{1}{2}$

In Exercises 65 and 66, find the time required for a $1000 investment to double at interest rate r, compounded continuously. Solve for t in the exponential equation $2000 = 1000e^{rt}$.

65. $r = 0.085$

66. $r = 0.12$

67. The demand equation for a certain product is given by

$$p = 500 - 0.5(e^{0.004x}).$$

Find the demand x for a price of (a) $p = \$350$ and (b) $p = \$300$.

68. The demand equation for a certain product is given by

$$p = 5000\left(1 - \frac{4}{4 + e^{-0.002x}}\right).$$

Find the demand x for a price of (a) $p = \$600$ and (b) $p = \$400$.

69. The yield V (in millions of cubic feet per acre) for a forest at age t years is given by

$$V = 6.7e^{-48.1/t}.$$

Find the time necessary to have a yield of (a) 1.3 million cubic feet and (b) 2 million cubic feet.

70. In a group project in learning theory, a mathematical model for the proportion P of correct responses after n trials was found to be

$$P = \frac{0.83}{1 + e^{-0.2n}}.$$

After how many trials will 60% of the responses be correct?

6.5 Applications of Exponential and Logarithmic Functions

The behavior of many physical, economic, and social phenomena can be described by exponential and logarithmic functions. In this section, we look at four basic types of applications: (1) Compound Interest, (2) Growth and Decay, (3) Logistics Models, and (4) Intensity Models. The problems presented in this section require the full range of solution techniques studied in this chapter.

Compound Interest

From Section 6.1, recall the following two compound interest formulas, where A is the account balance, P is the principal, r is the annual percentage rate, and t is the time in years.

n Compoundings per Year	*Continuous Compounding*
$A = P\left(1 + \dfrac{r}{n}\right)^{nt}$	$A = Pe^{rt}$

Exponential and Logarithmic Functions

EXAMPLE 1 *Doubling Time for an Investment at Quarterly Compounding*

An investment is made in a trust fund at an annual percentage rate of 9.5%, compounded quarterly. How long will it take for the investment to double in value?

SOLUTION

For quarterly compounding, we use the formula

$$A = P\left(1 + \frac{r}{4}\right)^{4t}.$$

Using $r = 0.095$, the time required for the investment to double is given by solving for t in the equation $2P = A$.

$$2P = P\left(1 + \frac{0.095}{4}\right)^{4t} \qquad 2P = A$$

$$2 = (1.02375)^{4t} \qquad \text{Divide both sides by } P$$

$$\ln 2 = \ln(1.02375)^{4t} \qquad \text{Take ln of both sides}$$

$$\ln 2 = 4t \ln(1.02375)$$

$$t = \frac{\ln 2}{4 \ln(1.02375)} \approx 7.4$$

Therefore, it will take approximately 7.4 years for the investment to double in value with quarterly compounding.

Try reworking Example 1 using continuous compounding. To do this you will need to solve the equation

$$2P = Pe^{0.095t}.$$

The solution is $t \approx 7.3$ years, which makes sense because the principal should double more quickly with continuous compounding than with quarterly compounding.

From Example 1, we see that the time required for an investment to double in value is independent of the amount invested. In general, the **doubling time** is as follows.

n Compoundings per Year	*Continuous Compounding*
$t = \dfrac{\ln 2}{n \ln[1 + (r/n)]}$	$t = \dfrac{\ln 2}{r}$

EXAMPLE 2 *Finding an Annual Percentage Rate*

An investment of $10,000 is compounded continuously. What annual percentage rate will produce a balance of $25,000 in ten years?

SOLUTION

We use the formula

$$A = Pe^{rt}$$

with $P = 10,000$, $A = 25,000$, and $t = 10$, and solve the following equation for r.

$$10,000e^{10r} = 25,000$$
$$e^{10r} = 2.5$$
$$10r = \ln 2.5$$
$$r = \frac{1}{10} \ln 2.5 \approx 0.0916$$

Thus, the annual percentage rate must be approximately 9.16%.

EXAMPLE 3 *The Effective Yield for an Investment*

A deposit is compounded continuously at an annual percentage rate of 7.5%. Find the **effective yield.** That is, find the simple interest rate that would yield the same balance at the end of one year.

SOLUTION

Using the formula $A = Pe^{rt}$ with $r = 0.075$ and $t = 1$, the balance at the end of one year is

$$A = Pe^{0.075(1)}$$
$$\approx P(1.0779)$$
$$= P(1 + 0.0779). \qquad\qquad A = P(1 + r)$$

Since the formula for simple interest after one year is

$$A = P(1 + r)$$

we conclude that the effective yield is approximately 7.79%.

Growth and Decay

The balance in an account earning *continuously compounded* interest is one example of a quantity that increases over time according to the **exponential growth model**

$$Q(t) = Ce^{kt}.$$

In this model, $Q(t)$ is the size of the population (balance, weight, and so forth) at any time t, C is the original population (when $t = 0$), and k is a constant determined by the rate of growth. If $k > 0$, the population *grows* (increases) over time, and if $k < 0$ it *decays* (decreases) over time. Example 9 of Section 6.1 is an example of population growth. Recall from Section 6.1 that we can remember this growth model as *(Then)* = *(Now)*(e^{kt}).

EXAMPLE 4 *Exponential Decay*

Radioactive iodine is a by-product of some types of nuclear reactors. Its **half-life** is 60 days. That is, after 60 days, a given amount of radioactive iodine will have decayed to half the original amount. Suppose a contained nuclear accident occurs and gives off an initial amount C of radioactive iodine.

(a) Write an equation for the amount of radioactive iodine present at any time t, following the accident.
(b) How long will it take for the radioactive iodine to decay to a level of 20% of the original amount?

SOLUTION

(a) We first need to find the rate k, in the exponential model $Q(t) = Ce^{kt}$. Knowing that half the original amount remains after $t = 60$ days, we obtain

$$Q(60) = Ce^{k(60)} = \frac{1}{2}C$$

$$e^{60k} = \frac{1}{2}$$

$$60k = -\ln 2$$

$$k = \frac{-\ln 2}{60} \approx -0.0116.$$

Thus, the exponential model is

$$Q(t) = Ce^{-0.0116t}.$$

(b) The time required to decay to 20% of the original amount is given by

$$Q(t) = Ce^{-0.0116t} = (0.2)C$$
$$e^{-0.0116t} = 0.2$$
$$-0.0116t = \ln 0.2$$

$$t = \frac{\ln 0.2}{-0.0116} \approx 139 \text{ days.}$$

In living organic material the ratio of radioactive carbon isotopes (Carbon 14) to the number of nonradioactive carbon isotopes (Carbon 12) is about 1 to 10^{12}. When organic material dies, its Carbon 12 content remains fixed, whereas its radioactive Carbon 14 begins to decay with a half-life of about 5700 years. To estimate the age of dead organic material, scientists use the following formula, which denotes the ratio of Carbon 14 to Carbon 12 present at any time t (in years).

$$R(t) = \frac{1}{10^{12}} 2^{-t/5700}$$

The graph of R is shown in Figure 6.17. Note that R decreases as the time t increases.

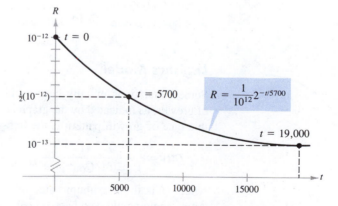

FIGURE 6.17

Exponential and Logarithmic Functions

EXAMPLE 5 Carbon Dating

Suppose that the Carbon 14/Carbon 12 ratio of a newly discovered fossil is

$$R = \frac{1}{10^{13}}.$$

Estimate the age of the fossil.

SOLUTION

In the carbon dating model, we substitute the given value of R to obtain

$$\frac{1}{10^{13}} = \frac{1}{10^{12}} 2^{-t/5700}$$

$$\frac{1}{10} = 2^{-t/5700}$$

$$10 = 2^{t/5700}$$

$$\ln 10 = \ln 2^{t/5700} = \frac{t}{5700} \ln 2$$

$$t = 5700 \frac{\ln 10}{\ln 2} \approx 19{,}000 \text{ years.} \qquad \textit{Nearest thousand}$$

Thus, we estimate the age of the fossil to be 19,000 years.

Remark: The carbon dating model in Example 5 assumed that the Carbon 14/Carbon 12 ratio was one part in 10,000,000,000,000. Suppose that an error in measurement occurred and the actual ratio was only one part in 8,000,000,000,000. The fossil age corresponding to the actual ratio would then be approximately 17,000 years. Try checking this result.

Logistics Model

Some populations initially have rapid growth, followed by a declining rate of growth, as indicated by the graph in Figure 6.18. One model for describing this type of growth pattern is the **logistics curve** given by the function

$$Q(t) = \frac{M}{1 + (M/Q_0 - 1)e^{-kt}}$$

where M is the maximum value of Q and Q_0 is the initial population size. An example would be a bacteria culture allowed to grow initially under ideal conditions, followed by less favorable conditions that inhibit growth.

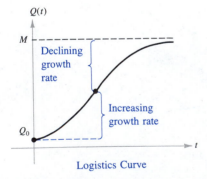

Logistics Curve

FIGURE 6.18

Applications of Exponential and Logarithmic Functions

EXAMPLE 6   Logistics Curve

On a college campus of 5000 students, one student returned from vacation with a contagious flu virus. The spread of the virus through the student body is given by

$$s(t) = \frac{5000}{1 + 4999e^{-0.8t}}$$

where $s(t)$ is the total number infected after t days. The college will cancel classes when 40% or more of the students are ill.

(a) How many are infected after five days?
(b) After how many days will the college cancel classes?

SOLUTION

(a) After five days, the number infected is

$$s(5) = \frac{5000}{1 + 4999e^{-0.8(5)}}$$

$$= \frac{5000}{1 + 4999e^{-4}}$$

$$\approx 54.$$

(b) In this case, the number infected is $(0.40)(5000) = 2000$. Therefore, we solve for t in the following equation.

$$2000 = \frac{5000}{1 + 4999e^{-0.8t}}$$

$$2000 + 9{,}998{,}000e^{-0.8t} = 5000$$

$$e^{-0.8t} = \frac{3000}{9{,}998{,}000} = \frac{3}{9998}$$

$$\ln(e^{-0.8t}) = \ln \frac{3}{9998}$$

$$-0.8t = \ln \frac{3}{9998}$$

$$t = \frac{\ln(3/9998)}{-0.8} \approx 10.1$$

Hence, after ten days, at least 40% of the students will be infected, and the college will cancel classes.

Exponential and Logarithmic Functions

Intensity Model

Sound and shock waves can be measured by the **intensity model**

$$S = K \log_{10} \frac{I}{I_0}$$

where I is the intensity of the stimulus wave, I_0 is the **threshold intensity** (the smallest value of I that can be detected by the listening device), and K determines the units in which S is measured. Sound heard by the human ear is measured in decibels. One **decibel** is considered the smallest detectable difference in the loudness of two sounds.

For earthquakes, the shock is measured by units on the *Richter Scale*, as demonstrated in Example 7.

EXAMPLE 7 Magnitude of Earthquakes

On the Richter Scale, the magnitude R of an earthquake of intensity I is given by

$$R = \log_{10} \frac{I}{I_0}$$

where $I_0 = 1$ is the minimum intensity used for comparison. Find the intensity per unit of area for the following earthquakes. (Intensity is a measure of the wave energy of an earthquake.)

(a) San Francisco in 1906, $R = 8.3$
(b) Mexico City in 1978, $R = 7.85$
(c) Predicted in 1990, $R = 6.3$

SOLUTION

(a) Since $I_0 = 1$ and $R = 8.3$, we have

$$8.3 = \log_{10} I$$
$$I = 10^{8.3} \approx 199,526,000.$$

(b) For Mexico City, we have $7.85 = \log_{10} I$, and

$$I = 10^{7.85} \approx 70,795,000.$$

(c) For $R = 6.3$, we have $6.3 = \log_{10} I$, and

$$I = 10^{6.3} \approx 1,995,000.$$

Note that an increase of two units on the Richter Scale (from 6.3 to 8.3) represents an intensity change by a factor of

$$\frac{199,526,000}{1,995,000} \approx 100.$$

EXERCISES 6.5

In Exercises 1–10, complete the table for a savings account in which interest is compounded continuously.

	Initial investment	Annual % rate	Effective yield	Time to double	Amount after 10 years
1.	$1000	12%			
2.	$20,000	$10\frac{1}{2}$%			
3.	$750			$7\frac{3}{4}$ yr	
4.	$10,000			5 yr	
5.	$500				$1,292.85
6.	$2000		4.5%		
7.		11%			$19,205.00
8.		8%			$20,000.00
9.	$5000		8.33%		
10.	$250		12.19%		

In Exercises 11 and 12, determine the principal P which must be invested at rate r, compounded monthly, so that $500,000 will be available for retirement in t years.

11. $r = 7\frac{1}{2}$%, $t = 20$

12. $r = 12$%, $t = 40$

In Exercises 13 and 14, determine the time necessary for $1000 to double if it is invested at interest rate r compounded (a) annually, (b) monthly, (c) daily, and (d) continuously.

13. $r = 11$%

14. $r = 10\frac{1}{2}$%

15. Complete the following table for the time t necessary for P dollars to triple if interest is compounded continuously at rate r.

r	2%	4%	6%	8%	10%	12%
t						

16. Repeat Exercise 15 for interest that is compounded annually.

In Exercises 17–20, consider making monthly deposits of P dollars into a savings account at an annual interest rate r, compounded monthly. Find the balance A after t years given that

$$A = \frac{P(e^{rt} - 1)}{e^{r/12} - 1}.$$

17. $P = 50, $r = 7$%, $t = 20$ years

18. $P = 75, $r = 9$%, $t = 25$ years

19. $P = 100, $r = 10$%, $t = 40$ years

20. $P = 20, $r = 6$%, $t = 50$ years

21. The population P of a city is given by

$$P = 105,300e^{0.015t}$$

where t is the time in years, with $t = 0$ corresponding to 1985. According to this model, in what year will the city have a population of 150,000?

22. The population P of a city is given by

$$P = 2500e^{kt}$$

where t is the time in years, with $t = 0$ corresponding to the year 1980. In 1935, the population was 3350. Find the value of k and use this result to predict the population in the year 2000.

23. Assume that the world population at time t is given by $P = P_0e^{rt}$. How long will it take the world population to double? To triple?

24. The number of bacteria N in a culture is given by the model

$$N = 100e^{kt}$$

where t is the time in hours, with $t = 0$ corresponding to the time when $N = 100$. When $t = 5$, $N = 300$. How long does it take the population to double in size?

In Exercises 25–30, complete the table for the given radio-active isotope.

Isotope	Half-life (years)	Initial quantity	Amount after 1000 years	Amount after 10,000 years
25. Ra^{226}	1,620	10 g		
26. Ra^{226}	1,620		1.5 g	
27. C^{14}	5,730			2 g
28. C^{14}	5,730	3 g		
29. Pu^{230}	24,360		2.1 g	
30. Pu^{230}	24,360			0.4 g

31. What percentage of a present amount of radioactive radium (Ra^{226}) will remain after 100 years?

32. Find the half-life of a radioactive material if, after one year, 99.57% of the initial amount remains.

In Exercises 33–36, find the constant k such that the exponential function $y = Ce^{kt}$ passes through the given points on the graph.

33.

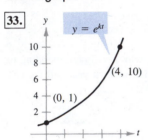

$y = e^{kt}$

(4, 10)

(0, 1)

34.

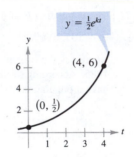

$y = \frac{1}{2}e^{kt}$

(4, 6)

(0, $\frac{1}{2}$)

35.

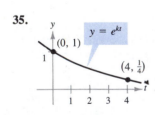

$y = e^{kt}$

(0, 1)

(4, $\frac{1}{4}$)

36.

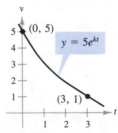

(0, 5)

$y = 5e^{kt}$

(3, 1)

37. The sales S (in thousands of units) of a new product after it has been on the market t years are given by

$$S(t) = 100(1 - e^{kt}).$$

(a) Find S as a function of t if 15,000 units have been sold after one year.

(b) How many units will be sold after five years?

38. The management at a factory has found that the maximum number of units a worker can produce in a day is 30. The learning curve for the number of units N produced per day after a new employee has worked t days is given by

$$N = 30(1 - e^{kt}).$$

After 20 days on the job, a particular worker produced 19 units.

(a) Find the learning curve for this worker (that is, find the value of k).

(b) How many days should pass before this worker is producing 25 units per day?

39. A certain lake is stocked with 500 fish, and the fish population increases according to the logistics curve

$$p(t) = \frac{10,000}{1 + 10e^{-t/5}}$$

where t is measured in months.

(a) Find $p(5)$.

(b) After how many months will the fish population be 2000?

40. A conservation organization releases 100 animals of an endangered species into a game preserve. The organization believes that the preserve has a carrying capacity of 1000 animals and that the growth of the herd will be modeled by the logistics curve

$$p(t) = \frac{1000}{1 + 9e^{-kt}}$$

where t is measured in years.

(a) Find k if the herd size is 134 after two years.

(b) Find $p(5)$.

41. The intensity level β, in decibels, of a sound wave is defined by

$$\beta(I) = 10 \log_{10} \frac{I}{I_0}$$

where I_0 is an intensity of 10^{-16} watts per square centimeter, corresponding roughly to the faintest sound that can be heard. Determine $\beta(I)$ for the following conditions.

(a) $I = 10^{-14}$ watts per centimeter (whisper)

(b) $I = 10^{-9}$ watts per centimeter (busy street corner)

(c) $I = 10^{-6.5}$ watts per centimeter (air hammer)

(d) $I = 10^{-4}$ watts per centimeter (threshold of pain)

42. Due to the installation of noise suppression materials, the noise level in an auditorium was reduced from 93 to 80 decibels. Find the percentage decrease in the intensity levels of the noise because of the installation of these materials.

In Exercises 43 and 44, use the Richter Scale (Example 7) for measuring the magnitude of earthquakes.

43. Find the magnitude R of an earthquake of intensity I (let $I_0 = 1$).
    (a) $I = 80,500,000$ (b) $I = 48,275,000$

44. Find the intensity I of an earthquake measuring R on the Richter Scale (let $I_0 = 1$).
    (a) Columbia in 1906, $R = 8.6$
    (b) Los Angeles in 1971, $R = 6.7$

In Exercises 45–48, use the acidity model given by $pH = -\log_{10}[H^+]$, where acidity (pH) is a measure of the hydrogen ion concentration $[H^+]$ (measured in moles of hydrogen per liter) of a solution.

45. Find the pH if $[H^+] = 2.3 \times 10^{-5}$.
46. Find the pH if $[H^+] = 11.3 \times 10^{-6}$.
47. Compute $[H^+]$ for a solution in which pH $= 5.8$.
48. Compute $[H^+]$ for a solution in which pH $= 3.2$.

In Exercises 49 and 50, use **Newton's Law of Cooling**, which states that the rate of change in the temperature of an object is proportional to the difference between its temperature and the temperature of its environment. If $T(t)$ is the temperature of the object at time t in minutes, T_0 is the initial temperature, and T_e is the constant temperature of the environment, then

$$T(t) = T_e + (T_0 - T_e)e^{-kt}.$$

49. An object in a room at 70° F cools from 350° F to 150° F in 45 minutes.
    (a) Find the temperature of the object as a function of time.
    (b) Find the temperature after it has cooled for one hour.
    (c) Find the time necessary for the object to cool to 80° F.

50. A thermometer is taken from a room at 72° F to the outdoors, where the temperature is 20° F. The reading on the thermometer drops to 48° F after one minute. Determine the reading after five minutes.

CHAPTER 6 REVIEW EXERCISES

In Exercises 1–16, sketch the graph of the function.

1. $f(x) = 6^x$
2. $f(x) = 0.3^x$
3. $g(x) = 6^{-x}$
4. $g(x) = 0.3^{-x}$
5. $h(x) = e^{-x/2}$
6. $h(x) = 2 - e^{-x/2}$
7. $f(x) = e^{x+2}$
8. $s(t) = 4e^{-2/t}, \quad t > 0$
9. $f(x) = \ln(x - 3)$
10. $f(x) = \ln|x|$
11. $f(x) = \ln x + 3$
12. $f(x) = \frac{1}{4}\ln x$
13. $g(x) = \log_2(-x)$
14. $g(x) = \log_5 x$
15. $h(x) = \ln(e^{x-1})$
16. $f(x) = e^{\ln x^2}$

In Exercises 17–24, use the properties of logarithms to write the expression as a sum, difference, or multiple of logarithms.

17. $\log_5 5x^2$
18. $\log_7 \dfrac{\sqrt{x}}{4}$
19. $\log_{10} \dfrac{5\sqrt{y}}{x^2}$
20. $\ln \left| \dfrac{x - 1}{x + 1} \right|$
21. $\ln \left| \dfrac{x^2 + 1}{x} \right|$
22. $\ln \sqrt{\dfrac{x^2 + 1}{x^4}}$
23. $\ln[(x^2 + 1)(x - 1)]$
24. $\ln \sqrt[5]{\dfrac{4x^2 - 1}{4x^2 + 1}}$

In Exercises 25–32, write the expression as the logarithm of a single quantity.

25. $\log_{10} 5 - 2 \log_{10}(x + 4)$

26. $6 \log_8 z + \frac{1}{2} \log_8(z^2 + 4)$

27. $\frac{1}{2} \ln|2x - 1| - 2 \ln|x + 1|$

28. $5 \ln|x - 2| - \ln|x + 2| - 3 \ln|x|$

29. $2(\ln x + \frac{1}{3} \ln \sqrt{x})$

30. $\frac{1}{2} \ln(x^2 + 4x) - \ln 2 - \ln x$

31. $\ln 3 + \frac{1}{3} \ln(4 - x^2) - \ln x$

32. $3[\ln x - 2 \ln(x^2 + 1)] + 2 \ln 5$

In Exercises 33–40, determine whether the statement or equation is true or false.

33. The domain of the function $f(x) = \ln|x|$ is the set of all real numbers.

34. The range of the function $g(x) = e^{-x}$ is the set of all positive real numbers.

35. $\ln(x + y) = \ln x + \ln y$

36. $\dfrac{\ln x}{\ln y} = \ln x - \ln y$

37. $\ln \sqrt{x^4 + 2x^2} = \ln(|x| \sqrt{x^2 + 2})$

38. $\log_b b^{2x} = 2x$

39. $\dfrac{e^{2x} - 1}{e^x - 1} = e^x + 1$

40. $e^{x-1} = \dfrac{e^x}{e}$

41. A solution of a certain drug contained 500 units per milliliter when prepared. It was analyzed after 40 days and found to contain 300 units per milliliter. Assuming that the rate of decomposition is proportional to the amount present, the equation giving the amount A after t days is

$A = 500e^{-0.013t}$.

Use this model to find A when $t = 60$.

42. The number of miles s of roads cleared of snow is approximated by the model

$s = 25 - \dfrac{13 \ln(h/12)}{\ln 3}$, $\quad 2 \le h \le 15$

where h is the depth of the snow in inches. Use this model to find s when $h = 10$ inches.

In Exercises 43 and 44, find the probability of waiting less than t units of time until the next occurrence of an event if that probability is approximated by the model

$F(t) = 1 - e^{-t/\lambda}$

where λ is the average time between successive occurrences of the event.

43. The average time between incoming calls at a switchboard is three minutes. If a call has just come in, find the probability that the next call will be within
(a) $\frac{1}{2}$ minute   (b) 2 minutes   (c) 5 minutes.

44. Trucks arrive at a terminal at an average of 3 per hour (therefore $\lambda = 20$ minutes). If a truck has just arrived, find the probability that the next arrival will be within
(a) 10 minutes   (b) 30 minutes   (c) 1 hour.

45. A certain automobile gets 28 miles per gallon of gasoline for speeds up to 50 miles per hour. Over 50 miles per hour, the number of miles per gallon drops at the rate of 12% for each 10 miles per hour. If s is the speed and y is the number of miles per gallon, then

$y = 28e^{0.6-0.012s}$, $s \ge 50$.

Use this function to complete the following table.

Speed	50	55	60	65	70
Miles per gallon					

46. Find the balance after 25 years if $25,000 is invested at 10% compounded
(a) monthly   (b) daily   (c) continuously.

In Exercises 47–50, find the exponential function $y = Ce^{kt}$ that passes through the two points.

47.

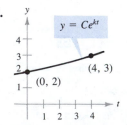

48.

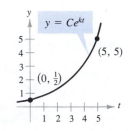

49.

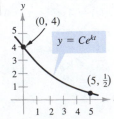

50.

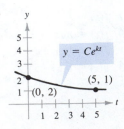

51. The demand equation for a certain product is given by

$$p = 500 - 0.5e^{0.004x}.$$

Find the demand x for a price of (a) $p = \$450$ and (b) $p = \$400$.

52. In a typing class, the average number of words per minute typed after t weeks of lessons was found to be

$$N = \frac{157}{1 + 5.4e^{-0.12t}}.$$

Find the time necessary to type (a) 50 words per minute and (b) 75 words per minute.

53. A deposit of \$750 is made in a savings account for which the interest is compounded continuously. The balance will double in $7\frac{3}{4}$ years.
(a) What is the annual percentage rate for this account?
(b) Find the balance in the account after ten years.
(c) Find the effective yield.

54. A deposit of \$10,000 is made in a savings account for which the interest is compounded continuously. The balance will double in five years.
(a) What is the annual percentage rate for this account?
(b) Find the balance after one year.
(c) Find the effective yield.

55. Complete the following table for the function $f(x) = e^{-x}$.

x	0	1	2	5	10	15
e^{-x}	1	0.36788				

This table demonstrates that for $k > 0$, e^{-kx} approaches 0 as x increases without bound.

56. Use the result of Exercise 55 to determine what each function ($k > 0$) approaches as x increases.
(a) $f(x) = 30(1 - e^{-0.025x})$

(b) $f(x) = \dfrac{50}{2 + 3e^{-0.2x}}$

(c) $f(x) = (100 - a)e^{-kx} + a$

(d) $f(x) = 5000\left(1 - \dfrac{4}{4 + e^{-0.002x}}\right)$

57. In calculus it can be shown that

$$e^x \approx 1 + x + \frac{x^2}{2} + \frac{x^3}{6} + \frac{x^4}{24}.$$

Use this equation to approximate the following, and compare the results to those obtained with a calculator.
(a) e (b) $e^{1/2}$ (c) $e^{-1/2}$

CHAPTER 7

Trigonometry

7.1 Angles and Their Measure

As derived from the Greek language, the word **trigonometry** means "measurement of triangles." Initially, trigonometry dealt with relationships among the sides and angles of triangles. As such, it was used in the development of astronomy, navigation, and surveying.

With the advent of calculus in the seventeenth century, and a resulting expansion of knowledge in the physical sciences, a different perspective arose—one that viewed the classic trigonometric relationships as *functions* with the set of real numbers as their domains. Consequently, the applications of trigonometry expanded to include a vast number of physical phenomena involving rotations, or vibrations. These include sound waves, light rays, planetary orbits, vibrating strings, pendulums, and orbits of atomic particles.

Our approach to trigonometry incorporates *both* perspectives, starting with angles and their measure.

Angles

An **angle** is determined by rotating a ray (half-line) about its endpoint. The starting position of the ray is called the **initial side** of the angle, and the position after rotation is called the **terminal side,** as shown in Figure 7.1.

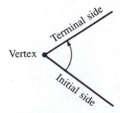

FIGURE 7.1

368

Angles and Their Measures

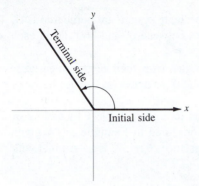

Standard Position of an Angle

FIGURE 7.2

The endpoint of the ray is called the **vertex** of the angle. This perception of an angle fits nicely into a coordinate system in which the origin is the vertex and the initial side coincides with the positive *x*-axis. Such an angle is said to be in **standard position,** as shown in Figure 7.2. **Positive angles** are generated by counterclockwise rotation, and **negative angles** by clockwise rotation, as shown in Figure 7.3.

To label angles in trigonometry, we use the Greek letters α (alpha), β (beta), and θ (theta), as well as upper-case letters *A*, *B*, and *C*. In Figure 7.4, note that angles α and β have the same initial and terminal sides. Such angles are called **coterminal.**

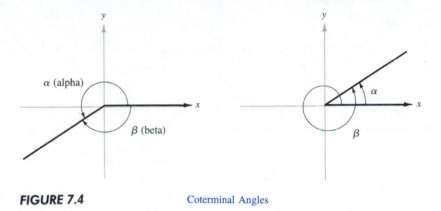

FIGURE 7.4   Coterminal Angles

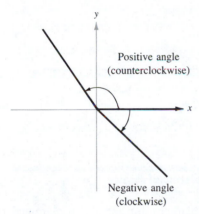

FIGURE 7.3

Degree Measure

The **measure of an angle** is determined by the amount of rotation from the initial to the terminal side. The most common unit of angle measure is the **degree,** denoted by the symbol °. A measure of **one degree (1°)** is equivalent to 1/360 of a complete revolution about the vertex. To measure angles, it is convenient to mark degrees on the circumference of a circle, as shown in Figure 7.5. Thus, a full revolution (counterclockwise) corresponds to 360°, a half revolution to 180°, and a quarter revolution to 90°.

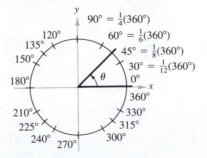

FIGURE 7.5   Degree Measure of an Angle

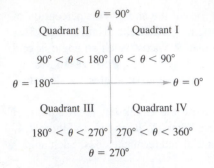

Quadrant Location of Angles

FIGURE 7.6

Recall that the four quadrants in a coordinate system are numbered I, II, III, and IV. Figure 7.6 shows which angles between 0° and 360° lie in each of the four quadrants.

Figure 7.7 shows several common angles with their degree measures. Note that we classify angles between 0° and 90° as **acute** and angles between 90° and 180° as **obtuse.**

We can find an angle that is coterminal to a given angle θ by adding or subtracting 360° (one revolution), as demonstrated in Example 1. (A given angle θ has many coterminal angles. For instance, $\theta = 30°$ is coterminal with $30° \pm n(360°)$, where n is an integer.)

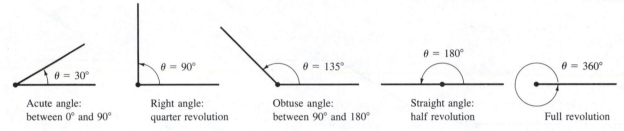

| Acute angle: between 0° and 90° | Right angle: quarter revolution | Obtuse angle: between 90° and 180° | Straight angle: half revolution | Full revolution |

FIGURE 7.7

EXAMPLE 1 *Sketching and Finding Coterminal Angles*

Sketch each of the following angles in standard position and find a coterminal angle for each.

(a) $\theta = 390°$ (b) $\theta = 135°$ (c) $\theta = -120°$

SOLUTION

(a) For the positive angle $\theta = 390°$, we subtract 360° and obtain the coterminal angle

$$390° - 360° = 30°.$$

Thus, the terminal side of θ lies in Quadrant I. Its sketch is shown in Figure 7.8(a).

(b) Again subtracting 360°, an angle coterminal with 135° is

$$135° - 360° = -225°.$$

Since $90° < \theta < 180°$, the terminal side of θ lies in Quadrant II. Moreover, since $\theta = 135°$ is 45° less than 180°, it follows that θ lies 45° up from the horizontal axis. A sketch is shown in Figure 7.8(b).

(a)

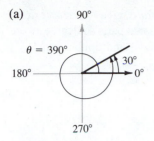

(b)

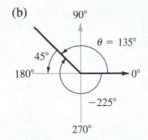

(c)

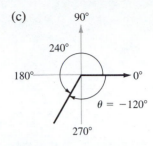

FIGURE 7.8

(c) For the negative angle $\theta = -120°$, we add $360°$ to obtain the coterminal angle

$$\theta = -120° + 360° = 240°$$

as shown in Figure 7.8(c).

Remark: We often abbreviate the phrase "the terminal side of θ lies in a quadrant" by simply saying that "θ lies in a quadrant."

Two *positive* angles α and β are said to be **complementary** (or complements of each other) if their sum is $90°$. For example, $30°$ and $60°$ are complementary angles because $30° + 60° = 90°$. Two positive angles are **supplementary** (or supplements of each other) if their sum is $180°$. For example, $50°$ and $130°$ are supplementary angles because $50° + 130° = 180°$.

EXAMPLE 2 *Complementary and Supplementary Angles*

If possible, find the complementary angle and the supplementary angle for each of the following angles.

(a) $72°$ (b) $148°$

SOLUTION

(a) The complement of $\theta = 72°$ is

$$90° - \theta = 90° - 72° = 18°.$$

The supplement of $\theta = 72°$ is

$$180° - \theta = 180° - 72° = 108°.$$

(b) Because $\theta = 148°$ is greater than $90°$, it has no complement. (Remember, we use only *positive* angles for complements.) The supplement of $\theta = 148°$ is

$$180° - \theta = 180° - 148° = 32°.$$

Trigonometry

With calculators it is convenient to use *decimal* degrees to denote fractional parts of degrees. Historically, however, fractional parts of degrees were expressed in *minutes* and *seconds*, using the prime (′) and double prime (″) notations, respectively. That is,

$$1' = \text{one minute} = \frac{1}{60}(1°)$$

$$1'' = \text{one second} = \frac{1}{60}(1') = \frac{1}{3600}(1°).$$

Consequently, an angle of 64 degrees, 32 minutes, and 47 seconds is represented by $\theta = 64° \ 32' \ 47''$.

Many calculators have special keys for converting an angle in degrees, minutes, and seconds (D° M′ S″) into decimal degree form, and conversely. If your calculator does not have these special keys, you can use the techniques demonstrated in the next two examples to make the conversions.

EXAMPLE 3 *Converting an Angle from D° M′ S″ to Decimal Form*

Convert 152° 15′ 29″ to decimal degree form.

SOLUTION

Since

$$1' = \left(\frac{1}{60}\right)° \quad \text{and} \quad 1'' = \left(\frac{1}{60}\right)\left(\frac{1}{60}\right)° = \left(\frac{1}{3600}\right)°$$

we have

$$152° \ 15' \ 29'' = 152° + \left(\frac{15}{60}\right)° + \left(\frac{29}{3600}\right)°$$
$$\approx 152° + 0.25° + 0.00806°$$
$$= 152.25806°.$$

Radian Measure

A second way to measure angles is in terms of radians. This type of measure is especially useful in calculus. To define a radian we use a **central angle** of a circle, one whose vertex is the center of the circle, as shown in Figure 7.9.

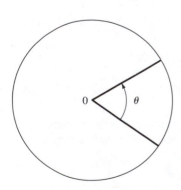

Central Angle θ

FIGURE 7.9

Definition of a Radian

One **radian** is the measure of a central angle θ that subtends (intercepts) an arc s equal in length to the radius r of the circle. [See Figure 7.10(a).]

Angles and Their Measures

(a)

(b)

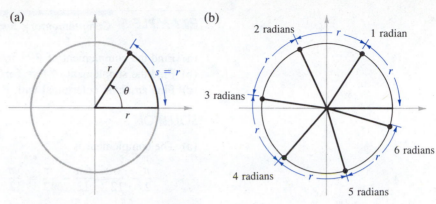

arc length = radius when $\theta = 1$ radian

FIGURE 7.10

Since the circumference of a circle is $2\pi r$, it follows that a central angle of one full revolution (counterclockwise) corresponds to an arc length of $s = 2\pi r$. Moreover, because each radian intercepts an arc of length r, we conclude that one full revolution corresponds to an angle of $(2\pi r)/r = 2\pi$ radians. Note that since $2\pi \approx 6.28$, there are just over six radii lengths in a full circle, as shown in Figure 7.10(b).

In general, the radian measure of a central angle θ is obtained by dividing the arc length s by r. That is,

$$\frac{s}{r} = \theta$$

where θ *is measured in radians*. Because the units of measure for s and r are the same, this ratio is unitless—it is simply a real number.

EXAMPLE 4   *Radian Measure of an Angle*

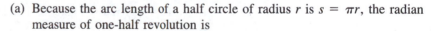

(a) Because the arc length of a half circle of radius r is $s = \pi r$, the radian measure of one-half revolution is

$$\frac{\pi r}{r} = \pi. \qquad\qquad \textit{One-half revolution}$$

(b) The radian measure of one-quarter revolution is

$$\frac{\pi r/2}{r} = \frac{\pi}{2}. \qquad\qquad \textit{One-quarter revolution}$$

Figure 7.11 shows these two angles together with the radian measure of some other common angles.

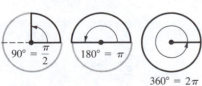

$360° = 2\pi$

Radian Measure for Several Common Angles

FIGURE 7.11

EXAMPLE 5 *Complementary, Supplementary, and Coterminal Angles*

(a) Find the complement of $\theta = \pi/12$.
(b) Find the supplement of $\theta = 5\pi/6$.
(c) Find an angle coterminal with $\theta = 17\pi/6$.

SOLUTION

(a) The complement is

$$\frac{\pi}{2} - \frac{\pi}{12} = \frac{6\pi}{12} - \frac{\pi}{12} = \frac{5\pi}{12}.$$

(b) The supplement is

$$\pi - \frac{5\pi}{6} = \frac{6\pi}{6} - \frac{5\pi}{6} = \frac{\pi}{6}.$$

(c) In radian measure, a coterminal angle is found by adding or subtracting 2π. For a positive angle, we subtract to obtain

$$\frac{17\pi}{6} - 2\pi = \frac{17\pi}{6} - \frac{12\pi}{6} = \frac{5\pi}{6}.$$

Remark: Note that when no units of angle measure are specified, *radian measure is implied.* For instance, if we write $\theta = \pi$ or $\theta = 2$, we mean $\theta = \pi$ radians or $\theta = 2$ radians.

Conversion of Angle Measure

Since 2π radians is the measure of an angle of one complete revolution, degrees and radians are related by the equations

$$360° = 2\pi \text{ rad} \qquad \text{or} \qquad 180° = \pi \text{ rad}.$$

From the latter equation, we obtain

$$1° = \frac{\pi}{180} \text{ rad} \qquad \text{and} \qquad 1 \text{ rad} = \left(\frac{180}{\pi}\right)°$$

which lead to the following conversion rules.

Conversions: Degrees $\Longleftrightarrow$ Radians

1. To convert degrees to radians, multiply degrees by $\dfrac{\pi \text{ rad}}{180°}$.

2. To convert radians to degrees, multiply radians by $\dfrac{180°}{\pi \text{ rad}}$.

To apply these two conversion rules, you simply need to remember the basic relationship π rad $= 180°$, as demonstrated in Examples 6, 7, and 8.

EXAMPLE 6 *Converting from Degrees to Radians*

Convert the following angles from degree to radian measure.

(a) $135°$ (b) $540°$ (c) $-270°$

SOLUTION

(a) $135° = (135 \text{ deg})\left(\dfrac{\pi \text{ rad}}{180 \text{ deg}}\right) = \dfrac{3\pi}{4} \text{ rad}$

(b) $540° = (540 \text{ deg})\left(\dfrac{\pi \text{ rad}}{180 \text{ deg}}\right) = 3\pi \text{ rad}$

(c) $-270° = (-270 \text{ deg})\left(\dfrac{\pi \text{ rad}}{180 \text{ deg}}\right) = -\dfrac{3\pi}{2} \text{ rad}$

EXAMPLE 7 *Converting from Radians to Degrees*

Convert the following radian measures to degree measures.

(a) $-\dfrac{\pi}{2}$ (b) $\dfrac{9\pi}{2}$ (c) 2

SOLUTION

(a) $-\dfrac{\pi}{2} \text{ rad} = \left(-\dfrac{\pi}{2} \text{ rad}\right)\left(\dfrac{180 \text{ deg}}{\pi \text{ rad}}\right) = -90°$

(b) $\dfrac{9\pi}{2} \text{ rad} = \left(\dfrac{9\pi}{2} \text{ rad}\right)\left(\dfrac{180 \text{ deg}}{\pi \text{ rad}}\right) = 810°$

(c) $2 \text{ rad} = (2 \text{ rad})\left(\dfrac{180 \text{ deg}}{\pi \text{ rad}}\right) = \dfrac{360}{\pi} \text{ deg} \approx 114.59°$

Remark: If you have a calculator with a "radian-to-degree" conversion key, try using it to verify the result shown in part (c) of Example 7.

EXAMPLE 8 *Converting an Angle to D° M′ S″ Form*

Convert 0.86492 radians to D° M′ S″ form.

SOLUTION

First we convert to decimal degrees.

$$0.86492 = 0.86492\left(\frac{180°}{\pi}\right) \approx 49.55627°$$

Then we have

$$
\begin{aligned}
49.55627° &= 49° + 0.55627° \\
&= 49° + (0.55627)(60') &&\text{\textit{Because 1° = 60'}} \\
&= 49° + 33.3762' \\
&= 49° + 33' + 0.3762' \\
&= 49° + 33' + (0.3762)(60'') &&\text{\textit{Because 1' = 60''}} \\
&= 49° + 33' + 22.572''.
\end{aligned}
$$

Consequently,

$$0.86492 \text{ radian} \approx 49° \ 33' \ 23''.$$

Applications

The *radian measure* formula, $\theta = s/r$, can be used to measure arc length along a circle. Specifically, for a circle of radius r, a central angle θ subtends an arc of length s given by

$$s = r\theta \qquad\qquad \textit{Length of circular arc}$$

where θ is measured in radians.

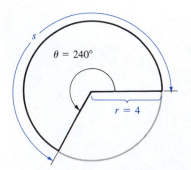

FIGURE 7.12

EXAMPLE 9 *Finding Arc Length*

A circle has a radius of 4 inches. Find the length of the arc cut off (subtended) by a central angle of 240°, as shown in Figure 7.12.

SOLUTION

To use the formula $s = r\theta$, we must first convert 240° to radian measure.

$$240° = (240 \text{ deg})\left(\frac{\pi \text{ rad}}{180 \text{ deg}}\right) = \frac{4\pi}{3} \text{ rad}$$

Then, using a radius of $r = 4$ inches, we find the arc length to be

$$s = r\theta = 4\left(\frac{4\pi}{3}\right) = \frac{16\pi}{3} \approx 16.76 \text{ inches.}$$

Note that the units for $r\theta$ are determined by the units for r because θ is given in radian measure and therefore has no units.

The formula for the length of a circular arc can be used to analyze the motion of a particle moving at a *constant speed* along a circular path. Assume that the particle is moving at a constant speed along a circular path (of radius r). If s is the length of the arc traveled in time t, then we say that the **speed** of the particle is

$$\text{speed} = \frac{\text{distance}}{\text{time}} = \frac{s}{t}.$$

Moreover, if θ is the angle (in radian measure) corresponding to the arc length s, then the **angular speed** of the particle is

$$\text{angular speed} = \frac{\theta}{t}.$$

EXAMPLE 10 *Finding the Speed of an Object*

The second hand on a clock is 4 inches long, as shown in Figure 7.13. Find the speed of the tip of this second hand.

SOLUTION

The time required for the second hand to make one full revolution is

$t = 60$ seconds $= 1$ minute.

Moreover, the distance traveled by the tip of the second hand in one revolution is

$$s = 2\pi(\text{radius}) = 2\pi(4) = 8\pi \text{ inches.}$$

Therefore, the speed of the tip of the second hand is

$$\text{speed} = \frac{s}{t} = \frac{8\pi \text{ inches}}{60 \text{ seconds}} \approx 0.419 \text{ in/sec.}$$

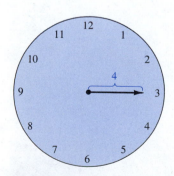

FIGURE 7.13

Trigonometry

EXAMPLE 11 Finding Angular Speed and Speed

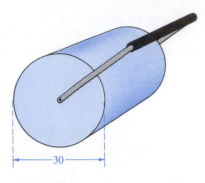

FIGURE 7.14

A lawn roller that is 30 inches in diameter makes 1.2 revolutions per second, as shown in Figure 7.14.

(a) Find the angular speed of the roller in radians per second.
(b) How fast is the roller moving across the lawn?

SOLUTION

(a) Since each revolution generates 2π radians, it follows that the roller turns $(1.2)(2\pi) = 2.4\pi$ radians per second. Thus, the angular speed is

$$\text{angular speed} = \frac{\theta}{t} = \frac{2.4\pi \text{ radians}}{1 \text{ second}} = 2.4\pi \text{ rad/sec.}$$

(b) To find the speed of the roller, we use the fact that its diameter is 30 inches. Thus, its radius is 15 inches and we have $s = 2\pi r = 2\pi(15) = 30\pi$. Since the roller makes 1.2 revolutions per second, its speed is

$$\text{speed} = \left(\frac{1.2 \text{ rev}}{1 \text{ sec}}\right)\left(\frac{30\pi \text{ in}}{1 \text{ rev}}\right) = 36\pi \text{ in/sec} \approx 113.1 \text{ in/sec.}$$

WARM UP

Solve for x.

1. $x + 135 = 180$

2. $790 = 720 + x$

3. $\pi = \frac{5\pi}{6} + x$

4. $2\pi - x = \frac{5\pi}{3}$

5. $\frac{45}{180} = \frac{x}{\pi}$

6. $\frac{240}{180} = \frac{x}{\pi}$

7. $\frac{\pi}{180} = \frac{x}{20}$

8. $\frac{180}{\pi} = \frac{330}{x}$

9. $\frac{x}{60} = \frac{3}{4}$

10. $\frac{x}{3600} = 0.0125$

EXERCISES 7.1

In Exercises 1–4, determine the quadrant in which the given angle lies.

1. (a) 130°   (b) 285°

2. (a) 8.3°   (b) 257° 30′

3. (a) −132° 50′   (b) −336°

4. (a) −260°   (b) −3.4°

In Exercises 5–10, determine the quadrant in which the given angle lies. (The angle measure is given in radians.)

5. (a) $\dfrac{\pi}{5}$ (b) $\dfrac{7\pi}{5}$

6. (a) $\dfrac{5\pi}{4}$ (b) $\dfrac{7\pi}{4}$

7. (a) $-\dfrac{\pi}{12}$ (b) $-\dfrac{11\pi}{9}$

8. (a) −1   (b) −2

9. (a) 3.5   (b) 2.25

10. (a) 5.63   (b) −2.25

In Exercises 11–16, sketch the given angle in standard position.

11. (a) 30°   (b) 150°

12. (a) −270°   (b) −120°

13. (a) 405°   (b) −480°

14. (a) $\dfrac{5\pi}{4}$ (b) $\dfrac{2\pi}{3}$

15. (a) $-\dfrac{7\pi}{4}$ (b) $-\dfrac{5\pi}{2}$

16. (a) $\dfrac{11\pi}{6}$ (b) 7π

In Exercises 17–20, determine two coterminal angles (one positive and one negative) for the given angle. Give the answers in degrees.

17. (a)   (b)

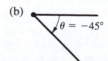 $\theta = 36°$ 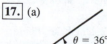 $\theta = -45°$

18. (a) 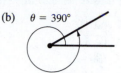 $\theta = -120°$ (b) $\theta = 390°$

19. (a) $\theta = 300°$ (b) $\theta = 740°$

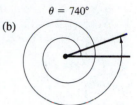

20. (a) $\theta = -420°$ (b) $\theta = 230°$

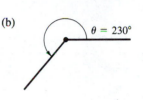

In Exercises 21–24, determine two coterminal angles (one positive and one negative) for the given angle. Give the answers in radians.

21. (a) $\theta = \dfrac{\pi}{9}$ (b) $\theta = \dfrac{4\pi}{3}$

22. (a) $\theta = \dfrac{11\pi}{6}$ (b) $\theta = -\dfrac{7\pi}{6}$

23. (a) $\theta = -\dfrac{9\pi}{4}$ (b) $\theta = -\dfrac{2\pi}{15}$

24. (a) $\theta = \dfrac{8\pi}{9}$ (b) $\theta = \dfrac{8\pi}{45}$

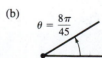

In Exercises 25–28, express the given angle in radian measure as a multiple of π. (Do not use a calculator.)

25. (a) 30°   (b) 150°

26. (a) 315°   (b) 120°

27. (a) −20°   (b) −240°

28. (a) −270°   (b) 144°

In Exercises 29–32, express the given angle in degree measure. (Do not use a calculator.)

29. (a) $\dfrac{3\pi}{2}$ (b) $\dfrac{7\pi}{6}$

30. (a) $-\dfrac{7\pi}{12}$ (b) $\dfrac{\pi}{9}$

31. (a) $\dfrac{7\pi}{3}$ (b) $-\dfrac{11\pi}{30}$

32. (a) $\dfrac{11\pi}{6}$ (b) $\dfrac{34\pi}{15}$

In Exercises 33–40, convert the angle from degrees to radian measure. List your answers to three decimal places.

33. 115°   **34.** 87.4°

35. −216.35°   **36.** −48.27°

37. 532°   **38.** 0.54°

39. −0.83°   **40.** 345°

In Exercises 41–48, convert the angle from radian to degree measure. List your answers to three decimal places.

41. $\dfrac{\pi}{7}$ **42.** $\dfrac{5\pi}{11}$

43. $\dfrac{15\pi}{8}$ **44.** 6.5π

45. -4.2π **46.** 4.8

47. −2   **48.** −0.57

In Exercises 49–52, convert the angle measurement to decimal form.

49. (a) 54° 45′   (b) −128° 30′

50. (a) 245° 10′   (b) 2° 12′

51. (a) 85° 18′ 30″   (b) 330° 25″

52. (a) −135° 36″   (b) −408° 16′ 25″

In Exercises 53–56, convert the angle measurement to D° M′ S″ form.

53. (a) 240.6°   (b) −145.8°

54. (a) −345.12°   (b) 0.45

55. (a) 2.5   (b) −3.58

56. (a) −0.355   (b) 0.7865

In Exercises 57–60, find the radian measure of the central angle using the given radius and arc length.

57. $r = 10$ inches, $s = 4$ inches

58. $r = 16$ feet, $s = 10$ feet

59. $r = 14.5$ cm, $s = 25$ cm

60. $r = 80$ km, $s = 160$ km

In Exercises 61–64, find the length of the arc on the circle of radius r subtended by the central angle θ.

61. $r = 15$ inches, $\theta = \pi$ radians

62. $r = 9$ feet, $\theta = \dfrac{\pi}{3}$

63. $r = 6$ m, $\theta = 2$ radians

64. $r = 40$ cm, $\theta = \dfrac{3\pi}{4}$

In Exercises 65–68, find the distance between the two cities of given latitudes. Assume that Earth is a sphere of radius 4000 miles and that the cities are on the same meridian (one city is due north of the other).

65. Dallas   32° 47′ 9″ N
Omaha   41° 15′ 42″ N

66. San Francisco 37° 46′ 39″ N
Seattle   47° 36′ 32″ N

67. Miami   25° 46′ 37″ N
Erie   42° 7′ 15″ N

68. Johannesburg, South Africa   26° 10′ S
Jerusalem, Israel   31° 47′ N

69. Assuming that Earth is a sphere of radius 4000 miles, what is the difference in latitude of two cities, one of which is 325 miles due north of the other?

70. The pointer on a voltmeter is 2 inches in length (see figure). Find the angle through which the pointer rotates when it moves 1/2 inch on the scale.

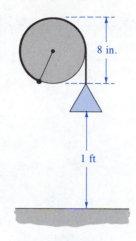

8 in.

1 ft

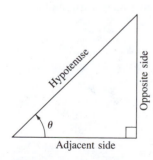

10

15   5

20   0

2

Volts

FIGURE FOR 70

FIGURE FOR 71

71. An electric hoist is being used to lift a piece of equipment (see figure). The diameter of the drum on the hoist is 8 inches and the equipment must be raised one foot. Find the number of degrees through which the drum must rotate.

72. Find the number of degrees through which a wheel will turn in 25 milliseconds (25/1000 seconds) for an angular speed of 3 radians per second.

73. A car is moving at the rate of 50 miles per hour, and the diameter of each of its wheels is 2.5 feet.
   (a) Find the number of revolutions per minute that the wheels are rotating.
   (b) Find the *angular speed* of the wheels.

74. Repeat Exercise 73 for the wheels on a truck that have a diameter of 3 feet.

75. A 2-inch-diameter pulley on an electric motor that runs at 1700 revolutions per minute is connected by a belt to a 4-inch-diameter pulley on a saw arbor.
   (a) Find the angular speed (radians per minute) of each pulley.
   (b) Find the revolutions per minute of the saw.

76. How long will it take a pulley rotating at 12 radians per second to make 100 complete revolutions?

7.2 Right Triangles and Trigonometric Functions

Our first look at the trigonometric functions is from a *right triangle* perspective. Note that the three sides of the right triangle shown in Figure 7.15 are labeled the **hypotenuse,** the **opposite side** (the side opposite the angle θ), and the **adjacent side** (the side adjacent to the angle θ). Using the lengths of these three sides, we can form six ratios that define the six trigonometric functions of the acute angle θ:

sine	**cosecant**
cosine	**secant**
tangent	**cotangent.**

These six functions are normally abbreviated as sin, csc, cos, sec, tan, and cot, respectively. In the following definition it is important to see that $0° < \theta < 90°$ and that for such angles the value of each of the six trigonometric functions is *positive*.

Hypotenuse

Opposite side

θ

Adjacent side

FIGURE 7.15

Right Triangle Definition of Trigonometric Functions

Let θ be an *acute* angle of a right triangle. Then the six trigonometric functions *of the angle* θ are defined as follows.

$$\sin \theta = \frac{\text{opp}}{\text{hyp}} \qquad \csc \theta = \frac{\text{hyp}}{\text{opp}}$$

$$\cos \theta = \frac{\text{adj}}{\text{hyp}} \qquad \sec \theta = \frac{\text{hyp}}{\text{adj}}$$

$$\tan \theta = \frac{\text{opp}}{\text{adj}} \qquad \cot \theta = \frac{\text{adj}}{\text{opp}}$$

The abbreviations opp, adj, and hyp represent the lengths of the three sides of the right triangle.

opp = the length of the side *opposite* θ
adj = the length of the side *adjacent to* θ
hyp = the length of the *hypotenuse*

Remark: As an aid to memorizing these definitions, note that the functions in the second column are the *reciprocals* of the corresponding members of the first column.

EXAMPLE 1 *Evaluating Trigonometric Functions*

Find the values of the six trigonometric functions of θ, as shown in Figure 7.16.

SOLUTION

By the Pythagorean Theorem, it follows that

$$\text{hyp} = \sqrt{3^2 + 4^2} = \sqrt{25} = 5.$$

Thus, we have adj = 3, opp = 4, and hyp = 5, so that the six trigonometric functions of θ have the following values.

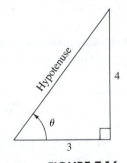

FIGURE 7.16

$$\sin \theta = \frac{\text{opp}}{\text{hyp}} = \frac{4}{5} \qquad \csc \theta = \frac{\text{hyp}}{\text{opp}} = \frac{5}{4}$$

$$\cos \theta = \frac{\text{adj}}{\text{hyp}} = \frac{3}{5} \qquad \sec \theta = \frac{\text{hyp}}{\text{adj}} = \frac{5}{3}$$

$$\tan \theta = \frac{\text{opp}}{\text{adj}} = \frac{4}{3} \qquad \cot \theta = \frac{\text{adj}}{\text{opp}} = \frac{3}{4}$$

In Example 1, we were given the lengths of the sides of the right triangle, but we were not given the angle θ. A much more common problem in trigonometry is to be asked to find the trigonometric functions for a *given* acute angle θ. To do this, we construct a right triangle having θ as one of its angles. This procedure is demonstrated in Examples 2 and 3.

EXAMPLE 2 *Evaluating Trigonometric Functions of 45°*

Find the value of sin 45°, cos 45°, and tan 45°.

SOLUTION

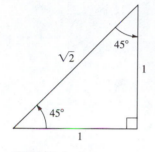

FIGURE 7.17

To begin, we construct a right triangle having 45° as one of its acute angles, as shown in Figure 7.17. We arbitrarily choose the length of the adjacent side to be 1. From geometry, we know that the other acute angle is also 45°, and therefore the triangle is isosceles. Hence, the length of the opposite side is also 1. Then, using the Pythagorean Theorem, we find the length of the hypotenuse to be

$$\text{hyp} = \sqrt{1^2 + 1^2} = \sqrt{2}.$$

Finally, we have the following:

$$\sin 45° = \frac{\text{opp}}{\text{hyp}} = \frac{1}{\sqrt{2}} = \frac{\sqrt{2}}{2}$$

$$\cos 45° = \frac{\text{adj}}{\text{hyp}} = \frac{1}{\sqrt{2}} = \frac{\sqrt{2}}{2}$$

$$\tan 45° = \frac{\text{opp}}{\text{adj}} = \frac{1}{1} = 1.$$

EXAMPLE 3 *Evaluating Trigonometric Functions of 30° and 60°*

Use the equilateral triangle shown in Figure 7.18 to find the value of sin 60°, cos 60°, sin 30°, and cos 30°.

SOLUTION

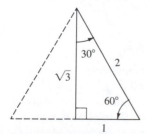

FIGURE 7.18

To begin, try using the Pythagorean Theorem and the 60°-60°-60° equilateral triangle to verify the lengths of the sides given in Figure 7.18. For $\theta = 60°$, we have adj = 1, opp = $\sqrt{3}$, and hyp = 2. Therefore,

$$\sin 60° = \frac{\text{opp}}{\text{hyp}} = \frac{\sqrt{3}}{2} \quad \text{and} \quad \cos 60° = \frac{\text{adj}}{\text{hyp}} = \frac{1}{2}.$$

For $\theta = 30°$, we have adj = $\sqrt{3}$, opp = 1, and hyp = 2. Therefore,

$$\sin 30° = \frac{\text{opp}}{\text{hyp}} = \frac{1}{2} \quad \text{and} \quad \cos 30° = \frac{\text{adj}}{\text{hyp}} = \frac{\sqrt{3}}{2}.$$

Because the angles 30°, 45°, and 60° occur frequently in trigonometry, we suggest that you learn to construct the triangles shown in Figures 7.17 and 7.18. The values of the sine, cosine, and tangent of 30°, 45°, and 60° are summarized in the following table.

Sine, Cosine, and Tangent of Special Angles

$\sin 30° = \dfrac{1}{2}$	$\cos 30° = \dfrac{\sqrt{3}}{2}$	$\tan 30° = \dfrac{1}{\sqrt{3}}$
$\sin 45° = \dfrac{\sqrt{2}}{2}$	$\cos 45° = \dfrac{\sqrt{2}}{2}$	$\tan 45° = 1$
$\sin 60° = \dfrac{\sqrt{3}}{2}$	$\cos 60° = \dfrac{1}{2}$	$\tan 60° = \sqrt{3}$

Trigonometric Identities

In the preceding table, note that for the angles 30° and 60°, we have $\sin 30° = 1/2 = \cos 60°$. This occurs because 30° and 60° are complementary angles, and, in general, it can be shown that *cofunctions of complementary angles are equal*. That is, if θ is an acute angle, then the following relationships are true.

$$\sin(90° - \theta) = \cos \theta$$
$$\tan(90° - \theta) = \cot \theta$$
$$\sec(90° - \theta) = \csc \theta$$
$$\cos(90° - \theta) = \sin \theta$$
$$\cot(90° - \theta) = \tan \theta$$
$$\csc(90° - \theta) = \sec \theta$$

For instance, since 10° and 80° are complementary angles, it follows that $\sin 10° = \cos 80°$ and $\tan 10° = \cot 80°$.

In trigonometry, a great deal of time is spent studying relationships between trigonometric functions. You will need to memorize many of these relationships (identities). We begin with some basic identities that are easily established from the definitions of the six trigonometric functions.

Right Triangles and Trigonometric Functions

Fundamental Trigonometric Identities

Reciprocal Identities

$$\sin \theta = \frac{1}{\csc \theta} \qquad \sec \theta = \frac{1}{\cos \theta} \qquad \tan \theta = \frac{1}{\cot \theta}$$

$$\csc \theta = \frac{1}{\sin \theta} \qquad \cos \theta = \frac{1}{\sec \theta} \qquad \cot \theta = \frac{1}{\tan \theta}$$

Tangent and Cotangent Identities

$$\tan \theta = \frac{\sin \theta}{\cos \theta} \qquad \cot \theta = \frac{\cos \theta}{\sin \theta}$$

Pythagorean Identities

$$\sin^2 \theta + \cos^2 \theta = 1 \qquad 1 + \tan^2 \theta = \sec^2 \theta$$
$$1 + \cot^2 \theta = \csc^2 \theta$$

Remark: We use $\sin^2 \theta$ to represent $(\sin \theta)^2$.

In the next two examples, we show how trigonometric identities can be used to find exact values of trigonometric functions.

EXAMPLE 4 *Applying Trigonometric Identities*

Let θ be the acute angle such that $\sin \theta = 0.6$. Find the values of (a) $\cos \theta$ and (b) $\tan \theta$.

SOLUTION

(a) To find the value of $\cos \theta$, we use the Pythagorean Identity

$$\sin^2 \theta + \cos^2 \theta = 1.$$

Thus, we have

$$(0.6)^2 + \cos^2 \theta = 1$$
$$\cos^2 \theta = 1 - (0.6)^2 = 0.64$$
$$\cos \theta = \sqrt{0.64} = 0.8.$$

Note that we choose the positive square root because θ is given to be an acute angle.

(b) Now, knowing the sine and cosine of θ, we find the tangent of θ to be

$$\tan \theta = \frac{\sin \theta}{\cos \theta} = \frac{0.6}{0.8} = 0.75.$$

Try using the triangle shown in Figure 7.19 to check these results.

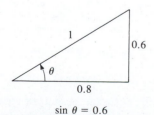

$\sin \theta = 0.6$

FIGURE 7.19

Remark: The triangle in Figure 7.19 was obtained from the fact that

$$\sin \theta = 0.6 = \frac{0.6}{1} = \frac{\text{opp}}{\text{hyp}}.$$

Thus, opp = 0.6, hyp = 1, and, by the Pythagorean Theorem, it follows that adj $= \sqrt{1^2 - (0.6)^2} = \sqrt{0.64} = 0.8$.

EXAMPLE 5 **Applying Trigonometric Identities**

Let θ be an acute angle such that $\tan \theta = 3$. Find the values of (a) cot θ and (b) sec θ.

SOLUTION

(a) Using the reciprocal identity

$$\cot \theta = \frac{1}{\tan \theta}$$

we have

$$\cot \theta = \frac{1}{3}.$$

(b) Using the Pythagorean Identity $1 + \tan^2 \theta = \sec^2 \theta$, we have

$$\sec^2 \theta = 1 + 3^2 = 10$$
$$\sec \theta = \sqrt{10}.$$

Try using the triangle shown in Figure 7.20 to check these results.

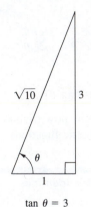

$\tan \theta = 3$

FIGURE 7.20

Evaluating Trigonometric Functions with a Calculator

A scientific calculator can be used to obtain decimal approximations of the values of the trigonometric functions of any angle. With a calculator, we need to set a switch to the desired *mode* of measurement (degrees or radians). For instance, to find the value of cos 28°, we can use the following keystroke sequence.

Degree Mode: 28 $\boxed{\text{cos}}$ *Display* 0.8829476

Similarly, we can find the value of $\tan(\pi/12)$ as follows.

Radian Mode: π $\boxed{\div}$ 12 $\boxed{=}$ $\boxed{\text{tan}}$ *Display* 0.2679492

Most calculators do not have keys for the cosecant, secant, or cotangent functions. To evaluate these functions, we use the $\boxed{1/x}$ key with their respective reciprocal functions sine, cosine, and tangent. For example, to evaluate $\csc(\pi/8)$, we use the fact that

$$\csc\frac{\pi}{8} = \frac{1}{\sin(\pi/8)}$$

and enter the following keystroke sequence.

Radian Mode: π ÷ 8 = **sin** **1/x**   *Display* 2.6131259

EXAMPLE 6   **Using a Calculator to Evaluate Trigonometric Functions**

Use a calculator to evaluate each of the following.

(a) $\sin 76.4°$ (b) $\cot 1.5$ (c) $\sec(5°\ 40'\ 12'')$

SOLUTION

Function	*Mode*	*Keystrokes*	*Display*
(a) $\sin 76.4°$	Degree	76.4 **sin**	0.9719610
(b) $\cot 1.5$	Radian	1.5 **tan** **1/x**	0.0709148

(c) Converting first to decimal form, we have

$$5°\ 40'\ 12'' = 5° + \left(\frac{40}{60}\right)^{\circ} + \left(\frac{12}{3600}\right)^{\circ}$$

$$= 5.67°.$$

Hence, it follows that

$$\sec(5°\ 40'\ 12'') = \sec 5.67° = \frac{1}{\cos 5.67°} \approx 1.00492.$$

Applications Involving Right Triangles

Many applications of trigonometry involve a process called **solving right triangles.** In this type of application, we are usually given two sides of a right triangle and asked to find one of its acute angles, *or* we are given one side and one of the acute angles and asked to find one of the other sides.

EXAMPLE 7   **Using Trigonometry to Solve a Right Triangle**

A surveyor is standing 50 feet from the base of a large tree, as shown in Figure 7.21. The surveyor measures the angle of elevation to the top of the tree as 71.5°. How tall is the tree?

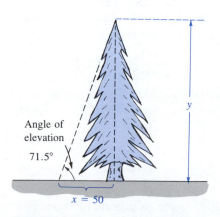

Angle of elevation

71.5°

y

$x = 50$

FIGURE 7.21

SOLUTION

From Figure 7.21, we see that

$$\tan 71.5° = \frac{\text{opp}}{\text{adj}} = \frac{y}{x}$$

where $x = 50$ and y is the height of the tree. Thus, we can determine the height of the tree to be

$$y = x \tan 71.5° \approx 50(2.98868) \approx 149.4 \text{ feet.}$$

Note in Example 7 that we were given one side and one of the acute angles of a right triangle and were asked to find the opposite side. In Example 8, we are given two sides and are asked to find one of the acute angles.

EXAMPLE 8 *Using Trigonometry to Solve a Right Triangle*

A person is standing 200 yards from a river. Rather than walk directly to the river, the person walks 400 yards along a straight path to the river's edge. Find the acute angle θ between this path and the river's edge, as indicated in Figure 7.22.

SOLUTION

From Figure 7.22, we see that the sine of the angle θ is given by

$$\sin \theta = \frac{\text{opp}}{\text{hyp}} = \frac{200}{400} = \frac{1}{2}.$$

Now, we recognize that $\theta = 30°$.

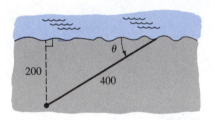

FIGURE 7.22

In Example 8, we were able to recognize that the acute angle that satisfies the equation $\sin \theta = 1/2$ is $\theta = 30°$. Suppose, however, that we were given the equation $\sin \theta = 0.6$ and asked to find the acute angle θ. Since

$$\sin 30° = \frac{1}{2} = 0.5000 \quad \text{and} \quad \sin 45° = \frac{1}{\sqrt{2}} \approx 0.7071$$

we know that θ lies somewhere between 30° and 45°. A more precise value of θ can be found using the |INV| key on a calculator. To do this, we use the following keystroke sequence.

Degree Mode: .6 |INV| |sin| *Display* 36.8699

Thus, we conclude that if $\sin \theta = 0.6$, then $\theta \approx 36.87°$.

Remark: Instead of an inverse key |INV|, some calculators have a second function key |2nd f|. In Section 7.7, we will explain the concepts involved in the use of the |INV| key.

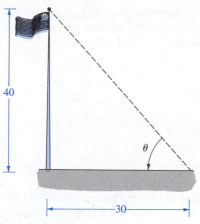

FIGURE 7.23

EXAMPLE 9 *Using Trigonometry to Solve a Right Triangle*

A 40-foot flagpole casts a 30-foot shadow, as shown in Figure 7.23. Find θ, the angle of elevation of the sun.

SOLUTION

Figure 7.23 shows that the *opposite* and *adjacent* sides are known. Thus,

$$\tan \theta = \frac{\text{opp}}{\text{adj}} = \frac{40}{30}.$$

With a calculator in degree mode we use the keystrokes

40 $\div$ 30 = INV tan

to obtain $\theta \approx 53.13°$.

WARM UP

Find the distance between each pair of points.

1. (3, 8), (1, 4)

2. (5, 2), (2, −7)

3. (−4, 0), (2, 8)

4. (−3, −3), (0, 0)

Perform the indicated operations. (Round your answers to two decimal places.)

5. 0.300×4.125

6. 7.30×43.50

7. $\dfrac{151.5}{2.40}$

8. $\dfrac{3740}{28.0}$

9. $\dfrac{19,500}{0.007}$

10. $\dfrac{(10.5)(3401)}{1240}$

EXERCISES 7.2

In Exercises 1–8, find the exact value of the six trigonometric functions of the angle θ given in the accompanying figure. (Use the Pythagorean Theorem to find the third side of the triangle.)

1.

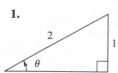

2.

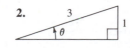

3.

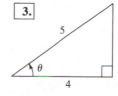

4.

5.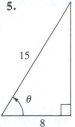

6.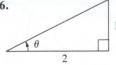

7.

8.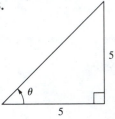

In Exercises 9–16, sketch a right triangle corresponding to the trigonometric function of the acute angle θ, and find the other five trigonometric functions of θ.

9. $\sin \theta = \frac{2}{3}$ **10.** $\cot \theta = 5$

11. $\sec \theta = 2$ **12.** $\cos \theta = \frac{5}{7}$

13. $\tan \theta = 3$ **14.** $\csc \theta = \frac{17}{4}$

15. $\cot \theta = \frac{3}{2}$ **16.** $\sin \theta = \frac{3}{8}$

In Exercises 17–20, use the given function values to find the indicated trigonometric functions.

17. $\sin 60° = \dfrac{\sqrt{3}}{2}, \quad \cos 60° = \dfrac{1}{2}$

   (a) $\tan 60°$ (b) $\sin 30°$
   (c) $\cos 30°$ (d) $\cot 60°$

18. $\sin 30° = \dfrac{1}{2}, \quad \tan 30° = \dfrac{\sqrt{3}}{3}$

   (a) $\csc 30°$ (b) $\cot 60°$
   (c) $\cos 30°$ (d) $\cot 30°$

19. $\csc \theta = 3, \quad \sec \theta = \dfrac{3\sqrt{2}}{4}$

   (a) $\sin \theta$ (b) $\cos \theta$
   (c) $\tan \theta$ (d) $\sec(90° - \theta)$

20. $\sec \theta = 5, \quad \tan \theta = 2\sqrt{6}$

   (a) $\cos \theta$ (b) $\cot \theta$
   (c) $\cot(90° - \theta)$ (d) $\sin \theta$

In Exercises 21–24, evaluate the given trigonometric function by memory or by constructing an appropriate triangle for the special angles 30°, 45°, and 60°.

21. (a) $\cos 60°$ (b) $\tan 30°$
22. (a) $\csc 30°$ (b) $\sin 45°$
23. (a) $\cot 45°$ (b) $\cos 45°$
24. (a) $\sin 60°$ (b) $\csc 45°$

In Exercises 25–34, use a calculator to evaluate each function. Round your answers to four decimal places. (Be sure the calculator is in the correct mode.)

25. (a) $\sin 10°$ (b) $\cos 80°$
26. (a) $\tan 23.5°$ (b) $\cot 66.5°$
27. (a) $\sec 42° \, 12'$ (b) $\csc 48° \, 7'$
28. (a) $\cos 16° \, 18'$ (b) $\sin 73° \, 56'$
29. (a) $\sin 16.35°$ (b) $\csc 16.35°$
30. (a) $\cos 4° \, 50' \, 15''$ (b) $\sec 4° \, 50' \, 15''$
31. (a) $\cot \dfrac{\pi}{16}$ (b) $\tan \dfrac{\pi}{16}$
32. (a) $\sec 0.75$ (b) $\cos 0.75$
33. (a) $\csc 1$ (b) $\sec\left(\dfrac{\pi}{2} - 1\right)$
34. (a) $\tan \dfrac{1}{2}$ (b) $\cot\left(\dfrac{\pi}{2} - \dfrac{1}{2}\right)$

In Exercises 35–40, find the value of θ in degrees ($0° < \theta < 90°$) and radians ($0 < \theta < \pi/2$) without a calculator.

35. (a) $\sin \theta = \dfrac{1}{2}$ (b) $\csc \theta = 2$
36. (a) $\cos \theta = \dfrac{\sqrt{2}}{2}$ (b) $\tan \theta = 1$
37. (a) $\sec \theta = 2$ (b) $\cot \theta = 1$
38. (a) $\tan \theta = \sqrt{3}$ (b) $\cos \theta = \dfrac{1}{2}$
39. (a) $\csc \theta = \dfrac{2\sqrt{3}}{3}$ (b) $\sin \theta = \dfrac{\sqrt{2}}{2}$
40. (a) $\cot \theta = \dfrac{\sqrt{3}}{3}$ (b) $\sec \theta = \sqrt{2}$

In Exercises 41–44, find the value of θ in degrees ($0° < \theta < 90°$) and radians ($0 < \theta < \pi/2$) by using the inverse key on a calculator.

41. (a) $\sin \theta = 0.8191$ (b) $\cos \theta = 0.0175$
42. (a) $\cos \theta = 0.9848$ (b) $\cos \theta = 0.8746$
43. (a) $\tan \theta = 1.1920$ (b) $\tan \theta = 0.4663$
44. (a) $\sin \theta = 0.3746$ (b) $\cos \theta = 0.3746$

45. Solve for y.

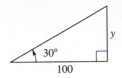

46. Solve for x.

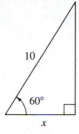

47. Solve for x.

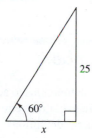

48. Solve for r.

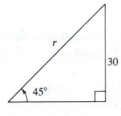

49. Solve for r.

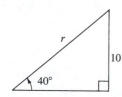

50. Solve for x.

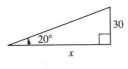

51. Solve for y.

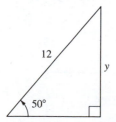

52. Solve for r.

53. A 6-foot person standing 12 feet from a streetlight casts an 8-foot shadow (see figure). What is the height of the streetlight?

54. A guy wire is stretched from a broadcasting tower at a point 200 feet above the ground to an anchor 125 feet from the base (see figure). How long is the wire?

FIGURE FOR 53

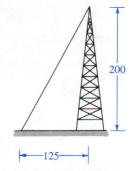

FIGURE FOR 54

55. A 20-foot ladder leaning against the side of a house makes a 75° angle with the ground (see figure). How far up the side of the house does the ladder reach?

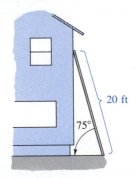

FIGURE FOR 55

56. A biologist wants to know the width w of a river in order to properly set instruments for studying the pollutants in the water. From point A, the biologist walks downstream 100 feet and sights to point C. From this sighting, it is determined that $\theta = 50°$ (see figure). How wide is the river?

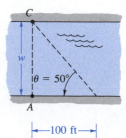

FIGURE FOR 56

57. From a 150-foot observation tower on the coast, a Coast Guard officer sights a boat in difficulty. The angle of depression of the boat is 4° (see figure). How far is the boat from the shoreline?

FIGURE FOR 57

58. A ramp $17\frac{1}{2}$ feet in length rises to a loading platform that is $3\frac{1}{3}$ feet off the ground (see figure). Find the angle θ that the ramp makes with the ground.

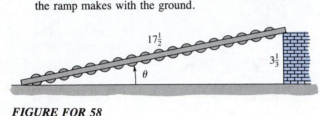

FIGURE FOR 58

In Exercises 59–64, determine whether the statement is true or false, and give reasons.

59. $\sin 60° \csc 60° = 1$

60. $\sec 30° = \csc 60°$

61. $\sin 45° + \cos 45° = 1$

62. $\cot^2 10° - \csc^2 10° = -1$

63. $\dfrac{\sin 60°}{\sin 30°} = \sin 2°$

64. $\tan[(0.8)^2] = \tan^2(0.8)$

7.3 Trigonometric Functions of Real Numbers

In Section 7.2, we restricted the evaluation of trigonometric functions to acute angles. We now extend our study of trigonometry to *any* angle (or, equivalently, any real number), starting with the following general definitions.

Definition of Trigonometric Functions of Any Angle

Let θ be an angle in standard position with (x, y) any point (except the origin) on the terminal side of θ and $r = \sqrt{x^2 + y^2}$, as shown in Figure 7.24.

$$\sin \theta = \frac{y}{r} \qquad \csc \theta = \frac{r}{y}, \quad y \neq 0$$

$$\cos \theta = \frac{x}{r} \qquad \sec \theta = \frac{r}{x}, \quad x \neq 0$$

$$\tan \theta = \frac{y}{x}, \quad x \neq 0 \qquad \cot \theta = \frac{x}{y}, \quad y \neq 0$$

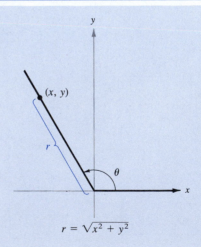

FIGURE 7.24

$$r = \sqrt{x^2 + y^2}$$

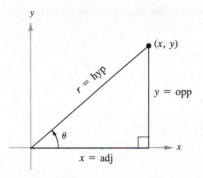

FIGURE 7.25

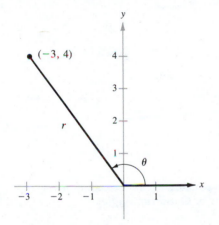

FIGURE 7.26

Remark: Since $r = \sqrt{x^2 + y^2}$ *cannot* be zero, it follows that the sine and cosine functions are defined for any real value of θ. However, if $x = 0$, the tangent and secant of θ are undefined. For example, the tangent of 90° is undefined. Similarly, if $y = 0$, the cotangent and cosecant of θ are undefined.

If θ is an *acute* angle, then these definitions coincide with those given in the previous section. To see this, note in Figure 7.25 that for an acute angle θ, $x = $ adj, $y = $ opp, and $r = $ hyp.

EXAMPLE 1 *Evaluating Trigonometric Functions*

Let $(-3, 4)$ be a point on the terminal side of θ. Find the sine, cosine, and tangent of θ.

SOLUTION

Referring to Figure 7.26, we see that $x = -3$, $y = 4$, and

$$r = \sqrt{x^2 + y^2} = \sqrt{(-3)^2 + 4^2} = \sqrt{25} = 5.$$

Thus, we have

$$\sin \theta = \frac{y}{r} = \frac{4}{5}$$

$$\cos \theta = \frac{x}{r} = \frac{-3}{5} = -\frac{3}{5}$$

$$\tan \theta = \frac{y}{x} = \frac{4}{-3} = -\frac{4}{3}.$$

The *signs* of the trigonometric functions in the four quadrants can be determined easily from the definitions of the functions. For instance, since $\cos \theta = x/r$, it follows that $\cos \theta$ is positive wherever $x > 0$, which is in Quadrants I and IV. (Remember, r is always positive.) In a similar manner we can verify the results shown in Figure 7.27.

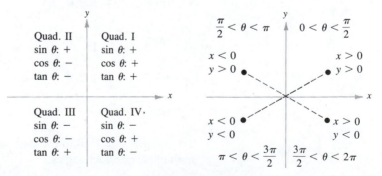

FIGURE 7.27   Signs of Trigonometric Functions

EXAMPLE 2 *Evaluating Trigonometric Functions*

Given $\tan \theta = -5/4$ and $\cos \theta > 0$, find $\sin \theta$ and $\sec \theta$.

SOLUTION

Note that θ lies in Quadrant IV because that is the only quadrant in which the tangent is negative and the cosine is positive. Moreover, using

$$\tan \theta = \frac{y}{x} = -\frac{5}{4}$$

and the fact that y is negative in Quadrant IV, we can conclude that $y = -5$ and $x = 4$. Hence, $r = \sqrt{25 + 16} = \sqrt{41}$ and we have

$$\sin \theta = \frac{y}{r} = \frac{-5}{\sqrt{41}} \approx -0.7809$$

$$\sec \theta = \frac{r}{x} = \frac{\sqrt{41}}{4} \approx 1.6008.$$

EXAMPLE 3 *Trigonometric Functions of Quadrant Angles*

Evaluate the sine function at the four quadrant angles 0, $\pi/2$, π, and $3\pi/2$.

SOLUTION

To begin, we choose a point on the terminal side of each angle, as shown in Figure 7.28. For each of the four given points, $r = 1$, and we have

$$\sin 0 = \frac{y}{r} = \frac{0}{1} = 0 \qquad\qquad (x, y) = (1, 0)$$

$$\sin \frac{\pi}{2} = \frac{y}{r} = \frac{1}{1} = 1 \qquad\qquad (x, y) = (0, 1)$$

$$\sin \pi = \frac{y}{r} = \frac{0}{1} = 0 \qquad\qquad (x, y) = (-1, 0)$$

$$\sin \frac{3\pi}{2} = \frac{y}{r} = \frac{-1}{1} = -1. \qquad\qquad (x, y) = (0, -1)$$

Try using Figure 7.28 to evaluate some of the other trigonometric functions at the four quadrant angles.

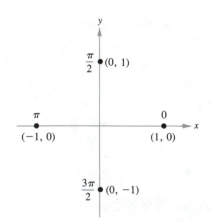

FIGURE 7.28

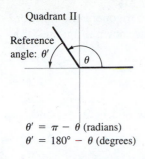

$\theta' = \pi - \theta$ (radians)
$\theta' = 180° - \theta$ (degrees)

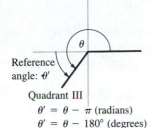

$\theta' = \theta - \pi$ (radians)
$\theta' = \theta - 180°$ (degrees)

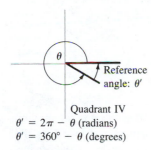

$\theta' = 2\pi - \theta$ (radians)
$\theta' = 360° - \theta$ (degrees)

FIGURE 7.29

Reference Angles

The values of the trigonometric functions of angles greater than 90° (or less than 0°) can be determined from their values at corresponding acute angles called **reference angles.**

Definition of Reference Angles

Let θ be an angle in standard position. Its **reference angle** is the acute angle θ' formed by the terminal side of θ and the horizontal axis.

Figure 7.29 shows the reference angle for θ in Quadrants II, III, and IV.

EXAMPLE 4 *Finding Reference Angles*

Find the reference angle θ' for each of the following.

(a) $\theta = 300°$ (b) $\theta = 2.3$ (c) $\theta = -135°$

SOLUTION

(a) Since $\theta = 300°$ lies in Quadrant IV, the angle it makes with the x-axis is

$$\theta' = 360° - 300° = 60°. \qquad \textit{Degrees}$$

(b) Since $\theta = 2.3$ lies between $\pi/2 \approx 1.5708$ and $\pi \approx 3.1416$, it follows that θ is in Quadrant II and its reference angle is

$$\theta' = \pi - 2.3 \approx 0.8416. \qquad \textit{Radians}$$

(c) First, we determine that $-135°$ is coterminal with $225°$, which lies in Quadrant III. Hence, the reference angle is

$$\theta' = 225° - 180° = 45°. \qquad \textit{Degrees}$$

Figure 7.30 shows each angle θ and its reference angle θ'.

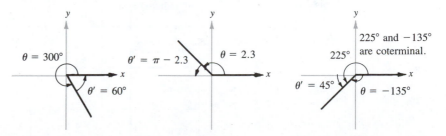

FIGURE 7.30   (a) θ in Quadrant IV   (b) θ in Quadrant II   (c) θ in Quadrant III

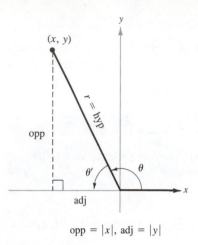

opp = $|x|$, adj = $|y|$

FIGURE 7.31

To see how a reference angle is used to *evaluate* a trigonometric function, consider the point (x, y) on the terminal side of θ, as shown in Figure 7.31. By definition, we know that

$$\sin \theta = \frac{y}{r} \quad \text{and} \quad \tan \theta = \frac{y}{x}.$$

For the right triangle with acute angle θ' and sides of lengths $|x|$ and $|y|$, we have

$$\sin \theta' = \frac{\text{opp}}{\text{hyp}} = \frac{|y|}{r} \quad \text{and} \quad \tan \theta' = \frac{\text{opp}}{\text{adj}} = \frac{|y|}{|x|}.$$

Thus, it follows that $\sin \theta$ and $\sin \theta'$ are equal, *except possibly in sign*. The same is true for $\tan \theta$ and $\tan \theta'$ *and* for the other four trigonometric functions. In all cases, the sign of the function value can be determined by the quadrant in which θ lies.

Use the following steps to evaluate trigonometric functions of any angle.

Evaluating Trigonometric Functions of Any Angle

To find the value of a trigonometric function of any angle θ,
1. Determine the function value for the associated reference angle θ'.
2. Depending on the quadrant in which θ lies, prefix the appropriate sign to the function value.

By using reference angles and the special angles discussed in the previous section, we can greatly extend our scope of *exact* trigonometric values. For instance, knowing the function values of 30° means that we know the function values of all angles for which 30° is a reference angle. For convenience, we provide the following table of the exact values of the trigonometric functions of special angles and quadrant angles. You should memorize these values.

TABLE 7.1

θ (degrees)	0°	30°	45°	60°	90°	180°	270°
θ (radians)	0	$\dfrac{\pi}{6}$	$\dfrac{\pi}{4}$	$\dfrac{\pi}{3}$	$\dfrac{\pi}{2}$	π	$\dfrac{3\pi}{2}$
$\sin \theta$	0	$\dfrac{1}{2}$	$\dfrac{\sqrt{2}}{2}$	$\dfrac{\sqrt{3}}{2}$	1	0	-1
$\cos \theta$	1	$\dfrac{\sqrt{3}}{2}$	$\dfrac{\sqrt{2}}{2}$	$\dfrac{1}{2}$	0	-1	0
$\tan \theta$	0	$\dfrac{\sqrt{3}}{3}$	1	$\sqrt{3}$	undef.	0	undef.

EXAMPLE 5 *Trigonometric Functions of Nonacute Angles*

Evaluate the following.

(a) $\cos \dfrac{4\pi}{3}$ (b) $\tan(-210°)$ (c) $\csc \dfrac{11\pi}{4}$

SOLUTION

(a) Since $\theta = 4\pi/3$ lies in Quadrant III, the reference angle is $\theta' = (4\pi/3) - \pi = \pi/3$, as shown in Figure 7.32(a). Moreover, the cosine is negative in Quadrant III, so that

$$\cos \frac{4\pi}{3} = (-)\cos \frac{\pi}{3} = -\frac{1}{2}. \qquad \textit{Special angle, } \pi/3$$

(b) Since $-210° + 360° = 150°$, it follows that $-210°$ is coterminal with the second-quadrant angle $150°$. Therefore, the reference angle is $\theta' = 180° - 150° = 30°$, as shown in Figure 7.32(b). Finally, since the tangent is negative in Quadrant II, we have

$$\tan(-210°) = (-)\tan 30° = -\frac{\sqrt{3}}{3}. \qquad \textit{Special angle, } 30°$$

(c) Since $(11\pi/4) - 2\pi = 3\pi/4$, it follows that $11\pi/4$ is coterminal with the second-quadrant angle $3\pi/4$. Therefore, the reference angle is $\theta' = \pi - (3\pi/4) = \pi/4$, as shown in Figure 7.32(c). Because the cosecant is positive in Quadrant II, we have

$$\csc \frac{11\pi}{4} = (+)\csc \frac{\pi}{4} = \frac{1}{\sin(\pi/4)} = \sqrt{2}. \quad \textit{Special angle, } \pi/4$$

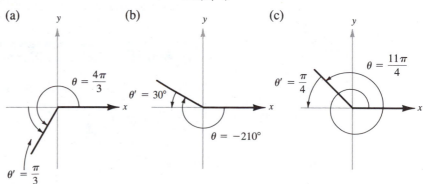

FIGURE 7.32

The fundamental trigonometric identities listed in the previous section (for an acute angle θ) are also valid when θ is any angle. In Example 6, we show how these identities can be used to find the exact value of trigonometric functions.

EXAMPLE 6 *Using Identities to Evaluate Trigonometric Functions*

Let θ be an angle in Quadrant II such that $\sin \theta = 1/3$. Find (a) $\cos \theta$ and (b) $\tan \theta$.

SOLUTION

(a) Since $\sin \theta = 1/3$, we can use the Pythagorean Identity $\sin^2 \theta + \cos^2 \theta = 1$ to obtain

$$\left(\frac{1}{3}\right)^2 + \cos^2 \theta = 1$$

$$\cos^2 \theta = 1 - \frac{1}{9} = \frac{8}{9}.$$

Since $\cos \theta < 0$ in Quadrant II, we use the negative root

$$\cos \theta = -\frac{\sqrt{8}}{\sqrt{9}} = -\frac{2\sqrt{2}}{3}.$$

(b) Using the result from part (a) and the trigonometric identity $\tan \theta = \sin \theta / \cos \theta$, we obtain

$$\tan \theta = \frac{1/3}{-2\sqrt{2}/3} = -\frac{1}{2\sqrt{2}} = -\frac{\sqrt{2}}{4}.$$

Scientific calculators can be used to approximate the values of trigonometric functions of any angle, as demonstrated in Example 7. (Note that some calculators restrict the magnitude of the angles they will accept.)

EXAMPLE 7 *Evaluating Trigonometric Functions with a Calculator*

Use a calculator to approximate the following values.

(a) $\cot 410°$ (b) $\sin(-7)$ (c) $\tan \dfrac{14\pi}{5}$

SOLUTION

Function	Mode	Keystrokes	Display
(a) $\cot 410°$	Degree	410 `tan` `1/x`	0.8390996
(b) $\sin(-7)$	Radian	7 `+/-` `sin`	-0.6569866
(c) $\tan \dfrac{14\pi}{5}$	Radian	14 `×` π `÷` 5 `=` `tan`	-0.7265425

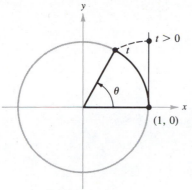

Positive numbers

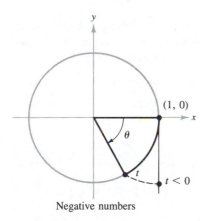

Negative numbers

FIGURE 7.33

Trigonometric Functions of Real Numbers

Many applications of trigonometric functions do not involve angles. For instance, in Section 7.8 we will look at the harmonic motion model

$$f(t) = a \sin \omega t$$

where a and ω are constants and t is a real number representing the time. (ω is the lower-case Greek letter omega.)

To define a trigonometric function of a real number (rather than an angle), we let t represent any real number. Then we imagine that the real number line is wrapped around a *unit circle*, as shown in Figure 7.33. Note that positive numbers correspond to a counterclockwise wrapping, and negative numbers correspond to a clockwise wrapping.

As the real line is wrapped around the unit circle, each real number t will correspond with a central angle θ. Moreover, since the circle has a radius of 1, the arc subtended by the angle θ will have a length of t. The point is that *if θ is measured in radians*, then $t = \theta$. Thus, we define the sin t to be

$$\sin t = \sin(t \text{ radians}).$$

Similarly, cos t = cos(t radians), tan t = tan(t radians), and so on.

EXAMPLE 8   *Evaluating Trigonometric Functions of Real Numbers*

Evaluate $f(t) = \sin t$ for (a) $t = 1$ and (b) $t = 3\pi/2$.

SOLUTION

(a) Using a calculator set in *radian mode*, we have

$$f(1) = \sin 1 \approx 0.841471.$$

(b) From Table 7.1, we have

$$f\left(\frac{3\pi}{2}\right) = \sin \frac{3\pi}{2} = -1.$$

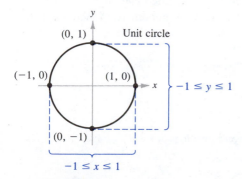

FIGURE 7.34

The *domain* of the sine and cosine functions is the set of all real numbers. To determine the *range* of these two functions, consider the unit circle shown in Figure 7.34. Since $r = 1$, it follows that sin $t = y$ and cos $t = x$. Moreover, because (x, y) is on the unit circle we know that $-1 \leq y \leq 1$ and $-1 \leq x \leq 1$, and it follows that the values of the sine and cosine also range between -1 and 1. That is,

$$\overset{-1 \leq y \leq 1}{-1 \leq \sin t \leq 1} \quad \text{and} \quad \overset{-1 \leq x \leq 1}{-1 \leq \cos t \leq 1}.$$

Suppose we add 2π to each value of t in the interval $[0, 2\pi]$, thus completing a second revolution around the unit circle, as shown in Figure

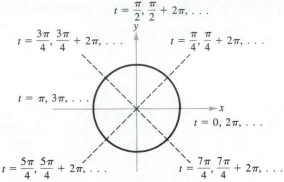

FIGURE 7.35

Repeated Revolutions
on the Unit Circle

7.35. The values of $\sin(t + 2\pi)$ and $\cos(t + 2\pi)$ correspond to those of $\sin t$ and $\cos t$. Similar results can be obtained for repeated revolutions (positive or negative) on the unit circle. This leads to the general result

$$\sin(t + 2\pi n) = \sin t \qquad \text{and} \qquad \cos(t + 2\pi n) = \cos t$$

for any integer n and real number t. Functions that behave in such a repetitive (or cyclic) manner are called **periodic.**

Definition of a Periodic Function

A function f is **periodic** if there exists a positive real number c such that

$$f(t + c) = f(t)$$

for all t in the domain of f. The least number c for which f is periodic is called the **period** of f.

From this definition it follows that the sine and cosine functions are periodic and have a period of 2π. The other four trigonometric functions are also periodic, and we will say more about that in Section 7.5.

Recall from Section 3.5 that a function f is called *even* if $f(-t) = f(t)$, and is called *odd* if $f(-t) = -f(t)$. Of the six trigonometric functions, two are even and four are odd, as stated in the following theorem.

Even and Odd Trigonometric Functions

The cosine and secant functions are *even*.

$$\cos(-t) = \cos t \qquad \sec(-t) = \sec t$$

The sine, cosecant, tangent, and cotangent functions are *odd*.

$$\sin(-t) = -\sin t \qquad \csc(-t) = -\csc t$$
$$\tan(-t) = -\tan t \qquad \cot(-t) = -\cot t$$

At this point, we have completed our introduction to basic trigonometry. We have measured angles in both degrees and radians. We have defined the six trigonometric functions from a right triangle perspective and as functions of real numbers. In our remaining work with trigonometry we will continue to rely on both perspectives. For instance, in the next three sections on graphing techniques we will think of the trigonometric functions as functions of real numbers. Later, in Chapter 9, we will look at applications involving angles and triangles.

For your convenience we have included on the inside cover of this text a summary of basic trigonometry.

WARM UP

Evaluate the trigonometric functions from memory.

1. $\sin 30°$

2. $\tan 45°$

3. $\cos \dfrac{\pi}{4}$

4. $\cot \dfrac{\pi}{3}$

5. $\sec \dfrac{\pi}{6}$

6. $\csc \dfrac{\pi}{4}$

Use the given trigonometric function of an *acute* angle θ to find the values of the remaining trigonometric functions.

7. $\tan \theta = \dfrac{3}{2}$

8. $\cos \theta = \dfrac{2}{3}$

9. $\sin \theta = \dfrac{1}{5}$

10. $\sec \theta = 3$

EXERCISES 7.3

In Exercises 1–4, determine the exact value of the six trigonometric functions of the given angle θ.

1. (a)

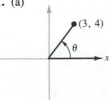

(b)

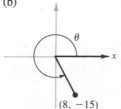

2. (a)

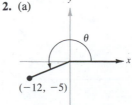

(b)

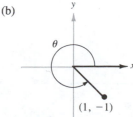

3. (a)

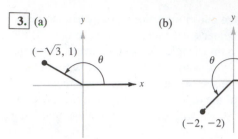

(b)

(−√3, 1)

θ

(−2, −2)

4. (a)

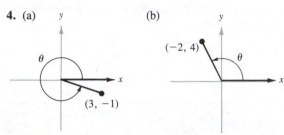

(b)

(−2, 4)

θ

θ

(3, −1)

In Exercises 5–8, the given point is on the terminal side of an angle in standard position. Determine the exact value of the six trigonometric functions of the angle.

5. (a) (7, 24)   (b) (7, −24)

6. (a) (8, 15)   (b) (−9, −40)

7. (a) (−4, 10)   (b) (3, −5)

8. (a) (−5, −2)   (b) (−3/2, 3)

In Exercises 9–12, use the two similar triangles in the accompanying figure to find (a) the unknown sides of the triangles and (b) the six trigonometric functions of the angles α_1 and α_2.

9. $a_1 = 3$, $b_1 = 4$, $a_2 = 9$

10. $b_1 = 12$, $c_1 = 13$, $c_2 = 26$

11. $a_1 = 1$, $c_1 = 2$, $b_2 = 5$

12. $b_1 = 4$, $a_2 = 4$, $b_2 = 10$

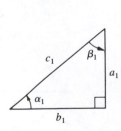

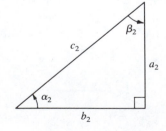

c_1 β_1 a_1

α_1

b_1

c_2 β_2 a_2

α_2

b_2

FIGURE FOR 9–12

In Exercises 13–16, determine the quadrant in which θ lies.

13. (a) $\sin \theta < 0$ and $\cos \theta < 0$
     (b) $\sin \theta > 0$ and $\cos \theta < 0$

14. (a) $\sin \theta > 0$ and $\cos \theta > 0$
     (b) $\sin \theta < 0$ and $\cos \theta > 0$

15. (a) $\sin \theta > 0$ and $\tan \theta < 0$
     (b) $\cos \theta > 0$ and $\tan \theta < 0$

16. (a) $\sec \theta > 0$ and $\cot \theta < 0$
     (b) $\csc \theta < 0$ and $\tan \theta > 0$

In Exercises 17–26, find the value (if possible) of the six trigonometric functions of θ.

17. θ lies in Quadrant II, $\sin \theta = \frac{3}{5}$

18. θ lies in Quadrant III, $\cos \theta = -\frac{4}{5}$

19. $\sin \theta < 0$, $\tan \theta = -\frac{15}{8}$

20. $\tan \theta < 0$, $\cos \theta = \frac{8}{17}$

21. $\sin \theta > 0$, $\sec \theta = -2$

22. $\frac{\pi}{2} \leq \theta \leq \frac{3\pi}{2}$, $\cot \theta$ is undefined

23. $\sin \theta = 0$, $\sec \theta = -1$

24. $\pi \leq \theta \leq 2\pi$, $\tan \theta$ is undefined

25. The terminal side of θ is in Quadrant III and lies on the line $y = 2x$.

26. The terminal side of θ is in Quadrant IV and lies on the line $4x + 3y = 0$.

In Exercises 27–34, find the reference angle θ′, and draw a sketch.

27. (a) $\theta = 203°$ (b) $\theta = 127°$

28. (a) $\theta = 309°$ (b) $\theta = 226°$

29. (a) $\theta = -245°$ (b) $\theta = -72°$

30. (a) $\theta = -145°$ (b) $\theta = -239°$

31. (a) $\theta = \frac{2\pi}{3}$ (b) $\theta = \frac{7\pi}{6}$

32. (a) $\theta = \frac{7\pi}{4}$ (b) $\theta = \frac{8\pi}{9}$

33. (a) $\theta = 3.5$ (b) $\theta = 5.8$

34. (a) $\theta = \frac{11\pi}{3}$ (b) $\theta = -\frac{7\pi}{10}$

In Exercises 35–44, evaluate the sine, cosine, and tangent of the given angles without using a calculator.

35. (a) 225°   (b) −225°

36. (a) 300°   (b) 330°

37. (a) $750°$ (b) $510°$

38. (a) $-405°$ (b) $-120°$

39. (a) $\dfrac{4\pi}{3}$ (b) $\dfrac{2\pi}{3}$

40. (a) $\dfrac{\pi}{4}$ (b) $\dfrac{5\pi}{4}$

41. (a) $-\dfrac{\pi}{6}$ (b) $\dfrac{5\pi}{6}$

42. (a) $-\dfrac{\pi}{2}$ (b) $\dfrac{\pi}{2}$

43. (a) $\dfrac{11\pi}{4}$ (b) $-\dfrac{13\pi}{6}$

44. (a) $\dfrac{10\pi}{3}$ (b) $\dfrac{17\pi}{3}$

In Exercises 45–52, use a calculator to evaluate the given trigonometric functions to four decimal places. (Be sure the calculator is set in the correct mode.)

45. (a) $\sin 10°$ (b) $\csc 10°$

46. (a) $\sec 225°$ (b) $\sec 135°$

47. (a) $\tan \dfrac{\pi}{9}$ (b) $\tan \dfrac{10\pi}{9}$

48. (a) $\cot 1.35$ (b) $\tan 1.35$

49. (a) $\cos(-110°)$ (b) $\cos 250°$

50. (a) $\sin(-0.65)$ (b) $\sin 5.63$

51. (a) $\tan 240°$ (b) $\cot 210°$

52. (a) $\csc 2.62$ (b) $\csc 150°$

In Exercises 53–58, find two values of θ that satisfy the given equation. List your answers in degrees ($0° \le \theta < 360°$) and radians ($0 \le \theta < 2\pi$). Do not use a calculator.

53. (a) $\sin \theta = \frac{1}{2}$ (b) $\sin \theta = -\frac{1}{2}$

54. (a) $\cos \theta = \dfrac{\sqrt{2}}{2}$ (b) $\cos \theta = -\dfrac{\sqrt{2}}{2}$

55. (a) $\csc \theta = \dfrac{2\sqrt{3}}{3}$ (b) $\cot \theta = -1$

56. (a) $\sec \theta = 2$ (b) $\sec \theta = -2$

57. (a) $\tan \theta = 1$ (b) $\cot \theta = -\sqrt{3}$

58. (a) $\sin \theta = \dfrac{\sqrt{3}}{2}$ (b) $\sin \theta = -\dfrac{\sqrt{3}}{2}$

In Exercises 59 and 60, use a calculator to approximate two values of θ ($0° \le \theta < 360°$) that satisfy the given equation. Round your answers to two decimal places.

59. (a) $\sin \theta = 0.8191$ (b) $\sin \theta = -0.2589$

60. (a) $\cos \theta = 0.8746$ (b) $\cos \theta = -0.2419$

In Exercises 61–64, use a calculator to approximate *two* values of θ ($0 \le \theta < 2\pi$) that satisfy the given equation. Round your answers to three decimal places.

61. (a) $\cos \theta = 0.9848$ (b) $\cos \theta = -0.5890$

62. (a) $\sin \theta = 0.0175$ (b) $\sin \theta = -0.6691$

63. (a) $\tan \theta = 1.192$ (b) $\tan \theta = -8.144$

64. (a) $\cot \theta = 5.671$ (b) $\cot \theta = -1.280$

In Exercises 65–68, evaluate the expression without using a calculator.

65. $\sin^2 2 + \cos^2 2$ **66.** $\tan^2 20° - \sec^2 20°$

67. $2 \sin^2 \dfrac{7\pi}{6} - \sin \dfrac{7\pi}{6} - 1$ **68.** $\sec^2 \dfrac{3\pi}{4} - 2 \tan \dfrac{3\pi}{4} - 2$

69. The average daily temperature (in degrees Fahrenheit) for a certain city is given by

$$T = 45 - 23 \cos\left[\frac{2\pi}{365}(t - 32)\right]$$

where t is the time in days, with $t = 1$ corresponding to January 1. Find the average temperature on (a) January 1, (b) July 4 ($t = 185$), and (c) October 18 ($t = 291$).

70. A company that produces a product with seasonal demands forecasts monthly sales over the next two years to be

$$S = 23.1 + 0.442t + 4.3 \sin\left(\frac{\pi t}{6}\right)$$

where S is measured in thousands of units and t is the time in months, with $t = 1$ representing January, 1988. Predict the sales for the following months.
(a) February, 1988   (b) February, 1989
(c) September, 1988   (d) September, 1989

71. The displacement from equilibrium of an oscillating weight suspended by a spring is given by

$$y(t) = \frac{1}{4} \cos 6t$$

where y is the displacement in feet and t is time in seconds. Find the displacement for (a) $t = 0$, (b) $t = 1/4$, and (c) $t = 1/2$.

72. Repeat Exercise 71 for the model given by

$$y(t) = \frac{1}{4}e^{-t} \cos 6t.$$

This model includes the damping effect of friction.

7.4 Graphs of Sine and Cosine

In this section we look at techniques for sketching the graphs of the sine and cosine functions. To accommodate the familiar xy-coordinate system, we use the variable x in place of θ or t. For example, we will write $y = \sin x$ and $y = \cos x$.

In Section 7.3 we discovered that the *domain* of the sine function is the set of all real numbers, and the *range* is the interval $[-1, 1]$. Moreover, the sine function has a *period* of 2π which implies that

$$\sin(x + 2\pi n) = \sin x$$

for all integers n.

Using this information, reference angles, and the function values from Table 7.1 (in the previous section), we can plot several points for $y = \sin x$ as x varies over the interval $[0, 2\pi]$. Then, by connecting these points with a smooth curve, we obtain the solid portion of the graph shown in Figure 7.36. This solid portion of the graph is called a **sine wave** or one **cycle** of the sine curve. The gray portion of the graph indicates that the basic sine wave repeats indefinitely to the right and left.

A similar analysis of the cosine function shows that its domain is also the set of all real numbers, its range is the interval $[-1, 1]$, and it has a period of 2π. The graph of the cosine function is shown in Figure 7.37.

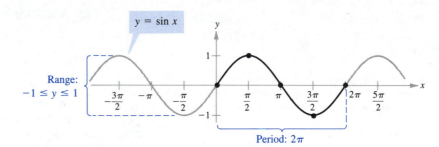

FIGURE 7.36

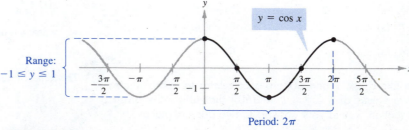

FIGURE 7.37

Note from Figures 7.36 and 7.37 that the sine graph is symmetric with respect to the *origin*, whereas the cosine graph is symmetric with respect to the *y-axis*. These properties of symmetry follow from the fact that the sine function is odd:

$$\sin(-x) = -\sin x \qquad\qquad \textit{Origin symmetry}$$

whereas the cosine function is even:

$$\cos(-x) = \cos x. \qquad\qquad \textit{y-axis symmetry}$$

Key Points on Graphs of Sine and Cosine Functions

To help you construct the graphs of the basic sine and cosine functions, we note five **key points** in one period of each graph: the intercepts, maximum points, and minimum points. For the sine function, the key points are

$$(0, 0), \quad \left(\frac{\pi}{2}, 1\right), \quad (\pi, 0), \quad \left(\frac{3\pi}{2}, -1\right), \quad \text{and} \quad (2\pi, 0).$$

For the cosine function, the key points are

$$(0, 1), \quad \left(\frac{\pi}{2}, 0\right), \quad (\pi, -1), \quad \left(\frac{3\pi}{2}, 0\right), \quad \text{and} \quad (2\pi, 1).$$

Note how the *x*-coordinates of these points divide the period of sin *x* and cos *x* into *four* equal parts, as indicated in Figure 7.38.

In Example 1, note how we use these key points to sketch the graph of $y = 2 \sin x$.

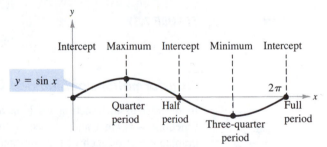

Period: 2π

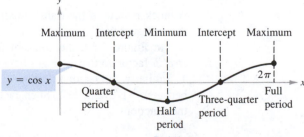

FIGURE 7.38

Period: 2π

EXAMPLE 1 Using Key Points to Sketch a Sine Curve

Sketch the graph of $y = 2 \sin x$ on the interval $[-\pi, 4\pi]$.

SOLUTION

In this case, note that $y = 2 \sin x = 2(\sin x)$ indicates that the y-values for the key points will have twice the magnitude of the graph of $y = \sin x$. Thus, the key points for $y = 2 \sin x$ are

$$(0, 0), \quad \left(\frac{\pi}{2}, 2\right), \quad (\pi, 0), \quad \left(\frac{3\pi}{2}, -2\right), \quad \text{and} \quad (2\pi, 0).$$

By connecting these key points with a smooth curve and extending the curve in both directions over the interval $[-\pi, 4\pi]$, we obtain the graph shown in Figure 7.39.

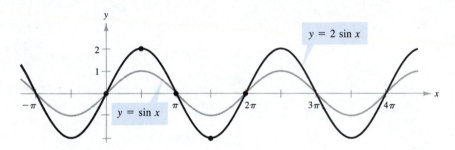

FIGURE 7.39

Transformations of Sine and Cosine Curves

In the rest of this section we look at variations in the graphs of the basic functions $y = \sin x$ and $y = \cos x$. In particular, we want to investigate the graphic effect of each of the constants a, b, and c in equations of the forms

$$y = a \sin(bx - c) \qquad \text{and} \qquad y = a \cos(bx - c).$$

A quick review of the transformations studied in Section 3.5 should help in this investigation.

In Figure 7.39, the constant factor 2 in $y = 2 \sin x$ acts as a vertical *stretch* factor so that y ranges between -2 and 2 instead of between -1 and 1. Similarly, the factor $1/2$ in $y = \frac{1}{2} \sin x$ *shrinks* the y-values so that y ranges between $-1/2$ and $1/2$. Such factors are referred to as the **amplitudes** of the functions.

Definition of Amplitude of Sine and Cosine Curves

The **amplitude** of $y = a \sin x$ and $y = a \cos x$ is the largest value of y and is given by

$$\text{amplitude} = |a|.$$

EXAMPLE 2 Using Amplitude to Sketch Graphs

On the same coordinate axes, sketch graphs of

$$y = \frac{1}{2} \cos x \quad \text{and} \quad y = 3 \cos x.$$

SOLUTION

Since the amplitude of $y = \frac{1}{2} \cos x$ is 1/2, the maximum value is 1/2 and the minimum value is $-1/2$. For one cycle, $0 \le x \le 2\pi$, the key points are

$$\left(0, \frac{1}{2}\right), \quad \left(\frac{\pi}{2}, 0\right), \quad \left(\pi, -\frac{1}{2}\right), \quad \left(\frac{3\pi}{2}, 0\right), \quad \text{and} \quad \left(2\pi, \frac{1}{2}\right).$$

A similar analysis shows that the amplitude of $y = 3 \cos x$ is 3, and the key points are

$$(0, 3), \quad \left(\frac{\pi}{2}, 0\right), \quad (\pi, -3), \quad \left(\frac{3\pi}{2}, 0\right), \quad \text{and} \quad (2\pi, 3).$$

The graphs of these two functions are shown in Figure 7.40.

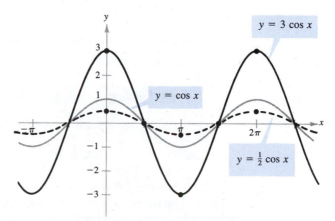

FIGURE 7.40   Amplitude Determines Vertical Stretch or Shrink

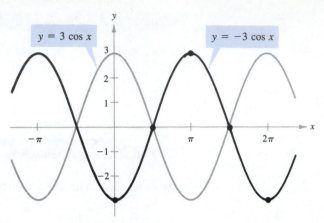

$y = 3 \cos x$

$y = -3 \cos x$

FIGURE 7.41   Reflection in the *x*-Axis

We know from Section 3.5 that the graph of $y = -f(x)$ is a **reflection** (in the *x*-axis) of the graph of $y = f(x)$. For instance, as shown in Figure 7.41, the graph of $y = -3 \cos x$ is a reflection of the graph of $y = 3 \cos x$.

Next, we consider the effect of the *positive* constant *b* on the graphs of

$$y = a \sin bx$$

and

$$y = a \cos bx.$$

Because $y = a \sin x$ completes one cycle from $x = 0$ to $x = 2\pi$, it follows that $y = a \sin bx$ completes one cycle from $bx = 0$ to $bx = 2\pi$, that is,

$$bx = 0 \quad \Longrightarrow \quad x = 0$$

$$bx = 2\pi \quad \Longrightarrow \quad x = \frac{2\pi}{b}.$$

Consequently, $y = a \sin bx$ completes one cycle from $x = 0$ to $x = 2\pi/b$, and hence its *period* is $2\pi/b$.

Period of Sine and Cosine Functions

Let *b* be a positive real number. The **period** of $y = a \sin bx$ and $y = a \cos bx$ is $2\pi/b$.

Note that if $0 < b < 1$, the period of $y = a \sin bx$ is greater than 2π and represents a *horizontal stretching* of the graph of $y = a \sin x$. Similarly, if $b > 1$, the period of $y = a \sin bx$ is less than 2π and represents a *horizontal shrinking* of the graph of $y = a \sin x$.

If b is negative, we use the identities $\sin(-x) = -\sin x$ and $\cos(-x) = \cos x$ to rewrite the function. For example, the period of

$$y = \sin(-2x)$$
$$= -\sin 2x$$

is $2\pi/2 = \pi$, and the period of

$$y = \cos(-3x)$$
$$= \cos 3x$$

is $2\pi/3$.

EXAMPLE 3 Horizontal Stretching

Sketch the graph of

$$y = \sin \frac{x}{2}.$$

SOLUTION

The amplitude is 1. Moreover, since $b = 1/2$, the period is

$$\frac{2\pi}{b} = \frac{2\pi}{1/2} = 4\pi.$$

Now, dividing the interval $[0, 4\pi]$ into four equal parts, we obtain the following key points on the graph:

$$(0, 0), \quad (\pi, 1), \quad (2\pi, 0), \quad (3\pi, -1), \quad \text{and} \quad (4\pi, 0).$$

The graph is shown in Figure 7.42.

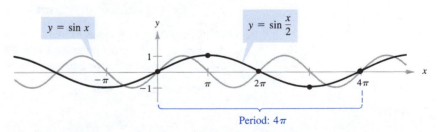

FIGURE 7.42

EXAMPLE 4 *Horizontal Shrinking*

Sketch the graph of

$$y = \sin 3x.$$

SOLUTION

The amplitude is 1, and since $b = 3$, the period is

$$\frac{2\pi}{b} = \frac{2\pi}{3}.$$

Dividing the interval $[0, 2\pi/3]$ into four equal parts, we obtain the following key points on the graph:

$$(0, 0), \quad \left(\frac{\pi}{6}, 1\right), \quad \left(\frac{\pi}{3}, 0\right), \quad \left(\frac{\pi}{2}, -1\right), \quad \text{and} \quad \left(\frac{2\pi}{3}, 0\right).$$

The graph is shown in Figure 7.43.

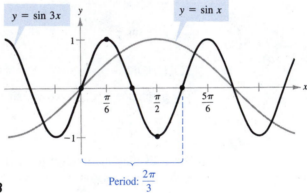

FIGURE 7.43

Our fourth graphic transformation is the horizontal shift caused by the constant c in the general equations

$$y = a \sin(bx - c) \quad \text{and} \quad y = a \cos(bx - c).$$

Comparing $y = a \sin bx$ with $y = a \sin(bx - c)$, we find that the graph of $y = a \sin(bx - c)$ completes one cycle from $bx - c = 0$ to $bx - c = 2\pi$. By solving for x, we find the interval for one cycle to be

Left Endpoint   Right Endpoint

$$\frac{c}{b} \leq x \leq \frac{c}{b} + \frac{2\pi}{b}.$$

Period

Graphs of Sine and Cosine

This implies that the period of $y = a \sin(bx - c)$ is $2\pi/b$, and the graph of $y = a \sin bx$ is shifted by an amount c/b. We call the number c/b the **phase shift.** We summarize these results as follows.

Graphs of the Sine and Cosine Functions

The graphs of $y = a \sin(bx - c)$ and $y = a \cos(bx - c)$ have the following characteristics. (Assume $b > 0$.)

   amplitude $= |a|$
   period $= 2\pi/b$

The **phase shift** and resulting interval for one cycle are solutions to the equations

$$bx - c = 0 \qquad \text{and} \qquad bx - c = 2\pi.$$

Note how we use this information to sketch graphs of the sine and cosine functions in Examples 5 and 6.

EXAMPLE 5   *Using Amplitude, Period, and Shift to Sketch Graphs*

Sketch the graph of

$$y = \frac{1}{2} \sin\left(x - \frac{\pi}{3}\right).$$

SOLUTION

The amplitude is $1/2$ and the period is 2π. By solving the equations

$$x - \frac{\pi}{3} = 0 \qquad \text{and} \qquad x - \frac{\pi}{3} = 2\pi$$

$$x = \frac{\pi}{3} \qquad\qquad\qquad x = \frac{7\pi}{3}$$

we see that the interval $[\pi/3, 7\pi/3]$ corresponds to one cycle of the graph. Dividing this interval into four equal parts produces the following key points:

$$\left(\frac{\pi}{3}, 0\right), \quad \left(\frac{5\pi}{6}, \frac{1}{2}\right), \quad \left(\frac{4\pi}{3}, 0\right), \quad \left(\frac{11\pi}{6}, -\frac{1}{2}\right), \quad \text{and} \quad \left(\frac{7\pi}{3}, 0\right).$$

The graph is shown in Figure 7.44.

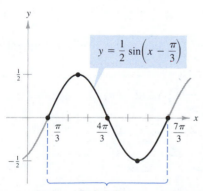

FIGURE 7.44

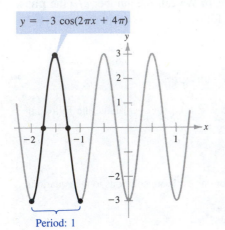

$y = -3 \cos(2\pi x + 4\pi)$

Period: 1

FIGURE 7.45

EXAMPLE 6   Using Amplitude, Period, and Shift to Sketch Graphs

Sketch the graph of $y = -3 \cos(2\pi x + 4\pi)$.

SOLUTION

For this function the amplitude is 3 and the period is $2\pi/2\pi = 1$. By solving the equations

$$2\pi x + 4\pi = 0 \qquad \text{and} \qquad 2\pi x + 4\pi = 2\pi$$
$$x = -2 \qquad\qquad\qquad x = -1$$

we see that one cycle corresponds to the interval $[-2, -1]$. Dividing this interval into four equal parts produces the following key points:

$$(-2, -3), \quad \left(-\frac{7}{4}, 0\right), \quad \left(-\frac{3}{2}, 3\right), \quad \left(-\frac{5}{4}, 0\right), \quad \text{and} \quad (-1, -3).$$

The graph is shown in Figure 7.45.

Our final graphic transformation is the *vertical shift* caused by the constant d in the equations

$$y = d + a \sin(bx - c) \qquad \text{and} \qquad y = d + a \cos(bx - c).$$

The shift is d units upward for $d > 0$ and downward for $d < 0$. In other words, the graph oscillates about the horizontal line $y = d$ instead of the x-axis.

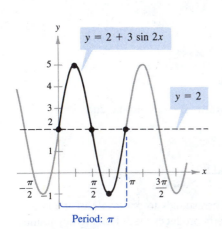

$y = 2 + 3 \sin 2x$

$y = 2$

Period: π

FIGURE 7.46

EXAMPLE 7   A Vertical Translation

Sketch the graph of $y = 2 + 3 \sin 2x$.

SOLUTION

The amplitude is 3 and the period is π. The key points over the interval $[0, \pi]$ are

$$(0, 2), \quad \left(\frac{\pi}{4}, 5\right), \quad \left(\frac{\pi}{2}, 2\right), \quad \left(\frac{3\pi}{4}, -1\right), \quad \text{and} \quad (\pi, 2).$$

The graph is shown in Figure 7.46.

For our last example in this section, we reverse the situation and show how to determine an equation for a given graph.

EXAMPLE 8 *Finding an Equation for a Given Graph*

Find the amplitude, period, and phase shift for the sine function whose graph is shown in Figure 7.47. Write an equation for this graph.

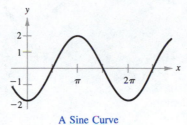

FIGURE 7.47   A Sine Curve

SOLUTION

The amplitude for this sine curve is 2. The period is 2π, and there is a right phase shift of $\pi/2$. Thus, we can write the following equation.

$$y = 2 \sin\left(x - \frac{\pi}{2}\right)$$

It is of interest to note that the graph of $y = \cos x$ corresponds to a *left* shift of $\pi/2$ units of the graph of $y = \sin x$, as shown in Figure 7.48. Similarly, the graph of $y = \sin x$ is a *right* shift of the graph of $y = \cos x$. This is consistent with the cofunction identities discussed in Section 7.2. That is,

$$\sin\left(\frac{\pi}{2} - x\right) = \cos x \qquad \text{and} \qquad \cos\left(\frac{\pi}{2} - x\right) = \sin x.$$

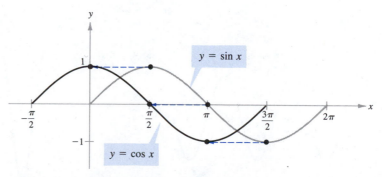

FIGURE 7.48   Shift the graph of $y = \sin x$ to the left $\frac{\pi}{2}$ units to obtain the graph of $y = \cos x$.

WARM UP

Simplify the expressions.

1. $\dfrac{2\pi}{1/3}$

2. $\dfrac{2\pi}{4\pi}$

Solve for x.

3. $2x - \dfrac{\pi}{3} = 0$

4. $2x - \dfrac{\pi}{3} = 2\pi$

5. $3\pi x + 6\pi = 0$

6. $3\pi x + 6\pi = 2\pi$

Evaluate the trigonometric functions from memory.

7. $\sin \dfrac{\pi}{2}$

8. $\sin \pi$

9. $\cos 0$

10. $\cos \dfrac{\pi}{2}$

EXERCISES 7.4

In Exercises 1–14, determine the period and amplitude of the given function.

1. $y = 2 \sin 2x$

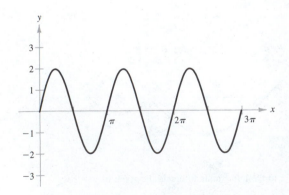

2. $y = 3 \cos 3x$

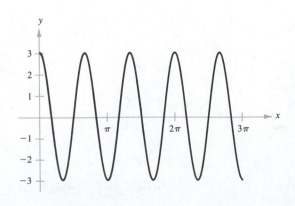

3. $y = \dfrac{3}{2} \cos \dfrac{x}{2}$

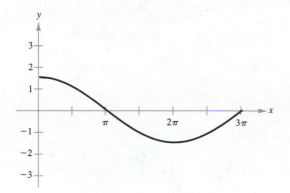

6. $y = \dfrac{5}{2} \cos \dfrac{\pi x}{2}$

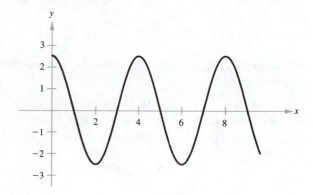

4. $y = -2 \sin \dfrac{x}{3}$

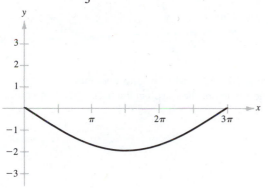

7. $y = 2 \sin x$

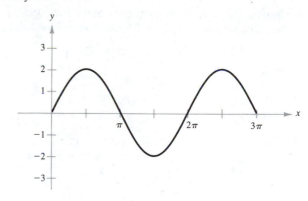

5. $y = \dfrac{1}{2} \sin \pi x$

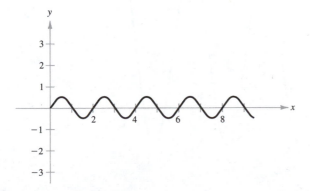

8. $y = -\cos \dfrac{2x}{3}$

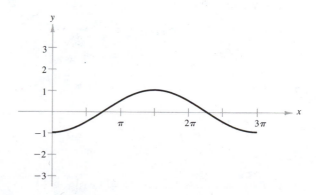

9. $y = -2 \sin 10x$

10. $y = \frac{1}{3} \sin 8x$

11. $y = \frac{1}{2} \cos \frac{2x}{3}$

12. $y = \frac{5}{2} \cos \frac{x}{4}$

13. $y = 3 \sin 4\pi x$

14. $y = \frac{2}{3} \cos \frac{\pi x}{10}$

In Exercises 15–22, describe the relationship between the graphs of f and g.

15. $f(x) = \sin x$
$g(x) = \sin(x - \pi)$

16. $f(x) = \cos x$
$g(x) = \cos(x + \pi)$

17. $f(x) = \cos 2x$
$g(x) = -\cos 2x$

18. $f(x) = \sin 3x$
$g(x) = \sin(-3x)$

19. $f(x) = \cos x$
$g(x) = \cos 2x$

20. $f(x) = \sin x$
$g(x) = \sin 3x$

21. $f(x) = \sin x$
$g(x) = 2 + \sin x$

22. $f(x) = \cos 4x$
$g(x) = -2 + \cos 4x$

In Exercises 23–30, sketch the graphs of the two functions on the same coordinate plane. (Include two full periods.)

23. $f(x) = -2 \sin x$
$g(x) = 4 \sin x$

24. $f(x) = \sin x$
$g(x) = \sin \frac{x}{3}$

25. $f(x) = \cos x$
$g(x) = 1 + \cos x$

26. $f(x) = 2 \cos 2x$
$g(x) = -\cos 4x$

27. $f(x) = -\frac{1}{2} \sin \frac{x}{2}$
$g(x) = 3 - \frac{1}{2} \sin \frac{x}{2}$

28. $f(x) = 4 \sin \pi x$
$g(x) = 4 \sin \pi x - 3$

29. $f(x) = 2 \cos x$
$g(x) = 2 \cos(x + \pi)$

30. $f(x) = -\cos x$
$g(x) = -\cos(x - \pi)$

In Exercises 31–56, sketch the graph of the given function. (Include two full periods.)

31. $y = -2 \sin 6x$

32. $y = -3 \cos 4x$

33. $y = \cos 2\pi x$

34. $y = \frac{3}{2} \sin \frac{\pi x}{4}$

35. $y = -\sin \frac{2\pi x}{3}$

36. $y = 10 \cos \frac{\pi x}{6}$

37. $y = 2 - \sin \frac{2\pi x}{3}$

38. $y = 2 \cos x - 3$

39. $y = \sin\left(x - \frac{\pi}{4}\right)$

40. $y = \frac{1}{2} \sin(x - \pi)$

41. $y = 3 \cos(x + \pi)$

42. $y = 4 \cos\left(x + \frac{\pi}{4}\right)$

43. $y = 3 \cos(x + \pi) - 3$

44. $y = 4 \cos\left(x + \frac{\pi}{4}\right) + 4$

45. $y = \frac{2}{3} \cos\left(\frac{x}{2} - \frac{\pi}{4}\right)$

46. $y = -3 \cos(6x + \pi)$

47. $y = -2 \sin(4x + \pi)$

48. $y = -4 \sin\left(\frac{2}{3}x - \frac{\pi}{3}\right)$

49. $y = \cos\left(2\pi x - \frac{\pi}{2}\right) + 1$

50. $y = 3 \cos\left(\frac{\pi x}{2} + \frac{\pi}{2}\right) - 2$

51. $y = -0.1 \sin\left(\frac{\pi x}{10} + \pi\right)$

52. $y = 5 \sin(\pi - 2x) + 10$

53. $y = 5 \cos(\pi - 2x) + 2$

54. $y = \frac{1}{100} \sin 120\pi t$

55. $y = \frac{1}{10} \cos 60\pi x$

56. $y = -3 + 5 \cos \frac{\pi t}{12}$

In Exercises 57–60, find a, b, and c so that the graph of the function matches the graph in the accompanying figure.

57. $y = a \sin(bx - c)$

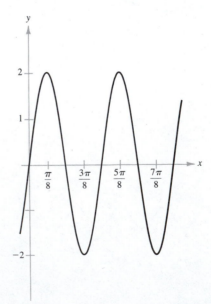

58. $y = a \sin(bx - c)$

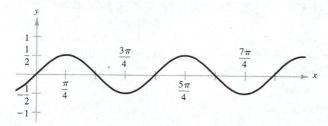

59. $y = a \cos(bx - c)$

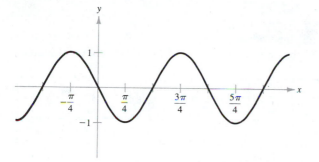

60. $y = a \sin(bx - c)$

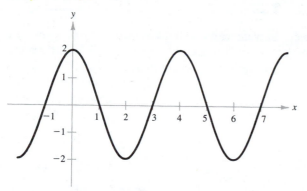

In Exercises 61–64, sketch the graph of f and g on the same coordinate axes and show that $f(x) = g(x)$ for all x. (Include two full periods.)

61. $f(x) = \sin x$

$g(x) = \cos\left(x - \dfrac{\pi}{2}\right)$

62. $f(x) = \sin x$

$g(x) = -\cos\left(x + \dfrac{\pi}{2}\right)$

63. $f(x) = \cos x$

$g(x) = -\sin\left(x - \dfrac{\pi}{2}\right)$

64. $f(x) = \cos x$

$g(x) = -\cos(x - \pi)$

65. For a person at rest, the velocity v (in liters per second) of air flow during a respiratory cycle is

$$v = 0.85 \sin \frac{\pi t}{3}$$

where t is the time in seconds. (Inhalation occurs when $v > 0$, and exhalation occurs when $v < 0$.)
(a) Find the time for one full respiratory cycle.
(b) Find the number of cycles per minute.
(c) Sketch the graph of the velocity function.

66. After exercising for a few minutes, a person has a respiratory cycle for which the velocity of air flow is approximated by

$$v = 1.75 \sin \frac{\pi t}{2}.$$

Use this model to repeat Exercise 65.

67. When tuning a piano, a technician strikes a tuning fork for the A above middle C and sets up wave motion that can be approximated by

$$y = 0.001 \sin 880\pi t$$

where t is the time in seconds.
(a) What is the period p of this function?
(b) The frequency f is given by $f = 1/p$. What is the frequency of this note?
(c) Sketch the graph of this function.

7.5 Graphs of Other Trigonometric Functions

In this section we continue our discussion of the graphs of the trigonometric functions, starting with the graph of the tangent function. Recall from Section 7.3 that the tangent function is odd, that is, $\tan(-x) = -\tan x$. Consequently, the graph of

$$y = \tan x$$

is symmetric with respect to the origin. We know also from the identity

$$\tan x = \frac{\sin x}{\cos x}$$

that the tangent is undefined when $\cos x = 0$. Two such values are $x = \pm\pi/2 \approx \pm 1.5708$. We examine this in more detail in the following table.

x	$-\dfrac{\pi}{2}$	-1.57	-1.5	-1	0	1	1.5	1.57	$\dfrac{\pi}{2}$
$\tan x$	undef.	-1255.8	-14.1	-1.56	0	1.56	14.1	1255.8	undef.

tan x approaches $-\infty$ as x approaches $-\pi/2$ from the right

tan x approaches ∞ as x approaches $\pi/2$ from the left

As indicated in this table, $\tan x$ increases without bound as x approaches $\pi/2$ from the left, and decreases without bound as x approaches $-\pi/2$ from the right. Thus, the graph of $y = \tan x$ has *vertical asymptotes* at $x = \pi/2$ and $-\pi/2$, as shown in Figure 7.49. Note in the graph that the period of the tangent function is π. Consequently, the tangent function has vertical asymptotes when $x = \pi/2 \pm n\pi$. The domain of the tangent function is the set of

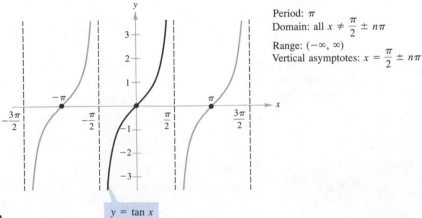

Period: π
Domain: all $x \neq \dfrac{\pi}{2} \pm n\pi$

Range: $(-\infty, \infty)$
Vertical asymptotes: $x = \dfrac{\pi}{2} \pm n\pi$

$y = \tan x$

FIGURE 7.49

all real numbers other than $x = \pi/2 \pm n\pi$, and the range is the set of all real numbers. Note from Figure 7.49 that the tangent function is increasing between each pair of consecutive asymptotes and that an x-intercept occurs at the midpoint of each cycle.

Sketching the graph of a function with the form

$$y = a \tan(bx - c)$$

is similar to sketching the graph of $y = a \sin(bx - c)$ in that we locate key points which identify the intercepts and asymptotes. Two consecutive asymptotes can be found by solving the equations

$$bx - c = -\frac{\pi}{2} \quad \text{and} \quad bx - c = \frac{\pi}{2}.$$

The midpoint between two consecutive asymptotes is an x-intercept of the graph. After plotting the asymptotes and the x-intercept, plot a few additional points between the two asymptotes and sketch one cycle. Additional cycles to the right or left can easily be sketched.

Remark: The period of the function $y = a \tan(bx - c)$ is the distance between two consecutive asymptotes. The amplitude of a tangent function is not defined.

EXAMPLE 1 *Sketching the Graph of a Tangent Function*

Sketch the graph of

$$y = \tan \frac{x}{2}.$$

SOLUTION

From the equations

$$\frac{x}{2} = -\frac{\pi}{2} \quad \text{and} \quad \frac{x}{2} = \frac{\pi}{2}$$

$$x = -\pi \qquad\qquad x = \pi$$

we see that two consecutive asymptotes occur at $x = -\pi$ and $x = \pi$. Between these two asymptotes, we plot a few points, as shown in the table, including the x-intercept, and complete the graph, as shown in Figure 7.50.

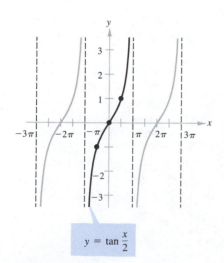

$$y = \tan \frac{x}{2}$$

FIGURE 7.50

x	$-\dfrac{\pi}{2}$	0	$\dfrac{\pi}{2}$
$\tan \dfrac{x}{2}$	-1	0	1

EXAMPLE 2 *Sketching the Graph of a Tangent Function*

Sketch the graph of $y = -3 \tan 2x$.

SOLUTION

By solving the equations

$$2x = -\frac{\pi}{2} \quad \text{and} \quad 2x = \frac{\pi}{2}$$

$$x = -\frac{\pi}{4} \qquad\qquad x = \frac{\pi}{4}$$

we see that two consecutive asymptotes occur at $x = -\pi/4$ and $x = \pi/4$. Between these two asymptotes, we plot a few points, as shown in the following table, and complete one cycle.

x	$-\dfrac{\pi}{8}$	0	$\dfrac{\pi}{8}$
$-3 \tan 2x$	3	0	-3

Four cycles of the graph are shown in Figure 7.51.

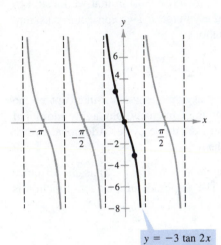

$y = -3 \tan 2x$

FIGURE 7.51

The graph of the cotangent function is similar to the graph of the tangent function. It also has a period of π. However, from the identity

$$y = \cot x = \frac{\cos x}{\sin x}$$

we can see that the cotangent function has vertical asymptotes at $x = n\pi$, because $\sin x$ is zero at these x-values. The graph of the cotangent function is shown in Figure 7.52.

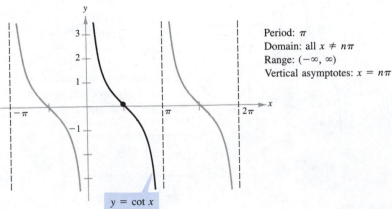

Period: π
Domain: all $x \neq n\pi$
Range: $(-\infty, \infty)$
Vertical asymptotes: $x = n\pi$

$y = \cot x$

FIGURE 7.52

EXAMPLE 3 *Sketching the Graph of a Cotangent Function*

Sketch the graph of

$$y = 2 \cot \frac{x}{3}.$$

SOLUTION

To locate two consecutive vertical asymptotes of the graph, we solve

$$\frac{x}{3} = 0 \qquad \text{and} \qquad \frac{x}{3} = \pi$$

$$x = 0 \qquad\qquad\qquad x = 3\pi.$$

Then, between these two asymptotes we plot the points shown in the following table, and complete one cycle of the graph. (Note that the period is 3π, the distance between consecutive asymptotes.)

x	$\dfrac{3\pi}{4}$	$\dfrac{3\pi}{2}$	$\dfrac{9\pi}{4}$
$2 \cot \dfrac{x}{3}$	2	0	−2

Three cycles of the graph are shown in Figure 7.53.

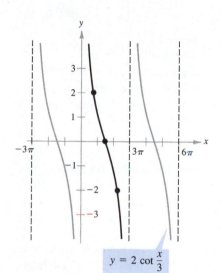

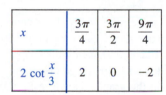

$y = 2 \cot \dfrac{x}{3}$

FIGURE 7.53

Graphs of the Reciprocal Functions

The graphs of the two remaining trigonometric functions can be obtained from the graphs of the sine and cosine functions using the reciprocal identities

$$\csc x = \frac{1}{\sin x} \qquad \text{and} \qquad \sec x = \frac{1}{\cos x}.$$

For instance, at a given value for x, the y-coordinate for $\sec x$ is the reciprocal of the y-coordinate for $\cos x$. Of course, when $\cos x = 0$, the reciprocal does not exist. Near such values for x, the behavior of the secant function is similar to that of the tangent function. In other words, the graphs of

$$\tan x = \frac{\sin x}{\cos x} \qquad \text{and} \qquad \sec x = \frac{1}{\cos x}$$

have vertical asymptotes at $x = (\pi/2) + n\pi$, because the cosine is zero at these x-values. Similarly,

$$\cot x = \frac{\cos x}{\sin x} \qquad \text{and} \qquad \csc x = \frac{1}{\sin x}$$

have vertical asymptotes where $\sin x = 0$, that is, at $x = n\pi$.

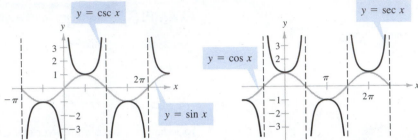

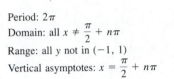

Period: 2π
Domain: all $x \neq n\pi$
Range: all y not in $(-1, 1)$
Vertical asymptotes: $x = n\pi$

Period: 2π
Domain: all $x \neq \dfrac{\pi}{2} + n\pi$
Range: all y not in $(-1, 1)$
Vertical asymptotes: $x = \dfrac{\pi}{2} + n\pi$

FIGURE 7.54

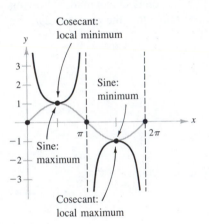

FIGURE 7.55

To sketch the graph of a secant or cosecant function we suggest that you first make a sketch of its reciprocal function. For instance, to sketch the graph of $y = \csc x$, first sketch the graph of $y = \sin x$, then take reciprocals of the y-coordinates to obtain points on the graph of $y = \csc x$. We use this procedure to obtain the graphs shown in Figure 7.54.

In comparing the graphs of the secant and cosecant functions with those of the sine and cosine functions, note that the "hills" and "valleys" are interchanged. For example, a hill (or maximum point) on the sine curve corresponds to a valley (a local minimum) on the cosecant curve. Similarly, a valley (or minimum point) on the sine curve corresponds to a hill (a local maximum) on the cosecant curve, as shown in Figure 7.55.

EXAMPLE 4 *Sketching the Graph of a Cosecant Function*

Sketch the graph of

$$y = 2 \csc\left(x + \frac{\pi}{4}\right).$$

SOLUTION

We begin by sketching the graph of

$$2 \sin\left(x + \frac{\pi}{4}\right).$$

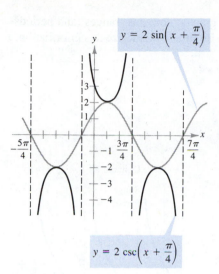

$y = 2 \sin\left(x + \dfrac{\pi}{4}\right)$

$y = 2 \csc\left(x + \dfrac{\pi}{4}\right)$

FIGURE 7.56

For this function, the amplitude is 2 and the period is 2π. By solving the equations

$$x + \frac{\pi}{4} = 0 \qquad \text{and} \qquad x + \frac{\pi}{4} = 2\pi$$

$$x = -\frac{\pi}{4} \qquad\qquad\qquad x = \frac{7\pi}{4}$$

we see that one cycle of the sine function corresponds to the interval from $x = -\pi/4$ to $x = 7\pi/4$. The graph of this sine function is represented by the gray curve in Figure 7.56. Because the sine function is zero at the endpoints of this interval, the corresponding cosecant function

$$y = 2 \csc\left(x + \frac{\pi}{4}\right)$$

$$= 2\left(\frac{1}{\sin[x + (\pi/4)]}\right)$$

has vertical asymptotes at $x = -\pi/4$ and $7\pi/4$. The graph of the cosecant function is represented by the solid curve.

EXAMPLE 5 *Sketching the Graph of a Secant Function*

Sketch the graph of $y = \sec 2x$.

SOLUTION

We begin by sketching the graph of $\cos 2x$ as indicated by the gray curve in Figure 7.57. Then, we form the graph of $y = \sec 2x$ as the solid curve in the figure. Note that the x-intercepts of $\cos 2x$

$$\left(\frac{\pi}{4}, 0\right), \quad \left(\frac{3\pi}{4}, 0\right), \quad \left(\frac{5\pi}{4}, 0\right), \quad \ldots$$

correspond to the vertical asymptotes

$$x = \frac{\pi}{4}, \quad x = \frac{3\pi}{4}, \quad x = \frac{5\pi}{4}, \quad \ldots$$

of the graph of $y = \sec 2x$.

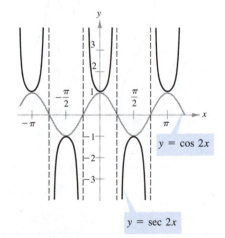

$y = \cos 2x$

$y = \sec 2x$

FIGURE 7.57

In Figure 7.58, we summarize the graphs, domains, ranges, and periods of the six basic trigonometric functions. Be sure to memorize this information.

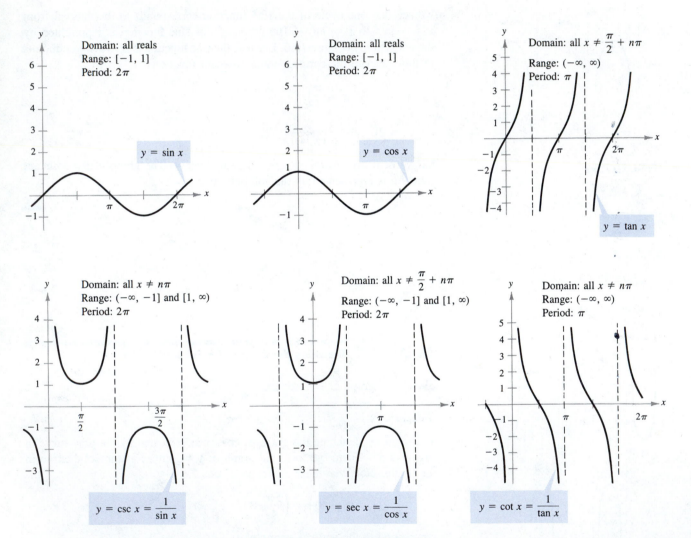

Graphs of the Six Trigonometric Functions

FIGURE 7.58

WARM UP

Evaluate the trigonometric functions from memory.

1. $\tan 0$

2. $\cos \dfrac{\pi}{4}$

3. $\tan \dfrac{\pi}{4}$

4. $\cot \dfrac{\pi}{2}$

5. $\sin \pi$

6. $\cos \dfrac{\pi}{2}$

Sketch the graph of each function. (Include two full periods.)

7. $y = -2 \cos 2x$

8. $y = 3 \sin \dfrac{x}{4}$

9. $y = \dfrac{3}{2} \sin 2\pi x$

10. $y = -2 \cos \dfrac{\pi x}{2}$

EXERCISES 7.5

In Exercises 1–8, match the trigonometric function with the correct graph and give the period of the function. [The graphs are labeled (a)–(h).]

1. $y = \sec 2x$

2. $y = \tan 3x$

3. $y = \tan \dfrac{x}{2}$

4. $y = 2 \csc \dfrac{x}{2}$

5. $y = \cot \pi x$

6. $y = \dfrac{1}{2} \sec \pi x$

7. $y = -\sec x$

8. $y = -2 \csc 2\pi x$

(c)
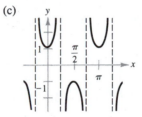

(d)
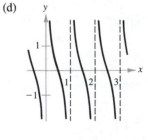

(a)
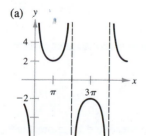

(b)
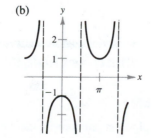

(e)
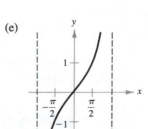

(f)
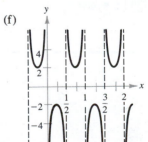

(g)

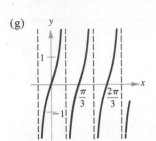

(h)

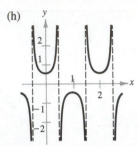

In Exercises 9–36, sketch the graph of the function through two periods.

9. $y = \tan 2x$

10. $y = -\tan 2x$

11. $y = \tan \dfrac{x}{3}$

12. $y = -3 \tan \pi x$

13. $y = -2 \sec 4x$

14. $y = 2 \sec 4x$

15. $y = -\sec \pi x$

16. $y = \sec \pi x$

17. $y = \sec \pi x - 1$

18. $y = -2 \sec 4x + 2$

19. $y = \csc \dfrac{x}{2}$

20. $y = \csc \dfrac{x}{3}$

21. $y = \cot \dfrac{x}{2}$

22. $y = 3 \cot \dfrac{\pi x}{2}$

23. $y = \frac{1}{2} \sec 2x$

24. $y = -\frac{1}{2} \tan x$

25. $y = \tan\left(x - \dfrac{\pi}{4}\right)$

26. $y = \sec(x + \pi)$

27. $y = \dfrac{1}{4} \csc\left(x + \dfrac{\pi}{4}\right)$

28. $y = -\csc(4x - \pi)$

29. $y = \dfrac{1}{4} \cot\left(x - \dfrac{\pi}{2}\right)$

30. $y = 2 \cot\left(x + \dfrac{\pi}{2}\right)$

31. $y = 2 \sec(2x - \pi)$

32. $y = \dfrac{1}{3} \sec\left(\dfrac{\pi x}{2} + \dfrac{\pi}{2}\right)$

33. $y = \tan \dfrac{\pi x}{4}$

34. $y = 0.1 \tan\left(\dfrac{\pi x}{4} + \dfrac{\pi}{4}\right)$

35. $y = \csc(\pi - x)$

36. $y = \sec(\pi - x)$

37. A plane flying at an altitude of 6 miles over level ground will pass directly over a radar antenna (see figure). Let d be the ground distance from the antenna to the point directly under the plane and let x be the angle of elevation to the plane from the antenna. Write d as a function of x, $0 < x < \pi/2$.

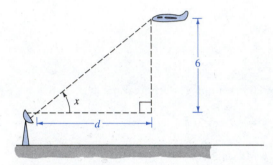

FIGURE FOR 37

38. A television camera is on a reviewing platform 100 feet from the street on which a parade will be passing from left to right (see figure). Express the distance d from the camera to a particular unit in the parade as a function of the angle x and sketch the graph of the function over the appropriate domain. (Consider x as negative when a unit in the parade approaches from the left.)

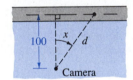

FIGURE FOR 38

39. Use the graph of $f(x) = \sec x$ to verify that the secant is an even function.

40. Use the graph of $f(x) = \tan x$ to verify that the tangent is an odd function.

7.6 Additional Graphing Techniques

Addition of Ordinates

The behavior of some physical phenomena can be represented by more than one trigonometric function, or by a combination of algebraic and trigonometric functions. We can use a technique called **addition of ordinates** (y-values)

Additional Graphing Techniques

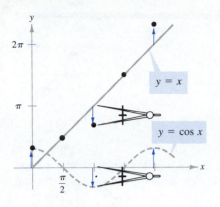

FIGURE 7.59

to sketch the graphs of functions like

$$y = \sin x - \cos 2x \quad \text{or} \quad y = x + \cos x.$$

For example, the graph of $y = x + \cos x$ can be obtained by first making sketches of $y = x$ and $y = \cos x$ on the same set of axes and then geometrically adding the ordinates (*y*-values) for several representative *x*-values. This addition of ordinates is aided by the use of a compass or ruler to measure the vertical displacements, as shown in Figure 7.59.

As with previous trigonometric graphs, the *key points* to plot are those for which one or both functions have an intercept, maximum point, or minimum point.

EXAMPLE 1 *Graphing by Addition of Ordinates*

Use addition of ordinates to sketch the graph of

$$y = x + \cos x.$$

SOLUTION

First, we make sketches (shown in gray) of the graphs of

$$y = x \quad \text{and} \quad y = \cos x$$

on the same coordinate plane. For $y = \cos x$, the key points are

$$(0, 1), \quad \left(\frac{\pi}{2}, 0\right), \quad (\pi, -1), \quad \left(\frac{3\pi}{2}, 0\right), \quad \text{and} \quad (2\pi, 1).$$

The points lying on the graph of $y = x$ directly above (or below) these five points are

$$(0, 0), \quad \left(\frac{\pi}{2}, \frac{\pi}{2}\right), \quad (\pi, \pi), \quad \left(\frac{3\pi}{2}, \frac{3\pi}{2}\right), \quad \text{and} \quad (2\pi, 2\pi).$$

We geometrically add the *y*-values of the two functions at these key points

$$(0, 1), \quad \left(\frac{\pi}{2}, \frac{\pi}{2}\right), \quad (\pi, -1 + \pi), \quad \left(\frac{3\pi}{2}, \frac{3\pi}{2}\right), \quad (2\pi, 1 + 2\pi)$$

and plot the results. Then we connect the resulting points by a smooth curve, obtaining the graph shown in Figure 7.60.

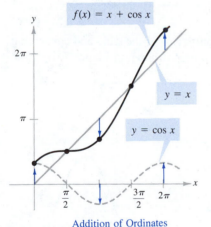

Addition of Ordinates

FIGURE 7.60

Note that the function in Figure 7.60 is not periodic as we defined the concept. However, the next example involves two trigonometric functions whose resulting difference is a periodic function.

EXAMPLE 2 *Graphing by Addition of Ordinates*

Sketch the graph of $y = \sin x - \cos 2x$.

SOLUTION

In this case, we make sketches (shown in gray) of the graphs of

$$y = \sin x \quad \text{and} \quad y = -\cos 2x$$

noting that their respective periods are 2π and π. From the shorter period (the one for $-\cos 2x$), we consider the key points given in the following table.

x	0	$\dfrac{\pi}{4}$	$\dfrac{\pi}{2}$	$\dfrac{3\pi}{4}$	π	$\dfrac{5\pi}{4}$	$\dfrac{3\pi}{2}$	$\dfrac{7\pi}{4}$	2π
$-\cos 2x$	-1	0	1	0	-1	0	1	0	-1
$\sin x$	0	$\dfrac{\sqrt{2}}{2}$	1	$\dfrac{\sqrt{2}}{2}$	0	$-\dfrac{\sqrt{2}}{2}$	-1	$-\dfrac{\sqrt{2}}{2}$	0
$\sin x - \cos 2x$	-1	$\dfrac{\sqrt{2}}{2}$	2	$\dfrac{\sqrt{2}}{2}$	-1	$-\dfrac{\sqrt{2}}{2}$	0	$-\dfrac{\sqrt{2}}{2}$	-1

By plotting the points indicated in the fourth row of the table and connecting them with a smooth curve, we obtain the graph shown in Figure 7.61. Note that the function has a period of 2π.

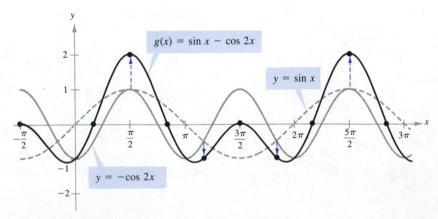

FIGURE 7.61　　　　　Addition of Ordinates

Additional Graphing Techniques

EXAMPLE 3 *Graphing by Addition of Ordinates*

Sketch the graph of $y = \sin x + 2 \cos x$.

SOLUTION

In this case, we make sketches (shown in gray) of the graphs of

$$y = \sin x \quad \text{and} \quad y = 2 \cos x$$

noting that both have periods of 2π. Then we consider the key points given in the following table.

x	0	$\dfrac{\pi}{2}$	π	$\dfrac{3\pi}{2}$	2π
$2 \cos x$	2	0	-2	0	2
$\sin x$	0	1	0	-1	0
$\sin x + 2 \cos x$	2	1	-2	-1	2

By plotting the points indicated in the fourth row of the table and connecting them with a smooth curve, we obtain the graph shown in Figure 7.62. Note that this function has a period of 2π.

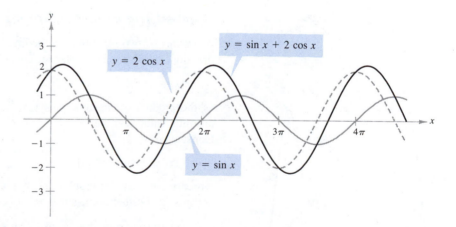

FIGURE 7.62

One of the simplest uses of addition of ordinates occurs in vertical shifts like those in the graphs of

$$y = 2 + \sin 3x \qquad \text{and} \qquad y = 3 + \cos\left(x - \frac{\pi}{4}\right).$$

We know from Sections 3.5 and 7.4 that adding a constant does not change the shape (or period) of a trigonometric graph—it only changes its vertical location. For example, using the graph of $y = \sin 3x$, we can easily sketch the graph of

$$y = 2 + \sin 3x$$

by adding 2 to each ordinate, as shown in Figure 7.63.

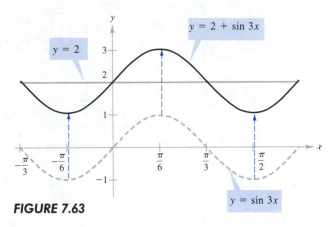

FIGURE 7.63

Damped Trigonometric Graphs

A *product* of two functions can be graphed using properties of the individual functions involved. For instance, consider the function

$$f(x) = x \sin x$$

as the product of the functions $y = x$ and $y = \sin x$. Using properties of absolute value and the fact that $|\sin x| \leq 1$, we have $0 \leq |x| \, |\sin x| \leq |x|$. Consequently,

$$-|x| \leq x \sin x \leq |x|$$

which means that the graph of $f(x) = x \sin x$ lies between the lines $y = -x$ and $y = x$. Furthermore, since

$$f(x) = x \sin x = \pm x \qquad \text{at} \qquad x = \frac{\pi}{2} + n\pi$$

$$f(x) = x \sin x = 0 \qquad \text{at} \qquad x = n\pi$$

the graph of f touches the line $y = -x$ or the line $y = x$ at $x = (\pi/2) + n\pi$ and has x-intercepts at $x = n\pi$. A sketch of f is shown in Figure 7.64.

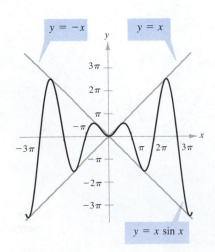

FIGURE 7.64

In the function $f(x) = x \sin x$ the factor x is called the **damping factor.** By changing the damping factor, we can change the graph significantly. For example, look in Figure 7.65 at the graphs of

$$y = \frac{1}{x} \sin x \qquad \text{and} \qquad y = e^{-x} \sin 3x.$$

We show another instance of a damping factor in Example 4.

EXAMPLE 4   Damped Cosine Wave

Sketch the graph of $f(x) = 2^{-x/2} \cos x$.

SOLUTION

Consider $f(x)$ as the product of the two functions

$$y = 2^{-x/2} \qquad \text{and} \qquad y = \cos x$$

each of which has the set of real numbers as its domain. For any real number x, we know that $2^{-x/2} \geq 0$ and $|\cos x| \leq 1$. Therefore, $|2^{-x/2}|\,|\cos x| \leq 2^{-x/2}$, which means that

$$-2^{-x/2} \leq 2^{-x/2} \cos x \leq 2^{-x/2}.$$

Furthermore, since

$$f(x) = 2^{-x/2} \cos x = \pm 2^{-x/2} \qquad \text{at} \qquad x = n\pi$$

and

$$f(x) = 2^{-x/2} \cos x = 0 \qquad \text{at} \qquad x = \frac{\pi}{2} + n\pi$$

the graph of f touches the curves $y = -2^{-x/2}$ and $y = 2^{-x/2}$ at $x = n\pi$ and has intercepts at $x = (\pi/2) + n\pi$. A sketch is shown in Figure 7.66.

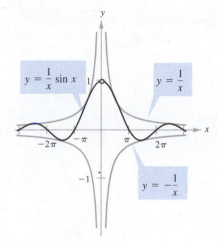

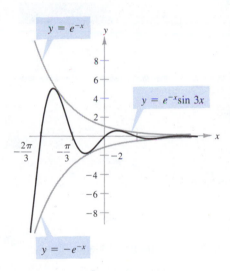

FIGURE 7.65

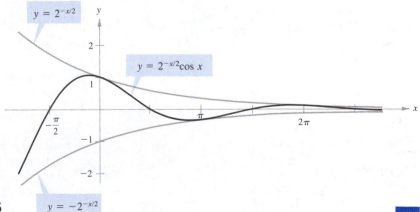

FIGURE 7.66

WARM UP

Find the x-values in the interval $[0, 2\pi]$ for which $f(x)$ is -1, 0, or 1.

1. $f(x) = \sin x$

2. $f(x) = \cos x$

3. $f(x) = \sin 2x$

4. $f(x) = \cos \dfrac{x}{2}$

Sketch the graph of each function.

5. $y = |x|$

6. $y = e^{-x}$

7. $y = \sin \pi x$

8. $y = \cos 2x$

Evaluate $f(x)$ when $x = 0$, $\pi/6$, $\pi/4$, $\pi/3$, and $\pi/2$.

9. $f(x) = x \cos x$

10. $f(x) = x + \sin x$

EXERCISES 7.6

In Exercises 1–26, use addition of ordinates to sketch the graph of the function.

1. $y = 2 - 2 \sin \dfrac{x}{2}$

2. $y = -3 + \cos x$

3. $y = 4 - 2 \cos \pi x$

4. $y = 5 - \frac{1}{2} \sin 2\pi x$

5. $y = -1 + \cot x$

6. $y = 2 + \tan \pi x$

7. $y = 1 + \csc x$

8. $y = 1 - \sec x$

9. $y = x + \sin x$

10. $y = x + \cos x$

11. $y = \frac{1}{2}x - 2 \cos x$

12. $y = 2x - \sin x$

13. $y = \sin x + \cos x$

14. $y = \cos x + \cos 2x$

15. $y = 2 \sin x + \sin 2x$

16. $y = 2 \sin x + \cos 2x$

17. $y = \cos x - \cos \dfrac{x}{2}$

18. $y = \sin x - \dfrac{1}{2} \sin \dfrac{x}{2}$

19. $y = \sin x + \frac{1}{3} \sin 5x$

20. $y = \cos x - \frac{1}{4} \cos 2x$

21. $y = -3 + \cos x + 2 \sin 2x$

22. $y = \sin \pi x + \sin \dfrac{\pi x}{2}$

23. $y = -x + \sin \dfrac{\pi x}{2}$

24. $y = -\dfrac{x}{2} - \dfrac{1}{2} \sin \dfrac{\pi x}{4}$

25. $y = \sin x + \cos \left(x + \dfrac{\pi}{2} \right)$

26. $y = \sin x - \cos \left(x + \dfrac{\pi}{2} \right)$

In Exercises 27–34, sketch the graph of the function.

27. $y = x \cos x$

28. $y = |x| \sin x$

29. $y = |x| \cos x$

30. $y = 2^{-x/4} \cos \pi x$

31. $y = e^{-x^2/2} \sin x$

32. $y = e^{-t} \cos t$

33. $y = \sin^2 x$

34. $y = \cos^2 \dfrac{\pi x}{2}$

35. The monthly sales S (in thousands of units) of a seasonal product is approximated by

$$S = 74.50 + 43.75 \sin \frac{\pi t}{6}$$

where t is the time in months, with $t = 1$ corresponding to January. Sketch the graph of this sales function over one year.

36. The function

$$P = 100 - 20 \cos \frac{5\pi t}{3}$$

approximates the blood pressure P (in millimeters of mercury) for a person at rest, where the time t is measured in seconds. Sketch the graph of this function over a 10-second interval of time.

37. Suppose that the population of a certain predator at time t (in months) in a given region is estimated to be

$$P = 10,000 + 3000 \sin \frac{2\pi t}{24}$$

and the population of its primary food source (its prey) is estimated to be

$$p = 15,000 + 5000 \cos \frac{2\pi t}{24}.$$

Sketch both of these functions on the same graph and explain the oscillations in the size of each population.

38. An object weighing W pounds is suspended from the ceiling by a steel spring (see figure). The weight is pulled downward (positive direction) from its equilibrium position and released. The resulting motion of the weight is described by the function

$$y = \frac{1}{2} e^{-t/4} \cos 4t$$

where y is distance in feet and t is the time in seconds. Sketch the graph of the function.

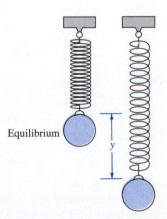

Equilibrium

FIGURE FOR 38

39. Use a calculator to evaluate the function

$$f(x) = \frac{1 - \cos x}{x}$$

at several points in the interval $[-1, 1]$, and then use these points to sketch the graph of f. This function is undefined when $x = 0$. From your graph, estimate the value that $f(x)$ is approaching as x approaches 0.

x	-0.5	-0.4	-0.3	-0.2	-0.1
$\dfrac{1 - \cos x}{x}$					

x	0.1	0.2	0.3	0.4	0.5
$\dfrac{1 - \cos x}{x}$					

7.7 *Inverse Trigonometric Functions*

Up to this point, much of our time has been spent evaluating trigonometric functions at specified angles or real numbers. However, in Section 7.2 we introduced the *inverse* problem: *Given the value of sin x, find x.* There we used the calculator key INV or 2nd f and promised to later explain the functions involved. We now investigate these **inverse trigonometric functions.**

Recall from Section 3.7 that, in order for a function to have an inverse, it must be one-to-one. From Figure 7.67 it is obvious that $y = \sin x$ is not one-to-one because different values of x yield the same y-value. However, if we restrict the domain to the interval $-\pi/2 \le x \le \pi/2$ (corresponding to the solid portion of the graph in Figure 7.67), the following properties hold.

1. On the interval $[-\pi/2, \pi/2]$, the function $y = \sin x$ is increasing.
2. On the interval $[-\pi/2, \pi/2]$, $y = \sin x$ takes on its full range of values, $-1 \le \sin x \le 1$.
3. On the interval $[-\pi/2, \pi/2]$, $y = \sin x$ is a one-to-one function.

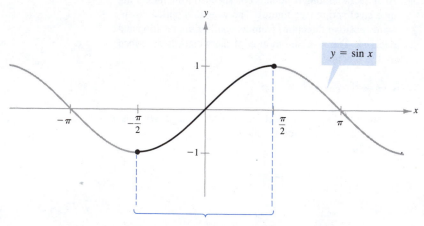

FIGURE 7.67

sin *x* is one-to-one on this interval.

Thus, on the restricted domain $-\pi/2 \le x \le \pi/2$, $y = \sin x$ has a unique inverse called the **inverse sine function.** It is denoted by

$$y = \arcsin x \qquad \text{or} \qquad y = \sin^{-1} x.$$

The notation $\sin^{-1} x$ is consistent with the inverse function notation $f^{-1}(x)$ used in Section 3.7. The arcsin x notation (read as "the arc sine of x") comes from the association of a central angle with its subtended *arc length* on a unit circle. Thus, arcsin x means the angle (or arc) whose sine is x. The values of arcsin x lie in the interval $-\pi/2 \le \arcsin x \le \pi/2$. The graph of $y = \arcsin x$ is shown in Example 2.

> ## Definition of Inverse Sine Function
>
> The **inverse sine function** is defined by
>
> $$y = \arcsin x \quad \text{if and only if} \quad \sin y = x$$
>
> where $-1 \le x \le 1$ and $-\pi/2 \le y \le \pi/2$. The domain of $y = \arcsin x$ is $[-1, 1]$ and the range is $[-\pi/2, \pi/2]$.

Remark: When evaluating the inverse sine function, it helps to remember the phrase "the arcsine of x is the angle (or number) whose sine is x."

Both notations, $\arcsin x$ and $\sin^{-1} x$, are commonly used in mathematics, so remember that $\sin^{-1} x$ denotes the *inverse* sine function rather than $1/\sin x$.

EXAMPLE 1   Evaluating the Inverse Sine Function

Find the values of the following (if possible).

(a) $\arcsin\left(-\dfrac{1}{2}\right)$ (b) $\sin^{-1}\dfrac{\sqrt{3}}{2}$ (c) $\sin^{-1} 2$

SOLUTION

(a) By definition, $y = \arcsin(-1/2)$ implies that

$$\sin y = -\frac{1}{2}, \quad \text{for } -\frac{\pi}{2} \le y \le \frac{\pi}{2}.$$

Because $\sin(-\pi/6) = -1/2$, we conclude that $y = -\pi/6$ and

$$\arcsin\left(-\frac{1}{2}\right) = -\frac{\pi}{6}.$$

(b) By definition, $y = \sin^{-1}(\sqrt{3}/2)$ implies that

$$\sin y = \frac{\sqrt{3}}{2}, \quad \text{for } -\frac{\pi}{2} \le y \le \frac{\pi}{2}.$$

Because $\sin(\pi/3) = \sqrt{3}/2$, we conclude that $y = \pi/3$ and

$$\sin^{-1}\frac{\sqrt{3}}{2} = \frac{\pi}{3}.$$

(c) It is not possible to evaluate $y = \sin^{-1} x$ when $x = 2$ because there is no angle whose sine is 2. Remember that the domain of the inverse sine function is $[-1, 1]$.

From Section 3.7 we know that graphs of inverse functions are reflections of each other in the line $y = x$. In Example 2 we demonstrate this result for the inverse sine function.

EXAMPLE 2 *Graphing the Arcsine Function*

Sketch a graph of $y = \arcsin x$.

SOLUTION

By definition, the equations

$$y = \arcsin x \qquad \text{and} \qquad \sin y = x$$

are equivalent. Hence, their graphs are the same. By assigning values to y in the second equation, we can make the following table of values.

y	$-\dfrac{\pi}{2}$	$-\dfrac{\pi}{4}$	$-\dfrac{\pi}{6}$	0	$\dfrac{\pi}{6}$	$\dfrac{\pi}{4}$	$\dfrac{\pi}{2}$
$x = \sin y$	-1	$-\dfrac{\sqrt{2}}{2}$	$-\dfrac{1}{2}$	0	$\dfrac{1}{2}$	$\dfrac{\sqrt{2}}{2}$	1

The resulting graph for $y = \arcsin x$ is shown in Figure 7.68. Note that it is the reflection (in line $y = x$) of the solid part of Figure 7.67. Be sure you see that Figure 7.68 shows the *entire* graph of the inverse sine function. Remember that the range of $y = \arcsin x$ is the closed interval $[-\pi/2, \pi/2]$.

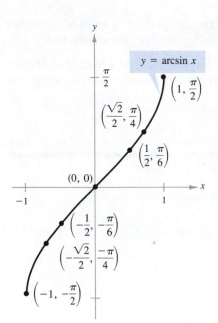

FIGURE 7.68

The cosine function is decreasing on the interval $0 \le x \le \pi$, as shown in Figure 7.69. Consequently, on this interval the cosine has an inverse

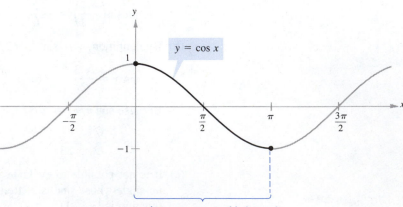

FIGURE 7.69

cos x is one-to-one on this interval.

function, which we call the **inverse cosine function** and denote by

$$y = \arccos x$$

or

$$y = \cos^{-1} x.$$

Similarly, we can define an **inverse tangent function** by restricting the domain of $y = \tan x$ to the interval $(-\pi/2, \pi/2)$. In the following list, we summarize the definitions of the three most common inverse trigonometric functions. The remaining three are discussed in the exercise set. (We summarize the graphs, domains, and ranges of *all six* inverse trigonometric functions in the Appendix.)

Definition of the Inverse Trigonometric Functions

Function	Domain	Range
$y = \textbf{arcsin } x$ if and only if $\sin y = x$	$-1 \le x \le 1$	$-\dfrac{\pi}{2} \le y \le \dfrac{\pi}{2}$
$y = \textbf{arccos } x$ if and only if $\cos y = x$	$-1 \le x \le 1$	$0 \le y \le \pi$
$y = \textbf{arctan } x$ if and only if $\tan y = x$	$-\infty < x < \infty$	$-\dfrac{\pi}{2} < y < \dfrac{\pi}{2}$

The graphs of these three inverse trigonometric functions are shown in Figure 7.70.

Domain: $[-1, 1]$
Range: $[-\pi/2, \pi/2]$

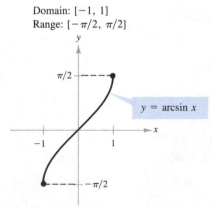

Domain: $[-1, 1]$
Range: $[0, \pi]$

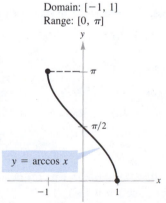

Domain: $(-\infty, \infty)$
Range: $(-\pi/2, \pi/2)$

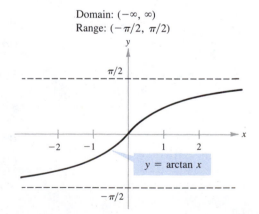

FIGURE 7.70

EXAMPLE 3 *Evaluating Inverse Trigonometric Functions*

Evaluate the following.

(a) $\arccos \dfrac{\sqrt{2}}{2}$ 　　　　　(b) $\arccos(-1)$ 　　　　　(c) $\arctan 0$

SOLUTION

(a) Because $\cos(\pi/4) = \sqrt{2}/2$, it follows that

$$\arccos \frac{\sqrt{2}}{2} = \frac{\pi}{4}.$$

(b) Because $\cos \pi = -1$, it follows that

$$\arccos(-1) = \pi.$$

(c) Because $\tan 0 = 0$, it follows that

$$\arctan 0 = 0.$$

In Example 3, we were able to evaluate the given inverse trigonometric function without a calculator. In the next example, a calculator is necessary to evaluate the functions.

EXAMPLE 4 *Evaluating Inverse Trigonometric Functions*

If possible, find the value of each of the following.

(a) $\arctan(-8.45)$ 　　　　　(b) $\arcsin 0.2447$ 　　　　　(c) $\arccos 2$

SOLUTION

Function	*Mode*	*Keystrokes*	*Display*
(a) $\arctan(-8.45)$	Radian	8.45 +/− INV tan	−1.453001
(b) $\arcsin 0.2447$	Radian	0.2447 INV sin	0.2472103
(c) $\arccos 2$	Radian	2 INV cos	ERROR

Note that the *error* in part (c) occurs because the domain of the inverse cosine function is $[-1, 1]$.

Remark: In Example 4, had we set the calculator to degree mode, the display would have been in degrees rather than radians. This convention is peculiar to calculators. By definition, the values of inverse trigonometric functions are always *in radians*.

Compositions of Trigonometric and Inverse Trigonometric Functions

Recall from Section 3.7 that inverse functions possess the properties

$$f(f^{-1}(x)) = x$$

and

$$f^{-1}(f(x)) = x.$$

The inverse trigonometric versions of these properties are given in the following list.

Inverse Properties

If $-1 \le x \le 1$ and $-\pi/2 \le y \le \pi/2$, then

$$\sin(\arcsin x) = x \qquad \text{and} \qquad \arcsin(\sin y) = y.$$

If $-1 \le x \le 1$ and $0 \le y \le \pi$, then

$$\cos(\arccos x) = x \qquad \text{and} \qquad \arccos(\cos y) = y.$$

If $-\pi/2 < y < \pi/2$, then

$$\tan(\arctan x) = x \qquad \text{and} \qquad \arctan(\tan y) = y.$$

Remark: Keep in mind that these inverse properties do not apply for arbitrary values of x and y. For instance,

$$\arcsin\left(\sin \frac{3\pi}{2}\right) = \arcsin(-1)$$

$$= -\frac{\pi}{2} \neq \frac{3\pi}{2}.$$

In other words, the property $\arcsin(\sin y) = y$ is not valid for values of y outside the interval $[-\pi/2, \pi/2]$.

EXAMPLE 5 Using Inverse Properties

If possible, find the value of the following.

(a) $\tan[\arctan(-5)]$ (b) $\arcsin\left(\sin\dfrac{5\pi}{3}\right)$ (c) $\cos(\cos^{-1}\pi)$

SOLUTION

(a) Since -5 lies in the domain of the arctan x, the inverse property applies, and we have

$$\tan[\arctan(-5)] = -5.$$

(b) In this case, $5\pi/3$ does not lie within the range of the arcsine function, $-\pi/2 \le x \le \pi/2$. However, $5\pi/3$ is coterminal with

$$\frac{5\pi}{3} - 2\pi = -\frac{\pi}{3}$$

which does lie in the range of the arcsine function, and we have

$$\arcsin\left(\sin\frac{5\pi}{3}\right) = \arcsin\left[\sin\left(-\frac{\pi}{3}\right)\right] = -\frac{\pi}{3}.$$

(c) The expression $\cos(\cos^{-1}\pi)$ is not defined because $\cos^{-1}\pi$ is not defined. Remember that the domain of the inverse cosine function is $[-1, 1]$.

As with the trigonometric functions, much of the work with the inverse trigonometric functions can be done by *exact* calculations rather than by calculator approximations. Exact calculations help to increase our understanding of the inverse functions by relating them to the triangle definitions of the trigonometric functions.

In Example 6, we show how to use right triangles to find exact values of functions of inverse functions. Then, in Example 7, we show how to use triangles to convert a trigonometric expression into an algebraic one. This conversion technique is used frequently in calculus.

EXAMPLE 6 Evaluating Functions of Inverse Trigonometric Functions

Find the exact value of the following.

(a) $\tan\left(\arccos\dfrac{2}{3}\right)$ (b) $\cos\left[\arcsin\left(-\dfrac{3}{5}\right)\right]$

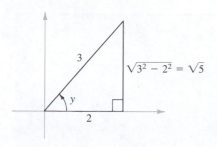

FIGURE 7.71

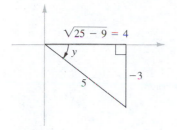

FIGURE 7.72

SOLUTION

(a) If we let $y = \arccos(2/3)$, then $\cos y = 2/3$. Because $\cos y$ is positive, y is a *first* quadrant angle. We can sketch and label y as shown in Figure 7.71. Consequently,

$$\tan\left(\arccos \frac{2}{3}\right) = \tan y$$

$$= \frac{\text{opp}}{\text{adj}} = \frac{\sqrt{5}}{2}.$$

(b) Let $y = \arcsin(-3/5)$. Then $\sin y = -3/5$. Because $\sin y$ is negative, y is a *fourth* quadrant angle. We can sketch and label y as shown in Figure 7.72. Consequently,

$$\cos\left[\arcsin\left(-\frac{3}{5}\right)\right] = \cos y$$

$$= \frac{\text{adj}}{\text{hyp}} = \frac{4}{5}.$$

EXAMPLE 7 *Some Problems from Calculus*

Write each of the following as an algebraic expression in x.

(a) $\sin(\arccos 3x), \quad 0 \le x \le \frac{1}{3}$ \qquad (b) $\cot(\arccos 3x), \quad 0 \le x \le \frac{1}{3}$

SOLUTION

Let $y = \arccos 3x$. Then we have $\cos y = 3x$. Since

$$\cos y = \frac{3x}{1} = \frac{\text{adj}}{\text{hyp}}$$

we can sketch a right triangle with acute angle y, as shown in Figure 7.73. From this triangle, we can easily convert each expression to algebraic form.

(a) $\sin(\arccos 3x) = \sin y = \dfrac{\text{opp}}{\text{hyp}} = \sqrt{1 - 9x^2}, \quad 0 \le x \le \dfrac{1}{3}$

(b) $\cot(\arccos 3x) = \cot y = \dfrac{\text{adj}}{\text{opp}} = \dfrac{3x}{\sqrt{1 - 9x^2}}, \quad 0 \le x \le \dfrac{1}{3}$

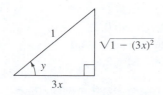

FIGURE 7.73

Remark: In Example 7, a similar argument can be made for x-values lying in the interval $[-1/3, 0]$.

WARM UP

Evaluate the trigonometric functions from memory.

1. $\sin\left(-\dfrac{\pi}{2}\right)$

2. $\cos \pi$

3. $\tan\left(-\dfrac{\pi}{4}\right)$

4. $\sin \dfrac{\pi}{4}$

Find an angle x in the interval $[-\pi/2, \pi/2]$ that has the same sine value as the given value.

5. $\sin 2\pi$

6. $\sin \dfrac{5\pi}{6}$

Find an angle x in the interval $[0, \pi]$ that has the same cosine value as the given value.

7. $\cos 3\pi$

8. $\cos\left(-\dfrac{\pi}{4}\right)$

Find an angle x in the interval $(-\pi/2, \pi/2)$ that has the same tangent value as the given value.

9. $\tan 4\pi$

10. $\tan \dfrac{3\pi}{4}$

EXERCISES 7.7

In Exercises 1–16, evaluate the given expression without using a calculator.

1. $\arcsin \frac{1}{2}$

2. $\arcsin 0$

3. $\arccos \frac{1}{2}$

4. $\arccos 0$

5. $\arctan \dfrac{\sqrt{3}}{3}$

6. $\arctan(-1)$

7. $\arccos\left(-\dfrac{\sqrt{3}}{2}\right)$

8. $\arcsin\left(-\dfrac{\sqrt{2}}{2}\right)$

9. $\arctan(-\sqrt{3})$

10. $\text{arccot}(-\sqrt{3})$

11. $\arccos(-\frac{1}{2})$

12. $\arcsin \dfrac{\sqrt{2}}{2}$

13. $\arcsin \dfrac{\sqrt{3}}{2}$

14. $\arccos 1$

15. $\arctan 0$

16. $\arctan\left(-\dfrac{\sqrt{3}}{3}\right)$

In Exercises 17–28, use a calculator to approximate the given value. [Round your answers to two decimal places.]

17. $\arccos 0.28$

18. $\arcsin 0.45$

19. $\arcsin(-0.75)$

20. $\arccos(-0.8)$

21. $\arctan(-2)$

22. $\arctan 15$

23. $\arcsin 0.31$

24. $\arccos 0.26$

25. $\arccos(-0.41)$

26. $\arcsin(-0.125)$

27. $\arctan 0.92$

28. $\arctan 2.8$

In Exercises 29–34, use the properties of inverse trigonometric functions to evaluate the given expression.

29. $\sin(\arcsin 0.3)$

30. $\tan(\arctan 25)$

31. $\cos[\arccos(-0.1)]$

32. $\sin[\arcsin(-0.2)]$

33. $\arcsin(\sin 3\pi)$

34. $\arccos\left(\cos \dfrac{7\pi}{2}\right)$

In Exercises 35–42, find the exact value of the given expression without using a calculator. [*Hint:* Make a sketch of a right triangle, as illustrated in Example 6.]

35. $\sin(\arctan \frac{3}{4})$

36. $\sec(\arcsin \frac{4}{5})$

37. $\cos(\arctan 2)$

38. $\sin\left(\arccos \frac{\sqrt{5}}{5}\right)$

39. $\cos(\arcsin \frac{5}{13})$

40. $\csc[\arctan(-\frac{5}{12})]$

41. $\sec[\arctan(-\frac{3}{5})]$

42. $\tan[\arcsin(-\frac{5}{6})]$

In Exercises 43–52, write an algebraic expression that is equivalent to the given expression. [*Hint:* Sketch a right triangle, as demonstrated in Example 7.]

43. $\cot(\arctan x)$

44. $\sin(\arctan x)$

45. $\cos(\arcsin 2x)$

46. $\sec(\arctan 3x)$

47. $\sin(\arccos x)$

48. $\cot\left(\arctan \frac{1}{x}\right)$

49. $\tan\left(\arccos \frac{x}{3}\right)$

50. $\sec[\arcsin (x - 1)]$

51. $\csc\left(\arctan \frac{x}{\sqrt{2}}\right)$

52. $\cos\left(\arcsin \frac{x - h}{r}\right)$

In Exercises 53–56, fill in the blanks.

53. $\arctan \dfrac{9}{x} = \arcsin(\underline{\quad\quad})$

54. $\arcsin \dfrac{\sqrt{36 - x^2}}{x} = \arccos(\underline{\quad\quad}), \quad |x| \le 6$

55. $\arccos \dfrac{3}{\sqrt{x^2 - 2x + 10}} = \arcsin(\underline{\quad\quad})$

56. $\arccos \dfrac{x - 2}{2} = \arctan(\underline{\quad\quad}), \quad |x - 2| \le 2$

In Exercises 57–60, sketch the graph of the function.

57. $f(x) = \arcsin(x - 1)$

58. $f(x) = \arctan x + \dfrac{\pi}{2}$

59. $f(x) = \arccos 2x$

60. $f(x) = \arccos \dfrac{x}{4}$

61. A photographer is taking a picture of a four-foot painting hung in an art gallery. The camera lens is one foot below the lower edge of the painting (see figure). The angle β subtended by the camera lens x feet from the painting is given by

$$\beta = \arctan \frac{4x}{x^2 + 5}.$$

Find β when (a) $x = 3$ feet and (b) $x = 6$ feet.

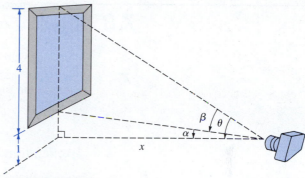

FIGURE FOR 61

62. In calculus, it is shown that the area of the region bounded by the graphs of $y = 0$, $y = 1/(x^2 + 1)$, $x = a$, and $x = b$ is given by

Area $= \arctan b - \arctan a$

(see figure). Find the area for the following values of a and b.

(a) $a = 0$, $b = 1$
(b) $a = -1$, $b = 1$
(c) $a = 0$, $b = 3$
(d) $a = -1$, $b = 3$

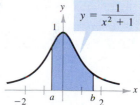

FIGURE FOR 62

63. Define the inverse cotangent function by restricting the domain of the cotangent to the interval $(0, \pi)$.

64. Define the inverse secant function by restricting the domain of the secant to the intervals $[0, \pi/2)$ and $(\pi/2, \pi]$.

65. Use the result of Exercise 64 to evaluate the following.
(a) arcsec $\sqrt{2}$ \qquad (b) arcsec 1

66. Define the inverse cosecant function by restricting the domain of the cosecant to the intervals $[-\pi/2, 0)$ and $(0, \pi/2]$.

In Exercises 67–72, prove the identity.

67. $\arcsin(-x) = -\arcsin x$

68. $\arctan(-x) = -\arctan x$

69. $\arccos(-x) = \pi - \arccos x$

70. $\arctan x + \arctan \dfrac{1}{x} = \dfrac{\pi}{2}, \quad x > 0$

71. $\arcsin x + \arccos x = \dfrac{\pi}{2}$

72. $\arcsin x = \arctan \dfrac{x}{\sqrt{1 - x^2}}$

7.8 Applications of Trigonometry

In keeping with our twofold perspective of trigonometry, this section includes both right triangle applications and applications that emphasize the periodic nature of the trigonometric functions.

Applications Involving Right Triangles

In this section we will denote the three angles of a right triangle by the letters A, B, and C (where C is the right angle), and the lengths of the sides opposite these angles by the letters a, b, and c (where c is the hypotenuse).

In Section 7.2, we introduced the problem of solving a right triangle. We review that procedure in the first four examples.

EXAMPLE 1 *Solving a Right Triangle, Given One Acute Angle and One Side*

Solve the right triangle having $A = 34.2°$ and $b = 19.4$, as shown in Figure 7.74.

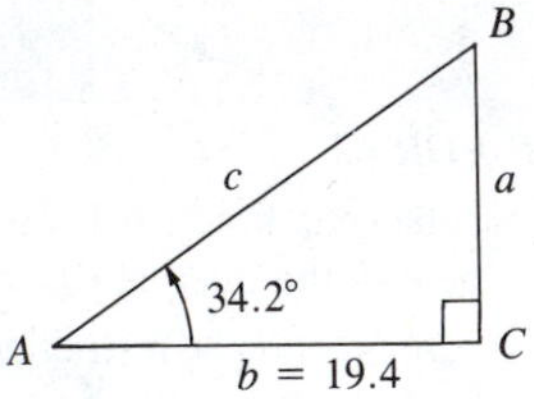

SOLUTION

Since $C = 90°$, it follows that $A + B = 90°$ and

$$B = 90° - 34.2° = 55.8°.$$

To solve for a we use the fact that

$$\tan A = \frac{\text{opp}}{\text{adj}} = \frac{a}{b} \implies a = b \tan A.$$

Thus, we have

$$a = 19.4 \tan 34.2° \approx 13.18.$$

Similarly, to solve for c we use the fact that

$$\cos A = \frac{\text{adj}}{\text{hyp}} = \frac{b}{c} \implies c = \frac{b}{\cos A}.$$

Thus, we have

$$c = \frac{19.4}{\cos 34.2°} \approx 23.46.$$

FIGURE 7.74

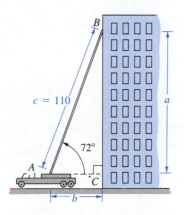

FIGURE 7.75

EXAMPLE 2   *Finding a Side of a Right Triangle*

A safety regulation states that the maximum angle of elevation for a rescue ladder is 72°. If a fire department's longest ladder is 110 feet, what is the maximum safe rescue height?

SOLUTION

A sketch is shown in Figure 7.75. From the equation

$$\sin A = \frac{a}{c}$$

it follows that

$$a = c \sin A = 110(\sin 72°) \approx 104.6 \text{ feet.}$$

In Example 2, we used the term **angle of elevation** to represent the angle from the horizontal upward to an object. For objects that lie below the horizontal, it is common to use the term **angle of depression,** as shown in Figure 7.76.

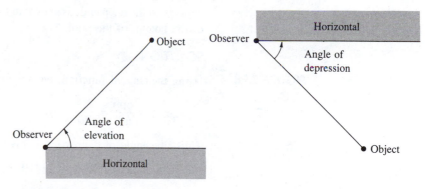

FIGURE 7.76

EXAMPLE 3   *Finding a Side of a Right Triangle*

From a point 200 feet from the base of a building, the angle of elevation to the *bottom* of a smokestack is 35°, while the angle of elevation to the *top* is 53°, as shown in Figure 7.77. Find the height, *s*, of the smokestack alone.

SOLUTION

Note from Figure 7.77 that this problem involves two right triangles. In the smaller right triangle, we use the fact that $\tan 35° = a/200$ to conclude that the height of the building is

$$a = 200 \tan 35°.$$

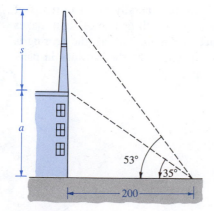

FIGURE 7.77

Now, from the larger right triangle, we use the equation

$$\tan 53° = \frac{a + s}{200}$$

to conclude that $a + s = 200 \tan 53°$. Hence, the height of the smokestack is

$$s = 200 \tan 53° - a = 200 \tan 53° - 200 \tan 35° \approx 125.4 \text{ feet.}$$

In Examples 1 through 3 we found the lengths of the sides of a right triangle, given an acute angle and the length of one of the sides. We can also find the angles of a right triangle given only the lengths of two sides, as demonstrated in Example 4.

EXAMPLE 4   *Finding an Acute Angle of a Right Triangle*

A swimming pool is 20 meters long and 12 meters wide. The bottom of the pool is slanted so that the water depth is 1.3 meters at the shallow end and 4 meters at the deep end, as shown in Figure 7.78. Find the angle of depression of the bottom of the pool.

SOLUTION

Using the tangent function, we see that

$$\tan A = \frac{\text{opp}}{\text{adj}} = \frac{2.7}{20} = 0.135.$$

Thus, the angle of depression is given by

$$A = \arctan 0.135 \approx 0.13419 \text{ (radians)} \approx 7.69°.$$

In surveying and navigation, directions are generally given in terms of **bearings**. A bearing measures the acute angle a path or line of sight makes with a fixed north-south line. For instance, in Figure 7.79(a), the bearing is S 35° E, meaning *35 degrees east of south*. Similarly, the bearings in parts (b) and (c) are N 80° W and N 45° E, respectively.

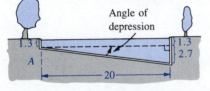

Angle of depression

FIGURE 7.78

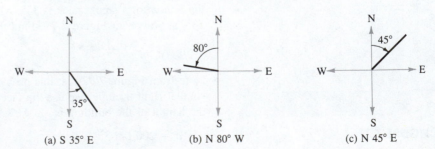

FIGURE 7.79   (a) S 35° E   (b) N 80° W   (c) N 45° E

EXAMPLE 5 Finding Directions in Terms of Bearings

A ship leaves port at noon and heads due west at 20 knots (nautical miles per hour). At 2 P.M., to avoid a storm, it changes course to N 54° W, as shown in Figure 7.80. Find the ship's bearing and distance from the port of departure at 3 P.M.

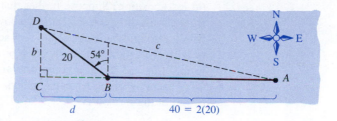

FIGURE 7.80

SOLUTION

In triangle BCD, we have $B = 90° - 54° = 36°$. The two sides of this triangle are determined as follows.

$$\sin B = \frac{b}{20}$$

$$b = 20 \sin 36°$$

$$\cos B = \frac{d}{20}$$

$$d = 20 \cos 36°$$

Now, in triangle ACD, we can determine angle A as follows.

$$\tan A = \frac{b}{d + 40}$$

$$= \frac{20 \sin 36°}{20 \cos 36° + 40} \approx 0.2092494$$

$$A \approx \arctan 0.2092494 \approx 0.2062732 \text{ radians} \approx 11.82°$$

The angle with the north-south line is $90° - 11.82° = 78.18°$. Therefore, the bearing of the ship is

N 78.18° W. *Bearing*

Finally, from triangle ACD, we have $\sin A = b/c$ which yields

$$c = \frac{b}{\sin A} = \frac{20 \sin 36°}{\sin(\arctan 0.2092494)}$$

$$c \approx 57.4 \text{ nautical miles.}$$ *Distance from port*

Harmonic Motion

The periodic nature of the trigonometric functions is useful for describing the motion of a point on an object that vibrates, oscillates, rotates, or is moved by wave motion.

For example, consider a ball that is bobbing up and down on the end of a spring, as shown in Figure 7.81. Suppose that 10 centimeters is the maximum distance the ball moves vertically upward or downward from its equilibrium (at rest) position. Suppose further that the time it takes for the ball to move from its maximum displacement above zero to its maximum displacement below zero and back again is $t = 4$ seconds. Assuming the ideal conditions of perfect elasticity and no friction or air resistance, the ball would continue to move up and down in a uniform and regular manner.

From this spring we can conclude that the period (time for one complete cycle) of the motion is

period = 4 seconds

and that its amplitude (maximum displacement from equilibrium) is

amplitude = 10 centimeters.

Motion of this nature can be described by a sine or cosine function, and is called **simple harmonic motion.**

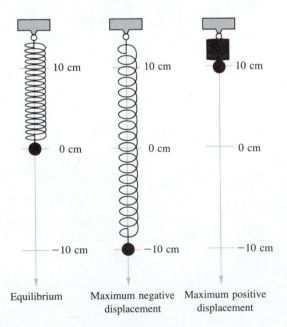

FIGURE 7.81   Simple Harmonic Motion

> ### Definition of Simple Harmonic Motion
>
> A point that moves on a coordinate line is said to be in **simple harmonic motion** if its distance d from the origin at time t is given by either
>
> $$d = a \sin \omega t \quad \text{or} \quad d = a \cos \omega t$$
>
> where a and ω are real numbers such that $\omega > 0$. The motion has **amplitude** $|a|$, **period** $2\pi/\omega$, and **frequency** $\omega/2\pi$.

EXAMPLE 6 Simple Harmonic Motion

Write the equation for the simple harmonic motion of the ball described in Figure 7.81, where the period is 4 seconds. What is the frequency for this motion?

SOLUTION

Since the spring is at equilibrium ($d = 0$) when $t = 0$, we use the equation

$$d = a \sin \omega t.$$

Moreover, since the maximum displacement from zero is 10 and the period is 4, we have

$$\text{amplitude} = |a| = 10$$

$$\text{period} = \frac{2\pi}{\omega} = 4 \implies \omega = \frac{\pi}{2}.$$

Consequently, the equation of motion is

$$d = 10 \sin \frac{\pi}{2} t.$$

Note that the choice of $a = 10$ or $a = -10$ depends on whether the ball initially moves up or down. The frequency is given by

$$\text{frequency} = \frac{\omega}{2\pi}$$

$$= \frac{\pi/2}{2\pi}$$

$$= \frac{1}{4} \text{ cycle per second.}$$

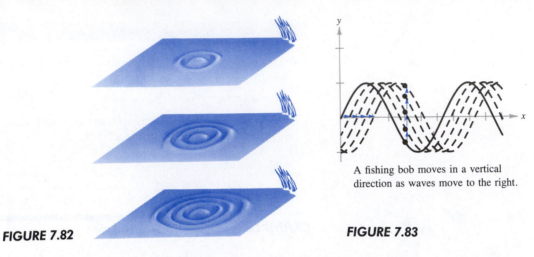

A fishing bob moves in a vertical direction as waves move to the right.

FIGURE 7.82   **FIGURE 7.83**

One clear illustration of the relation between sine waves and harmonic motion is seen in the wave motion resulting from dropping a stone into a calm pool of water. The waves move outward in roughly the shape of sine (or cosine) waves, as shown in Figure 7.82. As an example, suppose you are fishing and your fishing bob is attached so that it does not move horizontally. As the waves move outward from the dropped stone, your fishing bob will move up and down in simple harmonic motion, as shown in Figure 7.83.

EXAMPLE 7 Simple Harmonic Motion

Given the equation for simple harmonic motion

$$d = 6 \cos \frac{3\pi}{4} t$$

find (a) the maximum displacement, (b) the frequency, (c) the value of d when $t = 4$, and (d) the least positive value of t for which $d = 0$.

SOLUTION

The given equation has the form $d = a \cos \omega t$, with $a = 6$ and $\omega = 3\pi/4$.

(a) The maximum displacement (from the point of equilibrium) is given by the amplitude. Thus, the maximum displacement is 6.

(b) The frequency is

$$\text{frequency} = \frac{\omega}{2\pi} = \frac{3\pi/4}{2\pi} = \frac{3}{8} \text{ cycle per unit of time.}$$

(c) When $t = 4$, the position is given by

$$d = 6 \cos\left[\frac{3\pi}{4}(4)\right] = 6 \cos 3\pi = 6(-1) = -6.$$

(d) To find the least positive value of t for which $d = 0$, we solve the equation

$$d = 6 \cos \frac{3\pi}{4}t = 0$$

to obtain

$$\frac{3\pi}{4}t = \frac{\pi}{2}, \frac{3\pi}{2}, \frac{5\pi}{2}, \ldots$$

$$t = \frac{2}{3}, 2, \frac{10}{3}, \ldots.$$

Thus, the least positive value of t is $t = 2/3$.

Many other physical phenomena can be characterized by wave motion. These include electromagnetic waves such as radio waves, television waves, and microwaves. Radio waves transmit sound in two different ways. For an AM station, the *amplitude* of the wave is modified to carry sound (AM stands for **amplitude modulation**). See Figure 7.84(a). An FM radio signal has its *frequency* modified in order to carry sound, hence the term **frequency modulation**. See Figure 7.84(b).

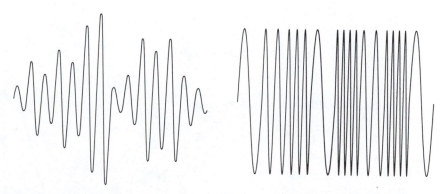

(a) AM: Amplitude Modulation   (b) FM: Frequency Modulation

Radio Waves

FIGURE 7.84

WARM UP

Evaluate the expressions. (Round your answers to two decimal places.)

1. $20 \sin 25°$

2. $42 \tan 62°$

3. $\arcsin 0.8723$

4. $\arctan 2.8703$

Solve for x. [Round your answers to two decimal places.]

5. $\cos 22° = \dfrac{x + 13 \sin 22°}{13 \sin 54°}$

6. $\tan 36° = \dfrac{x + 85 \tan 18°}{85}$

Find the amplitude and period of each function.

7. $f(x) = -4 \sin 2x$

8. $f(x) = \frac{1}{2} \sin \pi x$

9. $f(x) = 3 \cos 3\pi x$

10. $f(x) = 0.2 \cos \dfrac{x}{4}$

EXERCISES 7.8

In Exercises 1–10, solve the right triangle shown in the figure. (Round your answers to two decimal places.)

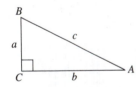

FIGURE FOR 1–10

1. $A = 20°, \quad b = 10$
2. $B = 54°, \quad c = 15$
3. $B = 71°, \quad b = 24$
4. $A = 8.4°, \quad a = 40.5$
5. $A = 12° \, 15', \quad c = 430.5$
6. $B = 65° \, 12', \quad a = 14.2$
7. $a = 6, \quad b = 10$
8. $a = 25, \quad c = 35$
9. $b = 16, \quad c = 52$
10. $b = 1.32, \quad c = 9.45$

11. A ladder of length 16 feet leans against the side of a house (see figure). Find the height h of the top of the ladder if the angle of elevation of the ladder is 74°.

12. The length of the shadow of a tree is 125 feet when the angle of elevation of the sun is 33° (see figure). Approximate the height h of the tree.

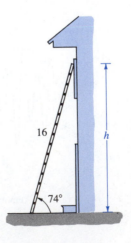

FIGURE FOR 11

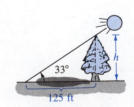

FIGURE FOR 12

13. An isosceles triangle has two angles of 52° (see figure). The base of the triangle is 4 inches. Find the altitude of the triangle.

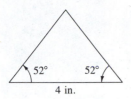

FIGURE FOR 13

14. The height of the top of an outdoor basketball backboard is $12\frac{1}{2}$ feet, and the backboard casts a shadow $17\frac{1}{3}$ feet long (see figure). Find the angle of elevation of the sun.

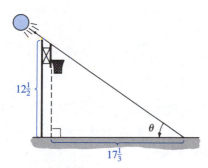

FIGURE FOR 14

15. An amateur radio operator erects a 75-foot vertical tower for his antenna. Find the angle of elevation to the top of the tower at a point on level ground 50 feet from the base.

16. Find the angle of depression from the top of a lighthouse 250 feet above water level to the water line of a ship 2 miles offshore.

17. A spacecraft is traveling in a circular orbit 100 miles above the surface of the earth (see figure). Find the angle of depression from the spacecraft to the horizon. Assume that the radius of the earth is 4000 miles.

FIGURE FOR 17

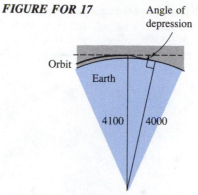

18. A sign on the roadway at the top of a mountain indicates that for the next 4 miles the grade is 12.5° (see figure). Find the change in elevation for a car descending the mountain.

FIGURE FOR 18

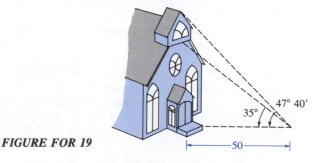

19. From a point 50 feet in front of a church, the angles of elevation to the base of the steeple and the top of the steeple are 35° and 47° 40′, respectively (see figure). Find the height of the steeple.

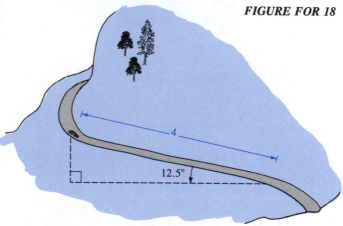

FIGURE FOR 19

20. From a point 100 feet in front of a public library, the angles of elevation to the base of a flagpole and to the top of the pole are 28° and 39° 45′, respectively. The flagpole is mounted on the front of the library's roof (see figure). Find the height of the pole.

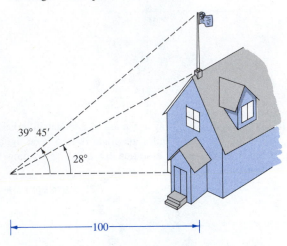

39° 45′

28°

|←————————100————————→|

FIGURE FOR 20

21. An airplane flying at 550 miles per hour has a bearing of N 52° E. After flying 1.5 hours, how far north and how far east has the plane traveled from its point of departure?

22. A ship leaves port at noon and has a bearing of S 27° W. If the ship is sailing at 20 knots, how many nautical miles south and how many nautical miles west has the ship traveled by 6 P.M.?

23. A ship is 45 miles east and 30 miles south of port. If the captain wants to travel directly to port, what bearing should be taken?

24. A plane is 120 miles north and 85 miles east of an airport. If the pilot wants to fly directly to the airport, what bearing should be taken?

25. A surveyor wishes to find the distance across a swamp (see figure). The bearing from *A* to *B* is N 32° W. The surveyor walks 50 yards from *A*, and at the point *C* the bearing to *B* is N 68° W. (a) Find the bearing from *A* to *C*. (b) Find the distance from *A* to *B*.

26. Two fire towers are 20 miles apart, tower *A* being due north of tower *B*. A fire is spotted from the towers, and its bearings from *A* and *B* are S 14° E and N 34° E (see figure). Find the distance *d* of the fire from the line segment *AB*. [*Hint:* Use the fact that $d = 20/(\cot 14° + \cot 34°)$.]

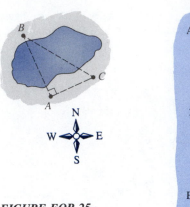

FIGURE FOR 25

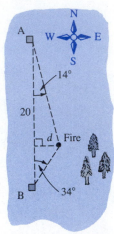

14°

20

d \ Fire

34°

FIGURE FOR 26

27. An observer in a lighthouse 300 feet above sea level spots two ships directly offshore. The angles of depression to the ships are 4° and 6.5° (see figure). How far apart are the ships?

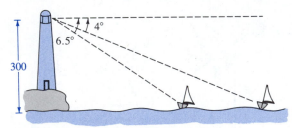

4°

6.5°

300

(Note: Angles are not drawn to scale.)

FIGURE FOR 27

28. A passenger in an airplane flying at 30,000 feet sees two towns directly to the left of the plane. The angles of depression to the towns are 28° and 55° (see figure). How far apart are the towns?

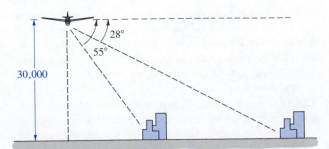

28°

55°

30,000

FIGURE FOR 28

29. A plane is observed approaching your home, and you assume that it is traveling at approximately 550 miles per hour. If the angle of elevation of the plane is 16° at one time and one minute later the angle is 57°, approximate the altitude.

30. In traveling across flat land, you notice a mountain directly in front of you. Its angle of elevation (to the peak) is 3.5°. After you drive 13 miles closer to the mountain, the angle of elevation is 9°. Approximate the height of the mountain.

31. A regular pentagon is inscribed in a circle of radius 25 inches. Find the length of the sides of the pentagon.

32. Repeat Exercise 31 using a hexagon.

33. Use the accompanying figure to find the distance y across the flat sides of the hexagonal nut as a function of r.

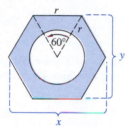

FIGURE FOR 33

34. The accompanying figure shows a circular sheet of diameter 25 cm, containing 12 equally spaced bolt holes. Determine the straight-line distance between the centers of the bolt heads.

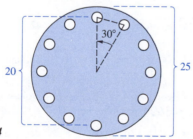

FIGURE FOR 34

In Exercises 35 and 36, find the lengths of all the unknown members of the given truss.

35.

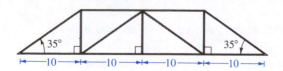

36.

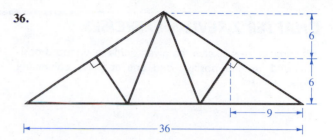

In Exercises 37–40, for the simple harmonic motion described by the given trigonometric function, find (a) the maximum displacement, (b) the frequency, and (c) the least positive value of t for which $d = 0$.

37. $d = 4 \cos 8\pi t$ **38.** $d = \frac{1}{2} \cos 20\pi t$

39. $d = \frac{1}{16} \sin 120\pi t$ **40.** $d = \frac{1}{64} \sin 792\pi t$

41. A point on the end of a tuning fork moves in simple harmonic motion described by $d = a \sin \omega t$. Find ω given that the tuning fork for middle C has a frequency of 264 vibrations per second.

42. A buoy oscillates in simple harmonic motion as waves move past. At a given time it is noted that the buoy moves a total of 3.5 feet from its low point to its high point, and that it returns to the high point every 10 seconds. Write an equation describing the motion of the buoy if, at $t = 0$, it is at its high point.

CHAPTER 7 REVIEW EXERCISES

In Exercises 1–4, sketch the given angle in standard position, and list one positive and one negative coterminal angle.

1. $\dfrac{11\pi}{4}$ **2.** $-405°$

3. $-110°$ **4.** $\dfrac{2\pi}{9}$

In Exercises 5–8, convert the angle measurement to decimal form. Round each answer to two decimal places.

5. $135°\ 16'\ 45''$ **6.** $-234°\ 50''$
7. $5°\ 22'\ 53''$ **8.** $280°\ 8'\ 50''$

In Exercises 9–12, convert the angle measurement to D° M′ S″ form.

9. $135.27°$ **10.** $25.1°$
11. $-85.15°$ **12.** $-327.85°$

In Exercises 13–16, convert the angle measurement from radians to degrees. Round each answer to two decimal places.

13. $\dfrac{5\pi}{7}$ **14.** $-\dfrac{3\pi}{5}$
15. -3.5 **16.** 1.75

In Exercises 17–20, convert the angle measurement from degrees to radians. Round each answer to four decimal places.

17. $480°$ **18.** $-16.5°$
19. $-33°\ 45'$ **20.** $84°\ 15'$

In Exercises 21–24, find the reference angle for the given angle.

21. $-\dfrac{6\pi}{5}$ **22.** $640°$
23. $252°$ **24.** $\dfrac{17\pi}{3}$

In Exercises 25–28, find the six trigonometric functions of the angle θ (in standard position) whose terminal side passes through the given point.

25. $(-7, 2)$ **26.** $(4, -8)$
27. $(-4, -6)$ **28.** $\left(\frac{2}{3}, \frac{5}{2}\right)$

In Exercises 29–32, use a right triangle to find the remaining five trigonometric functions of θ.

29. $\sec \theta = \frac{6}{5}$, $\tan \theta < 0$
30. $\tan \theta = -\frac{12}{5}$, $\sin \theta > 0$
31. $\sin \theta = \frac{3}{8}$, $\cos \theta < 0$
32. $\cos \theta = -\frac{2}{5}$, $\sin \theta > 0$

In Exercises 33–36, evaluate the given trigonometric functions without the use of a calculator.

33. $\sin \dfrac{5\pi}{3}$ **34.** $\cot\left(-\dfrac{5\pi}{6}\right)$

35. $\cos 495°$ **36.** $\csc \dfrac{3\pi}{2}$

In Exercises 37–40, use a calculator to evaluate the given trigonometric functions. Round each answer to two decimal places.

37. $\tan 33°$ **38.** $\csc 105°$
39. $\sec \dfrac{12\pi}{5}$ **40.** $\sin\left(-\dfrac{\pi}{9}\right)$

In Exercises 41–44, find two values of θ in degrees ($0° \le \theta < 360°$) and in radians ($0 \le \theta < 2\pi$) without using a calculator.

41. $\cos \theta = -\dfrac{\sqrt{2}}{2}$ **42.** $\sec \theta$ is undefined

43. $\csc \theta = -2$ **44.** $\tan \theta = \dfrac{\sqrt{3}}{3}$

In Exercises 45–48, find two values of θ in degrees ($0° \le \theta < 360°$) and in radians ($0 \le \theta < 2\pi$) by using a calculator.

45. $\sin \theta = 0.8387$ **46.** $\cot \theta = -1.5399$
47. $\sec \theta = -1.0353$ **48.** $\csc \theta = 11.4737$

In Exercises 49–52, use a right triangle to write an algebraic expression for the given expression.

49. $\sec[\arcsin(x - 1)]$ **50.** $\tan\left(\arccos \dfrac{x}{2}\right)$

51. $\sin\left(\arccos \dfrac{x^2}{4 - x^2}\right)$ **52.** $\csc(\arcsin 10x)$

In Exercises 53–70, sketch the graph of the given function.

53. $f(x) = 5 \sin \dfrac{2x}{5}$

54. $f(x) = 8 \cos\left(-\dfrac{x}{4}\right)$

55. $f(x) = -\dfrac{1}{4} \cos \dfrac{\pi x}{4}$

56. $f(x) = -\tan \dfrac{\pi x}{4}$

57. $g(t) = \dfrac{5}{2} \sin(\pi - \pi t)$

58. $g(t) = \sec\left(t - \dfrac{\pi}{4}\right)$

59. $h(t) = \csc\left(3t - \dfrac{\pi}{2}\right)$

60. $h(t) = 3 \csc\left(2t + \dfrac{\pi}{4}\right)$

61. $f(\theta) = \cot \dfrac{\pi \theta}{8}$

62. $E(t) = 110 \cos\left(120\pi t - \dfrac{\pi}{3}\right)$

63. $f(x) = \dfrac{x}{4} - \sin x$

64. $g(x) = 3\left(\sin \dfrac{\pi x}{3} + 1\right)$

65. $h(\theta) = \theta \sin \pi\theta$

66. $f(t) = 2.5e^{-t/4} \sin 2\pi t$

67. $f(x) = \arcsin \dfrac{x}{2}$

68. $f(x) = \arccos(x - \pi)$

69. $f(x) = \dfrac{\pi}{2} + \arctan x$

70. $f(x) = 2 \arccos x$

71. An observer 2.5 miles from the launch pad of a space shuttle measures the angle of elevation to the base of the vehicle to be 28° soon after liftoff (see figure). How high is the shuttle at that instant? Assume that the shuttle is still moving vertically.

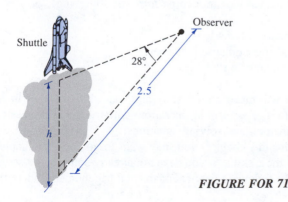

FIGURE FOR 71

72. From city A to city B, a plane flies 650 miles at a bearing that is N 48° E. From city B to city C the plane flies 810 miles at a bearing of S 65° E. Find the distance from A to C and the bearing from A to C.

73. A train travels 2.5 miles (horizontally) on a straight track with a grade of 1° 10' (see figure). What is the vertical rise of the train in that distance?

FIGURE FOR 73

74. In calculus it is shown that the sine and cosine functions can be represented by the infinite sums shown below, where x is in radians. Use the first four terms of the sum to approximate the given functional values and compare the result with that given by a calculator.

$$\sin x = x - \dfrac{x^3}{3!} + \dfrac{x^5}{5!} - \dfrac{x^7}{7!} + \cdots + \dfrac{(-1)^{n-1}x^{2n-1}}{(2n-1)!} + \cdots$$

$$\cos x = 1 - \dfrac{x^2}{2!} + \dfrac{x^4}{4!} - \dfrac{x^6}{6!} + \cdots + \dfrac{(-1)^{n-1}x^{2n-2}}{(2n-2)!} + \cdots$$

(a) $\sin 1$

(b) $\cos 1$

(c) $\sin \dfrac{1}{2}$

(d) $\cos(-1)$

(e) $\cos \dfrac{\pi}{4}$

(f) $\sin \dfrac{\pi}{6}$

C H A P T E R 8

Analytic Trigonometry

8.1 Applications of Fundamental Identities

In Chapter 7, we studied the basic definitions, properties, graphs, and applications of the individual trigonometric functions. In this chapter, we will study algebraic combinations of these functions. In particular, we will show how to use the fundamental identities to perform the following.

1. Simplify trigonometric expressions.
2. Develop additional trigonometric identities.
3. Solve trigonometric equations.

In our study, we will make use of many algebraic skills such as finding special products, factoring, performing operations with fractional expressions, rationalizing denominators, and solving equations.

As preparation for this chapter, you may want to go back to Chapter 1 for a quick review of these skills. If you have not already done so, we suggest that you memorize the following list of fundamental trigonometric identities.

Fundamental Identities

Reciprocal Identities

$$\sin x = \frac{1}{\csc x} \qquad \sec x = \frac{1}{\cos x} \qquad \tan x = \frac{1}{\cot x}$$

$$\csc x = \frac{1}{\sin x} \qquad \cos x = \frac{1}{\sec x} \qquad \cot x = \frac{1}{\tan x}$$

Tangent and Cotangent Identities

$$\tan x = \frac{\sin x}{\cos x} \qquad \cot x = \frac{\cos x}{\sin x}$$

Pythagorean Identities

$$\sin^2 x + \cos^2 x = 1 \qquad 1 + \tan^2 x = \sec^2 x \qquad 1 + \cot^2 x = \csc^2 x$$

Cofunction Identities

$$\sin\left(\frac{\pi}{2} - x\right) = \cos x \qquad \sec\left(\frac{\pi}{2} - x\right) = \csc x \qquad \tan\left(\frac{\pi}{2} - x\right) = \cot x$$

$$\csc\left(\frac{\pi}{2} - x\right) = \sec x \qquad \cos\left(\frac{\pi}{2} - x\right) = \sin x \qquad \cot\left(\frac{\pi}{2} - x\right) = \tan x$$

Negative Angle Identities

$$\sin(-x) = -\sin x \qquad \sec(-x) = \sec x \qquad \tan(-x) = -\tan x$$
$$\csc(-x) = -\csc x \qquad \cos(-x) = \cos x \qquad \cot(-x) = -\cot x$$

Remark: Pythagorean identities are sometimes used in radical form such as

$$\sin x = \pm\sqrt{1 - \cos^2 x} \qquad \text{or} \qquad \tan x = \pm\sqrt{\sec^2 x - 1}$$

where the sign depends on the choice of x.

EXAMPLE 1 *Using Identities to Evaluate a Function*

Use the fundamental identities to find $\csc x$, given that

$$\tan x = -6 \qquad \text{and} \qquad \sin x > 0.$$

SOLUTION

Since $\tan x < 0$ and $\sin x > 0$, it follows that x lies in Quadrant II. Using a reciprocal identity, we have

$$\cot x = \frac{1}{\tan x} = \frac{1}{-6}.$$

Analytic Trigonometry

From a Pythagorean identity, we obtain

$$\csc^2 x = 1 + \cot^2 x = 1 + \left(\frac{-1}{6}\right)^2 = \frac{37}{36}.$$

Finally, because $\csc x$ is positive when x is in Quadrant II, we choose the positive root and conclude that

$$\csc x = \frac{\sqrt{37}}{6}.$$

EXAMPLE 2 *Using Identities to Evaluate a Function*

Use the fundamental identities to find $\sin x$ and $\tan x$, given that

$$\sec x = -\frac{3}{2} \quad \text{and} \quad \tan x > 0.$$

SOLUTION

Because $\sec x < 0$ and $\tan x > 0$, it follows that x lies in Quadrant III. Moreover, by a reciprocal identity, we have

$$\cos x = \frac{1}{\sec x} = \frac{1}{-3/2} = -\frac{2}{3}.$$

Thus, by a Pythagorean identity, we obtain

$$\sin^2 x = 1 - \cos^2 x = 1 - \left(-\frac{2}{3}\right)^2 = 1 - \frac{4}{9} = \frac{5}{9}.$$

Because $\sin x$ is negative when x is in Quadrant III, we choose the negative root and obtain

$$\sin x = -\frac{\sqrt{5}}{3}.$$

Furthermore,

$$\tan x = \frac{\sin x}{\cos x} = \frac{-\sqrt{5}/3}{-2/3} = \frac{\sqrt{5}}{2}.$$

In the next five examples, we use algebraic techniques and the fundamental identities to factor and/or simplify trigonometric *expressions* such as

$$\cot x - \cos x \sin x, \quad \frac{1 - \sin x}{\cos^2 x}, \quad \text{and} \quad \frac{\tan x}{\sec x - 1}.$$

Applications of Fundamental Identities

EXAMPLE 3 *Simplifying a Trigonometric Expression*

Simplify the expression $\sin x \cos^2 x - \sin x$.

SOLUTION

In this case, we first factor out a common monomial factor and then use a fundamental identity.

$$
\begin{aligned}
\sin x \cos^2 x - \sin x &= \sin x(\cos^2 x - 1) \qquad && \textit{Monomial factor}\\
&= -\sin x(1 - \cos^2 x) \\
&= -\sin x(\sin^2 x) \qquad && \textit{Identity}\\
&= -\sin^3 x
\end{aligned}
$$

EXAMPLE 4 *Factoring Trigonometric Expressions*

Factor the following expressions.

(a) $\sec^2 \theta - 1$

(b) $4 \tan^2 \theta + \tan \theta - 3$

SOLUTION

(a) Here we have the difference of two squares, which factors as

$$\sec^2 \theta - 1 = (\sec \theta - 1)(\sec \theta + 1).$$

(b) This expression has the polynomial form, $ax^2 + bx + c$, and it factors as

$$4 \tan^2 \theta + \tan \theta - 3 = (4 \tan \theta - 3)(\tan \theta + 1).$$

On occasion, factoring or simplifying can best be done by first rewriting the expression in terms of just *one* trigonometric function or in terms of *sine and cosine alone*. This is demonstrated in the next three examples.

EXAMPLE 5 *Factoring a Trigonometric Expression*

Factor the expression $\csc^2 x - \cot x - 3$.

SOLUTION

As given, this expression cannot be factored, so we use the identity $\csc^2 x = 1 + \cot^2 x$ to rewrite the expression in terms of the cotangent alone. We then factor to obtain

$$\begin{aligned}
\csc^2 x - \cot x - 3 &= (1 + \cot^2 x) - \cot x - 3 && \textit{Identity}\\
&= \cot^2 x - \cot x - 2 && \textit{Combine terms}\\
&= (\cot x - 2)(\cot x + 1). && \textit{Factor} \quad\blacksquare
\end{aligned}$$

EXAMPLE 6 *Simplifying a Trigonometric Expression*

Simplify the expression $\sin t + \cot t \cos t$.

SOLUTION

Since this expression is not factorable as given, we convert all terms to sines and cosines to see what can be done.

$$\begin{aligned}
\sin t + \cot t \cos t &= \sin t + \left(\frac{\cos t}{\sin t}\right) \cos t \\
&= \frac{\sin^2 t + \cos^2 t}{\sin t} && \textit{Add fractions}\\
&= \frac{1}{\sin t} && \textit{Identity}\\
&= \csc t. && \textit{Identity} \quad\blacksquare
\end{aligned}$$

EXAMPLE 7 *Simplifying a Trigonometric Expression*

Simplify the function $f(x) = (\sin x + \cos x - 1)(\sin x + \cos x + 1)$.

SOLUTION

$$\begin{aligned}
f(x) &= (\sin x + \cos x - 1)(\sin x + \cos x + 1) \\
&= [(\sin x + \cos x) - 1][(\sin x + \cos x) + 1] \\
&= (\sin x + \cos x)^2 - 1 = \sin^2 x + 2 \sin x \cos x + \cos^2 x - 1 \\
&= 2 \sin x \cos x + 1 - 1 = 2 \sin x \cos x
\end{aligned}$$

Applications of Fundamental Identities

EXAMPLE 8 *Combining Fractional Expressions*

Perform the indicated addition and simplify the result.

$$\frac{\sin \theta}{1 + \cos \theta} + \frac{\cos \theta}{\sin \theta}$$

SOLUTION

Using the definition of addition of fractions

$$\frac{a}{b} + \frac{c}{d} = \frac{ad + bc}{bd}$$

we can write

$$\frac{\sin \theta}{1 + \cos \theta} + \frac{\cos \theta}{\sin \theta} = \frac{\sin \theta(\sin \theta) + \cos \theta(1 + \cos \theta)}{(1 + \cos \theta)\sin \theta}$$

$$= \frac{\sin^2 \theta + \cos^2 \theta + \cos \theta}{(1 + \cos \theta)\sin \theta}$$

$$= \frac{\cancel{1 + \cos \theta}}{\cancel{(1 + \cos \theta)}\sin \theta} \qquad \textit{Identity}$$

$$= \frac{1}{\sin \theta} \qquad \textit{Reduce}$$

$$= \csc \theta. \qquad \textit{Identity}$$

The last two examples of this section involve techniques for rewriting expressions into forms that are useful in calculus.

EXAMPLE 9 *Rewriting a Trigonometric Expression*

Rewrite the following expression so that it is *not* in fractional form.

$$\frac{1}{1 + \sin x}$$

SOLUTION

From the Pythagorean identity

$$\cos^2 x = 1 - \sin^2 x = (1 - \sin x)(1 + \sin x)$$

we can see that by multiplying both the numerator and the denominator by $(1 - \sin x)$ we will obtain a monomial denominator. That is,

$$\frac{1}{1 + \sin x} = \frac{1}{1 + \sin x} \cdot \frac{1 - \sin x}{1 - \sin x}$$

$$= \frac{1 - \sin x}{1 - \sin^2 x} \qquad\qquad \textit{Multiply}$$

$$= \frac{1 - \sin x}{\cos^2 x} \qquad\qquad \textit{Identity}$$

$$= \frac{1}{\cos^2 x} - \frac{\sin x}{\cos^2 x}$$

$$= \frac{1}{\cos^2 x} - \frac{\sin x}{\cos x} \cdot \frac{1}{\cos x}$$

$$= \sec^2 x - \tan x \sec x. \qquad\qquad \textit{Identity}$$

EXAMPLE 10 *Trigonometric Substitution*

Use the substitution $x = 2 \tan \theta$, $0 < \theta < \pi/2$ to express $\sqrt{4 + x^2}$ as a trigonometric function of θ.

SOLUTION

Letting $x = 2 \tan \theta$, we have

$$\sqrt{4 + x^2} = \sqrt{4 + (2 \tan \theta)^2}$$

$$= \sqrt{4(1 + \tan^2 \theta)}$$

$$= \sqrt{4 \sec^2 \theta} \qquad\qquad \textit{Pythagorean identity}$$

$$= 2 \sec \theta. \qquad\qquad \textit{sec } \theta > 0 \textit{ for } 0 < \theta < \frac{\pi}{2}$$

Figure 8.1 shows the right triangle illustration of the trigonometric substitution in Example 10. For $0 < \theta < \pi/2$, we have opp = x, adj = 2, and hyp = $\sqrt{4 + x^2}$. Thus, we can write

$$\sec \theta = \frac{\sqrt{4 + x^2}}{2} \implies \sqrt{4 + x^2} = 2 \sec \theta.$$

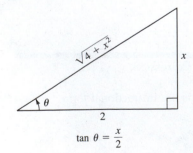

$$\tan \theta = \frac{x}{2}$$

FIGURE 8.1

Exercises

WARM UP

Use a right triangle to find the other five trigonometric functions of the acute angle θ.

1. $\tan \theta = \frac{3}{2}$

2. $\sec \theta = 3$

Determine the exact value of all six trigonometric functions of θ. Assume that the given point is on the terminal side of an angle θ, in standard position.

3. $(7, -3)$

4. $(-10, 5)$

Simplify the expressions.

5. $\sqrt{1 - \left(\dfrac{\sqrt{3}}{2}\right)^2}$

6. $\sqrt{\left(\dfrac{3}{4}\right)^2 + 1}$

7. $\sqrt{1 + \left(\dfrac{3}{8}\right)^2}$

8. $\sqrt{1 - \left(\dfrac{\sqrt{5}}{3}\right)^2}$

Perform the indicated operations and simplify.

9. $\dfrac{4}{1 + x} + \dfrac{x}{4}$

10. $\dfrac{3}{1 - x} - \dfrac{5}{1 + x}$

EXERCISES 8.1

In Exercises 1–14, use the fundamental identities to find the values of the other four trigonometric functions.

1. $\sin x = \dfrac{1}{2}, \quad \cos x = \dfrac{\sqrt{3}}{2}$

2. $\tan x = \dfrac{\sqrt{3}}{3}, \quad \cos x = -\dfrac{\sqrt{3}}{2}$

3. $\sec \theta = \sqrt{2}, \quad \sin \theta = -\dfrac{\sqrt{2}}{2}$

4. $\csc \theta = \dfrac{5}{3}, \quad \tan \theta = \dfrac{3}{4}$

5. $\tan x = \dfrac{5}{12}, \quad \sec x = -\dfrac{13}{12}$

6. $\cot \phi = -3, \quad \sin \phi = \dfrac{\sqrt{10}}{10}$

7. $\sec \phi = -1, \quad \sin \phi = 0$

8. $\cos\left(\dfrac{\pi}{2} - x\right) = \dfrac{3}{5}, \quad \cos x = \dfrac{4}{5}$

9. $\sin(-x) = -\dfrac{2}{3}, \quad \tan x = -\dfrac{2\sqrt{5}}{5}$

10. $\csc x = 5, \quad \cos x > 0$

11. $\tan \theta = 2, \quad \sin \theta < 0$

12. $\sec \theta = -3, \quad \tan \theta < 0$

13. $\sin \theta = -1, \quad \cot \theta = 0$

14. $\tan \theta$ is undefined, $\quad \sin \theta > 0$

In Exercises 15–20, match the trigonometric expression with one of the following.

(a) -1 \qquad (b) $\cos x$ \qquad (c) $\cot x$

(d) 1 \qquad (e) $-\tan x$ \qquad (f) $\sin x$

15. $\sec x \cos x$

16. $\dfrac{\sin(-x)}{\cos(-x)}$

17. $\tan^2 x - \sec^2 x$

18. $\dfrac{1 - \cos^2 x}{\sin x}$

19. $\cot x \sin x$

20. $\dfrac{\sin[(\pi/2) - x]}{\cos[(\pi/2) - x]}$

Analytic Trigonometry

In Exercises 21–26, match the trigonometric expression with one of the following.

(a) csc x   (b) tan x   (c) $\sin^2 x$

(d) sin x tan x   (e) $\sec^2 x$ (f) $\sec^2 x + \tan^2 x$

21. sin x sec x

22. $\cos^2 x(\sec^2 x - 1)$

23. $\dfrac{\sec^2 x - 1}{\sin^2 x}$

24. cot x sec x

25. $\sec^4 x - \tan^4 x$

26. $\dfrac{\cos^2[(\pi/2) - x]}{\cos x}$

In Exercises 27–40, use the fundamental identities to simplify the given expression.

27. $\tan \phi \csc \phi$

28. $\sin \phi(\csc \phi - \sin \phi)$

29. $\cos \beta \tan \beta$

30. $\sec \alpha \dfrac{\sin \alpha}{\tan \alpha}$

31. $\dfrac{\cot x}{\csc x}$

32. $\dfrac{\csc \theta}{\sec \theta}$

33. $\sec^2 x(1 - \sin^2 x)$

34. $\dfrac{1}{\tan^2 x + 1}$

35. $\dfrac{\sin(-x)}{\cos x}$

36. $\dfrac{\tan^2 \theta}{\sec^2 \theta}$

37. $\cos\left(\dfrac{\pi}{2} - x\right)\sec x$

38. $\cot\left(\dfrac{\pi}{2} - x\right)\cos x$

39. $\dfrac{\cos^2 y}{1 - \sin y}$

40. $\cos t(1 + \tan^2 t)$

In Exercises 41–48, factor each expression and use the fundamental identities to simplify the result.

41. $\tan^2 x - \tan^2 x \sin^2 x$

42. $\sec^2 x \tan^2 x + \sec^2 x$

43. $\sin^2 x \sec^2 x - \sin^2 x$

44. $\dfrac{\sec^2 x - 1}{\sec x - 1}$

45. $\tan^4 x + 2 \tan^2 x + 1$

46. $1 - 2 \cos^2 x + \cos^4 x$

47. $\sin^4 x - \cos^4 x$

48. $\csc^3 x - \csc^2 x - \csc x + 1$

In Exercises 49–52, perform the multiplication and use the fundamental identities to simplify the result.

49. $(\sin x + \cos x)^2$

50. $(\cot x + \csc x)(\cot x - \csc x)$

51. $(\sec x + 1)(\sec x - 1)$

52. $(3 - 3 \sin x)(3 + 3 \sin x)$

In Exercises 53–56, perform the addition and use the fundamental identities to simplify the result.

53. $\dfrac{1}{1 + \cos x} + \dfrac{1}{1 - \cos x}$

54. $\dfrac{1}{\sec x + 1} - \dfrac{1}{\sec x - 1}$

55. $\dfrac{\cos x}{1 + \sin x} + \dfrac{1 + \sin x}{\cos x}$

56. $\tan x - \dfrac{\sec^2 x}{\tan x}$

In Exercises 57–60, rewrite the given expression so that it is *not* in fractional form.

57. $\dfrac{\sin^2 y}{1 - \cos y}$

58. $\dfrac{5}{\tan x + \sec x}$

59. $\dfrac{3}{\sec x - \tan x}$

60. $\dfrac{\tan^2 x}{\csc x + 1}$

In Exercises 61–70, use the given trigonometric substitution to write the algebraic expression as a trigonometric expression involving θ, where $0 < \theta < \pi/2$.

61. $\sqrt{25 - x^2}, \quad x = 5 \sin \theta$

62. $\sqrt{16 - 4x^2}, \quad x = 2 \sin \theta$

63. $\sqrt{x^2 - 9}, \quad x = 3 \sec \theta$

64. $\sqrt{x^2 - 4}, \quad x = 2 \sec \theta$

65. $\sqrt{x^2 + 25}, \quad x = 5 \tan \theta$

66. $\sqrt{x^2 + 100}, \quad x = 10 \tan \theta$

67. $\sqrt{1 - (x - 1)^2}, \quad x - 1 = \sin \theta$

68. $\sqrt{1 - e^{2x}}, \quad e^x = \sin \theta$

69. $\sqrt{(9 + x^2)^3}, \quad x = 3 \tan \theta$

70. $\sqrt{(x^2 - 16)^3}, \quad x = 4 \sec \theta$

In Exercises 71 and 72, determine the values of θ, $0 \le \theta < 2\pi$, for which the equation is true.

71. $\sec \theta = \sqrt{1 + \tan^2 \theta}$ **72.** $\cos \theta = -\sqrt{1 - \sin^2 \theta}$

In Exercises 73 and 74, rewrite the given expression as a single logarithm and simplify.

73. $\ln|\cos \theta| - \ln|\sin \theta|$ **74.** $\ln|\cot t| + \ln(1 + \tan^2 t)$

In Exercises 75–78, determine whether the statement is true or false, and give a reason for your answer.

75. $\dfrac{\sin k\theta}{\cos k\theta} = \tan \theta, \quad k$ is constant

76. $5 \sec \theta = \dfrac{1}{5 \cos \theta}$

77. $\sin \theta \csc \theta = 1$ **78.** $\sin \theta \csc \phi = 1$

In Exercises 79–82, use a calculator to demonstrate that the identity is true for the given values of θ.

79. $\csc^2 \theta - \cot^2 \theta = 1$

(a) $\theta = 132°$ (b) $\theta = \dfrac{2\pi}{7}$

80. $\tan^2 \theta + 1 = \sec^2 \theta$

(a) $\theta = 346°$ (b) $\theta = 3.1$

81. $\cos\left(\dfrac{\pi}{2} - \theta\right) = \sin \theta$

(a) $\theta = 80°$ (b) $\theta = 0.8$

82. $\sin(-\theta) = -\sin \theta$

(a) $\theta = 250°$ (b) $\theta = \frac{1}{2}$

83. Express each of the other trigonometric functions of θ in terms of $\sin \theta$.

84. Express each of the other trigonometric functions of θ in terms of $\cos \theta$.

8.2 Verifying Trigonometric Identities

In the previous section, we showed how to rewrite trigonometric expressions in equivalent forms. In this section, we will demonstrate methods for proving (or verifying) trigonometric identities. And in Section 8.3, we will show how to solve trigonometric equations. The key to verifying identities and solving equations is the ability to use the fundamental identities and the rules of algebra to rewrite trigonometric expressions.

Before going on, let's review some distinctions among trigonometric expressions, equations, and identities. An *expression* has no equal sign. It is merely a combination of functions. When simplifying expressions, we use an equal sign only to indicate the equivalence of the original expression and the new form. An *equation* is a statement containing an equal sign that is true for a specific set of values. In this sense, it is really a *conditional* equation. For example, the equation

$$\sin x = -1$$

is true only for $x = (3\pi/2) \pm 2n\pi$. Hence, it is a conditional equation. On the other hand, an equation that is true for all real values in the domain of the variable is called an *identity*. For example, the familiar equation

$$\sin^2 x = 1 - \cos^2 x$$

is true for all real x; hence, it is an identity.

Though there are similarities, proving that a trigonometric equation is an identity is quite different from solving an equation. There is no well-defined set of rules to follow in verifying trigonometric identities, and the process is best learned by practice. However, the following guidelines should be helpful.

Guidelines for Verifying Trigonometric Identities

1. Work with one side of the equation at a time. It is often better to work with the more complicated side.
2. Look for opportunities to factor an expression, add fractions, square a binomial, or create a monomial denominator.
3. Look for opportunities to use the fundamental identities. Note which functions are in the final expression you want. Sines and cosines pair up well, as do secants and tangents, and cosecants and cotangents.
4. If the preceding guidelines do not help, try converting all terms to sines and cosines.
5. Do not just sit and stare at the problem. Try something! Even paths that lead to dead ends can give you insights.

Note how we use these guidelines in the examples in this section.

EXAMPLE 1 *Verifying a Trigonometric Identity*

Verify the identity

$$\frac{\sec^2 \theta - 1}{\sec^2 \theta} = \sin^2 \theta.$$

SOLUTION

Because the left side is more complicated, we will work with it.

$$\frac{\sec^2 \theta - 1}{\sec^2 \theta} = \frac{(\tan^2 \theta + 1) - 1}{\sec^2 \theta} \qquad \textit{Identity}$$

$$= \frac{\tan^2 \theta}{\sec^2 \theta} \qquad \textit{Simplify}$$

$$= \tan^2 \theta (\cos^2 \theta) \qquad \textit{Identity}$$

$$= \frac{\sin^2 \theta}{\cos^2 \theta} (\cos^2 \theta) \qquad \textit{Identity}$$

$$= \sin^2 \theta \qquad \textit{Simplify}$$

ALTERNATIVE SOLUTION

Sometimes it is helpful to separate a fraction into two parts. In this case, we have

$$\frac{\sec^2 \theta - 1}{\sec^2 \theta} = \frac{\sec^2 \theta}{\sec^2 \theta} - \frac{1}{\sec^2 \theta} \qquad \textit{Separate fractions}$$

$$= 1 - \cos^2 \theta \qquad \textit{Identity}$$

$$= \sin^2 \theta. \qquad \textit{Identity}$$

Remark: As you can see from Example 1, there can be more than one way to verify an identity. Your method may differ from that used by your instructor or fellow students. Here is a good chance to be creative and establish your own style, but try to be as efficient as possible.

EXAMPLE 2 Combining Fractions before Using Identities

Verify the identity

$$\frac{1}{1 - \sin \alpha} + \frac{1}{1 + \sin \alpha} = 2 \sec^2\alpha.$$

SOLUTION

Let's add the two fractions and see where we can go from there.

$$\frac{1}{1 - \sin \alpha} + \frac{1}{1 + \sin \alpha} = \frac{1 + \sin \alpha + 1 - \sin \alpha}{(1 - \sin \alpha)(1 + \sin \alpha)}$$

$$= \frac{2}{1 - \sin^2 \alpha} = \frac{2}{\cos^2 \alpha} = 2 \sec^2 \alpha$$

EXAMPLE 3 Verifying a Trigonometric Identity

Verify the identity

$$(\tan^2 x + 1)(\cos^2 x - 1) = -\tan^2 x.$$

SOLUTION

By multiplying the factors on the left side, we obtain

$$(\tan^2 x + 1)(\cos^2 x - 1) = \tan^2 x(\cos^2 x) - \tan^2 x + \cos^2 x - 1$$

$$= \frac{\sin^2 x}{\cos^2 x}(\cos^2 x) + \cos^2 x - \tan^2 x - 1$$

$$= (\sin^2 x + \cos^2 x) - \tan^2 x - 1$$

$$= 1 - \tan^2 x - 1 = -\tan^2 x.$$

ALTERNATIVE SOLUTION

By applying identities before multiplying, we obtain

$$(\tan^2 x + 1)(\cos^2 x - 1) = \sec^2 x(-\sin^2 x)$$

$$= -\frac{\sin^2 x}{\cos^2 x} = -\tan^2 x.$$

EXAMPLE 4 *Converting to Sines and Cosines*

Verify the identity

$$\tan x + \cot x = \sec x \csc x.$$

SOLUTION

In this case there appear to be no fractions to add, no products to find, and no opportunity to use one of the Pythagorean identities. Hence, we try converting the left side into sines and cosines to see what happens.

$$\tan x + \cot x = \frac{\sin x}{\cos x} + \frac{\cos x}{\sin x}$$

$$= \frac{\sin^2 x + \cos^2 x}{\cos x \sin x} \qquad \textit{Add fractions}$$

$$= \frac{1}{\cos x \sin x} \qquad \textit{sin}^2 x + \cos^2 x = 1$$

$$= \frac{1}{\cos x} \cdot \frac{1}{\sin x} \qquad \textit{Product of fractions}$$

$$= \sec x \csc x \qquad \textit{Reciprocal identities}$$

Recall from algebra that *rationalizing the denominator* is, on occasion, a powerful simplification technique. A related form of this technique works for simplifying trigonometric expressions as well.

EXAMPLE 5 *Verifying a Trigonometric Identity*

Verify the identity

$$\sec y + \tan y = \frac{\cos y}{1 - \sin y}.$$

SOLUTION

Let's work with the *right* side. Note that we can create a monomial denominator by multiplying the numerator and denominator by $(1 + \sin y)$.

$$\frac{\cos y}{1 - \sin y} = \frac{\cos y}{1 - \sin y}\left(\frac{1 + \sin y}{1 + \sin y}\right)$$

$$= \frac{\cos y + \cos y \sin y}{1 - \sin^2 y}$$

$$= \frac{\cos y + \cos y \sin y}{\cos^2 y} = \frac{\cos y}{\cos^2 y} + \frac{\cos y \sin y}{\cos^2 y}$$

$$= \frac{1}{\cos y} + \frac{\sin y}{\cos y} = \sec y + \tan y$$

So far in this section, we have been verifying trigonometric identities by working with one side of the equation and converting to the form given on the other side. On occasion it is practical to work with each side *separately*, to obtain one common form equivalent to both sides.

EXAMPLE 6 *Working with Each Side Separately*

Verify the identity

$$\frac{\cot^2 \theta}{1 + \csc \theta} = \frac{1 - \sin \theta}{\sin \theta}.$$

SOLUTION

Working with the left side, we have

$$\frac{\cot^2 \theta}{1 + \csc \theta} = \frac{\csc^2 \theta - 1}{1 + \csc \theta} \qquad\qquad \textit{cot}^2\ \theta = \textit{csc}^2\ \theta - 1$$

$$= \frac{(\csc \theta - 1)(\csc \theta + 1)}{1 + \csc \theta} \qquad\qquad \textit{Factor}$$

$$= \csc \theta - 1. \qquad\qquad \textit{Reduce}$$

Now, simplifying the right side, we have

$$\frac{1 - \sin \theta}{\sin \theta} = \frac{1}{\sin \theta} - \frac{\sin \theta}{\sin \theta} = \csc \theta - 1.$$

The identity is verified since both sides are equal to $\csc \theta - 1$.

Analytic Trigonometry

In the last example, we rewrite powers of trigonometric functions as more complicated sums of products of trigonometric functions. This is a common procedure used to simplify the operations of calculus.

EXAMPLE 7 An Example from Calculus

Verify the following identities.

(a) $\tan^4 x = \tan^2 x \sec^2 x - \tan^2 x$

(b) $\sin^3 x \cos^4 x = (\cos^4 x - \cos^6 x)\sin x$

SOLUTION

Note the use of the Pythagorean identities in the verifications.

(a) $\tan^4 x = (\tan^2 x)(\tan^2 x)$

$\qquad = \tan^2 x(\sec^2 x - 1)$

$\qquad = \tan^2 x \sec^2 x - \tan^2 x$

(b) $\sin^3 x \cos^4 x = \sin^2 x \cos^4 x \sin x$

$\qquad = (1 - \cos^2 x)\cos^4 x \sin x$

$\qquad = (\cos^4 x - \cos^6 x)\sin x$

WARM UP

Factor the expressions and, if possible, simplify the results.

1. (a) $x^2 - x^2 y^2$
   (b) $\sin^2 x - \sin^2 x \cos^2 x$

2. (a) $x^2 + x^2 y^2$
   (b) $\cos^2 x + \cos^2 x \tan^2 x$

3. (a) $x^4 - 1$
   (b) $\tan^4 x - 1$

4. (a) $z^3 + 1$
   (b) $\tan^3 x + 1$

5. (a) $x^3 - x^2 + x - 1$
   (b) $\cot^3 x - \cot^2 x + \cot x - 1$

6. (a) $x^4 - 2x^2 + 1$
   (b) $\sin^4 x - 2 \sin^2 x + 1$

Perform the indicated additions and subtractions and, if possible, simplify the results.

7. (a) $\dfrac{y^2}{x} - x$

   (b) $\dfrac{\csc^2 x}{\cot x} - \cot x$

8. (a) $1 - \dfrac{1}{x^2}$

   (b) $1 - \dfrac{1}{\sec^2 x}$

9. (a) $\dfrac{y}{1 + z} + \dfrac{1 + z}{y}$

   (b) $\dfrac{\sin x}{1 + \cos x} + \dfrac{1 + \cos x}{\sin x}$

10. (a) $\dfrac{y}{z} - \dfrac{z}{1 + y}$

    (b) $\dfrac{\tan x}{\sec x} - \dfrac{\sec x}{1 + \tan x}$

EXERCISES 8.2

In Exercises 1–60, verify the given identity.

1. $\sin t \csc t = 1$

2. $\tan y \cot y = 1$

3. $(1 + \sin \alpha)(1 - \sin \alpha) = \cos^2 \alpha$

4. $\cot^2 y(\sec^2 y - 1) = 1$

5. $\cos^2 \beta - \sin^2 \beta = 1 - 2 \sin^2 \beta$

6. $\cos^2 \beta - \sin^2 \beta = 2 \cos^2 \beta - 1$

7. $\tan^2 \theta + 4 = \sec^2 \theta + 3$

8. $2 - \sec^2 z = 1 - \tan^2 z$

9. $\sin^2 \alpha - \sin^4 \alpha = \cos^2 \alpha - \cos^4 \alpha$

10. $\cos x + \sin x \tan x = \sec x$

11. $\dfrac{\sec^2 x}{\tan x} = \sec x \csc x$

12. $\dfrac{\cot^3 t}{\csc t} = \cos t(\csc^2 t - 1)$

13. $\dfrac{\cot^2 t}{\csc t} = \csc t - \sin t$

14. $\dfrac{1}{\sin x} - \sin x = \dfrac{\cos^2 x}{\sin x}$

15. $\sin^{1/2} x \cos x - \sin^{5/2} x \cos x = \cos^3 x \sqrt{\sin x}$

16. $\sec^6 x(\sec x \tan x) - \sec^4 x(\sec x \tan x) = \sec^5 x \tan^3 x$

17. $\dfrac{1}{\sec x \tan x} = \csc x - \sin x$

18. $\dfrac{\sec \theta - 1}{1 - \cos \theta} = \sec \theta$

19. $\cos x + \sin x \tan x = \sec x$

20. $\sec x - \cos x = \sin x \tan x$

21. $\csc x - \sin x = \cos x \cot x$

22. $\dfrac{\sec x + \tan x}{\sec x - \tan x} = (\sec x + \tan x)^2$

23. $\dfrac{1}{\tan x} + \dfrac{1}{\cot x} = \tan x + \cot x$

24. $\dfrac{1}{\sin x} - \dfrac{1}{\csc x} = \csc x - \sin x$

25. $\dfrac{\cos \theta \cot \theta}{1 - \sin \theta} - 1 = \csc \theta$

26. $\dfrac{1 + \sin \theta}{\cos \theta} + \dfrac{\cos \theta}{1 + \sin \theta} = 2 \sec \theta$

27. $\dfrac{1}{\cot x + 1} + \dfrac{1}{\tan x + 1} = 1$

28. $\cos x - \dfrac{\cos x}{1 - \tan x} = \dfrac{\sin x \cos x}{\sin x - \cos x}$

29. $2 \sec^2 x - 2 \sec^2 x \sin^2 x - \sin^2 x - \cos^2 x = 1$

30. $\csc x(\csc x - \sin x) + \dfrac{\sin x - \cos x}{\sin x} + \cot x = \csc^2 x$

31. $2 + \cos^2 x - 3 \cos^4 x = \sin^2 x(2 + 3 \cos^2 x)$

32. $4 \tan^4 x + \tan^2 x - 3 = \sec^2 x(4 \tan^2 x - 3)$

33. $\csc^4 x - 2 \csc^2 x + 1 = \cot^4 x$

34. $\sin x(1 - 2 \cos^2 x + \cos^4 x) = \sin^5 x$

35. $\sec^4 \theta - \tan^4 \theta = 1 + 2 \tan^2 \theta$

36. $\csc^4 \theta - \cot^4 \theta = 2 \csc^2 \theta - 1$

37. $\dfrac{\sin \beta}{1 - \cos \beta} = \dfrac{1 + \cos \beta}{\sin \beta}$

38. $\dfrac{\cot \alpha}{\csc \alpha - 1} = \dfrac{\csc \alpha + 1}{\cot \alpha}$

39. $\dfrac{\tan^3 \alpha - 1}{\tan \alpha - 1} = \tan^2 \alpha + \tan \alpha + 1$

40. $\dfrac{\sin^3 \beta + \cos^3 \beta}{\sin \beta + \cos \beta} = 1 - \sin \beta \cos \beta$

41. $\cos\left(\dfrac{\pi}{2} - x\right)\csc x = 1$

42. $\dfrac{\cos[(\pi/2) - x]}{\sin[(\pi/2) - x]} = \tan x$

43. $\dfrac{\csc(-x)}{\sec(-x)} = -\cot x$

44. $(1 + \sin y)[1 + \sin(-y)] = \cos^2 y$

45. $\dfrac{\cos(-\theta)}{1 + \sin(-\theta)} = \sec \theta + \tan \theta$

46. $\dfrac{1 + \sec(-\theta)}{\sin(-\theta) + \tan(-\theta)} = -\csc \theta$

47. $\dfrac{\sin x \cos y + \cos x \sin y}{\cos x \cos y - \sin x \sin y} = \dfrac{\tan x + \tan y}{1 - \tan x \tan y}$

48. $\dfrac{\tan x + \tan y}{1 - \tan x \tan y} = \dfrac{\cot x + \cot y}{\cot x \cot y - 1}$

49. $\dfrac{\tan x + \cot y}{\tan x \cot y} = \tan y + \cot x$

50. $\dfrac{\cos x - \cos y}{\sin x + \sin y} + \dfrac{\sin x - \sin y}{\cos x + \cos y} = 0$

51. $\sqrt{\dfrac{1 + \sin \theta}{1 - \sin \theta}} = \dfrac{1 + \sin \theta}{|\cos \theta|}$

52. $\sqrt{\dfrac{1 - \cos \theta}{1 + \cos \theta}} = \dfrac{1 - \cos \theta}{|\sin \theta|}$

53. $\ln|\tan \theta| = \ln|\sin \theta| - \ln|\cos \theta|$

54. $\ln|\sec \theta| = -\ln|\cos \theta|$

55. $-\ln(1 + \cos \theta) = \ln(1 - \cos \theta) - 2 \ln|\sin \theta|$

56. $-\ln|\sec \theta + \tan \theta| = \ln|\sec \theta - \tan \theta|$

57. $\sin^2 x + \sin^2\left(\dfrac{\pi}{2} - x\right) = 1$

58. $\sec^2 y - \cot^2\left(\dfrac{\pi}{2} - y\right) = 1$

59. $\csc x \cos\left(\dfrac{\pi}{2} - x\right) = 1$

60. $\sec^2\left(\dfrac{\pi}{2} - x\right) - 1 = \cot^2 x$

In Exercises 61–64, explain why the equation is *not* an identity and find one value of the variable for which the equation is not true.

61. $\sin \theta = \sqrt{1 - \cos^2 \theta}$ **62.** $\tan \theta = \sqrt{\sec^2 \theta - 1}$

63. $\sqrt{\tan^2 x} = \tan x$

64. $\sqrt{\sin^2 x + \cos^2 x} = \sin x + \cos x$

8.3 Solving Trigonometric Equations

We now switch from *verifying* trigonometric identities to *solving* trigonometric equations. To see the difference, consider the following two equations.

$$\sin^2 x + \cos^2 x = 1 \qquad \text{and} \qquad \sin x = 1$$

The first equation is an identity because it is true for *all* real values of x. The second equation, however, is true only for *some* values of x. When we find these values, we say we are solving the equation.

To solve a trigonometric equation, we use standard algebraic techniques such as collecting like terms and factoring to isolate the trigonometric function involved in the equation. For example, in the equation

$$2 \sin x - 1 = 0$$

we isolate $\sin x$ as follows.

$$2 \sin x = 1$$

$$\sin x = \frac{1}{2}$$

Now, to solve for x, we note that the equation $\sin x = 1/2$ has solutions $x = \pi/6$ and $x = 5\pi/6$ in the interval $[0, 2\pi)$. Moreover, because $\sin x$ has a period of 2π, there are infinitely many other solutions, which can be written as

$$x = \frac{\pi}{6} + 2n\pi \qquad \text{and} \qquad x = \frac{5\pi}{6} + 2n\pi$$

where n is an integer, as shown in Figure 8.2. We call this the **general form** of the solution.

Another way to see that the equation $\sin x = 1/2$ has infinitely many solutions is indicated in Figure 8.3. For $0 \le x < 2\pi$, the solutions are $x = \pi/6$ and $x = 5\pi/6$. Any angles that are coterminal with $\pi/6$ or $5\pi/6$ will also be solutions of the equation.

Solving Trigonometric Equations

FIGURE 8.2

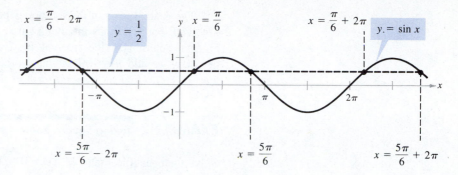

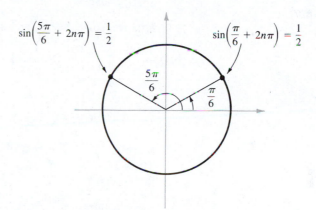

FIGURE 8.3

EXAMPLE 1 Collecting Like Terms

Solve the equation $\sin x + \sqrt{2} = -\sin x$.

SOLUTION

Collecting like terms and isolating $\sin x$, we obtain

$$\sin x + \sin x = -\sqrt{2}$$
$$2 \sin x = -\sqrt{2}$$
$$\sin x = -\frac{\sqrt{2}}{2}.$$

Since $\sin x$ has a period of 2π, we first find all solutions in the interval $[0, 2\pi)$. These are $x = 5\pi/4$ and $x = 7\pi/4$. Now we add $2n\pi$ to each of these solutions to get the general form

$$x = \frac{5\pi}{4} + 2n\pi \qquad \text{and} \qquad x = \frac{7\pi}{4} + 2n\pi$$

where n is an integer.

The general form of the solution depends upon the period of the function involved. For equations involving $\sin x$, $\cos x$, $\sec x$, or $\csc x$, we add $2n\pi$ to each solution in the interval $[0, 2\pi)$. For equations involving $\tan x$ or $\cot x$, we add $n\pi$. Unless stated otherwise, "solve the equation" means that we are to find the *general form* of the solution.

EXAMPLE 2 *Taking Square Roots*

Solve the equation $3 \tan^2 x - 1 = 0$.

SOLUTION

First we isolate $\tan^2 x$, and then we take the square root of both sides.

$$3 \tan^2 x = 1$$

$$\tan^2 x = \frac{1}{3}$$

$$\tan x = \pm \frac{1}{\sqrt{3}}$$

Since $\tan x$ has a period of π, we find all solutions in the interval $[0, \pi)$. These are $x = \pi/6$ and $x = 5\pi/6$. Finally, we add $n\pi$ to each and obtain the general form

$$x = \frac{\pi}{6} + n\pi \qquad \text{and} \qquad x = \frac{5\pi}{6} + n\pi$$

where n is an integer.

The equations in Examples 1 and 2 involved only one trigonometric function. When two or more functions occur in the same equation, we collect all terms to one side and try to separate the functions by factoring. This may produce factors that yield no solutions, as illustrated in Example 3.

EXAMPLE 3 *Factoring*

Solve the equation $\cot x \cos^2 x = 2 \cot x$.

SOLUTION

Collecting terms to one side and factoring produces

$$\cot x \cos^2 x - 2 \cot x = 0$$
$$\cot x (\cos^2 x - 2) = 0.$$

Solving Trigonometric Equations

By setting each of these factors to zero, we obtain the following.

$$\cot x = 0, \qquad \cos^2 x - 2 = 0$$

$$x = \frac{\pi}{2} \qquad\qquad \cos^2 x = 2$$

$$\cos x = \pm\sqrt{2}$$

No solution is obtained from $\cos x = \pm\sqrt{2}$ because $\pm\sqrt{2}$ are outside the range of the cosine function. Therefore, the general form of the solution is obtained by adding multiples of π to $x = \pi/2$, to get

$$x = \frac{\pi}{2} + n\pi$$

where n is an integer.

Equations of Quadratic Type

Many trigonometric equations are of quadratic type. Here are a couple of examples.

Quadratic in sin x		*Quadratic in sec x*
$2\sin^2 x - \sin x - 1 = 0$	and	$\sec^2 x - 3\sec x - 2 = 0$

To solve equations of this type, we factor the quadratic or, if that is not possible, we use the quadratic formula.

When working with an equation of quadratic type, be sure that the equation involves a *single* trigonometric function. For instance, we can rewrite the equation

$$2\sin^2 x + 3\cos x - 3 = 0$$

in terms of a single trigonometric function as follows.

$$2(1 - \cos^2 x) + 3\cos x - 3 = 0 \qquad \textit{Identity}$$
$$2\cos^2 x - 3\cos x + 1 = 0 \qquad \textit{Multiply by } -1$$

At this point, we can factor the equation to obtain

$$(2\cos x - 1)(\cos x - 1) = 0.$$

Finally, by setting each factor equal to zero, the solutions in the interval $[0, 2\pi)$ are determined to be $x = 0$, $x = \pi/3$, and $x = 5\pi/3$. The general solution is therefore

$$x = 2n\pi, \qquad x = \frac{\pi}{3} + 2n\pi, \qquad x = \frac{5\pi}{3} + 2n\pi$$

where n is an integer.

EXAMPLE 4 *Factoring an Equation of Quadratic Type*

Find all solutions of $2 \sin^2 x - \sin x - 1 = 0$ in the interval $[0, 2\pi)$.

SOLUTION

Treating the equation as a quadratic in $\sin x$ and factoring, we obtain

$$2 \sin^2 x - \sin x - 1 = 0$$
$$(2 \sin x + 1)(\sin x - 1) = 0.$$

Setting each factor to zero, we obtain the following solutions.

$$2 \sin x + 1 = 0, \qquad\qquad \sin x - 1 = 0$$
$$\sin x = -\frac{1}{2} \qquad\qquad \sin x = 1$$
$$x = \frac{7\pi}{6}, \frac{11\pi}{6} \qquad\qquad x = \frac{\pi}{2}$$

Remark: In Example 4, the general solution would be

$$x = \frac{7\pi}{6} + 2n\pi, \qquad x = \frac{11\pi}{6} + 2n\pi, \qquad x = \frac{\pi}{2} + 2n\pi$$

where n is an integer.

Sometimes we must square both sides of an equation to obtain a quadratic, as demonstrated in the next example. Because this procedure can introduce extraneous solutions, you should check any solutions in the original equation to see if they are valid or extraneous.

EXAMPLE 5 *Squaring and Converting to Quadratic Type*

Find all solutions of $\cos x + 1 = \sin x$ in the interval $[0, 2\pi)$.

SOLUTION

It is not immediately clear how to rewrite this equation in terms of a single trigonometric function. Let's see what happens when we square both sides of the equation.

$\cos x + 1 = \sin x$	*Given*
$\cos^2 x + 2 \cos x + 1 = \sin^2 x$	*Square both sides*
$\cos^2 x + 2 \cos x + 1 = 1 - \cos^2 x$	*Identity*
$2 \cos^2 x + 2 \cos x = 0$	*Collect terms*
$2 \cos x(\cos x + 1) = 0$	*Factor*

Setting each factor to zero produces the following.

$$2 \cos x = 0, \qquad\qquad \cos x + 1 = 0$$
$$\cos x = 0 \qquad\qquad\qquad \cos x = -1$$
$$x = \frac{\pi}{2}, \frac{3\pi}{2} \qquad\qquad\qquad x = \pi$$

Because we squared the original equation, we check for extraneous solutions. Of the three possible solutions, $x = 3\pi/2$ turns out to be extraneous. (Try checking this.) Thus, in the interval $[0, 2\pi)$, the only two solutions are $x = \pi/2$ and $x = \pi$.

Remark: In Example 5, the general solution would be

$$x = \frac{\pi}{2} + 2n\pi, \qquad x = \pi + 2n\pi$$

where n is an integer.

For trigonometric functions of *multiple angles*, extra care is needed to determine all possible solutions. We show how this is done in Example 6.

EXAMPLE 6   *Functions of Multiple Angles*

Find all solutions of $2 \cos 3t = 1$.

SOLUTION

Solving for $\cos 3t$, we have

$$2 \cos 3t = 1$$
$$\cos 3t = \frac{1}{2}.$$

In the interval $[0, 2\pi)$, we know that $3t = \pi/3$ and $3t = 5\pi/3$ so that, in general,

$$3t = \frac{\pi}{3} + 2n\pi \qquad \text{and} \qquad 3t = \frac{5\pi}{3} + 2n\pi.$$

Dividing this result by 3, we obtain the general form

$$t = \frac{\pi}{9} + \frac{2n\pi}{3} \qquad \text{and} \qquad t = \frac{5\pi}{9} + \frac{2n\pi}{3}$$

where n is an integer.

Analytic Trigonometry

EXAMPLE 7 *Functions of Multiple Angles*

Find all solutions of $3 \tan(x/2) + 3 = 0$.

SOLUTION

First we isolate $\tan(x/2)$, as follows.

$$3 \tan \frac{x}{2} = -3$$

$$\tan \frac{x}{2} = \frac{-3}{3} = -1$$

In the interval $[0, \pi)$, we know that $x/2 = 3\pi/4$ so that, in general, we have

$$\frac{x}{2} = \frac{3\pi}{4} + n\pi.$$

Multiplication by 2 yields the general form

$$x = \frac{3\pi}{2} + 2n\pi$$

where n is an integer.

Using Inverse Functions and a Calculator

So far in this section, we have chosen examples for which the solutions are special values like

$$0, \ \pm\frac{\pi}{6}, \ \pm\frac{\pi}{4}, \ \pm\frac{\pi}{3}, \ \pm\frac{\pi}{2}$$

and so on. In the remaining examples, we solve more general trigonometric equations by using inverse trigonometric functions or a calculator.

EXAMPLE 8 *Using Inverse Functions*

Find all solutions of $\sec^2 x - 2 \tan x = 4$.

SOLUTION

Collecting terms and substituting $(1 + \tan^2 x)$ for $\sec^2 x$, we have

$$1 + \tan^2 x - 2 \tan x - 4 = 0$$
$$\tan^2 x - 2 \tan x - 3 = 0$$
$$(\tan x - 3)(\tan x + 1) = 0.$$

Setting each factor equal to zero, we obtain two solutions in the interval $(-\pi/2, \pi/2)$. [Recall that the range of the inverse tangent function is $(-\pi/2, \pi/2)$.]

$$\tan x = 3, \qquad\qquad \tan x = -1$$

$$x = \arctan 3 \qquad\qquad x = \arctan(-1) = -\frac{\pi}{4}$$

Finally, by adding multiples of π (the period of the tangent), we obtain the general solution

$$x = \arctan 3 + n\pi$$

and

$$x = -\frac{\pi}{4} + n\pi$$

where n is an integer.

Remark: If you have access to a computer or calculator with plotting capabilities, you might want to check your solutions graphically. For instance, the graph of the function

$$f(x) = \tan^2 x - 2 \tan x - 3$$

shown in Figure 8.4, has the two x-intercepts

$$x = \arctan 3 \approx 1.2490$$

and

$$x = -\frac{\pi}{4} \approx -0.7854$$

in the interval $(-\pi/2, \pi/2)$.

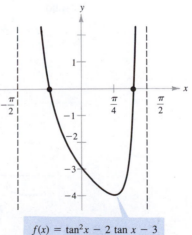

FIGURE 8.4

$$f(x) = \tan^2 x - 2 \tan x - 3$$

When using a calculator for arcsin x, arccos x, and arctan x, the displayed solution may need to be adjusted to obtain solutions in the desired interval, as shown in the next example.

EXAMPLE 9 *Using the Quadratic Formula*

Find all solutions of

$$\sin^2 t - 3 \sin t - 2 = 0$$

in the interval $[0, 2\pi)$.

SOLUTION

Since the left side of this equation is not factorable, we use the quadratic formula.

$$\sin t = \frac{-(-3) \pm \sqrt{(-3)^2 - 4(1)(-2)}}{2(1)} = \frac{3 \pm \sqrt{17}}{2}$$

$$\sin t \approx 3.561553 \quad \text{or} \quad -0.5615528$$

Because the range of the sine function is $[-1, 1]$, the equation $\sin t = 3.561553$ has no solution. To solve the equation $\sin t = -0.5615528$, we use a calculator and the inverse sine function as follows.

$$t \approx \arcsin(-0.5615528) \approx -0.5962613$$

Note that this solution is not in the interval $[0, 2\pi)$. To find the solutions in $[0, 2\pi)$, it is helpful to make a sketch of the graph of the sine function. From Figure 8.5, we see that the two solutions that lie in the interval $[0, 2\pi)$ are

$$t \approx \pi + 0.5962613 \approx 3.737854 \qquad \textit{Quadrant III}$$

and

$$t \approx 2\pi - 0.5962613 \approx 5.686924. \qquad \textit{Quadrant IV}$$

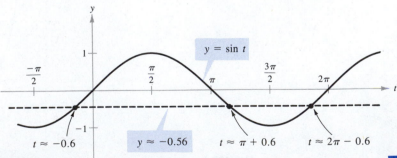

FIGURE 8.5

WARM UP

Find two values of θ $(0 \le \theta < 2\pi)$ that satisfy each equation.

1. $\cos \theta = -\frac{1}{2}$

2. $\sin \theta = \frac{\sqrt{3}}{2}$

3. $\cos \theta = \frac{\sqrt{2}}{2}$

4. $\sin \theta = -\frac{\sqrt{2}}{2}$

Find one value of θ $(0 \le \theta \le \pi)$ that satisfies each equation.

5. $\tan \theta = \sqrt{3}$

6. $\tan \theta = -1$

Solve for x.

7. $\frac{x}{3} + \frac{x}{5} = 1$

8. $2x(x + 3) - 5(x + 3) = 0$

9. $2x^2 - 4x - 5 = 0$

10. $\frac{1}{x} = \frac{x}{2x + 3}$

EXERCISES 8.3

In Exercises 1–6, verify that the given values of x are solutions of the equation.

1. $2 \cos x - 1 = 0$

   (a) $x = \frac{\pi}{3}$ (b) $x = \frac{5\pi}{3}$

2. $\csc x - 2 = 0$

   (a) $x = \frac{\pi}{6}$ (b) $x = \frac{5\pi}{6}$

3. $3 \tan^2 2x - 1 = 0$

   (a) $x = \frac{\pi}{12}$ (b) $x = \frac{5\pi}{12}$

4. $2 \cos^2 4x - 1 = 0$

   (a) $x = \frac{\pi}{16}$ (b) $x = \frac{3\pi}{16}$

5. $2 \sin^2 x - \sin x - 1 = 0$

   (a) $x = \frac{\pi}{2}$ (b) $x = \frac{7\pi}{6}$

6. $\sec^4 x - 4 \sec^2 x = 0$

   (a) $x = \frac{2\pi}{3}$ (b) $x = \frac{5\pi}{3}$

In Exercises 7–20, find all solutions of the equation. (Do not use a calculator.)

7. $2 \cos x + 1 = 0$
8. $2 \sin x - 1 = 0$
9. $\sqrt{3} \csc x - 2 = 0$
10. $\tan x + 1 = 0$
11. $2 \sin^2 x = 1$
12. $\tan^2 x = 3$
13. $3 \sec^2 x - 4 = 0$
14. $\csc^2 x - 2 = 0$
15. $\tan x(\tan x - 1) = 0$
16. $\sin^2 x = 3 \cos^2 x$
17. $\sin x(\sin x + 1) = 0$
18. $4 \sin^2 x - 3 = 0$
19. $\cos x(2 \cos x + 1) = 0$
20. $(3 \tan^2 x - 1)(\tan^2 x - 3) = 0$

In Exercises 21–40, find all solutions in the interval $[0, 2\pi)$. (Do not use a calculator.)

21. $\sec x \csc x - 2 \csc x = 0$
22. $\sec^2 x - \sec x - 2 = 0$
23. $2 \sin^2 x + 3 \sin x + 1 = 0$
24. $3 \tan^3 x - \tan x = 0$
25. $\cos^3 x = \cos x$
26. $4 \sin^3 x + 2 \sin^2 x - 2 \sin x - 1 = 0$
27. $2 \sin^2 x = 2 + \cos x$
28. $\csc^2 x = (1 + \sqrt{3}) - (1 - \sqrt{3})\cot x$
29. $2 \sec^2 x + \tan^2 x - 3 = 0$
30. $\sec^2 x = (1 + \sqrt{3}) - (1 - \sqrt{3})\tan x$
31. $2 \sin x + \csc x = 0$
32. $\csc x + \cot x = 1$
33. $\sin 2x = -\dfrac{\sqrt{3}}{2}$
34. $\tan 3x = 1$
35. $\cos \dfrac{x}{2} = \dfrac{\sqrt{2}}{2}$
36. $\sec 4x = 2$
37. $\dfrac{1 + \cos x}{1 - \cos x} = 0$
38. $\cos x + \sin x \tan x = 2$

39. $\dfrac{1 + \sin x}{\cos x} + \dfrac{\cos x}{1 + \sin x} = 4$
40. $\dfrac{\cos x \cot x}{1 - \sin x} = 3$

In Exercises 41–50, use a calculator to find all solutions in the interval $[0, 2\pi)$.

41. $2 \tan^2 x + 7 \tan x - 15 = 0$
42. $12 \cos^2 x + 5 \cos x - 3 = 0$
43. $12 \sin^2 x - 13 \sin x + 3 = 0$
44. $3 \tan^2 x + 4 \tan x - 4 = 0$
45. $6 \cos^2 x - 13 \cos x + 6 = 0$
46. $\sin^2 x + \sin x - 20 = 0$
47. $\tan^2 x - 8 \tan x + 13 = 0$
48. $2 \cos^2 x + 6 \cos x - 1 = 0$
49. $\sin^2 x + 2 \sin x - 1 = 0$
50. $4 \cos^2 x - 4 \cos x - 1 = 0$

51. The function $f(x) = \sin x + \cos x$ has maximum or minimum values when

$$\cos x - \sin x = 0.$$

Find all solutions of this equation in the interval $[0, 2\pi)$ and sketch a graph of the function f.

52. The function $f(x) = 2 \sin x + \cos 2x$ has maximum or minimum values when

$$2 \cos x - 4 \sin x \cos x = 0.$$

Find all solutions of this equation in the interval $[0, 2\pi)$ and sketch a graph of the function f.

53. A 5-pound weight is oscillating on the end of a spring, and the position of the weight relative to the point of equilibrium is given by

$$h(t) = \tfrac{1}{4}(\cos 8t - 3 \sin 8t)$$

where t is the time in seconds. Find the times when the weight is at the point of equilibrium $[h(t) = 0]$ for $0 \le t \le 1$.

54. The monthly sales (in thousands of units) of a seasonal product is approximated by

$$S = 74.50 + 43.75 \sin \frac{\pi t}{6}$$

where t is the time in months, with $t = 1$ corresponding to January. Determine the months when sales exceed 100,000 units.

55. A batted baseball leaves the bat at an angle of θ with the horizontal, with a velocity of $v_0 = 100$ feet per second, and is caught by an outfielder 300 feet from home plate (see figure). Find θ if the range of a projectile is given by

$$r = \tfrac{1}{32}v_0{}^2 \sin 2\theta.$$

FIGURE FOR 55

56. A gun with a muzzle velocity of 1200 feet per second is pointed at a target 1000 yards away (see figure). Neglecting air resistance, what should be the minimum angle of elevation of the gun? (Use the formula for range given in Exercise 55.)

FIGURE FOR 56

8.4 Sum and Difference Formulas

In this and the following section, we show the derivations and uses of several trigonometric identities (or formulas) that are useful in scientific applications.

We begin with six sum and difference formulas that express trigonometric functions of $(u \pm v)$ as functions of u and v alone.

Sum and Difference Formulas

Sine
$$\sin(u + v) = \sin u \cos v + \cos u \sin v$$
$$\sin(u - v) = \sin u \cos v - \cos u \sin v$$

Cosine
$$\cos(u + v) = \cos u \cos v - \sin u \sin v$$
$$\cos(u - v) = \cos u \cos v + \sin u \sin v$$

Tangent
$$\tan(u + v) = \frac{\tan u + \tan v}{1 - \tan u \tan v}$$

$$\tan(u - v) = \frac{\tan u - \tan v}{1 + \tan u \tan v}$$

Analytic Trigonometry

PROOF

We prove only the formulas for $\cos(u \pm v)$. In Figure 8.6 we let A be the point $(1, 0)$ and then use u and v to locate the points $B = (x_1, y_1)$, $C = (x_2, y_2)$, and $D = (x_3, y_3)$ on the unit circle. Thus, $x_i^2 + y_i^2 = 1$ for $i = 1, 2, 3$. For convenience, we assume that $0 < v < u < 2\pi$.

From Figure 8.7, note that arcs AC and BD have the same length. Hence, *line segments* AC and BD are also equal in length, which implies that

$$\sqrt{(x_2 - 1)^2 + (y_2 - 0)^2} = \sqrt{(x_3 - x_1)^2 + (y_3 - y_1)^2}$$

$$x_2^2 - 2x_2 + 1 + y_2^2 = x_3^2 - 2x_1x_3 + x_1^2 + y_3^2 - 2y_1y_3 + y_1^2$$

$$(x_2^2 + y_2^2) + 1 - 2x_2 = (x_3^2 + y_3^2) + (x_1^2 + y_1^2) - 2x_1x_3 - 2y_1y_3$$

$$1 + 1 - 2x_2 = 1 + 1 - 2x_1x_3 - 2y_1y_3$$

$$x_2 = x_3x_1 + y_3y_1.$$

Finally, by substituting the values $x_2 = \cos(u - v)$, $x_3 = \cos u$, $x_1 = \cos v$, $y_3 = \sin u$, and $y_1 = \sin v$, we obtain

$$\cos(u - v) = \cos u \cos v + \sin u \sin v.$$

The formula for $\cos(u + v)$ can be established by considering $u + v = u - (-v)$ and using the formula just derived to obtain

$$\begin{aligned}
\cos(u + v) &= \cos[u - (-v)] \\
&= \cos u \cos(-v) + \sin u \sin(-v) \\
&= \cos u \cos v - \sin u \sin v.
\end{aligned}$$

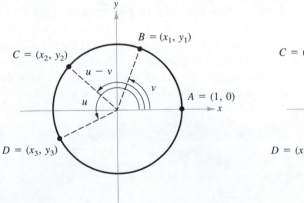

FIGURE 8.6

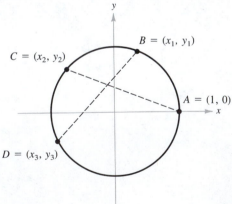

FIGURE 8.7

Remark: Note that $\sin(u + v) \neq \sin u + \sin v$. Similar statements can be made for $\cos(u + v)$ and $\tan(u + v)$.

Sum and Difference Formulas

Applications of Sum and Difference Formulas

In the remainder of this section, we show a variety of uses of sum and difference formulas. First, we show how sum and difference formulas can be used to find exact values of trigonometric functions involving sums or differences of special angles.

EXAMPLE 1 *Evaluating a Trigonometric Function*

Find the exact value of $\cos 75°$.

SOLUTION

To find the *exact* value of $\cos 75°$, we use the fact that

$$75° = 30° + 45°.$$

Consequently, the formula for $\cos(u + v)$ yields

$$\cos 75° = \cos(30° + 45°)$$
$$= \cos 30° \cos 45° - \sin 30° \sin 45°$$
$$= \frac{\sqrt{3}}{2}\left(\frac{\sqrt{2}}{2}\right) - \frac{1}{2}\left(\frac{\sqrt{2}}{2}\right)$$
$$= \frac{\sqrt{6} - \sqrt{2}}{4}.$$

Remark: Try checking the result obtained in Example 1 on your calculator. You will find that

$$\cos 75° = \frac{\sqrt{6} - \sqrt{2}}{4} \approx 0.259.$$

EXAMPLE 2   *Evaluating a Trigonometric Function*

Find the exact value of

$$\cos \frac{\pi}{12}.$$

SOLUTION

Using the fact that

$$\frac{\pi}{12} = \frac{\pi}{3} - \frac{\pi}{4}$$

together with the formula for $\cos(u - v)$, we obtain

$$\cos\frac{\pi}{12} = \cos\left(\frac{\pi}{3} - \frac{\pi}{4}\right)$$

$$= \cos\frac{\pi}{3}\cos\frac{\pi}{4} + \sin\frac{\pi}{3}\sin\frac{\pi}{4}$$

$$= \frac{1}{2}\left(\frac{\sqrt{2}}{2}\right) + \frac{\sqrt{3}}{2}\left(\frac{\sqrt{2}}{2}\right)$$

$$= \frac{\sqrt{2} + \sqrt{6}}{4}.$$

EXAMPLE 3 *Evaluating a Trigonometric Expression*

Find the exact value of $\sin 42° \cos 12° - \cos 42° \sin 12°$.

SOLUTION

Recognizing that this expression fits the formula for $\sin(u - v)$, we can write

$$\sin 42° \cos 12° - \cos 42° \sin 12° = \sin(42° - 12°)$$

$$= \sin 30°$$

$$= \frac{1}{2}.$$

EXAMPLE 4 *Evaluating a Trigonometric Expression*

Find the exact value of

$$\frac{\tan 80° + \tan 55°}{1 - \tan 80° \tan 55°}.$$

SOLUTION

From the formula for $\tan(u + v)$, we have

$$\frac{\tan 80° + \tan 55°}{1 - \tan 80° \tan 55°} = \tan(80° + 55°) = \tan 135° = -\tan 45° = -1.$$

EXAMPLE 5 An Application of a Difference Formula

Find $\cos(u - v)$ given that

$$\cos u = -\frac{15}{17}, \quad \pi < u < \frac{3\pi}{2}$$

and

$$\sin v = \frac{4}{5}, \quad 0 < v < \frac{\pi}{2}.$$

SOLUTION

Using the given values for $\cos u$ and $\sin v$, we can sketch angles u and v as shown in Figure 8.8. This implies that

$$\cos v = \frac{3}{5} \quad \text{and} \quad \sin u = -\frac{8}{17}.$$

Therefore,

$$\cos(u - v) = \cos u \cos v + \sin u \sin v$$

$$= \left(-\frac{15}{17}\right)\left(\frac{3}{5}\right) + \left(-\frac{8}{17}\right)\left(\frac{4}{5}\right)$$

$$= -\frac{77}{85}.$$

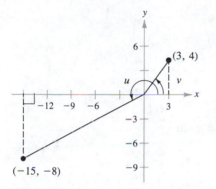

FIGURE 8.8

EXAMPLE 6 Proving a Cofunction Identity

Use the formula for $\cos(u - v)$ to prove the cofunction identity

$$\cos\left(\frac{\pi}{2} - x\right) = \sin x.$$

SOLUTION

Using the difference formula

$$\cos(u - v) = \cos u \cos v + \sin u \sin v$$

we have

$$\cos\left(\frac{\pi}{2} - x\right) = \cos\frac{\pi}{2}\cos x + \sin\frac{\pi}{2}\sin x$$

$$= (0)\cos x + (1)\sin x$$

$$= \sin x.$$

Analytic Trigonometry

Sum and difference formulas can be used to derive **reduction formulas** involving expressions like

$$\sin\left(\theta + \frac{n\pi}{2}\right) \qquad \text{and} \qquad \cos\left(\theta + \frac{n\pi}{2}\right)$$

where n is an integer. We show two such formulas in Example 7.

EXAMPLE 7 *Deriving Reduction Formulas*

Simplify the following expressions.

(a) $\cos\left(\theta - \dfrac{3\pi}{2}\right)$ \qquad\qquad (b) $\sin(\theta + 3\pi)$

SOLUTION

(a) Using the formula

$$\cos(u - v) = \cos u \cos v + \sin u \sin v$$

we have

$$\cos\left(\theta - \frac{3\pi}{2}\right) = \cos\theta\cos\frac{3\pi}{2} + \sin\theta\sin\frac{3\pi}{2}$$
$$= (\cos\theta)(0) + (\sin\theta)(-1)$$
$$= -\sin\theta.$$

(b) Using the formula

$$\sin(u + v) = \sin u \cos v + \cos u \sin v$$

we have

$$\sin(\theta + 3\pi) = \sin\theta\cos 3\pi + \cos\theta\sin 3\pi$$
$$= (\sin\theta)(-1) + (\cos\theta)(0)$$
$$= -\sin\theta.$$

EXAMPLE 8 *Solving a Trigonometric Equation*

Find all solutions of

$$\sin\left(x + \frac{\pi}{4}\right) + \sin\left(x - \frac{\pi}{4}\right) = -1$$

in the interval $[0, 2\pi)$.

SOLUTION

Using sum and difference formulas, we rewrite the given equation as

$$\sin x \cos \frac{\pi}{4} + \cos x \sin \frac{\pi}{4} + \sin x \cos \frac{\pi}{4} - \cos x \sin \frac{\pi}{4} = -1$$

$$2 \sin x \cos \frac{\pi}{4} = -1$$

$$2(\sin x)\left(\frac{\sqrt{2}}{2}\right) = -1$$

$$\sin x = -\frac{1}{\sqrt{2}}$$

$$\sin x = -\frac{\sqrt{2}}{2}.$$

Therefore, the only solutions in the interval $[0, 2\pi)$ are

$$x = \frac{5\pi}{4} \qquad \text{and} \qquad x = \frac{7\pi}{4}.$$

Our last example shows how a sum formula allows us to rewrite a trigonometric expression in a form that is useful in calculus.

EXAMPLE 9   *An Application from Calculus*

Verify that

$$\frac{\sin(x + h) - \sin x}{h} = (\cos x)\left(\frac{\sin h}{h}\right) - (\sin x)\left(\frac{1 - \cos h}{h}\right)$$

where $h \neq 0$.

SOLUTION

Using the formula for $\sin(u + v)$, we have

$$\frac{\sin(x + h) - \sin x}{h} = \frac{\sin x \cos h + \cos x \sin h - \sin x}{h}$$

$$= \frac{\cos x \sin h - \sin x(1 - \cos h)}{h}$$

$$= (\cos x)\left(\frac{\sin h}{h}\right) - (\sin x)\left(\frac{1 - \cos h}{h}\right).$$

WARM UP

Find sin θ.

1. $\tan \theta = \frac{1}{3}$ (θ in Quadrant I)

2. $\cot \theta = \frac{3}{5}$ (θ in Quadrant III)

3. $\cos \theta = \frac{3}{4}$ (θ in Quadrant IV)

4. $\sec \theta = -3$ (θ in Quadrant II)

Solve for *x*.

5. $\sin x = \frac{\sqrt{2}}{2}$

6. $\cos x = 0$

Simplify the expressions.

7. $\tan x \sec^2 x - \tan x$

8. $\dfrac{\cos x \csc x}{\tan x}$

9. $\dfrac{\cos x}{1 - \sin x} - \tan x$

10. $1 - 3 \cos^2 x + 2 \cos^4 x$

EXERCISES 8.4

In Exercises 1–10, determine the exact values of the sine, cosine, and tangent of the given angle.

1. $75° = 30° + 45°$

2. $15° = 45° - 30°$

3. $105° = 60° + 45°$

4. $165° = 135° + 30°$

5. $195° = 225° - 30°$

6. $255° = 300° - 45°$

7. $\dfrac{11\pi}{12} = \dfrac{3\pi}{4} + \dfrac{\pi}{6}$

8. $\dfrac{7\pi}{12} = \dfrac{\pi}{3} + \dfrac{\pi}{4}$

9. $\dfrac{17\pi}{12} = \dfrac{9\pi}{4} - \dfrac{5\pi}{6}$

10. $-\dfrac{\pi}{12} = \dfrac{\pi}{6} - \dfrac{\pi}{4}$

In Exercises 11–20, simplify the given expression.

11. $\cos 25° \cos 15° - \sin 25° \sin 15°$

12. $\sin 140° \cos 50° + \cos 140° \sin 50°$

13. $\sin 230° \cos 30° - \cos 230° \sin 30°$

14. $\cos 20° \cos 30° + \sin 20° \sin 30°$

15. $\dfrac{\tan 325° - \tan 86°}{1 + \tan 325° \tan 86°}$

16. $\dfrac{\tan 140° - \tan 60°}{1 + \tan 140° \tan 60°}$

17. $\sin 3 \cos 1.2 - \cos 3 \sin 1.2$

18. $\cos \dfrac{\pi}{7} \cos \dfrac{\pi}{5} - \sin \dfrac{\pi}{7} \sin \dfrac{\pi}{5}$

19. $\dfrac{\tan 2x + \tan x}{1 - \tan 2x \tan x}$

20. $\cos 3x \cos 2y + \sin 3x \sin 2y$

In Exercises 21–24, find the exact value of the trigonometric function given that

$$\sin u = \frac{5}{13}, \, 0 < u < \frac{\pi}{2} \quad \text{and} \quad \cos v = -\frac{3}{5}, \frac{\pi}{2} < v < \pi.$$

21. $\sin(u + v)$

22. $\cos(v - u)$

23. $\cos(v + u)$

24. $\sin(u - v)$

In Exercises 25–28, find the exact value of the trigonometric function given that

$$\sin u = \frac{7}{25}, \frac{\pi}{2} < u < \pi \quad \text{and} \quad \cos v = \frac{4}{5}, \frac{3\pi}{2} < v < 2\pi.$$

25. $\cos(u + v)$

26. $\sin(u + v)$

27. $\sin(v - u)$

28. $\cos(u - v)$

In Exercises 29–48, verify the given identity.

29. $\sin\left(\dfrac{\pi}{2} + x\right) = \cos x$

30. $\sin(3\pi - x) = \sin x$

31. $\cos\left(\dfrac{3\pi}{2} - x\right) = -\sin x$

32. $\cos(\pi + x) = -\cos x$

33. $\sin\left(\dfrac{\pi}{6} + x\right) = \dfrac{1}{2}(\cos x + \sqrt{3} \sin x)$

34. $\cos\left(\dfrac{5\pi}{4} - x\right) = -\dfrac{\sqrt{2}}{2}(\cos x + \sin x)$

35. $\cos(\pi - \theta) + \sin\left(\dfrac{\pi}{2} + \theta\right) = 0$

36. $\sin\left(\dfrac{3\pi}{2} + \theta\right) + \sin(\pi - \theta) = \sin\theta - \cos\theta$

37. $\tan(\pi + \theta) = \tan\theta$

38. $\tan\left(\dfrac{\pi}{4} - \theta\right) = \dfrac{1 - \tan\theta}{1 + \tan\theta}$

39. $\cos(x + y)\cos(x - y) = \cos^2 x - \sin^2 y$

40. $\sin(x + y)\sin(x - y) = \sin^2 x - \sin^2 y$

41. $\sin(x + y) + \sin(x - y) = 2\sin x \cos y$

42. $\cos(x + y) + \cos(x - y) = 2\cos x \cos y$

43. $\sin(x + y + z) = \sin x \cos y \cos z + \sin y \cos x \cos z + \sin z \cos x \cos y - \sin x \sin y \sin z$

44. $\cos(x + y + z) = \cos x \cos y \cos z - \cos x \sin y \sin z - \cos y \sin x \sin z - \cos z \sin x \sin y$

45. $\cos(n\pi + \theta) = (-1)^n\cos\theta,\quad n$ is an integer

46. $\sin(n\pi + \theta) = (-1)^n\sin\theta,\quad n$ is an integer

47. $a\sin B\theta + b\cos B\theta = \sqrt{a^2 + b^2}\,\sin(B\theta + C)$,

where $C = \arctan\dfrac{b}{a}$

48. $a\sin B\theta + b\cos B\theta = \sqrt{a^2 + b^2}\,\cos(B\theta - C)$,

where $C = \arctan\dfrac{a}{b}$

In Exercises 49–52, use the formulas given in Exercises 47 and 48 to write the given trigonometric expression in the following forms.

(a) $\sqrt{a^2 + b^2}\,\sin(B\theta + C)$ (b) $\sqrt{a^2 + b^2}\,\cos(B\theta - C)$

49. $\sin\theta + \cos\theta$

50. $3\sin 2\theta + 4\cos 2\theta$

51. $12\sin 3\theta + 5\cos 3\theta$

52. $\sin 2\theta - \cos 2\theta$

In Exercises 53 and 54, use the formulas given in Exercises 47 and 48 to write the given trigonometric expression in the form $a\sin B\theta + b\cos B\theta$.

53. $2\sin\left(\theta + \dfrac{\pi}{4}\right)$

54. $5\cos\left(\theta + \dfrac{3\pi}{4}\right)$

In Exercises 55 and 56, write the trigonometric expression as an algebraic expression in x. [*Hint*: See Examples 6 and 7 in Section 7.7.]

55. $\sin(\arcsin x + \arccos x)$

56. $\sin(\arctan 2x - \arccos x)$

In Exercises 57–62, find all solutions in the interval $[0, 2\pi)$.

57. $\sin\left(x + \dfrac{\pi}{3}\right) + \sin\left(x - \dfrac{\pi}{3}\right) = 1$

58. $\sin\left(x + \dfrac{\pi}{6}\right) - \sin\left(x - \dfrac{\pi}{6}\right) = \dfrac{1}{2}$

59. $\cos\left(x + \dfrac{\pi}{4}\right) + \cos\left(x - \dfrac{\pi}{4}\right) = 1$

60. $\cos\left(x + \dfrac{\pi}{4}\right) - \cos\left(x - \dfrac{\pi}{4}\right) = 1$

61. $\tan(x + \pi) + 2\sin(x + \pi) = 0$

62. $\tan(x + \pi) - \cos\left(x + \dfrac{\pi}{2}\right) = 0$

63. Show that

$$\dfrac{\cos(x + h) - \cos x}{h} = \cos x\left(\dfrac{\cos h - 1}{h}\right) - \sin x\left(\dfrac{\sin h}{h}\right).$$

64. A weight is attached to a spring suspended vertically from the ceiling. When a certain driving force is applied to the system, the weight moves vertically from its equilibrium position and this motion is described by the model

$$y = \tfrac{1}{3}\sin 2t + \tfrac{1}{4}\cos 2t$$

where y is the distance from equilibrium measured in feet and t is time in seconds.

(a) Write the model in the form

$$y = \sqrt{a^2 + b^2}\,\sin(Bt + C).$$

(See Exercise 47.)

(b) Find the amplitude of the oscillations of the weight.

(c) Find the frequency of the oscillations of the weight.

65. Use the difference formula

$$\cos(u - v) = \cos u \cos v + \sin u \sin v$$

together with the cofunction identity

$$\sin(u + v) = \cos\left[\dfrac{\pi}{2} - (u + v)\right] = \cos\left[\left(\dfrac{\pi}{2} - u\right) - v\right]$$

to prove the sum formula

$$\sin(u + v) = \sin u \cos v + \cos u \sin v.$$

66. Use the sum formula for $\sin(u + v)$ and $\cos(u + v)$ to prove the sum formula

$$\tan(u + v) = \dfrac{\tan u + \tan v}{1 - \tan u \tan v}.$$

Analytic Trigonometry

8.5 Multiple-Angle Formulas and Product-Sum Formulas

In this section we look at two other categories of trigonometric identities. The first category involves functions of multiple angles such as $\sin ku$ or $\cos ku$. The second category involves products of trigonometric functions such as $\sin u \cos v$.

Multiple-Angle Formulas

The most commonly used multiple-angle formulas are the double-angle formulas. They are used often enough so that you should memorize them.

Double-Angle Formulas

$$\sin 2u = 2 \sin u \cos u$$
$$\cos 2u = \cos^2 u - \sin^2 u = 2 \cos^2 u - 1 = 1 - 2 \sin^2 u$$
$$\tan 2u = \frac{2 \tan u}{1 - \tan^2 u}$$

PROOF

To prove the first formula, we let $v = u$ in the formula for $\sin(u + v)$, and obtain

$$\sin 2u = \sin(u + u)$$
$$= \sin u \cos u + \cos u \sin u$$
$$= 2 \sin u \cos u.$$

The other double-angle formulas can be proved in a similar way, and we leave their proofs for you to do.

Remark: Note that $\sin 2u \neq 2 \sin u$. Similar statements can be made for $\cos 2u$ and $\tan 2u$.

EXAMPLE 1 *Solving a Trigonometric Equation*

Find all solutions of $2 \cos x + \sin 2x = 0$.

SOLUTION

We begin by rewriting the equation so that it involves functions of x (rather than $2x$). Then we can factor and solve as usual.

$$2 \cos x + \sin 2x = 0 \qquad \text{\textit{Given}}$$
$$2 \cos x + 2 \sin x \cos x = 0 \qquad \text{\textit{Double-angle formula}}$$
$$2 \cos x (1 + \sin x) = 0 \qquad \text{\textit{Factor}}$$
$$\cos x = 0, \qquad\qquad 1 + \sin x = 0 \qquad \text{\textit{Set factors to zero}}$$
$$x = \frac{\pi}{2}, \frac{3\pi}{2} \qquad\qquad x = \frac{3\pi}{2} \qquad \text{\textit{Solutions in [0, 2\pi)}}$$

Therefore, the general solution is

$$x = \frac{\pi}{2} + 2n\pi \quad \cdot \quad \text{and} \quad x = \frac{3\pi}{2} + 2n\pi$$

where n is an integer.

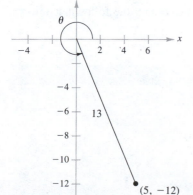

FIGURE 8.9

EXAMPLE 2 *Evaluating Functions Involving Double Angles*

Use the fact that

$$\cos \theta = \frac{5}{13}, \quad \frac{3\pi}{2} < \theta < 2\pi$$

to find $\sin 2\theta$, $\cos 2\theta$, and $\tan 2\theta$.

SOLUTION

From Figure 8.9, we can see that

$$\sin \theta = \frac{y}{r} = -\frac{12}{13}.$$

Consequently,

$$\sin 2\theta = 2 \sin \theta \cos \theta = 2\left(\frac{-12}{13}\right)\left(\frac{5}{13}\right) = -\frac{120}{169}$$

$$\cos 2\theta = 2 \cos^2 \theta - 1 = 2\left(\frac{25}{169}\right) - 1 = -\frac{119}{169}$$

$$\tan 2\theta = \frac{\sin 2\theta}{\cos 2\theta} = \frac{120}{119}.$$

The double-angle formulas are not restricted to angles 2θ and θ. Other *double* combinations like 4θ and 2θ or 6θ and 3θ are also valid. Here are a couple of examples.

$$\sin 4\theta = 2 \sin 2\theta \cos 2\theta$$
$$\cos 6\theta = \cos^2 3\theta - \sin^2 3\theta$$

By using double-angle formulas together with the sum formulas derived in the previous section, we can form other multiple-angle formulas. For instance, in the following example, we derive a *triple*-angle formula for $\sin 3x$.

EXAMPLE 3   *Deriving a Triple-Angle Formula*

Express $\sin 3x$ in terms of $\sin x$.

SOLUTION

Considering $3x = 2x + x$, we have

$$\begin{aligned}
\sin 3x &= \sin(2x + x) \\
&= \sin 2x \cos x + \cos 2x \sin x \\
&= 2 \sin x \cos x \cos x + (1 - 2 \sin^2 x)\sin x \\
&= 2 \sin x \cos^2 x + \sin x - 2 \sin^3 x \\
&= 2 \sin x(1 - \sin^2 x) + \sin x - 2 \sin^3 x \\
&= 2 \sin x - 2 \sin^3 x + \sin x - 2 \sin^3 x \\
&= 3 \sin x - 4 \sin^3 x.
\end{aligned}$$

The double-angle formulas can be used to obtain the following **power-reducing formulas.**

Power-Reducing Formulas

$$\sin^2 u = \frac{1 - \cos 2u}{2}$$

$$\cos^2 u = \frac{1 + \cos 2u}{2}$$

$$\tan^2 u = \frac{1 - \cos 2u}{1 + \cos 2u}$$

PROOF

The first two formulas can be verified by solving for $\sin^2 u$ and $\cos^2 u$, respectively, in the double-angle formulas

$$\cos 2u = 1 - 2\sin^2 u \quad \text{and} \quad \cos 2u = 2\cos^2 u - 1.$$

The third formula can be verified using the fact that

$$\tan^2 u = \frac{\sin^2 u}{\cos^2 u}.$$

Example 4 shows a typical power reduction that is used in calculus.

EXAMPLE 4 *Reducing the Power of a Trigonometric Function*

Rewrite $\sin^4 x$ as a sum involving first powers of the cosine of multiple angles.

SOLUTION

Note the repeated use of power-reducing formulas in the following procedure.

$$\sin^4 x = (\sin^2 x)^2 = \left(\frac{1 - \cos 2x}{2}\right)^2$$

$$= \frac{1}{4}(1 - 2\cos 2x + \cos^2 2x)$$

$$= \frac{1}{4}\left(1 - 2\cos 2x + \frac{1 + \cos 4x}{2}\right)$$

$$= \frac{1}{4} - \frac{1}{2}\cos 2x + \frac{1}{8} + \frac{1}{8}\cos 4x$$

$$= \frac{3}{8} - \frac{1}{2}\cos 2x + \frac{1}{8}\cos 4x$$

$$= \frac{1}{8}(3 - 4\cos 2x + \cos 4x)$$

We can derive some useful alternative forms of the power-reducing formulas by replacing u with $u/2$. The results are called **half-angle formulas.**

$$\sin\frac{u}{2} = \pm\sqrt{\frac{1 - \cos u}{2}} \qquad \cos\frac{u}{2} = \pm\sqrt{\frac{1 + \cos u}{2}}$$

The signs depend upon the quadrant in which $u/2$ lies.

$$\tan\frac{u}{2} = \frac{1 - \cos u}{\sin u} = \frac{\sin u}{1 + \cos u}$$

EXAMPLE 5 *Using a Half-Angle Formula*

Find the exact value of sin 105°.

SOLUTION

We begin by noting that 105° is half of 210°. Then, using the half-angle formula for $\sin(u/2)$ and the fact that 105° lies in Quadrant II, we have

$$\sin 105° = \sqrt{\frac{1 - \cos 210°}{2}} = \sqrt{\frac{1 - (-\cos 30°)}{2}}$$

$$= \sqrt{\frac{1 + (\sqrt{3}/2)}{2}} = \frac{\sqrt{2 + \sqrt{3}}}{2}.$$

Note that we chose the positive square root because sin θ is positive in Quadrant II.

Remark: Try using your calculator to verify the result obtained in Example 5. That is, evaluate

$$\sin 105° \quad \text{and} \quad \frac{\sqrt{2 + \sqrt{3}}}{2}$$

and you will see that both values are approximately 0.9659258.

EXAMPLE 6 *Solving a Trigonometric Equation*

Find all solutions of

$$2 - \sin^2 x = 2 \cos^2 \frac{x}{2}.$$

SOLUTION

$$2 - \sin^2 x = 2\left(\frac{1 + \cos x}{2}\right) \qquad \textit{Half-angle formula}$$

$$2 - \sin^2 x = 1 + \cos x$$

$$2 - (1 - \cos^2 x) = 1 + \cos x \qquad \textit{Pythagorean identity}$$

$$\cos^2 x - \cos x = 0 \qquad \textit{Simplify}$$

$$\cos x (\cos x - 1) = 0 \qquad \textit{Factor}$$

$$\cos x = 0, \qquad\qquad \cos x = 1 \qquad \textit{Set factors to zero}$$

$$x = \frac{\pi}{2}, \frac{3\pi}{2} \qquad\qquad x = 0 \qquad \textit{Solutions in } [0, 2\pi)$$

Therefore, the general solution is

$$x = 2n\pi, \qquad x = \frac{\pi}{2} + 2n\pi, \qquad \text{and} \qquad x = \frac{3\pi}{2} + 2n\pi$$

where n is an integer.

Product-Sum Formulas

Each of the following **product-to-sum formulas** are easily verified using the sum and difference formulas discussed in the preceding section.

Product-to-Sum Formulas

$$\sin u \sin v = \frac{1}{2}[\cos(u - v) - \cos(u + v)]$$

$$\cos u \cos v = \frac{1}{2}[\cos(u - v) + \cos(u + v)]$$

$$\sin u \cos v = \frac{1}{2}[\sin(u + v) + \sin(u - v)]$$

$$\cos u \sin v = \frac{1}{2}[\sin(u + v) - \sin(u - v)]$$

EXAMPLE 7 *Writing Products as Sums*

Rewrite $\cos 5x \sin 4x$ as a sum or difference.

SOLUTION

$$\cos 5x \sin 4x = \frac{1}{2}[\sin(5x + 4x) - \sin(5x - 4x)]$$

$$= \frac{1}{2} \sin 9x - \frac{1}{2} \sin x$$

Occasionally, it is useful to reverse the procedure and write a sum of trigonometric functions as a product. This can be accomplished with the following **sum-to-product formulas.**

Sum-to-Product Formulas

$$\sin x + \sin y = 2 \sin\left(\frac{x + y}{2}\right)\cos\left(\frac{x - y}{2}\right)$$

$$\sin x - \sin y = 2 \cos\left(\frac{x + y}{2}\right)\sin\left(\frac{x - y}{2}\right)$$

$$\cos x + \cos y = 2 \cos\left(\frac{x + y}{2}\right)\cos\left(\frac{x - y}{2}\right)$$

$$\cos x - \cos y = -2 \sin\left(\frac{x + y}{2}\right)\sin\left(\frac{x - y}{2}\right)$$

PROOF

To prove the first formula, we let $x = u + v$ and $y = u - v$. Then, substituting $u = (x + y)/2$ and $v = (x - y)/2$ in the product-to-sum formula

$$\sin u \cos v = \frac{1}{2}[\sin(u + v) + \sin(u - v)]$$

we get

$$\sin\left(\frac{x + y}{2}\right)\cos\left(\frac{x - y}{2}\right) = \frac{1}{2}(\sin x + \sin y)$$

or equivalently,

$$\sin x + \sin y = 2 \sin\left(\frac{x + y}{2}\right)\cos\left(\frac{x - y}{2}\right).$$

EXAMPLE 8 *Using a Sum-to-Product Formula*

Find the exact value of $\cos 195° + \cos 105°$.

SOLUTION

Using the appropriate sum-to-product formula, we get

$$\cos 195° + \cos 105° = 2 \cos\left(\frac{195° + 105°}{2}\right)\cos\left(\frac{195° - 105°}{2}\right)$$

$$= 2 \cos 150° \cos 45°$$

$$= 2\left(-\frac{\sqrt{3}}{2}\right)\left(\frac{\sqrt{2}}{2}\right)$$

$$= -\frac{\sqrt{6}}{2}.$$

EXAMPLE 9 *Solving a Trigonometric Equation*

Find all solutions of $\sin 5x + \sin 3x = 0$.

SOLUTION

$$\sin 5x + \sin 3x = 0 \qquad \qquad \textit{Given}$$

$$2 \sin\left(\frac{5x + 3x}{2}\right)\cos\left(\frac{5x - 3x}{2}\right) = 0 \qquad \textit{Sum-to-product formula}$$

$$2 \sin 4x \cos x = 0 \qquad \qquad \textit{Simplify}$$

$$\sin 4x = 0, \qquad \cos x = 0 \qquad \textit{Set factors to zero}$$

In the interval $[0, 2\pi)$, the solutions of $\sin 4x = 0$ are

$$x = 0, \quad \frac{\pi}{4}, \quad \frac{\pi}{2}, \quad \frac{3\pi}{4}, \quad \pi, \quad \frac{5\pi}{4}, \quad \frac{3\pi}{2}, \quad \frac{7\pi}{4}.$$

Moreover, the equation $\cos x = 0$ yields no additional solutions, and we can conclude that the solutions are of the form

$$x = \frac{n\pi}{4}$$

where n is an integer.

EXAMPLE 10 *Verifying a Trigonometric Identity*

Verify the identity

$$\frac{\sin t + \sin 3t}{\cos t + \cos 3t} = \tan 2t.$$

SOLUTION

Using appropriate sum-to-product formulas, we have

$$\frac{\sin t + \sin 3t}{\cos t + \cos 3t} = \frac{2 \sin 2t \cos(-t)}{2 \cos 2t \cos(-t)}$$

$$= \frac{\sin 2t}{\cos 2t}$$

$$= \tan 2t.$$

WARM UP

Factor the expressions.

1. $2 \sin x + \sin x \cos x$

2. $\cos^2 x - \cos x - 2$

Find all solutions in the interval $[0, 2\pi)$.

3. $\sin 2x = 0$

4. $\cos 2x = 0$

5. $\cos \dfrac{x}{2} = 0$

6. $\sin \dfrac{x}{2} = 0$

Simplify the expressions.

7. $\dfrac{1 - \cos(\pi/4)}{2}$

8. $\dfrac{1 + \cos(\pi/3)}{2}$

9. $\dfrac{2 \sin 3x \cos x}{2 \cos 3x \cos x}$

10. $(1 - 2 \sin^2 x)\cos x - 2 \sin x \cos x \sin x$

EXERCISES 8.5

In Exercises 1–6, use a double-angle formula to determine the exact values of the sine, cosine, and tangent of the given angle.

1. $90° = 2(45°)$

2. $180° = 2(90°)$

3. $60° = 2(30°)$

4. $120° = 2(60°)$

5. $\dfrac{2\pi}{3} = 2\left(\dfrac{\pi}{3}\right)$

6. $\dfrac{3\pi}{2} = 2\left(\dfrac{3\pi}{4}\right)$

In Exercises 7–12, find the exact values of $\sin 2u$, $\cos 2u$, and $\tan 2u$.

7. $\sin u = \dfrac{3}{5}, \quad 0 < u < \dfrac{\pi}{2}$

8. $\cos u = -\dfrac{2}{3}, \quad \dfrac{\pi}{2} < u < \pi$

9. $\tan u = \dfrac{1}{2}, \quad \pi < u < \dfrac{3\pi}{2}$

10. $\cot u = -4, \quad \dfrac{3\pi}{2} < u < 2\pi$

11. $\sec u = -\dfrac{5}{2}, \quad \dfrac{\pi}{2} < u < \pi$

12. $\csc u = 3, \quad \dfrac{\pi}{2} < u < \pi$

In Exercises 13–18, use half-angle formulas to determine the exact values of the sine, cosine, and tangent of the given angle.

13. $105° = \frac{1}{2}(210°)$

14. $165° = \frac{1}{2}(330°)$

15. $112° \ 30' = \frac{1}{2}(225°)$

16. $67° \ 30' = \frac{1}{2}(135°)$

17. $\dfrac{\pi}{8} = \dfrac{1}{2}\left(\dfrac{\pi}{4}\right)$

18. $\dfrac{\pi}{12} = \dfrac{1}{2}\left(\dfrac{\pi}{6}\right)$

In Exercises 19–24, find the exact values of $\sin(u/2)$, $\cos(u/2)$, and $\tan(u/2)$ by using the half-angle formulas and the given information.

19. $\sin u = \dfrac{5}{13}, \quad \dfrac{\pi}{2} < u < \pi$

20. $\cos u = \dfrac{3}{5}, \quad 0 < u < \dfrac{\pi}{2}$

21. $\tan u = -\dfrac{5}{8}, \quad \dfrac{3\pi}{2} < u < 2\pi$

22. $\cot u = 3, \quad \pi < u < \dfrac{3\pi}{2}$

23. $\csc u = -\dfrac{5}{3}, \quad \pi < u < \dfrac{3\pi}{2}$

24. $\sec u = -\dfrac{7}{2}, \quad \dfrac{\pi}{2} < u < \pi$

In Exercises 25–28, use the half-angle formulas to simplify the given expression.

25. $\sqrt{\dfrac{1 - \cos 6x}{2}}$

26. $\sqrt{\dfrac{1 + \cos 4x}{2}}$

27. $-\sqrt{\dfrac{1 - \cos 8x}{1 + \cos 8x}}$

28. $-\sqrt{\dfrac{1 - \cos(x - 1)}{2}}$

In Exercises 29 and 30, use the power-reducing formulas to write each expression in terms of the first power of the cosine.

29. (a) $\cos^4 x$
(b) $\sin^2 x \cos^4 x$

30. (a) $\cos^6 x$
(b) $\sin^2 x \cos^2 x$

In Exercises 31–40, rewrite the given product as a sum.

31. $6 \sin \dfrac{\pi}{4} \cos \dfrac{\pi}{4}$

32. $4 \sin \dfrac{\pi}{3} \cos \dfrac{5\pi}{6}$

33. $\sin 5\theta \cos 3\theta$

34. $3 \sin 2\alpha \sin 3\alpha$

35. $5 \cos(-5\beta) \cos 3\beta$

36. $\cos 2\theta \cos 4\theta$

37. $\sin(x + y) \sin(x - y)$

38. $\sin(x + y) \cos(x - y)$

39. $\sin(\theta + \pi) \cos(\theta - \pi)$

40. $10 \cos 75° \cos 15°$

In Exercises 41–50, express the given sum (or difference) as a product.

41. $\sin 60° + \sin 30°$

42. $\cos 120° + \cos 30°$

43. $\cos \dfrac{3\pi}{4} - \cos \dfrac{\pi}{4}$

44. $\sin 5\theta - \sin 3\theta$

45. $\cos 6x + \cos 2x$

46. $\sin x + \sin 5x$

47. $\sin(\alpha + \beta) - \sin(\alpha - \beta)$

48. $\cos\left(\theta + \dfrac{\pi}{2}\right) - \cos\left(\theta - \dfrac{\pi}{2}\right)$

49. $\cos(\phi + 2\pi) + \cos \phi$

50. $\sin\left(x + \dfrac{\pi}{2}\right) + \sin\left(x - \dfrac{\pi}{2}\right)$

In Exercises 51–72, verify the given identity.

51. $\csc 2\theta = \dfrac{\csc \theta}{2 \cos \theta}$

52. $\sec 2\theta = \dfrac{\sec^2 \theta}{2 - \sec^2 \theta}$

53. $\cos^2 2\alpha - \sin^2 2\alpha = \cos 4\alpha$

54. $\sin \dfrac{\alpha}{3} \cos \dfrac{\alpha}{3} = \dfrac{1}{2} \sin \dfrac{2\alpha}{3}$

55. $(\sin x + \cos x)^2 = 1 + \sin 2x$

56. $\cos^4 x - \sin^4 x = \cos 2x$

57. $\cos 3\beta = \cos^3 \beta - 3 \sin^2 \beta \cos \beta$

58. $\sin 4\beta = 4 \sin \beta \cos \beta(1 - 2 \sin^2 \beta)$

59. $1 + \cos 10y = 2 \cos^2 5y$

60. $\dfrac{\cos 3\beta}{\cos \beta} = 1 - 4 \sin^2 \beta$

61. $\sec \dfrac{u}{2} = \pm \sqrt{\dfrac{2 \tan u}{\tan u + \sin u}}$

62. $\tan \dfrac{u}{2} = \csc u - \cot u$

63. $\dfrac{\cos 4x + \cos 2x}{\sin 4x + \sin 2x} = \cot 3x$

64. $\dfrac{\cos 3x - \cos x}{\sin 3x - \sin x} = -\tan 2x$

65. $\dfrac{\cos 4x - \cos 2x}{2 \sin 3x} = -\sin x$

66. $\dfrac{\sin x \pm \sin y}{\cos x + \cos y} = \tan \dfrac{x \pm y}{2}$

67. $\dfrac{\sin x \pm \sin y}{\cos x - \cos y} = -\cot \dfrac{x \mp y}{2}$

68. $\dfrac{\sin x + \sin y}{\sin x - \sin y} = \dfrac{\tan[(x + y)/2]}{\tan[(x - y)/2]}$

69. $\dfrac{\cos t + \cos 3t}{\sin 3t - \sin t} = \cot t$

70. $\dfrac{\sin 6t - \sin 2t}{\cos 2t + \cos 6t} = \tan 2t$

71. $\sin^2 4x - \sin^2 2x = \sin 2x \sin 6x$

72. $\sin\left(\dfrac{\pi}{6} + x\right) + \sin\left(\dfrac{\pi}{6} - x\right) = \cos x$

In Exercises 73–84, find all solutions in the interval $[0, 2\pi)$.

73. $4 \sin x \cos x = 1$

74. $\sin 2x \sin x = \cos x$

75. $\cos 2x = \cos x$

76. $\cos 3x - \cos x = 0$

77. $\sin 4x + 2 \sin 2x = 0$

78. $\tan\left(x + \dfrac{\pi}{4}\right) - \tan\left(x - \dfrac{\pi}{4}\right) = 4$

79. $(\sin 2x + \cos 2x)^2 = 1$

80. $\cos 2x - \cos 6x = 0$

81. $\sin 6x + \sin 2x = 0$

82. $\sin^2 3x - \sin^2 x = 0$

83. $\dfrac{\cos 2x}{\sin 3x - \sin x} = 1$

84. $\sin 2x + \sin 4x + \sin 6x = 0$

In Exercises 85 and 86, sketch the graph by using the power-reducing formulas.

85. $f(x) = \sin^2 x$ **86.** $f(x) = \cos^2 x$

In Exercises 87 and 88, write the trigonometric expression as an algebraic expression in x.

87. $\sin(2 \arcsin x)$ **88.** $\cos(2 \arccos x)$

In Exercises 89 and 90, verify the identity for the complementary angles ϕ and θ.

89. $\sin(\phi - \theta) = \cos 2\theta$ **90.** $\cos(\phi - \theta) = \sin 2\theta$

In Exercises 91–93, prove the product-to-sum formulas.

91. $\cos u \cos v = \frac{1}{2}[\cos(u - v) + \cos(u + v)]$

92. $\sin u \cos v = \frac{1}{2}[\sin(u + v) + \sin(u - v)]$

93. $\cos u \sin v = \frac{1}{2}[\sin(u + v) - \sin(u - v)]$

CHAPTER 8 REVIEW EXERCISES

In Exercises 1–10, simplify the given expression.

1. $\dfrac{1}{\cot^2 x + 1}$ **2.** $\dfrac{\sin 2\alpha}{\cos^2 \alpha - \sin^2 \alpha}$

3. $\dfrac{\sin^2 \alpha - \cos^2 \alpha}{\sin^2 \alpha - \sin \alpha \cos \alpha}$ **4.** $\dfrac{\sin^3 \beta + \cos^3 \beta}{\sin \beta + \cos \beta}$

5. $\cos^2 \beta + \cos^2 \beta \tan^2 \beta$

6. $\dfrac{\sin \theta}{1 + \cos \theta} + \dfrac{1 + \cos \theta}{\sin \theta}$

7. $\tan^2 \theta(\csc^2 \theta - 1)$ **8.** $\dfrac{2 \tan(x + 1)}{1 - \tan^2(x + 1)}$

9. $1 - 4 \sin^2 x \cos^2 x$ **10.** $\sqrt{\dfrac{1 - \cos^2 x}{1 + \cos x}}$

In Exercises 11–40, verify the given identity.

11. $\tan x(1 - \sin^2 x) = \frac{1}{2} \sin 2x$

12. $\cos^3 x \sin^2 x = (\sin^2 x - \sin^4 x) \cos x$

13. $\sin^5 x \cos^2 x = (\cos^2 x - 2 \cos^4 x + \cos^6 x) \sin x$

14. $\sin^4 2x = \frac{1}{8}(\cos 8x - 4 \cos 4x + 3)$

15. $\sin^2 x \cos^4 x = \frac{1}{16}(1 - \cos 4x + 2 \sin^2 2x \cos 2x)$

16. $\sin 3x \cos 2x = \frac{1}{2}(\sin 5x + \sin x)$

17. $\sin 3\theta \sin \theta = \frac{1}{2}(\cos 2\theta - \cos 4\theta)$

18. $\sqrt{1 - \cos x} = \dfrac{|\sin x|}{\sqrt{1 + \cos x}}$

19. $\sqrt{\dfrac{1 - \sin \theta}{1 + \sin \theta}} = \dfrac{1 - \sin \theta}{|\cos \theta|}$

20. $\sin 4x = 8 \cos^3 x \sin x - 4 \cos x \sin x$

21. $\cos 3x = 4 \cos^3 x - 3 \cos x$

22. $\cos 4x = 8 \cos^4 x - 8 \cos^2 x + 1$

23. $\sin\left(x - \dfrac{3\pi}{2}\right) = \cos x$

24. $\cos\left(x + \dfrac{\pi}{2}\right) = -\sin x$

25. $\dfrac{\sec x - 1}{\tan x} = \tan \dfrac{x}{2}$

26. $\dfrac{2 \cos 3x}{\sin 4x - \sin 2x} = \csc x$

27. $\dfrac{\cos 3x - \cos x}{\sin 3x - \sin x} = -\tan 2x$

28. $1 - \cos 2x = 2 \sin^2 x$

29. $\sin(\pi - x) = \sin x$

30. $\cot\left(\dfrac{\pi}{2} - x\right) = \tan x$

31. $2 \sin y \cos y \sec 2y = \tan 2y$

32. $\cot \dfrac{u}{2} = \pm\sqrt{\dfrac{\sec u + 1}{\sec u - 1}}$

33. $\sin \dfrac{\theta}{2} + \cos \dfrac{\theta}{2} = \pm\sqrt{1 + \sin \theta}$

34. $\cos^2 5x - \cos^2 x = -\sin 4x \sin 6x$

35. $\tan^2 x = \dfrac{1 - \cos 2x}{1 + \cos 2x}$

36. $\dfrac{\sin(\alpha + \beta)}{\cos \alpha \cos \beta} = \tan \alpha + \tan \beta$

37. $\sin 2x + \sin 4x - \sin 6x = 4 \sin x \sin 2x \sin 3x$

38. $\sin 2x + \sin 4x + \sin 6x = 4 \cos x \cos 2x \sin 3x$

39. $1 + \cos 2x + \cos 4x + \cos 6x = 4 \cos x \cos 2x \cos 3x$

40. $1 - \cos 2x + \cos 4x - \cos 6x = 4 \sin x \cos 2x \sin 3x$

In Exercises 41–44, find the exact value of the trigonometric function by using the sum, difference, or half-angle formulas.

41. $\sin \dfrac{5\pi}{12} = \sin\left(\dfrac{2\pi}{3} - \dfrac{\pi}{4}\right)$

42. $\cos 285° = \cos(225° + 60°)$

43. $\cos(157° \, 30') = \cos \dfrac{315°}{2}$

44. $\sin \dfrac{3\pi}{8} = \sin\left[\dfrac{1}{2}\left(\dfrac{3\pi}{4}\right)\right]$

In Exercises 45–50, find the exact value of the trigonometric function given that

$\sin u = \dfrac{3}{4}$ and $\cos v = -\dfrac{5}{13}$ (u and v in Quadrant II).

45. $\sin(u + v)$ **46.** $\tan(u + v)$

47. $\cos(u - v)$ **48.** $\sin 2v$

49. $\cos \dfrac{u}{2}$ **50.** $\tan 2v$

In Exercises 51–60, find all solutions of the given equation in the interval $[0, 2\pi)$.

51. $\sin x - \tan x = 0$ **52.** $\csc x - 2 \cot x = 0$

53. $\dfrac{1 + \sin x}{\cos x} + \dfrac{\cos x}{1 + \sin x} = 4$

54. $\cos x = \cos \dfrac{x}{2}$

55. $\sin 2x + \sqrt{2} \sin x = 0$

56. $\cos 4x - 7 \cos 2x = 8$

57. $\cos^2 x + \sin x = 1$ **58.** $\sin 4x - \sin 2x = 0$

59. $\tan^3 x - \tan^2 x + 3 \tan x - 3 = 0$

60. $\sin x + \sin 3x + \sin 5x = 0$

In Exercises 61 and 62, write the trigonometric expression as a product.

61. $\cos 3\theta + \cos 2\theta$

62. $\sin\left(x + \dfrac{\pi}{4}\right) - \sin\left(x - \dfrac{\pi}{4}\right)$

In Exercises 63 and 64, write the trigonometric product as a sum or difference.

63. $\sin 3\alpha \sin 2\alpha$ **64.** $\cos \dfrac{x}{2} \cos \dfrac{x}{4}$

65. The function $f(x) = x \sin x$ has maximum or minimum values when

$$\tan x + x = 0.$$

Approximate the solutions of this equation in the interval $[-\pi, \pi]$ and sketch a graph of the function. [Use the Bisection Method discussed in Section 4.6 and list your approximations to three-decimal-place accuracy.]

66. Write $\cos(2 \arccos 2x)$ as an algebraic expression in x.

67. A standing wave on a string of given length is modeled by the equation

$$y = A\left(\cos\left[2\pi\left(\dfrac{t}{T} - \dfrac{x}{\lambda}\right)\right] + \cos\left[2\pi\left(\dfrac{t}{T} + \dfrac{x}{\lambda}\right)\right]\right).$$

Use the trigonometric identities for the cosine of the sum and difference of two angles to verify that the following equation is an equivalent model for the standing wave

$$y = 2A \cos \dfrac{2\pi t}{T} \cos \dfrac{2\pi x}{\lambda}.$$

CHAPTER 9

Additional Applications of Trigonometry

9.1 Law of Sines

In Chapter 7 we looked at techniques for solving right triangles. In this section and the next, we will solve triangles that have no right angles. Such triangles are called **oblique.** As standard notation, we label the vertices of a triangle as A, B, and C, and their opposite sides as a, b, and c, as shown in Figure 9.1.

To solve an oblique triangle, we need to know the measure of at least one side and any two other parts of the triangle—either two sides, two angles, or one angle and one side. This breaks down into the following four possible cases.

1. Two angles and any side (AAS or ASA).
2. Two sides and an angle opposite one of them (SSA).
3. Three sides (SSS).
4. Two sides and their included angle (SAS).

The first two cases can be solved using what is called the **Law of Sines,** while the last two cases require the **Law of Cosines** (to be discussed in Section 9.2).

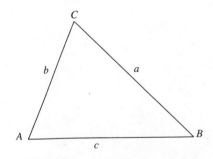

FIGURE 9.1

506

Law of Sines

If ABC is a triangle with sides a, b, and c, then

$$\frac{a}{\sin A} = \frac{b}{\sin B} = \frac{c}{\sin C}.$$

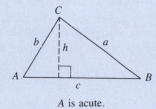 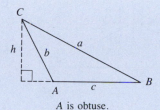

A is acute.   A is obtuse.

Oblique Triangles

FIGURE 9.2

PROOF

Let h be the altitude of either triangle shown in Figure 9.2. Then, we have

$$\sin A = \frac{h}{b} \quad \text{or} \quad h = b \sin A$$

$$\sin B = \frac{h}{a} \quad \text{or} \quad h = a \sin B.$$

Equating these two values of h, we have

$$a \sin B = b \sin A \quad \text{or} \quad \frac{a}{\sin A} = \frac{b}{\sin B}.$$

Note that $\sin A \neq 0$ and $\sin B \neq 0$ because no angle of a triangle can have a measure of $0°$ or $180°$. By constructing an altitude from vertex B to extended side AC, we can show that

$$\frac{a}{\sin A} = \frac{c}{\sin C}.$$

Hence, the Law of Sines is established.

Remark: The Law of Sines can also be written in the form

$$\frac{\sin A}{a} = \frac{\sin B}{b} = \frac{\sin C}{c}.$$

Additional Applications of Trigonometry

When using a calculator with the Law of Sines, remember to store all intermediate calculations. By not rounding until the final result, you minimize the round-off error.

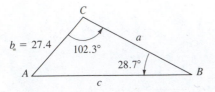

FIGURE 9.3

EXAMPLE 1 *Given Two Angles and One Side—AAS*

Given a triangle with $C = 102.3°$, $B = 28.7°$, and $b = 27.4$, as shown in Figure 9.3, find the remaining angle and sides.

SOLUTION

The third angle of the triangle is

$$A = 180° - B - C = 180° - 28.7° - 102.3° = 49.0°.$$

By the Law of Sines, we have

$$\frac{a}{\sin 49°} = \frac{b}{\sin 28.7°} = \frac{c}{\sin 102.3°}.$$

Because $b = 27.4$, we obtain

$$a = \frac{27.4}{\sin 28.7°}(\sin 49°) \approx 43.06$$

and

$$c = \frac{27.4}{\sin 28.7°}(\sin 102.3°) \approx 55.75.$$

Note that the ratio $(27.4)/(\sin 28.7°)$ occurs in both solutions, and you can save time by storing this result for repeated use.

When solving triangles, a careful sketch is useful as a quick test for the feasibility of an answer. Remember that the longest side lies opposite the largest angle, and the shortest side lies opposite the smallest angle of a triangle.

EXAMPLE 2 *Given Two Angles and One Side—ASA*

A pole tilts *toward* the sun at an 8° angle from vertical, and it casts a 22-foot shadow. The angle of elevation from the tip of the shadow to the top of the pole is 43°. How tall is the pole?

Law of Sines

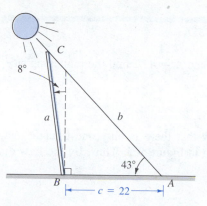

FIGURE 9.4

SOLUTION

From Figure 9.4, note that $A = 43°$, and $B = 90° + 8° = 98°$. Thus, the third angle is

$$C = 180° - A - B = 180° - 43° - 98° = 39°.$$

By the Law of Sines, we have

$$\frac{a}{\sin 43°} = \frac{c}{\sin 39°}.$$

Because $c = 22$ feet, the length of the pole is

$$a = \frac{22}{\sin 39°}(\sin 43°) \approx 23.84 \text{ feet.}$$

Remark: For practice, try reworking Example 2 for a pole that tilts *away from* the sun under the same conditions.

In Examples 1 and 2 we saw that two angles (whose sum is less than 180°) and one side determine a unique triangle. However, if two sides and one opposite angle are given, three possible situations can occur: (1) no such triangle exists, (2) one such triangle exists, or (3) two distinct triangles may satisfy the conditions. The possibilities in this *ambiguous* case (SSA) are summarized in the following table.

The Ambiguous Case (SSA) (Given: a, b, and A)

	A Is Acute				A Is Obtuse	
Sketch ($h = b \sin A$)						
Necessary Condition	$a < h$	$a = h$	$a > b$	$h < a < b$	$a \le b$	$a > b$
Triangles Possible	None	One	One	Two	None	One

In order to determine which of the possibilities hold for a given pair of sides and opposite angle, we suggest that you make a sketch, as demonstrated in the next three examples.

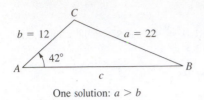

One solution: $a > b$

FIGURE 9.5

EXAMPLE 3 *Single Solution Case—SSA*

Given a triangle with $a = 22$, $b = 12$, and $A = 42°$, find the remaining side and angles.

SOLUTION

Since A is acute and $a > b$, we know that there is only one triangle that satisfies the given conditions, as shown in Figure 9.5. Thus, by the Law of Sines, we have

$$\frac{22}{\sin 42°} = \frac{12}{\sin B}$$

which implies that

$$\sin B = 12\left(\frac{\sin 42°}{22}\right) \approx 0.3649803$$

$$B \approx 21.41°.$$

Now, we can determine that $C \approx 180° - 42° - 21.41° = 116.59°$, and the remaining side is given by

$$\frac{c}{\sin 116.59°} = \frac{22}{\sin 42°}$$

$$c = \sin 116.59°\left(\frac{22}{\sin 42°}\right) \approx 29.40.$$

EXAMPLE 4 *No Solution Case—SSA*

Show that there is no triangle that satisfies either of the following conditions.

(a) $a = 15$, $b = 25$, $A = 85°$

(b) $a = 15.2$, $b = 20$, $A = 110°$

SOLUTION

(a) We begin by making the sketch shown in Figure 9.6. From this figure it appears that no triangle is formed. We can verify this using the Law of Sines, as follows.

$$\frac{a}{\sin A} = \frac{b}{\sin B}$$

$$\frac{15}{\sin 85°} = \frac{25}{\sin B}$$

$$\sin B = 25\left(\frac{\sin 85°}{15}\right) \approx 1.660 > 1$$

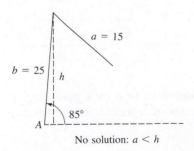

No solution: $a < h$

FIGURE 9.6

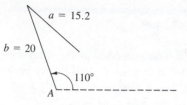

No solution: $a < b$ and $A > 90°$

FIGURE 9.7

This contradicts the fact that $|\sin B| \leq 1$. Hence, no triangle can be formed having sides $a = 15$ and $b = 25$ and an angle of $A = 85°$.

(b) Because A is obtuse and $a = 15.2$ is less than $b = 20$, we conclude that there is *no solution*, as shown in Figure 9.7. Try using the Law of Sines to verify this.

EXAMPLE 5 *Two-Solution Case—SSA*

Find two triangles for which $a = 12$, $b = 31$, and $A = 20.5°$.

SOLUTION

To begin, we note that

$$h = b \sin A = 31(\sin 20.5°) \approx 10.86.$$

Hence, $h < a < b$ and we conclude that there are two possible triangles. By the Law of Sines, we obtain

$$\frac{a}{\sin A} = \frac{b}{\sin B}$$

which implies that

$$\sin B = b\left(\frac{\sin A}{a}\right) = 31\left(\frac{\sin 20.5°}{12}\right) \approx 0.9047.$$

There are two angles $B_1 \approx 64.8°$ and $B_2 \approx 115.2°$ between $0°$ and $180°$ whose sine is 0.9047. For $B_1 \approx 64.8°$, we obtain

$$C \approx 180° - 20.5° - 64.8° = 94.7°$$

$$c = \frac{a}{\sin A}(\sin C) = \frac{12}{\sin 20.5°}(\sin 94.7°) \approx 34.15.$$

For $B_2 \approx 115.2°$, we obtain

$$C \approx 180° - 20.5° - 115.2° = 44.3°$$

$$c = \frac{a}{\sin A}(\sin C) = \frac{12}{\sin 20.5°}(\sin 44.3°) \approx 23.9.$$

The resulting triangles are shown in Figures 9.8 and 9.9.

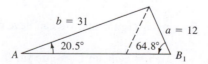

FIGURE 9.8

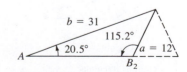

FIGURE 9.9

Additional Applications of Trigonometry

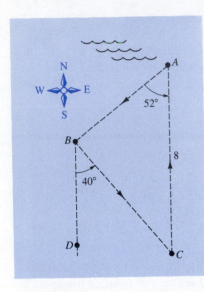

FIGURE 9.10

EXAMPLE 6   *An Application of the Law of Sines*

The course for a boat race starts at point A and proceeds in the direction S 52° W to point B, then in the direction S 40° E to point C, and finally back to A, as shown in Figure 9.10. The point C lies 8 kilometers directly south of point A. Approximate the total distance of the race course.

SOLUTION

Because lines BD and AC are parallel, it follows that $\angle BCA = \angle DBC$. Consequently, triangle ABC has the measures shown in Figure 9.11. For angle B, we have

$$B = 180° - 52° - 40° = 88°.$$

Thus, using the Law of Sines,

$$\frac{a}{\sin 52°} = \frac{b}{\sin 88°} = \frac{c}{\sin 40°},$$

we let $b = 8$ and obtain

$$a = \frac{8}{\sin 88°}(\sin 52°) \approx 6.308$$

$$c = \frac{8}{\sin 88°}(\sin 40°) \approx 5.145.$$

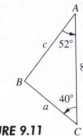

FIGURE 9.11

Finally, the total length of the course is approximately

$$\text{Length} \approx 8 + 6.308 + 5.145 = 19.453 \text{ kilometers.}$$

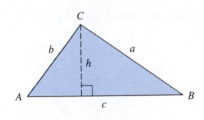

A is acute.

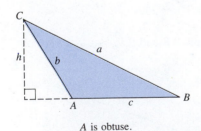

A is obtuse.

Oblique Triangles

FIGURE 9.12

The procedure used to prove the Law of Sines leads to a simple formula for the area of an oblique triangle. Referring to Figure 9.12, we note that each triangle has a height of

$$h = b \sin A.$$

Consequently, the area of each triangle is given by

$$\text{Area} = \frac{1}{2}(\text{base})(\text{height}) = \frac{1}{2}(c)(b \sin A) = \frac{1}{2}bc \sin A.$$

By similar arguments, we can develop the formulas

$$\text{Area} = \frac{1}{2}ab \sin C = \frac{1}{2}ac \sin B.$$

This leads to the following theorem.

Area of an Oblique Triangle

The area of any triangle is given by one-half the product of the lengths of two sides times the sine of their included angle. That is,

$$\text{Area} = \frac{1}{2}bc \sin A = \frac{1}{2}ab \sin C = \frac{1}{2}ac \sin B.$$

EXAMPLE 7 *Finding the Area of an Oblique Triangle*

Find the area of a triangular lot having two sides of lengths 90 meters and 52 meters and an included angle of 102°.

SOLUTION

Consider $a = 90$ m, $b = 52$ m, and angle $C = 102°$, as shown in Figure 9.13. Then the area of the triangle is

$$\text{Area} = \frac{1}{2}ab \sin C = \frac{1}{2}(90)(52)(\sin 102°)$$

$$\approx 2289 \text{ square meters.}$$

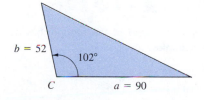

FIGURE 9.13

WARM UP

Solve the *right* triangle shown in the figure.

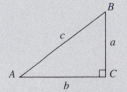

1. $a = 3$, $c = 6$ **2.** $a = 5$, $b = 5$
3. $b = 15$, $c = 17$ **4.** $A = 42°$, $a = 7.5$
5. $B = 10°$, $b = 4$ **6.** $B = 72°\ 15'$, $c = 150$

Find the altitude h of each triangle.

7.

8.

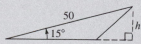

Solve for x.

9. $\dfrac{2}{\sin 30°} = \dfrac{9}{x}$

10. $\dfrac{100}{\sin 72°} = \dfrac{x}{\sin 60°}$

EXERCISES 9.1

In Exercises 1–16, find the remaining sides and angles of the triangle.

1.

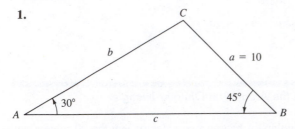

2.

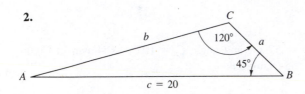

3.

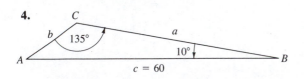

4.

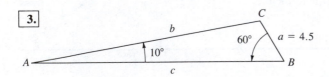

5. $A = 36°$, $a = 8$, $b = 5$ **6.** $A = 60°$, $a = 9$, $c = 10$

7. $A = 150°$, $C = 20°$, $a = 200$

8. $A = 24.3°$, $C = 54.6°$, $c = 2.68$

9. $A = 83° 20'$, $C = 54.6°$, $c = 18.1$

10. $A = 5° 40'$, $B = 8° 15'$, $b = 4.8$

11. $B = 15° 30'$, $a = 4.5$, $b = 6.8$

12. $C = 85° 20'$, $a = 35$, $c = 50$

13. $C = 145°$, $b = 4$, $c = 14$

14. $A = 100°$, $a = 125$, $c = 10$

15. $A = 110° 15'$, $a = 48$, $b = 16$

16. $B = 2° 45'$, $b = 6.2$, $c = 5.8$

In Exercises 17–22, solve the triangle, if possible. If two solutions exist, find both.

17. $A = 58°$, $a = 4.5$, $b = 12.8$

18. $A = 58°$, $a = 11.4$, $b = 12.8$

19. $A = 58°$, $a = 4.5$, $b = 5$

20. $A = 58°$, $a = 42.4$, $b = 50$

21. $A = 110°$, $a = 125$, $b = 200$

22. $A = 110°$, $a = 125$, $b = 100$

In Exercises 23 and 24, find a value for b such that the triangle has (a) one solution, (b) two solutions, and (c) no solution.

23. $A = 36°$, $a = 5$ **24.** $A = 60°$, $a = 10$

In Exercises 25–30, find the area of the triangle having the indicated sides and angles.

25. $C = 120°$, $a = 4$, $b = 6$

26. $B = 72° 30'$, $a = 105$, $c = 64$

27. $A = 43° 45'$, $b = 57$, $c = 85$

28. $A = 5° 15'$, $b = 4.5$, $c = 22$

29. $B = 130°$, $a = 62$, $c = 20$

30. $C = 84° 30'$, $a = 16$, $b = 20$

31. Find the length d of the brace required to support the streetlight shown in the figure.

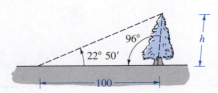

FIGURE FOR 31

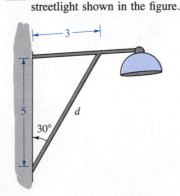

FIGURE FOR 32

32. Because of prevailing winds, a tree grew so that it was leaning 6° from the vertical. At a point 100 feet from the tree, the angle of elevation to the top of the tree is 22° 50′ (see figure). Find the height h of the tree.

33. A bridge is to be built across a small lake from B to C (see figure). The bearing from B to C is S 41° W. From a point A, 100 yards from B, the bearings to B and C are S 74° E and S 28° E, respectively. Find the distance from B to C.

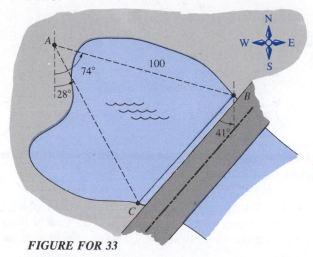

FIGURE FOR 33

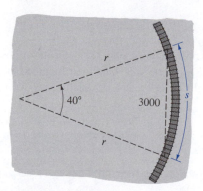

FIGURE FOR 34

34. The circular arc of a railroad curve has a chord of length 3000 feet, and a central angle of 40° (see figure). Find (a) the radius r of the circular arc and (b) the length s of the circular arc.

35. The angles of elevation to an airplane from two points A and B on level ground are 51° and 68°, respectively. A and B are 6 miles apart, and the airplane is between A and B in the same vertical plane.

36. The angles of elevation to an airplane from two points A and B on level ground are 51° and 68°, respectively. A and B are 2.5 miles apart, and the airplane is to the east of A and B in the same vertical plane. Find the altitude of the airplane.

37. Two fire towers A and B are 18.5 miles apart. The bearing from A to B is N 65° E. A fire is spotted by the ranger in each tower, and its bearings from A and B are N 28° E and N 16.5° W, respectively (see figure). Find the distance of the fire from each tower.

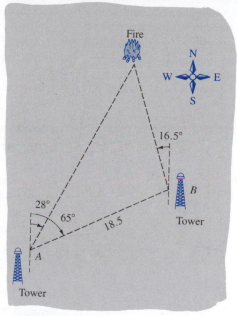

FIGURE FOR 37

38. A boat is sailing due east parallel to the shoreline at a speed of 10 miles per hour. At a given time the bearing to a lighthouse is S 72° E, and 15 minutes later the bearing is S 66° E (see figure). Find the distance from the boat to the shoreline if the lighthouse is at the shoreline.

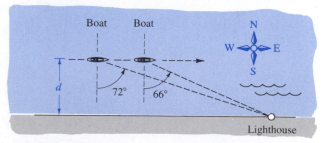

FIGURE FOR 38

39. A family is traveling due west on a road that passes a famous landmark. At a given time the bearing to the landmark is N 62° W, and after the family travels 5 miles farther the bearing is N 38° W. What is the closest the family will come to the landmark while on the road?

40. A rescue vehicle is located near an apartment complex, and its emergency light is turning at the rate of 30 revolutions per minute. One-eighth of a second after illuminating the nearest point on the apartment complex, the lightbeam reaches a point 50 feet along the apartments. One-eighth of a second later, it reaches a point 70 feet further along the apartment wall. How far is the rescue truck from the apartment complex?

41. The following information about a triangular parcel of land is given at a zoning board meeting: "One side is 450 feet long, and another is 120 feet long. The angle opposite the shorter side is 30°." Could this information be correct?

42. The angles of elevation, θ and ϕ, to an airplane are being continuously monitored at two observation points A and B which are two miles apart (see figure). Write an equation giving the distance d between the plane and point B in terms of θ and ϕ.

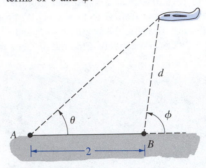

FIGURE FOR 42

9.2 Law of Cosines

Two cases remain in our list of conditions needed to solve an oblique triangle—SSS and SAS. In Section 9.1, we mentioned that the Law of Sines would not work in either of these cases. To see why, consider the three ratios given in the Law of Sines:

$$\frac{a}{\sin A} = \frac{b}{\sin B} = \frac{c}{\sin C}.$$

To use the Law of Sines we must know at least one side and its opposite angle. If we are given three sides (SSS), or two sides and their included angle (SAS), none of the above ratios would be complete. In such cases we rely on the **Law of Cosines.**

Law of Cosines

If ABC is a triangle with sides a, b, and c, then the following equations are valid.

Standard Form	*Alternative Form*
$a^2 = b^2 + c^2 - 2bc \cos A$	$\cos A = \dfrac{b^2 + c^2 - a^2}{2bc}$
$b^2 = a^2 + c^2 - 2ac \cos B$	$\cos B = \dfrac{a^2 + c^2 - b^2}{2ac}$
$c^2 = a^2 + b^2 - 2ab \cos C$	$\cos C = \dfrac{a^2 + b^2 - c^2}{2ab}$

Law of Cosines

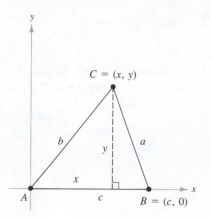

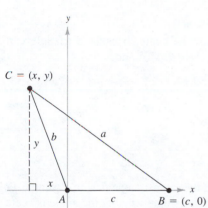

FIGURE 9.14

PROOF

In Figure 9.14, vertex B has coordinates $(c, 0)$. Furthermore, C has coordinates (x, y), where $x = b \cos A$ and $y = b \sin A$. Since a is the distance from vertex C to vertex B, it follows that

$$a = \sqrt{(x - c)^2 + (y - 0)^2}$$
$$a^2 = (b \cos A - c)^2 + (b \sin A)^2$$
$$= b^2 \cos^2 A - 2bc \cos A + c^2 + b^2 \sin^2 A$$
$$= b^2(\sin^2 A + \cos^2 A) + c^2 - 2bc \cos A.$$

Using the identity $\sin^2 A + \cos^2 A = 1$, we then obtain

$$a^2 = b^2 + c^2 - 2bc \cos A.$$

Similar arguments with angles B and C in standard position establish the other two equations.

Note that if $A = 90°$ in the right-hand triangle in Figure 9.14, then $\cos A = 0$ and the first form of the Law of Cosines becomes the Pythagorean Theorem:

$$a^2 = b^2 + c^2.$$

Thus, the Pythagorean Theorem is actually just a special case of the more general Law of Cosines.

EXAMPLE 1 *Given Three Sides of a Triangle—SSS*

Find the three angles of the triangle whose sides have lengths $a = 8.65$, $b = 19.2$, and $c = 13.7$.

SOLUTION

It is a good idea first to find the angle opposite the longest side—side b in this case (see Figure 9.15). Using the Law of Cosines, we find that

$$\cos B = \frac{a^2 + c^2 - b^2}{2ac}$$
$$= \frac{(8.65)^2 + (13.7)^2 - (19.2)^2}{2(8.65)(13.7)}$$
$$\approx -0.4477765.$$

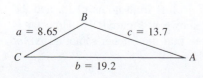

FIGURE 9.15

Since cos *B* is negative, we know *B* is an *obtuse* angle given by

$$B \approx 116.60°.$$

At this point we could use the Law of Cosines to find cos *A* and cos *C*. However, knowing that *B* ≈ 116.60°, it is simpler to use the Law of Sines to obtain

$$\frac{b}{\sin B} = \frac{a}{\sin A}$$

$$\sin A = a\left(\frac{\sin B}{b}\right)$$

$$\approx 8.65\left(\frac{\sin 116.60°}{19.2}\right)$$

$$\approx 0.4028351.$$

Since *B* is obtuse, we know that *A* must be acute, because a triangle can have, at most, one obtuse angle. Thus, *A* ≈ 23.76° and

$$C \approx 180° - 23.76° - 116.60° = 39.64°.$$

Do you see why it was wise to find the largest angle *first* in Example 1? Knowing the cosine of an angle, we can determine whether the angle is acute or obtuse. That is,

cos $\theta > 0$	for	$0° < \theta < 90°$	*(Acute)*
cos $\theta < 0$	for	$90° < \theta < 180°.$	*(Obtuse)*

So, in Example 1, once we found that *B* was obtuse, we subsequently knew that angles *A* and *C* were both acute. If the largest angle is acute, then the remaining two angles will be acute also.

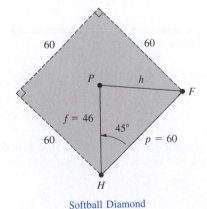

Softball Diamond

FIGURE 9.16

EXAMPLE 2 **Given Two Sides and the Included Angle—SAS**

The pitcher's mound on a softball field is 46 feet from home plate, and the distance between the bases is 60 feet, as shown in Figure 9.16. How far is the pitcher's mound from first base? (Note that the pitcher's mound is *not* halfway between home plate and second base.)

SOLUTION

In triangle HPF, we have $H = 45°$ (line HP bisects the right angle at H), $f = 46$, and $p = 60$. Using the Law of Cosines for this SAS case, we have

$$h^2 = f^2 + p^2 - 2fp \cos H$$
$$= 46^2 + 60^2 - 2(46)(60) \cos 45°$$
$$\approx 1812.8.$$

Therefore, the approximate distance from the pitcher's mound to first base is

$$h \approx \sqrt{1812.8} \approx 42.58 \text{ feet.}$$

EXAMPLE 3 **Given Two Sides and the Included Angle—SAS**

A ship travels 60 miles due east, then adjusts its course 15° northward, as shown in Figure 9.17. After traveling 80 miles in that direction, how far is the ship from its point of departure?

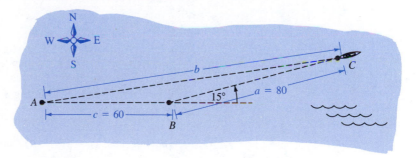

FIGURE 9.17

SOLUTION

We have $c = 60$, $B = 180° - 15° = 165°$, and $a = 80$. Consequently, by the Law of Cosines, we find that

$$b^2 = a^2 + c^2 - 2ac \cos B$$
$$= 80^2 + 60^2 - 2(80)(60) \cos 165°$$
$$\approx 19{,}273.$$

Therefore, the distance b is

$$b \approx \sqrt{19{,}273} \approx 138.8 \text{ miles.}$$

The Law of Cosines can be used to establish the following formula for the area of a triangle. This formula is credited to the Greek mathematician Heron (c. 100 B.C.).

Additional Applications of Trigonometry

Heron's Area Formula

Given any triangle with sides of length a, b, and c, the area of the triangle is

$$\text{Area} = \sqrt{s(s - a)(s - b)(s - c)}$$

where $s = (a + b + c)/2$.

PROOF

From the previous section, we know that

$$\text{Area} = \frac{1}{2}bc \sin A = \sqrt{\frac{1}{4}b^2c^2 \sin^2 A} = \sqrt{\frac{1}{4}b^2c^2(1 - \cos^2 A)}$$

$$= \sqrt{\left[\frac{1}{2}bc(1 + \cos A)\right]\left[\frac{1}{2}bc(1 - \cos A)\right]}.$$

Using the Law of Cosines, we can show that

$$\frac{1}{2}bc(1 + \cos A) = \frac{a + b + c}{2} \cdot \frac{-a + b + c}{2}$$

and

$$\frac{1}{2}bc(1 - \cos A) = \frac{a - b + c}{2} \cdot \frac{a + b - c}{2}.$$

(See Exercises 37 and 38.) Letting $s = (a + b + c)/2$, these two equations can be rewritten as

$$\frac{1}{2}bc(1 + \cos A) = s(s - a) \quad \text{and} \quad \frac{1}{2}bc(1 - \cos A) = (s - b)(s - c).$$

Thus, we conclude that

$$\text{Area} = \sqrt{s(s - a)(s - b)(s - c)}.$$

EXAMPLE 4   *Using Heron's Area Formula*

Find the area of the triangular region having sides of lengths $a = 47$ yards, $b = 58$ yards, and $c = 78.6$ yards.

SOLUTION

Since

$$s = \frac{1}{2}(a + b + c) = \frac{183.6}{2} = 91.8$$

Exercises

Heron's Formula yields

$$\text{Area} = \sqrt{s(s - a)(s - b)(s - c)}$$
$$= \sqrt{91.8(44.8)(33.8)(13.2)}$$
$$\approx 1,354.58 \text{ square yards.}$$

WARM UP

Simplify the expressions.

1. $\sqrt{(7 - 3)^2 + [1 - (-5)]^2}$

2. $\sqrt{[-2 - (-5)]^2 + (12 - 6)^2}$

Find the distance between the two points.

3. $(4, -2), (8, 10)$

4. $(1, 3), (7, 12)$

Find the area of each triangle.

5.

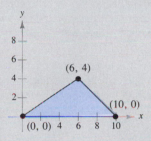

6.

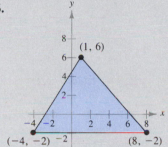

Find the remaining sides and angles of each triangle (if possible).

7. $A = 10°, C = 100°, b = 25$

8. $A = 20°, C = 90°, c = 100$

9. $B = 30°, b = 6.5, c = 15$

10. $A = 30°, b = 6.5, a = 10$

EXERCISES 9.2

In Exercises 1–14, use the Law of Cosines to solve the given triangle.

1.

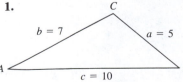

2.

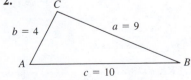

3.

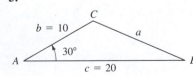

4.

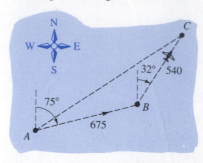

C

$b = 4.5$ $a = 10$

$110°$

A c B

5. $a = 9, b = 12, c = 15$ **6.** $a = 55, b = 25, c = 72$

7. $a = 75.4, b = 52, c = 52$

8. $a = 1.42, b = 0.75, c = 1.25$

9. $A = 120°, b = 3, c = 10$

10. $A = 55°, b = 3, c = 10$

11. $B = 8° 45', a = 25, c = 15$

12. $B = 75° 20', a = 6.2, c = 9.5$

13. $C = 125° 40', a = 32, b = 32$

14. $C = 15°, a = 6.25, b = 2.15$

In Exercises 15–20, use Heron's Formula to find the area of the triangle.

15. $a = 5, b = 7, c = 10$ **16.** $a = 2.5, b = 10.2, c = 9$

17. $a = 12, b = 15, c = 9$

18. $a = 75.4, b = 52, c = 52$

19. $a = 20, b = 20, c = 10$

20. $a = 4.25, b = 1.55, c = 3.00$

21. A boat race occurs along a triangular course marked by buoys A, B, and C. The race starts with the boats going 8000 feet in a northerly direction. The other two sides of the course lie to the east of the first side, and their lengths are 3500 feet and 6500 feet (see figure). Find the bearings for the last two legs of the race.

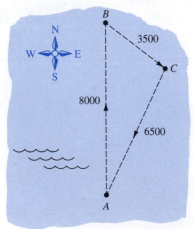

FIGURE FOR 21

22. A plane flies 675 miles from A to B with a bearing of N 75° E. Then it flies 540 miles from B to C with a bearing of N 32° E (see figure). Find the straight-line distance and bearing for the flight from C to A.

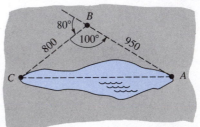

FIGURE FOR 22

23. Two ships leave a port at 9 A.M. One travels at a bearing of N 53° W at 12 miles per hour and the other at a bearing of S 67° W at 16 miles per hour. Approximately how far apart are they at noon that day?

24. A triangular parcel of land has 375 feet of frontage, and the two other boundaries have lengths of 250 feet and 300 feet. What angles does the frontage make with the two other boundaries?

25. A 100-foot vertical tower is to be erected on the side of a hill that makes an 8° angle with the horizontal (see figure). Find the lengths of each of the two guy wires that will be anchored 75 feet uphill and downhill from the base of the tower.

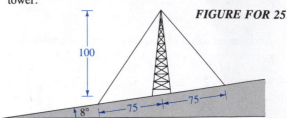

FIGURE FOR 25

26. To approximate the length of a marsh, a surveyor walks 950 feet from point A to point B, then turns 80° and walks 800 feet to point C (see figure). Approximate the length $\overline{AC}$ of the marsh.

FIGURE FOR 26

27. Determine the angle θ as shown on the streetlight in the figure.

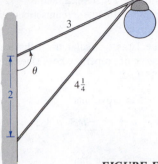

3

θ

2

$4\frac{1}{4}$

FIGURE FOR 27

28. In order to determine the distance between two aircraft, a tracking station continuously determines the distance to each aircraft and the angle α between them (see figure). Determine the distance a between the planes when $\alpha = 42°$, $b = 35$ miles, and $c = 20$ miles.

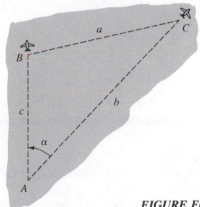

a

B

C

c

b

α

A

FIGURE FOR 28

29. If Q is the midpoint of the line segment $\overline{PR}$, find the lengths of the line segments $\overline{PQ}$, $\overline{QS}$, and $\overline{RS}$ on the truss rafter shown in the figure.

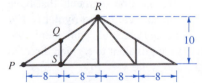

R

Q

10

P

S

8   8   8   8

FIGURE FOR 29

30. In a certain process with continuous paper, the paper passes across three rollers of radii 3 inches, 4 inches, and 6 inches (see figure). The centers of the 3-inch and 6-inch rollers are d inches apart, and the length of the arc in contact with the paper on the 4-inch roller is s inches. Complete the following table.

d (inches)	9	10	11	12	13	14	15	16
θ (degrees)								
s (inches)								

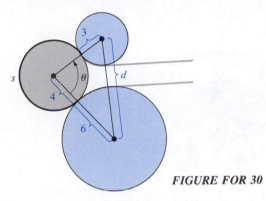

3

s

θ

d

4

6

FIGURE FOR 30

31. In a (square) baseball diamond with 90-foot sides the pitcher's mound is 60 feet from home plate.
   (a) How far is it from the pitcher's mound to third base?
   (b) When a runner is halfway from second to third, how far is the runner from the pitcher's mound?

32. On a certain map, Minneapolis is 6.5 inches due west of Albany, Phoenix is 8.5 inches from Minneapolis, and Phoenix is 14.5 inches from Albany.
   (a) Find the bearing of Minneapolis from Phoenix.
   (b) Find the bearing of Albany from Phoenix.

33. On a certain map, Orlando is 7 inches due south of Niagara Falls, Denver is 10.75 inches from Orlando, and Denver is 9.25 inches from Niagara Falls (see figure).
   (a) Find the bearing of Denver from Orlando.
   (b) Find the bearing of Denver from Niagara Falls.

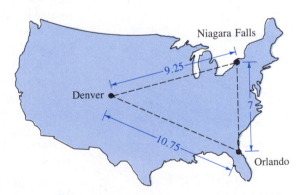

Niagara Falls

9.25

Denver

7

10.75

Orlando

FIGURE FOR 33

34. Let R and r be the radii of the circumscribed and inscribed circles of triangle ABC, respectively, and let $s = (a + b + c)/2$ (see figure). Prove the following.

(a) $2R = \dfrac{a}{\sin A} = \dfrac{b}{\sin B} = \dfrac{c}{\sin C}$

(b) $r = \sqrt{\dfrac{(s - a)(s - b)(s - c)}{s}}$

R = radius of the large circle

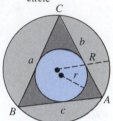

r = radius of the small circle

FIGURE FOR 34

In Exercises 35 and 36, use the results of Exercise 34.

35. Given the triangle with $a = 25$, $b = 55$, and $c = 72$, find the area of (a) the triangle, (b) the circumscribed circle, and (c) the inscribed circle.

36. Find the length of the largest circular track that can be built on a triangular piece of property whose sides are 200 feet, 250 feet, and 325 feet.

37. Use the Law of Cosines to prove that

$$\frac{1}{2}bc(1 + \cos A) = \frac{a + b + c}{2} \cdot \frac{-a + b + c}{2}.$$

38. Use the Law of Cosines to prove that

$$\frac{1}{2}bc(1 - \cos A) = \frac{a - b + c}{2} \cdot \frac{a + b - c}{2}.$$

9.3 Vectors

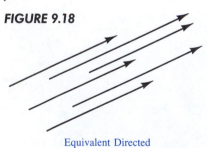

FIGURE 9.18

Equivalent Directed Line Segments

FIGURE 9.19

Many quantities in geometry and physics, such as area, time, and temperature, can be represented by a single real number.

Other quantities, such as force and velocity, involve both *magnitude* and *direction* and cannot be completely characterized by a single real number. To represent such a quantity, we use a **directed line segment,** as shown in Figure 9.18. The directed line segment $\overrightarrow{PQ}$ has **initial point** P and **terminal point** Q and we denote its **length** by $\|\overrightarrow{PQ}\|$. Two directed line segments that have the same length (or magnitude) and direction are called **equivalent.** For example, the directed line segments in Figure 9.19 are all equivalent. We call the set of all directed line segments that are equivalent to a given directed line segment $\overrightarrow{PQ}$, a **vector v in the plane,** and write $\mathbf{v} = \overrightarrow{PQ}$.

Remark: We denote vectors by lower-case, boldface letters such as $\mathbf{u}$, $\mathbf{v}$, and $\mathbf{w}$.

Be sure you see that a vector in the plane can be represented by many different directed line segments. This is illustrated in the following example.

Vectors

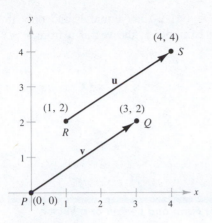

FIGURE 9.20

EXAMPLE 1 *Vector Representation by Directed Line Segments*

Let **u** be represented by the directed line segment from $P = (0, 0)$ to $Q = (3, 2)$, and let **v** be represented by the directed line segment from $R = (1, 2)$ to $S = (4, 4)$, as shown in Figure 9.20. Show that **u** = **v**.

SOLUTION

From the distance formula, we see that $\overrightarrow{PQ}$ and $\overrightarrow{RS}$ have the *same length*.

$$\|\overrightarrow{PQ}\| = \sqrt{(3 - 0)^2 + (2 - 0)^2} = \sqrt{13}$$

$$\|\overrightarrow{RS}\| = \sqrt{(4 - 1)^2 + (4 - 2)^2} = \sqrt{13}$$

Moreover, both line segments have the *same direction* since they are both directed toward the upper right on lines having a slope of 2/3. Thus, $\overrightarrow{PQ}$ and $\overrightarrow{RS}$ have the same length and direction, and we conclude that **u** = **v**.

The directed line segment whose initial point is the origin is often the most convenient representative of a set of equivalent directed line segments. We say that this representative of the vector **v** is in **standard position.**

A vector whose initial point is at the origin $(0, 0)$ can be uniquely represented by the coordinates of its terminal point (v_1, v_2). We call this the **component form of a vector v** and write

$$\mathbf{v} = \langle v_1, v_2 \rangle.$$

The coordinates v_1 and v_2 are called the **components** of **v**. If both the initial point and the terminal point lie at the origin, then **v** is called the **zero vector** and is denoted by $\mathbf{0} = \langle 0, 0 \rangle$.

To convert directed line segments to component form we use the following procedure.

Component Form of a Vector

The component form of the vector with initial point $P = (p_1, p_2)$ and terminal point $Q = (q_1, q_2)$ is

$$\overrightarrow{PQ} = \langle q_1 - p_1, q_2 - p_2 \rangle = \langle v_1, v_2 \rangle = \mathbf{v}.$$

The *length* (or magnitude) of **v** is given by

$$\|\mathbf{v}\| = \sqrt{(q_1 - p_1)^2 + (q_2 - p_2)^2} = \sqrt{v_1{}^2 + v_2{}^2}.$$

If $\|\mathbf{v}\| = 1$, then **v** is called a **unit vector.** Moreover, $\|\mathbf{v}\| = 0$ if and only if **v** is the zero vector **0**.

Two vectors $\mathbf{u} = \langle u_1, u_2 \rangle$ and $\mathbf{v} = \langle v_1, v_2 \rangle$ are **equal** if and only if $u_1 = v_1$ and $u_2 = v_2$. For instance, in Example 1, the vector $\mathbf{u}$ from $P = (0, 0)$ to $Q = (3, 2)$ is

$$\mathbf{u} = \overrightarrow{PQ} = \langle 3 - 0, 2 - 0 \rangle = \langle 3, 2 \rangle$$

and the vector $\mathbf{v}$ from $R = (1, 2)$ to $S = (4, 4)$ is

$$\mathbf{v} = \overrightarrow{RS} = \langle 4 - 1, 4 - 2 \rangle = \langle 3, 2 \rangle.$$

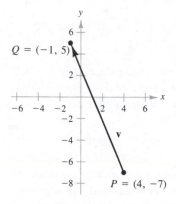

Component form of $\mathbf{v}$:

$$\mathbf{v} = \langle -5, 12 \rangle$$

FIGURE 9.21

EXAMPLE 2 *Finding the Component Form and Length of a Vector*

Find the component form and length of the vector $\mathbf{v}$ that has initial point $(4, -7)$ and terminal point $(-1, 5)$.

SOLUTION

We let $P = (4, -7) = (p_1, p_2)$ and $Q = (-1, 5) = (q_1, q_2)$. Then, the components of $\mathbf{v} = \langle v_1, v_2 \rangle$ are given by

$$v_1 = q_1 - p_1 = -1 - 4 = -5$$
$$v_2 = q_2 - p_2 = 5 - (-7) = 12.$$

Thus, $\mathbf{v} = \langle -5, 12 \rangle$ and the length of $\mathbf{v}$ is

$$\|\mathbf{v}\| = \sqrt{(-5)^2 + 12^2} = \sqrt{169} = 13$$

as shown in Figure 9.21.

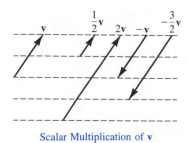

Scalar Multiplication of $\mathbf{v}$

FIGURE 9.22

Vector Operations

The two basic vector operations are **scalar multiplication** and **vector addition.** (In this text, we use the term **scalar** to mean a real number.) Geometrically, the product of a vector $\mathbf{v}$ and a scalar k is the vector that is k times as long as $\mathbf{v}$. If k is positive, then $k\mathbf{v}$ has the same direction as $\mathbf{v}$, and if k is negative, then $k\mathbf{v}$ has the opposite direction of $\mathbf{v}$, as shown in Figure 9.22.

To add two vectors geometrically, we position them (without changing length or direction) so that the initial point of one coincides with the terminal point of the other. The sum $\mathbf{u} + \mathbf{v}$ is formed by joining the initial point of

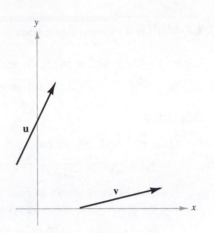

FIGURE 9.23   To find **u** + **v**,   move the initial point of **v** to the terminal
                                                 point of **u**.

the second vector **v** with the terminal point of the first vector **u**, as shown in
Figure 9.23. Since the vector **u** + **v** is the diagonal of a parallelogram having
u and **v** as its adjacent sides, we call this the **parallelogram law** for vector
addition.

Vector addition and scalar multiplication can also be defined using components of vectors.

Definition of Vector Addition and Scalar Multiplication

Let $\mathbf{u} = \langle u_1, u_2 \rangle$ and $\mathbf{v} = \langle v_1, v_2 \rangle$ be vectors and let k be a scalar (a real number).
Then the **sum** of **u** and **v** is the vector

$$\mathbf{u} + \mathbf{v} = \langle u_1 + v_1, u_2 + v_2 \rangle \qquad \textit{Sum}$$

and the **scalar multiple** of k times **u** is the vector

$$k\mathbf{u} = \langle k u_1, k u_2 \rangle. \qquad \textit{Scalar multiple}$$

The **negative** of $\mathbf{v} = \langle v_1, v_2 \rangle$ is

$$-\mathbf{v} = (-1)\mathbf{v} = \langle -v_1, -v_2 \rangle \qquad \textit{Negative}$$

and the **difference** of **u** and **v** is

$$\mathbf{u} - \mathbf{v} = \mathbf{u} + (-\mathbf{v}) = \langle u_1 - v_1, u_2 - v_2 \rangle. \qquad \textit{Difference}$$

To represent **u** − **v** graphically, we use directed line segments with the *same*
initial points. The difference **u** − **v** is the vector from the terminal point of
v to the terminal point of **u**, as shown in Figure 9.24.

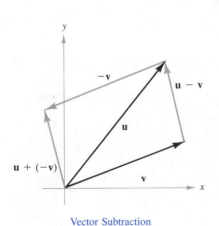

Vector Subtraction

FIGURE 9.24

(a)

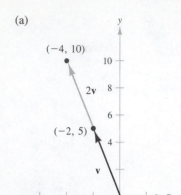

(b)

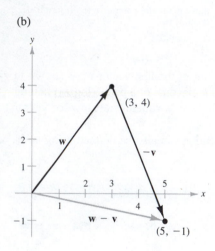

(c)

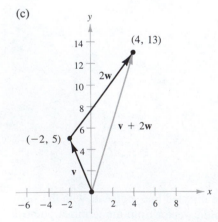

FIGURE 9.25

EXAMPLE 3 *Vector Operations*

Let $\mathbf{v} = \langle -2, 5 \rangle$ and $\mathbf{w} = \langle 3, 4 \rangle$, and find the following vectors.

(a) $2\mathbf{v}$ (b) $\mathbf{w} - \mathbf{v}$ (c) $\mathbf{v} + 2\mathbf{w}$

SOLUTION

(a) Since $\mathbf{v} = \langle -2, 5 \rangle$, we have

$$2\mathbf{v} = \langle 2(-2), 2(5) \rangle = \langle -4, 10 \rangle.$$

A sketch of $2\mathbf{v}$ is shown in Figure 9.25(a).

(b) The difference of $\mathbf{w}$ and $\mathbf{v}$ is given by

$$\mathbf{w} - \mathbf{v} = \langle 3 - (-2), 4 - 5 \rangle = \langle 5, -1 \rangle.$$

A sketch of $\mathbf{w} - \mathbf{v}$ is shown in Figure 9.25(b).

(c) Since $2\mathbf{w} = \langle 6, 8 \rangle$, it follows that

$$\begin{aligned} \mathbf{v} + 2\mathbf{w} &= \langle -2, 5 \rangle + \langle 6, 8 \rangle \\ &= \langle -2 + 6, 5 + 8 \rangle \\ &= \langle 4, 13 \rangle. \end{aligned}$$

A sketch of $\mathbf{v} + 2\mathbf{w}$ is shown in Figure 9.25(c).

Vector addition and scalar multiplication share many of the properties of ordinary arithmetic.

Properties of Vector Addition and Scalar Multiplication

Let $\mathbf{u}$, $\mathbf{v}$, and $\mathbf{w}$ be vectors and let c and d be scalars. Then the following properties are true.

1. $\mathbf{u} + \mathbf{v} = \mathbf{v} + \mathbf{u}$
2. $(\mathbf{u} + \mathbf{v}) + \mathbf{w} = \mathbf{u} + (\mathbf{v} + \mathbf{w})$
3. $\mathbf{u} + \mathbf{0} = \mathbf{u}$
4. $\mathbf{u} + (-\mathbf{u}) = \mathbf{0}$
5. $c(d\mathbf{u}) = (cd)\mathbf{u}$
6. $(c + d)\mathbf{u} = c\mathbf{u} + d\mathbf{u}$
7. $c(\mathbf{u} + \mathbf{v}) = c\mathbf{u} + c\mathbf{v}$
8. $1(\mathbf{u}) = \mathbf{u}, \quad 0(\mathbf{u}) = \mathbf{0}$
9. $\|c\mathbf{v}\| = |c| \, \|\mathbf{v}\|$

Remark: Property 9 can be stated as follows: The length of the vector $c\mathbf{v}$ is the absolute value of c times the length of $\mathbf{v}$.

In many applications of vectors it is useful to find a unit vector which has the same direction as a given vector **v**. To do this, we divide **v** by its length to obtain

$$\mathbf{u} = \left(\frac{1}{\|\mathbf{v}\|}\right)\mathbf{v} = \frac{\mathbf{v}}{\|\mathbf{v}\|}.$$

Note that **u** is a scalar multiple of **v**. The vector **u** has length 1 and the same direction as **v**. We call **u** a **unit vector in the direction of v.**

EXAMPLE 4 *Finding a Unit Vector*

Find a unit vector in the direction of $\mathbf{v} = \langle -2, 5 \rangle$ and verify that the result has length 1.

SOLUTION

The unit vector in the direction of **v** is

$$\frac{\mathbf{v}}{\|\mathbf{v}\|} = \frac{\langle -2, 5 \rangle}{\sqrt{(-2)^2 + (5)^2}}$$

$$= \frac{1}{\sqrt{29}}\langle -2, 5 \rangle = \left\langle \frac{-2}{\sqrt{29}}, \frac{5}{\sqrt{29}} \right\rangle.$$

This vector has length 1 since

$$\sqrt{\left(\frac{-2}{\sqrt{29}}\right)^2 + \left(\frac{5}{\sqrt{29}}\right)^2} = \sqrt{\frac{4}{29} + \frac{25}{29}} = \sqrt{\frac{29}{29}} = 1.$$

Standard Unit Vectors

The unit vectors $\langle 1, 0 \rangle$ and $\langle 0, 1 \rangle$ are called the **standard unit vectors** and are denoted by

$$\mathbf{i} = \langle 1, 0 \rangle \qquad \text{and} \qquad \mathbf{j} = \langle 0, 1 \rangle$$

as shown in Figure 9.26. (Note that the lower-case letter **i** is written in boldface to distinguish it from the imaginary number $i = \sqrt{-1}$.) These vectors can be used to represent any vector $\mathbf{v} = \langle v_1, v_2 \rangle$ as follows.

$$\mathbf{v} = \langle v_1, v_2 \rangle = v_1\langle 1, 0 \rangle + v_2\langle 0, 1 \rangle = v_1\mathbf{i} + v_2\mathbf{j}$$

The scalars v_1 and v_2 are called the **horizontal** and **vertical components of v,** respectively. The vector sum $v_1\mathbf{i} + v_2\mathbf{j}$ is called a **linear combination** of the vectors **i** and **j**. Any vector in the plane can be expressed as a linear combination of the standard unit vectors **i** and **j**.

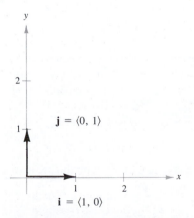

Standard Unit Vectors **i** and **j**

FIGURE 9.26

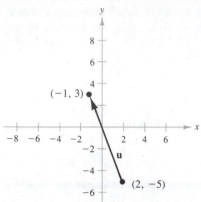

EXAMPLE 5 *Representing a Vector as a Linear Combination of Unit Vectors*

Let **u** be the vector with initial point $(2, -5)$ and terminal point $(-1, 3)$. Write **u** as a linear combination of the standard unit vectors **i** and **j**.

SOLUTION

$$\mathbf{u} = \langle -1 - 2, 3 + 5 \rangle$$
$$= \langle -3, 8 \rangle$$
$$= -3\mathbf{i} + 8\mathbf{j}$$

This result is shown graphically in Figure 9.27.

EXAMPLE 6 *Vector Operations*

Let $\mathbf{u} = -3\mathbf{i} + 8\mathbf{j}$ and $\mathbf{v} = 2\mathbf{i} - \mathbf{j}$. Find $2\mathbf{u} - 3\mathbf{v}$.

SOLUTION

$$2\mathbf{u} - 3\mathbf{v} = 2(-3\mathbf{i} + 8\mathbf{j}) - 3(2\mathbf{i} - \mathbf{j})$$
$$= -6\mathbf{i} + 16\mathbf{j} - 6\mathbf{i} + 3\mathbf{j}$$
$$= -12\mathbf{i} + 19\mathbf{j}$$

FIGURE 9.27

If **u** is a *unit vector* such that θ is the angle (measured counterclockwise) from the positive x-axis to **u**, then the terminal point of **u** lies on the unit circle and we have

$$\mathbf{u} = \langle \cos \theta, \sin \theta \rangle$$
$$= (\cos \theta)\mathbf{i} + (\sin \theta)\mathbf{j}$$

as shown in Figure 9.28. We call θ the **direction angle** of the vector **u**.

Suppose that **u** is a unit vector with direction angle θ. If **v** is any vector that makes an angle θ with the positive x-axis, then it has the same direction as **u** and we can write

$$\mathbf{v} = \|\mathbf{v}\|\langle \cos \theta, \sin \theta \rangle$$
$$= \|\mathbf{v}\|(\cos \theta)\mathbf{i} + \|\mathbf{v}\|(\sin \theta)\mathbf{j}.$$

For instance, the vector **v** of length 3 making an angle of 30° with the positive x-axis is given by

$$\mathbf{v} = 3(\cos 30°)\mathbf{i} + 3(\sin 30°)\mathbf{j} = \frac{3\sqrt{3}}{2}\mathbf{i} + \frac{3}{2}\mathbf{j}$$

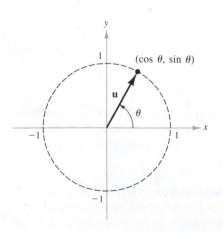

FIGURE 9.28

where $\|\mathbf{v}\| = 3$.

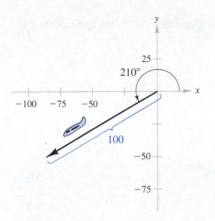

FIGURE 9.29

EXAMPLE 7 *Finding Component Form, Given Magnitude and Direction*

Find the component form of the vector that represents the velocity of an airplane descending at a speed of 100 miles per hour at an angle 30° below horizontal, as shown in Figure 9.29.

SOLUTION

The velocity vector **v** has magnitude of 100 and direction angle of $\theta = 210°$. Hence, the component form of **v** is

$$\mathbf{v} = \|\mathbf{v}\|(\cos \theta)\mathbf{i} + \|\mathbf{v}\|(\sin \theta)\mathbf{j}$$
$$= 100(\cos 210°)\mathbf{i} + 100(\sin 210°)\mathbf{j}$$
$$= 100\left(\frac{-\sqrt{3}}{2}\right)\mathbf{i} + 100\left(\frac{-1}{2}\right)\mathbf{j}$$
$$= -50\sqrt{3}\,\mathbf{i} - 50\mathbf{j}$$
$$= \langle -50\sqrt{3}, -50 \rangle.$$

(You should check to see that $\|\mathbf{v}\| = 100$.)

Since $\mathbf{v} = a\mathbf{i} + b\mathbf{j} = \|\mathbf{v}\| \cos \theta\mathbf{i} + \|\mathbf{v}\| \sin \theta\mathbf{j}$, it follows that the direction angle θ for **v** is determined from

$$\tan \theta = \frac{\sin \theta}{\cos \theta} = \frac{\|\mathbf{v}\| \sin \theta}{\|\mathbf{v}\| \cos \theta} = \frac{b}{a}.$$

We use this result in the next example.

EXAMPLE 8 *Finding Direction Angles of Vectors*

Find the direction angles of the following vectors.

(a) $\mathbf{u} = 3\mathbf{i} + 3\mathbf{j}$ (b) $\mathbf{v} = 3\mathbf{i} - 4\mathbf{j}$

SOLUTION

(a) The direction angle is given by

$$\tan \theta = \frac{b}{a} = \frac{3}{3} = 1.$$

Therefore, $\theta = 45°$, as shown in Figure 9.30.

(b) The direction angle is given by

$$\tan \theta = \frac{b}{a} = \frac{-4}{3}.$$

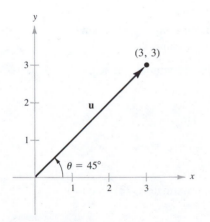

FIGURE 9.30

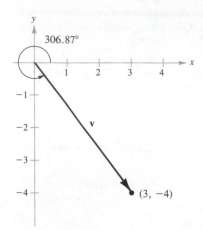

FIGURE 9.31

Moreover, since $\mathbf{v} = 3\mathbf{i} - 4\mathbf{j}$ lies in Quadrant IV, θ lies in Quadrant IV and its reference angle is

$$\theta' = \left| \arctan\left(-\frac{4}{3}\right) \right| \approx |-53.13°| = 53.13°.$$

Therefore, it follows that

$$\theta \approx 360° - 53.13° = 306.87°$$

as shown in Figure 9.31.

Applications of Vectors

Many applications of vectors involve the use of triangles and trigonometry in their solutions.

EXAMPLE 9 An Application

A force of 600 pounds is required to pull a boat and trailer up a ramp inclined at 15° from horizontal. Find the combined weight of the boat and trailer. (Assume no friction is involved.)

SOLUTION

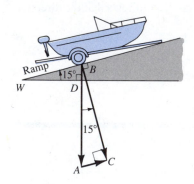

FIGURE 9.32

Based on Figure 9.32, we make the following observations.

$\|\overrightarrow{BA}\|$ = force of gravity = combined weight of boat and trailer

$\|\overrightarrow{BC}\|$ = force against ramp

$\|\overrightarrow{AC}\|$ = force required to move boat up ramp = 600 pounds
 (Note that AC is parallel to the ramp.)

By construction, triangles WBD and ABC are similar. Hence, angle ABC is 15°. Therefore, in triangle ABC we have

$$\sin 15° = \frac{\|\overrightarrow{AC}\|}{\|\overrightarrow{BA}\|} = \frac{600}{\|\overrightarrow{BA}\|}$$

$$\|\overrightarrow{BA}\| = \frac{600}{\sin 15°} \approx 2318.$$

Consequently, the combined weight is approximately 2318 pounds.

EXAMPLE 10 An Application

An airplane is traveling at a fixed altitude with a negligible wind factor. The airplane is headed N 30° W at a speed of 500 miles per hour, as shown in Figure 9.33. As the airplane reaches a certain point, it encounters a wind with a velocity of 70 miles per hour in the direction N 45° E. What are the resultant speed and direction of the airplane?

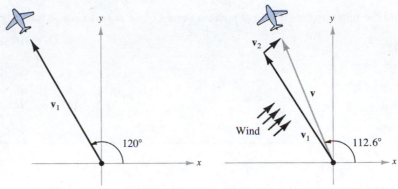

FIGURE 9.33

SOLUTION

Using Figure 9.33, we can represent the velocity of the airplane by the vector

$$\mathbf{v}_1 = 500 \langle \cos 120°, \sin 120° \rangle = \langle -250, 250\sqrt{3} \rangle$$

and the velocity of the wind by the vector

$$\mathbf{v}_2 = 70 \langle \cos 45°, \sin 45° \rangle = \langle 35\sqrt{2}, 35\sqrt{2} \rangle.$$

Thus, the velocity of the airplane is given by the vector

$$\mathbf{v} = \mathbf{v}_1 + \mathbf{v}_2 = \langle -250 + 35\sqrt{2}, 250\sqrt{3} + 35\sqrt{2} \rangle$$
$$\approx \langle -200.5, 482.5 \rangle$$

and the speed of the airplane is

$$\|\mathbf{v}\| = \sqrt{(-200.5)^2 + (482.5)^2} \approx 522.5 \text{ miles per hour.}$$

Finally, if θ is the direction angle of the flight path, we have

$$\tan \theta = \frac{482.5}{-200.5} \approx -2.4065$$

which implies that

$$\theta \approx 180° + \arctan(-2.4065) \approx 180° - 67.4° = 112.6°.$$

WARM UP

Find the distance between the two points.

1. $(-2, 6), (5, -15)$ **2.** $(0, 0), (-3, -7)$

Find an equation of the line passing through the two points.

3. $(3, 1), (-2, 4)$ **4.** $(-2, -3), (4, 5)$

Find the angle θ ($0 \leq \theta < 360°$) whose terminal side passes through the given point.

5. $(-2, 5)$ **6.** $(4, -3)$

Find the sine and cosine of θ.

7. $\theta = 30°$ **8.** $\theta = 120°$
9. $\theta = 300°$ **10.** $\theta = 210°$

EXERCISES 9.3

In Exercises 1–6, use the figure to sketch a graph of the indicated vector.

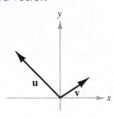

FIGURE FOR 1–6

1. $-\mathbf{u}$ **2.** $3\mathbf{v}$
3. $\mathbf{u} + \mathbf{v}$ **4.** $\mathbf{u} + 2\mathbf{v}$
5. $\mathbf{u} - \mathbf{v}$ **6.** $\mathbf{v} - \frac{1}{2}\mathbf{u}$

In Exercises 7–16, find the component form and the magnitude of the vector **v**.

7.
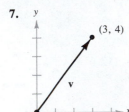

8.
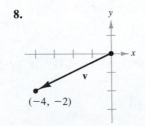

9.
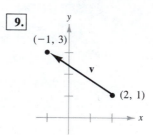

10.
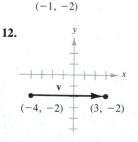

11.
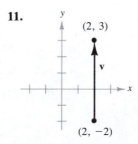

12.

13. Initial Point: $(-1, 5)$
Terminal Point: $(15, 2)$

14. Initial Point: $(1, 11)$
Terminal Point: $(9, 3)$

15. Initial Point: $(-3, -5)$
Terminal Point: $(5, -1)$

16. Initial Point: $(-3, 11)$ ·
Terminal Point: $(9, 40)$

In Exercises 17–26, find (a) $\mathbf{u} + \mathbf{v}$, (b) $\mathbf{u} - \mathbf{v}$, and (c) $2\mathbf{u} - 3\mathbf{v}$.

17. $\mathbf{u} = \langle 1, 2 \rangle$, $\mathbf{v} = \langle 3, 1 \rangle$ **18.** $\mathbf{u} = \langle 2, 3 \rangle$, $\mathbf{v} = \langle 4, 0 \rangle$
19. $\mathbf{u} = \langle -2, 3 \rangle$, $\mathbf{v} = \langle -2, 1 \rangle$
20. $\mathbf{u} = \langle 0, 1 \rangle$, $\mathbf{v} = \langle 0, -1 \rangle$
21. $\mathbf{u} = \langle 4, -2 \rangle$, $\mathbf{v} = \langle 0, 0 \rangle$
22. $\mathbf{u} = \langle 0, 0 \rangle$, $\mathbf{v} = \langle 2, 1 \rangle$
23. $\mathbf{u} = \mathbf{i} + \mathbf{j}$, $\mathbf{v} = 2\mathbf{i} - 3\mathbf{j}$
24. $\mathbf{u} = 2\mathbf{i} - \mathbf{j}$, $\mathbf{v} = -\mathbf{i} + \mathbf{j}$
25. $\mathbf{u} = 2\mathbf{i}$, $\mathbf{v} = \mathbf{j}$ **26.** $\mathbf{u} = 3\mathbf{j}$, $\mathbf{v} = 2\mathbf{i}$

In Exercises 27–34, sketch $\mathbf{v}$ and find its component form. (Assume θ is measured counterclockwise from the x-axis to the vector.)

27. $\|\mathbf{v}\| = 3$, $\theta = 0°$ **28.** $\|\mathbf{v}\| = 1$, $\theta = 45°$
29. $\|\mathbf{v}\| = 1$, $\theta = 150°$ **30.** $\|\mathbf{v}\| = \frac{5}{2}$, $\theta = 45°$
31. $\|\mathbf{v}\| = 3\sqrt{2}$, $\theta = 150°$ **32.** $\|\mathbf{v}\| = 8$, $\theta = 90°$
33. $\|\mathbf{v}\| = 2$, $\mathbf{v}$ in the direction $\mathbf{i} + 3\mathbf{j}$
34. $\|\mathbf{v}\| = 3$, $\mathbf{v}$ in the direction $3\mathbf{i} + 4\mathbf{j}$

In Exercises 35–40, find the component form of $\mathbf{v}$, and sketch the indicated vector operations geometrically, where $\mathbf{u} = 2\mathbf{i} - \mathbf{j}$ and $\mathbf{w} = \mathbf{i} + 2\mathbf{j}$.

35. $\mathbf{v} = \frac{3}{2}\mathbf{u}$ **36.** $\mathbf{v} = \mathbf{u} + \mathbf{w}$
37. $\mathbf{v} = \mathbf{u} + 2\mathbf{w}$ **38.** $\mathbf{v} = -\mathbf{u} + \mathbf{w}$
39. $\mathbf{v} = \frac{1}{2}(3\mathbf{u} + \mathbf{w})$ **40.** $\mathbf{v} = \mathbf{u} - 2\mathbf{w}$

In Exercises 41–44, find the component form of the sum of the vectors $\mathbf{u}$ and $\mathbf{v}$ with direction angles $\theta_\mathbf{u}$ and $\theta_\mathbf{v}$ respectively.

41. $\|\mathbf{u}\| = 5$, $\theta_\mathbf{u} = 0°$ **42.** $\|\mathbf{u}\| = 2$, $\theta_\mathbf{u} = 30°$
$\|\mathbf{v}\| = 5$, $\theta_\mathbf{v} = 90°$ $\|\mathbf{v}\| = 2$, $\theta_\mathbf{v} = 90°$
43. $\|\mathbf{u}\| = 20$, $\theta_\mathbf{u} = 45°$ **44.** $\|\mathbf{u}\| = 35$, $\theta_\mathbf{u} = 25°$
$\|\mathbf{v}\| = 50$, $\theta_\mathbf{v} = 180°$ $\|\mathbf{v}\| = 50$, $\theta_\mathbf{v} = 120°$

In Exercises 45–48, find a unit vector in the direction of the given vector.

45. $\mathbf{v} = 4\mathbf{i} - 3\mathbf{j}$ **46.** $\mathbf{v} = \mathbf{i} + \mathbf{j}$
47. $\mathbf{v} = 2\mathbf{j}$ **48.** $\mathbf{v} = \mathbf{i} - 2\mathbf{j}$

In Exercises 49–52, use the Law of Cosines to find the angle α between the given vectors. (Assume $0° \le \alpha \le 180°$.)

49. $\mathbf{v} = \mathbf{i} + \mathbf{j}$, $\mathbf{w} = 2(\mathbf{i} - \mathbf{j})$
50. $\mathbf{v} = 3\mathbf{i} + \mathbf{j}$, $\mathbf{w} = 2\mathbf{i} - \mathbf{j}$
51. $\mathbf{v} = \mathbf{i} + \mathbf{j}$, $\mathbf{w} = 3\mathbf{i} - \mathbf{j}$
52. $\mathbf{v} = \mathbf{i} + 2\mathbf{j}$, $\mathbf{w} = 2\mathbf{i} - \mathbf{j}$

53. Two forces, one of 35 pounds and the other of 50 pounds, act on the same object. The angle between the forces is 30°. Find the magnitude of the resultant (vector sum) of these forces.

54. Two forces, one of 100 pounds and the other of 150 pounds, act on the same object, at angles of 20° and 60°, respectively, with the positive x-axis. Find the direction and magnitude of the resultant (vector sum) of these forces.

55. Three forces of 75 pounds, 100 pounds, and 125 pounds act on the same object, at angles of 30°, 45°, and 120°, respectively, with the positive x-axis. Find the direction and magnitude of the resultant of these forces.

56. Three forces of 70 pounds, 40 pounds, and 60 pounds act on the same object, at angles of −30°, 45°, and 135°, respectively, with the positive x-axis. Find the direction and magnitude of the resultant of these forces.

57. A heavy implement is dragged 10 feet across the floor, using a force of 85 pounds. Find the work done if the direction of the force is 60° above the horizontal (see figure). (Use the formula for work, $W = FD$, where F is the horizontal component of force and D is the horizontal distance.)

58. To carry a 100-pound cylindrical weight, two people lift on the ends of short ropes that are tied to an eyelet on the top center of the cylinder. One of the ropes makes a 20° angle away from the vertical and the other a 30° angle (see figure). Find the vertical component of each person's force.

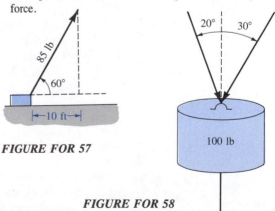

FIGURE FOR 57

FIGURE FOR 58

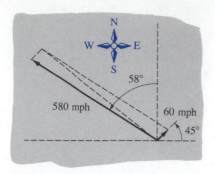

FIGURE FOR 59

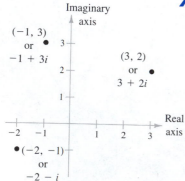

FIGURE FOR 60

59. An airplane's velocity with respect to the air is 580 miles per hour, and it is headed N 58° W. The wind, at the altitude of the plane, is from the southwest and has a velocity of 60 miles per hour (see figure). What is the true direction of the plane, and what is its speed with respect to the ground?

60. A tether ball weighing 1 pound is pulled outward from the pole by a horizontal force **u** until the rope makes a 30° angle with the pole (see figure). Determine the resulting tension in the rope and the magnitude of **u**.

61. An airplane is flying in the direction S 32° E, with an airspeed of 540 miles per hour. Because of the wind, its groundspeed and direction are 500 miles per hour and S 40° E, respectively. Find the direction and speed of the wind.

62. A ball is thrown into the air with an initial velocity of 80 feet per second, at an angle of 50° with the horizontal. Find the vertical and horizontal components of the velocity.

In Exercises 63 and 64, find the angle between the two forces for the given magnitude of their sum.

63. Force One: 45 lb
    Force Two: 60 lb
    Resultant Force: 90 lb

64. Force One: 3000 lb
    Force Two: 1000 lb
    Resultant Force: 3750 lb

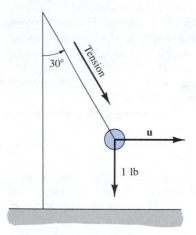

The Complex Plane

FIGURE 9.34

9.4 *Trigonometric Form of Complex Numbers*

In this section we develop the trigonometric form of a complex number. The usefulness of this form will not be fully apparent until we introduce DeMoivre's Theorem in Section 9.5.

Just as real numbers can be represented by points on the real number line, we can represent a complex number

$$z = a + bi$$

as the point (a, b) in a coordinate plane, called the **complex plane.** In this context, we call the horizontal axis the **real axis** and the vertical axis the **imaginary axis,** as shown in Figure 9.34.

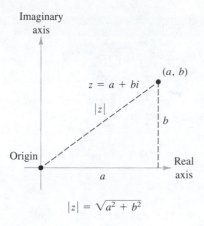

$$|z| = \sqrt{a^2 + b^2}$$

FIGURE 9.35

The **absolute value** of the complex number $a + bi$ is defined to be the distance between the origin $(0, 0)$ and the point (a, b), as shown in Figure 9.35.

Definition of the Absolute Value of a + bi

The **absolute value** of the complex number $z = a + bi$ is given by

$$|a + bi| = \sqrt{a^2 + b^2}.$$

EXAMPLE 1 *Finding the Absolute Value of a Complex Number*

Plot the points corresponding to the following complex numbers and find the absolute value of each.

(a) $z = -3i$ (b) $z = -2 + 5i$

SOLUTION

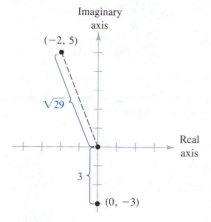

FIGURE 9.36

The points are shown in Figure 9.36.

(a) The complex number $z = 0 + (-3)i$ has an absolute value of

$$|z| = \sqrt{0^2 + (-3)^2} = \sqrt{9} = 3.$$

(b) The complex number $z = -2 + 5i$ has an absolute value of

$$|z| = \sqrt{(-2)^2 + 5^2} = \sqrt{29}.$$

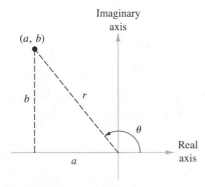

Complex Number: $a + bi$

FIGURE 9.37

In Section 2.5 we discussed how to add, subtract, multiply, and divide complex numbers. To work effectively with *powers* and *roots* of complex numbers, it is helpful to write complex numbers in **trigonometric form.** In Figure 9.37, consider the nonzero complex number $a + bi$. By letting θ be the angle from the positive x-axis (measured counterclockwise) to the line segment connecting the origin and the point (a, b), we can write

$$a = r \cos \theta \quad \text{and} \quad b = r \sin \theta$$

where $r = \sqrt{a^2 + b^2}$. Consequently, we have

$$a + bi = (r \cos \theta) + (r \sin \theta)i$$

from which we obtain the following **trigonometric form of a complex number.**

Trigonometric Form of a Complex Number

Let $z = a + bi$ be a complex number. The **trigonometric form** of z is

$$z = r(\cos \theta + i \sin \theta)$$

where $a = r \cos \theta$, $b = r \sin \theta$, $r = \sqrt{a^2 + b^2}$, and $\tan \theta = b/a$. The number r is called the **modulus** of z, and θ is called an **argument** of z.

Remark: The trigonometric form of a complex number is also called the **polar form.** Because there are infinitely many choices for θ, the trigonometric form of a complex number is not unique. Normally, we use θ values in the interval $0 \le \theta < 2\pi$, though on occasion we may use $\theta < 0$.

(a)

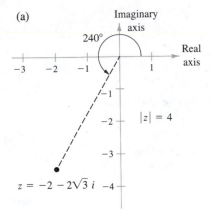

$z = -2 - 2\sqrt{3}\,i$

(b)

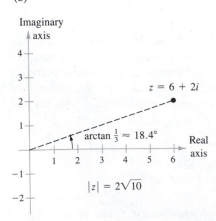

FIGURE 9.38

EXAMPLE 2 *Writing Complex Numbers in Trigonometric Form*

Write the following complex numbers in trigonometric form.

(a) $z = -2 - 2\sqrt{3}\,i$ (b) $z = 6 + 2i$

SOLUTION

(a) The absolute value of z is

$$r = |-2 - 2\sqrt{3}\,i| = \sqrt{(-2)^2 + (-2\sqrt{3})^2} = \sqrt{16} = 4$$

and the angle θ is given by

$$\tan \theta = \frac{b}{a} = \frac{-2\sqrt{3}}{-2} = \sqrt{3}.$$

Since $\tan 60° = \sqrt{3}$ and $z = -2 - 2\sqrt{3}\,i$ lies in Quadrant III, we choose θ to be $\theta = 180° + 60° = 240°$. Thus, the trigonometric form is

$$z = r(\cos \theta + i \sin \theta) = 4(\cos 240° + i \sin 240°).$$

See Figure 9.38(a).

(b) Here we have $r = |6 + 2i| = 2\sqrt{10}$ with θ given by

$$\tan \theta = \frac{2}{6} = \frac{1}{3} \qquad \text{\textit{θ in Quadrant I}}$$

$$\theta = \arctan \frac{1}{3} \approx 18.4°.$$

Therefore the trigonometric form of z is

$$z = r(\cos \theta + i \sin \theta)$$

$$= 2\sqrt{10}\left[\cos\left(\arctan \frac{1}{3}\right) + i \sin\left(\arctan \frac{1}{3}\right)\right]$$

$$\approx 2\sqrt{10}(\cos 18.4° + i \sin 18.4°).$$

See Figure 9.38(b).

EXAMPLE 3   *Writing a Complex Number in Standard Form*

Write the following complex number in standard form $a + bi$.

$$z = \sqrt{8}\left[\cos\left(-\frac{\pi}{3}\right) + i \sin\left(-\frac{\pi}{3}\right)\right]$$

SOLUTION

Since $\cos(-\pi/3) = 1/2$ and $\sin(-\pi/3) = -\sqrt{3}/2$, we can write

$$z = \sqrt{8}\left[\cos\left(-\frac{\pi}{3}\right) + i \sin\left(-\frac{\pi}{3}\right)\right]$$

$$= \sqrt{8}\left[\frac{1}{2} - \frac{\sqrt{3}}{2}i\right]$$

$$= 2\sqrt{2}\left[\frac{1}{2} - \frac{\sqrt{3}}{2}i\right]$$

$$= \sqrt{2} - \sqrt{6}\,i.$$

The trigonometric form adapts nicely to multiplication and division of complex numbers. Suppose we are given two complex numbers

$$z_1 = r_1(\cos \theta_1 + i \sin \theta_1) \qquad \text{and} \qquad z_2 = r_2(\cos \theta_2 + i \sin \theta_2).$$

The product of z_1 and z_2 is

$$z_1 z_2$$
$$= r_1 r_2 (\cos \theta_1 + i \sin \theta_1)(\cos \theta_2 + i \sin \theta_2)$$
$$= r_1 r_2 [(\cos \theta_1 \cos \theta_2 - \sin \theta_1 \sin \theta_2) + i(\sin \theta_1 \cos \theta_2 + \cos \theta_1 \sin \theta_2)].$$

Using the sum and difference formulas for cosine and sine, we can rewrite this equation as

$$z_1 z_2 = r_1 r_2 [\cos(\theta_1 + \theta_2) + i \sin(\theta_1 + \theta_2)].$$

This establishes the first part of the following rule. The second part is left to you (see Exercise 49).

Additional Applications of Trigonometry

Product and Quotient of Two Complex Numbers

Let $z_1 = r_1(\cos \theta_1 + i \sin \theta_1)$ and $z_2 = r_2(\cos \theta_2 + i \sin \theta_2)$ be complex numbers.

$$z_1 z_2 = r_1 r_2 [\cos(\theta_1 + \theta_2) + i \sin(\theta_1 + \theta_2)] \qquad \textit{Product}$$

$$\frac{z_1}{z_2} = \frac{r_1}{r_2}[\cos(\theta_1 - \theta_2) + i \sin(\theta_1 - \theta_2)], \quad z_2 \neq 0 \qquad \textit{Quotient}$$

Note that this rule says that to multiply two complex numbers we multiply moduli and add arguments, whereas to divide two complex numbers we divide moduli and subtract arguments.

EXAMPLE 4 Multiplying Complex Numbers in Trigonometric Form

Find the product of the following complex numbers.

$$z_1 = 2\left(\cos \frac{2\pi}{3} + i \sin \frac{2\pi}{3}\right)$$

$$z_2 = 8\left(\cos \frac{11\pi}{6} + i \sin \frac{11\pi}{6}\right)$$

SOLUTION

$$z_1 z_2 = 2\left(\cos \frac{2\pi}{3} + i \sin \frac{2\pi}{3}\right) \cdot 8\left(\cos \frac{11\pi}{6} + i \sin \frac{11\pi}{6}\right)$$

$$= 16\left[\cos\left(\frac{2\pi}{3} + \frac{11\pi}{6}\right) + i \sin\left(\frac{2\pi}{3} + \frac{11\pi}{6}\right)\right]$$

$$= 16\left[\cos \frac{5\pi}{2} + i \sin \frac{5\pi}{2}\right]$$

$$= 16\left[\cos \frac{\pi}{2} + i \sin \frac{\pi}{2}\right]$$

$$= 16[0 + i(1)] = 16i$$

Try checking this result by first converting to the standard forms $z_1 = -1 + \sqrt{3}\, i$ and $z_2 = 4\sqrt{3} - 4i$ and then multiplying.

EXAMPLE 5 Dividing Complex Numbers in Trigonometric Form

Find z_1/z_2 for the following two complex numbers.

$$z_1 = 24(\cos 300° + i \sin 300°)$$
$$z_2 = 8(\cos 75° + i \sin 75°)$$

SOLUTION

$$\frac{z_1}{z_2} = \frac{24(\cos 300° + i \sin 300°)}{8(\cos 75° + i \sin 75°)}$$

$$= \frac{24}{8}[\cos(300° - 75°) + i \sin(300° - 75°)]$$

$$= 3[\cos 225° + i \sin 225°]$$

$$= 3\left[\left(-\frac{\sqrt{2}}{2}\right) + i\left(-\frac{\sqrt{2}}{2}\right)\right]$$

$$= -\frac{3\sqrt{2}}{2} - \frac{3\sqrt{2}}{2}i$$

WARM UP

Write the complex numbers in standard form.

1. $-5 - \sqrt{-100}$

2. $7 + \sqrt{-54}$

3. $-4i + i^2$

4. $3i^3$

Perform the indicated operations and write the answers in standard form.

5. $(3 - 10i) - (-3 + 4i)$

6. $(2 + \sqrt{-50}) + (4 - \sqrt{2}\,i)$

7. $(4 - 2i)(-6 + i)$

8. $(3 - 2i)(3 + 2i)$

9. $\dfrac{1 + 4i}{1 - i}$

10. $\dfrac{3 - 5i}{2i}$

EXERCISES 9.4

In Exercises 1–4, express the complex number in trigonometric form.

1.
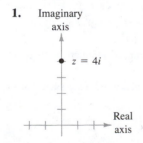
$z = 4i$

2.
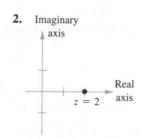
$z = 2$

3.
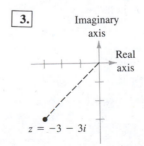
$z = -3 - 3i$

4.
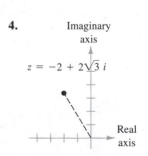
$z = -2 + 2\sqrt{3}\,i$

In Exercises 5–20, represent the complex numbers graphically, and find the trigonometric form of the number.

5. $3 - 3i$

7. $\sqrt{3} + i$

9. $-2(1 + \sqrt{3}\,i)$

11. $6i$

13. $-7 + 4i$

15. 7

17. $1 + 6i$

19. $-3 - i$

6. $-2 - 2i$

8. $-1 + \sqrt{3}\,i$

10. $\frac{5}{2}(\sqrt{3} - i)$

12. 4

14. $3 - i$

16. $-2i$

18. $2\sqrt{2} - i$

20. $1 + 3i$

In Exercises 21–30, represent the complex number graphically, and find the standard form of the number.

21. $2(\cos 150° + i \sin 150°)$

22. $5(\cos 135° + i \sin 135°)$

23. $\frac{3}{2}(\cos 300° + i \sin 300°)$

24. $\frac{3}{4}(\cos 315° + i \sin 315°)$

25. $3.75\left(\cos \dfrac{3\pi}{4} + i \sin \dfrac{3\pi}{4}\right)$

26. $8\left(\cos \dfrac{\pi}{12} + i \sin \dfrac{\pi}{12}\right)$

27. $4\left(\cos \dfrac{3\pi}{2} + i \sin \dfrac{3\pi}{2}\right)$

28. $6[\cos(230°\ 30') + i \sin(230°\ 30')]$

29. $3[\cos(18°\ 45') + i \sin(18°\ 45')]$

30. $7(\cos 0° + i \sin 0°)$

In Exercises 31–42, perform the indicated operation and leave the result in trigonometric form.

31. $[3(\cos 60° + i \sin 60°)][4(\cos 30° + i \sin 30°)]$

32. $[\frac{3}{2}(\cos 90° + i \sin 90°)][6(\cos 45° + i \sin 45°)]$

33. $[\frac{5}{3}(\cos 140° + i \sin 140°)][\frac{2}{3}(\cos 60° + i \sin 60°)]$

34. $[0.5(\cos 100° + i \sin 100°)][0.8(\cos 300° + i \sin 300°)]$

35. $[0.45(\cos 310° + i \sin 310°)][0.60(\cos 200° + i \sin 200°)]$

36. $(\cos 5° + i \sin 5°)(\cos 20° + i \sin 20°)$

37. $\dfrac{2(\cos 120° + i \sin 120°)}{4(\cos 40° + i \sin 40°)}$

38. $\dfrac{\cos 40° + i \sin 40°}{\cos 10° + i \sin 10°}$

39. $\dfrac{\cos(5\pi/3) + i \sin(5\pi/3)}{\cos \pi + i \sin \pi}$

40. $\dfrac{5[\cos(4.3) + i \sin(4.3)]}{4[\cos(2.1) + i \sin(2.1)]}$

41. $\dfrac{12(\cos 52° + i \sin 52°)}{3(\cos 110° + i \sin 110°)}$

42. $\dfrac{9(\cos 20° + i \sin 20°)}{5(\cos 75° + i \sin 75°)}$

In Exercises 43–48, (a) give the trigonometric form of the complex numbers, (b) perform the indicated operation using the trigonometric form, and (c) perform the indicated operation using the standard form and check your result with the answer in part (b).

43. $(2 + 2i)(1 - i)$

44. $(\sqrt{3} + i)(1 + i)$

45. $-2i(1 + i)$

46. $\dfrac{3 + 4i}{1 - \sqrt{3}\,i}$

47. $\dfrac{5}{2 + 3i}$

48. $\dfrac{4i}{-4 + 2i}$

49. Given two complex numbers $z_1 = r_1(\cos \theta_1 + i \sin \theta_1)$ and $z_2 = r_2(\cos \theta_2 + i \sin \theta_2)$, $z_2 \neq 0$, prove that

$$\frac{z_1}{z_2} = \frac{r_1}{r_2}[\cos(\theta_1 - \theta_2) + i \sin(\theta_1 - \theta_2)].$$

50. Show that the complex conjugate of

$$z = r(\cos \theta + i \sin \theta)$$

is

$$\bar{z} = r[\cos(-\theta) + i \sin(-\theta)].$$

51. Use the trigonometric form of z and $\bar{z}$ in Exercise 50 to find

(a) $z\bar{z}$ (b) $z/\bar{z}$, $z \neq 0$

52. Show that the negative of $z = r(\cos \theta + i \sin \theta)$ is $-z = r[\cos(\theta + \pi) + i \sin(\theta + \pi)]$.

53. Sketch the graph of all complex numbers z such that $|z| = 2$.

54. Sketch the graph of all complex numbers z such that the argument of each is $\theta = \pi/6$.

9.5 DeMoivre's Theorem and nth Roots

Our final look at complex numbers involves procedures for finding their powers and roots. Repeated use of the multiplication rule in the previous section yields

$$z = r(\cos \theta + i \sin \theta)$$

$$z^2 = r(\cos \theta + i \sin \theta)r(\cos \theta + i \sin \theta)$$
$$= r^2(\cos 2\theta + i \sin 2\theta)$$

$$z^3 = z^2(z) = r^2(\cos 2\theta + i \sin 2\theta)r(\cos \theta + i \sin \theta)$$
$$= r^3(\cos 3\theta + i \sin 3\theta).$$

Similarly,

$$z^4 = r^4(\cos 4\theta + i \sin 4\theta)$$
$$z^5 = r^5(\cos 5\theta + i \sin 5\theta)$$
$$\vdots$$

This pattern leads to the following important theorem.

DeMoivre's Theorem

If $z = r(\cos \theta + i \sin \theta)$ is a complex number and n is a positive integer, then

$$z^n = [r(\cos \theta + i \sin \theta)]^n = r^n(\cos n\theta + i \sin n\theta).$$

A complete proof of this theorem can be given using mathematical induction (see Section 12.4).

EXAMPLE 1 *Finding Powers of a Complex Number*

Use DeMoivre's Theorem to find $(-1 + \sqrt{3}\,i)^{12}$.

SOLUTION

We first convert to trigonometric form.

$$-1 + \sqrt{3}\,i = 2\left(\cos \frac{2\pi}{3} + i \sin \frac{2\pi}{3}\right)$$

Then, by DeMoivre's Theorem, we have

$$(-1 + \sqrt{3}\,i)^{12} = \left[2\left(\cos\frac{2\pi}{3} + i\,\sin\frac{2\pi}{3}\right)\right]^{12}$$

$$= 2^{12}\left[\cos(12)\frac{2\pi}{3} + i\,\sin(12)\frac{2\pi}{3}\right]$$

$$= 4096(\cos 8\pi + i\,\sin 8\pi)$$

$$= 4096(1 + 0) = 4096.$$

Are you surprised to see a real number as the answer?

Recall that a consequence of the Fundamental Theorem of Algebra (Section 4.5) is that a polynomial equation of degree n has n solutions in the complex number system. Hence, an equation like $x^6 = 1$ has six solutions, and in this particular case we can find the six solutions by factoring and using the quadratic formula.

$$x^6 - 1 = (x^3 - 1)(x^3 + 1)$$
$$= (x - 1)(x^2 + x + 1)(x + 1)(x^2 - x + 1) = 0$$

Consequently, the solutions are

$$x = \pm 1, \qquad x = \frac{-1 \pm \sqrt{3}\,i}{2}, \qquad \text{and} \qquad x = \frac{1 \pm \sqrt{3}\,i}{2}.$$

Each of these numbers is called a sixth root of 1. In general, we define the **nth root** of a complex number as follows.

Definition of nth Root of a Complex Number

The complex number $u = a + bi$ is an **nth root** of the complex number z if

$$z = u^n = (a + bi)^n.$$

To find a formula for an nth root of a complex number, we let u be an nth root of z, where

$$u = s(\cos \beta + i\,\sin \beta) \qquad \text{and} \qquad z = r(\cos \theta + i\,\sin \theta).$$

By DeMoivre's Theorem and the fact that $u^n = z$, we have

$$s^n(\cos n\beta + i\,\sin n\beta) = r(\cos \theta + i\,\sin \theta).$$

Now, taking the absolute value of both sides of this equation, it follows that $s^n = r$. Substituting back into the previous equation and dividing by r, we get

$$\cos n\beta + i\,\sin n\beta = \cos \theta + i\,\sin \theta.$$

Thus, it follows that

$$\cos n\beta = \cos \theta \quad \text{and} \quad \sin n\beta = \sin \theta.$$

Since both sine and cosine have a period of 2π, these last two equations have solutions if and only if the angles differ by a multiple of 2π. Consequently, there must exist an integer k such that

$$n\beta = \theta + 2\pi k$$

$$\beta = \frac{\theta + 2\pi k}{n}.$$

By substituting this value for β into the trigonometric form of u, we get the result stated in the following theorem.

nth Roots of a Complex Number

For a positive integer n, the complex number $z = r(\cos \theta + i \sin \theta)$ has exactly n distinct nth roots given by

$$\sqrt[n]{r}\left(\cos \frac{\theta + 2\pi k}{n} + i \sin \frac{\theta + 2\pi k}{n}\right)$$

where $k = 0, 1, 2, \ldots, n - 1$.

Remark: Note that when k exceeds $n - 1$ the roots begin to repeat. For instance, if $k = n$, the angle

$$\frac{\theta + 2\pi n}{n} = \frac{\theta}{n} + 2\pi$$

is coterminal with θ/n, which is also obtained when $k = 0$.

This formula for the nth roots of a complex number z has a nice geometrical interpretation, as shown in Figure 9.39. Note that because the nth roots of z all have the same magnitude $\sqrt[n]{r}$, they all lie on a circle of radius $\sqrt[n]{r}$ with center at the origin. Furthermore, the n roots are equally spaced along the circle, since successive nth roots have arguments that differ by $2\pi/n$.

We have already found the sixth roots of 1 by factoring and the quadratic formula. Now let's see how we can solve the same problem with the formula for nth roots.

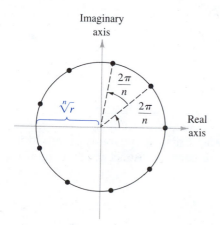

FIGURE 9.39

EXAMPLE 2   *Finding nth Roots of a Real Number*

Find all the sixth roots of 1.

SOLUTION

First we write 1 in the trigonometric form

$$1 = 1(\cos 0 + i \sin 0).$$

Then, by the *n*th root formula, with $n = 6$ and $r = 1$, the roots have the form

$$\sqrt[6]{1}\left(\cos \frac{0 + 2\pi k}{6} + i \sin \frac{0 + 2\pi k}{6}\right)$$

or simply $\cos(\pi k/3) + i \sin(\pi k/3)$. Thus, for $k = 0, 1, 2, 3, 4, 5$, the sixth roots are

$$\cos 0 + i \sin 0 = 1$$
$$\cos \frac{\pi}{3} + i \sin \frac{\pi}{3} = \frac{1}{2} + \frac{\sqrt{3}}{2} i$$
$$\cos \frac{2\pi}{3} + i \sin \frac{2\pi}{3} = -\frac{1}{2} + \frac{\sqrt{3}}{2} i$$
$$\cos \pi + i \sin \pi = -1$$
$$\cos \frac{4\pi}{3} + i \sin \frac{4\pi}{3} = -\frac{1}{2} - \frac{\sqrt{3}}{2} i$$
$$\cos \frac{5\pi}{3} + i \sin \frac{5\pi}{3} = \frac{1}{2} - \frac{\sqrt{3}}{2} i$$

as shown in Figure 9.40.

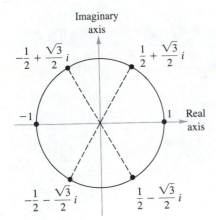

FIGURE 9.40

In Figure 9.40, notice that the roots obtained in Example 2 all have a magnitude of 1 and are equally spaced around this unit circle. Also, notice that the complex roots occur in conjugate pairs, as previously discussed in Section 4.5. We refer to the special case of the *n* distinct *n*th roots of 1 as the **nth roots of unity.**

EXAMPLE 3   *Finding the nth Roots of a Complex Number*

Find the three cube roots of $z = -2 + 2i$.

SOLUTION

Because z lies in the second quadrant, the trigonometric form for z is

$$z = -2 + 2i = \sqrt{8}(\cos 135° + i \sin 135°).$$

By our formula for nth roots, the cube roots have the form

$$\sqrt[6]{8}\left(\cos \frac{135° + 360°k}{3} + i \sin \frac{135° + 360°k}{3} \right).$$

Finally, for $k = 0, 1, 2$, we obtain the roots

$$\sqrt{2}(\cos 45° + i \sin 45°) = 1 + i$$
$$\sqrt{2}(\cos 165° + i \sin 165°) \approx -1.3660 + 0.3660\,i$$
$$\sqrt{2}(\cos 285° + i \sin 285°) \approx 0.3660 - 1.3660\,i.$$

The nth roots of a complex number can be useful for solving polynomial equations, as shown in Example 4.

EXAMPLE 4 *Finding the Roots of a Polynomial Equation*

Find all solutions to the equation $x^4 + 16 = 0$.

SOLUTION

The given equation can be written as

$$x^4 = -16 = 16(\cos \pi + i \sin \pi)$$

which means that we can solve the equation by finding the four fourth roots of -16. Each of these roots has the form

$$\sqrt[4]{16}\left(\cos \frac{\pi + 2\pi k}{4} + i \sin \frac{\pi + 2\pi k}{4} \right).$$

Finally, using $k = 0, 1, 2, 3$, we obtain the roots

$$2\left(\cos \frac{\pi}{4} + i \sin \frac{\pi}{4} \right) = 2\left(\frac{\sqrt{2}}{2} + \frac{\sqrt{2}}{2}\,i \right) = \sqrt{2} + \sqrt{2}\,i$$

$$2\left(\cos \frac{3\pi}{4} + i \sin \frac{3\pi}{4} \right) = 2\left(-\frac{\sqrt{2}}{2} + \frac{\sqrt{2}}{2}\,i \right) = -\sqrt{2} + \sqrt{2}\,i$$

$$2\left(\cos \frac{5\pi}{4} + i \sin \frac{5\pi}{4} \right) = 2\left(-\frac{\sqrt{2}}{2} - \frac{\sqrt{2}}{2}\,i \right) = -\sqrt{2} - \sqrt{2}\,i$$

$$2\left(\cos \frac{7\pi}{4} + i \sin \frac{7\pi}{4} \right) = 2\left(\frac{\sqrt{2}}{2} - \frac{\sqrt{2}}{2}\,i \right) = \sqrt{2} - \sqrt{2}\,i.$$

WARM UP

Simplify the expressions.

1. $\sqrt[3]{54}$

2. $\sqrt[4]{16 + 48}$

Write the complex numbers in trigonometric form.

3. $-5 + 5i$

4. $-3i$

5. -12

6. 12

Perform the indicated operations. Leave the results in trigonometric form.

7. $\left(\cos \dfrac{\pi}{4} + i \sin \dfrac{\pi}{4} \right)\left(\cos \dfrac{\pi}{2} + i \sin \dfrac{\pi}{2} \right)$

8. $\left(\cos \dfrac{\pi}{12} + i \sin \dfrac{\pi}{12} \right)\left(\cos \dfrac{5\pi}{6} + i \sin \dfrac{5\pi}{6} \right)$

9. $\dfrac{6[\cos(2\pi/3) + i \sin(2\pi/3)]}{3[\cos(\pi/6) + i \sin(\pi/6)]}$

10. $\dfrac{2(\cos 55° + i \sin 55°)}{3(\cos 10° + i \sin 10°)}$

EXERCISES 9.5

In Exercises 1–12, use DeMoivre's Theorem to find the indicated powers of the given complex number. Express the result in standard form.

1. $(1 + i)^5$

2. $(2 + 2i)^6$

3. $(-1 + i)^{10}$

4. $(1 - i)^{12}$

5. $2(\sqrt{3} + i)^7$

6. $4(1 - \sqrt{3}\,i)^3$

7. $[5(\cos 20° + i \sin 20°)]^3$

8. $[3(\cos 150° + i \sin 150°)]^4$

9. $\left(\cos \dfrac{5\pi}{4} + i \sin \dfrac{5\pi}{4} \right)^{10}$

10. $\left[2\left(\cos \dfrac{\pi}{2} + i \sin \dfrac{\pi}{2} \right) \right]^8$

11. $[5(\cos 3.2 + i \sin 3.2)]^4$

12. $(\cos 0 + i \sin 0)^{20}$

In Exercises 13–24, (a) use DeMoivre's Theorem to find the indicated roots, (b) represent each of the roots graphically, and (c) express each of the roots in standard form.

13. Square roots:
$9(\cos 120° + i \sin 120°)$

14. Square roots:
$16(\cos 60° + i \sin 60°)$

15. Fourth roots:
$16\left(\cos \dfrac{4\pi}{3} + i \sin \dfrac{4\pi}{3} \right)$

16. Fifth roots:
$32\left(\cos \dfrac{5\pi}{6} + i \sin \dfrac{5\pi}{6} \right)$

17. Square roots:
$-25i$

18. Fourth roots:
$625i$

19. Cube roots:
$-\dfrac{125}{2}(1 + \sqrt{3}\,i)$

20. Cube roots:
$-4\sqrt{2}(1 - i)$

21. Cube roots:
8

22. Fourth roots:
i

23. Fifth roots:
1

24. Cube roots:
1000

In Exercises 25–32, find all the solutions to the given equation and represent your solutions graphically.

25. $x^4 - i = 0$

26. $x^3 + 1 = 0$

27. $x^5 + 243 = 0$

28. $x^4 - 81 = 0$

29. $x^3 + 64i = 0$

30. $x^6 - 64i = 0$

31. $x^3 - (1 - i) = 0$

32. $x^4 + (1 + i) = 0$

CHAPTER 9 REVIEW EXERCISES

In Exercises 1–16, solve the triangle (if possible) using the three given parts. If two solutions are possible, list both.

1. $a = 5$, $b = 8$, $c = 10$
2. $a = 6$, $b = 9$, $C = 45°$
3. $A = 12°$, $B = 58°$, $a = 5$
4. $B = 110°$, $C = 30°$, $c = 10.5$
5. $B = 110°$, $a = 4$, $c = 4$
6. $a = 80$, $b = 60$, $c = 100$
7. $A = 75°$, $a = 2.5$, $b = 16.5$
8. $A = 130°$, $a = 50$, $b = 30$
9. $B = 115°$, $a = 7$, $b = 14.5$
10. $C = 50°$, $a = 25$, $c = 22$
11. $A = 15°$, $a = 5$, $b = 10$
12. $B = 150°$, $a = 64$, $b = 10$
13. $B = 150°$, $a = 10$, $c = 20$
14. $a = 2.5$, $b = 15.0$, $c = 4.5$
15. $B = 25°$, $a = 6.2$, $b = 4$
16. $B = 90°$, $a = 5$, $c = 12$

In Exercises 17–20, find the area of the triangle having the given parts.

17. $a = 4$, $b = 5$, $c = 7$ 18. $a = 15$, $b = 8$, $c = 10$
19. $A = 27°$, $b = 5$, $c = 8$ 20. $B = 80°$, $a = 4$, $c = 8$

21. Find the height of a tree that stands on a hillside of slope 32° (from the horizontal) if from a point 75 feet downhill the angle of elevation to the top of the tree is 48° (see figure).

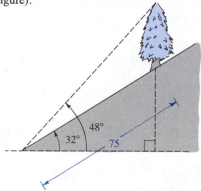

FIGURE FOR 21

22. To approximate the length of a marsh, a surveyor walks 450 meters from point A to point B. Then the surveyor turns 65° and walks 325 meters to point C. Approximate the length AC of the marsh (see figure).

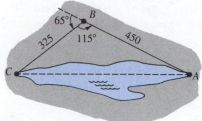

FIGURE FOR 22

23. From a certain distance, the angle of elevation of the top of a building is 17°. At a point 50 meters closer to the building, the angle of elevation is 31°. Approximate the height of the building.

24. Determine the width of a river that flows due east, if a tree on the opposite bank has a bearing of N 22° 30′ E and, after walking 400 feet downstream, a surveyor finds that the tree has a bearing of N 15° W.

25. Two planes leave an airport at approximately the same time. One is flying at 425 miles per hour at a bearing of N 5° W, and the other is flying at 530 miles per hour at a bearing of N 67° E (see figure). How far apart are the two planes after flying for 2 hours?

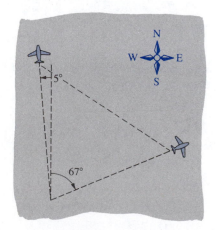

FIGURE FOR 25

26. The lengths of the diagonals of a parallelogram are 10 feet and 16 feet. Find the lengths of the sides of the parallelogram if the diagonals intersect at an angle of 28°.

In Exercises 27–30, find the component form of the vector **v** satisfying the given conditions.

27. Initial Point: (0, 10)
    Terminal Point: (7, 3)
28. Initial Point: (1, 5)
    Terminal Point: (15, 9)
29. $\|\mathbf{v}\| = 8$, $\theta = 120°$
30. $\|\mathbf{v}\| = 1/2$, $\theta = 225°$

In Exercises 31–34, find the component form of the required vector and sketch its graph given that $\mathbf{u} = 6\mathbf{i} - 5\mathbf{j}$ and $\mathbf{v} = 10\mathbf{i} + 3\mathbf{j}$.

31. $\dfrac{1}{\|\mathbf{u}\|}\mathbf{u}$

32. $3\mathbf{v}$

33. $4\mathbf{u} - 5\mathbf{v}$

34. $-\frac{1}{2}\mathbf{v}$

35. If two forces of 85 pounds and 50 pounds act on a single point, and the angle between them is 15°, find the magnitude of the resultant of the two forces.

36. A 100-pound weight is being supported by two ropes (see figure). Find the tension in each rope.

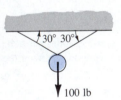

FIGURE FOR 36

37. Find the direction and magnitude of the resultant of the three forces shown in the figure.

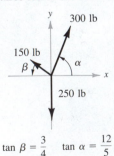

$\tan \beta = \dfrac{3}{4}$ $\tan \alpha = \dfrac{12}{5}$

FIGURE FOR 37

38. A 500-pound motorcycle is on a hill inclined at 12°. What force is required to keep the motorcycle from rolling back down the hill?

39. An airplane has an air speed of 450 miles per hour at a bearing of N 30° E. If the wind velocity is 20 miles per hour from the west, find the ground speed and the direction of the plane.

40. Two forces of 60 pounds and 100 pounds have a resultant force of magnitude 125 pounds. Find the angle between the two given forces.

In Exercises 41–44, find the trigonometric form of the complex number.

41. $5 - 5i$

42. $-3\sqrt{3} + 3i$

43. $5 + 12i$

44. -7

In Exercises 45 and 46, write the complex number in standard form.

45. $100(\cos 240° + i \sin 240°)$

46. $8\left(\cos \dfrac{5\pi}{6} + i \sin \dfrac{5\pi}{6}\right)$

In Exercises 47–50, (a) express the two given complex numbers in trigonometric form, and then (b) find z_1z_2 and z_1/z_2 using the trigonometric form.

47. $z_1 = -5$, $z_2 = 5i$

48. $z_1 = 2\sqrt{3} - 2i$, $z_2 = -10i$

49. $z_1 = -3(1 + i)$, $z_2 = 2(\sqrt{3} + i)$

50. $z_1 = 5i$, $z_2 = 2 - 2i$

In Exercises 51–54, use DeMoivre's Theorem to find the indicated power of the given complex number. Express the result in standard form.

51. $\left[5\left(\cos \dfrac{\pi}{12} + i \sin \dfrac{\pi}{12}\right)\right]^4$

52. $\left[2\left(\cos \dfrac{4\pi}{15} + i \sin \dfrac{4\pi}{15}\right)\right]^5$

53. $(2 + 3i)^6$

54. $(1 - i)^8$

In Exercises 55–58, use DeMoivre's Theorem to find the roots of the given complex numbers.

55. The sixth roots of $-729i$

56. The fourth roots of 256

57. The cube roots of -1

58. The fourth roots of $-1 + i$

CHAPTER 10

Systems of Equations and Inequalities

10.1 Systems of Equations

Up to this point in the text most problems have involved either a function of one variable or a single equation in two variables. However, many problems in science, business, and engineering involve two or more equations in two or more variables. To solve such problems, we need to find solutions of a **system of equations.** For example, consider the following problem.

A total of $12,000 is invested in two funds paying 9% and 11% simple interest, respectively. If the yearly interest is $1180, how much of the $12,000 is invested at each rate?

Letting x and y denote the amount in each fund, we can translate this problem into the following *system of two equations in two variables.*

$$
\begin{aligned}
x + \quad\ y &= 12{,}000 \qquad &\text{\textit{Equation 1}}\\
0.09x + 0.11y &= 1{,}180 \qquad &\text{\textit{Equation 2}}
\end{aligned}
$$

A solution of this system is an ordered pair that satisfies each equation in the system, and when we find the set of all solutions we say we are **solving the system** of equations.

Systems of Equations and Inequalities

Definition of a Solution of a System of Equations

The ordered pair (a, b) is a **solution of an equation** in the two variables x and y if the equation is true when a and b are substituted for x and y, respectively.

The ordered pair (a, b) is a **solution of a system** of equations in the two variables x and y if it is a solution of *each* equation in the system.

In this section we begin with the **method of substitution** for solving a system of equations. (Other methods will be considered later.)

EXAMPLE 1 *Solving a System of Two Equations in Two Variables*

Solve the following system of equations.

$$x + y = 4 \qquad \qquad \text{\textit{Equation 1}}$$
$$x - y = 2 \qquad \qquad \text{\textit{Equation 2}}$$

SOLUTION

Solving for y in Equation 1, we get

$$y = 4 - x.$$

Substituting $(4 - x)$ for y in Equation 2, we obtain a single-variable equation which we then solve for x.

$$x - (4 - x) = 2$$
$$x - 4 + x = 2$$
$$2x = 6$$
$$x = 3$$

Finally, by back-substituting $x = 3$ into the equation $y = 4 - x$, we obtain

$$y = 4 - x = 4 - 3 = 1.$$

Thus, the solution is the ordered pair $(3, 1)$. ▬▬

Remark: The term *back-substitution* implies that we work *backwards*. First we solve for one of the variables, and then we substitute that value back into one of the equations in the system to find the value of the other variable.

Because many steps are required to solve a system of equations, it is very easy to make errors in arithmetic. Thus, we suggest that you develop the habit of *checking your solution by substituting it into each equation in the original system*. In Example 1, we check the solution $x = 3$ and $y = 1$:

Equation 1: $3 + 1 = 4$ *Check solution in each equation in the original system.*
Equation 2: $3 - 1 = 2$

The method of substitution demonstrated in Example 1 has five steps.

Method of Substitution

To solve a system of two equations in two variables, use the following steps.

1. Solve one of the equations for one variable in terms of the other.
2. Substitute the expression found in Step 1 into the other equation to obtain an equation in one variable.
3. Solve the equation obtained in Step 2.
4. Back-substitute the solution in Step 3 into the expression obtained in Step 1 to find the value of the other variable.
5. Check your answer to see that it satisfies each of the original equations.

In Example 2, we apply this method to solve the "interest rate problem" introduced at the beginning of this section.

EXAMPLE 2 *Solving a System by Substitution: One-Solution Case*

Solve the following system of equations.

$$x + \quad y = 12{,}000 \qquad \textit{Equation 1}$$
$$0.09x + 0.11y = \quad 1{,}180 \qquad \textit{Equation 2}$$

SOLUTION

To begin, it is convenient to multiply both sides of the second equation by 100 to obtain $9x + 11y = 118{,}000$. This eliminates the need to work with decimals. Then the following steps are performed.

1. Solve for x in Equation 1.

$$x = 12{,}000 - y$$

2. Substitute this expression for x into Equation 2.

$$9(12{,}000 - y) + 11y = 118{,}000$$

3. Solve for y.

$$108{,}000 - 9y + 11y = 118{,}000$$
$$2y = 10{,}000$$
$$y = 5{,}000$$

4. Back-substitute the value $y = 5000$ to solve for x.

$$x = 12{,}000 - 5{,}000 = 7{,}000$$

Therefore, the solution is the ordered pair (7000, 5000). Try checking to see that $x = 7000$ and $y = 5000$ satisfies each of the original equations. ▄▄▄

Systems of Equations and Inequalities

Note that the equations in Examples 1 and 2 are linear. That is, the variables x and y occurred to the first power only. The method of substitution can also be used to solve systems in which one or both of the equations are nonlinear. We illustrate this in our next two examples.

EXAMPLE 3 *Solving a System by Substitution: Two-Solution Case*

Solve the following system of equations.

$$x^2 - 2x + y^2 = 1 \qquad \text{\textit{Equation 1}}$$
$$-x + y = -1 \qquad \text{\textit{Equation 2}}$$

SOLUTION

1. Solve for y in Equation 2.

$$-x + y = -1$$
$$y = x - 1$$

2. Substitute this expression for y into Equation 1.

$$x^2 - 2x + (x - 1)^2 = 1$$

3. Solve for x.

$$x^2 - 2x + x^2 - 2x + 1 = 1$$
$$2x^2 - 4x = 0$$
$$2x(x - 2) = 0$$
$$x = 0, 2$$

4. Back-substitute these values of x to solve for the corresponding values of y.

$$\text{For } x = 0: \quad y = 0 - 1 = -1$$
$$\text{For } x = 2: \quad y = 2 - 1 = 1$$

Thus, there are two solutions: $(0, -1)$ and $(2, 1)$. Try checking these in the original system to see that both work.

EXAMPLE 4 *Solving a System by Substitution: No-Solution Case*

Solve the following system of equations.

$$x^2 + y^2 = 1 \qquad \text{\textit{Equation 1}}$$
$$x^2 + y = 3 \qquad \text{\textit{Equation 2}}$$

SOLUTION

1. Solve for x^2 in Equation 2.

   $$x^2 = 3 - y$$

2. Substitute this expression for x^2 into Equation 1.

   $$(3 - y) + y^2 = 1$$

3. Solve for y.

   $$y^2 - y + 3 = 1$$
   $$y^2 - y + 2 = 0$$
   $$y = \frac{1 \pm \sqrt{1 - 8}}{2} \qquad \text{\textit{Use Quadratic Formula}}$$

Since the discriminant is negative, the equation $y^2 - y + 2 = 0$ has no (real) solution. Hence, this system has no (real) solution.

Graphical Approach to Finding Solutions

From Examples 2, 3, and 4, we can see that a system of two equations in two unknowns can have exactly one solution, more than one solution, or no solution. In practice, we can gain insight about the location and number of solutions of a system of equations by graphing each of the equations on the same coordinate plane. The solutions of the system correspond to the **points of intersection** of the graphs. For instance, in Figure 10.1(a), the two equations graph as two lines with a *single point* of intersection. The two equations in Example 3 graph as a circle and a line with *two points* of intersection, as shown in Figure 10.1(b). Moreover, the two equations in Example 4 graph as a circle and a parabola that happen to have *no points* of intersection, as shown in Figure 10.1(c).

FIGURE 10.1

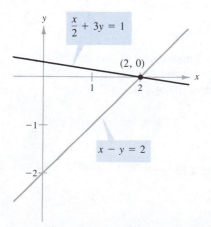

(a) One point of intersection

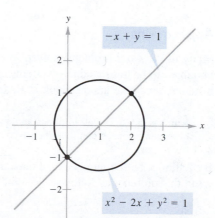

(b) Two points of intersection

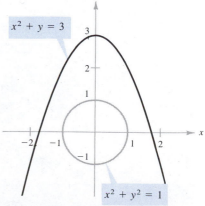

(c) No points of intersection

Occasionally, the graphical approach to solving a system of equations is easier than the method of substitution. This is illustrated in Example 5.

EXAMPLE 5 *Solving a System of Equations*

Solve the following system of equations.

$$y = \ln x \qquad \qquad \textit{Equation 1}$$
$$x + y = 1 \qquad \qquad \textit{Equation 2}$$

SOLUTION

The graph of each equation is shown in Figure 10.2. From this sketch it is clear that there is only one point of intersection. Also, it appears that $(1, 0)$ is the solution point, and we confirm this by checking these coordinates in *both* equations.

CHECK   Let $x = 1$ and $y = 0$.

$$0 = \ln 1 \qquad \qquad \textit{Equation 1 checks}$$
$$1 + 0 = 1 \qquad \qquad \textit{Equation 2 checks}$$

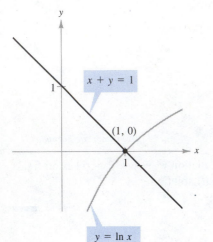

FIGURE 10.2

Remark: Example 5 shows us the value of a graphical approach to solving systems of equations in two variables. Notice what would have happened if we had tried only the substitution method in Example 5. By substituting $y = \ln x$ into $x + y = 1$, we obtain

$$x + \ln x = 1.$$

It would be difficult to solve this equation for x using standard algebraic techniques.

Applications

Many of the word problems solved in earlier sections of this text using one variable can be translated into systems of two equations in two variables. For instance, let's take another look at Example 10 of Section 2.2.

Kim has saved \$21.20 in dimes and quarters. If there are 119 coins in all, how many of each coin has she saved?

By letting x be the number of dimes and y the number of quarters, we can write this problem as follows.

$$x + y = 119$$
$$0.10x + 0.25y = 21.20$$

The solution can then be found by the method of substitution. Try solving this system. (You should find that the solution is $x = 57$ and $y = 62$.)

EXAMPLE 6 An Application of a System of Equations

One type of variety pack contains three candy bars of type A and five of type B. A second variety pack contains four bars of each type. The first pack costs $2.86 and the second costs $2.80. What is the cost of each type of candy bar?

SOLUTION

Using two variables, we let

$x =$ cost of bar A
$y =$ cost of bar B.

The problem can then be translated into the following system.

$$3x + 5y = 2.86 \qquad \text{\textit{Equation 1}}$$
$$4x + 4y = 2.80 \qquad \text{\textit{Equation 2}}$$

1. Solve for y in Equation 2.

$$4y = 2.80 - 4x$$
$$y = 0.70 - x$$

2. Substitute this expression for y into Equation 1.

$$3x + 5(0.70 - x) = 2.86$$

3. Solve for x.

$$-2x = -0.64$$
$$x = 0.32$$

4. Back-substitute the value $x = 0.32$ to solve for y.

$$y = 0.70 - 0.32 = 0.38$$

Therefore, bar A costs $0.32 and bar B costs $0.38.

The total cost C of producing x units of a product typically has two components—the initial cost and the cost per unit. When enough units have been sold so that the total revenue R equals the total cost, we say that sales have reached the **break-even point.** We can find this break-even point by setting C equal to R and solving for x. In other words, the break-even point corresponds to the point of intersection of the cost and revenue curves.

Systems of Equations and Inequalities

EXAMPLE 7 An Application: Break-Even Analysis

A small business invests $10,000 in equipment to produce a product. Each unit of the product costs $0.65 to produce and is sold for $1.20. How many items must be sold before the business breaks even?

SOLUTION

The total cost of producing x units is

$$C = \overbrace{0.65x}^{\substack{\text{Cost per} \\ \text{unit}}} + \overbrace{10,000}^{\substack{\text{Initial} \\ \text{cost}}}$$

and the revenue obtained by selling x units is

$$R = \overbrace{1.2x}^{\substack{\text{Price} \\ \text{per unit}}}.$$

Since the break-even point occurs when $R = C$, we have

$$1.2x = 0.65x + 10,000$$

$$0.55x = 10,000$$

$$x = \frac{10,000}{0.55} \approx 18,182 \text{ units.}$$

Note in Figure 10.3 that sales less than the break-even point correspond to an overall loss, while sales greater than the break-even point correspond to a profit.

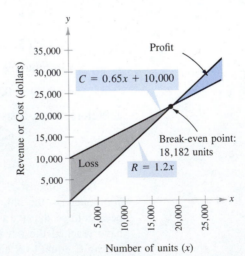

FIGURE 10.3

WARM UP

Sketch the graph of each equation.

1. $y = -\frac{1}{3}x + 6$

2. $y = 2(x - 3)$

3. $x^2 + y^2 = 4$

4. $y = 5 - (x - 3)^2$

Perform the indicated operations and simplify.

5. $(3x + 2y) - 2(x + y)$

6. $(-10u + 3v) + 5(2u - 8v)$

7. $x^2 + (x - 3)^2 + 6x$

8. $y^2 - (y + 1)^2 + 2y$

Solve the equations.

9. $3x + (x - 5) = 15 + 4$

10. $y^2 + (y - 2)^2 = 2$

EXERCISES 10.1

In Exercises 1–10, solve the system by the method of substitution.

1. $2x + y = 4$
$\quad\ -x + y = 1$

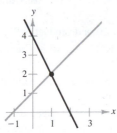

2. $x - \ y = -5$
$\quad\ x + 2y = \ \ 4$

5. $x + 3y = 15$
$\quad\ x^2 + y^2 = 25$

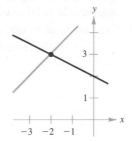

6. $x - y = 0$
$\quad\ x^3 - 5x + y = 0$

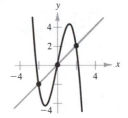

3. $x - y = -3$
$\quad\ x^2 - y = -1$

4. $3x - y = -2$
$\quad\ x^3 - y = 0$

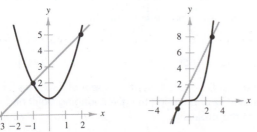

7. $x^2 - y = 0$
$\quad\ x^2 - 4x + y = 0$

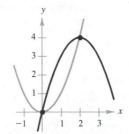

8. $y = -x^2 + 1$
$\quad\ y = x^4 - 2x^2 + 1$

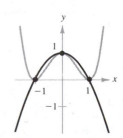

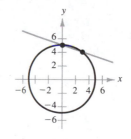

9. $x - 3y = -4$
$x^2 - y^3 = 0$

10. $y = x^3 - 3x^2 + 3$
$2x + y = 3$

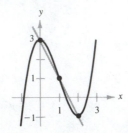

In Exercises 11–30, solve the system by the method of substitution.

11. $x - y = 0$
$5x - 3y = 10$

12. $x + 2y = 1$
$5x - 4y = -23$

13. $2x - y + 2 = 0$
$4x + y - 5 = 0$

14. $6x - 3y - 4 = 0$
$x + 2y - 4 = 0$

15. $30x - 40y - 33 = 0$
$10x + 20y - 21 = 0$

16. $1.5x + 0.8y = 2.3$
$0.3x - 0.2y = 0.1$

17. $\frac{1}{5}x + \frac{1}{2}y = 8$
$x + y = 20$

18. $\frac{1}{2}x + \frac{3}{4}y = 10$
$\frac{3}{2}x - y = 4$

19. $x - y = 0$
$2x + y = 0$

20. $x - 2y = 0$
$3x - y = 0$

21. $y = 2x$
$y = x^2 + 1$

22. $x + y = 4$
$x^2 - y = 2$

23. $3x - 7y + 6 = 0$
$x^2 - y^2 = 4$

24. $x^2 + y^2 = 25$
$2x + y = 10$

25. $x^2 + y^2 = 5$
$x - y = 1$

26. $y = x^3 - 2x^2 + x - 1$
$y = -x^2 + 3x - 1$

27. $y = x^4 - 2x^2 + 1$
$y = 1 - x^2$

28. $x^2 + y = 4$
$2x - y = 1$

29. $xy - 1 = 0$
$2x - 4y + 7 = 0$

30. $x - 2y = 1$
$y = \sqrt{x - 1}$

In Exercises 31–38, find all points of intersection of the graphs of the given pair of equations. [*Hint:* A graphical approach, as demonstrated in Example 5, may be helpful.]

31. $x + y = 4$
$x^2 + y^2 - 4x = 0$

32. $3x - 2y = 0$
$x^2 - y^2 = 4$

33. $2x - y + 3 = 0$
$x^2 + y^2 - 4x = 0$

34. $x - y + 3 = 0$
$x^2 - 4x + 7 = y$

35. $y = e^x$
$x - y + 1 = 0$

36. $x^2 + y^2 = 8$
$y = x^2$

37. $y = \cos x$
$y = \sin 2x$

38. $x - y = 3$
$x - y^2 = 1$

In Exercises 39–42, find x, y, and λ satisfying the given system. These systems arise in certain optimization problems in calculus, and λ is called a Lagrange Multiplier. [*Hint:* You can reduce each system to a system of two equations in two variables by solving for λ in the first equation and substituting into the second equation.]

39. $y + \lambda = 0$
$x + \lambda = 0$
$x + y - 10 = 0$

40. $2x + \lambda = 0$
$2y + \lambda = 0$
$x + y - 4 = 0$

41. $2x - 2x\lambda = 0$
$-2y + \lambda = 0$
$y - x^2 = 0$

42. $2 + 2y + 2\lambda = 0$
$2x + 1 + \lambda = 0$
$2x + y - 100 = 0$

In Exercises 43 and 44, find the sales necessary to break even ($R = C$) for the given cost C of x units, and the given revenue R obtained by selling x units. (Round your answer to the nearest whole unit.)

43. $C = 8650x + 250,000$, $R = 9950x$

44. $C = 5.5\sqrt{x} + 10,000$, $R = 3.29x$

45. A person is setting up a small business and has invested $16,000 to produce an item that will sell for $5.95. If each unit can be produced for $3.45, how many units must be sold to break even?

46. A person setting up a part-time business makes an initial investment of $5000. The unit cost of the product is $21.60, and the selling price is $34.10. How many units must be sold to break even?

47. Find two numbers whose sum is 20 and product is 96.

48. Find two numbers whose sum is 12 and whose sum of squares is 80.

49. What are the dimensions of a rectangle if its perimeter is 40 miles and its area is 96 square miles?

50. What are the dimensions of an isosceles right triangle whose hypotenuse is 2 inches long and area is 1 square inch?

In Exercises 51 and 52, find the initial velocity v_0 and the time t of travel of a projectile that is thrown from the point $(0, 0)$ at an angle of inclination θ and that lands at the given point. The path of the projectile is described by the model

$$x = (v_0 \cos \theta)t \quad \text{and} \quad y = (v_0 \sin \theta)t - 16t^2$$

where t is time in seconds, x and y are distances in feet.

51. $\theta = 45°$, $(144, 0)$

52. $\theta = 30°$, $(144, 0)$

In Exercises 53 and 54, use the model of Exercises 51 and 52 to find v_0 and θ using the given landing point and flight time of the projectile.

53. $(100, 0)$, $t = 4$

54. $(200, 0)$, $t = 4$

10.2 Systems of Linear Equations in Two Variables

Another method for solving a system of equations is called **elimination.** It is especially convenient for solving systems of *linear equations* such as

$$3x + 5y = 7 \qquad \textit{Equation 1}$$
$$-3x - 2y = -1. \qquad \textit{Equation 2}$$

To solve this system by elimination we can add the second equation to the first to obtain

$$0x + 3y = 6.$$

Note that the variable x has been "eliminated." Now, we solve for y to obtain $y = 2$ and back-substitute into Equation 1 to obtain

$$3x + 5(2) = 7$$
$$3x = -3$$
$$x = -1.$$

Thus, the solution is $(-1, 2)$.

For the given system, it was convenient to eliminate the variable x by adding the second equation to the first. However, we could have chosen to eliminate the variable y. To do this we would multiply both sides of the first equation by 2 and both sides of the second equation by 5 to obtain

$$6x + 10y = 14 \qquad \textit{New Equation 1}$$
$$-15x - 10y = -5. \qquad \textit{New Equation 2}$$

By adding the two equations, we obtain $-9x = 9$, which yields $x = -1$.

For a linear system of two equations in two variables, we give the following steps in the **method of elimination.**

The Method of Elimination

To solve a system of linear equations in two variables x and y, use the following steps.

1. Obtain coefficients for x (or y) that differ only in sign by multiplying all terms of one or both equations by suitably chosen constants.
2. Add the two equations to obtain a linear equation in (at most) one variable.
3. Solve the resulting linear equation in one variable.
4. Back-substitute this solution into either of the original equations to solve for the other variable.
5. Check your solution in *both* of the original equations.

EXAMPLE 1 *Solving a Linear System by Elimination*

Solve the following system of linear equations.

$$2x - 3y = -7 \qquad \text{\textit{Equation 1}}$$
$$3x + y = -5 \qquad \text{\textit{Equation 2}}$$

SOLUTION

1. We can obtain coefficients of y that differ only in sign by multiplying the second equation by 3.

$$9x + 3y = -15 \qquad \text{\textit{New Equation 2}}$$

2. Now we can eliminate y by adding the first equation to the new second equation as follows.

$$\begin{aligned} 2x - 3y &= -7 \\ 9x + 3y &= -15 \\ \hline 11x &= -22 \end{aligned}$$

3. Solve for x.

$$11x = -22$$
$$x = -2$$

4. Back-substitute (into Equation 1) to solve for y.

$$\begin{aligned} 2x - 3y &= -7 \\ 2(-2) - 3y &= -7 \\ -3y &= -3 \\ y &= 1 \end{aligned}$$

5. Check the solution $x = -2$, $y = 1$.

$$2(-2) - 3(1) = -4 - 3 = -7 \qquad \text{\textit{Equation 1 checks}}$$
$$3(-2) + 1 = -6 + 1 = -5 \qquad \text{\textit{Equation 2 checks}}$$

As we observed in Section 10.1, it is possible for a *general* system of equations to have exactly one solution, two or more solutions, or no solution. For a system of *linear* equations we can strengthen this result somewhat. Specifically, if a system of linear equations has two different solutions, then it must have an infinite number of solutions. To see why this is true, consider the following graphical interpretations of a system of two linear equations in two variables. (Remember that the graph of a linear equation is a straight line.)

Graphic Interpretation of Solutions

For a system of two linear equations in two variables, the number of solutions is given by one of the following.

Number of Solutions	*Graphical Interpretation*
1. Exactly one solution	The lines intersect at one point.
2. Infinitely many solutions	The lines are identical.
3. No solution	The lines are parallel.

Figure 10.4 graphically illustrates the three cases.

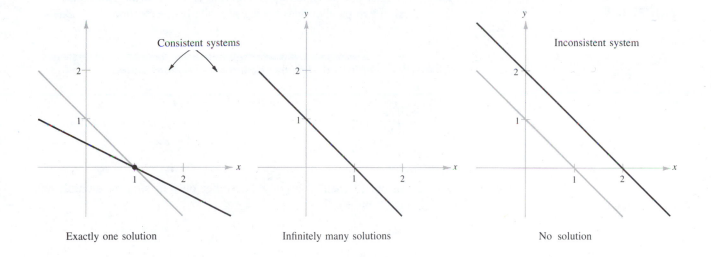

Exactly one solution   Infinitely many solutions   No solution

FIGURE 10.4

A system of two equations in two variables is called **consistent** if it has at least one solution, and **inconsistent** if it has no solution. Moreover, a consistent system is called **dependent** if it has an infinite number of solutions.

EXAMPLE 2   *Method of Elimination: No-Solution Case*

Solve the following system of linear equations.

$$x - 2y = 3 \qquad\qquad \textit{Equation 1}$$
$$-2x + 4y = 1 \qquad\qquad \textit{Equation 2}$$

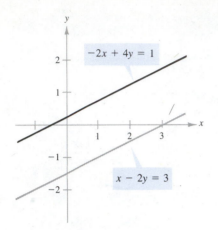

FIGURE 10.5

SOLUTION

1. We can obtain coefficients of x that differ only in sign by multiplying the first equation by 2.

$$2x - 4y = 6 \qquad \text{\textit{New Equation 1}}$$

2. Now, to eliminate x, we add this new first equation to the second equation.

$$
\begin{array}{r}
2x - 4y = 6 \\
-2x + 4y = 1 \\
\hline
0x + 0y = 7 \\
0 = 7
\end{array}
$$

There are no values for x or y for which $0 = 7$. Hence, the system is inconsistent and has no solution. Figure 10.5 shows that the two equations correspond to two parallel lines having no point of intersection. ▬

EXAMPLE 3   *Method of Elimination: Many-Solutions Case*

Solve the following system of linear equations.

$$
\begin{array}{ll}
2x - y = 1 & \qquad \text{\textit{Equation 1}} \\
4x - 2y = 2 & \qquad \text{\textit{Equation 2}}
\end{array}
$$

SOLUTION

1. We can obtain coefficients of x that differ only in sign by multiplying Equation 2 by $-1/2$.

$$-2x + y = -1 \qquad \text{\textit{New Equation 2}}$$

2. Now, to eliminate x, we add this new second equation to the first equation.

$$
\begin{array}{r}
2x - y = 1 \\
-2x + y = -1 \\
\hline
0x + 0y = 0
\end{array}
$$

The latter equation imposes no restrictions on x or y. This means that the solution set for the first equation is the same as for the second equation, or graphically that the two lines coincide, as shown in Figure 10.6. The linear equation $2x - y = 1$ (or $y = 2x - 1$) has an infinite solution set consisting of all ordered pairs (a, b) such that $b = 2a - 1$. Hence, every ordered pair of the form

$$(a, 2a - 1) \qquad \text{\textit{a is a real number}}$$

is a solution of the given system. For instance, if $a = 1$, then we see that $(1, 1)$ is a solution. Similarly, if $a = 2$, we see that $(2, 3)$ is a solution.

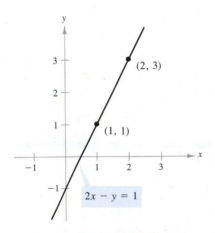

Infinite Number of Solutions

FIGURE 10.6

In Example 3, we could have reached the same conclusion by multiplying the first equation by 2 to obtain

$$4x - 2y = 2 \qquad \text{\textit{New Equation 1}}$$
$$4x - 2y = 2. \qquad \text{\textit{Equation 2}}$$

Because these two equations are identical, they obviously have the same solution set.

Applications

Systems of linear equations have many applications. Example 4 shows how to solve a problem involving speed with a system of linear equations.

EXAMPLE 4 An Application of a Linear System

An airplane flying into a headwind travels the 2000-mile flying distance between two cities in 4 hours and 24 minutes. On the return flight, the same distance is traveled in 4 hours. Find the ground speed of the plane and the speed of the wind, assuming that both remain constant.

SOLUTION

The two unknown quantities are the speeds of the wind and the plane. If r_1 is the speed of the plane and r_2 is the speed of the wind, then

$$r_1 - r_2 = \text{speed of the plane \textit{against} the wind}$$
$$r_1 + r_2 = \text{speed of the plane \textit{with} the wind}$$

as shown in Figure 10.7. Using the formula

$$\text{distance} = (\text{rate})(\text{time})$$

for these two speeds, we obtain the following equations.

$$2000 = (r_1 - r_2)\left(4 + \frac{24}{60}\right)$$
$$2000 = (r_1 + r_2)(4)$$

These two equations simplify as follows.

$$5000 = 11r_1 - 11r_2 \qquad \text{\textit{Equation 1}}$$
$$500 = r_1 + r_2 \qquad \text{\textit{Equation 2}}$$

By elimination, the solution is

$$r_1 = \frac{5250}{11} \approx 477.27 \text{ miles per hour}$$

$$r_2 = \frac{250}{11} \approx 22.73 \text{ miles per hour.}$$

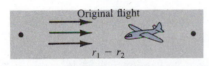

Original flight
$r_1 - r_2$

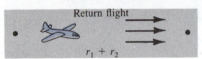

Return flight
$r_1 + r_2$

FIGURE 10.7

WARM UP

Sketch the graph of each equation.

1. $2x + y = 4$

2. $5x - 2y = 3$

In each case, find an equation of the line passing through the two points.

3. $(-1, 3), (4, 8)$

4. $(2, 6), (5, 1)$

Determine the slope of each line.

5. $3x + 6y = 4$

6. $7x - 4y = 10$

In each case, determine whether the lines represented by the pair of equations are parallel, perpendicular, or neither.

7. $2x - 3y = -10$
   $3x + 2y = 11$

8. $4x - 12y = 5$
   $-2x + 6y = 3$

9. $5x + y = 2$
   $3x + 2y = 1$

10. $x - 3y = 2$
   $6x + 2y = 4$

EXERCISES 10.2

In Exercises 1–10, solve the linear system by elimination.
Identify and label each line with the appropriate equation.

1. $2x + y = 4$
   $x - y = 2$

2. $x + 3y = 2$
   $-x + 2y = 3$

5. $x - y = 1$
   $-2x + 2y = 5$

6. $3x + 2y = 2$
   $6x + 4y = 14$

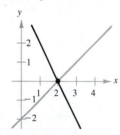

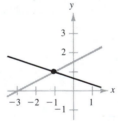

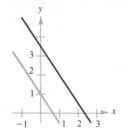

3. $x - y = 0$
   $3x - 2y = -1$

4. $2x - y = 2$
   $4x + 3y = 24$

7. $3x - 2y = 6$
   $-6x + 4y = -12$

8. $x - 2y = 5$
   $6x + 2y = 7$

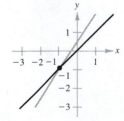

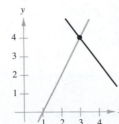

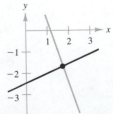

9. $9x - 3y = -1$
$3x + 6y = -5$

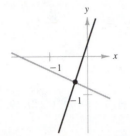

10. $5x + 3y = 18$
$2x - 7y = -1$

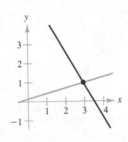

In Exercises 11–30, solve the system by elimination.

11. $x + 2y = 4$
$x - 2y = 1$

12. $3x - 5y = 2$
$2x + 5y = 13$

13. $2x + 3y = 18$
$5x - y = 11$

14. $x + 7y = 12$
$3x - 5y = 10$

15. $3x + 2y = 10$
$2x + 5y = 3$

16. $8r + 16s = 20$
$16r + 50s = 55$

17. $2u + v = 120$
$u + 2v = 120$

18. $5u + 6v = 24$
$3u + 5v = 18$

19. $6r - 5s = 3$
$10s - 12r = 5$

20. $1.8x + 1.2y = 4$
$9x + 6y = 3$

21. $\dfrac{x}{4} + \dfrac{y}{6} = 1$
$x - y = 3$

22. $\frac{2}{3}x + \frac{1}{6}y = \frac{2}{3}$
$4x + y = 4$

23. $\dfrac{x + 3}{4} + \dfrac{y - 1}{3} = 1$
$2x - y = 12$

24. $\dfrac{x - 1}{2} + \dfrac{y + 2}{3} = 4$
$x - 2y = 5$

25. $2.5x - 3y = 1.5$
$10x - 12y = 6$

26. $0.02x - 0.05y = -0.19$
$0.03x + 0.04y = 0.52$

27. $0.05x - 0.03y = 0.21$
$0.07x + 0.02y = 0.16$

28. $3b + 3m = 7$
$3b + 5m = 3$

29. $4b + 3m = 3$
$3b + 11m = 13$

30. $\dfrac{12}{x} - \dfrac{12}{y} = 7$ [*Hint:* Let $X = 1/x$ and $Y = 1/y$.
Then solve the linear system in X and Y.]
$\dfrac{3}{x} + \dfrac{4}{y} = 0$

In Exercises 31 and 32, the graphs of the two equations appear to be parallel. Yet, when the system is solved algebraically, we find that the system does have a solution. Find the solution and explain why it does not appear on the portion of the graph that is shown.

31. $200y - x = 200$
$199y - x = -198$

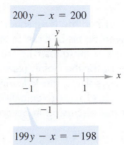

32. $25x - 24y = 0$
$13x - 12y = 120$

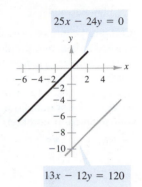

In Exercises 33–42, solve the problem using a system of linear equations.

33. Find two numbers whose sum is 20 and difference is 2.

34. Find two numbers whose sum is 35 and difference is 11.

35. An airplane flying into a headwind travels the 1800-mile flying distance between two cities in 3 hours and 36 minutes. On the return flight, the distance is traveled in 3 hours. Find the ground speed of the plane and the speed of the wind, assuming that both remain constant.

36. Two planes start from the same airport and fly in opposite directions. The second plane starts one-half hour after the first plane, but its speed is 50 miles per hour faster. Find the ground speed of each plane if 2 hours after the first plane starts its flight the planes are 2000 miles apart.

37. Ten gallons of a 30% acid solution are obtained by mixing a 20% solution with a 50% solution. How much of each must be used?

38. Ten pounds of mixed nuts sell for $5.95 per pound. The mixture is obtained from two kinds of nuts, with one variety priced at $4.29 per pound and the other at $6.55 per pound. How much of each must be used?

39. Five hundred tickets were sold for a certain performance of a play. The tickets for adults and children sold for $7.50 and $4.00, respectively, and the receipts for the performance were $3312.50. How many of each kind of ticket were sold?

40. A total of $12,000 is invested in two corporate bonds that pay 10.5% and 12% simple interest. The annual interest is $1380. How much is invested in each bond?

41. The perimeter of a rectangle is 40 feet and the length is 4 feet greater than the width. Find the dimensions of the rectangle.

42. The sum of the digits of a given two-digit number is 12. If the digits are reversed, the number is increased by 36. Find the number.

[Optional] In Exercises 43–46, (a) find the least squares regression line, and (b) plot the given points and sketch the least squares regression line on the same axes. The *least squares regression line*, $y = ax + b$, for the points (x_1, y_1), $(x_2, y_2), \ldots, (x_n, y_n)$ is obtained by solving the following system of linear equations for a and b. (If you are unfamiliar with summation notation, look at the discussion in Section 12.1.)

$$nb + \left(\sum_{i=1}^{n} x_i\right)a = \sum_{i=1}^{n} y_i$$

$$\left(\sum_{i=1}^{n} x_i\right)b + \left(\sum_{i=1}^{n} x_i^2\right)a = \sum_{i=1}^{n} x_i y_i$$

43. $(-2, 0), (0, 1), (2, 3)$

44. $(-3, 0), (-1, 1), (1, 1), (3, 2)$

45. $(0, 4), (1, 3), (1, 1), (2, 0)$

46. $(1, 0), (2, 0), (3, 0), (3, 1), (4, 1), (4, 2), (5, 2), (6, 2)$

10.3 Systems of Linear Equations in More than Two Variables

The method of elimination can be applied to a system of linear equations in more than two variables. In fact, this method easily adapts to computer use for solving linear systems with dozens of variables.

When using elimination to solve a system of linear equations, the goal is to rewrite the system in a form to which back-substitution can be applied. To see how this works, consider the following two systems of linear equations.

$$\begin{array}{rcr} x - 2y + 3z &=& 9 \\ -x + 3y &=& -4 \\ 2x - 5y + 5z &=& 17 \end{array} \qquad \begin{array}{rcl} x - 2y + 3z &=& 9 \\ y + 3z &=& 5 \\ z &=& 2 \end{array}$$

The system on the right is clearly easier to solve. We say that this system is in **row-echelon form,** which means that it follows a stair-step pattern and has leading coefficients of 1. To solve such a system, we use back-substitution, working from the bottom equation to the top equation, as demonstrated in Example 1.

EXAMPLE 1 *Using Back-Substitution to Solve a System in Row-Echelon Form*

Solve the following system of linear equations.

$$\begin{array}{rcl} x - 2y + 3z &=& 9 \qquad \textit{Equation 1} \\ y + 3z &=& 5 \qquad \textit{Equation 2} \\ z &=& 2 \qquad \textit{Equation 3} \end{array}$$

SOLUTION

From Equation 3, we already know the value of z. To solve for y, we substitute $z = 2$ into Equation 2 to obtain

$$y + 3(2) = 5 \qquad \textit{Substitute } z = 2$$
$$y = -1. \qquad \textit{Solve for } y$$

Finally, we substitute $y = -1$ and $z = 2$ into Equation 1 to obtain

$$x - 2(-1) + 3(2) = 9 \qquad \textit{Substitute } y = -1,\ z = 2$$
$$x = 1. \qquad \textit{Solve for } x$$

Thus, the solution is $x = 1$, $y = -1$, and $z = 2$.

Two systems of equations are called **equivalent** if they have precisely the same solution set. To solve a system that is not in row-echelon form, we first change it to an *equivalent* system that is in row-echelon form by using the following operations.

Operations That Lead to Equivalent Systems of Equations

Each of the following operations on a system of linear equations produces an *equivalent* system.

1. Interchange two equations.
2. Multiply an equation by a nonzero constant.
3. Add a multiple of an equation to another equation.

Rewriting a system of linear equations in row-echelon form usually involves a *chain* of equivalent systems, each of which is obtained by using one of the three basic operations. This process is called **Gaussian elimination,** after the German mathematician Carl Friedrich Gauss (1777–1855).

EXAMPLE 2   *Using Elimination to Solve a Linear System*

Solve the following system of linear equations.

$$\begin{array}{rcl} x - 2y + 3z &=& 9 \\ -x + 3y &=& -4 \\ 2x - 5y + 5z &=& 17 \end{array}$$

Systems of Equations and Inequalities

SOLUTION

Although there are several ways to begin, our goal is to develop a systematic procedure that can be applied to large systems. We work from the upper left corner, saving the x in the upper left position and eliminating the other x's from the first column.

$$\begin{aligned} x - 2y + 3z &= 9 \\ y + 3z &= 5 \\ 2x - 5y + 5z &= 17 \end{aligned}$$

> Adding the first equation to the second equation produces a new second equation.

$$\begin{aligned} x - 2y + 3z &= 9 \\ y + 3z &= 5 \\ -y - z &= -1 \end{aligned}$$

> Adding -2 times the first equation to the third equation produces a new third equation.

Now that all but the first x has been eliminated from the first column, we go to work on the second column. (We need to eliminate y from the third equation.)

$$\begin{aligned} x - 2y + 3z &= 9 \\ y + 3z &= 5 \\ 2z &= 4 \end{aligned}$$

> Adding the second equation to the third equation produces a new third equation.

Finally, we need a coefficient of 1 for z in the third equation.

$$\begin{aligned} x - 2y + 3z &= 9 \\ y + 3z &= 5 \\ z &= 2 \end{aligned}$$

> Multiplying the third equation by $1/2$ produces a new third equation.

This is the same system we solved in Example 1, and, as in that example, we conclude that the solution is

$$x = 1, \quad y = -1, \quad \text{and} \quad z = 2.$$

The solution can also be written as the **ordered triple** $(1, -1, 2)$.

In Example 2, we can check the solution $x = 1$, $y = -1$, and $z = 2$, as follows.

Equation 1: $\quad (1) - 2(-1) + 3(2) = \quad 9$
Equation 2: $\quad -(1) + 3(-1) \qquad\quad = -4$
Equation 3: $\quad 2(1) - 5(-1) + 5(2) = \quad 17$

Check solution in each equation of original system.

We now look at an inconsistent system—one that has no solution. The key to recognizing an inconsistent system is that at some stage in the elimination process, we obtain an absurdity such as $0 = 7$. This is demonstrated in Example 3.

EXAMPLE 3 *An Inconsistent System*

Solve the following system of linear equations.

$$x - 3y + z = 1$$
$$2x - y - 2z = 2$$
$$x + 2y - 3z = -1$$

SOLUTION

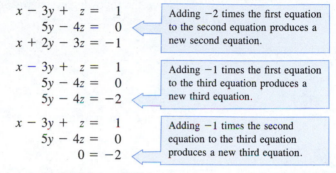

$$x - 3y + z = 1$$
$$5y - 4z = 0$$
$$x + 2y - 3z = -1$$

Adding -2 times the first equation to the second equation produces a new second equation.

$$x - 3y + z = 1$$
$$5y - 4z = 0$$
$$5y - 4z = -2$$

Adding -1 times the first equation to the third equation produces a new third equation.

$$x - 3y + z = 1$$
$$5y - 4z = 0$$
$$0 = -2$$

Adding -1 times the second equation to the third equation produces a new third equation.

Because the third "equation" is absurd, we conclude that this system has no solution. Moreover, because this system is equivalent to the original system, we can conclude that the original system also has no solution.

As with a system of linear equations in two variables, the solution(s) to a system of linear equations in more than two variables must fall into one and only one of the following categories.

1. There is exactly one solution.
2. There are infinitely many solutions.
3. There is no solution.

When a system of equations has no solution, we simply state that it is *inconsistent*. If a system has exactly one solution, we list the value of each variable. However, for systems that have infinitely many solutions, we encounter a certain awkwardness in listing the solutions. For example, we might give the solutions to a system in three variables as

$$(a, a + 1, 2a), \qquad \text{where } a \text{ is any real number.}$$

This means that for each real number a, we have a valid solution to the system. A few of the infinitely many possible solutions are found by letting $a = -1, 0, 1$, and 2 to obtain $(-1, 0, -2)$, $(0, 1, 0)$, $(1, 2, 2)$, and $(2, 3, 4)$, respectively. Now consider the solutions represented by

$$(b - 1, b, 2b - 2), \qquad \text{where } b \text{ is any real number.}$$

Systems of Equations and Inequalities

Here again a few possible solutions are $(-1, 0, -2)$, $(0, 1, 0)$, $(1, 2, 2)$, and $(2, 3, 4)$, found by letting $b = 0$, 1, 2, and 3, respectively. Note that both descriptions result in the same collection of solutions. Thus, when comparing descriptions of an infinite solution set, keep in mind that there is more than one way to describe the set.

EXAMPLE 4 *A System with Infinitely Many Solutions*

Solve the following system of linear equations.

$$\begin{array}{rcrcrcl} x & + & y & - & 3z & = & -1 \\ & & y & - & z & = & 0 \\ -x & + & 2y & & & = & 1 \end{array}$$

Equation 1
Equation 2
Equation 3

SOLUTION

We begin by rewriting the system in row-echelon form as follows.

$$\begin{array}{rcrcrcl} x & + & y & - & 3z & = & -1 \\ & & y & - & z & = & 0 \\ & & 3y & - & 3z & = & 0 \end{array}$$

Adding the first equation to the third equation produces a new third equation.

$$\begin{array}{rcrcrcl} x & + & y & - & 3z & = & -1 \\ & & y & - & z & = & 0 \\ & & & & 0 & = & 0 \end{array}$$

Adding -3 times the second equation to the third equation produces a new third equation.

This means that Equation 3 is *dependent* on Equations 1 and 2 in the sense that it gives us no additional information about the variables. Thus, the original system is equivalent to the system

$$\begin{array}{rcrcrcl} x & + & y & - & 3z & = & -1 \\ & & y & - & z & = & 0. \end{array}$$

In this last equation, we solve for y in terms of z to obtain $y = z$. Back-substituting for y into the previous equation, we find x in terms of z, as follows.

$$\begin{array}{rcl} x + z - 3z & = & -1 \\ x - 2z & = & -1 \\ x & = & 2z - 1 \end{array}$$

Finally, letting $z = a$, the solutions to the given system are all of the form

$$x = 2a - 1, \quad y = a, \quad z = a$$

where a is a real number. Thus, every ordered triple of the form

$$(2a - 1, a, a), \quad a \text{ is a real number}$$

is a solution of the system.

Nonsquare Systems

So far we have only considered **square** systems, for which the number of equations is equal to the number of variables. In a **nonsquare** system, the number of equations differs from the number of variables. In Chapter 11 we will prove that a system of linear equations cannot have a unique solution unless there are at least as many equations as there are variables in the system.

EXAMPLE 5 A System with Fewer Equations than Variables

Solve the following system of linear equations.

$$x - 2y + z = 2 \qquad\qquad \textit{Equation 1}$$
$$2x - y - z = 1 \qquad\qquad \textit{Equation 2}$$

SOLUTION

$$x - 2y + z = 2$$
$$3y - 3z = -3$$

Adding -2 times the first equation to the second equation produces a new second equation.

$$x - 2y + z = 2$$
$$y - z = -1$$

Multiplying the second equation by $1/3$ produces a new second equation.

Solving for y in terms of z, we get $y = z - 1$, and back-substitution into Equation 1 yields

$$x - 2(z - 1) + z = 2$$
$$x - 2z + 2 + z = 2$$
$$x = z.$$

Finally, by letting $z = a$, we have the solution

$$x = a, \qquad y = a - 1, \qquad \text{and} \qquad z = a$$

where a is a real number. Thus, every ordered triple of the form

$$(a, a - 1, a), \qquad a \text{ is a real number}$$

is a solution of the system.

Homogeneous Systems

A system of linear equations in which the constant term in each equation is zero is called **homogeneous.** For example, a homogeneous system of three linear equations in the three variables x, y, and z has the form

$$a_1 x + b_1 y + c_1 z = 0$$
$$a_2 x + b_2 y + c_2 z = 0$$
$$a_3 x + b_3 y + c_3 z = 0.$$

The *trivial* (or obvious) solution to this homogeneous system is $(0, 0, 0)$. This means that if all variables are given the value zero, then the equations must be satisfied. Often homogeneous systems will have nontrivial solutions also, and we can find these in the same way that we find solutions for nonhomogeneous systems, as demonstrated in Example 6.

EXAMPLE 6 *Solving a Homogeneous System of Linear Equations*

Solve the following system of linear equations.

$$x - y + 3z = 0 \qquad \textit{Equation 1}$$
$$2x + y + 3z = 0 \qquad \textit{Equation 2}$$
$$x \qquad + 2z = 0 \qquad \textit{Equation 3}$$

SOLUTION

$$x - y + 3z = 0$$
$$3y - 3z = 0$$
$$x \qquad + 2z = 0$$

Adding -2 times the first equation to the second equation produces a new second equation.

$$x - y + 3z = 0$$
$$3y - 3z = 0$$
$$y - z = 0$$

Adding -1 times the first equation to the third equation produces a new third equation.

$$x - y + 3z = 0$$
$$y - z = 0$$
$$y - z = 0$$

Multiplying the second equation by $1/3$ produces a new second equation.

$$x - y + 3z = 0$$
$$y - z = 0$$

Adding -1 times the second equation to the third equation eliminates the third equation.

Solving for y in terms of z, we get $y = z$, and back-substitution into Equation 1 yields $x = -2z$. Finally, by letting $z = a$, we obtain the solution

$$x = -2a, \qquad y = a, \qquad \text{and} \qquad z = a$$

where a is a real number. Thus, every ordered triple of the form

$$(-2a, a, a), \qquad a \text{ is a real number}$$

is a solution of the system. Note that this homogeneous system of linear equations has an infinite number of solutions, one of which is the trivial solution (given by $a = 0$).

Applications

We conclude this section with two applications involving systems of linear equations in three variables.

EXAMPLE 7 An Application: Moving Object

The height at time t of an object that is moving in a (vertical) line with constant acceleration a is given by the **position equation**

$$s = \frac{1}{2}at^2 + v_0 t + s_0.$$

The height s is measured in feet, t is measured in seconds, v_0 is the initial velocity (at time $t = 0$), and s_0 is the initial height. Find the values of a, v_0, and s_0, if $s = 52$ feet at 1 second, $s = 52$ feet at 2 seconds, and $s = 20$ feet at 3 seconds.

SOLUTION

By substituting the three values of t and s into the position equation, we obtain three linear equations in a, v_0, and s_0.

When $t = 1$: $\dfrac{1}{2}a(1^2) + v_0(1) + s_0 = 52$

When $t = 2$: $\dfrac{1}{2}a(2^2) + v_0(2) + s_0 = 52$

When $t = 3$: $\dfrac{1}{2}a(3^2) + v_0(3) + s_0 = 20$

By multiplying the first and third equations by 2, this system can be rewritten

$$
\begin{aligned}
a + 2v_0 + 2s_0 &= 104 \\
2a + 2v_0 + s_0 &= 52 \\
9a + 6v_0 + 2s_0 &= 40
\end{aligned}
$$

and we apply elimination as follows.

$$
\begin{aligned}
a + 2v_0 + 2s_0 &= 104 \\
-2v_0 - 3s_0 &= -156 \\
9a + 6v_0 + 2s_0 &= 40
\end{aligned}
$$

> Adding -2 times the first equation to the second equation produces a new second equation.

$$
\begin{aligned}
a + 2v_0 + 2s_0 &= 104 \\
-2v_0 - 3s_0 &= -156 \\
-12v_0 - 16s_0 &= -896
\end{aligned}
$$

> Adding -9 times the first equation to the third equation produces a new third equation.

$$
\begin{aligned}
a + 2v_0 + 2s_0 &= 104 \\
-2v_0 - 3s_0 &= -156 \\
2s_0 &= 40
\end{aligned}
$$

> Adding -6 times the second equation to the third equation produces a new third equation.

Systems of Equations and Inequalities

From the third equation, we find that $s_0 = 20$, so that back-substitution into the second equation yields

$$-2v_0 - 3(20) = -156$$
$$-2v_0 = -96$$
$$v_0 = 48.$$

Finally, by back-substituting $s_0 = 20$ and $v_0 = 48$ into the first equation, we have

$$a + 2(48) + 2(20) = 104$$
$$a = -32.$$

Thus, the position equation for this object is

$$s = -16t^2 + 48t + 20.$$

In the next example we show how to fit a parabola through three given points in the plane. This procedure can be generalized to fit an nth degree polynomial function to $n + 1$ points in the plane. The only restriction to the procedure is that (since we are trying to fit a *function* to the points) every point must have a distinct x-coordinate.

EXAMPLE 8 An Application: Curve-Fitting

Find a quadratic function

$$f(x) = ax^2 + bx + c$$

whose graph passes through the points $(-1, 3)$, $(1, 1)$, and $(2, 6)$.

SOLUTION

Since the graph of f passes through the points $(-1, 3)$, $(1, 1)$, and $(2, 6)$, we have

$$f(-1) = a(-1)^2 + b(-1) + c = 3$$
$$f(1) = a(1)^2 + b(1) + c = 1$$
$$f(2) = a(2)^2 + b(2) + c = 6.$$

This produces the following system of linear equations in the variables a, b, and c.

$$a - b + c = 3$$
$$a + b + c = 1$$
$$4a + 2b + c = 6$$

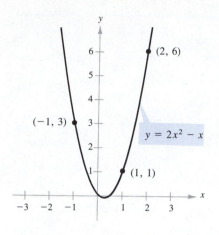

FIGURE 10.8

The solution to this system turns out to be

$$a = 2, \qquad b = -1, \qquad \text{and} \qquad c = 0.$$

Thus, the equation of the parabola passing through the three given points is

$$f(x) = 2x^2 - x$$

as shown in Figure 10.8.

EXAMPLE 9   An Application: Partial Fractions

Write the partial fraction decomposition for

$$\frac{3x + 4}{x^3 - 2x - 4}.$$

SOLUTION

Since

$$x^3 - 2x - 4 = (x - 2)(x^2 + 2x + 2)$$

we write

$$\frac{3x + 4}{x^3 - 2x - 4} = \frac{A}{x - 2} + \frac{Bx + C}{x^2 + 2x + 2}$$

$$3x + 4 = A(x^2 + 2x + 2) + (Bx + C)(x - 2) \qquad \textit{Basic equation}$$

$$3x + 4 = (A + B)x^2 + (2A - 2B + C)x + (2A - 2C)$$

By equating coefficients of like powers on opposite sides of the expanded equation, we obtain the following system of linear equations in A, B, and C.

$$
\begin{aligned}
A + B \qquad\quad &= 0 \\
2A - 2B + C &= 3 \\
2A \qquad\;\; - 2C &= 4
\end{aligned}
$$

The solution of this system is $A = 1$, $B = -1$, and $C = -1$. Therefore, the partial fraction decomposition is

$$\frac{3x + 4}{x^3 - 2x - 4} = \frac{1}{x - 2} + \frac{-x - 1}{x^2 + 2x + 2}$$

$$= \frac{1}{x - 2} - \frac{x + 1}{x^2 + 2x + 2}.$$

WARM UP

Solve each system of linear equations.

1. $x + y = 25$
$\quad\quad y = 10$

3. $x + y = 32$
$\quad x - y = 24$

2. $2x - 3y = \quad 4$
$\quad 6x \quad\quad = -12$

4. $2r - \quad s = \quad 5$
$\quad r + 2s = 10$

In each case, determine whether the ordered triple is a solution of the equation.

5. $5x - 3y + 4z = 2$
$\quad (-1, -2, 1)$

7. $2x - 5y + 3z = -9$
$\quad (a - 2, a + 1, a)$

6. $x - 2y + 12z = 9$
$\quad (6, 3, 2)$

8. $-5x + y + z = 21$
$\quad (a - 4, 4a + 1, a)$

Solve for x in terms of a.

9. $x + 2y - 3z = 4$
$\quad y = 1 - a, \quad z = a$

10. $x - 3y + 5z = 4$
$\quad y = 2a + 3, \quad z = a$

EXERCISES 10.3

In Exercises 1–26, solve the system of linear equations.

1. $x + y + z = 6$
$\quad 2x - y + z = 3$
$\quad 3x \quad\quad - z = 0$

2. $x + y + z = 2$
$\quad -x + 3y + 2z = 8$
$\quad 4x + y \quad\quad = 4$

3. $4x + y - 3z = \quad 11$
$\quad 2x - 3y + 2z = \quad 9$
$\quad x + y + z = -3$

4. $2x \quad\quad + 2z = 2$
$\quad 5x + 3y \quad\quad = 4$
$\quad 3y - 4z = 4$

5. $\quad\quad 6y + 4z = -12$
$\quad 3x + 3y \quad\quad = \quad 9$
$\quad 2x \quad\quad - 3z = \quad 10$

6. $2x + 4y + z = -4$
$\quad 2x - 4y + 6z = \quad 13$
$\quad 4x - 2y + z = \quad 6$

7. $3x - 2y + 4z = 1$
$\quad x + y - 2z = 3$
$\quad 2x - 3y + 6z = 8$

8. $5x - 3y + 2z = 3$
$\quad 2x + 4y - z = 7$
$\quad x - 11y + 4z = 3$

9. $3x + 3y + 5z = 1$
$\quad 3x + 5y + 9z = 0$
$\quad 5x + 9y + 17z = 0$

10. $2x + y + 3z = 1$
$\quad 2x + 6y + 8z = 3$
$\quad 6x + 8y + 18z = 5$

11. $x + 2y - 7z = -4$
$\quad 2x + y + z = 13$
$\quad 3x + 9y - 36z = -33$

12. $2x + y - 3z = 4$
$\quad 4x \quad\quad + 2z = 10$
$\quad -2x + 3y - 13z = -8$

13. $x \quad\quad + 4z = 13$
$\quad 4x - 2y + z = 7$
$\quad 2x - 2y - 7z = -19$

14. $4x - y + 5z = 11$
$\quad x + 2y - z = 5$
$\quad 5x - 8y + 13z = 7$

15. $x - 2y + 5z = 2$
$\quad 3x + 2y - z = -2$

16. $x - 3y + 2z = 18$
$\quad 5x - 13y + 12z = 80$

17. $2x - 3y + z = -2$
$\quad -4x + 9y = 7$

18. $2x + 3y + 3z = 7$
$\quad 4x + 18y + 15z = 44$

19. $x \quad\quad\quad + 3w = 4$
$\quad 2y - z - w = 0$
$\quad 3y \quad\quad - 2w = 1$
$\quad 2x - y + 4z \quad\quad = 5$

20. $x + y + z + w = 6$
$\quad 2x + 3y \quad\quad - w = 0$
$\quad -3x + 4y + z + 2w = 4$
$\quad x + 2y - z + w = 0$

21. $x \quad\quad + 4z = 1$
$\quad x + y + 10z = 10$
$\quad 2x - y + 2z = -5$

22. $3x - 2y - 6z = -4$
$\quad -3x + 2y + 6z = 1$
$\quad x - y - 5z = -3$

23. $4x + 3y + 17z = 0$
$\quad 5x + 4y + 22z = 0$
$\quad 4x + 2y + 19z = 0$

24. $2x + 3y \quad\quad = 0$
$\quad 4x + 3y - z = 0$
$\quad 8x + 3y + 3z = 0$

25. $5x + 5y - z = 0$
$\quad 10x + 5y + 2z = 0$
$\quad 5x + 15y - 9z = 0$

26. $12x + 5y + z = 0$
$\quad 12x + 4y - z = 0$

In Exercises 27–30, find the equation of the parabola

$$y = ax^2 + bx + c$$

that passes through the given points.

27. $(0, -4)$, $(1, 1)$, $(2, 10)$
28. $(0, 5)$, $(1, 6)$, $(2, 5)$
29. $(1, 0)$, $(3, 0)$, $(2, -1)$
30. $(1, 2)$, $(2, 1)$, $(3, -4)$

In Exercises 31–34, find the equation of the circle

$$x^2 + y^2 + Dx + Ey + F = 0$$

that passes through the given points.

31. $(0, 0)$, $(2, -2)$, $(4, 0)$
32. $(0, 0)$, $(0, 6)$, $(-3, 3)$
33. $(3, -1)$, $(-2, 4)$, $(6, 8)$
34. $(0, 0)$, $(0, 2)$, $(3, 0)$

In Exercises 35–38, find a, v_0, and s_0 in the position equation

$$s = \tfrac{1}{2}at^2 + v_0 t + s_0.$$

35. At $t = 1$ second, $s = 128$ feet
At $t = 2$ seconds, $s = 80$ feet
At $t = 3$ seconds, $s = 0$ feet

36. At $t = 1$ second, $s = 48$ feet
At $t = 2$ seconds, $s = 64$ feet
At $t = 3$ seconds, $s = 48$ feet

37. At $t = 1$ second, $s = 452$ feet
At $t = 3$ seconds, $s = 260$ feet
At $t = 4$ seconds, $s = 116$ feet

38. At $t = 2$ seconds, $s = 132$ feet
At $t = 3$ seconds, $s = 100$ feet
At $t = 4$ seconds, $s = 36$ feet

In Exercises 39–42, use a system of linear equations to decompose each rational fraction into partial fractions. (See Example 9 and Section 5.2.)

39. $\dfrac{1}{x^3 - x} = \dfrac{A}{x} + \dfrac{B}{x - 1} + \dfrac{C}{x + 1}$

40. $\dfrac{3}{x^2 + x - 2} = \dfrac{A}{x - 1} + \dfrac{B}{x + 2}$

41. $\dfrac{x^2 - 3x - 3}{x(x - 2)(x + 3)} = \dfrac{A}{x} + \dfrac{B}{x - 2} + \dfrac{C}{x + 3}$

42. $\dfrac{12}{x(x - 2)(x + 3)} = \dfrac{A}{x} + \dfrac{B}{x - 2} + \dfrac{C}{x + 3}$

43. A small company that manufactures products A and B has an order for 15 units of product A and 16 units of product B. The company has trucks of three different sizes that can haul the products, as shown in the following table.

Truck	Product A	Product B
Large	6	3
Medium	4	4
Small	0	3

How many trucks of each size are needed to deliver the order? (Give *two* possible solutions.)

44. A chemist needs 10 liters of a 25% acid solution. The solution is to be mixed from three solutions whose concentrations are 10%, 20%, and 50%, respectively. How many liters of each solution should the chemist use to satisfy the following?
(a) Use as little as possible of the 50% solution.
(b) Use as much as possible of the 50% solution.
(c) Use two liters of the 50% solution.

[*Optional*]   In Exercises 45–48, (a) find the least squares regression parabola, then (b) plot the given points and sketch the least squares parabola on the same axes. The least squares regression parabola, $y = ax^2 + bx + c$, for the points (x_1, y_1), (x_2, y_2), . . . , (x_n, y_n) is obtained by solving the following system of linear equations for a, b, and c.

$$nc + \left(\sum_{i=1}^{n} x_i\right)b + \left(\sum_{i=1}^{n} x_i^2\right)a = \sum_{i=1}^{n} y_i$$

$$\left(\sum_{i=1}^{n} x_i\right)c + \left(\sum_{i=1}^{n} x_i^2\right)b + \left(\sum_{i=1}^{n} x_i^3\right)a = \sum_{i=1}^{n} x_i y_i$$

$$\left(\sum_{i=1}^{n} x_i^2\right)c + \left(\sum_{i=1}^{n} x_i^3\right)b + \left(\sum_{i=1}^{n} x_i^4\right)a = \sum_{i=1}^{n} x_i^2 y_i$$

45. $(-4, 5)$, $(-2, 6)$, $(2, 6)$, $(4, 2)$
46. $(-2, 0)$, $(-1, 0)$, $(0, 1)$, $(1, 2)$, $(2, 5)$
47. $(0, 0)$, $(2, 2)$, $(3, 6)$, $(4, 12)$
48. $(0, 10)$, $(1, 9)$, $(2, 6)$, $(3, 0)$

10.4 Systems of Inequalities

The following statements are inequalities in two variables:

$$3x - 2y < 6 \quad \text{and} \quad 2x^2 + 3y^2 \geq 6.$$

An ordered pair (a, b) is a **solution of an inequality** in x and y if the inequality is true when a and b are substituted for x and y, respectively. The **graph** of an inequality is the collection of all solutions of the inequality. To sketch the graph of an inequality such as $3x - 2y < 6$, we begin by sketching the graph of the *corresponding equation* $3x - 2y = 6$. This graph is made with a dashed line for the strict inequalities $<$ or $>$ and a solid line for the inequalities $\leq$ or $\geq$. The graph of the equation will normally separate the plane into two or more regions. In each such region, one of the following must be true.

1. All points in the region are solutions of the inequality.
2. No points in the region are solutions of the inequality.

Thus, we can determine whether the points in an entire region satisfy the inequality by simply testing *one* point in the region.

Sketching the Graph of an Inequality in Two Variables

1. Replace the inequality sign by an equal sign, and sketch the graph of the resulting equation. (Use a dashed line for $<$ or $>$ and a solid line for $\leq$ or $\geq$.)
2. Test one point in each of the regions formed by the graph in Step 1. If the point satisfies the inequality, then shade the entire region to denote that every point in the region satisfies the inequality.

EXAMPLE 1 Sketching the Graph of an Inequality

Sketch the graph of the inequality $y \geq x^2 - 1$.

SOLUTION

The graph of the corresponding *equation* $y = x^2 - 1$ is a parabola, as shown in Figure 10.9. By testing a point *above* the parabola $(0, 0)$, and a point *below* the parabola $(0, -2)$, we see that the points that satisfy the inequality are those lying above (or on) the parabola.

The inequality given in Example 1 is a nonlinear inequality in two variables. In this section, however, we will work primarily with **linear inequalities** of the form

$$ax + by < c \qquad ax + by \leq c$$
$$ax + by > c \qquad ax + by \geq c.$$

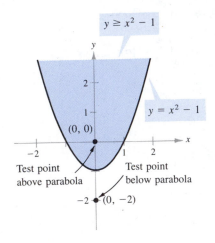

$y \geq x^2 - 1$

$y = x^2 - 1$

$(0, 0)$

Test point above parabola

Test point below parabola

$(0, -2)$

FIGURE 10.9

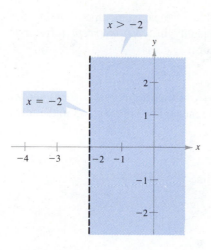

FIGURE 10.10

The graph of each of these linear inequalities is a half-plane lying on one side of the line $ax + by = c$.

The simplest linear inequalities are those corresponding to horizontal or vertical lines, as shown in Example 2.

EXAMPLE 2 *Sketching the Graph of a Linear Inequality*

Sketch the graphs of the following linear inequalities.

(a) $x > -2$ (b) $y \leq 3$

SOLUTION

(a) The graph of the corresponding equation $x = -2$ is a vertical line. The points that satisfy the inequality $x > -2$ are those lying to the right of this line, as shown in Figure 10.10.

(b) The graph of the corresponding equation $y = 3$ is a horizontal line. The points that satisfy the inequality $y \leq 3$ are those lying below (or on) this line, as shown in Figure 10.11.

EXAMPLE 3 *Sketching the Graph of a Linear Inequality*

Sketch the graph of $x - y < 2$.

SOLUTION

The graph of the corresponding equation $x - y = 2$ is a line, as shown in Figure 10.12. Since the origin $(0, 0)$ satisfies the inequality, the graph consists of the half-plane lying above the line. (Try checking a point below the line. Regardless of which point you choose, you will see that it does not satisfy the inequality.)

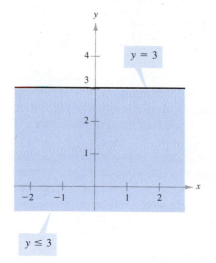

FIGURE 10.11

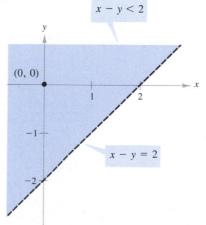

FIGURE 10.12

For a linear inequality in two variables, we can sometimes simplify the graphing procedure by writing the inequality in *slope-intercept* form. For instance, by writing $x - y < 2$ in the form

$$y > x - 2$$

we can see that the solution points lie *above* the line $y = x - 2$, as shown in Figure 10.12. Similarly, by writing the inequality $3x - 2y > 5$ in the form

$$y < \frac{3}{2}x - \frac{5}{2}$$

we see that the solutions lie *below* the line $y = \frac{3}{2}x - \frac{5}{2}$.

Systems of Inequalities

Many practical problems in business, science, and engineering involve systems of linear inequalities. Here are two examples of such systems.

$$\begin{aligned} 2x - y &\le 5 \\ x + 2y &> 2 \end{aligned} \qquad \begin{aligned} x + y &\le 12 \\ 3x - 4y &\le 15 \\ x &\ge 0 \\ y &\ge 0 \end{aligned}$$

A **solution** of a system of inequalities in x and y is a point (x, y) that satisfies each inequality in the system. For instance, $(2, 4)$ is a solution of the system on the right because $x = 2$ and $y = 4$ satisfy each of the four inequalities in the system.

To sketch the graph of a system of inequalities in two variables, we first sketch the graph of each individual inequality (on the same coordinate system) and then find the region that is *common* to every graph in the system. For systems of linear inequalities, it is helpful to find the *vertices* of the solution region, as shown in the following example.

EXAMPLE 4 *Solving a System of Inequalities*

Sketch the graph (and label the vertices) of the solution set of the following system.

$$\begin{aligned} x - y &< 2 \\ x &> -2 \\ y &\le 3 \end{aligned}$$

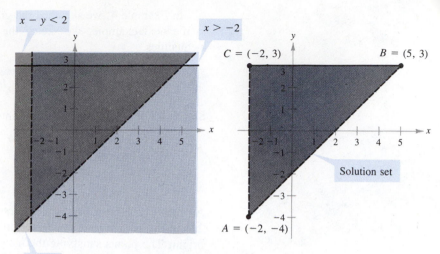

FIGURE 10.13

SOLUTION

We already sketched the graph of each inequality in Examples 2 and 3. The triangular region common to all three graphs can be found by superimposing the graphs on the same coordinate plane, as shown in Figure 10.13. To find the vertices of the region, we solve the three systems of corresponding equations obtained by taking *pairs* of equations representing the boundaries of the individual regions.

Vertex A: $(-2, -4)$
Obtained by solving the system

$$x - y = 2$$
$$x = -2$$

Vertex B: $(5, 3)$
Obtained by solving the system

$$x - y = 2$$
$$y = 3$$

Vertex C: $(-2, 3)$
Obtained by solving the system

$$x = -2$$
$$y = 3$$

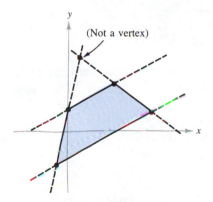

Boundary lines can intersect at a point that is not a vertex.

FIGURE 10.14

For the triangular region shown in Figure 10.13 each point of intersection of a pair of boundary lines corresponds to a vertex. With more complicated regions, two border lines can sometimes intersect at a point that is not a vertex of the region, as shown in Figure 10.14. In order to keep track of which points of intersection are actually vertices of the region, we suggest that you make a careful sketch of the region and refer to your sketch as you find each point of intersection.

In Example 4, we sketched the graph of a system of *linear* inequalities. In the next example, we sketch the graph of a more general system of inequalities.

EXAMPLE 5   Solving a System of Inequalities

Sketch the region containing all points that satisfy the following system.

$$x^2 + y^2 \le 25$$
$$7x - y \le 25$$

SOLUTION

As shown in Figure 10.15, the points that satisfy the inequality $x^2 + y^2 \le 25$ are the points lying inside (or on) the circle of radius 5 centered at the origin. The points satisfying the inequality $7x - y \le 25$ (or $y \ge 7x - 25$) are the points lying on or above the line given by $y = 7x - 25$. To find the points of intersection of the circle and the line, we solve the following system of corresponding equations:

$$x^2 + y^2 = 25$$
$$7x - y = 25.$$

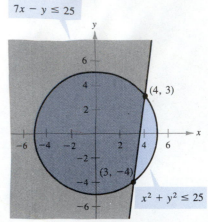

FIGURE 10.15

Using the method of substitution, we find the solutions (vertices) to be $(3, -4)$ and $(4, 3)$, as shown in Figure 10.15.

When solving a system of inequalities, you should be aware that the system might have no solution. For instance, the system

$$x + y > 3$$
$$x + y < -1$$

has no solution points, because the quantity $(x + y)$ cannot be both less than -1 and greater than 3, as shown in Figure 10.16.

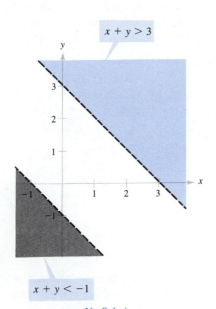

No Solution

FIGURE 10.16

Systems of Inequalities

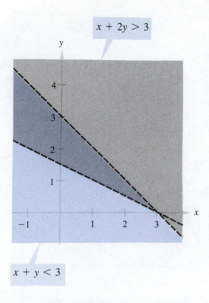

$x + 2y > 3$

$x + y < 3$

Unbounded Region

FIGURE 10.17

Another possibility is that the solution set of a system of inequalities can be unbounded. For instance, the solution set of

$$x + y < 3$$
$$x + 2y > 3$$

forms an *infinite wedge,* as shown in Figure 10.17.

Our last example in this section shows how a system of linear inequalities can arise in an applied problem.

EXAMPLE 6 An Application of a System of Inequalities

The liquid portion of a diet is to provide at least 300 calories, 36 units of vitamin A, and 90 units of vitamin C daily. A cup of dietary drink X provides 60 calories, 12 units of vitamin A, and 10 units of vitamin C. A cup of dietary drink Y provides 60 calories, 6 units of vitamin A, and 30 units of vitamin C. Set up a system of linear inequalities that describes the minimum daily requirements for calories and vitamins.

SOLUTION

We let

$$x = \text{number of cups of dietary drink } X$$
$$y = \text{number of cups of dietary drink } Y.$$

Then, to meet the minimum daily requirements, the following inequalities must be satisfied.

$$
\begin{aligned}
\text{For calories:} \quad & 60x + 60y \geq 300 \\
\text{For vitamin A:} \quad & 12x + 6y \geq 36 \\
\text{For vitamin C:} \quad & 10x + 30y \geq 90 \\
& x \geq 0 \\
& y \geq 0
\end{aligned}
$$

The last two inequalities are included because x and y cannot be negative. (More is said about this application in Example 3 in Section 10.5.)

WARM UP

Identify the graph of each equation.

1. $x + y = 3$

3. $y = x^2 - 4$

5. $x^2 + y^2 = 9$

2. $4x - y = 8$

4. $y = -x^2 + 1$

6. $\dfrac{x^2}{4} + \dfrac{y^2}{9} = 1$

Solve each system of equations.

7. $\begin{aligned} x + 2y &= 3 \\ 4x - 7y &= -3 \end{aligned}$

9. $\begin{aligned} x^2 + y &= 5 \\ 2x - 4y &= 0 \end{aligned}$

8. $\begin{aligned} 2x - 3y &= 4 \\ x + 5y &= 2 \end{aligned}$

10. $\begin{aligned} x^2 + y^2 &= 13 \\ x + y &= 5 \end{aligned}$

EXERCISES 10.4

In Exercises 1–8, match the inequality with its graph. [The graphs are labeled (a)–(h).]

1. $x > 3$

3. $2x + 3y \le 6$

5. $x^2 + y^2 < 4$

7. $xy > 2$

2. $y \le 2$

4. $2x - y \ge -2$

6. $(x - 2)^2 + (y - 3)^2 > 4$

8. $y \le 4 - x^2$

(a)

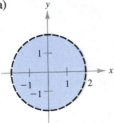

(b)

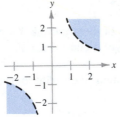

(c)

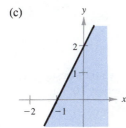

(d)

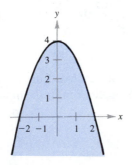

(e)

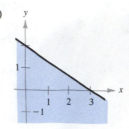

(f)

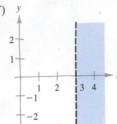

(g)

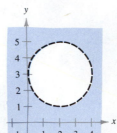

(h)

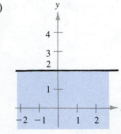

In Exercises 9–20, sketch the graph of the inequality.

9. $x \ge 2$

11. $y \ge -1$

13. $y < 2 - x$

10. $x \le 4$

12. $y \le 3$

14. $y > 2x - 4$

15. $2y - x \geq 4$ **16.** $5x + 3y \geq -15$

17. $(x + 1)^2 + (y - 2)^2 < 9$

18. $y^2 - x < 0$

19. $y \leq \dfrac{1}{1 + x^2}$ **20.** $y < \ln x$

In Exercises 21–40, sketch the graph of the solution of the system of inequalities.

21. $\begin{aligned} x + y &\leq 1 \\ -x + y &\leq 1 \\ y &\geq 0 \end{aligned}$ **22.** $\begin{aligned} 3x + 2y &< 6 \\ x \quad\;\; &> 0 \\ y &> 0 \end{aligned}$

23. $\begin{aligned} x + y &\leq 5 \\ x \quad &\geq 2 \\ y &\geq 0 \end{aligned}$ **24.** $\begin{aligned} 2x + y &\geq 2 \\ x \quad &\leq 2 \\ y &\leq 1 \end{aligned}$

25. $\begin{aligned} -3x + 2y &< 6 \\ x + 4y &> -2 \\ 2x + y &< 3 \end{aligned}$ **26.** $\begin{aligned} x - 7y &> -36 \\ 5x + 2y &> 5 \\ 6x - 5y &> 6 \end{aligned}$

27. $\begin{aligned} 2x + y &> 2 \\ 6x + 3y &< 2 \end{aligned}$ **28.** $\begin{aligned} x - 2y &< -6 \\ 5x - 3y &> -9 \end{aligned}$

29. $\begin{aligned} x &\geq 1 \\ x - 2y &\leq 3 \\ 3x + 2y &\geq 9 \\ x + y &\leq 6 \end{aligned}$ **30.** $\begin{aligned} x - y^2 &> 0 \\ x - y &< 2 \end{aligned}$

31. $\begin{aligned} x^2 + y^2 &\leq 9 \\ x^2 + y^2 &\geq 1 \end{aligned}$ **32.** $\begin{aligned} x^2 + y^2 &\leq 25 \\ 4x - 3y &\leq 0 \end{aligned}$

33. $\begin{aligned} x &> y^2 \\ x &< y + 2 \end{aligned}$ **34.** $\begin{aligned} x &< 2y - y^2 \\ 0 &< x + y \end{aligned}$

35. $\begin{aligned} y &\leq \sqrt{3}x + 1 \\ y &\geq x + 1 \end{aligned}$ **36.** $\begin{aligned} y &< -x^2 + 2x + 3 \\ y &> x^2 - 4x + 3 \end{aligned}$

37. $\begin{aligned} y &< x^3 - 2x + 1 \\ y &> -2x \\ x &\leq 1 \end{aligned}$ **38.** $\begin{aligned} y &\geq x^4 - 2x^2 + 1 \\ y &\leq 1 - x^2 \end{aligned}$

39. $\begin{aligned} x^2 y &\geq 1 \\ 0 &< x \leq 4 \\ y &\leq 4 \end{aligned}$ **40.** $\begin{aligned} y &\leq e^{-x^2/2} \\ y &\geq 0 \\ -2 &\leq x \leq 2 \end{aligned}$

In Exercises 41–46, derive a set of inequalities to describe the region.

41. Rectangular region with vertices at $(2, 1)$, $(5, 1)$, $(5, 7)$, and $(2, 7)$

42. Parallelogram region with vertices at $(0, 0)$, $(4, 0)$, $(1, 4)$, and $(5, 4)$

43. Triangular region with vertices at $(0, 0)$, $(5, 0)$, and $(2, 3)$

44. Triangular region with vertices at $(-1, 0)$, $(1, 0)$, and $(0, 1)$

45. Sector of a circle   **46.** Sector of a circle

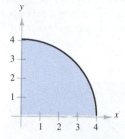
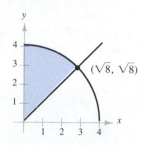

47. A furniture company can sell all the tables and chairs it produces. Each table requires 1 hour in the assembly center and $1\frac{1}{3}$ hours in the finishing center. Each chair requires $1\frac{1}{2}$ hours in the assembly center and $1\frac{1}{2}$ hours in the finishing center. The company's assembly center is available 12 hours per day, and its finishing center is available 15 hours per day. If x is the number of tables produced per day and y is the number of chairs, find a system of inequalities describing all possible production levels. Sketch the graph of the system.

48. A store sells two models of a certain brand of computer. Because of the demand, it is necessary to stock twice as many units of model A as units of model B. The cost to the store for the two models is $800 and $1200, respectively. The management does not want more than $20,000 in computer inventory at any one time, and it wants at least four model A computers and two model B computers in inventory at all times. Devise a system of inequalities describing all possible inventory levels, and sketch the graph of the system.

49. A person plans to invest a total of at most $20,000 in two different interest-bearing accounts. Each account is to contain at least $5000. Moreover, one account should have at least twice the amount that is in the other account. Find a system of inequalities to describe the various amounts that can be deposited in each account, and sketch the graph of the system.

50. Two types of tickets are to be sold for a concert. One type costs $15 per ticket and the other type costs $25 per ticket. The promoter of the concert must sell at least 15,000 tickets including 8000 of the $15 tickets and 4000 of the $25 tickets. Moreover, the gross receipts must total at least $275,000 in order for the concert to be held. Find a system of inequalities describing the different numbers of tickets that can be sold, and sketch the graph of the system.

51. An indoor running track is to be constructed with a space for body-building equipment inside the track (see figure). The inside track must be at least 125 meters long, and the body-building space must have an area of at least 500 square meters. Find a system of inequalities describing the various sizes of the track, and sketch the graph of the system.

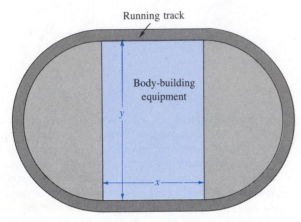

Running track

Body-building equipment

y

x

FIGURE FOR 51

52. A dietitian is asked to design a special diet supplement using two different foods. Each ounce of food X contains 20 units of calcium, 15 units of iron, and 10 units of vitamin B. Each ounce of food Y contains 10 units of calcium, 10 units of iron, and 20 units of vitamin B. The minimum daily requirements in the diet are 300 units of calcium, 150 units of iron, and 200 units of vitamin B. Find a system of inequalities describing the different amounts of food X and food Y that can be used in the diet, and sketch the graph of the system.

10.5 *Linear Programming*

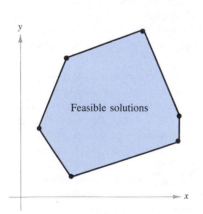

Feasible solutions

The objective function has its optimal value at one of the vertices of the region determined by the constraints.

FIGURE 10.18

Many applications in business and economics involve a process called **optimization.** In such problems we may be required to find the minimum cost, the maximum profit, or the minimum use of resources. In this section we discuss one type of optimization problem called **linear programming.**

A two-dimensional linear programming problem consists of a linear **objective function** and a system of linear inequalities called **constraints.** The objective function gives the quantity that is to be maximized (or minimized), and the constraints determine the set of **feasible solutions.**

For example, consider a linear programming problem in which we are asked to maximize the value of

$$C = ax + by \qquad \textit{Objective function}$$

subject to the set of constraints indicated in Figure 10.18. Because every point in the region satisfies each constraint, it is not clear how we should go about finding the point that yields a maximum value of C. Fortunately, it can be shown that if there is an optimal solution, it must occur at one of the vertices of the region. In other words, *we can find the maximum value by testing C at each of the vertices,* as illustrated in Example 1.

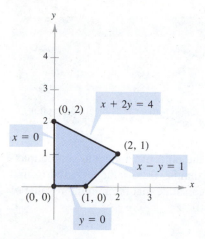

FIGURE 10.19

EXAMPLE 1 *Solving a Linear Programming Problem*

Find the maximum value of

$$C = 3x + 2y \qquad\qquad \textit{Objective function}$$

subject to the following constraints.

$$
\begin{aligned}
x &\geq 0 \qquad\qquad \textit{Constraints}\\
y &\geq 0\\
x + 2y &\leq 4\\
x - y &\leq 1
\end{aligned}
$$

SOLUTION

The constraints form the boundaries for the region shown in Figure 10.19. At the four vertices of this region, the objective function has the following values.

$$
\begin{array}{llll}
\text{At } (0, 0): & C = 3(0) + 2(0) = 0 \\
\text{At } (1, 0): & C = 3(1) + 2(0) = 3 \\
\text{At } (2, 1): & C = 3(2) + 2(1) = 8 & \textit{(Maximum value of C)} \\
\text{At } (0, 2): & C = 3(0) + 2(2) = 4
\end{array}
$$

Thus, the maximum value of C is 8. This occurs when $x = 2$ and $y = 1$.

Remark: In Example 1, try testing some of the interior points in the region. You will see that the corresponding values of C are less than 8.

To see why the maximum value of the objective function in Example 1 must occur at a vertex, consider writing the objective function in the form

$$y = -\frac{3}{2}x + \frac{C}{2}.$$

Of the infinitely many lines represented by this equation, we want the one that has the largest C-value, while still intersecting the region determined by the constraints. In other words, of all the lines whose slope is $-3/2$, we want the one that has the largest y-intercept *and* intersects the given region, as shown in Figure 10.20. It should be clear that such a line will pass through one (or more) of the vertices of the region.

We outline the steps used in Example 1 as follows.

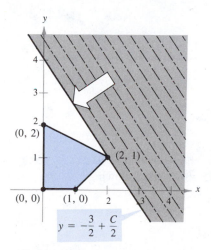

FIGURE 10.20

Systems of Equations and Inequalities

Solving a Linear Programming Problem

To solve a linear programming problem, use the following steps.

1. Sketch the region corresponding to the feasible solution set for the system of constraints.
2. Find the vertices of the region.
3. Test the objective function at each of the vertices and select the values of the variables that optimize the objective function.

These guidelines will work whether the objective function is to be maximized *or* minimized. For instance, in Example 1 the same test used to find the maximum value of C can be used to conclude that the minimum value of C is 0, and this occurs at the vertex $(0, 0)$.

EXAMPLE 2 *Solving a Linear Programming Problem*

Find the maximum value of the objective function

$$C = 4x + 6y \qquad \text{\textit{Objective function}}$$

where $x \geq 0$ and $y \geq 0$, subject to the constraints

$$-x + y \leq 11$$
$$x + y \leq 27 \qquad \text{\textit{Constraints}}$$
$$2x + 5y \leq 90.$$

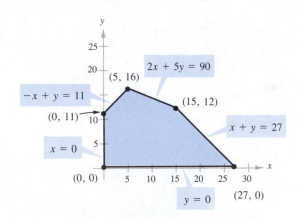

FIGURE 10.21

SOLUTION

The region bounded by the constraints is shown in Figure 10.21. By testing the objective function at each vertex, we obtain the following.

At $(0, 0)$: $C = 4(0) + 6(0) = 0$
At $(0, 11)$: $C = 4(0) + 6(11) = 66$
At $(5, 16)$: $C = 4(5) + 6(16) = 116$
At $(15, 12)$: $C = 4(15) + 6(12) = 132$ *(Maximum value of C)*
At $(27, 0)$: $C = 4(27) + 6(0) = 108$

Thus, the maximum value of C is 132, and this occurs when $x = 15$ and $y = 12$.

EXAMPLE 3 An Application of Linear Programming

In Example 6 in the preceding section, we set up a system of linear equations to describe a dietary supplement problem. Suppose in that example that dietary drink X costs \$0.12 per cup and drink Y costs \$0.15 per cup. How many cups of each drink should be consumed each day to minimize the cost and still meet the stated daily requirements?

SOLUTION

Recall from Example 6 that x is the number of cups of dietary drink X and y is the number of cups of dietary drink Y. Moreover, to meet the minimum daily requirements, the following inequalities must be satisfied.

For calories:	$60x + 60y \geq 300$	*Constraints*
For vitamin A:	$12x + 6y \geq 36$	
For vitamin C:	$10x + 30y \geq 90$	
	$x \geq 0$	
	$y \geq 0$	

The cost C is given by

$$C = 0.12x + 0.15y. \qquad \textit{Objective function}$$

The graph of the region corresponding to the constraints is shown in Figure 10.22. To determine the minimum cost, we test C at each vertex of the region as follows.

At $(0, 6)$: $C = 0.12(0) + 0.15(6) = 0.90$
At $(1, 4)$: $C = 0.12(1) + 0.15(4) = 0.72$
At $(3, 2)$: $C = 0.12(3) + 0.15(2) = 0.66$ *(Minimum cost)*
At $(9, 0)$: $C = 0.12(9) + 0.15(0) = 1.08$

Thus, the minimum cost is \$0.66 per day, and this occurs when three cups of drink X and two cups of drink Y are consumed each day.

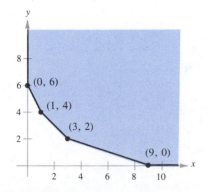

FIGURE 10.22

EXAMPLE 4 *An Application of Linear Programming*

A manufacturer wants to maximize the profit for two products. The first product yields a profit of $1.50 per unit, and the second product yields a profit of $2.00 per unit. Market tests and available resources have indicated the following constraints.

1. The combined production level should not exceed 1200 units per month.
2. The demand for product II is, at most, half of that for product I.
3. The production level of product I can exceed three times the production level of product II by, at most, 600 units.

SOLUTION

If we let x be the number of units of product I and y the number of units of product II, then the objective function (for the combined profit) is given by

$$P = 1.5x + 2y. \qquad \textit{Objective function}$$

The three constraints translate into the following linear inequalities.

1. $x + y \leq 1200$ $\implies$ $x + y \leq 1200$

2. $\quad y \leq \dfrac{1}{2}x$ $\implies$ $x - 2y \geq 0$

3. $\quad x \leq 3y + 600$ $\implies$ $x - 3y \leq 600$

Since neither x nor y can be negative, we also have the two additional constraints of $x \geq 0$ and $y \geq 0$. Figure 10.23 shows the region determined by the constraints. To find the maximum profit, we test the value of P at the vertices of the region.

At (0, 0): $P = 1.5(0) + 2(0) = 0$
At (800, 400): $P = 1.5(800) + 2(400) = \$2000$ *(Maximum profit)*
At (1050, 150): $P = 1.5(1050) + 2(150) = \$1875$
At (600, 0): $P = 1.5(600) + 2(0) = \$900$

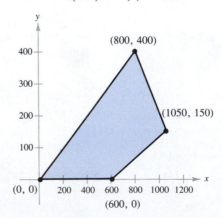

FIGURE 10.23

Thus, the maximum profit is $2000, and it occurs when the monthly production consists of 800 units of product I and 400 units of product II.

WARM UP

Sketch the graph of each linear equation.

1. $y + x = 3$

2. $y - x = 12$

3. $x = 0$

4. $y = 4$

In each case, find the point of intersection of the two lines.

5. $x + y = 4$
$\quad x \quad\;\; = 0$

6. $x + 2y = 12$
$\qquad\quad y = 0$

7. $\;\; x + \;\, y = 4$
$\;\; 2x + 3y = 9$

8. $\;\; x + 2y = 12$
$\;\; 2x + \;\, y = 9$

Sketch the graph of each inequality.

9. $2x + 3y \geq 18$

10. $4x + 3y \geq 12$

EXERCISES 10.5

In Exercises 1–16, find the minimum and maximum values of the given objective function, subject to the indicated constraints.

1. Objective function:

$C = 3x + 2y$

Constraints:

$x \geq 0$
$y \geq 0$
$x + 3y \leq 15$
$4x + y \leq 16$

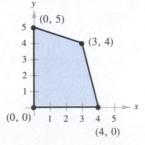

2. Objective function:

$C = 4x + 3y$

Constraints:

$x \geq 0$
$2x + 3y \geq 6$
$3x - 2y \leq 9$
$x + 5y \leq 20$

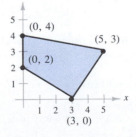

3. Objective function:

$C = 5x + 0.5y$

Constraints:

(See Exercise 1.)

4. Objective function:

$C = x + 6y$

Constraints:

(See Exercise 2.)

5. Objective function:

$C = 10x + 7y$

Constraints:

$$0 \le x \le 60$$
$$0 \le y \le 45$$
$$5x + 6y \le 420$$

6. Objective function:

$C = 50x + 35y$

Constraints:

$$x \ge 0$$
$$y \ge 0$$
$$8x + 9y \le 7200$$
$$8x + 9y \ge 5400$$

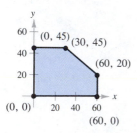

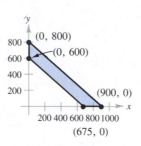

7. Objective function:

$C = 25x + 30y$

Constraints:

(See Exercise 5.)

8. Objective function:

$C = 16x + 18y$

Constraints:

(See Exercise 6.)

9. Objective function:

$C = 4x + 5y$

Constraints:

$$x \ge 0$$
$$y \ge 0$$
$$4x + 3y \ge 27$$
$$x + y \ge 8$$
$$3x + 5y \ge 30$$

10. Objective function:

$C = 4x + 5y$

Constraints:

$$x \ge 0$$
$$y \ge 0$$
$$2x + 2y \le 10$$
$$x + 2y \le 6$$

11. Objective function:

$C = 2x + 7y$

Constraints:

(See Exercise 9.)

12. Objective function:

$C = 2x - y$

Constraints:

(See Exercise 10.)

13. Objective function:

$C = 4x + y$

Constraints:

$$x \ge 0$$
$$y \ge 0$$
$$x + 2y \le 40$$
$$x + y \ge 30$$
$$2x + 3y \ge 72$$

14. Objective function:

$C = x$

Constraints:

$$x \ge 0$$
$$y \ge 0$$
$$2x + 3y \le 60$$
$$2x + y \le 28$$
$$4x + y \le 48$$

15. Objective function:

$C = x + 4y$

Constraints:

(See Exercise 13.)

16. Objective function:

$C = y$

Constraints:

(See Exercise 14.)

17. A merchant plans to sell two models of home computers at costs of $250 and $400, respectively. The $250 model yields a profit of $45 and the $400 model yields a profit of $50. The merchant estimates that the total monthly demand will not exceed 250 units. Find the number of units of each model that should be stocked in order to maximize profit. Assume that the merchant does not want to invest more than $70,000 in computer inventory.

18. A fruit grower has 150 acres of land available to raise two crops, *A* and *B*. It takes one day to trim an acre of crop *A* and two days to trim an acre of crop *B*, and there are 240 days per year available for trimming. It takes 0.3 day to pick an acre of crop *A* and 0.1 day to pick an acre of crop *B*, and there are 30 days per year available for picking. Find the number of acres of each fruit that should be planted to maximize profit, assuming that the profit is $140 per acre for crop *A* and $235 per acre for crop *B*.

19. A farmer mixes two brands of cattle feed. Brand *X* costs $25 per bag and contains 2 units of nutritional element *A*, 2 units of element *B*, and 2 units of element *C*. Brand *Y* costs $20 per bag and contains 1 unit of nutritional element *A*, 9 units of element *B*, and 3 units of element *C*. Find the number of bags of each brand that should be mixed to produce a mixture having a minimum cost per bag. The minimum requirements of nutrients *A*, *B*, and *C* are 12 units, 36 units, and 24 units, respectively.

20. Two gasolines, type *A* and type *B*, have octane ratings of 80 and 92, respectively. Type *A* costs $0.83 per gallon and type *B* costs $0.98 per gallon. Determine the blend of minimum cost with an octane rating of at least 90. [*Hint:* Let *x* be the fraction of each gallon that is type *A* and *y* be the fraction that is type *B*.]

In Exercises 21–24, the given linear programming problem has an unusual characteristic. Sketch a graph of the solution region for the problem and describe the unusual characteristic. (In each problem, the objective function is to be maximized.)

21. Objective function:

$$C = 2.5x + y$$

Constraints:

$$x \geq 0$$
$$y \geq 0$$
$$3x + 5y \leq 15$$
$$5x + 2y \leq 10$$

22. Objective function:

$$C = x + y$$

Constraints:

$$x \geq 0$$
$$y \geq 0$$
$$-x + y \leq 1$$
$$-x + 2y \leq 4$$

23. Objective function:

$$C = -x + 2y$$

Constraints:

$$x \geq 0$$
$$y \geq 0$$
$$x \qquad \leq 10$$
$$x + y \leq 7$$

24. Objective function:

$$C = x + y$$

Constraints:

$$x \geq 0$$
$$y \geq 0$$
$$-x + y \leq 0$$
$$-3x + y \geq 3$$

In Exercises 25 and 26, determine t-values such that the given objective function has a maximum value at the indicated vertex.

25. Objective function:

$$C = x + ty$$

Constraints:

$$x \geq 0$$
$$y \geq 0$$
$$x \leq 1$$
$$y \leq 1$$

(a) $(0, 0)$ (b) $(1, 0)$
(c) $(1, 1)$ (d) $(0, 1)$

26. Objective function:

$$C = 3x + ty$$

Constraints:

$$x \geq 0$$
$$y \geq 0$$
$$x + 2y \geq 4$$
$$x - y \leq 1$$

(a) $(0, 0)$ (b) $(1, 0)$
(c) $(2, 1)$ (d) $(0, 2)$
[Note that for $t = 2$, this problem is equivalent to that given in Example 1.]

CHAPTER 10 REVIEW EXERCISES

In Exercises 1–20, solve the system of equations.

1. $x + y = 2$
$x - y = 0$

2. $2x = 3(y - 1)$
$y = x$

3. $x^2 - y^2 = 9$
$x - y = 1$

4. $x^2 + y^2 = 169$
$3x + 2y = 39$

5. $y = 2x^2$
$y = x^4 - 2x^2$

6. $x = y + 3$
$x = y^2 + 1$

7. $y^2 - 2y + x = 0$
$x + y = 0$

8. $y = 2x^2 - 4x + 1$
$y = x^2 - 4x + 3$

9. $2x - y = 2$
$6x + 8y = 39$

10. $40x + 30y = 24$
$20x - 50y = -14$

11. $0.2x + 0.3y = 0.14$
$0.4x + 0.5y = 0.20$

12. $12x + 42y = -17$
$30x - 18y = 19$

13. $3x - 2y = 0$
$3x + 2(y + 5) = 10$

14. $7x + 12y = 63$
$2x + 3y = 15$

15. $x + 2y + 6z = 4$
$-3x + 2y - z = -4$
$4x + 2z = 16$

16. $2x + 3y + 4z = 21$
$5x + y - 2z = -17$
$8x + 9y + z = -12$

17. $x - 2y + z = -6$
$2x - 3y = -7$
$-x + 3y - 3z = 11$

18. $2x + 6z = -9$
$3x - 2y + 11z = -16$
$3x - y + 7z = -11$

19. $2x + 5y - 19z = 34$
$3x + 8y - 31z = 54$

20. $2x + y + z + 2w = -1$
$5x - 2y + z - 3w = 0$
$-x + 3y + 2z + 2w = 1$
$3x + 2y + 3z - 5w = 12$

In Exercises 21–24, solve the problem using a system of linear equations.

21. Find three positive numbers whose sum is 88 if the second number is twice the first and the third number is two-thirds the first.

22. An inheritance of $20,000 was divided among three investments yielding $1780 in interest per year. The interest rates for the three investments were 7%, 9%, and 11%. Find the amount placed in each investment if the second and third were $3000 and $1000 less than the first, respectively.

23. A mixture of 6 parts of chemical A, 8 parts of chemical B, and 13 parts of chemical C is required to kill a certain destructive crop insect. Commercial spray X contains 1, 2, and 2 parts, respectively, of these chemicals. Commercial spray Y contains only chemical C. Commercial spray Z contains chemicals A, B, and C in equal amounts. How much of each type of commercial spray is needed to get the desired mixture?

24. Two planes leave Pittsburgh and Philadelphia at the same time, each going to the other city. Because of the wind, one plane flies 25 miles per hour faster than the other. Find the ground speed of each plane if the cities are 275 miles apart and the planes pass one another (at different altitudes) after 40 minutes of flying time.

In Exercises 25–32, sketch a graph of the solution of the system of inequalities.

25. $x + 2y \le 160$
$3x + y \le 180$
$x \ge 0$
$y \ge 0$

26. $2x + 3y \le 24$
$2x + y \le 16$
$x \ge 0$
$y \ge 0$

27. $3x + 2y \ge 24$
$x + 2y \ge 12$
$2 \le x \le 15$
$y \le 15$

28. $2x + y \ge 16$
$x + 3y \ge 18$
$0 \le x \le 25$
$0 \le y \le 25$

29. $y < x + 1$
$y > x^2 - 1$

30. $y \le 6 - 2x - x^2$
$y \ge x + 6$

31. $2x - 3y \ge 0$
$2x - y \le 8$
$y \ge 0$

32. $x^2 + y^2 \le 9$
$(x - 3)^2 + y^2 \le 9$

In Exercises 33 and 34, determine a system of inequalities that models the description and sketch a graph of the solution of the system.

33. A Pennsylvania fruit grower has 1500 bushels of apples that are to be divided between markets in Harrisburg and Philadelphia. These two markets need at least 400 bushels and 600 bushels, respectively.

34. A warehouse operator has 24,000 square feet of floor space in which to store two products. Each unit of product I requires 20 square feet of floor space and costs $12 per day to store. Each unit of product II requires 30 square feet of floor space and costs $8 per day to store. The total storage cost per day cannot exceed $12,400.

In Exercises 35–38, find the required optimum value of the objective function subject to the indicated constraints.

35.

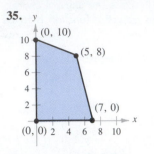

Maximize: $C = 3x + 4y$

36.

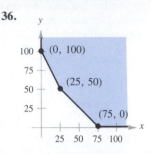

Minimize: $C = 10x + 7y$

37. Minimize:

$C = 1.75x + 2.25y$

Constraints:

$2x + y \ge 25$
$3x + 2y \ge 45$
$x \ge 0$
$y \ge 0$

38. Maximize:

$C = 50x + 70y$

Constraints:

$x + 2y \le 1500$
$5x + 2y \le 3500$
$x \ge 0$
$y \ge 0$

39. A manufacturer produces two products A and B yielding profits of $18 and $24, respectively. Each product must go through three processes with the required times per unit as shown in the following table.

Process	Hours for product A	Hours for product B	Hours available per day
I	4	2	24
II	1	2	9
III	1	1	8

Find the daily production level for each unit so that the profit is maximized.

CHAPTER 11

Matrices and Determinants

11.1 Matrices and Systems of Linear Equations

In mathematics we always look for valid shortcuts to solving problems. In this section we look at a streamlined technique for solving systems of linear equations. This technique involves the use of a rectangular array of real numbers called a **matrix.**

Definition of a Matrix

If m and n are positive integers, then an $m \times n$ **matrix** (read "m by n") is a rectangular array

$$\begin{bmatrix} a_{11} & a_{12} & a_{13} & \cdots & a_{1n} \\ a_{21} & a_{22} & a_{23} & \cdots & a_{2n} \\ a_{31} & a_{32} & a_{33} & \cdots & a_{3n} \\ \vdots & \vdots & \vdots & & \vdots \\ a_{m1} & a_{m2} & a_{m3} & \cdots & a_{mn} \end{bmatrix} \left.\vphantom{\begin{matrix}a\\a\\a\\a\\a\end{matrix}}\right\} m \text{ rows}$$

$$\underbrace{\phantom{a_{11} \quad a_{12} \quad a_{13} \quad \cdots \quad a_{1n}}}_{n \text{ columns}}$$

in which each **entry,** a_{ij}, of the matrix is a number. An $m \times n$ matrix has m **rows** (horizontal lines) and n **columns** (vertical lines).

Remark: The plural of matrix is *matrices*. If each entry of a matrix is a *real* number, then the matrix is called a **real matrix.** *In this text, we work only with real matrices.*

The entry in the ith row and jth column is denoted by the *double subscript* notation a_{ij}. We call i the **row subscript** because it gives the position in the horizontal lines, and j the **column subscript** because it gives the position in the vertical lines.

An $m \times n$ matrix is said to be of **order** $m \times n$. If $m = n$, the matrix is called **square** of order n. For a square matrix, the entries $a_{11}, a_{22}, a_{33}, \ldots$ are called the **main diagonal** entries.

EXAMPLE 1 *Examples of Matrices*

The following matrices have the indicated orders.

(a) Order: 1×1

$$[2]$$

(b) Order: 2×2

$$\begin{bmatrix} 0 & 0 \\ 0 & 0 \end{bmatrix}$$

(c) Order: 1×4

$$\begin{bmatrix} 1 & -3 & 0 & \frac{1}{2} \end{bmatrix}$$

(d) Order: 3×2

$$\begin{bmatrix} e & \pi \\ 2 & \sqrt{2} \\ -7 & 4 \end{bmatrix}$$

A matrix derived from a system of linear equations (each written in standard form with the constant term on the right) is called the **augmented matrix** of the system. Moreover, the matrix derived from the coefficients of the system (but which does not include the constant terms) is called the **coefficient matrix** of the system. Here is an example.

System	Augmented Matrix	Coefficient Matrix
$\begin{aligned} x - 4y + 3z &= 5 \\ -x + 3y - z &= -3 \\ 2x \quad\quad - 4z &= 6 \end{aligned}$	$\begin{bmatrix} 1 & -4 & 3 & \vdots & 5 \\ -1 & 3 & -1 & \vdots & -3 \\ 2 & 0 & -4 & \vdots & 6 \end{bmatrix}$	$\begin{bmatrix} 1 & -4 & 3 \\ -1 & 3 & -1 \\ 2 & 0 & -4 \end{bmatrix}$

Remark: Note the use of 0 for the missing y-variable in the third equation, and also note the fourth column of constant terms in the augmented matrix.

When forming either the coefficient matrix or the augmented matrix of a system, you should begin by vertically aligning the variables in the equations.

Given System	*Line Up Variables*	*Form Augmented Matrix*
$x + 3y = 9$	$x + 3y \quad\;\; = 9$	
$-y + 4z = -2$	$-y + 4z = -2$	
$x - 5z = 0$	$x \qquad - 5z = 0$	

$$\begin{bmatrix} 1 & 3 & 0 & \vdots & 9 \\ 0 & -1 & 4 & \vdots & -2 \\ 1 & 0 & -5 & \vdots & 0 \end{bmatrix}$$

Elementary Row Operations

In Section 10.3 we studied three operations that can be used on a system of linear equations to produce an equivalent system.

1. Interchange two equations.
2. Multiply an equation by a nonzero constant.
3. Add a multiple of an equation to another equation.

In matrix terminology these three operations correspond to **elementary row operations.** An elementary row operation on an augmented matrix of a given system of linear equations produces a new augmented matrix corresponding to a new (but equivalent) system of linear equations. Two matrices are said to be **row-equivalent** if one can be obtained from the other by a sequence of elementary row operations.

Elementary Row Operations

1. Interchange two rows.
2. Multiply a row by a nonzero constant.
3. Add a multiple of a row to another row.

Although elementary row operations are simple to perform, they involve a lot of arithmetic. Because it is easy to make a mistake, we suggest that you get in the habit of noting the elementary row operation performed in each step so that you can go back and check your work.

EXAMPLE 2 *Elementary Row Operations*

(a) Interchange the first and second rows.

Original Matrix	*New Row Equivalent Matrix*
$\begin{bmatrix} 0 & 1 & 3 & 4 \\ -1 & 2 & 0 & 3 \\ 2 & -3 & 4 & 1 \end{bmatrix}$	$\begin{matrix} R_2 \\ R_1 \end{matrix} \begin{bmatrix} -1 & 2 & 0 & 3 \\ 0 & 1 & 3 & 4 \\ 2 & -3 & 4 & 1 \end{bmatrix}$

Matrices and Determinants

(b) Multiply the first row by 1/2.

<u>Original Matrix</u>   <u>New Row Equivalent Matrix</u>

$$\begin{bmatrix} 2 & -4 & 6 & -2 \\ 1 & 3 & -3 & 0 \\ 5 & -2 & 1 & 2 \end{bmatrix} \quad \tfrac{1}{2}R_1 \rightarrow \quad \begin{bmatrix} 1 & -2 & 3 & -1 \\ 1 & 3 & -3 & 0 \\ 5 & -2 & 1 & 2 \end{bmatrix}$$

(c) Add -2 times the first row to the third row.

<u>Original Matrix</u>   <u>New Row Equivalent Matrix</u>

$$\begin{bmatrix} 1 & 2 & -4 & 3 \\ 0 & 3 & -2 & -1 \\ 2 & 1 & 5 & -2 \end{bmatrix} \quad \substack{\\ \\ -2R_1 + R_3 \rightarrow} \quad \begin{bmatrix} 1 & 2 & -4 & 3 \\ 0 & 3 & -2 & -1 \\ 0 & -3 & 13 & -8 \end{bmatrix}$$

Note that we write the elementary row operation beside the row that we are *changing*.

In Section 10.3 we used Gaussian elimination with back-substitution to solve a system of linear equations. We now demonstrate the matrix version of Gaussian elimination. The two methods are essentially the same. The basic difference is that with matrices we do not need to keep writing the variables.

EXAMPLE 3 *Using Elementary Row Operations to Solve a System*

<u>Linear System</u>   <u>Associated Augmented Matrix</u>

$$\begin{aligned} x - 2y + 3z &= 9 \\ -x + 3y &= -4 \\ 2x - 5y + 5z &= 17 \end{aligned} \qquad \begin{bmatrix} 1 & -2 & 3 & \vdots & 9 \\ -1 & 3 & 0 & \vdots & -4 \\ 2 & -5 & 5 & \vdots & 17 \end{bmatrix}$$

Add the first equation to the second equation.

Add the first row to the second row ($R_1 + R_2$).

$$\begin{aligned} x - 2y + 3z &= 9 \\ y + 3z &= 5 \\ 2x - 5y + 5z &= 17 \end{aligned} \qquad R_1 + R_2 \rightarrow \qquad \begin{bmatrix} 1 & -2 & 3 & \vdots & 9 \\ 0 & 1 & 3 & \vdots & 5 \\ 2 & -5 & 5 & \vdots & 17 \end{bmatrix}$$

Add -2 times the first equation to the third equation.

Add -2 times the first row to the third row ($-2R_1 + R_3$).

$$\begin{aligned} x - 2y + 3z &= 9 \\ y + 3z &= 5 \\ -y - z &= -1 \end{aligned} \qquad -2R_1 + R_3 \rightarrow \qquad \begin{bmatrix} 1 & -2 & 3 & \vdots & 9 \\ 0 & 1 & 3 & \vdots & 5 \\ 0 & -1 & -1 & \vdots & -1 \end{bmatrix}$$

Add the second equation to the third equation.

$$\begin{aligned} x - 2y + 3z &= 9 \\ y + 3z &= 5 \\ 2z &= 4 \end{aligned}$$

Add the second row to the third row $(R_2 + R_3)$.

$$R_2 + R_3 \rightarrow \begin{bmatrix} 1 & -2 & 3 & \vdots & 9 \\ 0 & 1 & 3 & \vdots & 5 \\ 0 & 0 & 2 & \vdots & 4 \end{bmatrix}$$

Multiply the third equation by 1/2.

$$\begin{aligned} x - 2y + 3z &= 9 \\ y + 3z &= 5 \\ z &= 2 \end{aligned}$$

Multiply the third row by 1/2.

$$\tfrac{1}{2} R_3 \rightarrow \begin{bmatrix} 1 & -2 & 3 & \vdots & 9 \\ 0 & 1 & 3 & \vdots & 5 \\ 0 & 0 & 1 & \vdots & 2 \end{bmatrix}$$

At this point, we can use back-substitution to find that the solution is $x = 1$, $y = -1$, and $z = 2$, as we did in Section 10.3.

The last matrix in Example 3 is said to be in **row-echelon form.** The term *echelon* refers to the stair-step pattern formed by the nonzero elements of the matrix. To be in this form, a matrix must have the following properties.

Definition of Row-Echelon Form and Reduced Row-Echelon Form

A matrix in **row-echelon form** has the following properties.

1. All rows consisting entirely of zeros occur at the bottom of the matrix.
2. For each row that does not consist entirely of zeros, the first nonzero entry is 1 (called a **leading 1**).
3. For two successive (nonzero) rows, the leading 1 in the higher row is farther to the left than the leading 1 in the lower row.

A matrix in *row-echelon form* is in **reduced row-echelon form** if every column that has a leading 1 has zeros in every position above and below its leading 1.

EXAMPLE 4 *Row-Echelon Form*

The following matrices are in row-echelon form.

(a) $\begin{bmatrix} 1 & 2 & -1 & 4 \\ 0 & 1 & 0 & 3 \\ 0 & 0 & 1 & -2 \end{bmatrix}$
(b) $\begin{bmatrix} 0 & 1 & 0 & 5 \\ 0 & 0 & 1 & 3 \\ 0 & 0 & 0 & 0 \end{bmatrix}$

(c) $\begin{bmatrix} 1 & -5 & 2 & -1 & 3 \\ 0 & 0 & 1 & 3 & -2 \\ 0 & 0 & 0 & 1 & 4 \\ 0 & 0 & 0 & 0 & 1 \end{bmatrix}$
(d) $\begin{bmatrix} 1 & 0 & 0 & -1 \\ 0 & 1 & 0 & 2 \\ 0 & 0 & 1 & 3 \\ 0 & 0 & 0 & 0 \end{bmatrix}$

The matrices in (b) and (d) also happen to be in *reduced* row-echelon form. The following matrices are not in row-echelon form.

$$\text{(e)} \begin{bmatrix} 1 & 2 & -3 & 4 \\ 0 & 2 & 1 & -1 \\ 0 & 0 & 1 & -3 \end{bmatrix} \qquad \text{(f)} \begin{bmatrix} 1 & 2 & -1 & 2 \\ 0 & 0 & 0 & 0 \\ 0 & 1 & 2 & -4 \end{bmatrix}$$

Every matrix is row equivalent to a matrix in row-echelon form. For instance, in Example 4, we can change the matrix in part (e) to row-echelon form by multiplying its second row by 1/2.

Guidelines for using Gaussian elimination with back-substitution to solve a system of linear equations are summarized as follows.

Gaussian Elimination with Back-Substitution

1. Write the augmented matrix of the system of linear equations.
2. Use elementary row operations to rewrite the augmented matrix in row-echelon form.
3. Write the system of linear equations corresponding to the matrix in row-echelon form, and use back-substitution to find the solution.

Gaussian elimination with back-substitution works well for solving systems of linear equations with a computer. For this algorithm, the order in which the elementary row operations are performed is important. We operate from *left to right by columns*, using elementary row operations to obtain zeros in all entries directly below the leading 1's.

EXAMPLE 5 *Gaussian Elimination with Back-Substitution*

Solve the following system.

$$\begin{aligned} y + z - 2w &= -3 \\ x + 2y - z &= 2 \\ 2x + 4y + z - 3w &= -2 \\ x - 4y - 7z - w &= -19 \end{aligned}$$

SOLUTION

The augmented matrix for this system is

$$\begin{bmatrix} 0 & 1 & 1 & -2 & \vdots & -3 \\ 1 & 2 & -1 & 0 & \vdots & 2 \\ 2 & 4 & 1 & -3 & \vdots & -2 \\ 1 & -4 & -7 & -1 & \vdots & -19 \end{bmatrix}$$

Matrices and Systems of Linear Equations

We begin by obtaining a leading 1 in the upper left corner and then proceed to obtain zeros elsewhere in the first column.

$$\begin{matrix} \curvearrowright R_2 \\ \searrow R_1 \end{matrix} \begin{bmatrix} 1 & 2 & -1 & 0 & \vdots & 2 \\ 0 & 1 & 1 & -2 & \vdots & -3 \\ 2 & 4 & 1 & -3 & \vdots & -2 \\ 1 & -4 & -7 & -1 & \vdots & -19 \end{bmatrix}$$

First column has leading 1 in upper left corner.

$$\begin{matrix} \\ \\ -2R_1 + R_3 \rightarrow \\ -R_1 + R_4 \rightarrow \end{matrix} \begin{bmatrix} 1 & 2 & -1 & 0 & \vdots & 2 \\ 0 & 1 & 1 & -2 & \vdots & -3 \\ 0 & 0 & 3 & -3 & \vdots & -6 \\ 0 & -6 & -6 & -1 & \vdots & -21 \end{bmatrix}$$

First column has zeros below its leading 1.

Now that the first column is in the desired form, we change the second, third, and fourth columns as follows.

$$\begin{matrix} \\ \\ \\ 6R_2 + R_4 \rightarrow \end{matrix} \begin{bmatrix} 1 & 2 & -1 & 0 & \vdots & 2 \\ 0 & 1 & 1 & -2 & \vdots & -3 \\ 0 & 0 & 3 & -3 & \vdots & -6 \\ 0 & 0 & 0 & -13 & \vdots & -39 \end{bmatrix}$$

Second column has zeros below its leading 1.

$$\begin{matrix} \\ \\ \frac{1}{3}R_3 \rightarrow \\ \\ \end{matrix} \begin{bmatrix} 1 & 2 & -1 & 0 & \vdots & 2 \\ 0 & 1 & 1 & -2 & \vdots & -3 \\ 0 & 0 & 1 & -1 & \vdots & -2 \\ 0 & 0 & 0 & -13 & \vdots & -39 \end{bmatrix}$$

Third column has zeros below its leading 1.

$$\begin{matrix} \\ \\ \\ -\frac{1}{13}R_4 \rightarrow \end{matrix} \begin{bmatrix} 1 & 2 & -1 & 0 & \vdots & 2 \\ 0 & 1 & 1 & -2 & \vdots & -3 \\ 0 & 0 & 1 & -1 & \vdots & -2 \\ 0 & 0 & 0 & 1 & \vdots & 3 \end{bmatrix}$$

Fourth column has a leading 1.

The matrix is now in row-echelon form, and the corresponding system of linear equations is

$$\begin{aligned} x + 2y - z \quad\quad &= 2 \\ y + z - 2w &= -3 \\ z - w &= -2 \\ w &= 3. \end{aligned}$$

Using back-substitution, we can determine that the solution is

$$x = -1, \quad y = 2, \quad z = 1, \quad \text{and} \quad w = 3.$$

We can now check for errors in our elementary row operations by substituting these values in each equation in the *original* system. (If they don't check, then we know that we made an error in the back-substitution or one of the elementary row operations.)

Matrices and Determinants

When solving a system of linear equations, remember that it is possible for the system to have no solution. If, in the elimination process, you obtain a row with zeros except for the last entry, it is unnecessary to continue the elimination process. You can simply conclude that the system is inconsistent.

EXAMPLE 6 A System with No Solution

Solve the following system.

$$
\begin{aligned}
x - y + 2z &= 4 \\
x \quad\;\; + z &= 6 \\
2x - 3y + 5z &= 4 \\
3x + 2y - z &= 1
\end{aligned}
$$

SOLUTION

To the augmented matrix for this system we apply Gaussian elimination as follows.

$$
\begin{bmatrix}
1 & -1 & 2 & \vdots & 4 \\
1 & 0 & 1 & \vdots & 6 \\
2 & -3 & 5 & \vdots & 4 \\
3 & 2 & -1 & \vdots & 1
\end{bmatrix}
\begin{array}{l}
-R_1 + R_2 \rightarrow \\
-2R_1 + R_3 \rightarrow \\
-3R_1 + R_4 \rightarrow
\end{array}
\begin{bmatrix}
1 & -1 & 2 & \vdots & 4 \\
0 & 1 & -1 & \vdots & 2 \\
0 & -1 & 1 & \vdots & -4 \\
0 & 5 & -7 & \vdots & -11
\end{bmatrix}
$$

$$
R_2 + R_3 \rightarrow
\begin{bmatrix}
1 & -1 & 2 & \vdots & 4 \\
0 & 1 & -1 & \vdots & 2 \\
0 & 0 & 0 & \vdots & -2 \\
0 & 5 & -7 & \vdots & -11
\end{bmatrix}
$$

Note that the third row of this matrix consists of zeros except for the last entry. This means that the original system of linear equations is *inconsistent*. You can see why this is true by converting back to a system of linear equations.

$$
\begin{aligned}
x - y + 2z &= 4 \\
y - z &= 2 \\
0 &= -2 \\
5y - 7z &= -11
\end{aligned}
$$

The third equation means that there must be three real numbers x, y, and z such that $0x + 0y + 0z = -2$. Since this is not possible, it follows that the system has no solution.

Gauss-Jordan Elimination

With Gaussian elimination, we apply elementary row operations to a matrix to obtain a (row-equivalent) row-echelon form. A second method of elimination called **Gauss-Jordan elimination,** after Carl Gauss and Wilhelm Jordan (1842–1899), continues the reduction process until a *reduced* row-echelon form is obtained. We demonstrate this procedure in the following example.

EXAMPLE 7   *Gauss-Jordan Elimination*

Use Gauss-Jordan elimination to solve the system

$$\begin{aligned} x - 2y + 3z &= 9 \\ -x + 3y \phantom{{}+ 3z} &= -4 \\ 2x - 5y + 5z &= 17. \end{aligned}$$

SOLUTION

In Example 3 we used Gaussian elimination to obtain the following row-echelon form.

$$\left[\begin{array}{ccc:c} 1 & -2 & 3 & 9 \\ 0 & 1 & 3 & 5 \\ 0 & 0 & 1 & 2 \end{array}\right]$$

Now, rather than using back-substitution, we apply elementary row operations until we obtain a matrix in reduced row-echelon form. To do this, we must produce zeros above each of the leading 1's, as follows.

$$\begin{array}{c} 2R_2 + R_1 \rightarrow \\ \\ \\ \end{array} \left[\begin{array}{ccc:c} 1 & 0 & 9 & 19 \\ 0 & 1 & 3 & 5 \\ 0 & 0 & 1 & 2 \end{array}\right]$$

Second column has zeros above its leading 1.

$$\begin{array}{c} -9R_3 + R_1 \rightarrow \\ -3R_3 + R_2 \rightarrow \\ \\ \end{array} \left[\begin{array}{ccc:c} 1 & 0 & 0 & 1 \\ 0 & 1 & 0 & -1 \\ 0 & 0 & 1 & 2 \end{array}\right]$$

Third column has zeros above its leading 1.

Now, converting back to a system of linear equations, we have

$$\begin{aligned} x \phantom{{}+{}} &= 1 \\ y \phantom{{}+{}} &= -1 \\ z &= 2. \end{aligned}$$

The beauty of Gauss-Jordan elimination is that, from the reduced row-echelon form, we can simply read the solution.

It is worth noting that the row-echelon form for a matrix is not unique. That is, two different sequences of elementary row operations may yield different row-echelon forms. For instance, the following sequence of elementary row operations on the matrix in Example 3 produces a slightly different row-echelon form.

Matrices and Determinants

$$\begin{bmatrix} 1 & -2 & 3 & \vdots & 9 \\ -1 & 3 & 0 & \vdots & -4 \\ 2 & -5 & 5 & \vdots & 17 \end{bmatrix} \quad \begin{matrix} \curvearrowright R_2 \\ \searrow R_1 \end{matrix} \begin{bmatrix} -1 & 3 & 0 & \vdots & -4 \\ 1 & -2 & 3 & \vdots & 9 \\ 2 & -5 & 5 & \vdots & 17 \end{bmatrix}$$

$$-R_1 \rightarrow \begin{bmatrix} 1 & -3 & 0 & \vdots & 4 \\ 1 & -2 & 3 & \vdots & 9 \\ 2 & -5 & 5 & \vdots & 17 \end{bmatrix}$$

$$\begin{matrix} -R_1 + R_2 \rightarrow \\ -2R_1 + R_3 \rightarrow \end{matrix} \begin{bmatrix} 1 & -3 & 0 & \vdots & 4 \\ 0 & 1 & 3 & \vdots & 5 \\ 0 & 1 & 5 & \vdots & 9 \end{bmatrix}$$

$$-R_2 + R_3 \rightarrow \begin{bmatrix} 1 & -3 & 0 & \vdots & 4 \\ 0 & 1 & 3 & \vdots & 5 \\ 0 & 0 & 2 & \vdots & 4 \end{bmatrix}$$

$$\tfrac{1}{2} R_3 \rightarrow \begin{bmatrix} 1 & -3 & 0 & \vdots & 4 \\ 0 & 1 & 3 & \vdots & 5 \\ 0 & 0 & 1 & \vdots & 2 \end{bmatrix}$$

However, the *reduced* row-echelon form for a given matrix *is* unique. You should try applying Gauss-Jordan elimination to this matrix to see that you obtain the same reduced row-echelon form as in Example 7.

The elimination procedures described in this section employ an algorithmic approach that is easily adapted to computer use. However, the procedure makes no effort to avoid fractional coefficients. For instance, if the system given in Example 7 had been listed as

$$\begin{aligned} 2x - 5y + 5z &= 17 \\ x - 2y + 3z &= 9 \\ -x + 3y &= -4 \end{aligned}$$

our procedure would then have required multiplying the first row by 1/2, which would have introduced fractions in the first row. For hand computations, fractions can sometimes be avoided by judiciously choosing the order in which the elementary row operations are applied.

The next example demonstrates how Gauss-Jordan elimination can be used to solve a system with an infinite number of solutions.

EXAMPLE 8 *A System with an Infinite Number of Solutions*

Solve the following system of linear equations.

$$\begin{aligned} 2x + 4y - 2z &= 0 \\ 3x + 5y &= 1 \end{aligned}$$

SOLUTION

Using Gauss-Jordan elimination, the augmented matrix reduces as follows.

$$\begin{bmatrix} 2 & 4 & -2 & \vdots & 0 \\ 3 & 5 & 0 & \vdots & 1 \end{bmatrix} \qquad \tfrac{1}{2}R_1 \rightarrow \begin{bmatrix} 1 & 2 & -1 & \vdots & 0 \\ 3 & 5 & 0 & \vdots & 1 \end{bmatrix}$$

$$-3R_1 + R_2 \rightarrow \begin{bmatrix} 1 & 2 & -1 & \vdots & 0 \\ 0 & -1 & 3 & \vdots & 1 \end{bmatrix}$$

$$-R_2 \rightarrow \begin{bmatrix} 1 & 2 & -1 & \vdots & 0 \\ 0 & 1 & -3 & \vdots & -1 \end{bmatrix}$$

$$-2R_2 + R_1 \rightarrow \begin{bmatrix} 1 & 0 & 5 & \vdots & 2 \\ 0 & 1 & -3 & \vdots & -1 \end{bmatrix}$$

The corresponding system of equations is

$$x \qquad + 5z = \quad 2$$
$$y - 3z = -1.$$

Solving for x and y in terms of z, we have $x = -5z + 2$ and $y = 3z - 1$. Then, letting $z = a$, the solution set has the form

$$(-5a + 2, \ 3a - 1, \ a), \quad \text{where } a \text{ is a real number.}$$ ▬

We have looked at two elimination methods for solving a system of linear equations. Which is better? To some degree, it depends on personal preference. For hand computations, Gaussian elimination with back-substitution is often preferred. However, we will encounter other applications in which Gauss-Jordan elimination is better. Thus, you should know both methods.

WARM UP

Evaluate the expressions.

1. $2(-1) - 3(5) + 7(2)$ **2.** $-4(-3) + 6(7) + 8(-3)$ **3.** $11(\tfrac{1}{2}) - 7(-\tfrac{3}{2}) - 5(2)$ **4.** $\tfrac{2}{3}(\tfrac{1}{2}) + \tfrac{4}{3}(-\tfrac{1}{3})$

In each case, determine whether $x = 1$, $y = 3$, and $z = -1$ is a solution.

5. $\begin{aligned} 4x - 2y + 3z &= -5 \\ x + 3y - \ z &= 11 \\ -x + 2y \qquad &= \ 5 \end{aligned}$ **6.** $\begin{aligned} -x + 2y + \ z &= \ 4 \\ 2x \qquad - 3z &= \ 5 \\ 3x + 5y - 2z &= 21 \end{aligned}$

Use back-substitution to solve the systems of linear equations.

7. $\begin{aligned} 2x - 3y &= 4 \\ y &= 2 \end{aligned}$ **8.** $\begin{aligned} 5x + 4y &= \ 0 \\ y &= -3 \end{aligned}$ **9.** $\begin{aligned} x - 3y + \ z &= 0 \\ y - 3z &= 8 \\ z &= 2 \end{aligned}$ **10.** $\begin{aligned} 2x - 5y + 3z &= -2 \\ y - 4z &= \ 0 \\ z &= \ 1 \end{aligned}$

Matrices and Determinants

EXERCISES 11.1

In Exercises 1–6, determine the order of the given matrix.

1. $\begin{bmatrix} 4 & -2 \\ 7 & 0 \\ 0 & 8 \end{bmatrix}$

2. $\begin{bmatrix} 5 & -3 & 8 & 7 \end{bmatrix}$

3. $\begin{bmatrix} -9 \\ 2 \\ 36 \\ 11 \\ 3 \end{bmatrix}$

4. $\begin{bmatrix} 11 & 0 & 8 & 5 & 5 \\ -3 & 7 & 15 & 0 & 10 \\ 0 & 6 & 3 & 3 & 9 \\ 12 & 4 & 16 & 9 & 0 \\ 1 & 1 & 6 & 7 & 8 \end{bmatrix}$

5. $\begin{bmatrix} 33 & 45 \\ -9 & 20 \end{bmatrix}$

6. $[4]$

In Exercises 7–10, determine whether the matrix is in row-echelon form. If it is, determine if it is also in reduced row-echelon form.

7. $\begin{bmatrix} 1 & 0 & 0 & 0 \\ 0 & 1 & 1 & 5 \\ 0 & 0 & 0 & 0 \end{bmatrix}$

8. $\begin{bmatrix} 1 & 0 & 2 & 1 \\ 0 & 1 & -3 & 10 \\ 0 & 0 & 1 & 0 \end{bmatrix}$

9. $\begin{bmatrix} 2 & 0 & 4 & 0 \\ 0 & -1 & 3 & 6 \\ 0 & 0 & 1 & 5 \end{bmatrix}$

10. $\begin{bmatrix} 1 & 3 & 0 & 0 \\ 0 & 0 & 1 & 8 \\ 0 & 0 & 0 & 0 \end{bmatrix}$

11. Perform the indicated *sequence* of elementary row operations to write the given matrix in reduced row-echelon form.

$$\begin{bmatrix} 1 & 2 & 3 \\ 2 & -1 & -4 \\ 3 & 1 & -1 \end{bmatrix}$$

(a) Add (-2) times Row 1 to Row 2. (Only Row 2 should change.)

(b) Add (-3) times Row 1 to Row 3. (Only Row 3 should change.)

(c) Add (-1) times Row 2 to Row 3.

(d) Multiply Row 2 by $(-\frac{1}{5})$.

(e) Add (-2) times Row 2 to Row 1.

12. Perform the indicated *sequence* of elementary row operations to write the given matrix in reduced row-echelon form.

$$\begin{bmatrix} 7 & 1 \\ 0 & 2 \\ -3 & 4 \\ 4 & 1 \end{bmatrix}$$

(a) Add Row 3 to Row 4. (Only Row 4 should change.)

(b) Interchange Rows 1 and 4. (Note that the first element in the matrix is now 1, and it was obtained without introducing fractions.)

(c) Add (3) times Row 1 to Row 3.

(d) Add (-7) times Row 1 to Row 4.

(e) Multiply Row 2 by $\frac{1}{2}$.

(f) Add the appropriate multiple of Row 2 to Rows 1, 3, and 4.

In Exercises 13–16, write the matrix in row-echelon form. Remember that the row-echelon form for a given matrix is not unique.

13. $\begin{bmatrix} 1 & 1 & 0 & 5 \\ -2 & -1 & 2 & -10 \\ 3 & 6 & 7 & 14 \end{bmatrix}$

14. $\begin{bmatrix} 1 & 2 & -1 & 3 \\ 3 & 7 & -5 & 14 \\ -2 & -1 & -3 & 8 \end{bmatrix}$

15. $\begin{bmatrix} 1 & -1 & -1 & 1 \\ 5 & -4 & 1 & 8 \\ -6 & 8 & 18 & 0 \end{bmatrix}$

16. $\begin{bmatrix} 1 & -3 & 0 & -7 \\ -3 & 10 & 1 & 23 \\ 4 & -10 & 2 & -24 \end{bmatrix}$

In Exercises 17–20, write the matrix in *reduced* row-echelon form.

17. $\begin{bmatrix} 3 & 3 & 3 \\ -1 & 0 & -4 \\ 2 & 4 & -2 \end{bmatrix}$

18. $\begin{bmatrix} 1 & 3 & 2 \\ 5 & 15 & 9 \\ 2 & 6 & 10 \end{bmatrix}$

19. $\begin{bmatrix} 1 & 2 & 3 & -5 \\ 1 & 2 & 4 & -9 \\ -2 & -4 & -4 & 3 \\ 4 & 8 & 11 & -14 \end{bmatrix}$

20. $\begin{bmatrix} 1 & -3 \\ -1 & 8 \\ 0 & 4 \\ -2 & 10 \end{bmatrix}$

In Exercises 21–24, write the system of linear equations represented by the augmented matrix.

21. $\left[\begin{array}{cc:c} 4 & 3 & 8 \\ 1 & -2 & 3 \end{array}\right]$

22. $\left[\begin{array}{cc:c} 9 & -4 & 0 \\ 6 & 1 & -4 \end{array}\right]$

23. $\left[\begin{array}{ccc:c} 1 & 0 & 2 & -10 \\ 0 & 3 & -1 & 5 \\ 4 & 2 & 0 & 3 \end{array}\right]$

24. $\left[\begin{array}{cccc:c} 5 & 8 & 2 & 0 & -1 \\ -2 & 15 & 5 & 1 & 9 \\ 1 & 6 & -7 & 0 & -3 \end{array}\right]$

In Exercises 25–28, determine the augmented matrix for the given system of linear equations.

25. $\begin{aligned} 4x - 5y &= -2 \\ -x + 8y &= 10 \end{aligned}$

26. $\begin{aligned} 8x + 3y &= 25 \\ 3x - 9y &= 12 \end{aligned}$

27. $\begin{aligned} x + 10y - 3z &= 2 \\ 5x - 3y + 4z &= 0 \\ 2x + 4y &= 6 \end{aligned}$

28. $\begin{aligned} 9w - 3x + 20y + z &= 13 \\ 12w - 8y &= 5 \end{aligned}$

In Exercises 29–50, solve the system of equations. Use Gaussian elimination with back-substitution or Gauss-Jordan elimination.

29. $\begin{aligned} x + 2y &= 7 \\ 2x + y &= 8 \end{aligned}$

30. $\begin{aligned} 2x + 6y &= 16 \\ 2x + 3y &= 7 \end{aligned}$

31. $\begin{aligned} -3x + 5y &= -22 \\ 3x + 4y &= 4 \\ 4x - 8y &= 32 \end{aligned}$

32. $\begin{aligned} x + 2y &= 0 \\ x + y &= 6 \\ 3x - 2y &= 8 \end{aligned}$

33. $\begin{aligned} 8x - 4y &= 7 \\ 5x + 2y &= 1 \end{aligned}$

34. $\begin{aligned} 2x - y &= -0.1 \\ 3x + 2y &= 1.6 \end{aligned}$

35. $\begin{aligned} -x + 2y &= 1.5 \\ 2x - 4y &= 3 \end{aligned}$

36. $\begin{aligned} x - 3y &= 5 \\ -2x + 6y &= -10 \end{aligned}$

37. $\begin{aligned} x - 3z &= -2 \\ 3x + y - 2z &= 5 \\ 2x + 2y + z &= 4 \end{aligned}$

38. $\begin{aligned} 2x - y + 3z &= 24 \\ 2y - z &= 14 \\ 7x - 5y &= 6 \end{aligned}$

39. $\begin{aligned} x + y - 5z &= 3 \\ x - 2z &= 1 \\ 2x - y - z &= 0 \end{aligned}$

40. $\begin{aligned} 2x + 3z &= 3 \\ 4x - 3y + 7z &= 5 \\ 8x - 9y + 15z &= 9 \end{aligned}$

41. $\begin{aligned} x + 2y + z &= 8 \\ 3x + 7y + 6z &= 26 \end{aligned}$

42. $\begin{aligned} 4x + 12y - 7z - 20w &= 22 \\ 3x + 9y - 5z - 28w &= 30 \end{aligned}$

43. $\begin{aligned} 3x + 3y + 12z &= 6 \\ x + y + 4z &= 2 \\ 2x + 5y + 20z &= 10 \\ -x + 2y + 8z &= 4 \end{aligned}$

44. $\begin{aligned} 2x + 10y + 2z &= 6 \\ x + 5y + 2z &= 6 \\ x + 5y + z &= 3 \\ -3x - 15y - 3z &= -9 \end{aligned}$

45. $\begin{aligned} 2x + y - z + 2w &= -6 \\ 3x + 4y + w &= 1 \\ x + 5y + 2z + 6w &= -3 \\ 5x + 2y - z - w &= 3 \end{aligned}$

46. $\begin{aligned} x + 2y + 2z + 4w &= 11 \\ 3x + 6y + 5z + 12w &= 30 \end{aligned}$

47. $\begin{aligned} x + 2y &= 0 \\ -x - y &= 0 \end{aligned}$

48. $\begin{aligned} x + 2y &= 0 \\ 2x + 4y &= 0 \end{aligned}$

49. $\begin{aligned} x + y + z &= 0 \\ 2x + 3y + z &= 0 \\ 3x + 5y + z &= 0 \end{aligned}$

50. $\begin{aligned} x + 2y + z + 3w &= 0 \\ x - y + w &= 0 \\ y - z + 2w &= 0 \end{aligned}$

In Exercises 51–56, solve the word problem by using a system of linear equations and matrices.

51. A small corporation borrowed $1,500,000 to expand its product line. Some of the money was borrowed at 8%, some at 9%, and some at 12%. How much was borrowed at each rate if the annual interest was $133,000 and the amount borrowed at 8% was 4 times the amount borrowed at 12%?

52. A grocer wishes to mix three kinds of nuts costing $3.50, $4.50, and $6.00 per pound, to obtain 50 pounds of a mixture costing $4.95 per pound. How many pounds of each variety should the grocer use if half of the mixture is composed of the two cheapest varieties?

53. Find a, b, and c for the quadratic function $f(x) = ax^2 + bx + c$, such that $f(1) = 1$, $f(-3) = 17$, and $f(2) = -\frac{1}{2}$. [*Hint:* See Example 8 in Section 10.3.]

54. Find a, b, c, and d for the cubic function $f(x) = ax^3 + bx^2 + cx + d$, such that $f(1) = 3$, $f(2) = 19$, $f(-1) = -11$, and $f(-2) = -33$. [*Hint:* See Example 8 in Section 10.3.]

55. Find D, E, and F such that $(1, 1)$, $(3, 3)$, and $(4, 2)$ are solution points of the equation $x^2 + y^2 + Dx + Ey + F = 0$. [*Hint:* See Example 8 in Section 10.3.]

56. The sum of three positive numbers is 33. Find the three numbers if the second is 3 greater than the first and the third is 4 times the first.

11.2 Operations with Matrices

In Section 11.1 we used matrices to solve systems of linear equations. Matrices, however, can do much more than that. There is a rich mathematical theory of matrices, and its applications are numerous. This section and the next introduce some fundamentals of matrix theory.

It is standard mathematical convention to represent matrices in any of the following three ways.

1. A matrix can be denoted by an upper-case letter such as

   $A, B, C, \ldots$.

2. A matrix can be denoted by a representative element enclosed in brackets, such as

   $[a_{ij}], [b_{ij}], [c_{ij}], \ldots$.

3. A matrix can be denoted by a rectangular array of numbers

$$
A = [a_{ij}] = \begin{bmatrix} a_{11} & a_{12} & a_{13} & \cdots & a_{1n} \\ a_{21} & a_{22} & a_{23} & \cdots & a_{2n} \\ a_{31} & a_{32} & a_{33} & \cdots & a_{3n} \\ \vdots & \vdots & \vdots & & \vdots \\ a_{m1} & a_{m2} & a_{m3} & \cdots & a_{mn} \end{bmatrix}.
$$

As mentioned in Section 11.1, the matrices in this text are *real matrices*. That is, their entries are real numbers.

Two matrices are said to be **equal** if their corresponding entries are equal.

Definition of Equality of Matrices

Two matrices $A = [a_{ij}]$ and $B = [b_{ij}]$ are **equal** if they have the same order $(m \times n)$ and

$\quad a_{ij} = b_{ij}$

for $1 \leq i \leq m$ and $1 \leq j \leq n$.

EXAMPLE 1 *Equality of Matrices*

Solve for a_{11}, a_{12}, a_{21}, and a_{22} in the following matrix equation.

$$
\begin{bmatrix} a_{11} & a_{12} \\ a_{21} & a_{22} \end{bmatrix} = \begin{bmatrix} 2 & -1 \\ -3 & 0 \end{bmatrix}
$$

SOLUTION

Because two matrices are equal only if corresponding entries are equal, we conclude that

$$a_{11} = 2, \quad a_{12} = -1, \quad a_{21} = -3, \quad \text{and} \quad a_{22} = 0.$$

Matrix Addition and Scalar Multiplication

We **add** two matrices (of the same order) by adding their corresponding entries.

Definition of Matrix Addition

If $A = [a_{ij}]$ and $B = [b_{ij}]$ are matrices of order $m \times n$, then their **sum** is the $m \times n$ matrix given by

$$A + B = [a_{ij} + b_{ij}].$$

The sum of two matrices of different orders is undefined.

EXAMPLE 2 Addition of Matrices

(a) $\begin{bmatrix} -1 & 2 \\ 0 & 1 \end{bmatrix} + \begin{bmatrix} 1 & 3 \\ -1 & 2 \end{bmatrix} = \begin{bmatrix} -1 + 1 & 2 + 3 \\ 0 - 1 & 1 + 2 \end{bmatrix} = \begin{bmatrix} 0 & 5 \\ -1 & 3 \end{bmatrix}$

(b) $\begin{bmatrix} 0 & 1 & -2 \\ 1 & 2 & 3 \end{bmatrix} + \begin{bmatrix} 0 & 0 & 0 \\ 0 & 0 & 0 \end{bmatrix} = \begin{bmatrix} 0 & 1 & -2 \\ 1 & 2 & 3 \end{bmatrix}$

(c) $\begin{bmatrix} 1 \\ -3 \\ -2 \end{bmatrix} + \begin{bmatrix} -1 \\ 3 \\ 2 \end{bmatrix} = \begin{bmatrix} 0 \\ 0 \\ 0 \end{bmatrix}$

(d) The sum of

$$A = \begin{bmatrix} 2 & 1 & 0 \\ 4 & 0 & -1 \\ 3 & -2 & 2 \end{bmatrix} \quad \text{and} \quad B = \begin{bmatrix} 0 & 1 \\ -1 & 3 \\ 2 & 4 \end{bmatrix}$$

is undefined.

When working with matrices, we usually refer to numbers as **scalars.** In this text, scalars will always be real numbers. We multiply a matrix A by a scalar c by multiplying each entry in A by c.

Matrices and Determinants

Definition of Scalar Multiplication

If $A = [a_{ij}]$ is an $m \times n$ matrix and c is a scalar, then the **scalar multiple** of A by c is the $m \times n$ matrix given by

$$cA = [ca_{ij}].$$

We use $-A$ to represent the scalar product $(-1)A$. Moreover, if A and B are of the same order, $A - B$ represents the sum of A and $(-1)B$. That is,

$$A - B = A + (-1)B. \qquad \text{\textit{Subtraction of matrices}}$$

EXAMPLE 3 Scalar Multiplication and Matrix Subtraction

For the matrices

$$A = \begin{bmatrix} 1 & 2 & 4 \\ -3 & 0 & -1 \\ 2 & 1 & 2 \end{bmatrix} \quad \text{and} \quad B = \begin{bmatrix} 2 & 0 & 0 \\ 1 & -4 & 3 \\ -1 & 3 & 2 \end{bmatrix}$$

find the following.

(a) $3A$

(b) $-B$

(c) $3A - B$

SOLUTION

(a) $3A = 3\begin{bmatrix} 1 & 2 & 4 \\ -3 & 0 & -1 \\ 2 & 1 & 2 \end{bmatrix} = \begin{bmatrix} 3(1) & 3(2) & 3(4) \\ 3(-3) & 3(0) & 3(-1) \\ 3(2) & 3(1) & 3(2) \end{bmatrix} = \begin{bmatrix} 3 & 6 & 12 \\ -9 & 0 & -3 \\ 6 & 3 & 6 \end{bmatrix}$

(b) $-B = (-1)\begin{bmatrix} 2 & 0 & 0 \\ 1 & -4 & 3 \\ -1 & 3 & 2 \end{bmatrix} = \begin{bmatrix} -2 & 0 & 0 \\ -1 & 4 & -3 \\ 1 & -3 & -2 \end{bmatrix}$

(c) $3A - B = \begin{bmatrix} 3 & 6 & 12 \\ -9 & 0 & -3 \\ 6 & 3 & 6 \end{bmatrix} - \begin{bmatrix} 2 & 0 & 0 \\ 1 & -4 & 3 \\ -1 & 3 & 2 \end{bmatrix} = \begin{bmatrix} 1 & 6 & 12 \\ -10 & 4 & -6 \\ 7 & 0 & 4 \end{bmatrix}$

Remark: It is often convenient to rewrite the scalar multiple cA by factoring c out of every entry in the matrix. For instance, in the following example, the scalar $1/2$ has been factored out of the matrix.

$$\begin{bmatrix} \frac{1}{2} & -\frac{3}{2} \\ \frac{5}{2} & \frac{1}{2} \end{bmatrix} = \frac{1}{2}\begin{bmatrix} 1 & -3 \\ 5 & 1 \end{bmatrix}$$

The properties of matrix addition and scalar multiplication are similar to those of addition and multiplication of real numbers, and we summarize them in the following list.

Properties of Matrix Addition and Scalar Multiplication

If A, B, and C are $m \times n$ matrices and c and d are scalars, then the following properties are true.

1. $A + B = B + A$	Commutative Property of Addition
2. $A + (B + C) = (A + B) + C$	Associative Property of Addition
3. $(cd)A = c(dA)$	Associative Property of Scalar Multiplication
4. $1A = A$	Scalar Identity
5. $c(A + B) = cA + cB$	Distributive Property
6. $(c + d)A = cA + dA$	Distributive Property

Note that the associative property for matrix addition allows us to write expressions like $A + B + C$ without ambiguity because the same sum occurs no matter how the matrices are grouped. In other words, we obtain the same sum whether we group $A + B + C$ as $(A + B) + C$ or as $A + (B + C)$. This same reasoning applies to sums of four or more matrices.

EXAMPLE 4 Addition of More than Two Matrices

By adding corresponding entries, we can obtain the following sum of four matrices.

$$\begin{bmatrix} 1 \\ 2 \\ -3 \end{bmatrix} + \begin{bmatrix} -1 \\ -1 \\ 2 \end{bmatrix} + \begin{bmatrix} 0 \\ 1 \\ 4 \end{bmatrix} + \begin{bmatrix} 2 \\ -3 \\ -2 \end{bmatrix} = \begin{bmatrix} 2 \\ -1 \\ 1 \end{bmatrix}$$

One important property of the addition of real numbers is that the number 0 is the additive identity. That is, $c + 0 = c$ for any real number c. For matrices, a similar property holds. That is, if A is an $m \times n$ matrix and O is the $m \times n$ **zero matrix** consisting entirely of zeros, then

$$A + O = A.$$

In other words, O is the **additive identity** for the set of all $m \times$

For example, the following matrix is the additive identity for th

2×3 matrices.

$$O = \begin{bmatrix} 0 & 0 & 0 \\ 0 & 0 & 0 \end{bmatrix}$$ *Zero 2×3 matrix*

Similarly, the additive identity for the set of all 3×4 matrices is

$$O = \begin{bmatrix} 0 & 0 & 0 & 0 \\ 0 & 0 & 0 & 0 \\ 0 & 0 & 0 & 0 \end{bmatrix}.$$ *Zero 3×4 matrix*

The algebra of real numbers and the algebra of matrices have similarities.* For example, compare the following solutions.

Real Numbers *(Solve for x.)*	*m × n Matrices* *(Solve for X.)*
$x + a = b$	$X + A = B$
$x + a + (-a) = b + (-a)$	$X + A + (-A) = B + (-A)$
$x + 0 = b - a$	$X + O = B - A$
$x = b - a$	$X = B - A$

The process of solving a matrix equation is demonstrated in Example 5.

EXAMPLE 5 Solving a Matrix Equation

Solve for X in the equation $3X + A = B$, where

$$A = \begin{bmatrix} 1 & -2 \\ 0 & 3 \end{bmatrix} \quad \text{and} \quad B = \begin{bmatrix} -3 & 4 \\ 2 & 1 \end{bmatrix}.$$

SOLUTION

We begin by solving the given equation for X to obtain

$$3X = B - A \implies X = \frac{1}{3}(B - A).$$

Now, using the given matrices A and B, we have

$$X = \frac{1}{3}\left(\begin{bmatrix} -3 & 4 \\ 2 & 1 \end{bmatrix} - \begin{bmatrix} 1 & -2 \\ 0 & 3 \end{bmatrix} \right) = \frac{1}{3}\begin{bmatrix} -4 & 6 \\ 2 & -2 \end{bmatrix} = \begin{bmatrix} -\frac{4}{3} & 2 \\ \frac{2}{3} & -\frac{2}{3} \end{bmatrix}.$$

*There are also some important differences, which we will discuss later.

27. $x + 10y - 3z = 2$
$5x - 3y + 4z = 0$
$2x + 4y = 6$

28. $9w - 3x + 20y + z = 13$
$12w - 8y = 5$

In Exercises 29–50, solve the system of equations. Use Gaussian elimination with back-substitution or Gauss-Jordan elimination.

29. $x + 2y = 7$
$2x + y = 8$

30. $2x + 6y = 16$
$2x + 3y = 7$

31. $-3x + 5y = -22$
$3x + 4y = 4$
$4x - 8y = 32$

32. $x + 2y = 0$
$x + y = 6$
$3x - 2y = 8$

33. $8x - 4y = 7$
$5x + 2y = 1$

34. $2x - y = -0.1$
$3x + 2y = 1.6$

35. $-x + 2y = 1.5$
$2x - 4y = 3$

36. $x - 3y = 5$
$-2x + 6y = -10$

37. $x - 3z = -2$
$3x + y - 2z = 5$
$2x + 2y + z = 4$

38. $2x - y + 3z = 24$
$2y - z = 14$
$7x - 5y = 6$

39. $x + y - 5z = 3$
$x - 2z = 1$
$2x - y - z = 0$

40. $2x + 3z = 3$
$4x - 3y + 7z = 5$
$8x - 9y + 15z = 9$

41. $x + 2y + z = 8$
$3x + 7y + 6z = 26$

42. $4x + 12y - 7z - 20w = 22$
$3x + 9y - 5z - 28w = 30$

43. $3x + 3y + 12z = 6$
$x + y + 4z = 2$
$2x + 5y + 20z = 10$
$-x + 2y + 8z = 4$

44. $2x + 10y + 2z = 6$
$x + 5y + 2z = 6$
$x + 5y + z = 3$
$-3x - 15y - 3z = -9$

45. $2x + y - z + 2w = -6$
$3x + 4y + w = 1$
$x + 5y + 2z + 6w = -3$
$5x + 2y - z - w = 3$

46. $x + 2y + 2z + 4w = 11$
$3x + 6y + 5z + 12w = 30$

47. $x + 2y = 0$
$-x - y = 0$

48. $x + 2y = 0$
$2x + 4y = 0$

49. $x + y + z = 0$
$2x + 3y + z = 0$
$3x + 5y + z = 0$

50. $x + 2y + z + 3w = 0$
$x - y + w = 0$
$y - z + 2w = 0$

In Exercises 51–56, solve the word problem by using a system of linear equations and matrices.

51. A small corporation borrowed $1,500,000 to expand its product line. Some of the money was borrowed at 8%, some at 9%, and some at 12%. How much was borrowed at each rate if the annual interest was $133,000 and the amount borrowed at 8% was 4 times the amount borrowed at 12%?

52. A grocer wishes to mix three kinds of nuts costing $3.50, $4.50, and $6.00 per pound, to obtain 50 pounds of a mixture costing $4.95 per pound. How many pounds of each variety should the grocer use if half of the mixture is composed of the two cheapest varieties?

53. Find a, b, and c for the quadratic function $f(x) = ax^2 + bx + c$, such that $f(1) = 1$, $f(-3) = 17$, and $f(2) = -\frac{1}{2}$. [*Hint:* See Example 8 in Section 10.3.]

54. Find a, b, c, and d for the cubic function $f(x) = ax^3 + bx^2 + cx + d$, such that $f(1) = 3$, $f(2) = 19$, $f(-1) = -11$, and $f(-2) = -33$. [*Hint:* See Example 8 in Section 10.3.]

55. Find D, E, and F such that $(1, 1)$, $(3, 3)$, and $(4, 2)$ are solution points of the equation $x^2 + y^2 + Dx + Ey + F = 0$. [*Hint:* See Example 8 in Section 10.3.]

56. The sum of three positive numbers is 33. Find the three numbers if the second is 3 greater than the first and the third is 4 times the first.

Matrices and Determinants

11.2 Operations with Matrices

In Section 11.1 we used matrices to solve systems of linear equations. Matrices, however, can do much more than that. There is a rich mathematical theory of matrices, and its applications are numerous. This section and the next introduce some fundamentals of matrix theory.

It is standard mathematical convention to represent matrices in any of the following three ways.

1. A matrix can be denoted by an upper-case letter such as

   $A, B, C, \ldots$.

2. A matrix can be denoted by a representative element enclosed in brackets, such as

   $[a_{ij}], [b_{ij}], [c_{ij}], \ldots$.

3. A matrix can be denoted by a rectangular array of numbers

$$A = [a_{ij}] = \begin{bmatrix} a_{11} & a_{12} & a_{13} & \cdots & a_{1n} \\ a_{21} & a_{22} & a_{23} & \cdots & a_{2n} \\ a_{31} & a_{32} & a_{33} & \cdots & a_{3n} \\ \vdots & \vdots & \vdots & & \vdots \\ a_{m1} & a_{m2} & a_{m3} & \cdots & a_{mn} \end{bmatrix}.$$

As mentioned in Section 11.1, the matrices in this text are *real matrices*. That is, their entries are real numbers.

Two matrices are said to be **equal** if their corresponding entries are equal.

Definition of Equality of Matrices

Two matrices $A = [a_{ij}]$ and $B = [b_{ij}]$ are **equal** if they have the same order $(m \times n)$ and

$a_{ij} = b_{ij}$

for $1 \leq i \leq m$ and $1 \leq j \leq n$.

EXAMPLE 1 *Equality of Matrices*

Solve for a_{11}, a_{12}, a_{21}, and a_{22} in the following matrix equation.

$$\begin{bmatrix} a_{11} & a_{12} \\ a_{21} & a_{22} \end{bmatrix} = \begin{bmatrix} 2 & -1 \\ -3 & 0 \end{bmatrix}$$

SOLUTION

Because two matrices are equal only if corresponding entries are equal, we conclude that

$$a_{11} = 2, \quad a_{12} = -1, \quad a_{21} = -3, \quad \text{and} \quad a_{22} = 0.$$

Matrix Addition and Scalar Multiplication

We **add** two matrices (of the same order) by adding their corresponding entries.

Definition of Matrix Addition

If $A = [a_{ij}]$ and $B = [b_{ij}]$ are matrices of order $m \times n$, then their **sum** is the $m \times n$ matrix given by

$$A + B = [a_{ij} + b_{ij}].$$

The sum of two matrices of different orders is undefined.

EXAMPLE 2 **Addition of Matrices**

(a) $\begin{bmatrix} -1 & 2 \\ 0 & 1 \end{bmatrix} + \begin{bmatrix} 1 & 3 \\ -1 & 2 \end{bmatrix} = \begin{bmatrix} -1+1 & 2+3 \\ 0-1 & 1+2 \end{bmatrix} = \begin{bmatrix} 0 & 5 \\ -1 & 3 \end{bmatrix}$

(b) $\begin{bmatrix} 0 & 1 & -2 \\ 1 & 2 & 3 \end{bmatrix} + \begin{bmatrix} 0 & 0 & 0 \\ 0 & 0 & 0 \end{bmatrix} = \begin{bmatrix} 0 & 1 & -2 \\ 1 & 2 & 3 \end{bmatrix}$

(c) $\begin{bmatrix} 1 \\ -3 \\ -2 \end{bmatrix} + \begin{bmatrix} -1 \\ 3 \\ 2 \end{bmatrix} = \begin{bmatrix} 0 \\ 0 \\ 0 \end{bmatrix}$

(d) The sum of

$$A = \begin{bmatrix} 2 & 1 & 0 \\ 4 & 0 & -1 \\ 3 & -2 & 2 \end{bmatrix} \quad \text{and} \quad B = \begin{bmatrix} 0 & 1 \\ -1 & 3 \\ 2 & 4 \end{bmatrix}$$

is undefined.

When working with matrices, we usually refer to numbers as **scalars.** In this text, scalars will always be real numbers. We multiply a matrix A by a scalar c by multiplying each entry in A by c.

Matrices and Determinants

Definition of Scalar Multiplication

If $A = [a_{ij}]$ is an $m \times n$ matrix and c is a scalar, then the **scalar multiple** of A by c is the $m \times n$ matrix given by

$$cA = [ca_{ij}].$$

We use $-A$ to represent the scalar product $(-1)A$. Moreover, if A and B are of the same order, $A - B$ represents the sum of A and $(-1)B$. That is,

$$A - B = A + (-1)B. \qquad \textit{Subtraction of matrices}$$

EXAMPLE 3 *Scalar Multiplication and Matrix Subtraction*

For the matrices

$$A = \begin{bmatrix} 1 & 2 & 4 \\ -3 & 0 & -1 \\ 2 & 1 & 2 \end{bmatrix} \quad \text{and} \quad B = \begin{bmatrix} 2 & 0 & 0 \\ 1 & -4 & 3 \\ -1 & 3 & 2 \end{bmatrix}$$

find the following.

(a) $3A$ 　　　　　　　　(b) $-B$ 　　　　　　　　(c) $3A - B$

SOLUTION

(a) $3A = 3\begin{bmatrix} 1 & 2 & 4 \\ -3 & 0 & -1 \\ 2 & 1 & 2 \end{bmatrix} = \begin{bmatrix} 3(1) & 3(2) & 3(4) \\ 3(-3) & 3(0) & 3(-1) \\ 3(2) & 3(1) & 3(2) \end{bmatrix} = \begin{bmatrix} 3 & 6 & 12 \\ -9 & 0 & -3 \\ 6 & 3 & 6 \end{bmatrix}$

(b) $-B = (-1)\begin{bmatrix} 2 & 0 & 0 \\ 1 & -4 & 3 \\ -1 & 3 & 2 \end{bmatrix} = \begin{bmatrix} -2 & 0 & 0 \\ -1 & 4 & -3 \\ 1 & -3 & -2 \end{bmatrix}$

(c) $3A - B = \begin{bmatrix} 3 & 6 & 12 \\ -9 & 0 & -3 \\ 6 & 3 & 6 \end{bmatrix} - \begin{bmatrix} 2 & 0 & 0 \\ 1 & -4 & 3 \\ -1 & 3 & 2 \end{bmatrix} = \begin{bmatrix} 1 & 6 & 12 \\ -10 & 4 & -6 \\ 7 & 0 & 4 \end{bmatrix}$

Remark: It is often convenient to rewrite the scalar multiple cA by factoring c out of every entry in the matrix. For instance, in the following example, the scalar $1/2$ has been factored out of the matrix.

$$\begin{bmatrix} \frac{1}{2} & -\frac{3}{2} \\ \frac{5}{2} & \frac{1}{2} \end{bmatrix} = \frac{1}{2}\begin{bmatrix} 1 & -3 \\ 5 & 1 \end{bmatrix}$$

The properties of matrix addition and scalar multiplication are similar to those of addition and multiplication of real numbers, and we summarize them in the following list.

Properties of Matrix Addition and Scalar Multiplication

If A, B, and C are $m \times n$ matrices and c and d are scalars, then the following properties are true.

1. $A + B = B + A$ *Commutative Property of Addition*
2. $A + (B + C) = (A + B) + C$ *Associative Property of Addition*
3. $(cd)A = c(dA)$ *Associative Property of Scalar Multiplication*
4. $1A = A$ *Scalar Identity*
5. $c(A + B) = cA + cB$ *Distributive Property*
6. $(c + d)A = cA + dA$ *Distributive Property*

Note that the associative property for matrix addition allows us to write expressions like $A + B + C$ without ambiguity because the same sum occurs no matter how the matrices are grouped. In other words, we obtain the same sum whether we group $A + B + C$ as $(A + B) + C$ or as $A + (B + C)$. This same reasoning applies to sums of four or more matrices.

EXAMPLE 4   *Addition of More than Two Matrices*

By adding corresponding entries, we can obtain the following sum of four matrices.

$$\begin{bmatrix} 1 \\ 2 \\ -3 \end{bmatrix} + \begin{bmatrix} -1 \\ -1 \\ 2 \end{bmatrix} + \begin{bmatrix} 0 \\ 1 \\ 4 \end{bmatrix} + \begin{bmatrix} 2 \\ -3 \\ -2 \end{bmatrix} = \begin{bmatrix} 2 \\ -1 \\ 1 \end{bmatrix}$$

One important property of the addition of real numbers is that the number 0 is the additive identity. That is, $c + 0 = c$ for any real number c. For matrices, a similar property holds. That is, if A is an $m \times n$ matrix and O is the $m \times n$ **zero matrix** consisting entirely of zeros, then

$$A + O = A.$$

Matrices and Determinants

In other words, O is the **additive identity** for the set of all $m \times n$ matrices. For example, the following matrix is the additive identity for the set of all 2×3 matrices.

$$O = \begin{bmatrix} 0 & 0 & 0 \\ 0 & 0 & 0 \end{bmatrix} \qquad \textit{Zero 2 × 3 matrix}$$

Similarly, the additive identity for the set of all 3×4 matrices is

$$O = \begin{bmatrix} 0 & 0 & 0 & 0 \\ 0 & 0 & 0 & 0 \\ 0 & 0 & 0 & 0 \end{bmatrix}. \qquad \textit{Zero 3 × 4 matrix}$$

The algebra of real numbers and the algebra of matrices have many similarities.* For example, compare the following solutions.

Real Numbers	*m × n Matrices*
(Solve for x.)	*(Solve for X.)*

$$
\begin{array}{cc}
x + a = b & X + A = B \\
x + a + (-a) = b + (-a) & X + A + (-A) = B + (-A) \\
x + 0 = b - a & X + O = B - A \\
x = b - a & X = B - A
\end{array}
$$

The process of solving a matrix equation is demonstrated in Example 5.

EXAMPLE 5 Solving a Matrix Equation

Solve for X in the equation $3X + A = B$, where

$$A = \begin{bmatrix} 1 & -2 \\ 0 & 3 \end{bmatrix} \quad \text{and} \quad B = \begin{bmatrix} -3 & 4 \\ 2 & 1 \end{bmatrix}.$$

SOLUTION

We begin by solving the given equation for X to obtain

$$3X = B - A \implies X = \frac{1}{3}(B - A).$$

Now, using the given matrices A and B, we have

$$X = \frac{1}{3}\left(\begin{bmatrix} -3 & 4 \\ 2 & 1 \end{bmatrix} - \begin{bmatrix} 1 & -2 \\ 0 & 3 \end{bmatrix} \right) = \frac{1}{3} \begin{bmatrix} -4 & 6 \\ 2 & -2 \end{bmatrix} = \begin{bmatrix} -\frac{4}{3} & 2 \\ \frac{2}{3} & -\frac{2}{3} \end{bmatrix}.$$

---

*There are also some important differences, which we will discuss later.

Operations with Matrices

Matrix Multiplication

The third basic matrix operation is **matrix multiplication.** At first glance the definition may seem unusual. You will see later, however, that this definition of the product of two matrices has many practical applications.

Definition of Matrix Multiplication

If $A = [a_{ij}]$ is an $m \times n$ matrix and $B = [b_{ij}]$ is an $n \times p$ matrix, then the **product** AB is an $m \times p$ matrix

$$AB = [c_{ij}]$$

where $c_{ij} = a_{i1}b_{1j} + a_{i2}b_{2j} + a_{i3}b_{3j} + \cdots + a_{in}b_{nj}$.

This definition indicates a *row-by-column* multiplication, where the entry in the ith row and jth column of the product AB is obtained by multiplying the entries in the ith row of A by the corresponding entries in the jth column of B and then adding the results. The following example illustrates the process.

EXAMPLE 6 **Finding the Product of Two Matrices**

Find the product AB where

$$A = \begin{bmatrix} -1 & 3 \\ 4 & -2 \\ 5 & 0 \end{bmatrix} \quad \text{and} \quad B = \begin{bmatrix} -3 & 2 \\ -4 & 1 \end{bmatrix}.$$

SOLUTION

First note that the product AB is defined because the number of columns of A is equal to the number of rows of B. Moreover, the product AB has order 3×2 and will take the form

$$\begin{bmatrix} -1 & 3 \\ 4 & -2 \\ 5 & 0 \end{bmatrix}\begin{bmatrix} -3 & 2 \\ -4 & 1 \end{bmatrix} = \begin{bmatrix} c_{11} & c_{12} \\ c_{21} & c_{22} \\ c_{31} & c_{32} \end{bmatrix}.$$

To find c_{11} (the entry in the first row and first column of the product), we multiply corresponding entries in the first row of A and the first column of B. That is,

$$c_{11} = (-1)(-3) + (3)(-4) = -9$$

$$\begin{bmatrix} -1 & 3 \\ 4 & -2 \\ 5 & 0 \end{bmatrix}\begin{bmatrix} -3 & 2 \\ -4 & 1 \end{bmatrix} = \begin{bmatrix} -9 & c_{12} \\ c_{21} & c_{22} \\ c_{31} & c_{32} \end{bmatrix}.$$

Matrices and Determinants

Similarly, to find c_{12}, we multiply corresponding entries in the first row of A and the second column of B to obtain

$$c_{12} = (-1)(2) + (3)(1) = 1$$

$$\begin{bmatrix} -1 & 3 \\ 4 & -2 \\ 5 & 0 \end{bmatrix} \begin{bmatrix} -3 & 2 \\ -4 & 1 \end{bmatrix} = \begin{bmatrix} -9 & 1 \\ c_{21} & c_{22} \\ c_{31} & c_{32} \end{bmatrix}.$$

Continuing this pattern produces the following results.

$$\begin{aligned} c_{21} &= (4)(-3) + (-2)(-4) = -4 \\ c_{22} &= (4)(2) + (-2)(1) = 6 \\ c_{31} &= (5)(-3) + (0)(-4) = -15 \\ c_{32} &= (5)(2) + (0)(1) = 10. \end{aligned}$$

Thus, the product is

$$AB = \begin{bmatrix} -1 & 3 \\ 4 & -2 \\ 5 & 0 \end{bmatrix} \begin{bmatrix} -3 & 2 \\ -4 & 1 \end{bmatrix} = \begin{bmatrix} -9 & 1 \\ -4 & 6 \\ -15 & 10 \end{bmatrix}.$$

Be sure you understand that for the product of two matrices to be defined, the number of columns of the first matrix must equal the number of rows of the second matrix. That is,

$$\begin{array}{ccc} A & B & = & AB. \\ m \times n & n \times p & & m \times p \end{array}$$

equal

order of AB

The general pattern for matrix multiplication is as follows. To obtain the element in the ith row and the jth column of the product AB, use the ith row of A and the jth column of B.

$$\begin{bmatrix} a_{11} & a_{12} & a_{13} & \cdots & a_{1n} \\ a_{21} & a_{22} & a_{23} & \cdots & a_{2n} \\ \vdots & \vdots & \vdots & & \vdots \\ a_{i1} & a_{i2} & a_{i3} & \cdots & a_{in} \\ \vdots & \vdots & \vdots & & \vdots \\ a_{m1} & a_{m2} & a_{m3} & \cdots & a_{mn} \end{bmatrix} \begin{bmatrix} b_{11} & b_{12} & \cdots & b_{1j} & \cdots & b_{1p} \\ b_{21} & b_{22} & \cdots & b_{2j} & \cdots & b_{2p} \\ b_{31} & b_{32} & \cdots & b_{3j} & \cdots & b_{3p} \\ \vdots & \vdots & & \vdots & & \vdots \\ b_{n1} & b_{n2} & \cdots & b_{nj} & \cdots & b_{np} \end{bmatrix} = \begin{bmatrix} c_{11} & c_{12} & \cdots & c_{1j} & \cdots & c_{1p} \\ c_{21} & c_{22} & \cdots & c_{2j} & \cdots & c_{2p} \\ \vdots & \vdots & & \vdots & & \vdots \\ c_{i1} & c_{i2} & \cdots & c_{ij} & \cdots & c_{ip} \\ \vdots & \vdots & & \vdots & & \vdots \\ c_{m1} & c_{m2} & \cdots & c_{mj} & \cdots & c_{mp} \end{bmatrix}$$

$$a_{i1}b_{1j} + a_{i2}b_{2j} + a_{i3}b_{3j} + \cdots + a_{in}b_{nj} = c_{ij}$$

Operations with Matrices

EXAMPLE 7 Matrix Multiplication

(a) $\begin{bmatrix} 1 & 0 & 3 \\ 2 & -1 & -2 \end{bmatrix} \begin{bmatrix} -2 & 4 & 2 \\ 1 & 0 & 0 \\ -1 & 1 & -1 \end{bmatrix} = \begin{bmatrix} -5 & 7 & -1 \\ -3 & 6 & 6 \end{bmatrix}$

         2 × 3   3 × 3   2 × 3

(b) $\begin{bmatrix} 3 & 4 \\ -2 & 5 \end{bmatrix} \begin{bmatrix} 1 & 0 \\ 0 & 1 \end{bmatrix} = \begin{bmatrix} 3 & 4 \\ -2 & 5 \end{bmatrix}$

     2 × 2   2 × 2   2 × 2

(c) $\begin{bmatrix} 1 & 2 \\ 1 & 1 \end{bmatrix} \begin{bmatrix} -1 & 2 \\ 1 & -1 \end{bmatrix} = \begin{bmatrix} 1 & 0 \\ 0 & 1 \end{bmatrix}$

     2 × 2   2 × 2   2 × 2

(d) $[1 \quad -2 \quad -3] \begin{bmatrix} 2 \\ -1 \\ 1 \end{bmatrix} = [1]$

     1 × 3   3 × 1   1 × 1

(e) $\begin{bmatrix} 2 \\ -1 \\ 1 \end{bmatrix} [1 \quad -2 \quad -3] = \begin{bmatrix} 2 & -4 & -6 \\ -1 & 2 & 3 \\ 1 & -2 & -3 \end{bmatrix}$

   3 × 1   1 × 3   3 × 3

(f) The product AB for

$$A = \begin{bmatrix} -2 & 1 \\ 1 & -3 \\ 1 & 4 \end{bmatrix} \quad \text{and} \quad B = \begin{bmatrix} -2 & 3 & 1 & 4 \\ 0 & 1 & -1 & 2 \\ 2 & -1 & 0 & 1 \end{bmatrix}$$

           3 × 2   3 × 4

is not defined (nor is the product BA).

Remark: In parts (d) and (e) of Example 7, note the difference between the two products. Matrix multiplication is not, in general, commutative. That is, for most matrices, $AB \neq BA$.

Properties of Matrix Multiplication

If A, B, and C are matrices and c is a scalar, then the following properties are true.

1. $A(BC) = (AB)C$ *Associative Property of Multiplication*
2. $A(B + C) = AB + AC$ *Distributive Property*
3. $(A + B)C = AC + BC$ *Distributive Property*
4. $c(AB) = (cA)B = A(cB)$

The $n \times n$ matrix that consists of 1's on its main diagonal and 0's elsewhere is called the **identity matrix of order** n and is denoted by

$$I_n = \begin{bmatrix} 1 & 0 & 0 & \cdots & 0 \\ 0 & 1 & 0 & \cdots & 0 \\ 0 & 0 & 1 & \cdots & 0 \\ \vdots & \vdots & \vdots & & \vdots \\ 0 & 0 & 0 & \cdots & 1 \end{bmatrix}. \qquad \textit{Identity matrix}$$

Note that an identity matrix must be *square*. When the order is understood to be n, we often denote I_n simply by I. If A is an $n \times n$ matrix, then the identity matrix has the property that

$$AI_n = A \qquad \text{and} \qquad I_nA = A.$$

For example,

$$\begin{bmatrix} 3 & -2 & 5 \\ 1 & 0 & 4 \\ -1 & 2 & -3 \end{bmatrix} \begin{bmatrix} 1 & 0 & 0 \\ 0 & 1 & 0 \\ 0 & 0 & 1 \end{bmatrix} = \begin{bmatrix} 3 & -2 & 5 \\ 1 & 0 & 4 \\ -1 & 2 & -3 \end{bmatrix}$$

and

$$\begin{bmatrix} 1 & 0 & 0 \\ 0 & 1 & 0 \\ 0 & 0 & 1 \end{bmatrix} \begin{bmatrix} 3 & -2 & 5 \\ 1 & 0 & 4 \\ -1 & 2 & -3 \end{bmatrix} = \begin{bmatrix} 3 & -2 & 5 \\ 1 & 0 & 4 \\ -1 & 2 & -3 \end{bmatrix}.$$

Applications

EXAMPLE 8 An Application of Matrix Multiplication

Two softball teams submit equipment lists to their sponsors.

	Women's Team	*Men's Team*
Bats	12	15
Balls	45	38
Gloves	15	17

Each bat costs \$18, each ball costs \$3, and each glove costs \$25. Use matrices to find the total cost of equipment for each team.

SOLUTION

The equipment lists can be written in matrix form as

$$E = \begin{bmatrix} 12 & 15 \\ 45 & 38 \\ 15 & 17 \end{bmatrix}$$

and the cost per item can be written in matrix form as

$$C = [18 \quad 3 \quad 25].$$

The total cost of equipment for each team is given by the product

$$18(12) + 3(45) + 25(15) = 726 \quad \text{(Women's team)}$$

$$CE = [18 \quad 3 \quad 25] \begin{bmatrix} 12 & 15 \\ 45 & 38 \\ 15 & 17 \end{bmatrix} = [\,726 \quad 809\,].$$

$$18(15) + 3(38) + 25(17) = 809 \quad \text{(Men's team)}$$

Thus, the total cost of equipment for the women's team is \$726, and the total cost of equipment for the men's team is \$809.

Another useful application of matrix multiplication is in representing a system of linear equations. Note how the system

$$a_{11}x_1 + a_{12}x_2 + a_{13}x_3 = b_1$$
$$a_{21}x_1 + a_{22}x_2 + a_{23}x_3 = b_2$$
$$a_{31}x_1 + a_{32}x_2 + a_{33}x_3 = b_3$$

can be written as the matrix equation $AX = B$, where A is the *coefficient matrix* of the system. That is, we can write

$$\begin{bmatrix} a_{11} & a_{12} & a_{13} \\ a_{21} & a_{22} & a_{23} \\ a_{31} & a_{32} & a_{33} \end{bmatrix} \begin{bmatrix} x_1 \\ x_2 \\ x_3 \end{bmatrix} = \begin{bmatrix} b_1 \\ b_2 \\ b_3 \end{bmatrix}.$$

$$A \qquad\qquad X \quad = \quad B$$

EXAMPLE 9   *Solving a System of Linear Equations*

Solve the matrix equation $AX = B$ for X, where

Coefficient matix   Constant matrix

$$A = \begin{bmatrix} 1 & -2 & 1 \\ 0 & 1 & 2 \\ 2 & 3 & -2 \end{bmatrix} \quad \text{and} \quad B = \begin{bmatrix} -4 \\ 4 \\ 2 \end{bmatrix}.$$

SOLUTION

As a system of linear equations, $AX = B$ looks like

$$x_1 - 2x_2 + x_3 = -4$$
$$x_2 + 2x_3 = 4$$
$$2x_1 + 3x_2 - 2x_3 = 2.$$

Matrices and Determinants

Using Gauss-Jordan elimination on the augmented matrix of this system, we obtain

$$\begin{bmatrix} 1 & 0 & 0 & \vdots & -1 \\ 0 & 1 & 0 & \vdots & 2 \\ 0 & 0 & 1 & \vdots & 1 \end{bmatrix}.$$

Thus, the solution of the system of linear equations is $x_1 = -1$, $x_2 = 2$, and $x_3 = 1$, and the solution of the matrix equation is

$$X = \begin{bmatrix} x_1 \\ x_2 \\ x_3 \end{bmatrix} = \begin{bmatrix} -1 \\ 2 \\ 1 \end{bmatrix}.$$

WARM UP

Evaluate the expressions.

1. $-3\left(-\frac{5}{6}\right) + 10\left(-\frac{3}{4}\right)$

2. $-22\left(\frac{5}{2}\right) + 6(8)$

Determine whether the matrices are in *reduced row-echelon form.*

3. $\begin{bmatrix} 0 & 1 & 0 & -5 \\ 1 & 0 & 3 & 2 \\ 0 & 0 & 1 & 0 \end{bmatrix}$

4. $\begin{bmatrix} 1 & 0 & 0 & 2 & 3 \\ 0 & 0 & 0 & 0 & 0 \\ 0 & 1 & 1 & 3 & 10 \end{bmatrix}$

Write the augmented matrix for each system of linear equations.

5. $\begin{aligned} -5x + 10y &= 12 \\ 7x - 3y &= 0 \\ -x + 7y &= 25 \end{aligned}$

6. $\begin{aligned} 10x + 15y - 9z &= 42 \\ 6x - 5y &= 0 \end{aligned}$

Solve the systems of linear equations represented by the augmented matrices.

7. $\begin{bmatrix} 1 & 0 & \vdots & 0 \\ 0 & 1 & \vdots & 2 \end{bmatrix}$

8. $\begin{bmatrix} 1 & 0 & -1 & \vdots & 2 \\ 0 & 1 & 1 & \vdots & 3 \end{bmatrix}$

9. $\begin{bmatrix} 1 & 2 & 1 & \vdots & 0 \\ 0 & 0 & 1 & \vdots & -1 \\ 0 & 0 & 0 & \vdots & 0 \end{bmatrix}$

10. $\begin{bmatrix} 1 & -1 & 0 & \vdots & 3 \\ 0 & 1 & -2 & \vdots & 1 \\ 0 & 0 & 1 & \vdots & -1 \end{bmatrix}$

EXERCISES 11.2

In Exercises 1–4, find x and y.

1. $\begin{bmatrix} x & -2 \\ 7 & y \end{bmatrix} = \begin{bmatrix} -4 & -2 \\ 7 & 22 \end{bmatrix}$

2. $\begin{bmatrix} -5 & x \\ y & 8 \end{bmatrix} = \begin{bmatrix} -5 & 13 \\ 12 & 8 \end{bmatrix}$

3. $\begin{bmatrix} 16 & 4 & 5 & 4 \\ -3 & 13 & 15 & 6 \\ 0 & 2 & 4 & 0 \end{bmatrix} = \begin{bmatrix} 16 & 4 & 2x+1 & 4 \\ -3 & 13 & 15 & 3x \\ 0 & 2 & 3y-5 & 0 \end{bmatrix}$

4. $\begin{bmatrix} x+2 & 8 & -3 \\ 1 & 2y & 2x \\ 7 & -2 & y+2 \end{bmatrix} = \begin{bmatrix} 2x+6 & 8 & -3 \\ 1 & 18 & -8 \\ 7 & -2 & 11 \end{bmatrix}$

In Exercises 5–10, find (a) A + B, (b) A − B, (c) 3A, and (d) 3A − 2B.

5. $A = \begin{bmatrix} 1 & -1 \\ 2 & -1 \end{bmatrix}$, $B = \begin{bmatrix} 2 & -1 \\ -1 & 8 \end{bmatrix}$

6. $A = \begin{bmatrix} 1 & 2 \\ 2 & 1 \end{bmatrix}$, $B = \begin{bmatrix} -3 & -2 \\ 4 & 2 \end{bmatrix}$

7. $A = \begin{bmatrix} 6 & -1 \\ 2 & 4 \\ -3 & 5 \end{bmatrix}$, $B = \begin{bmatrix} 1 & 4 \\ -1 & 5 \\ 1 & 10 \end{bmatrix}$

8. $A = \begin{bmatrix} 2 & 1 & 1 \\ -1 & -1 & 4 \end{bmatrix}$, $B = \begin{bmatrix} 2 & -3 & 4 \\ -3 & 1 & -2 \end{bmatrix}$

9. $A = \begin{bmatrix} 2 & 2 & -1 & 0 & 1 \\ 1 & 1 & -2 & 0 & -1 \end{bmatrix}$,

$B = \begin{bmatrix} 1 & 1 & -1 & 1 & 0 \\ -3 & 4 & 9 & -6 & -7 \end{bmatrix}$

10. $A = \begin{bmatrix} 3 \\ 2 \\ -1 \end{bmatrix}$, $B = \begin{bmatrix} -4 \\ 6 \\ 2 \end{bmatrix}$

In Exercises 11–16, find (a) AB, (b) BA, and if possible (c) A^2. [Note: $A^2 = AA$.]

11. $A = \begin{bmatrix} 1 & 2 \\ 4 & 2 \end{bmatrix}$, $B = \begin{bmatrix} 2 & -1 \\ -1 & 8 \end{bmatrix}$

12. $A = \begin{bmatrix} 2 & -1 \\ 1 & 4 \end{bmatrix}$, $B = \begin{bmatrix} 0 & 0 \\ 3 & -3 \end{bmatrix}$

13. $A = \begin{bmatrix} 3 & -1 \\ 1 & 3 \end{bmatrix}$, $B = \begin{bmatrix} 1 & -3 \\ 3 & 1 \end{bmatrix}$

14. $A = \begin{bmatrix} 1 & -1 \\ 1 & 1 \end{bmatrix}$, $B = \begin{bmatrix} 1 & 3 \\ -3 & 1 \end{bmatrix}$

15. $A = \begin{bmatrix} 1 & -1 & 7 \\ 2 & -1 & 8 \\ 3 & 1 & -1 \end{bmatrix}$, $B = \begin{bmatrix} 1 & 1 & 2 \\ 2 & 1 & 1 \\ 1 & -3 & 2 \end{bmatrix}$

16. $A = \begin{bmatrix} 3 & 2 & 1 \end{bmatrix}$, $B = \begin{bmatrix} 2 \\ 3 \\ 0 \end{bmatrix}$

In Exercises 17–24, find AB, if possible.

17. $A = \begin{bmatrix} 2 & 1 \\ -3 & 4 \\ 1 & 6 \end{bmatrix}$, $B = \begin{bmatrix} 0 & -1 & 0 \\ 4 & 0 & 2 \\ 8 & -1 & 7 \end{bmatrix}$

18. $A = \begin{bmatrix} 0 & -1 & 0 \\ 4 & 0 & 2 \\ 8 & -1 & 7 \end{bmatrix}$, $B = \begin{bmatrix} 2 & 1 \\ -3 & 4 \\ 1 & 6 \end{bmatrix}$

19. $A = \begin{bmatrix} -1 & 3 \\ 4 & -5 \\ 0 & 2 \end{bmatrix}$, $B = \begin{bmatrix} 1 & 2 \\ 0 & 7 \end{bmatrix}$

20. $A = \begin{bmatrix} 1 & 0 & 0 \\ 0 & 4 & 0 \\ 0 & 0 & -2 \end{bmatrix}$, $B = \begin{bmatrix} 3 & 0 & 0 \\ 0 & -1 & 0 \\ 0 & 0 & 5 \end{bmatrix}$

21. $A = \begin{bmatrix} 5 & 0 & 0 \\ 0 & -8 & 0 \\ 0 & 0 & 7 \end{bmatrix}$, $B = \begin{bmatrix} \frac{1}{5} & 0 & 0 \\ 0 & -\frac{1}{8} & 0 \\ 0 & 0 & \frac{1}{2} \end{bmatrix}$

22. $A = \begin{bmatrix} 0 & 0 & 5 \\ 0 & 0 & -3 \\ 0 & 0 & 4 \end{bmatrix}$, $B = \begin{bmatrix} 6 & -11 & 4 \\ 8 & 16 & 4 \\ 0 & 0 & 0 \end{bmatrix}$

23. $A = \begin{bmatrix} 6 \\ -2 \\ 1 \\ 6 \end{bmatrix}$, $B = \begin{bmatrix} 10 & 12 \end{bmatrix}$

24. $A = \begin{bmatrix} 1 & 0 & 3 & -2 & 4 \\ 6 & 13 & 8 & -17 & 10 \end{bmatrix}$, $B = \begin{bmatrix} 1 & 6 \\ 4 & 2 \end{bmatrix}$

In Exercises 25–28, solve for X given

$A = \begin{bmatrix} -2 & -1 \\ 1 & 0 \\ 3 & -4 \end{bmatrix}$ and $B = \begin{bmatrix} 0 & 3 \\ 2 & 0 \\ -4 & -1 \end{bmatrix}$.

25. $X = 3A - 2B$

26. $2X = 2A - B$

27. $2X + 3A = B$

28. $2A + 4B = -2X$

In Exercises 29–32, find matrices A, X, and B such that the given system of linear equations can be written as the matrix equation AX = B. Solve the system of equations.

29. $-x + y = 4$
$-2x + y = 0$

30. $2x + 3y = 5$
$x + 4y = 10$

31. $x - 2y + 3z = 9$
$-x + 3y - z = -6$
$2x - 5y + 5z = 17$

32. $x + y - 3z = -1$
$-x + 2y = 1$
$-y + z = 0$

33. If a, b, and c are real numbers such that $c \neq 0$ and $ac = bc$, then $a = b$. (See Section 1.1.) However, if A, B, and C are matrices such that $AC = BC$, then A is *not* necessarily equal to B. Illustrate this using the following matrices.

$$A = \begin{bmatrix} 1 & 2 & 3 \\ 0 & 5 & 4 \\ 3 & -2 & 1 \end{bmatrix}, \quad B = \begin{bmatrix} 4 & -6 & 3 \\ 5 & 4 & 4 \\ -1 & 0 & 1 \end{bmatrix},$$

$$C = \begin{bmatrix} 0 & 0 & 0 \\ 0 & 0 & 0 \\ 4 & -2 & 3 \end{bmatrix}$$

34. If a and b are real numbers such that $ab = 0$, then $a = 0$ or $b = 0$. (See Section 1.1.) However, if A and B are matrices such that $AB = O$, then it is *not* necessarily true that $A = O$ or $B = O$. Illustrate this using the following matrices.

$$A = \begin{bmatrix} 3 & 3 \\ 4 & 4 \end{bmatrix}, \quad B = \begin{bmatrix} 1 & -1 \\ -1 & 1 \end{bmatrix}$$

35. If A and B are real numbers, then the following equations are true. If A and B are $n \times n$ matrices, are they true? Give reasons for your answers.
(a) $(A + B)(A - B) = A^2 - B^2$
(b) $(A + B)(A + B) = A^2 + 2AB + B^2$

36. A certain corporation has four factories, each of which manufactures two products. The number of units of product i produced at factory j in one day is represented by a_{ij} in the matrix

$$A = \begin{bmatrix} 100 & 90 & 70 & 30 \\ 40 & 20 & 60 & 60 \end{bmatrix}.$$

Find the production levels if production is increased by 10%. (*Hint:* Since an increase of 10% corresponds to 100% + 10%, multiply the given matrix by 1.10.)

37. A fruit grower raises two crops, which are shipped to three outlets. The number of units of product i that are shipped to outlet j is represented by a_{ij} in the matrix

$$A = \begin{bmatrix} 100 & 75 & 75 \\ 125 & 150 & 100 \end{bmatrix}.$$

The profit per unit is represented by the matrix

$B = [\$3.75 \quad \$7.00]$.

Find the product BA, and state what each entry of the product represents.

38. The matrix

$$P = \begin{matrix} & \text{To } R & \text{To } D & \text{To } I \\ \text{From } R & \\ \text{From } D & \\ \text{From } I & \end{matrix} \begin{bmatrix} 0.75 & 0.15 & 0.10 \\ 0.20 & 0.60 & 0.20 \\ 0.30 & 0.40 & 0.30 \end{bmatrix}$$

is called a stochastic matrix. Each entry p_{ij} ($i \neq j$) represents the proportion of the voting population that changes from party i to party j, and p_{ii} represents the proportion that remains loyal to the party from one election to the next. Find P^2. (This matrix gives the transition probabilities from the first election to the third.)

[Optional] In Exercises 39–42, find $f(A) = a_0 I_n + a_1 A + a_2 A^2 + \cdots + a_n A^n$.

39. $f(x) = x^2 - 5x + 2$, $A = \begin{bmatrix} 2 & 0 \\ 4 & 5 \end{bmatrix}$

40. $f(x) = x^2 - 7x + 6$, $A = \begin{bmatrix} 5 & 4 \\ 1 & 2 \end{bmatrix}$

41. $f(x) = x^3 - 10x^2 + 31x - 30$, $A = \begin{bmatrix} 3 & 1 & 4 \\ 0 & 2 & 6 \\ 0 & 0 & 5 \end{bmatrix}$

42. $f(x) = x^2 - 10x + 24$, $A = \begin{bmatrix} 8 & -4 \\ 2 & 2 \end{bmatrix}$

11.3 The Inverse of a Matrix

This section further develops the algebra of matrices to include the solution of matrix equations involving matrix multiplication. To begin, consider the real number equation $ax = b$. To solve this equation for x, we multiply both sides of the equation by a^{-1} (provided $a \neq 0$).

$$ax = b$$
$$(a^{-1}a)x = a^{-1}b$$
$$(1)x = a^{-1}b$$
$$x = a^{-1}b$$

The number a^{-1} is called the *multiplicative inverse* of a because it has the property that $a^{-1}a = 1$. The definition of a multiplicative inverse of a matrix is similar.

Definition of an Inverse of a Matrix

Let A be a square matrix of order n. If there exists a matrix A^{-1} such that

$$AA^{-1} = I_n = A^{-1}A$$

then A^{-1} is called the **inverse** of A.

Remark: The symbol A^{-1} is read "A inverse."

If a matrix A has an inverse, then A is called **invertible** (or **nonsingular**); otherwise, A is called **singular.** A nonsquare matrix cannot have an inverse. To see this, note that if A is of order $m \times n$ and B is of order $n \times m$ (where $m \neq n$), then the products AB and BA are of different orders and could therefore not be equal to each other. Indeed, not all square matrices possess inverses (see Example 4). If, however, a matrix does possess an inverse, then that inverse is unique.

EXAMPLE 1 The Inverse of a Matrix

Show that B is the inverse of A, where

$$A = \begin{bmatrix} -1 & 2 \\ -1 & 1 \end{bmatrix} \quad \text{and} \quad B = \begin{bmatrix} 1 & -2 \\ 1 & -1 \end{bmatrix}.$$

Matrices and Determinants

SOLUTION

Using the definition of an inverse matrix, we can show that B is the inverse of A by showing that $AB = I = BA$ as follows.

$$AB = \begin{bmatrix} -1 & 2 \\ -1 & 1 \end{bmatrix} \begin{bmatrix} 1 & -2 \\ 1 & -1 \end{bmatrix} = \begin{bmatrix} -1+2 & 2-2 \\ -1+1 & 2-1 \end{bmatrix} = \begin{bmatrix} 1 & 0 \\ 0 & 1 \end{bmatrix}$$

$$BA = \begin{bmatrix} 1 & -2 \\ 1 & -1 \end{bmatrix} \begin{bmatrix} -1 & 2 \\ -1 & 1 \end{bmatrix} = \begin{bmatrix} -1+2 & 2-2 \\ -1+1 & 2-1 \end{bmatrix} = \begin{bmatrix} 1 & 0 \\ 0 & 1 \end{bmatrix}$$

Remark: Recall that it is not always true that $AB = BA$, even if both products are defined. However, if A and B are both square matrices and $AB = I_n$, then it can be shown that $BA = I_n$. Hence, in Example 1, we needed only to check that $AB = I_2$.

The following example shows how to use a system of equations to find the inverse of a matrix.

EXAMPLE 2   *Finding the Inverse of a Matrix*

Find the inverse of the matrix

$$A = \begin{bmatrix} 1 & 4 \\ -1 & -3 \end{bmatrix}.$$

SOLUTION

To find the inverse of A, we try to solve the matrix equation $AX = I$ for X.

$$\overset{A}{\begin{bmatrix} 1 & 4 \\ -1 & -3 \end{bmatrix}} \overset{X}{\begin{bmatrix} x_{11} & x_{12} \\ x_{21} & x_{22} \end{bmatrix}} = \overset{I}{\begin{bmatrix} 1 & 0 \\ 0 & 1 \end{bmatrix}}$$

$$\begin{bmatrix} x_{11} + 4x_{21} & x_{12} + 4x_{22} \\ -x_{11} - 3x_{21} & -x_{12} - 3x_{22} \end{bmatrix} = \begin{bmatrix} 1 & 0 \\ 0 & 1 \end{bmatrix}$$

Equating corresponding entries, we obtain the following two systems of linear equations.

$$\begin{aligned} x_{11} + 4x_{21} &= 1 \\ -x_{11} - 3x_{21} &= 0 \end{aligned} \qquad \begin{aligned} x_{12} + 4x_{22} &= 0 \\ -x_{12} - 3x_{22} &= 1 \end{aligned}$$

From the first system we find that $x_{11} = -3$ and $x_{21} = 1$, and from the second system we find that $x_{12} = -4$ and $x_{22} = 1$. Therefore, the inverse of A is

$$X = A^{-1} = \begin{bmatrix} -3 & -4 \\ 1 & 1 \end{bmatrix}.$$

Try using matrix multiplication to check this result.

In Example 2, note that the two systems of linear equations have the *same coefficient matrix A*. Rather than solve the two systems represented by

$$\begin{bmatrix} 1 & 4 & \vdots & 1 \\ -1 & -3 & \vdots & 0 \end{bmatrix}$$

and

$$\begin{bmatrix} 1 & 4 & \vdots & 0 \\ -1 & -3 & \vdots & 1 \end{bmatrix}$$

separately, we can solve them *simultaneously* by **adjoining** the identity matrix to the coefficient matrix to obtain

$$\begin{matrix} A & & I \end{matrix}$$
$$\begin{bmatrix} 1 & 4 & \vdots & 1 & 0 \\ -1 & -3 & \vdots & 0 & 1 \end{bmatrix}.$$

Then, applying Gauss-Jordan elimination to this matrix, we can solve *both* systems with a single elimination process as follows.

$$\begin{bmatrix} 1 & 4 & \vdots & 1 & 0 \\ -1 & -3 & \vdots & 0 & 1 \end{bmatrix} \quad \begin{matrix} R_1 + R_2 \rightarrow \end{matrix} \begin{bmatrix} 1 & 4 & \vdots & 1 & 0 \\ 0 & 1 & \vdots & 1 & 1 \end{bmatrix}$$
$$\begin{matrix} -4R_2 + R_1 \rightarrow \end{matrix} \begin{bmatrix} 1 & 0 & \vdots & -3 & -4 \\ 0 & 1 & \vdots & 1 & 1 \end{bmatrix}$$

Thus, from the "doubly augmented" matrix $[A \vdots I]$ we obtained the matrix $[I \vdots A^{-1}]$.

$$\begin{matrix} A & & I & & I & & A^{-1} \end{matrix}$$
$$\begin{bmatrix} 1 & 4 & \vdots & 1 & 0 \\ -1 & -3 & \vdots & 0 & 1 \end{bmatrix} \Longrightarrow \begin{bmatrix} 1 & 0 & \vdots & -3 & -4 \\ 0 & 1 & \vdots & 1 & 1 \end{bmatrix}.$$

This procedure (or algorithm) works for an arbitrary square matrix that has an inverse.

Finding the Inverse of a Matrix by Gauss-Jordan Elimination

Let A be a square matrix of order n.

1. Write the $n \times 2n$ matrix that consists of the given matrix A on the left and the $n \times n$ identity matrix I on the right to obtain $[A \vdots I]$. Note that we separate the matrices A and I by a dotted line. We call this process **adjoining** the matrices A and I.
2. If possible, row reduce A to I using elementary row operations on the *entire* matrix $[A \vdots I]$. The result will be the matrix $[I \vdots A^{-1}]$. If this is not possible, then A is not invertible.
3. Check your work by multiplying to see that $AA^{-1} = I = A^{-1}A$.

Matrices and Determinants

EXAMPLE 3 *Finding the Inverse of a Matrix*

Find the inverse of the following matrix.

$$A = \begin{bmatrix} 1 & -1 & 0 \\ 1 & 0 & -1 \\ 6 & -2 & -3 \end{bmatrix}$$

SOLUTION

We begin by adjoining the identity matrix to A to form the matrix

$$[A \vdots I] = \begin{bmatrix} 1 & -1 & 0 & \vdots & 1 & 0 & 0 \\ 1 & 0 & -1 & \vdots & 0 & 1 & 0 \\ 6 & -2 & -3 & \vdots & 0 & 0 & 1 \end{bmatrix}.$$

Now, using elementary row operations, we rewrite this matrix in the form $[I \vdots A^{-1}]$ as follows.

$$\begin{bmatrix} 1 & -1 & 0 & \vdots & 1 & 0 & 0 \\ 1 & 0 & -1 & \vdots & 0 & 1 & 0 \\ 6 & -2 & -3 & \vdots & 0 & 0 & 1 \end{bmatrix}$$

$$\begin{matrix} \\ -R_1 + R_2 \rightarrow \\ -6R_1 + R_3 \rightarrow \end{matrix} \begin{bmatrix} 1 & -1 & 0 & \vdots & 1 & 0 & 0 \\ 0 & 1 & -1 & \vdots & -1 & 1 & 0 \\ 0 & 4 & -3 & \vdots & -6 & 0 & 1 \end{bmatrix}$$

$$\begin{matrix} R_2 + R_1 \rightarrow \\ \\ -4R_2 + R_3 \rightarrow \end{matrix} \begin{bmatrix} 1 & 0 & -1 & \vdots & 0 & 1 & 0 \\ 0 & 1 & -1 & \vdots & -1 & 1 & 0 \\ 0 & 0 & 1 & \vdots & -2 & -4 & 1 \end{bmatrix}$$

$$\begin{matrix} R_3 + R_1 \rightarrow \\ R_3 + R_2 \rightarrow \\ \end{matrix} \begin{bmatrix} 1 & 0 & 0 & \vdots & -2 & -3 & 1 \\ 0 & 1 & 0 & \vdots & -3 & -3 & 1 \\ 0 & 0 & 1 & \vdots & -2 & -4 & 1 \end{bmatrix}$$

Therefore, the matrix A is invertible and its inverse is

$$A^{-1} = \begin{bmatrix} -2 & -3 & \vdots & 1 \\ -3 & -3 & \vdots & 1 \\ -2 & -4 & \vdots & 1 \end{bmatrix}.$$

Try confirming this by multiplying A and A^{-1} to obtain I.

The process shown in Example 3 applies to any $n \times n$ matrix A. If A has an inverse, this process will find it. On the other hand, if A does not have an inverse, the process will tell us that. The next example shows what happens when we apply the process to a singular matrix.

EXAMPLE 4 A Singular Matrix

Find the inverse of A, if it exists.

$$A = \begin{bmatrix} 1 & 2 & 0 \\ 3 & -1 & 2 \\ -2 & 3 & -2 \end{bmatrix}$$

SOLUTION

We adjoin the identity matrix to A to form

$$[A \vdots I] = \begin{bmatrix} 1 & 2 & 0 & \vdots & 1 & 0 & 0 \\ 3 & -1 & 2 & \vdots & 0 & 1 & 0 \\ -2 & 3 & -2 & \vdots & 0 & 0 & 1 \end{bmatrix}$$

and apply Gauss-Jordan elimination as follows.

$$\begin{bmatrix} 1 & 2 & 0 & \vdots & 1 & 0 & 0 \\ 3 & -1 & 2 & \vdots & 0 & 1 & 0 \\ -2 & 3 & -2 & \vdots & 0 & 0 & 1 \end{bmatrix} \begin{matrix} \\ -3R_1 + R_2 \rightarrow \\ 2R_1 + R_3 \rightarrow \end{matrix} \begin{bmatrix} 1 & 2 & 0 & \vdots & 1 & 0 & 0 \\ 0 & -7 & 2 & \vdots & -3 & 1 & 0 \\ 0 & 7 & -2 & \vdots & 2 & 0 & 1 \end{bmatrix}$$

$$\begin{matrix} \\ \\ R_2 + R_3 \rightarrow \end{matrix} \begin{bmatrix} 1 & 2 & 0 & \vdots & 1 & 0 & 0 \\ 0 & -7 & 2 & \vdots & -3 & 1 & 0 \\ 0 & 0 & 0 & \vdots & -1 & 1 & 1 \end{bmatrix}$$

Because the "A portion" of the matrix has a row of zeros, we conclude that it is not possible to rewrite the matrix $[A \vdots I]$ in the form $[I \vdots A^{-1}]$. This means that A has no inverse.

Using Gauss-Jordan elimination to find the inverse of a matrix works well (even as a computer technique) for matrices of order 3×3 or greater. For 2×2 matrices, however, many people prefer to use a formula for the inverse, rather than find the inverse by Gauss-Jordan elimination. This simple formula, which *only* works for 2×2 matrices, is explained as follows. If A is a 2×2 matrix given by

$$A = \begin{bmatrix} a & b \\ c & d \end{bmatrix}$$

then A is invertible if and only if $ad - bc \neq 0$. Moreover, if $ad - bc \neq 0$, then the inverse is given by

$$A^{-1} = \frac{1}{ad - bc} \begin{bmatrix} d & -b \\ -c & a \end{bmatrix}.$$

Try verifying this inverse by multiplication.

Remark: The denominator $ad - bc$ is called the **determinant** of the 2×2 matrix A. We will study determinants in detail in the next two sections.

EXAMPLE 5 Finding the Inverse of a 2 × 2 Matrix

If possible, find the inverses of the following matrices.

(a) $A = \begin{bmatrix} 3 & -1 \\ -2 & 2 \end{bmatrix}$
(b) $B = \begin{bmatrix} 3 & -1 \\ -6 & 2 \end{bmatrix}$

SOLUTION

(a) For the matrix A, we apply the formula for the inverse of a 2×2 matrix to obtain $ad - bc = (3)(2) - (-1)(-2) = 4$. Since this quantity is not zero, the inverse is formed by interchanging the entries on the main diagonal and changing the sign of the other two entries as follows.

$$A^{-1} = \frac{1}{4}\begin{bmatrix} 2 & 1 \\ 2 & 3 \end{bmatrix} = \begin{bmatrix} \frac{1}{2} & \frac{1}{4} \\ \frac{1}{2} & \frac{3}{4} \end{bmatrix}$$

(b) For the matrix B, we have $ad - bc = (3)(2) - (-1)(-6) = 0$, which means that B is not invertible.

Systems of Equations

We know that a system of linear equations can have exactly one solution, an infinite number of solutions, or no solution. For *square* systems (those having the same number of equations as variables), we can use the following theorem to determine whether the system has a unique solution.

Systems of Equations with Unique Solutions

If A is an invertible matrix, then the system of linear equations represented by $AX = B$ has a unique solution given by

$$X = A^{-1}B.$$

Solving a system of linear equations by finding the inverse of the coefficient matrix is not very efficient. That is, it is usually more work to find A^{-1} and then multiply by B than simply to solve the system using Gaussian elimination with back-substitution. One case in which you might consider using an inverse matrix as a computational technique would be with *several* systems of linear equations, all of which have the same coefficient matrix A. In such a case, you could find the inverse matrix once and then solve each system by computing the product $A^{-1}B$. This is demonstrated in Example 6.

EXAMPLE 6 Solving a System of Equations Using an Inverse

Use an inverse matrix to solve the following systems.

(a) $2x + 3y + z = -1$
$\quad 3x + 3y + z = 1$
$\quad 2x + 4y + z = -2$

(b) $2x + 3y + z = 4$
$\quad 3x + 3y + z = 8$
$\quad 2x + 4y + z = 5$

(c) $2x + 3y + z = 0$
$\quad 3x + 3y + z = 0$
$\quad 2x + 4y + z = 0$

SOLUTION

We first note that the coefficient matrix for each system is

$$A = \begin{bmatrix} 2 & 3 & 1 \\ 3 & 3 & 1 \\ 2 & 4 & 1 \end{bmatrix}.$$

Using Gauss-Jordan elimination, we find A^{-1} to be

$$A^{-1} = \begin{bmatrix} -1 & 1 & 0 \\ -1 & 0 & 1 \\ 6 & -2 & -3 \end{bmatrix}.$$

To solve each system, we use matrix multiplication as follows.

(a) $X = A^{-1}B = \begin{bmatrix} -1 & 1 & 0 \\ -1 & 0 & 1 \\ 6 & -2 & -3 \end{bmatrix} \begin{bmatrix} -1 \\ 1 \\ -2 \end{bmatrix} = \begin{bmatrix} 2 \\ -1 \\ -2 \end{bmatrix}$

The solution is $x = 2$, $y = -1$, and $z = -2$.

(b) $X = A^{-1}B = \begin{bmatrix} -1 & 1 & 0 \\ -1 & 0 & 1 \\ 6 & -2 & -3 \end{bmatrix} \begin{bmatrix} 4 \\ 8 \\ 5 \end{bmatrix} = \begin{bmatrix} 4 \\ 1 \\ -7 \end{bmatrix}$

The solution is $x = 4$, $y = 1$, and $z = -7$.

(c) $X = A^{-1}B = \begin{bmatrix} -1 & 1 & 0 \\ -1 & 0 & 1 \\ 6 & -2 & -3 \end{bmatrix} \begin{bmatrix} 0 \\ 0 \\ 0 \end{bmatrix} = \begin{bmatrix} 0 \\ 0 \\ 0 \end{bmatrix}$

The solution is trivial: $x = 0$, $y = 0$, and $z = 0$.

WARM UP

Perform the indicated matrix operations.

1. $4 \begin{bmatrix} 1 & 6 \\ 0 & -4 \\ 12 & 2 \end{bmatrix}$

2. $\frac{1}{2} \begin{bmatrix} 11 & 10 & 48 \\ 1 & 0 & 16 \\ 0 & 2 & 8 \end{bmatrix}$

3. $\begin{bmatrix} 1 & -10 & 3 \\ 4 & 1 & 0 \end{bmatrix} - 2 \begin{bmatrix} 3 & -4 & 8 \\ 0 & 7 & 1 \end{bmatrix}$

4. $\begin{bmatrix} 5 & 20 \\ -7 & 15 \end{bmatrix} - 3 \begin{bmatrix} 6 & 3 \\ 4 & -2 \end{bmatrix}$

5. $\begin{bmatrix} 1 & -2 \\ -1 & 3 \end{bmatrix} \begin{bmatrix} 3 & 2 \\ 1 & 1 \end{bmatrix}$

6. $\begin{bmatrix} 1 & 0 \\ 0 & 1 \end{bmatrix} \begin{bmatrix} 6 & 5 \\ 3 & -2 \end{bmatrix}$

7. $\begin{bmatrix} 2 & 0 & 0 \\ 0 & -1 & 0 \\ 0 & 0 & 3 \end{bmatrix} \begin{bmatrix} \frac{1}{2} & 0 & 0 \\ 0 & -1 & 0 \\ 0 & 0 & \frac{1}{3} \end{bmatrix}$

8. $\begin{bmatrix} 1 & -1 & 0 \\ 1 & 0 & -1 \\ 6 & -2 & -3 \end{bmatrix} \begin{bmatrix} -2 & -3 & 1 \\ -3 & -3 & 1 \\ -2 & -4 & 1 \end{bmatrix}$

Rewrite the matrices in reduced row-echelon form.

9. $\begin{bmatrix} 3 & -2 & 1 & 0 \\ 4 & -3 & 0 & 1 \end{bmatrix}$

10. $\begin{bmatrix} 1 & 1 & 2 & 1 & 0 & 0 \\ -1 & 0 & 3 & 0 & 1 & 0 \\ 1 & 2 & 8 & 0 & 0 & 1 \end{bmatrix}$

EXERCISES 11.3

In Exercises 1–4, show that B is the inverse of A.

1. $A = \begin{bmatrix} 1 & 2 \\ 3 & 4 \end{bmatrix}$, $B = \begin{bmatrix} -2 & 1 \\ \frac{3}{2} & -\frac{1}{2} \end{bmatrix}$

2. $A = \begin{bmatrix} 1 & -1 \\ 2 & 3 \end{bmatrix}$, $B = \begin{bmatrix} \frac{3}{5} & \frac{1}{5} \\ -\frac{2}{5} & \frac{1}{5} \end{bmatrix}$

3. $A = \begin{bmatrix} -2 & 2 & 3 \\ 1 & -1 & 0 \\ 0 & 1 & 4 \end{bmatrix}$, $B = \frac{1}{3} \begin{bmatrix} -4 & -5 & 3 \\ -4 & -8 & 3 \\ 1 & 2 & 0 \end{bmatrix}$

4. $A = \begin{bmatrix} 2 & -17 & 11 \\ -1 & 11 & -7 \\ 0 & 3 & -2 \end{bmatrix}$, $B = \begin{bmatrix} 1 & 1 & 2 \\ 2 & 4 & -3 \\ 3 & 6 & -5 \end{bmatrix}$

In Exercises 5–30, find the inverse of the matrix (if it exists).

5. $\begin{bmatrix} 2 & 0 \\ 0 & 3 \end{bmatrix}$

6. $\begin{bmatrix} 1 & 2 \\ 3 & 7 \end{bmatrix}$

7. $\begin{bmatrix} 1 & -2 \\ 2 & -3 \end{bmatrix}$

8. $\begin{bmatrix} -7 & 33 \\ 4 & -19 \end{bmatrix}$

9. $\begin{bmatrix} -1 & 1 \\ -2 & 1 \end{bmatrix}$

10. $\begin{bmatrix} 11 & 1 \\ -1 & 0 \end{bmatrix}$

11. $\begin{bmatrix} 2 & 4 \\ 4 & 8 \end{bmatrix}$

12. $\begin{bmatrix} 2 & 3 \\ 1 & 4 \end{bmatrix}$

13. $\begin{bmatrix} 2 & 7 & 1 \\ -3 & -9 & 2 \end{bmatrix}$

14. $\begin{bmatrix} -2 & 5 \\ 6 & -15 \\ 0 & 1 \end{bmatrix}$

15. $\begin{bmatrix} 1 & 1 & 1 \\ 3 & 5 & 4 \\ 3 & 6 & 5 \end{bmatrix}$

16. $\begin{bmatrix} 1 & 2 & 2 \\ 3 & 7 & 9 \\ -1 & -4 & -7 \end{bmatrix}$

17. $\begin{bmatrix} 1 & 2 & -1 \\ 3 & 7 & -10 \\ -5 & -7 & -15 \end{bmatrix}$

18. $\begin{bmatrix} 10 & 5 & -7 \\ -5 & 1 & 4 \\ 3 & 2 & -2 \end{bmatrix}$

19. $\begin{bmatrix} 1 & -2 & -1 & -2 \\ 3 & -5 & -2 & -3 \\ 2 & -5 & -2 & -5 \\ -1 & 4 & 4 & 11 \end{bmatrix}$

20. $\begin{bmatrix} 4 & 8 & -7 & 14 \\ 2 & 5 & -4 & 6 \\ 0 & 2 & 1 & -7 \\ 3 & 6 & -5 & 10 \end{bmatrix}$

21. $\begin{bmatrix} 1 & 1 & 2 \\ 3 & 1 & 0 \\ -2 & 0 & 3 \end{bmatrix}$

22. $\begin{bmatrix} 3 & 2 & 2 \\ 2 & 2 & 2 \\ -4 & 4 & 3 \end{bmatrix}$

23. $\begin{bmatrix} 0.1 & 0.2 & 0.3 \\ -0.3 & 0.2 & 0.2 \\ 0.5 & 0.4 & 0.4 \end{bmatrix}$

24. $\begin{bmatrix} 2 & 0 & 0 \\ 0 & 3 & 0 \\ 0 & 0 & 5 \end{bmatrix}$

25. $\begin{bmatrix} -8 & 0 & 0 & 0 \\ 0 & 1 & 0 & 0 \\ 0 & 0 & 4 & 0 \\ 0 & 0 & 0 & -5 \end{bmatrix}$

26. $\begin{bmatrix} 1 & 3 & -2 & 0 \\ 0 & 2 & 4 & 6 \\ 0 & 0 & -2 & 1 \\ 0 & 0 & 0 & 5 \end{bmatrix}$

27. $\begin{bmatrix} 1 & 0 & 0 \\ 3 & 4 & 0 \\ 2 & 5 & 5 \end{bmatrix}$

28. $\begin{bmatrix} 1 & 0 & 0 \\ 3 & 0 & 0 \\ 2 & 5 & 5 \end{bmatrix}$

29. $\begin{bmatrix} 1 & 0 & 3 & 0 \\ 0 & 2 & 0 & 4 \\ 1 & 0 & 3 & 0 \\ 0 & 2 & 0 & 4 \end{bmatrix}$

30. $\begin{bmatrix} 1/a & 0 \\ 0 & a \end{bmatrix}, \quad a \neq 0$

31. Use an inverse matrix to solve the following systems (see Exercise 9).

(a) $\begin{aligned} -x + y &= 4 \\ -2x + y &= 0 \end{aligned}$ (b) $\begin{aligned} -x + y &= -3 \\ -2x + y &= 5 \end{aligned}$

(c) $\begin{aligned} -x + y &= 20 \\ -2x + y &= 10 \end{aligned}$ (d) $\begin{aligned} -x + y &= 0 \\ -2x + y &= 7 \end{aligned}$

32. Use an inverse matrix to solve the following systems (see Exercise 12).

(a) $\begin{aligned} 2x + 3y &= 5 \\ x + 4y &= 10 \end{aligned}$ (b) $\begin{aligned} 2x + 3y &= 0 \\ x + 4y &= 3 \end{aligned}$

(c) $\begin{aligned} 2x + 3y &= 4 \\ x + 4y &= 2 \end{aligned}$ (d) $\begin{aligned} 2x + 3y &= 1 \\ x + 4y &= -2 \end{aligned}$

33. Use an inverse matrix to solve the following systems (see Exercise 19).

(a) $\begin{aligned} x_1 - 2x_2 - x_3 - 2x_4 &= 0 \\ 3x_1 - 5x_2 - 2x_3 - 3x_4 &= 1 \\ 2x_1 - 5x_2 - 2x_3 - 5x_4 &= -1 \\ -x_1 + 4x_2 + 4x_3 + 11x_4 &= 2 \end{aligned}$

(b) $\begin{aligned} x_1 - 2x_2 - x_3 - 2x_4 &= 1 \\ 3x_1 - 5x_2 - 2x_3 - 3x_4 &= -2 \\ 2x_1 - 5x_2 - 2x_3 - 5x_4 &= 0 \\ -x_1 + 4x_2 + 4x_3 + 11x_4 &= -3 \end{aligned}$

34. Use an inverse matrix to solve the following systems (see Exercise 22).

(a) $\begin{aligned} 3x + 2y + 2z &= 0 \\ 2x + 2y + 2z &= 5 \\ -4x + 4y + 3z &= 2 \end{aligned}$

(b) $\begin{aligned} 3x + 2y + 2z &= -1 \\ 2x + 2y + 2z &= 2 \\ -4x + 4y + 3z &= 0 \end{aligned}$

11.4 *The Determinant of a Matrix*

Every *square* matrix can be associated with a real number called its **determinant.** Determinants have many uses, and several will be discussed in this chapter.

Historically, the use of determinants arose from the recognition of special patterns that occur in the solution of systems of linear equations. For instance, the general solution to the system

$$a_{11}x_1 + a_{12}x_2 = b_1$$
$$a_{21}x_1 + a_{22}x_2 = b_2$$

can be shown to be

$$x_1 = \frac{b_1 a_{22} - b_2 a_{12}}{a_{11} a_{22} - a_{21} a_{12}} \quad \text{and} \quad x_2 = \frac{b_2 a_{11} - b_1 a_{21}}{a_{11} a_{22} - a_{21} a_{12}}$$

provided that $a_{11}a_{22} - a_{21}a_{12} \neq 0$. Note that both fractions have the same denominator, $a_{11}a_{22} - a_{21}a_{12}$. This quantity is called the **determinant** of the coefficient matrix A.

Matrices and Determinants

Definition of the Determinant of a 2 × 2 Matrix

The **determinant** of the matrix

$$A = \begin{bmatrix} a_{11} & a_{12} \\ a_{21} & a_{22} \end{bmatrix}$$

is given by

$$\det(A) = |A| = a_{11}a_{22} - a_{21}a_{12}.$$

Remark: In this text $\det(A)$ and $|A|$ are used interchangeably to represent the determinant of A. Although vertical bars are used also to denote the absolute value of a real number, the context will show which use is intended.

A convenient method for remembering the formula for the determinant of a 2 × 2 matrix is shown in the following diagram.

$$|A| = \begin{vmatrix} a_{11} & a_{12} \\ a_{21} & a_{22} \end{vmatrix} = a_{11}a_{22} - a_{21}a_{12}$$

Note that the determinant is given by the difference of the products of the two diagonals of the matrix.

EXAMPLE 1 The Determinant of a Matrix of Order 2

Find the determinants of the following matrices.

(a) $A = \begin{bmatrix} 2 & -3 \\ 1 & 2 \end{bmatrix}$
(b) $B = \begin{bmatrix} 2 & 1 \\ 4 & 2 \end{bmatrix}$
(c) $C = \begin{bmatrix} 0 & 3 \\ 2 & 4 \end{bmatrix}$

SOLUTION

(a) $|A| = \begin{vmatrix} 2 & -3 \\ 1 & 2 \end{vmatrix} = 2(2) - 1(-3) = 4 + 3 = 7$

(b) $|B| = \begin{vmatrix} 2 & 1 \\ 4 & 2 \end{vmatrix} = 2(2) - 4(1) = 4 - 4 = 0$

(c) $|C| = \begin{vmatrix} 0 & 3 \\ 2 & 4 \end{vmatrix} = 0(4) - 2(3) = 0 - 6 = -6$

Remark: In Example 1, note that the determinant of a matrix can be positive, zero, or negative.

The determinant of a matrix of order 1 is defined simply as the entry of the matrix. For instance, if $A = [-2]$, then $\det(A) = -2$. To define the determinant of a matrix of orders higher than 2, it is convenient to introduce the notions of **minors** and **cofactors.**

Definition of Minors and Cofactors of a Matrix

If A is a square matrix, then the **minor** M_{ij} of the entry a_{ij} is the determinant of the matrix obtained by deleting the ith row and jth column of A. The **cofactor** C_{ij} is given by

$$C_{ij} = (-1)^{i+j} M_{ij}.$$

For example, if A is a 3×3 matrix, then the minors of a_{21} and a_{22} are as shown in the following diagram.

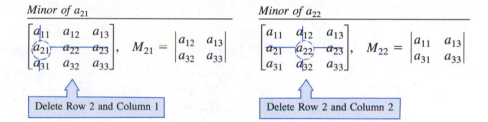

The minors and cofactors of a matrix differ at most in sign. To obtain the cofactor C_{ij}, first find the minor M_{ij}, and then multiply by $(-1)^{i+j}$. The value of $(-1)^{i+j}$ is given by the following checkerboard pattern of $+$'s and $-$'s.

Sign Pattern for Cofactors

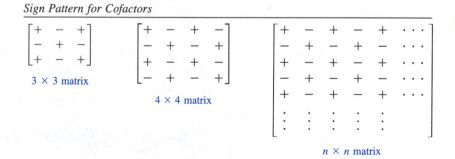

Note that *odd* positions (where $i + j$ is odd) have negative signs, and *even* positions (where $i + j$ is even) have positive signs.

Matrices and Determinants

EXAMPLE 2 Finding the Minors and Cofactors of a Matrix

Find all the minors and cofactors of

$$A = \begin{bmatrix} 0 & 2 & 1 \\ 3 & -1 & 2 \\ 4 & 0 & 1 \end{bmatrix}.$$

SOLUTION

To find the minor M_{11}, we delete the first row and first column of A and evaluate the determinant of the resulting matrix.

$$\begin{bmatrix} 0 & 2 & 1 \\ 3 & -1 & 2 \\ 4 & 0 & 1 \end{bmatrix}, \quad M_{11} = \begin{vmatrix} -1 & 2 \\ 0 & 1 \end{vmatrix} = -1(1) - 0(2) = -1$$

Similarly, to find M_{12}, we delete the first row and second column.

$$\begin{bmatrix} 0 & 2 & 1 \\ 3 & -1 & 2 \\ 4 & 0 & 1 \end{bmatrix}, \quad M_{12} = \begin{vmatrix} 3 & 2 \\ 4 & 1 \end{vmatrix} = 3(1) - 4(2) = -5$$

Continuing this pattern, we obtain the following minors.

$$\begin{array}{lll} M_{11} = -1 & M_{12} = -5 & M_{13} = 4 \\ M_{21} = 2 & M_{22} = -4 & M_{23} = -8 \\ M_{31} = 5 & M_{32} = -3 & M_{33} = -6 \end{array}$$

Now, to find the cofactors, we combine our checkerboard pattern of signs with these minors to obtain

$$\begin{array}{lll} C_{11} = -1 & C_{12} = 5 & C_{13} = 4 \\ C_{21} = -2 & C_{22} = -4 & C_{23} = 8 \\ C_{31} = 5 & C_{32} = 3 & C_{33} = -6. \end{array}$$

Having defined the minors and cofactors of a matrix, we are ready for a general definition of the determinant of a matrix. The definition given below is called **inductive** because it uses determinants of matrices of order $n - 1$ to define the determinant of a matrix of order n.

Definition of the Determinant of a Matrix

If A is a square matrix (of order 2 or greater), then the determinant of A is the sum of the entries in any row (or column) of A multiplied by their respective cofactors. For instance, expanding along the first row yields

$$|A| = a_{11}C_{11} + a_{12}C_{12} + \cdots + a_{1n}C_{1n}.$$

Remark: Try checking that for 2×2 matrices this definition yields $|A| = a_{11}a_{22} - a_{12}a_{21}$ as previously defined.

When this definition is used to evaluate a determinant, we say that we are **expanding by cofactors.** This procedure is demonstrated for a 3×3 matrix in Example 3.

EXAMPLE 3   *The Determinant of a Matrix of Order 3*

Find the determinant of

$$A = \begin{bmatrix} 0 & 2 & 1 \\ 3 & -1 & 2 \\ 4 & 0 & 1 \end{bmatrix}.$$

SOLUTION

Note that this matrix is the same as that given in Example 2. There we found the cofactors of the entries in the first row to be

$$C_{11} = -1, \quad C_{12} = 5, \quad \text{and} \quad C_{13} = 4.$$

Therefore, by the definition of a determinant, we have the following.

$$|A| = a_{11}C_{11} + a_{12}C_{12} + a_{13}C_{13} \qquad \textit{First row expansion}$$
$$= 0(-1) + 2(5) + 1(4) = 14.$$

In Example 3 we found the determinant by expanding by the cofactors in the first row. We could have used any row or column. For instance, we could have expanded along the second row to obtain

$$|A| = a_{21}C_{21} + a_{22}C_{22} + a_{23}C_{23} \qquad \textit{Second row expansion}$$
$$= 3(-2) + (-1)(-4) + 2(8) = 14$$

or along the first column to obtain

$$|A| = a_{11}C_{11} + a_{21}C_{21} + a_{31}C_{31} \qquad \textit{First column expansion}$$
$$= 0(-1) + 3(-2) + 4(5) = 14.$$

Try some other possibilities to see that the determinant of A can be evaluated by expanding by cofactors along *any* row or column.

When expanding by cofactors we do not need to find cofactors of zero entries because zero times its cofactor is zero.

$$a_{ij}C_{ij} = (0)C_{ij} = 0$$

Thus, the row (or column) containing the most zeros is usually the best choice for expansion by cofactors. This is demonstrated in the next example.

EXAMPLE 4 *The Determinant of a Matrix of Order 4*

Find the determinant of

$$A = \begin{bmatrix} 1 & -2 & 3 & 0 \\ -1 & 1 & 0 & 2 \\ 0 & 2 & 0 & 3 \\ 3 & 4 & 0 & 2 \end{bmatrix}.$$

SOLUTION

Inspecting this matrix, we see that three of the entries in the third column are zeros. Thus, we can eliminate some of the work in the expansion by using the third column:

$$|A| = 3(C_{13}) + 0(C_{23}) + 0(C_{33}) + 0(C_{43}).$$

Because C_{23}, C_{33}, and C_{43} have zero coefficients, we need only find the cofactor C_{13}. To do this, we delete the first row and third column of A and evaluate the determinant of the resulting matrix.

$$C_{13} = (-1)^{1+3} \begin{vmatrix} -1 & 1 & 2 \\ 0 & 2 & 3 \\ 3 & 4 & 2 \end{vmatrix}$$

$$= \begin{vmatrix} -1 & 1 & 2 \\ 0 & 2 & 3 \\ 3 & 4 & 2 \end{vmatrix}$$

Expanding by cofactors in the second row yields the following.

$$C_{13} = (0)(-1)^3 \begin{vmatrix} 1 & 2 \\ 4 & 2 \end{vmatrix} + (2)(-1)^4 \begin{vmatrix} -1 & 2 \\ 3 & 2 \end{vmatrix} + (3)(-1)^5 \begin{vmatrix} -1 & 1 \\ 3 & 4 \end{vmatrix}$$

$$= 0 + 2(1)(-8) + 3(-1)(-7)$$

$$= 5$$

Thus, we obtain

$$|A| = 3C_{13} = 3(5) = 15.$$

There is an alternative method that is commonly used for evaluating the determinant of a 3×3 matrix A. (This method *only* works for 3×3 matrices.) To apply this method, copy the first and second columns of A to form fourth and fifth columns. The determinant of A is then obtained by adding the products of three diagonals and subtracting the products of three diagonals, as shown in the following diagram.

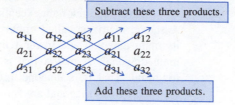

Thus, the determinant of the 3×3 matrix A is given by the following.

$$|A| = a_{11}a_{22}a_{33} + a_{12}a_{23}a_{31} + a_{13}a_{21}a_{32}$$
$$- a_{31}a_{22}a_{13} - a_{32}a_{23}a_{11} - a_{33}a_{21}a_{12}$$

EXAMPLE 5 **The Determinant of a Matrix of Order 3**

Find the determinant of

$$A = \begin{bmatrix} 0 & 2 & 1 \\ 3 & -1 & 2 \\ 4 & -4 & 1 \end{bmatrix}.$$

SOLUTION

Since A is a 3×3 matrix, we can use the alternate procedure for finding $|A|$. We begin by recopying the first two columns and then computing the six diagonal products as follows.

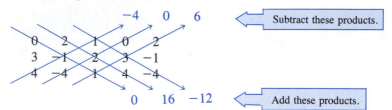

Now, by adding the lower three products and subtracting the upper three products, we find the determinant of A to be

$$|A| = 0 + 16 - 12 - (-4) - 0 - 6 = 2.$$

Remark: The diagonal process illustrated in Example 5 is valid *only* for matrices of order 3. For matrices of higher orders, another method must be used.

Triangular Matrices

Evaluating determinants of matrices of order 4 or higher can be tedious. There is, however, an important exception: the determinant of a **triangular** matrix. A square matrix is called **upper triangular** if it has all zero entries below its main diagonal and **lower triangular** if it has all zero entries above its main diagonal. A matrix that is both upper and lower triangular is called **diagonal.** That is, a diagonal matrix is one in which all entries above and below the main diagonal are zero.

Upper Triangular Matrix

$$\begin{bmatrix} a_{11} & a_{12} & a_{13} & \cdots & a_{1n} \\ 0 & a_{22} & a_{23} & \cdots & a_{2n} \\ 0 & 0 & a_{33} & \cdots & a_{3n} \\ \vdots & \vdots & \vdots & & \vdots \\ 0 & 0 & 0 & \cdots & a_{nn} \end{bmatrix}$$

Lower Triangular Matrix

$$\begin{bmatrix} a_{11} & 0 & 0 & \cdots & 0 \\ a_{21} & a_{22} & 0 & \cdots & 0 \\ a_{31} & a_{32} & a_{33} & \cdots & 0 \\ \vdots & \vdots & \vdots & & \vdots \\ a_{n1} & a_{n2} & a_{n3} & \cdots & a_{nn} \end{bmatrix}$$

To find the determinant of any order triangular matrix, we simply form the product of the entries on the main diagonal. It is easy to see that this procedure is valid for triangular matrices of order 2 or 3. For instance, the determinant of

$$A = \begin{bmatrix} 2 & 3 & -1 \\ 0 & -1 & 2 \\ 0 & 0 & 3 \end{bmatrix}$$

can be found by expanding by the third row to obtain

$$|A| = 0 \begin{vmatrix} 3 & -1 \\ -1 & 2 \end{vmatrix} - 0 \begin{vmatrix} 2 & -1 \\ 0 & 2 \end{vmatrix} + 3 \begin{vmatrix} 2 & 3 \\ 0 & -1 \end{vmatrix} = 3(2)(-1)$$

which is the product of the entries on the main diagonal.

EXAMPLE 6 *The Determinant of a Triangular Matrix*

Find the determinants of the following matrices.

(a) $A = \begin{bmatrix} 2 & 0 & 0 & 0 \\ 4 & -2 & 0 & 0 \\ -5 & 6 & 1 & 0 \\ 1 & 5 & 3 & 3 \end{bmatrix}$

(b) $B = \begin{bmatrix} -1 & 0 & 0 & 0 & 0 \\ 0 & 3 & 0 & 0 & 0 \\ 0 & 0 & 2 & 0 & 0 \\ 0 & 0 & 0 & 4 & 0 \\ 0 & 0 & 0 & 0 & -2 \end{bmatrix}$

SOLUTION

(a) The determinant of this triangular matrix is given by

$$|A| = (2)(-2)(1)(3) = -12.$$

(b) This *diagonal* matrix is both upper and lower triangular. Therefore, its determinant is given by

$$|B| = (-1)(3)(2)(4)(-2) = 48.$$

WARM UP

Perform the indicated matrix operations.

1. $\begin{bmatrix} 1 & -2 \\ 0 & 3 \end{bmatrix} + \begin{bmatrix} 2 & 7 \\ 4 & -3 \end{bmatrix}$

2. $\begin{bmatrix} -2 & 5 \\ 3 & -2 \end{bmatrix} - \begin{bmatrix} 0 & -3 \\ 1 & 2 \end{bmatrix}$

3. $3 \begin{bmatrix} 3 & -4 & 2 \\ 1 & 0 & -1 \\ 0 & 1 & -2 \end{bmatrix}$

4. $4 \begin{bmatrix} 0 & 2 & 3 \\ -1 & 2 & 3 \\ -2 & 1 & -2 \end{bmatrix}$

Perform the indicated arithmetic operations.

5. $[(1)(3) + (-3)(2)] - [(1)(4) + (3)(5)]$

6. $[(4)(4) + (-1)(-3)] - [(-1)(2) + (-2)(7)]$

7. $\dfrac{4(7) - 1(-2)}{(-5)(-2) - 3(4)}$

8. $\dfrac{3(6) - 2(7)}{6(-5) - 2(1)}$

9. $-5(-1)^2[6(-2) - 7(-3)]$

10. $4(-1)^3[3(6) - 2(7)]$

EXERCISES 11.4

In Exercises 1–20, find the determinant of the matrix.

1. $[5]$

2. $[-8]$

3. $\begin{bmatrix} 2 & 1 \\ 3 & 4 \end{bmatrix}$

4. $\begin{bmatrix} -3 & 1 \\ 5 & 2 \end{bmatrix}$

5. $\begin{bmatrix} 5 & 2 \\ -6 & 3 \end{bmatrix}$

6. $\begin{bmatrix} 2 & -2 \\ 4 & 3 \end{bmatrix}$

7. $\begin{bmatrix} -7 & 6 \\ \frac{1}{2} & 3 \end{bmatrix}$

8. $\begin{bmatrix} 4 & -3 \\ 0 & 0 \end{bmatrix}$

9. $\begin{bmatrix} 2 & 6 \\ 0 & 3 \end{bmatrix}$

10. $\begin{bmatrix} 2 & -3 \\ -6 & 9 \end{bmatrix}$

11. $\begin{bmatrix} 2 & -1 & 0 \\ 4 & 2 & 1 \\ 4 & 2 & 1 \end{bmatrix}$

12. $\begin{bmatrix} -2 & 2 & 3 \\ 1 & -1 & 0 \\ 0 & 1 & 4 \end{bmatrix}$

13. $\begin{bmatrix} 3 & 2 & 2 \\ 2 & 2 & 2 \\ -4 & 4 & 3 \end{bmatrix}$

14. $\begin{bmatrix} 0.1 & 0.2 & 0.3 \\ -0.3 & 0.2 & 0.2 \\ 0.5 & 0.4 & 0.4 \end{bmatrix}$

15. $\begin{bmatrix} 1 & 4 & -2 \\ 3 & 6 & -6 \\ -2 & 1 & 4 \end{bmatrix}$

16. $\begin{bmatrix} 2 & 3 & 1 \\ 0 & 5 & -2 \\ 0 & 0 & -2 \end{bmatrix}$

17. $\begin{bmatrix} 6 & 3 & -7 \\ 0 & 0 & 0 \\ 4 & -6 & 3 \end{bmatrix}$

18. $\begin{bmatrix} 1 & 1 & 2 \\ 3 & 1 & 0 \\ -2 & 0 & 3 \end{bmatrix}$

19. $\begin{bmatrix} x & y & 1 \\ -2 & -2 & 1 \\ 1 & 5 & 1 \end{bmatrix}$

20. $\begin{bmatrix} 3 - \lambda & 2 \\ 4 & 1 - \lambda \end{bmatrix}$

In Exercises 21–24, find (a) all minors and (b) cofactors for the given matrix.

21. $\begin{bmatrix} 3 & 4 \\ 2 & -5 \end{bmatrix}$

22. $\begin{bmatrix} 11 & 0 \\ -3 & 2 \end{bmatrix}$

23. $\begin{bmatrix} 3 & -2 & 8 \\ 3 & 2 & -6 \\ -1 & 3 & 6 \end{bmatrix}$

24. $\begin{bmatrix} -2 & 9 & 4 \\ 7 & -6 & 0 \\ 6 & 7 & -6 \end{bmatrix}$

In Exercises 25–30, find the determinant of the matrix by the method of expansion by cofactors. Expand using the indicated row or column.

25. $\begin{bmatrix} -3 & 2 & 1 \\ 4 & 5 & 6 \\ 2 & -3 & 1 \end{bmatrix}$
(a) Row 1
(b) Column 2

26. $\begin{bmatrix} -3 & 4 & 2 \\ 6 & 3 & 1 \\ 4 & -7 & -8 \end{bmatrix}$
(a) Row 2
(b) Column 3

27. $\begin{bmatrix} 5 & 0 & -3 \\ 0 & 12 & 4 \\ 1 & 6 & 3 \end{bmatrix}$
(a) Row 2
(b) Column 2

28. $\begin{bmatrix} 10 & -5 & 5 \\ 30 & 0 & 10 \\ 0 & 10 & 1 \end{bmatrix}$
(a) Row 3
(b) Column 1

29. $\begin{bmatrix} 6 & 0 & -3 & 5 \\ 4 & 13 & 6 & -8 \\ -1 & 0 & 7 & 4 \\ 8 & 6 & 0 & 2 \end{bmatrix}$
(a) Row 2
(b) Column 2

30. $\begin{bmatrix} 10 & 8 & 3 & -7 \\ 4 & 0 & 5 & -6 \\ 0 & 3 & 2 & 7 \\ 1 & 0 & -3 & 2 \end{bmatrix}$
(a) Row 3
(b) Column 1

In Exercises 31–40, find the determinant of the matrix.

31. $\begin{bmatrix} 1 & 4 & -2 \\ 3 & 2 & 0 \\ -1 & 4 & 3 \end{bmatrix}$

32. $\begin{bmatrix} 2 & -1 & 3 \\ 1 & 4 & 4 \\ 1 & 0 & 2 \end{bmatrix}$

33. $\begin{bmatrix} 2 & 4 & 6 \\ 0 & 3 & 1 \\ 0 & 0 & -5 \end{bmatrix}$

34. $\begin{bmatrix} -3 & 0 & 0 \\ 7 & 11 & 0 \\ 1 & 2 & 2 \end{bmatrix}$

35. $\begin{bmatrix} 3 & 6 & -5 & 4 \\ -2 & 0 & 6 & 0 \\ 1 & 1 & 2 & 2 \\ 0 & 3 & -1 & -1 \end{bmatrix}$

36. $\begin{bmatrix} 2 & 6 & 6 & 2 \\ 2 & 7 & 3 & 6 \\ 1 & 5 & 0 & 1 \\ 3 & 7 & 0 & 7 \end{bmatrix}$

37. $\begin{bmatrix} 5 & 3 & 0 & 6 \\ 4 & 6 & 4 & 12 \\ 0 & 2 & -3 & 4 \\ 0 & 1 & -2 & 2 \end{bmatrix}$

38. $\begin{bmatrix} 1 & 4 & 3 & 2 \\ -5 & 6 & 2 & 1 \\ 0 & 0 & 0 & 0 \\ 3 & -2 & 1 & 5 \end{bmatrix}$

39. $\begin{bmatrix} 3 & 2 & 4 & -1 & 5 \\ -2 & 0 & 1 & 3 & 2 \\ 1 & 0 & 0 & 4 & 0 \\ 6 & 0 & 2 & -1 & 0 \\ 3 & 0 & 5 & 1 & 0 \end{bmatrix}$

40. $\begin{bmatrix} 5 & 2 & 0 & 0 & -2 \\ 0 & 1 & 4 & 3 & 2 \\ 0 & 0 & 2 & 6 & 3 \\ 0 & 0 & 3 & 4 & 1 \\ 0 & 0 & 0 & 0 & 2 \end{bmatrix}$

In Exercises 41 and 42, solve for x.

41. $\begin{vmatrix} x-1 & 2 \\ 3 & x-2 \end{vmatrix} = 0$

42. $\begin{vmatrix} x-2 & -1 \\ -3 & x \end{vmatrix} = 0$

[*Optional*] In Exercises 43–50, evaluate the determinant of the matrix where the entries are functions. Determinants of this type occur in calculus.

43. $\begin{bmatrix} 4u & -1 \\ -1 & 2v \end{bmatrix}$

44. $\begin{bmatrix} 3x^2 & -3y^2 \\ 1 & 1 \end{bmatrix}$

45. $\begin{bmatrix} \cos x & \sin x \\ -\sin x & \cos x \end{bmatrix}$

46. $\begin{bmatrix} \sec x & \tan x \\ \sec x \tan x & \sec^2 x \end{bmatrix}$

47. $\begin{bmatrix} e^{2x} & e^{3x} \\ 2e^{2x} & 3e^{3x} \end{bmatrix}$

48. $\begin{bmatrix} e^{-x} & xe^{-x} \\ -e^{-x} & (1-x)e^{-x} \end{bmatrix}$

49. $\begin{bmatrix} x & \ln x \\ 1 & 1/x \end{bmatrix}$

50. $\begin{bmatrix} x & x \ln x \\ 1 & 1 + \ln x \end{bmatrix}$

11.5 Properties of Determinants

Which of the following two determinants is easier to evaluate?

$$|A| = \begin{vmatrix} 1 & -2 & 3 & 1 \\ 4 & -6 & 3 & 2 \\ -2 & 4 & -9 & -3 \\ 3 & -6 & 9 & 2 \end{vmatrix} \quad \text{or} \quad |B| = \begin{vmatrix} 1 & -2 & 3 & 1 \\ 0 & 2 & -9 & -2 \\ 0 & 0 & -3 & -1 \\ 0 & 0 & 0 & -1 \end{vmatrix}$$

From what you now know about the determinant of a triangular matrix, it is clear that the second determinant is *much* easier to evaluate. Its determinant is simply the product of the entries on the main diagonal. That is,

$$|B| = (1)(2)(-3)(-1) = 6.$$

On the other hand, using expansion by cofactors (the only technique discussed so far) to evaluate the first determinant is messy. For instance, if we expand by cofactors across the first row, we have the following.

$$|A| = 1\begin{vmatrix} -6 & 3 & 2 \\ 4 & -9 & -3 \\ -6 & 9 & 2 \end{vmatrix} + 2\begin{vmatrix} 4 & 3 & 2 \\ -2 & -9 & -3 \\ 3 & 9 & 2 \end{vmatrix} + 3\begin{vmatrix} 4 & -6 & 2 \\ -2 & 4 & -3 \\ 3 & -6 & 2 \end{vmatrix} - 1\begin{vmatrix} 4 & -6 & 3 \\ -2 & 4 & -9 \\ 3 & -6 & 9 \end{vmatrix}$$

Evaluating the determinants of these four 3×3 matrices produces

$$|A| = (1)(-60) + (2)(39) + (3)(-10) - (1)(-18) = 6.$$

It is not coincidental that these two determinants have the same value. In fact, we obtained matrix B by performing elementary row operations on matrix A. (Try verifying this.) In this section, we consider the effect of elementary row (and column) operations on the value of a determinant. Here is an example.

EXAMPLE 1   The Effect of Elementary Row Operations on a Determinant

(a) The matrix B was obtained from A by interchanging the rows of A.

$$|A| = \begin{vmatrix} 2 & -3 \\ 1 & 4 \end{vmatrix} = 11 \quad \text{and} \quad |B| = \begin{vmatrix} 1 & 4 \\ 2 & -3 \end{vmatrix} = -11$$

(b) The matrix B was obtained from A by adding -2 times the first row of A to the second row of A.

$$|A| = \begin{vmatrix} 1 & -3 \\ 2 & -4 \end{vmatrix} = 2 \quad \text{and} \quad |B| = \begin{vmatrix} 1 & -3 \\ 0 & 2 \end{vmatrix} = 2$$

(c) The matrix B was obtained from A by multiplying the first row of A by $1/2$.

$$|A| = \begin{vmatrix} 2 & -8 \\ -2 & 9 \end{vmatrix} = 2 \quad \text{and} \quad |B| = \begin{vmatrix} 1 & -4 \\ -2 & 9 \end{vmatrix} = 1$$

In Example 1, we see that interchanging two rows of the matrix changed the sign of its determinant. Adding a multiple of one row to another did not change the determinant. Finally, multiplying a row by a nonzero constant multiplied the determinant by that same constant. The following theorem generalizes these observations.

Elementary Row Operations and Determinants

Let A and B be square matrices.

1. If B is obtained from A by interchanging two rows of A, then

$$|B| = -|A|.$$

2. If B is obtained from A by adding a multiple of a row of A to another row of A, then

$$|B| = |A|.$$

3. If B is obtained from A by multiplying a row of A by a nonzero constant c, then

$$|B| = c|A|.$$

Remark: Note that the third property allows us to take a common factor out of a row. For instance,

$$\begin{vmatrix} 2 & 4 \\ 1 & 3 \end{vmatrix} = 2\begin{vmatrix} 1 & 2 \\ 1 & 3 \end{vmatrix}.$$

Factor 2 out of the first row.

This theorem provides a practical way to evaluate determinants (in fact, this method works well with computers). To find the determinant of a matrix A, we use elementary row operations to obtain a triangular matrix B that is row equivalent to A. At each step in the elimination process we incorporate the effect of the elementary row operation on the determinant. Finally, we find the determinant of B by forming the product of the entries on its main diagonal. This process is demonstrated in the next example.

EXAMPLE 2 *Evaluating a Determinant Using Elementary Row Operations*

Find the determinant of

$$A = \begin{bmatrix} 2 & -3 & 10 \\ 1 & 2 & -2 \\ 0 & 1 & -3 \end{bmatrix}.$$

Properties of Determinants

SOLUTION

Using elementary row operations, we rewrite A in triangular form as follows.

$$\begin{vmatrix} 2 & -3 & 10 \\ 1 & 2 & -2 \\ 0 & 1 & -3 \end{vmatrix} = -\begin{vmatrix} 1 & 2 & -2 \\ 2 & -3 & 10 \\ 0 & 1 & -3 \end{vmatrix}$$

Interchange the first two rows.

$$= -\begin{vmatrix} 1 & 2 & -2 \\ 0 & -7 & 14 \\ 0 & 1 & -3 \end{vmatrix}$$

Adding -2 times the first row to the second row produces a new second row.

$$= 7\begin{vmatrix} 1 & 2 & -2 \\ 0 & 1 & -2 \\ 0 & 1 & -3 \end{vmatrix}$$

Factor -7 out of the second row.

$$= 7\begin{vmatrix} 1 & 2 & -2 \\ 0 & 1 & -2 \\ 0 & 0 & -1 \end{vmatrix}$$

Adding -1 times the second row to the third row produces a new third row.

Now, because the final matrix is triangular, we may conclude that the determinant is

$$|A| = 7(1)(1)(-1) = -7.$$

Determinants and Elementary Column Operations

Although the preceding theorem was stated in terms of elementary *row* operations, the theorem remains valid if the word "row" is replaced by the word "column." Operations performed on the columns of a matrix (rather than the rows) are called **elementary column operations,** and two matrices are called **column equivalent** if one can be obtained from the other by elementary column operations. This is illustrated as follows.

$$\begin{vmatrix} 2 & 1 & -3 \\ 4 & 0 & 1 \\ 0 & 0 & 2 \end{vmatrix} = -\begin{vmatrix} 1 & 2 & -3 \\ 0 & 4 & 1 \\ 0 & 0 & 2 \end{vmatrix} \qquad \begin{vmatrix} 2 & 3 & -5 \\ 4 & 1 & 0 \\ -2 & 4 & -3 \end{vmatrix} = 2\begin{vmatrix} 1 & 3 & -5 \\ 2 & 1 & 0 \\ -1 & 4 & -3 \end{vmatrix}$$

Interchange the first and second columns.

Factor 2 out of the first column.

In evaluating a determinant it is occasionally convenient to use elementary column operations, as shown in Example 3.

EXAMPLE 3 Evaluating a Determinant Using Elementary Column Operations

Find the determinant of

$$A = \begin{bmatrix} -1 & 2 & 2 \\ 3 & -6 & 4 \\ 5 & -10 & -3 \end{bmatrix}.$$

SOLUTION

Because the first two columns of A are multiples of each other, we can obtain a column of zeros by adding two times the first column to the second column, as follows.

$$\begin{vmatrix} -1 & 2 & 2 \\ 3 & -6 & 4 \\ 5 & -10 & -3 \end{vmatrix} = \begin{vmatrix} -1 & 0 & 2 \\ 3 & 0 & 4 \\ 5 & 0 & -3 \end{vmatrix}$$

At this point we need not continue to rewrite the matrix in triangular form. Since there is an entire column of zeros, we simply conclude that the determinant is zero. The validity of this conclusion follows from an expansion by cofactors along the second column.

$$|A| = (0)C_{12} + (0)C_{22} + (0)C_{32} = 0$$

Example 3 shows that if one column of a matrix is a scalar multiple of another column, we can immediately conclude that the determinant of the matrix is zero. This is one of three conditions listed below that yield a determinant of zero.

Conditions That Yield a Zero Determinant

If A is a square matrix and any one of the following conditions is true, then $|A| = 0$.

1. An entire row (or an entire column) is zero.
2. Two rows (or two columns) are equal.
3. One row (or column) is a multiple of another row (or column).

Recognizing the conditions listed in this theorem can make evaluating a determinant much easier. For instance, consider the following three evaluations.

Properties of Determinants

$$\begin{vmatrix} 0 & 0 & 0 \\ 2 & 4 & -5 \\ 3 & -5 & 2 \end{vmatrix} = 0 \qquad \begin{vmatrix} 1 & -2 & 4 \\ 0 & 1 & 2 \\ 1 & -2 & 4 \end{vmatrix} = 0 \qquad \begin{vmatrix} 1 & 2 & -3 \\ 2 & -1 & -6 \\ -2 & 0 & 6 \end{vmatrix} = 0$$

First row has all zeros. | First and third rows are the same. | Third column is a multiple of the first column.

Do not conclude that the three conditions listed in this theorem are the *only* conditions that produce a determinant of zero. The theorem is often used indirectly. That is, you can begin with a matrix that does not satisfy any of the three conditions listed in the theorem, and through elementary row or column operations obtain a matrix that does satisfy one of the conditions. Then you may conclude that the original matrix has a determinant of zero. This process is demonstrated in Example 4.

EXAMPLE 4   A Matrix with a Zero Determinant

Find the determinant of

$$A = \begin{bmatrix} 1 & 4 & 1 \\ 2 & -1 & 0 \\ 0 & 18 & 4 \end{bmatrix}.$$

SOLUTION

Adding -2 times the first row to the second row produces

$$|A| = \begin{vmatrix} 1 & 4 & 1 \\ 2 & -1 & 0 \\ 0 & 18 & 4 \end{vmatrix} = \begin{vmatrix} 1 & 4 & 1 \\ 0 & -9 & -2 \\ 0 & 18 & 4 \end{vmatrix}.$$

Because the second and third rows are multiples of each other, we conclude that the determinant is zero.

In Example 4 we could have obtained a matrix with a row of zeros by performing an additional elementary row operation (adding two times the second row to the third row). This is true in general. That is, a square matrix has a determinant of zero if and only if it is row (or column) equivalent to a matrix that has at least one row (or column) consisting entirely of zeros.

We have now surveyed several methods for evaluating determinants. Of these, the best method for computers is to use elementary row operations to write the matrix in triangular form. When evaluating a determinant *by hand*, however, you can sometimes save steps by using elementary row (or column)

Matrices and Determinants

operations to create a row or column having zeros in all but one position and then using cofactor expansion to reduce the order of the matrix by one. For instance, to evaluate the 4×4 determinant

$$|A| = \begin{vmatrix} 1 & -2 & 3 & 0 \\ -1 & 1 & 0 & 2 \\ 0 & 2 & 0 & 3 \\ 3 & 4 & 0 & -2 \end{vmatrix}$$

we would expand by cofactors down the third column to obtain the 3×3 determinant

$$|A| = 3(-1)^4 \begin{vmatrix} -1 & 1 & 2 \\ 0 & 2 & 3 \\ 3 & 4 & -2 \end{vmatrix}.$$

This idea is illustrated further in the next example.

EXAMPLE 5   *Evaluating a Determinant*

Evaluate the determinant of

$$A = \begin{bmatrix} 2 & 0 & 1 & 3 & -2 \\ -2 & 1 & 3 & 2 & -1 \\ 1 & 0 & -1 & 2 & 3 \\ 3 & -1 & 2 & 4 & -3 \\ 1 & 1 & 3 & 2 & 0 \end{bmatrix}$$

SOLUTION

Because the second column of this matrix already has two zeros, we choose it for cofactor expansion. Two additional zeros can be created in the second column by adding the second row to the fourth row, and then adding -1 times the second row to the fifth row.

$$|A| = \begin{vmatrix} 2 & 0 & 1 & 3 & -2 \\ -2 & 1 & 3 & 2 & -1 \\ 1 & 0 & -1 & 2 & 3 \\ 3 & -1 & 2 & 4 & -3 \\ 1 & 1 & 3 & 2 & 0 \end{vmatrix} = \begin{vmatrix} 2 & 0 & 1 & 3 & -2 \\ -2 & 1 & 3 & 2 & -1 \\ 1 & 0 & -1 & 2 & 3 \\ 1 & 0 & 5 & 6 & -4 \\ 3 & 0 & 0 & 0 & 1 \end{vmatrix}$$

$$= (1)(-1)^4 \begin{vmatrix} 2 & 1 & 3 & -2 \\ 1 & -1 & 2 & 3 \\ 1 & 5 & 6 & -4 \\ 3 & 0 & 0 & 1 \end{vmatrix}$$

(Note that we have now reduced the problem of finding the determinant of a 5×5 matrix to finding the determinant of a 4×4 matrix.) Because we

Properties of Determinants

already have two zeros in the fourth row, it is chosen for the next cofactor expansion. Adding -3 times the fourth column to the first column produces the following.

$$|A| = \begin{vmatrix} 2 & 1 & 3 & -2 \\ 1 & -1 & 2 & 3 \\ 1 & 5 & 6 & -4 \\ 3 & 0 & 0 & 1 \end{vmatrix} = \begin{vmatrix} 8 & 1 & 3 & -2 \\ -8 & -1 & 2 & 3 \\ 13 & 5 & 6 & -4 \\ 0 & 0 & 0 & 1 \end{vmatrix}$$

$$= (1)(-1)^8 \begin{vmatrix} 8 & 1 & 3 \\ -8 & -1 & 2 \\ 13 & 5 & 6 \end{vmatrix}$$

Finally, adding the second row to the first row, and then expanding by cofactors along the first row, yields

$$|A| = \begin{vmatrix} 8 & 1 & 3 \\ -8 & -1 & 2 \\ 13 & 5 & 6 \end{vmatrix} = \begin{vmatrix} 0 & 0 & 5 \\ -8 & -1 & 2 \\ 13 & 5 & 6 \end{vmatrix}$$

$$= 5(-1)^4 \begin{vmatrix} -8 & -1 \\ 13 & 5 \end{vmatrix}$$

$$= 5(-40 + 13) = -135.$$

Determinants and the Inverse of a Matrix

We saw in Section 11.3 that some square matrices are not invertible. However, it can be difficult to tell simply by inspection whether a matrix possesses an inverse. For instance, can you tell which of the following two matrices is invertible?

$$A = \begin{bmatrix} 0 & 2 & -1 \\ 3 & -2 & 1 \\ 3 & 2 & -1 \end{bmatrix} \quad \text{or} \quad B = \begin{bmatrix} 0 & 2 & -1 \\ 3 & -2 & 1 \\ 3 & 2 & 1 \end{bmatrix}$$

The following theorem shows how determinants can be used to classify square matrices as invertible or noninvertible.

Determinant of an Invertible Matrix

A square matrix A is invertible (nonsingular) if and only if

$$|A| \neq 0.$$

Matrices and Determinants

EXAMPLE 6 *Classifying Square Matrices as Singular or Nonsingular*

Which of the following matrices possesses an inverse?

(a) $\begin{bmatrix} 0 & 2 & -1 \\ 3 & -2 & 1 \\ 3 & 2 & -1 \end{bmatrix}$
(b) $\begin{bmatrix} 0 & 2 & -1 \\ 3 & -2 & 1 \\ 3 & 2 & 1 \end{bmatrix}$

SOLUTION

(a) Because

$$\begin{vmatrix} 0 & 2 & -1 \\ 3 & -2 & 1 \\ 3 & 2 & -1 \end{vmatrix} = 0$$

we conclude that this matrix has no inverse (it is singular).

(b) Because

$$\begin{vmatrix} 0 & 2 & -1 \\ 3 & -2 & 1 \\ 3 & 2 & 1 \end{vmatrix} = -12 \neq 0$$

we conclude that this matrix has an inverse (it is nonsingular).

WARM UP

Write the matrices in row-echelon form.

1. $\begin{bmatrix} 2 & -6 \\ 5 & 20 \end{bmatrix}$

2. $\begin{bmatrix} -7 & 21 \\ 3 & 5 \end{bmatrix}$

3. $\begin{bmatrix} 1 & 3 & 4 \\ 0 & 1 & 1 \\ 2 & 4 & 6 \end{bmatrix}$

4. $\begin{bmatrix} 4 & 8 & 16 \\ 3 & -1 & 2 \\ -2 & 10 & 12 \end{bmatrix}$

Evaluate the determinants.

5. $\begin{vmatrix} 4 & -3 \\ -2 & 1 \end{vmatrix}$

6. $\begin{vmatrix} 10 & -20 \\ -1 & 2 \end{vmatrix}$

7. $\begin{vmatrix} 4 & 0 \\ -3 & -2 \end{vmatrix}$

8. $\begin{vmatrix} x & x^2 \\ 1 & 2x \end{vmatrix}$

9. $\begin{vmatrix} 4 & 0 & -2 \\ 3 & 1 & 2 \\ -8 & 0 & 6 \end{vmatrix}$

10. $\begin{vmatrix} 3 & 2 & 5 \\ 0 & 0 & -4 \\ -6 & 1 & 1 \end{vmatrix}$

EXERCISES 11.5

In Exercises 1–14, state the property of determinants that verifies the equation.

1. $\begin{vmatrix} 2 & -6 \\ 1 & -3 \end{vmatrix} = 0$

2. $\begin{vmatrix} -4 & 5 \\ 12 & -15 \end{vmatrix} = 0$

3. $\begin{vmatrix} 1 & 4 & 2 \\ 0 & 0 & 0 \\ 5 & 6 & -7 \end{vmatrix} = 0$

4. $\begin{vmatrix} -4 & 3 & 2 \\ 8 & 0 & 0 \\ -4 & 3 & 2 \end{vmatrix} = 0$

5. $\begin{vmatrix} 1 & 3 & 4 \\ -7 & 2 & -5 \\ 6 & 1 & 2 \end{vmatrix} = -\begin{vmatrix} 1 & 4 & 3 \\ -7 & -5 & 2 \\ 6 & 2 & 1 \end{vmatrix}$

6. $\begin{vmatrix} 1 & 3 & 4 \\ -2 & 2 & 0 \\ 1 & 6 & 2 \end{vmatrix} = -\begin{vmatrix} 1 & 6 & 2 \\ -2 & 2 & 0 \\ 1 & 3 & 4 \end{vmatrix}$

7. $\begin{vmatrix} 5 & 10 & 15 \\ 2 & -3 & 4 \\ 2 & -7 & 1 \end{vmatrix} = 5\begin{vmatrix} 1 & 2 & 3 \\ 2 & -3 & 4 \\ 2 & -7 & 1 \end{vmatrix}$

8. $\begin{vmatrix} 1 & 8 & -3 \\ 3 & -12 & 6 \\ 7 & 4 & 9 \end{vmatrix} = 12\begin{vmatrix} 1 & 2 & -1 \\ 3 & -3 & 2 \\ 7 & 1 & 3 \end{vmatrix}$

9. $\begin{vmatrix} 5 & 0 & 10 \\ 25 & -30 & 40 \\ -15 & 5 & 20 \end{vmatrix} = 5^3\begin{vmatrix} 1 & 0 & 2 \\ 5 & -6 & 8 \\ -3 & 1 & 4 \end{vmatrix}$

10. $\begin{vmatrix} 6 & 0 & 0 \\ 0 & 6 & 0 \\ 0 & 0 & 6 \end{vmatrix} = 6^3\begin{vmatrix} 1 & 0 & 0 \\ 0 & 1 & 0 \\ 0 & 0 & 1 \end{vmatrix}$

11. $\begin{vmatrix} 2 & -3 \\ 8 & 7 \end{vmatrix} = \begin{vmatrix} 2 & -3 \\ 0 & 19 \end{vmatrix}$

12. $\begin{vmatrix} 1 & -3 \\ 5 & 2 \end{vmatrix} = \begin{vmatrix} 1 & -3 \\ 0 & 17 \end{vmatrix}$

13. $\begin{vmatrix} 3 & 2 & 4 \\ -2 & 1 & 5 \\ 5 & -7 & -20 \end{vmatrix} = \begin{vmatrix} 7 & 2 & -6 \\ 0 & 1 & 0 \\ -9 & -7 & 15 \end{vmatrix}$

14. $\begin{vmatrix} 5 & 4 & 2 \\ 2 & -3 & 4 \\ 7 & 6 & 3 \end{vmatrix} = \begin{vmatrix} 1 & 10 & -6 \\ 2 & -3 & 4 \\ 7 & 6 & 3 \end{vmatrix}$

In Exercises 15–30, use elementary row (or column) operations as aids for evaluating the determinant.

15. $\begin{vmatrix} 1 & 2 & 5 \\ 1 & 4 & 2 \\ 0 & 3 & -4 \end{vmatrix}$

16. $\begin{vmatrix} 1 & 7 & -3 \\ 1 & 3 & 1 \\ 4 & 8 & 1 \end{vmatrix}$

17. $\begin{vmatrix} 1 & 1 & 1 \\ 2 & -1 & -2 \\ 1 & -2 & -1 \end{vmatrix}$

18. $\begin{vmatrix} 2 & -1 & -1 \\ 1 & 3 & 2 \\ 1 & 1 & 3 \end{vmatrix}$

19. $\begin{vmatrix} 3 & -1 & -3 \\ -1 & -4 & -2 \\ 3 & -1 & -1 \end{vmatrix}$

20. $\begin{vmatrix} 4 & 3 & -2 \\ 5 & 4 & 1 \\ -2 & 3 & 4 \end{vmatrix}$

21. $\begin{vmatrix} 3 & 8 & -7 \\ 0 & -5 & 4 \\ 8 & 1 & 6 \end{vmatrix}$

22. $\begin{vmatrix} 5 & -8 & 0 \\ 9 & 7 & 4 \\ -8 & 7 & 1 \end{vmatrix}$

23. $\begin{vmatrix} 4 & -8 & 5 & 0 \\ 8 & -5 & 3 & 0 \\ 8 & 5 & 2 & 0 \\ 1 & 7 & -5 & 1 \end{vmatrix}$

24. $\begin{vmatrix} 4 & -7 & 9 & 1 \\ 6 & 2 & 7 & 0 \\ 3 & 6 & -3 & 3 \\ 0 & 7 & 4 & -1 \end{vmatrix}$

25. $\begin{vmatrix} 9 & -4 & 2 & 5 \\ 2 & 7 & 6 & -5 \\ 4 & 1 & -2 & 0 \\ 7 & 3 & 4 & 10 \end{vmatrix}$

26. $\begin{vmatrix} 1 & -2 & 7 & 9 \\ 3 & -4 & 5 & 5 \\ 3 & 6 & 1 & -1 \\ 4 & 5 & 3 & 2 \end{vmatrix}$

27. $\begin{vmatrix} 0 & -3 & 8 & 2 \\ 8 & 1 & -1 & 6 \\ -4 & 6 & 0 & 9 \\ -7 & 0 & 0 & 14 \end{vmatrix}$

28. $\begin{vmatrix} 1 & -1 & 8 & 4 \\ 2 & 6 & 0 & -4 \\ 2 & 0 & 2 & 6 \\ 0 & 2 & 8 & 0 \end{vmatrix}$

29. $\begin{vmatrix} 3 & -2 & 4 & 3 & 1 \\ -1 & 0 & 2 & 1 & 0 \\ 5 & -1 & 0 & 3 & 2 \\ 4 & 7 & -8 & 0 & 0 \\ 1 & 2 & 3 & 0 & 2 \end{vmatrix}$

30. $\begin{vmatrix} 4 & 2 & -1 & 0 & 3 \\ 0 & 1 & 1 & 2 & -3 \\ 0 & 0 & -2 & 8 & 12 \\ 0 & 0 & 0 & 5 & 13 \\ 0 & 0 & 0 & 0 & 3 \end{vmatrix}$

In Exercises 31–36, use a determinant to determine whether the matrix is invertible.

31. $\begin{bmatrix} 5 & 4 \\ 10 & 8 \end{bmatrix}$

32. $\begin{bmatrix} 3 & -6 \\ 4 & 2 \end{bmatrix}$

33. $\begin{bmatrix} 14 & 7 & 0 \\ 2 & 3 & 0 \\ 1 & -5 & 2 \end{bmatrix}$

34. $\begin{bmatrix} 1 & 0 & 4 \\ 0 & 6 & 3 \\ 2 & -1 & 4 \end{bmatrix}$

35. $\begin{bmatrix} \frac{1}{2} & \frac{3}{2} & 2 \\ \frac{2}{3} & -\frac{1}{3} & 0 \\ 1 & 1 & 1 \end{bmatrix}$

36. $\begin{bmatrix} 2 & -1 & 6 \\ 1 & -3 & 4 \\ 4 & -2 & 12 \end{bmatrix}$

37. Show that $\begin{vmatrix} 1 & x & x^2 \\ 1 & y & y^2 \\ 1 & z & z^2 \end{vmatrix} = (y - x)(z - x)(z - y).$

38. Show that

$$\begin{vmatrix} a_{11} & 0 & 0 & \cdots & 0 \\ a_{21} & a_{22} & 0 & \cdots & 0 \\ \vdots & \vdots & & & \vdots \\ a_{n1} & a_{n2} & a_{n3} & \cdots & a_{nn} \end{vmatrix} = a_{11}a_{22}a_{23} \cdots a_{nn}.$$

39. For any values of x and y, what is the relationship between the determinants

$$\begin{vmatrix} a & b \\ c & d \end{vmatrix} \quad \text{and} \quad \begin{vmatrix} a & b & x \\ c & d & y \\ 0 & 0 & 1 \end{vmatrix}?$$

40. Find square matrices A and B to demonstrate that
$$|A + B| \neq |A| + |B|.$$

In Exercises 41–44, verify the equation.

41. $\begin{vmatrix} w & x \\ y & z \end{vmatrix} = -\begin{vmatrix} y & z \\ w & x \end{vmatrix}$

42. $\begin{vmatrix} w & cx \\ y & cz \end{vmatrix} = c\begin{vmatrix} w & x \\ y & z \end{vmatrix}$

43. $\begin{vmatrix} w & x \\ y & z \end{vmatrix} = \begin{vmatrix} w & x + cw \\ y & z + cy \end{vmatrix}$

44. $\begin{vmatrix} w & x \\ cw & cx \end{vmatrix} = 0$

11.6 Applications of Determinants and Matrices

Our focus in much of Chapters 10 and 11 has been on methods for solving systems of linear equations. In the first part of this section, we look at yet another method, called Cramer's Rule. Then we look at three applications: one involving area, one involving lines in the plane, and one involving cryptography.

Cramer's Rule

Cramer's Rule, named after Gabriel Cramer (1704–1752), gives a simple rule for solving a system of n linear equations in n variables. This rule can be applied only to systems of linear equations that have unique solutions. (For such systems, the determinant of the coefficient matrix is nonzero.)

To see how Cramer's Rule arises, let's look at the solution of a system involving two linear equations in two unknowns.

$$a_{11}x_1 + a_{12}x_2 = b_1$$
$$a_{21}x_1 + a_{22}x_2 = b_2$$

Multiplying the first equation by $-a_{21}$, the second by a_{11}, and adding the results produces the following.

$$-a_{21}a_{11}x_1 - a_{21}a_{12}x_2 = -a_{21}b_1$$
$$\underline{a_{11}a_{21}x_1 + a_{11}a_{22}x_2 = \quad a_{11}b_2}$$
$$(a_{11}a_{22} - a_{21}a_{12})x_2 = a_{11}b_2 - a_{21}b_1$$

Applications of Determinants and Matrices

Solving for x_2 (provided $a_{11}a_{22} - a_{21}a_{12} \neq 0$) produces

$$x_2 = \frac{a_{11}b_2 - a_{21}b_1}{a_{11}a_{22} - a_{21}a_{12}}.$$

In a similar way, we can solve for x_1 to obtain

$$x_1 = \frac{a_{22}b_1 - a_{12}b_2}{a_{11}a_{22} - a_{21}a_{12}}.$$

Finally, recognizing that the numerator and denominator for both x_1 and x_2 can be represented as determinants, we have

$$x_1 = \frac{\begin{vmatrix} b_1 & a_{12} \\ b_2 & a_{22} \end{vmatrix}}{\begin{vmatrix} a_{11} & a_{12} \\ a_{21} & a_{22} \end{vmatrix}}, \qquad x_2 = \frac{\begin{vmatrix} a_{11} & b_1 \\ a_{21} & b_2 \end{vmatrix}}{\begin{vmatrix} a_{11} & a_{12} \\ a_{21} & a_{22} \end{vmatrix}}, \qquad a_{11}a_{22} - a_{21}a_{12} \neq 0.$$

The denominator for both x_1 and x_2 is simply the determinant of the coefficient matrix A. The determinant forming the numerator for x_1 can be obtained from A by replacing its first column by the column representing the constants of the system. The determinant forming the numerator for x_2 can be obtained in a similar way. We denote these two determinants by $|A_1|$ and $|A_2|$ as follows.

$$|A_1| = \begin{vmatrix} b_1 & a_{12} \\ b_2 & a_{22} \end{vmatrix} \qquad \text{and} \qquad |A_2| = \begin{vmatrix} a_{11} & b_1 \\ a_{21} & b_2 \end{vmatrix}$$

Thus, we have

$$x_1 = \frac{|A_1|}{|A|} \qquad \text{and} \qquad x_2 = \frac{|A_2|}{|A|}.$$

This determinant form of the solution is called **Cramer's Rule.**

EXAMPLE 1 *Using Cramer's Rule*

Use Cramer's Rule to solve the following system of linear equations.

$$4x_1 - 2x_2 = 10$$
$$3x_1 - 5x_2 = 11$$

SOLUTION

First we find the determinant of the coefficient matrix.

$$|A| = \begin{vmatrix} 4 & -2 \\ 3 & -5 \end{vmatrix} = -14$$

Matrices and Determinants

Since $|A| \neq 0$, we know that the system has a unique solution, and applying Cramer's Rule produces the following.

$$x_1 = \frac{|A_1|}{|A|} = \frac{\begin{vmatrix} 10 & -2 \\ 11 & -5 \end{vmatrix}}{-14} = \frac{-28}{-14} = 2$$

$$x_2 = \frac{|A_2|}{|A|} = \frac{\begin{vmatrix} 4 & 10 \\ 3 & 11 \end{vmatrix}}{-14} = \frac{14}{-14} = -1$$

Thus, the solution is $x_1 = 2$ and $x_2 = -1$.

Cramer's Rule generalizes easily to systems of n linear equations in n variables. The value of each variable is given as the quotient of two determinants. The denominator is the determinant of the coefficient matrix, and the numerator is the determinant of the matrix formed by replacing the column corresponding to the variable (being solved for) with the column representing the constants. For example, the solution for x_3 in the system

$$a_{11}x_1 + a_{12}x_2 + a_{13}x_3 = b_1$$
$$a_{21}x_1 + a_{22}x_2 + a_{23}x_3 = b_2$$
$$a_{31}x_1 + a_{32}x_2 + a_{33}x_3 = b_3$$

is given by

$$x_3 = \frac{|A_3|}{|A|} = \frac{\begin{vmatrix} a_{11} & a_{12} & b_1 \\ a_{21} & a_{22} & b_2 \\ a_{31} & a_{32} & b_3 \end{vmatrix}}{\begin{vmatrix} a_{11} & a_{12} & a_{13} \\ a_{21} & a_{22} & a_{23} \\ a_{31} & a_{32} & a_{33} \end{vmatrix}}.$$

Cramer's Rule

If a system of n linear equations in n variables has a coefficient matrix with a nonzero determinant $|A|$, then the solution to the system is given by

$$x_1 = \frac{|A_1|}{|A|}, \qquad x_2 = \frac{|A_2|}{|A|}, \qquad \ldots, \qquad x_n = \frac{|A_n|}{|A|}$$

where the ith column of A_i is the column of constants in the system of equations.

EXAMPLE 2 *Using Cramer's Rule*

Use Cramer's Rule to solve the following system of linear equations for x.

$$
\begin{aligned}
-x + 2y - 3z &= 1 \\
2x \qquad\;\; + z &= 0 \\
3x - 4y + 4z &= 2
\end{aligned}
$$

SOLUTION

The determinant of the coefficient matrix is

$$
|A| = \begin{vmatrix} -1 & 2 & -3 \\ 2 & 0 & 1 \\ 3 & -4 & 4 \end{vmatrix} = 10.
$$

Since $|A| \neq 0$, we know that the solution is unique, and Cramer's Rule may be applied to solve for x as follows.

$$
x = \frac{\begin{vmatrix} 1 & 2 & -3 \\ 0 & 0 & 1 \\ 2 & -4 & 4 \end{vmatrix}}{10} = \frac{(-1)^5 \begin{vmatrix} 1 & 2 \\ 2 & -4 \end{vmatrix}}{10} = \frac{(-1)(-8)}{10} = \frac{4}{5}.
$$

Remark: Try applying Cramer's Rule in Example 2 to solve for y and z. You will see that the solution is $y = -3/2$ and $z = -8/5$.

Because of the many calculations with determinants, Cramer's Rule is less efficient than Gaussian elimination. Moreover, if the number of equations is not the same as the number of variables, Cramer's Rule does not apply.

Area and Equations of Lines

Determinants have many applications in analytic geometry. Two are presented here. The first gives a formula for the area of a triangle in the xy-plane.

Area of a Triangle in the xy-Plane

The area of the triangle whose vertices are (x_1, y_1), (x_2, y_2), and (x_3, y_3) is given by

$$
\text{area} = \pm \frac{1}{2} \begin{vmatrix} x_1 & y_1 & 1 \\ x_2 & y_2 & 1 \\ x_3 & y_3 & 1 \end{vmatrix}
$$

where the sign $(\pm)$ is chosen to give a positive area.

Matrices and Determinants

EXAMPLE 3 *Finding the Area of a Triangle*

Find the area of the triangle whose vertices are (1, 0), (2, 2), and (4, 3), as shown in Figure 11.1.

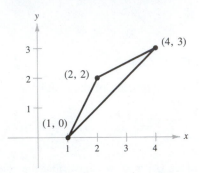

FIGURE 11.1

SOLUTION

It is unnecessary to know the relative position of the three vertices. We simply evaluate the determinant

$$\frac{1}{2}\begin{vmatrix} 1 & 0 & 1 \\ 2 & 2 & 1 \\ 4 & 3 & 1 \end{vmatrix} = -\frac{3}{2}$$

and conclude that the area of the triangle is 3/2.

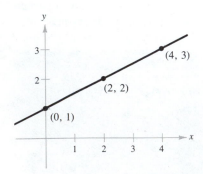

FIGURE 11.2

Suppose the three points in Example 3 had been on the same line. What would have happened had we applied our area formula to three such points? The answer is that the determinant would have been zero. Consider, for instance, the collinear points (0, 1), (2, 2), and (4, 3), as shown in Figure 11.2. The determinant giving the area of the "triangle" having these three points as vertices is

$$\frac{1}{2}\begin{vmatrix} 0 & 1 & 1 \\ 2 & 2 & 1 \\ 4 & 3 & 1 \end{vmatrix} = 0.$$

We generalize this result as follows.

Test for Collinear Points in the xy-Plane

Three points (x_1, y_1), (x_2, y_2), and (x_3, y_3) are collinear if and only if

$$\begin{vmatrix} x_1 & y_1 & 1 \\ x_2 & y_2 & 1 \\ x_3 & y_3 & 1 \end{vmatrix} = 0.$$

The following determinant form for the equation of the line passing through two points in the *xy*-plane is a corollary to the test for collinear points.

Two-Point Form of the Equation of a Line

The equation of the line passing through the distinct points (x_1, y_1) and (x_2, y_2) is given by

$$\begin{vmatrix} x & y & 1 \\ x_1 & y_1 & 1 \\ x_2 & y_2 & 1 \end{vmatrix} = 0.$$

EXAMPLE 4   *Finding the Equation of the Line Passing Through Two Points*

Find the equation of the line passing through the points $(2, 4)$ and $(-1, 3)$.

SOLUTION

Applying the determinant formula for the equation of the line passing through these two points produces

$$\begin{vmatrix} x & y & 1 \\ 2 & 4 & 1 \\ -1 & 3 & 1 \end{vmatrix} = 0.$$

To evaluate this determinant, we expand by cofactors along the top row to obtain

$$x\begin{vmatrix} 4 & 1 \\ 3 & 1 \end{vmatrix} - y\begin{vmatrix} 2 & 1 \\ -1 & 1 \end{vmatrix} + 1\begin{vmatrix} 2 & 4 \\ -1 & 3 \end{vmatrix} = x - 3y + 10.$$

Therefore, the equation of the line is

$$x - 3y + 10 = 0.$$

Matrices and Determinants

Cryptography

A **cryptogram** is a message written according to a secret code. (The Greek word "kryptos" means "hidden.") We describe next a method for using matrix multiplication to **encode** and **decode** messages.

We begin by assigning a number to each letter in the alphabet (with 0 assigned to a blank space) as follows.

0 = _	9 = I	18 = R
1 = A	10 = J	19 = S
2 = B	11 = K	20 = T
3 = C	12 = L	21 = U
4 = D	13 = M	22 = V
5 = E	14 = N	23 = W
6 = F	15 = O	24 = X
7 = G	16 = P	25 = Y
8 = H	17 = Q	26 = Z

Then the message is converted to numbers and partitioned into **uncoded row matrices,** each having n entries, as demonstrated in Example 5.

EXAMPLE 5 *Forming Uncoded Row Matrices*

Write the uncoded row matrices of order 1×3 for the message

MEET ME MONDAY.

SOLUTION

Partitioning the message (including blank spaces, but ignoring other punctuation) into groups of three produces the following uncoded row matrices.

$$[13 \quad 5 \quad 5] \quad [20 \quad 0 \quad 13] \quad [5 \quad 0 \quad 13] \quad [15 \quad 14 \quad 4] \quad [1 \quad 25 \quad 0]$$
$$\text{M} \quad \text{E} \quad \text{E} \quad \text{T} \quad \_ \quad \text{M} \quad \text{E} \quad \_ \quad \text{M} \quad \text{O} \quad \text{N} \quad \text{D} \quad \text{A} \quad \text{Y} \quad \_$$

Note that a blank space is used to fill out the last uncoded row matrix.

To **encode** a message we choose an $n \times n$ invertible matrix A and multiply the uncoded row matrices (on the right) by A to obtain **coded row matrices.** This process is demonstrated in Example 6.

EXAMPLE 6 *Encoding a Message*

Use the matrix

$$A = \begin{bmatrix} 1 & -2 & 2 \\ -1 & 1 & 3 \\ 1 & -1 & -4 \end{bmatrix}$$

to encode the message MEET ME MONDAY.

SOLUTION

The coded row matrices are obtained by multiplying each of the uncoded row matrices found in Example 5 by the matrix A as follows.

Uncoded Row Matrix	Encoding Matrix A		Coded Row Matrix

$$[13 \quad 5 \quad 5] \begin{bmatrix} 1 & -2 & 2 \\ -1 & 1 & 3 \\ 1 & -1 & -4 \end{bmatrix} = [13 \ -26 \ 21]$$

$$[20 \quad 0 \quad 13] \begin{bmatrix} 1 & -2 & 2 \\ -1 & 1 & 3 \\ 1 & -1 & -4 \end{bmatrix} = [33 \ -53 \ -12]$$

$$[5 \quad 0 \quad 13] \begin{bmatrix} 1 & -2 & 2 \\ -1 & 1 & 3 \\ 1 & -1 & -4 \end{bmatrix} = [18 \ -23 \ -42]$$

$$[15 \quad 14 \quad 4] \begin{bmatrix} 1 & -2 & 2 \\ -1 & 1 & 3 \\ 1 & -1 & -4 \end{bmatrix} = [5 \ -20 \ 56]$$

$$[1 \quad 25 \quad 0] \begin{bmatrix} 1 & -2 & 2 \\ -1 & 1 & 3 \\ 1 & -1 & -4 \end{bmatrix} = [-24 \ 23 \ 77]$$

Thus, the sequence of coded row matrices is

$$[13 \ -26 \ 21] \quad [33 \ -53 \ -12] \quad [18 \ -23 \ -42] \quad [5 \ -20 \ 56] \quad [-24 \ 23 \ 77].$$

Finally, removing the matrix notation produces the following cryptogram.

$$13 \ -26 \ 21 \ 33 \ -53 \ -12 \ 18 \ -23 \ -42 \ 5 \ -20 \ 56 \ -24 \ 23 \ 77$$

Matrices and Determinants

For those who do not know the matrix A, decoding the cryptogram found in Example 6 is difficult. But for an authorized receiver who knows the matrix A, decoding is simple. The receiver need only multiply the coded row matrices by A^{-1} to retrieve the uncoded row matrices. In other words, if

$$X = [x_1 \quad x_2 \quad \ldots \quad x_n]$$

is an uncoded $1 \times n$ matrix, then $Y = XA$ is the corresponding encoded matrix. The receiver of the encoded matrix can decode Y by multiplying on the right by A^{-1} to obtain

$$YA^{-1} = (XA)A^{-1} = X.$$

This procedure is demonstrated in Example 7.

EXAMPLE 7   *Decoding a Message*

Use the inverse of the matrix

$$A = \begin{bmatrix} 1 & -2 & 2 \\ -1 & 1 & 3 \\ 1 & -1 & -4 \end{bmatrix}$$

to decode the following cryptogram.

$$13 \ -26 \ 21 \ 33 \ -53 \ -12 \ 18 \ -23 \ -42 \ 5 \ -20 \ 56 \ -24 \ 23 \ 77$$

SOLUTION

We begin by using Gauss-Jordan elimination to find A^{-1}.

$$[A \vdots I] \qquad\qquad\qquad [I \vdots A^{-1}]$$

$$\begin{bmatrix} 1 & -2 & 2 & \vdots & 1 & 0 & 0 \\ -1 & 1 & 3 & \vdots & 0 & 1 & 0 \\ 1 & -1 & -4 & \vdots & 0 & 0 & 1 \end{bmatrix} \Longrightarrow \begin{bmatrix} 1 & 0 & 0 & \vdots & -1 & -10 & -8 \\ 0 & 1 & 0 & \vdots & -1 & -6 & -5 \\ 0 & 0 & 1 & \vdots & 0 & -1 & -1 \end{bmatrix}$$

Now, to decode the message, we partition the message into groups of three to form the following coded row matrices.

$$[13 \ -26 \ 21] \quad [33 \ -53 \ -12] \quad [18 \ -23 \ -42] \quad [5 \ -20 \ 56] \quad [-24 \ 23 \ 77]$$

Then we multiply each coded row matrix by A^{-1} (on the right) to obtain the decoded row matrices.

Coded Row Matrix	Decoding Matrix A^{-1}	Decoded Row Matrix

$$[13 \ -26 \ 21] \begin{bmatrix} -1 & -10 & -8 \\ -1 & -6 & -5 \\ 0 & -1 & -1 \end{bmatrix} = [13 \ 5 \ 5]$$

$$[33 \ -53 \ -12] \begin{bmatrix} -1 & -10 & -8 \\ -1 & -6 & -5 \\ 0 & -1 & -1 \end{bmatrix} = [20 \ 0 \ 13]$$

$$[18 \ -23 \ -42] \begin{bmatrix} -1 & -10 & -8 \\ -1 & -6 & -5 \\ 0 & -1 & -1 \end{bmatrix} = [5 \ 0 \ 13]$$

$$[5 \ -20 \ 56] \begin{bmatrix} -1 & -10 & -8 \\ -1 & -6 & -5 \\ 0 & -1 & -1 \end{bmatrix} = [15 \ 14 \ 4]$$

$$[-24 \ 23 \ 77] \begin{bmatrix} -1 & -10 & -8 \\ -1 & -6 & -5 \\ 0 & -1 & -1 \end{bmatrix} = [1 \ 25 \ 0]$$

Thus, the sequence of decoded row matrices is

$$[13 \ 5 \ 5] \quad [20 \ 0 \ 13] \quad [5 \ 0 \ 13] \quad [15 \ 14 \ 4] \quad [1 \ 25 \ 0]$$

and the message is

13 5 5 20 0 13 5 0 13 15 14 4 1 25 0.

M E E T _ M E _ M O N D A Y _

WARM UP

Solve the systems of equations. Use Gaussian elimination with back-substitution or Gauss-Jordan elimination.

1. $\begin{aligned} x - 3y &= -2 \\ x + y &= 2 \end{aligned}$

2. $\begin{aligned} -x + 3y &= 5 \\ 4x - y &= 2 \end{aligned}$

3. $\begin{aligned} x + 2y - z &= 7 \\ -y - z &= 4 \\ 4x \quad - z &= 16 \end{aligned}$

4. $\begin{aligned} 3x \quad + 6z &= 0 \\ -2x + y \quad &= 5 \\ y + 2z &= 3 \end{aligned}$

Evaluate the determinants.

5. $\begin{vmatrix} 10 & 8 \\ -6 & -4 \end{vmatrix}$

6. $\begin{vmatrix} -7 & 14 \\ 2 & 3 \end{vmatrix}$

7. $\begin{vmatrix} 1 & 0 & -2 \\ 0 & 1 & 0 \\ -2 & 0 & 1 \end{vmatrix}$

8. $\begin{vmatrix} 0 & 3 & 1 \\ 5 & -2 & 1 \\ 1 & 6 & 1 \end{vmatrix}$

9. $\begin{vmatrix} 0 & -2 & 1 & 0 \\ -2 & 0 & 5 & -1 \\ 1 & 5 & 0 & 2 \\ 0 & -1 & 2 & 0 \end{vmatrix}$

10. $\begin{vmatrix} 1 & 0 & -2 & 1 \\ 0 & 2 & 5 & 1 \\ 3 & -3 & 2 & 1 \\ 0 & 0 & 4 & 1 \end{vmatrix}$

Matrices and Determinants

EXERCISES 11.6

In Exercises 1–20, use Cramer's Rule to solve (if possible) the system of equations.

1. $x + 2y = 5$
   $-x + y = 1$

2. $2x - y = -10$
   $3x + 2y = -1$

3. $3x + 4y = -2$
   $5x + 3y = 4$

4. $18x + 12y = 13$
   $30x + 24y = 23$

5. $20x + 8y = 11$
   $12x - 24y = 21$

6. $13x - 6y = 17$
   $26x - 12y = 8$

7. $-0.4x + 0.8y = 1.6$
   $2x - 4y = 5$

8. $-0.4x + 0.8y = 1.6$
   $0.2x + 0.3y = 2.2$

9. $3x + 6y = 5$
   $6x + 14y = 11$

10. $3x + 2y = 1$
    $2x + 10y = 6$

11. $4x - y + z = -5$
    $2x + 2y + 3z = 10$
    $5x - 2y + 6z = 1$

12. $4x - 2y + 3z = -2$
    $2x + 2y + 5z = 16$
    $8x - 5y - 2z = 4$

13. $3x + 4y + 4z = 11$
    $4x - 4y + 6z = 11$
    $6x - 6y = 3$

14. $14x - 21y - 7z = 10$
    $-4x + 2y - 2z = 4$
    $56x - 21y + 7z = 5$

15. $3x + 3y + 5z = 1$
    $3x + 5y + 9z = 2$
    $5x + 9y + 17z = 4$

16. $2x + 3y + 5z = 4$
    $3x + 5y + 9z = 7$
    $5x + 9y + 17z = 13$

17. $5x - 3y + 2z = 2$
    $2x + 2y - 3z = 3$
    $x - 7y + 8z = -4$

18. $3x + 2y + 5z = 4$
    $4x - 3y - 4z = 1$
    $-8x + 2y + 3z = 0$

19. $7x - 3y + 2w = 41$
    $-2x + y - w = -13$
    $4x + z - 2w = 12$
    $-x + y - w = -8$

20. $2x + 5y + w = 11$
    $x + 4y + 2z - 2w = -7$
    $2x - 2y + 5z + w = 3$
    $x - 3w = 1$

In Exercises 21–26, use a determinant to find the area of the triangle with the given vertices.

21. $(0, 0)$, $(3, 1)$, $(1, 5)$
22. $(0, 0)$, $(5, -2)$, $(4, 5)$
23. $(-2, -3)$, $(2, -3)$, $(0, 4)$
24. $(-2, 1)$, $(3, -1)$, $(1, 6)$
25. $(0, \frac{1}{2})$, $(\frac{5}{2}, 0)$, $(4, 3)$
26. $(-4, -5)$, $(6, -1)$, $(6, 10)$

In Exercises 27–32, use a determinant to determine if the points are collinear.

27. $(3, -1)$, $(0, -3)$, $(12, 5)$
28. $(-3, -5)$, $(6, 1)$, $(10, 2)$
29. $(2, -\frac{1}{2})$, $(-4, 4)$, $(6, -3)$
30. $(0, 1)$, $(4, -2)$, $(-8, 7)$
31. $(0, 2)$, $(1, 2.4)$, $(-1, 1.6)$
32. $(2, 3)$, $(3, 3.5)$, $(-1, 2)$

In Exercises 33–38, use a determinant to find an equation of the line through the given points (x_1, y_1) and (x_2, y_2).

33. $(0, 0)$, $(5, 3)$
34. $(0, 0)$, $(-2, 2)$
35. $(-4, 3)$, $(2, 1)$
36. $(10, 7)$, $(-2, -7)$
37. $(-\frac{1}{2}, 3)$, $(\frac{5}{2}, 1)$
38. $(\frac{2}{3}, 4)$, $(6, 12)$

In Exercises 39–42, write a cryptogram for each message (see Example 6) using the matrix

$$A = \begin{bmatrix} 1 & 2 & 2 \\ 3 & 7 & 9 \\ -1 & -4 & -7 \end{bmatrix}.$$

39. LANDING SUCCESSFUL

40. BEAM ME UP SCOTTY

41. HAPPY BIRTHDAY

42. OPERATION OVERLORD

In Exercises 43 and 44, decode the cryptogram by using the inverse of matrix A from Exercises 39–42 (see Example 7).

43. 20 17 −15 −12 −56 −104 1 −25 −65 62 143 181

44. 13 −9 −59 61 112 106 −17 −73 −131 11 24 29
65 144 172

CHAPTER 11 REVIEW EXERCISES

In Exercises 1–12, use matrices and elementary row operations to solve the system of equations.

1. $5x + 4y = 2$
$-x + y = -22$

2. $2x - 5y = 2$
$3x - 7y = 1$

3. $0.2x - 0.1y = 0.07$
$0.4x - 0.5y = -0.01$

4. $2x + y = 0.3$
$3x - y = -1.3$

5. $-x + y + 2z = 1$
$2x + 3y + z = -2$
$5x + 4y + 2z = 4$

6. $2x + 3y + z = 10$
$2x - 3y - 3z = 22$
$4x - 2y + 3z = -2$

7. $2x + 3y + 3z = 3$
$6x + 6y + 12z = 13$
$12x + 9y - z = 2$

8. $4x + 4y + 4z = 5$
$4x - 2y - 8z = 1$
$5x + 3y + 8z = 6$

9. $2x + y + 2z = 4$
$2x + 2y = 5$
$2x - y + 6z = 2$

10. $3x + 21y - 29z = -1$
$2x + 15y - 21z = 0$

11. $x + 2y + 6z = 1$
$2x + 5y + 15z = 4$
$3x + y + 3z = -6$

12. $x + 2y + w = 3$
$-3y + 3z = 0$
$4x + 4y + z + 2w = 0$
$2x + z = 3$

In Exercises 13–20, perform the indicated matrix operations (if possible).

13. $\begin{bmatrix} 2 & 1 & 0 \\ 0 & 5 & -4 \end{bmatrix} - 3\begin{bmatrix} 5 & 3 & -6 \\ 0 & -2 & 5 \end{bmatrix}$

14. $-2\begin{bmatrix} 1 & 2 \\ 5 & -4 \\ 6 & 0 \end{bmatrix} + 8\begin{bmatrix} 7 & 1 \\ 1 & 2 \\ 1 & 4 \end{bmatrix}$

15. $\begin{bmatrix} 1 & 2 \\ 5 & -4 \\ 6 & 0 \end{bmatrix}\begin{bmatrix} 6 & -2 & 8 \\ 4 & 0 & 0 \end{bmatrix}$

16. $\begin{bmatrix} 1 & 5 & 6 \\ 2 & -4 & 0 \end{bmatrix}\begin{bmatrix} 6 & -2 & 8 \\ 4 & 0 & 0 \end{bmatrix}$

17. $\begin{bmatrix} 1 & 5 & 6 \\ 2 & -4 & 0 \end{bmatrix}\begin{bmatrix} 6 & 4 \\ -2 & 0 \\ 8 & 0 \end{bmatrix}$

18. $\begin{bmatrix} 4 \\ 6 \end{bmatrix} \begin{bmatrix} 6 & -2 \end{bmatrix}$

19. $\begin{bmatrix} 1 & 3 & 2 \\ 0 & 2 & -4 \\ 0 & 0 & 3 \end{bmatrix}\begin{bmatrix} 4 & -3 & 2 \\ 0 & 3 & -1 \\ 0 & 0 & 2 \end{bmatrix}$

20. $\begin{bmatrix} 2 & 1 \\ 6 & 0 \end{bmatrix}\left(\begin{bmatrix} 4 & 2 \\ -3 & 1 \end{bmatrix} + \begin{bmatrix} -2 & 4 \\ 0 & 4 \end{bmatrix}\right)$

In Exercises 21–24, solve for X given

$$A = \begin{bmatrix} -4 & 0 \\ 1 & -5 \\ -3 & 2 \end{bmatrix} \text{ and } B = \begin{bmatrix} 1 & 2 \\ -2 & 1 \\ 4 & 4 \end{bmatrix}.$$

21. $X = 3A - 2B$

22. $6X = 4A + 3B$

23. $3X + 2A = B$

24. $2A - 5B = 3X$

In Exercises 25–28, evaluate the determinant.

25. $\begin{vmatrix} 50 & -30 \\ 10 & 5 \end{vmatrix}$

26. $\begin{vmatrix} 8 & 5 \\ 2 & -4 \end{vmatrix}$

27. $\begin{vmatrix} 3 & 0 & -4 & 0 \\ 0 & 8 & 1 & 2 \\ 6 & 1 & 8 & 2 \\ 0 & 3 & -4 & 1 \end{vmatrix}$

28. $\begin{vmatrix} -5 & 6 & 0 & 0 \\ 0 & 1 & -1 & 2 \\ -3 & 4 & -5 & 1 \\ 1 & 6 & 0 & 3 \end{vmatrix}$

In Exercises 29–32, find the inverse of the matrix (if it exists)

29. $\begin{bmatrix} 2 & 6 \\ 3 & -6 \end{bmatrix}$

30. $\begin{bmatrix} 3 & -10 \\ 4 & 2 \end{bmatrix}$

31. $\begin{bmatrix} 2 & 0 & 3 \\ -1 & 1 & 1 \\ 2 & -2 & 1 \end{bmatrix}$

32. $\begin{bmatrix} 1 & 4 & 6 \\ 2 & -3 & 1 \\ -1 & 18 & 16 \end{bmatrix}$

33. Write the system of linear equations represented by the matrix equation

$$\begin{bmatrix} 5 & 4 \\ -1 & 1 \end{bmatrix} \begin{bmatrix} x \\ y \end{bmatrix} = \begin{bmatrix} 2 \\ -22 \end{bmatrix}.$$

34. Write the matrix equation $AX = B$ for the following system of linear equations.

$$2x + 3y + z = 10$$
$$2x - 3y - 3z = 22$$
$$4x - 2y + 3z = -2$$

In Exercises 35–40, solve the system of linear equations using (a) the inverse of the coefficient matrix and (b) Cramer's Rule.

35. $x + 2y = -1$
$3x + 4y = -5$

36. $x + 3y = 23$
$-6x + 2y = -18$

37. $-3x - 3y - 4z = 2$
$y + z = -1$
$4x + 3y + 4z = -1$

38. $x - 3y - 2z = 8$
$-2x + 7y + 3z = -19$
$x - y - 3z = 3$

39. $x + 3y + 2z = 2$
$-2x - 5y - z = 10$
$2x + 4y = -12$

40. $2x + 4y = -12$
$3x + 4y - 2z = -14$
$-x + y + 2z = -6$

In Exercises 41–44, use a determinant to find the area of the triangle with the given vertices.

41. $(1, 0), (5, 0), (5, 8)$

42. $(-4, 0), (4, 0), (0, 6)$

43. $(1, 2), (4, -5), (3, 2)$

44. $\left(\frac{3}{2}, 1\right), \left(4, -\frac{1}{2}\right), (4, 2)$

In Exercises 45–48, use a determinant to find an equation of the line through the given points.

45. $(-4, 0), (4, 4)$

46. $(2, 5), (6, -1)$

47. $\left(-\frac{5}{2}, 3\right), \left(\frac{7}{2}, 1\right)$

48. $(-0.8, 0.2), (0.7, 3.2)$

49. If A is a 3×3 matrix and $|A| = 2$, then what is the value of $|4A|$? Give the reason for your answer.

50. Prove that

$$\begin{vmatrix} a_{11} & a_{12} & a_{13} \\ a_{21} & a_{22} & a_{23} \\ a_{31} + c_1 & a_{32} + c_2 & a_{33} + c_3 \end{vmatrix}$$

$$= \begin{vmatrix} a_{11} & a_{12} & a_{13} \\ a_{21} & a_{22} & a_{23} \\ a_{31} & a_{32} & a_{33} \end{vmatrix} + \begin{vmatrix} a_{11} & a_{12} & a_{13} \\ a_{21} & a_{22} & a_{23} \\ c_1 & c_2 & c_3 \end{vmatrix}.$$

CHAPTER 12

Sequences, Series, and Probability

12.1 Sequences and Summation Notation

In this chapter, we look at several special topics in algebra that are extensions of concepts studied earlier in the text.

Sequences

In mathematics, the word "sequence" is used in much the same way as it is in ordinary English. When we say that a collection of objects is listed "in sequence," we usually mean that the collection is ordered so that it has a first member, second member, third member, and so on. Two examples are

$$1, 3, 5, 7, \ldots \qquad \text{and} \qquad \frac{1}{2}, \frac{1}{4}, \frac{1}{8}, \frac{1}{16}, \ldots$$

Mathematically, we can think of a sequence as a *function* whose domain is the set of positive integers. Even though a sequence is a function, we usually represent a sequence by subscript notation, rather than the standard function notation. For instance, we write the terms of the sequence

$$f(1), f(2), f(3), f(4), \ldots, f(n), \ldots$$

as

$$a_1, \quad a_2, \quad a_3, \quad a_4, \quad \ldots, \quad a_n \ldots.$$

663

Note that 1 corresponds to a_1, 2 corresponds to a_2, and so on. We call a_n the **nth term** of the sequence and denote the entire sequence by $\{a_n\}$.

Definition of a Sequence

An **infinite sequence** $\{a_n\}$ is a function whose domain is the set of positive integers. The function values

$$a_1, a_2, a_3, \ldots, a_n, \ldots$$

are called the **terms** of the sequence. An *infinite sequence* is often referred to simply as a **sequence.**

Remark: On occasion it is convenient to begin subscripting a sequence with zero instead of one so that the terms of the sequence become a_0, a_1, a_2, a_3, . . . , a_n, ...

EXAMPLE 1 *Finding Terms in a Sequence*

(a) The first four terms of the sequence whose *n*th term is $a_n = 3 + (-1)^n$ are as follows.

$$a_1 = 3 + (-1)^1 = 3 - 1 = 2$$
$$a_2 = 3 + (-1)^2 = 3 + 1 = 4$$
$$a_3 = 3 + (-1)^3 = 3 - 1 = 2$$
$$a_4 = 3 + (-1)^4 = 3 + 1 = 4$$

(b) The first four terms of the sequence whose *n*th term is $a_n = 2n/(1 + n)$ are as follows.

$$a_1 = \frac{2(1)}{1 + 1} = \frac{2}{2} = 1$$

$$a_2 = \frac{2(2)}{1 + 2} = \frac{4}{3}$$

$$a_3 = \frac{2(3)}{1 + 3} = \frac{6}{4} = \frac{3}{2}$$

$$a_4 = \frac{2(4)}{1 + 4} = \frac{8}{5}$$

The terms of a sequence need not all be positive, as demonstrated in Example 2.

EXAMPLE 2 A Sequence with Alternating Signs

Find the first four terms and the thirty-seventh term of the sequence whose nth term is

$$a_n = \frac{(-1)^n}{2n - 1}.$$

SOLUTION

$$a_1 = \frac{(-1)^1}{2(1) - 1} = -1$$

$$a_2 = \frac{(-1)^2}{2(2) - 1} = \frac{1}{3}$$

$$a_3 = \frac{(-1)^3}{2(3) - 1} = -\frac{1}{5}$$

$$a_4 = \frac{(-1)^4}{2(4) - 1} = \frac{1}{7}$$

$$a_{37} = \frac{(-1)^{37}}{2(37) - 1} = -\frac{1}{73}$$

Remark: Try finding the first four terms of the sequence whose nth term is

$$a_n = \frac{(-1)^{n+1}}{2n - 1}.$$

How do they differ from the first four terms of the sequence in Example 2?

Some very important sequences in mathematics involve terms that are defined with special types of products, called **factorials.**

Definition of Factorial

If n is a positive integer, then n **factorial** is defined by

$$n! = 1 \cdot 2 \cdot 3 \cdot 4 \cdots (n - 1) \cdot n.$$

As a special case, we define zero factorial to be $0! = 1$.

Sequences, Series, and Probability

Here are some values of $n!$.

$0! = 1$
$1! = 1$
$2! = 1 \cdot 2 = 2$
$3! = 1 \cdot 2 \cdot 3 = 6$
$4! = 1 \cdot 2 \cdot 3 \cdot 4 = 24$
$5! = 1 \cdot 2 \cdot 3 \cdot 4 \cdot 5 = 120$

The value of n does not have to be very large before the value of $n!$ becomes huge. For instance, $10! = 3,628,800$. Many scientific calculators have a factorial key, denoted by $\boxed{x!}$.

Factorials follow the same conventions for order of operation as do exponents. For instance,

$$2n! = 2(n!) = 2(1 \cdot 2 \cdot 3 \cdot 4 \cdots n)$$

whereas

$$(2n)! = 1 \cdot 2 \cdot 3 \cdot 4 \cdots n \cdot (n + 1) \cdots (2n).$$

EXAMPLE 3 *Finding Terms of a Sequence Involving Factorials*

List the first five terms of the sequence whose nth term is

$$a_n = \frac{2^n}{n!}.$$

Begin with $n = 0$.

SOLUTION

$$a_0 = \frac{2^0}{0!} = \frac{1}{1} = 1$$

$$a_1 = \frac{2^1}{1!} = \frac{2}{1} = 2$$

$$a_2 = \frac{2^2}{2!} = \frac{4}{2} = 2$$

$$a_3 = \frac{2^3}{3!} = \frac{8}{6} = \frac{4}{3}$$

$$a_4 = \frac{2^4}{4!} = \frac{16}{24} = \frac{2}{3}$$

EXAMPLE 4 *Finding Terms of a Sequence Involving Factorials*

List the first four terms of the sequence whose nth term is

$$a_n = \frac{(2n)!}{2^n n!}.$$

Begin with $n = 1$.

SOLUTION

$$a_1 = \frac{2!}{2^1(1!)} = \frac{1 \cdot 2}{2(1)} = \frac{1 \cdot \cancel{2}}{\cancel{2}} = 1$$

$$a_2 = \frac{4!}{2^2(2!)} = \frac{1 \cdot 2 \cdot 3 \cdot 4}{2 \cdot 2(1 \cdot 2)} = \frac{1 \cdot \cancel{2} \cdot 3 \cdot \cancel{4}}{\cancel{2} \cdot \cancel{4}} = 1 \cdot 3$$

$$a_3 = \frac{6!}{2^3(3!)} = \frac{1 \cdot 2 \cdot 3 \cdot 4 \cdot 5 \cdot 6}{2 \cdot 2 \cdot 2(1 \cdot 2 \cdot 3)} = \frac{1 \cdot \cancel{2} \cdot 3 \cdot \cancel{4} \cdot 5 \cdot \cancel{6}}{\cancel{2} \cdot \cancel{4} \cdot \cancel{6}} = 1 \cdot 3 \cdot 5$$

$$a_4 = \frac{8!}{2^4(4!)} = \frac{1 \cdot 2 \cdot 3 \cdot 4 \cdot 5 \cdot 6 \cdot 7 \cdot 8}{2 \cdot 2 \cdot 2 \cdot 2(1 \cdot 2 \cdot 3 \cdot 4)} = \frac{1 \cdot \cancel{2} \cdot 3 \cdot \cancel{4} \cdot 5 \cdot \cancel{6} \cdot 7 \cdot \cancel{8}}{\cancel{2} \cdot \cancel{4} \cdot \cancel{6} \cdot \cancel{8}}$$

$$= 1 \cdot 3 \cdot 5 \cdot 7$$

Notice the cancellation that took place in Example 4. It often happens in work with factorials that appropriate cancellation will greatly simplify an expression. For instance, the following types of cancellations are common.

$$\frac{n!}{(n-1)!} = \frac{1 \cdot 2 \cdot 3 \cdots (n-1) \cdot n}{1 \cdot 2 \cdot 3 \cdots (n-1)} = n$$

$$\frac{(2n-1)!}{(2n+1)!} = \frac{1 \cdot 2 \cdot 3 \cdots (2n-2)(2n-1)}{1 \cdot 2 \cdot 3 \cdots (2n-2)(2n-1)(2n)(2n+1)} = \frac{1}{2n(2n+1)}$$

It is important to realize that simply listing the first few terms is not sufficient to define a sequence—the nth term *must be given*. To see this, consider the following sequences, both of which have the *same first three terms*.

$$\frac{1}{2}, \frac{1}{4}, \frac{1}{8}, \frac{1}{16}, \cdots, \frac{1}{2^n}, \cdots$$

$$\frac{1}{2}, \frac{1}{4}, \frac{1}{8}, \frac{1}{15}, \cdots, \frac{6}{(n+1)(n^2-n+6)}, \cdots$$

Sequences, Series, and Probability

Sigma Notation

In the next two sections of this chapter we will work with the sum

$$a_1 + a_2 + a_3 + a_4 + \cdots + a_n$$

of the first n terms of the sequence $\{a_n\}$. A convenient shorthand notation for such a sum involves **sigma notation.** This name comes from the use of the upper-case Greek letter sigma, written as Σ.

Definition of Sigma Notation

If $\{a_n\}$ is a sequence, then the sum of the first n terms of the sequence is represented by

$$\sum_{i=1}^{n} a_i = a_1 + a_2 + a_3 + a_4 + \cdots + a_n$$

where i is called the **index of summation,** n is the **upper limit of summation,** and 1 is the **lower limit of summation.**

EXAMPLE 5 Sigma Notation for Sums

Find the following sums.

(a) $\displaystyle\sum_{i=1}^{5} i$
(b) $\displaystyle\sum_{k=3}^{6} (1 + k^2)$
(c) $\displaystyle\sum_{i=0}^{8} \frac{1}{i!}$

SOLUTION

(a) By letting i take on successive values from 1 to 5, we obtain

$$\sum_{i=1}^{5} i = 1 + 2 + 3 + 4 + 5 = 15.$$

(b) By letting k take on successive values from 3 to 6, we obtain

$$\begin{aligned}
\sum_{k=3}^{6} (1 + k^2) &= (1 + 3^2) + (1 + 4^2) + (1 + 5^2) + (1 + 6^2) \\
&= 10 + 17 + 26 + 37 \\
&= 90.
\end{aligned}$$

(c) By letting i take on successive values from 0 to 8, we obtain

$$\sum_{i=0}^{8} \frac{1}{i!} = \frac{1}{0!} + \frac{1}{1!} + \frac{1}{2!} + \frac{1}{3!} + \frac{1}{4!} + \frac{1}{5!} + \frac{1}{6!} + \frac{1}{7!} + \frac{1}{8!}$$

$$= 1 + 1 + \frac{1}{2} + \frac{1}{6} + \frac{1}{24} + \frac{1}{120} + \frac{1}{720} + \frac{1}{5040} + \frac{1}{40320}$$

$$\approx 2.71828.$$

Remark: Note that the lower index of a summation does not have to be 1. Also note that the index does not have to be the letter i. For instance, in part (b), the letter k is the index.

The following properties of sums are useful.

Properties of Sums

1. $\displaystyle\sum_{i=1}^{n} ca_i = c \sum_{i=1}^{n} a_i,$ c is any constant

2. $\displaystyle\sum_{i=1}^{n} (a_i + b_i) = \sum_{i=1}^{n} a_i + \sum_{i=1}^{n} b_i$

3. $\displaystyle\sum_{i=1}^{n} (a_i - b_i) = \sum_{i=1}^{n} a_i - \sum_{i=1}^{n} b_i$

PROOF

Each of these properties follows directly from the associative property of addition, the commutative property of addition, and the distributive property of multiplication over addition. For example, note the use of the distributive property in the proof of Property 1.

$$\sum_{i=1}^{n} ca_i = ca_1 + ca_2 + ca_3 + \cdots + ca_n$$

$$= c(a_1 + a_2 + a_3 + \cdots + a_n) = c \sum_{i=1}^{n} a_i$$

Variations in the upper and lower limits of summation can produce quite different-looking sigma notations for *the same sum.* For example, consider the following two sums.

$$\sum_{i=1}^{5} 3(2^i) = 3 \sum_{i=1}^{5} 2^i = 3(2^1 + 2^2 + 2^3 + 2^4 + 2^5)$$

$$\sum_{i=0}^{4} 3(2^{i+1}) = 3 \sum_{i=0}^{4} 2^{i+1} = 3(2^1 + 2^2 + 2^3 + 2^4 + 2^5)$$

One of the most important types of sequences used in mathematics is the sequence obtained by taking successive sums of terms of a given sequence. Consider the sequence

$$a_1, a_2, a_3, a_4, \ldots, a_n, \ldots$$

Sequences, Series, and Probability

By finding the successive sums

$$S_1 = a_1$$
$$S_2 = a_1 + a_2$$
$$S_3 = a_1 + a_2 + a_3$$
$$\vdots$$
$$S_n = a_1 + a_2 + a_3 + a_4 + \cdots + a_n$$

we can form a new sequence $\{S_n\}$, which is called the **sequence of partial sums.** In particular, S_n is called the ***n*th partial sum** of the sequence $\{a_n\}$. If there is a number S such that the value of S_n approaches S as n increases without bound, then we write

$$S = a_1 + a_2 + a_3 + a_4 + \cdots + a_n + \cdots = \sum_{i=1}^{\infty} a_i.$$

The expression $a_1 + a_2 + a_3 + a_4 + \cdots$ is called an **infinite series,** and the number S is called the sum of the infinite series.

EXAMPLE 6   *Finding a Sequence of Partial Sums*

For the sequence whose *n*th term is $a_n = 3/10^n$, find the first five partial sums, and use the results to find the sum of the infinite series

$$\sum_{i=1}^{\infty} \frac{3}{10^n}.$$

SOLUTION

Since the terms of the given sequence are

$$\frac{3}{10^1}, \frac{3}{10^2}, \frac{3}{10^3}, \frac{3}{10^4}, \frac{3}{10^5}, \cdots$$

$$0.3, 0.03, 0.003, 0.0003, 0.00003, \ldots$$

it follows that the first five partial sums are

$$S_1 = 0.3$$
$$S_2 = 0.3 + 0.03 = 0.33$$
$$S_3 = 0.3 + 0.03 + 0.003 = 0.333$$
$$S_4 = 0.3 + 0.03 + 0.003 + 0.0003 = 0.3333$$
$$S_5 = 0.3 + 0.03 + 0.003 + 0.0003 + 0.00003 = 0.33333.$$

From these results it appears that $S_n \rightarrow 0.3\overline{3}$ as $n \rightarrow \infty$, which suggests that

$$\sum_{i=1}^{\infty} \frac{3}{10^n} = 0.3\overline{3} = \frac{1}{3}.$$

WARM UP

1. Find $f(2)$ for $f(n) = \dfrac{2n}{n^2 + 1}$.

2. Find $f(3)$ for $f(n) = \dfrac{4}{3(n + 1)}$.

Factor the expressions.

3. $4n^2 - 1$

5. $n^2 - 3n + 2$

4. $4n^2 - 8n + 3$

6. $n^2 + 3n + 2$

Perform the indicated operations and/or simplify.

7. $\left(\frac{2}{3}\right)\left(\frac{3}{4}\right)\left(\frac{4}{5}\right)\left(\frac{5}{6}\right)$

8. $\dfrac{2 \cdot 4 \cdot 6 \cdot 8}{2^4}$

9. $\dfrac{1}{2 \cdot 2} + \dfrac{1}{2 \cdot 3} + \dfrac{1}{2 \cdot 4}$

10. $\dfrac{1}{1 \cdot 2} + \dfrac{1}{2 \cdot 3} + \dfrac{1}{3 \cdot 4}$

EXERCISES 12.1

In Exercises 1–20, write the first five terms of the indicated sequence. (Assume n begins with 1.)

1. $a_n = 2^n$

2. $a_n = \dfrac{n}{n + 1}$

3. $a_n = \left(-\dfrac{1}{2}\right)^n$

4. $a_n = \dfrac{n^2 - 1}{n^2 + 2}$

5. $a_n = \dfrac{3^n}{n!}$

6. $a_n = 5 - \dfrac{1}{n} + \dfrac{1}{n^2}$

7. $a_n = \dfrac{(-1)^n}{n^2}$

8. $a_n = \dfrac{3n!}{(n - 1)!}$

9. $a_n = \dfrac{n + 1}{n}$

10. $a_n = \dfrac{1}{n^{3/2}}$

11. $a_n = \dfrac{n^2 - 1}{n + 1}$

12. $a_n = 1 + (-1)^n$

13. $a_n = \dfrac{1 + (-1)^n}{n}$

14. $a_n = \dfrac{n!}{n}$

15. $a_n = 3 - \dfrac{1}{2^n}$

16. $a_n = \dfrac{3^n}{4^n}$

17. $a_n = (-1)^n\left(\dfrac{n}{n + 1}\right)$

18. $a_n = \dfrac{3n^2 - n + 4}{2n^2 + 1}$

19. $a_1 = 3$ and $a_{k+1} = 2(a_k - 1)$

20. $a_1 = 4$ and $a_{k+1} = \left(\dfrac{k + 1}{2}\right)a_k$

In Exercises 21–26, simplify the ratio of factorials.

21. $\dfrac{10!}{8!}$

22. $\dfrac{25!}{23!}$

23. $\dfrac{(n + 1)!}{n!}$

24. $\dfrac{(n + 2)!}{n!}$

25. $\dfrac{(2n - 1)!}{(2n + 1)!}$

26. $\dfrac{(2n + 2)!}{(2n)!}$

In Exercises 27–40, write an expression for the nth term of the sequence. (Assume n begins with 1.)

27. $1, 4, 7, 10, 13, \ldots$

28. $3, 7, 11, 15, 19, \ldots$

29. $0, 3, 8, 15, 24, \ldots$

30. $1, \frac{1}{4}, \frac{1}{9}, \frac{1}{16}, \frac{1}{25}, \ldots$

31. $\frac{2}{3}, \frac{3}{4}, \frac{4}{5}, \frac{5}{6}, \frac{6}{7}, \ldots$

32. $\frac{2}{1}, \frac{3}{3}, \frac{4}{5}, \frac{5}{7}, \frac{6}{9}, \ldots$

33. $\frac{1}{2}, \frac{-1}{4}, \frac{1}{8}, \frac{-1}{16}, \ldots$

34. $\frac{1}{3}, \frac{2}{9}, \frac{4}{27}, \frac{8}{81}, \ldots$

35. $1 + \frac{1}{1}, 1 + \frac{1}{2}, 1 + \frac{1}{3}, 1 + \frac{1}{4}, 1 + \frac{1}{5}, \ldots$

36. $1 + \frac{1}{2}, 1 + \frac{3}{4}, 1 + \frac{7}{8}, 1 + \frac{15}{16}, 1 + \frac{31}{32}, \ldots$

37. $1, \frac{1}{2}, \frac{1}{6}, \frac{1}{24}, \frac{1}{120}, \ldots$

38. $2, -4, 6, -8, 10, \ldots$

Sequences, Series, and Probability

39. $1, -1, 1, -1, 1, \ldots$

40. $1, x, \dfrac{x^2}{2}, \dfrac{x^3}{6}, \dfrac{x^4}{24}, \dfrac{x^5}{120}, \ldots$

In Exercises 41–50, find the given sum.

41. $\displaystyle\sum_{i=1}^{5} (2i + 1)$ **42.** $\displaystyle\sum_{i=1}^{6} 2i$

43. $\displaystyle\sum_{k=0}^{3} \dfrac{1}{k^2 + 1}$ **44.** $\displaystyle\sum_{j=3}^{5} \dfrac{1}{j}$

45. $\displaystyle\sum_{k=1}^{4} 10$ **46.** $\displaystyle\sum_{k=1}^{5} c, \quad c$ is constant

47. $\displaystyle\sum_{i=1}^{4} [(i - 1)^2 + (i + 1)^3]$

48. $\displaystyle\sum_{k=2}^{5} (k + 1)(k - 3)$

49. $\displaystyle\sum_{i=1}^{4} (x^2 + 2i)$ **50.** $\displaystyle\sum_{j=1}^{4} (-2)^{j-1}$

In Exercises 51–60, use sigma notation to write the given sum.

51. $\dfrac{1}{3(1)} + \dfrac{1}{3(2)} + \dfrac{1}{3(3)} + \cdots + \dfrac{1}{3(9)}$

52. $\dfrac{5}{1 + 1} + \dfrac{5}{1 + 2} + \dfrac{5}{1 + 3} + \cdots + \dfrac{5}{1 + 15}$

53. $[2(\frac{1}{8}) + 3] + [2(\frac{2}{8}) + 3] + \cdots + [2(\frac{8}{8}) + 3]$

54. $[1 - (\frac{1}{6})^2] + [1 - (\frac{2}{6})^2] + \cdots + [1 - (\frac{6}{6})^2]$

55. $3 - 9 + 27 - 81 + 243 - 729$

56. $1 - \frac{1}{2} + \frac{1}{4} - \frac{1}{8} + \cdots - \frac{1}{128}$

57. $\dfrac{1}{1^2} - \dfrac{1}{2^2} + \dfrac{1}{3^2} - \dfrac{1}{4^2} + \cdots - \dfrac{1}{20^2}$

58. $\dfrac{1}{1 \cdot 3} + \dfrac{1}{2 \cdot 4} + \dfrac{1}{3 \cdot 5} + \cdots + \dfrac{1}{10 \cdot 12}$

59. $\frac{1}{4} + \frac{3}{8} + \frac{7}{16} + \frac{15}{32} + \frac{31}{64}$

60. $\frac{1}{2} + \frac{2}{4} + \frac{6}{8} + \frac{24}{16} + \frac{120}{32} + \frac{720}{64}$

In Exercises 61 and 62, use the following definition of the arithmetic mean $\bar{x}$ of a set of n measurements $x_1, x_2, x_3, \ldots, x_n$.

$$\bar{x} = \frac{1}{n} \sum_{i=1}^{n} x_i$$

61. Prove that $\displaystyle\sum_{i=1}^{n} (x_i - \bar{x}) = 0$.

62. Prove that $\displaystyle\sum_{i=1}^{n} (x_i - \bar{x})^2 = \sum_{i=1}^{n} x_i^2 - \frac{1}{n}\left(\sum_{i=1}^{n} x_i\right)^2$.

63. A deposit of $5000 is made in an account that earns 8% interest compounded quarterly. The balance in the account after n quarters is given by

$$A_n = 5000\left(1 + \frac{0.08}{4}\right)^n, \quad n = 1, 2, 3, \ldots.$$

  (a) Compute the first eight terms of this sequence.
  (b) Find the balance in this account after 10 years by computing the fortieth term of the sequence.

64. A deposit of $100 is made *each* month in an account that earns 12% interest compounded monthly. The balance in the account after n months is given by

$$A_n = 100(101)[(1.01)^n - 1], \quad n = 1, 2, 3, \ldots.$$

  (a) Compute the first six terms of this sequence.
  (b) Find the balance after 5 years by computing the sixtieth term of the sequence.
  (c) Find the balance after 20 years by computing the two hundred fortieth term of the sequence.

12.2 Arithmetic Sequences

In this section we look at a special type of sequence called an **arithmetic sequence** (or **arithmetic progression**). In an arithmetic sequence, consecutive terms have a common difference. For instance, in the sequence defined by $a_n = 2 + 3n$, consecutive terms have a common difference of 3.

$$a_1 = 2 + 3 = 5$$

$$8 - 5 = 3$$

$$a_2 = 2 + 6 = 8$$

$$11 - 8 = 3$$

$$a_3 = 2 + 9 = 11$$

$$14 - 11 = 3$$

$$a_4 = 2 + 12 = 14$$

Definition of an Arithmetic Sequence

A sequence $\{a_n\}$ is **arithmetic** if each pair of consecutive terms differs by the same amount

$$d = a_{i+1} - a_i.$$

The number d is called the **common difference** in the sequence.

Remark: Note from this definition that we can write $a_{i+1} = a_i + d$. This is called a **recursive** formula because it defines a given term by reference to the preceding term.

EXAMPLE 1 Finding the Common Difference in an Arithmetic Sequence

Which of the sequences with the following nth terms is arithmetic? If the sequence is arithmetic, find the common difference.

(a) $a_n = 3 - 5n$ (b) $b_n = -2 + 3^n$

SOLUTION

(a) For the sequence given by $a_n = 3 - 5n$ the difference between two consecutive terms is

$$\begin{aligned} a_{i+1} - a_i &= [3 - 5(i + 1)] - [3 - 5(i)] \\ &= 3 - 5i - 5 - 3 + 5i \\ &= -5. \end{aligned}$$

Thus, the sequence *is* arithmetic, and the common difference is $d = -5$.

(b) For the sequence given by $b_n = -2 + 3^n$,

$$1, 7, 25, 79, \ldots$$

we see that the difference between the first two terms is 6, whereas the difference between the second and third terms is 18. Since these two differences are not the same, we conclude that the sequence is *not* arithmetic.

Remark: In Example 1(a), be sure you see that the common difference in an arithmetic sequence can be negative.

If you know the first term a_1 of an arithmetic sequence and the common difference d, you can form the other terms by repeatedly adding d.

$$
\begin{aligned}
a_1 &= a_1 & &= a_1 + (0)d = a_1 + (1 - 1)d \\
a_2 &= a_1 + d & &= a_1 + (1)d = a_1 + (2 - 1)d \\
a_3 &= a_1 + d + d & &= a_1 + (2)d = a_1 + (3 - 1)d \\
a_4 &= a_1 + d + d + d & &= a_1 + (3)d = a_1 + (4 - 1)d \\
a_5 &= a_1 + d + d + d + d &= a_1 + (4)d = a_1 + (5 - 1)d
\end{aligned}
$$

This leads to the following rule.

The nth Term of an Arithmetic Sequence

The nth term of an arithmetic sequence whose first term is a_1 and whose common difference is d is given by

$$a_n = a_1 + (n - 1)d.$$

EXAMPLE 2 *Finding the Terms of an Arithmetic Sequence*

Write the first four terms of the arithmetic sequence whose first term is -1 and whose common difference is 4. What is the twentieth term of the sequence?

SOLUTION

Since $a_1 = -1$ and $d = 4$, we have

$$
\begin{aligned}
a_1 &= -1 \\
a_2 &= -1 + 1(4) = 3 \\
a_3 &= -1 + 2(4) = 7 \\
a_4 &= -1 + 3(4) = 11.
\end{aligned}
$$

The twentieth term of the sequence is

$$a_{20} = a_1 + (20 - 1)d = -1 + 19(4) = 75.$$

EXAMPLE 3   *Finding the nth Term of an Arithmetic Sequence*

Find the sixty-fourth term of the arithmetic sequence whose first three terms are -1, 11, and 23.

SOLUTION

For this sequence we find that $a_1 = -1$ and $a_2 = 11$ so that $d = 11 - (-1) = 12$. Therefore, to find the sixty-fourth term of the sequence we let $n = 64$ and write

$$a_{64} = a_1 + (n - 1)d = -1 + 63(12) = -1 + 756 = 755.$$

EXAMPLE 4   *Finding the nth Term of an Arithmetic Sequence*

Find the thirty-eighth term of the arithmetic sequence whose first term is 8 and whose ith term is given by

$$a_{i+1} = a_i - 7.$$

SOLUTION

From the formula $a_{i+1} = a_i - 7$, we can see that

$$d = a_{i+1} - a_i = (a_i - 7) - a_i = -7.$$

Since we are given that $a_1 = 8$, the thirty-eighth term of the sequence is

$$a_{38} = a_1 + (n - 1)d = 8 + 37(-7) = 8 - 259 = -251.$$

EXAMPLE 5   *Finding the nth Term of an Arithmetic Sequence*

The seventh term of an arithmetic sequence is 55, and the twenty-second term is 145. Find the eighteenth term.

Sequences, Series, and Probability

SOLUTION

Using the nth-term formula, we have the following.

$$a_7 = a_1 + 6d \qquad \text{and} \qquad a_{22} = a_1 + 21d$$
$$55 = a_1 + 6d \qquad \qquad \qquad 145 = a_1 + 21d$$

Solving this system of linear equations for d produces the following.

$$
\begin{aligned}
a_1 + 21d &= 145 \\
-a_1 - 6d &= -55 \\
\hline
15d &= 90 \\
d &= 6
\end{aligned}
$$

Back-substituting the value $d = 6$ into the first equation yields $a_1 = 19$. Thus, the eighteenth term is

$$a_{18} = a_1 + (n - 1)d = 19 + 17(6) = 121.$$

Arithmetic Mean

Recall that $(a + b)/2$ is the midpoint between the two numbers a and b on the real number line. As a result, the terms

$$a, \frac{a + b}{2}, b$$

have a common difference. We call $(a + b)/2$ the **arithmetic mean** of the numbers a and b. We can generalize this concept by finding k numbers m_1, $m_2, m_3, \ldots, m_k$ between a and b such that the terms

$$a, m_1, m_2, m_3, \ldots, m_k, b$$

have a common difference. This process is referred to as **inserting k arithmetic means** between a and b.

EXAMPLE 6 *Inserting Arithmetic Means Between Two Numbers*

Insert three arithmetic means between 4 and 15.

SOLUTION

We need to find three numbers m_1, m_2, and m_3 such that the terms

$$4, m_1, m_2, m_3, 15$$

have a common difference. In this case we have $a_1 = 4$, $n = 5$, and $a_5 = 15$. Therefore,

$$a_5 = 15 = a_1 + (n - 1)d = 4 + 4d.$$

Since $15 = 4 + 4d$, we find that $d = 11/4$, and the three arithmetic means are

$$m_1 = a_1 + d = 4 + \frac{11}{4} = \frac{27}{4}$$

$$m_2 = m_1 + d = \frac{27}{4} + \frac{11}{4} = \frac{38}{4}$$

$$m_3 = m_2 + d = \frac{38}{4} + \frac{11}{4} = \frac{49}{4}.$$

We now develop a formula for finding the *nth partial sum* of an arithmetic sequence. For example, the arithmetic sequence

$$1, 3, 5, 7, 9, \ldots, 2n - 1, \ldots$$

gives rise to the following partial sums.

$$S_1 = 1 = 1^2$$
$$S_2 = 1 + 3 = 4 = 2^2$$
$$S_3 = 1 + 3 + 5 = 9 = 3^2$$
$$S_4 = 1 + 3 + 5 + 7 = 16 = 4^2$$
$$S_5 = 1 + 3 + 5 + 7 + 9 = 25 = 5^2$$

Judging from this pattern, the *n*th partial sum

$$S_n = 1 + 3 + 5 + 7 + 9 + \cdots + 2n - 1 = \sum_{i=1}^{n} (2i - 1)$$

appears to be n^2. We can verify this observation using the following formula for finding the *n*th partial sum of an arithmetic sequence.

The nth Partial Sum of an Arithmetic Sequence

The *n*th partial sum of the arithmetic sequence $\{a_n\}$ with common difference d is given by either of the following formulas.

1. $S_n = \dfrac{n}{2}(a_1 + a_n)$

2. $S_n = \dfrac{n}{2}[2a_1 + (n - 1)d]$

Sequences, Series, and Probability

PROOF

We begin by generating the terms of the arithmetic sequence in two ways. In the first way, we repeatedly add d to the first term to obtain

$$S_n = a_1 + a_2 + a_3 + \cdots + a_{n-2} + a_{n-1} + a_n$$
$$= a_1 + [a_1 + d] + [a_1 + 2d] + \cdots + [a_1 + (n-1)d].$$

In the second way, we repeatedly subtract d from the nth term to obtain

$$S_n = a_n + a_{n-1} + a_{n-2} + \cdots + a_3 + a_2 + a_1$$
$$= a_n + [a_n - d] + [a_n - 2d] + \cdots + [a_n - (n-1)d].$$

If we add these two versions of S_n, the multiples of d cancel and we obtain

$$\overbrace{2S_n = (a_1 + a_n) + (a_1 + a_n) + (a_1 + a_n) + \cdots + (a_1 + a_n)}^{n \text{ terms}}$$
$$= n(a_1 + a_n).$$

Thus, we have

$$S_n = \frac{n}{2}(a_1 + a_n).$$

But since $a_n = a_1 + (n-1)d$, we can also write

$$S_n = \frac{n}{2}[2a_1 + (n-1)d].$$

EXAMPLE 7 *Finding the nth Partial Sum of an Arithmetic Sequence*

Verify the formula

$$S_n = 1 + 3 + 5 + \cdots + (2n-1) = n^2.$$

SOLUTION

Using the first formula for the nth partial sum of an arithmetic sequence, we can write

$$S_n = 1 + 3 + 5 + \cdots + (2n-1)$$

$$= \frac{n}{2}(a_1 + a_n)$$

$$= \frac{n}{2}[1 + (2n-1)]$$

$$= \frac{n}{2}(2n)$$

$$= n^2.$$

EXAMPLE 8 *Finding the nth Partial Sum of an Arithmetic Sequence*

Find the sum of the first 99 terms of the arithmetic sequence whose nth term is

$$a_n = 7 + \frac{n}{2}.$$

SOLUTION

Using $n = 99$, $a_1 = 7 + \frac{1}{2}$, and $a_{99} = 7 + \frac{99}{2}$, we find the sum of the first 99 terms of the sequence to be

$$S_{99} = \frac{n}{2}(a_1 + a_n) = \frac{99}{2}\left[\left(7 + \frac{1}{2}\right) + \left(7 + \frac{99}{2}\right)\right]$$

$$= \frac{99}{2}\left(\frac{15}{2} + \frac{113}{2}\right)$$

$$= \frac{99}{2}\left(\frac{128}{2}\right)$$

$$= 3168.$$

EXAMPLE 9 *Finding the nth Partial Sum of an Arithmetic Sequence*

Find the sum of the first 150 terms of the arithmetic sequence

$$5, 16, 27, 38, 49, \ldots.$$

SOLUTION

For this sequence, we have $a_1 = 5$ and $d = 16 - 5 = 11$. Therefore, for $n = 150$, the sum of the first 150 terms is

$$S_{150} = \frac{n}{2}[2a_1 + (n - 1)d] = \frac{150}{2}[2(5) + 149(11)]$$

$$= 75(10 + 1639)$$

$$= 123{,}675.$$

EXAMPLE 10 *An Application to Business*

A small business sells \$10,000 worth of products during its first year. The owner of the business has set a goal of increasing annual sales by \$7500 each year for 9 years. Assuming that this goal is met, find the total sales during the first 10 years this business is in operation.

Sequences, Series, and Probability

SOLUTION

The annual sales form an arithmetic sequence in which $a_1 = 10,000$ and $d = 7500$. We can find the total sales for the first 10 years as follows.

$$S_{10} = \frac{n}{2}[2a_1 + (n-1)d] = \frac{10}{2}[20,000 + 9(7500)]$$
$$= 5(87,500)$$
$$= \$437,500$$

WARM UP

Find the sums.

1. $\sum_{i=1}^{6} (2i - 1)$

2. $\sum_{i=1}^{10} (4i + 2)$

Find the distance between the two real numbers.

3. $\frac{5}{2}, 8$

4. $\frac{4}{3}, \frac{14}{3}$

5. Find $f(3)$ for $f(n) = 10 + (n-1)4$.

6. Find $f(10)$ for $f(n) = 1 + (n-1)\frac{1}{3}$.

Evaluate the expressions.

7. $\frac{11}{2}(1 + 25)$

8. $\frac{16}{2}(4 + 16)$

9. $\frac{20}{2}[2(5) + (12-1)3]$

10. $\frac{8}{2}[2(-3) + (15-1)5]$

EXERCISES 12.2

In Exercises 1–10, determine whether the sequence is arithmetic. If it is, find the common difference.

1. 4, 7, 10, 13, 16, . . .   **2.** 10, 8, 6, 4, 2, . . .

3. 1, 2, 4, 8, 16, . . .   **4.** $3, \frac{5}{2}, 2, \frac{3}{2}, 1, \ldots$

5. $\frac{9}{4}, 2, \frac{7}{4}, \frac{3}{2}, \frac{5}{4}, \ldots$ **6.** $-12, -8, -4, 0, 4, \ldots$

7. $\frac{1}{3}, \frac{2}{3}, \frac{4}{3}, \frac{8}{3}, \frac{16}{3}, \ldots$

8. ln 1, ln 2, ln 3, ln 4, ln 5, . . .

9. 5.3, 5.7, 6.1, 6.5, 6.9, . . .

10. $1^2, 2^2, 3^2, 4^2, 5^2, \ldots$

In Exercises 11–18, write the first five terms of the specified sequence. Determine whether the sequence is arithmetic, and if it is, find the common difference.

11. $a_n = 5 + 3n$

12. $a_n = (2^n)n$

13. $a_n = \dfrac{1}{n+1}$

14. $a_n = 1 + (n-1)4$

15. $a_n = 100 - 3n$

16. $a_n = 2^{n-1}$

17. $a_1 = 1, a_2 = 1, a_n = a_{n-1} + a_{n-2}, \quad n \geq 3$

18. $a_n = n \sin \dfrac{n\pi}{2}$

In Exercises 19–26, find a_n for the given arithmetic sequence.

19. $a_1 = 1, d = 3, n = 10$

20. $a_1 = 15, d = 4, n = 25$

21. $a_1 = 100, d = -8, n = 8$

22. $a_1 = 0$, $d = -\frac{2}{3}$, $n = 12$

23. $a_1 = x$, $d = 2x$, $n = 50$

24. $a_1 = -y$, $d = 5y$, $n = 10$

25. $4, \frac{3}{2}, -1, -\frac{7}{2}, \ldots$, $n = 10$

26. $10, 5, 0, -5, -10, \ldots$, $n = 50$

In Exercises 27–36, write the first five terms of the arithmetic sequence.

27. $a_1 = 5$, $d = 6$ **28.** $a_1 = 5$, $d = -\frac{3}{4}$

29. $a_1 = -2.6$, $d = -0.4$

30. $a_1 = 6$, $a_{k+1} = a_k + 12$

31. $a_1 = \frac{3}{2}$, $a_{k+1} = a_k - \frac{1}{4}$

32. $a_5 = 28$, $a_{10} = 53$

33. $a_1 = 2$, $a_{12} = 46$ **34.** $a_4 = 16$, $a_{10} = 46$

35. $a_8 = 26$, $a_{12} = 42$ **36.** $a_3 = 1.9$, $a_{15} = -1.7$

In Exercises 37–44, find the *n*th partial sum of the arithmetic sequence.

37. $8, 20, 32, 44, \ldots$, $n = 10$

38. $2, 8, 14, 20, \ldots$, $n = 25$

39. $-6, -2, 2, 6, \ldots$, $n = 50$

40. $0.5, 0.9, 1.3, 1.7, \ldots$, $n = 10$

41. $40, 37, 34, 31, \ldots$, $n = 10$

42. $1.50, 1.45, 1.40, 1.35, \ldots$, $n = 20$

43. $a_1 = 100$, $a_{25} = 220$, $n = 25$

44. $a_1 = 15$, $a_{100} = 307$, $n = 100$

In Exercises 45–54, find the indicated sum.

45. $\displaystyle\sum_{n=1}^{50} n$ **46.** $\displaystyle\sum_{n=1}^{100} 2n$

47. $\displaystyle\sum_{n=1}^{100} 5n$ **48.** $\displaystyle\sum_{n=51}^{100} n - \sum_{n=1}^{50} n$

49. $\displaystyle\sum_{n=1}^{500} (n + 3)$ **50.** $\displaystyle\sum_{n=1}^{250} (1000 - n)$

51. $\displaystyle\sum_{n=1}^{20} (2n + 5)$ **52.** $\displaystyle\sum_{n=1}^{100} \frac{n + 4}{2}$

53. $\displaystyle\sum_{n=0}^{50} (1000 - 5n)$ **54.** $\displaystyle\sum_{n=0}^{100} \frac{8 - 3n}{16}$

55. Find the sum of the first 100 odd integers.

56. Find the sum of the integers from -10 to 50.

57. A person accepts a position with a company and will receive a salary of $27,500 for the first year. The person is guaranteed a raise of $1500 per year for the first five years.

  (a) What will the salary be during the sixth year of employment?

  (b) How much will the company have paid the person by the end of the six years?

58. A freely falling object will fall 16 feet during the first second, 48 feet during the second second, 80 feet during the third second, 112 feet during the fourth second, and so on. What is the total distance the object will fall in 10 seconds if this pattern continues?

59. Determine the seating capacity of an auditorium with 30 rows of seats if there are 20 seats in the first row, 24 seats in the second row, 28 seats in the third row, and so on.

60. As a farmer bales a field of hay, each trip around the field gets shorter, since he is getting closer to the center. Suppose on the first round there were 267 bales and on the second round there were 253 bales. Assume that the decrease will be the same on each round and that there are 11 more trips. How many bales of hay will the farmer get from the field?

In Exercises 61–64, insert *k* arithmetic means between the given pair of numbers.

61. $5, 17$, $k = 2$ **62.** $24, 56$, $k = 3$

63. $3, 6$, $k = 3$ **64.** $2, 5$, $k = 4$

12.3 Geometric Sequences and Series

In this section we look at another important type of sequence called a **geometric sequence** (or **geometric progression**), characterized by the fact that consecutive terms have a common *ratio*. For instance, in the sequence defined by $a_n = 3(2^{n-1})$, the consecutive terms have a common ratio of 2.

$$a_1 = 3(2^0) = 3$$
$$a_2 = 3(2^1) = 6$$
$$a_3 = 3(2^2) = 12$$
$$a_4 = 3(2^3) = 24$$

$$\frac{6}{3} = 2$$
$$\frac{12}{6} = 2$$
$$\frac{24}{12} = 2$$

Definition of a Geometric Sequence

A sequence $\{a_n\}$ is **geometric** if each pair of consecutive terms has the same (nonzero) ratio

$$r = \frac{a_{i+1}}{a_i}, \qquad r \neq 0.$$

The number r is called the **common ratio** of the sequence.

Remark: Note that this definition yields the *recursive* formula $a_{i+1} = ra_i$.

EXAMPLE 1 *Finding the Common Ratio of a Geometric Sequence*

Determine which of the following sequences are geometric.

(a) $a_n = \dfrac{2}{3^n}$　　　　　(b) $1, -1, 1, -1, 1, \ldots$　　　　　(c) $a_n = n!$

SOLUTION

(a) For the sequence given by $a_n = 2/3^n$, the ratio of consecutive terms is

$$r = \frac{a_{i+1}}{a_i} = \frac{2/3^{i+1}}{2/3^i} = \frac{2}{3^{i+1}}\left(\frac{3^i}{2}\right) = \frac{1}{3}.$$

Thus, the sequence *is* geometric.

(b) For this sequence the ratio of consecutive terms is -1. Thus, the sequence *is* geometric.

(c) For the sequence

$$1, 2, 6, 24, 120, 720, \ldots$$

given by $a_n = n!$, the ratio of the first two terms is 2 and the ratio of the third term to the second is 3. Since these two ratios differ, we conclude that the sequence is *not* geometric.

Each successive term in a geometric sequence can be obtained by multiplying the preceding term by r.

$$a_1, a_1r, a_1r^2, a_1r^3, a_1r^4, \ldots, a_1r^{n-1}, \ldots$$

$$\downarrow \quad \downarrow \quad \downarrow \quad \downarrow \quad \downarrow \qquad \qquad \downarrow$$

$$a_1, \ a_2, \ a_3, \ a_4, \ a_5, \ \ldots, \quad a_n, \quad \ldots$$

From this, we can see that the nth term of a geometric sequence has the following form.

The nth Term of a Geometric Sequence

The nth term of a geometric sequence whose first term is a_1 and whose common ratio is r is given by

$$a_n = a_1 r^{n-1}.$$

EXAMPLE 2   Finding the Terms of a Geometric Sequence

Write the first five terms of the geometric sequence whose first term is $a_1 = 3$ and whose common ratio is $r = -2$.

SOLUTION

Starting with 3, we successively multiply by -2 to obtain the following.

$$a_1 = 3$$
$$a_2 = 3(-2)^1 = -6$$
$$a_3 = 3(-2)^2 = 12$$
$$a_4 = 3(-2)^3 = -24$$
$$a_5 = 3(-2)^4 = 48$$

EXAMPLE 3 *Finding a Term of a Geometric Sequence*

Find the fifteenth term of the geometric sequence whose first term is 20 and whose common ratio is 1.05.

SOLUTION

Since $a_1 = 20$ and $r = 1.05$, the fifteenth term $(n = 15)$ is

$$a_{15} = a_1 r^{n-1} = 20(1.05)^{14} \approx 39.599.$$

EXAMPLE 4 *Finding a Term of a Geometric Sequence*

Find the twelfth term of the following geometric sequence

$$5, -15, 45, \ldots.$$

SOLUTION

The common ratio of this sequence is $r = -15/5 = -3$. Therefore, since the first term is $a_1 = 5$, we determine the twelfth term $(n = 12)$ to be

$$a_{12} = a_1 r^{n-1} = 5(-3)^{11} = 5(-177,147) = -885,735.$$

EXAMPLE 5 *Finding a Term of a Geometric Sequence*

The fourth term of a geometric sequence is 125, and the tenth term is 125/64. Find the fourteenth term.

SOLUTION

Using the nth-term formula for a geometric sequence, we can write the following two equations:

$$a_4 = a_1 r^3 = 125$$

and

$$a_{10} = a_1 r^9 = \frac{125}{64}.$$

Solving for a_1 in the first equation, we have $a_1 = 125/r^3$. Then, by substituting in the second equation, we obtain

$$\left(\frac{125}{r^3}\right)r^9 = \frac{125}{64}$$

$$r^6 = \frac{1}{64}$$

$$r = \frac{1}{2}.$$

By back-substituting $r = 1/2$ into the equation $a_1 r^3 = 125$, we obtain

$$a_1\left(\frac{1}{2}\right)^3 = 125$$

$$a_1 = 125(2^3) = 1000.$$

Finally, by the nth-term formula, the fourteenth term is

$$a_{14} = a_1 r^{n-1} = 1000\left(\frac{1}{2}\right)^{13} = \frac{1000}{8192} = \frac{125}{1024}.$$

Geometric Series

An infinite series that is formed from a geometric sequence is called an **infinite geometric series** (or simply a geometric series). For example, the geometric sequence

$$a_1, a_1 r, a_1 r^2, a_1 r^3, a_1 r^4, \ldots$$

gives rise to the following geometric series:

$$a_1 + a_1 r + a_1 r^2 + a_1 r^3 + a_1 r^4 + \cdots.$$

The formula for the nth partial sum of a geometric sequence

$$S_n = a_1 + a_1 r + a_1 r^2 + \cdots + a_1 r^{n-1} = \sum_{i=1}^{n} a_1 r^{i-1}$$

is given as follows.

The nth Partial Sum of a Geometric Sequence

The nth partial sum of the geometric sequence $\{a_n\}$ with common ratio $r \neq 1$ is given by

$$S_n = a_1\left(\frac{1 - r^n}{1 - r}\right).$$

Sequences, Series, and Probability

PROOF

We begin by writing out the *n*th partial sum.

$$S_n = a_1 + a_1 r + a_1 r^2 + \cdots + a_1 r^{n-2} + a_1 r^{n-1}$$

Multiplication by *r* yields

$$r S_n = a_1 r + a_1 r^2 + a_1 r^3 + \cdots + a_1 r^{n-1} + a_1 r^n.$$

Subtracting the second equation from the first gives us

$$S_n - r S_n = a_1 - a_1 r^n.$$

Therefore, $S_n(1 - r) = a_1(1 - r^n)$, and since $r \neq 1$, we have

$$S_n = a_1 \left(\frac{1 - r^n}{1 - r} \right).$$

When using the formula for the *n*th partial sum of a geometric sequence, be careful to check that the index begins at $i = 1$. If the index begins at $i = 0$, you must adjust the formula for the *n*th partial sum, as demonstrated in the following example.

EXAMPLE 6 *Finding the nth Partial Sum of a Geometric Sequence*

Find the following sums.

(a) $\displaystyle\sum_{i=1}^{12} 4(0.3)^{i-1}$

(b) $\displaystyle\sum_{i=0}^{10} 10\left(-\frac{1}{2}\right)^i$

SOLUTION

(a) By writing out a few terms, we have

$$\sum_{i=1}^{12} 4(0.3)^{i-1} = 4 + 4(0.3) + 4(0.3)^2 + \cdots + 4(0.3)^{11}.$$

Now, since $a_1 = 4$, $r = 0.3$, and $n = 12$, we apply the formula for the *n*th partial sum of a geometric sequence to obtain

$$\sum_{i=1}^{12} 4(0.3)^{i-1} = a_1 \left(\frac{1 - r^n}{1 - r} \right) = 4 \left(\frac{1 - (0.3)^{12}}{1 - 0.3} \right) \approx 5.714.$$

(b) By writing out a few terms, we have

$$\sum_{i=0}^{10} 10\left(-\frac{1}{2}\right)^i = 10 - 10\left(\frac{1}{2}\right) + 10\left(\frac{1}{2}\right)^2 - \cdots + 10\left(\frac{1}{2}\right)^{10}.$$

Geometric Sequences and Series

Now, we see that the *first term* is $a_1 = 10$ and $r = -1/2$. Moreover, by starting with $i = 0$ and ending with $i = 10$, we are adding $n = 11$ terms, which means that the partial sum is

$$\sum_{i=0}^{10} 10\left(-\frac{1}{2}\right)^i = a_1\left(\frac{1 - r^n}{1 - r}\right) = 10\left(\frac{1 - (-1/2)^{11}}{1 + (1/2)}\right) \approx 6.670.$$

EXAMPLE 7   An Application: Compound Interest

A deposit of \$50 is made the first day of each month in a savings account that pays 12% compounded monthly. What is the balance at the end of two years?

SOLUTION

The formula for compound interest is

$$A = P\left(1 + \frac{r}{n}\right)^{tn}$$

where A is the balance of the account, P is the initial deposit, r is the annual percentage rate, n is the number of compoundings per year, and t is the time (in years). To find the balance in the account after 24 months, it is helpful to consider each of the 24 deposits separately. For example, the first deposit will gain interest for a full 24 months, and its balance will be

$$A_{24} = 50\left(1 + \frac{0.12}{12}\right)^{24}.$$

The second deposit will gain interest for 23 months, and its balance will be

$$A_{23} = 50\left(1 + \frac{0.12}{12}\right)^{23}.$$

The last (twenty-fourth) deposit will gain interest for only 1 month, and its balance will be

$$A_1 = 50\left(1 + \frac{0.12}{12}\right)^1 = 50(1.01).$$

Finally, the total balance in the account will be the sum of the balances of the 24 deposits.

$$S_{24} = A_1 + A_2 + A_3 + \cdots + A_{23} + A_{24}$$

Using the formula for the nth partial sum of a geometric sequence, with $A_1 = 50(1.01)$ and $r = 1.01$, we have

$$S_{24} = 50(1.01)\left(\frac{1 - (1.01)^{24}}{1 - 1.01}\right) = \$1362.16.$$

The formula for the *n*th partial sum of a geometric sequence involves the sum of a *finite* number of terms. Occasionally, we can add all the terms of an *infinite* geometric sequence and still obtain a finite sum. Specifically, if the common ratio *r* has the property that $|r| < 1$, then it can be shown that r^n becomes arbitrarily close to zero as *n* increases without bound. Consequently,

$$S_n \rightarrow \frac{a_1(1 - 0)}{1 - r} = \frac{a_1}{1 - r} \quad \text{as} \quad n \rightarrow \infty.$$

The number $a_1/(1 - r)$ is called the **sum of the infinite geometric series**

$$\sum_{n=1}^{\infty} a_1 r^{n-1} = a_1 + a_1 r + a_1 r^2 + a_1 r^3 + \cdots + a_1 r^{n-1} + \cdots$$

as summarized in the following statement.

Sum of an Infinite Geometric Series

If $|r| < 1$, then the infinite geometric series

$$a_1 + a_1 r + a_1 r^2 + a_1 r^3 + \cdots + a_1 r^{n-1} + \cdots$$

has the sum

$$S = \frac{a_1}{1 - r}.$$

EXAMPLE 8 *Finding the Sum of an Infinite Geometric Series*

Find the sum of the following infinite geometric series.

$$\sum_{n=1}^{\infty} 4(-0.6)^{n-1}$$

SOLUTION

Since $a_1 = 4$, $r = -0.6$, and $|r| < 1$, we have

$$\sum_{n=1}^{\infty} 4(-0.6)^{n-1} = \frac{a_1}{1 - r}$$

$$= \frac{4}{1 - (-0.6)} = 2.5.$$

Geometric Sequences and Series

When we introduced the real number system in Section 1.1, we noted that if a real number has a decimal representation that indefinitely repeats a block of digits, then the real number must be rational. For example, the repeating decimals

$$0.1111. . . = 0.\overline{1} \quad \text{and} \quad 5.2135135135. . . = 5.2\overline{135}$$

represent rational numbers. To write a repeating decimal as the ratio of two integers, we can use a geometric series, as shown in the following example.

EXAMPLE 9 *Writing a Repeating Decimal as a Ratio of Two Integers*

Write the following repeating decimals as the ratio of two integers.

(a) $0.2\overline{2}$

(b) $2.1\overline{425}$

SOLUTION

(a) As an infinite geometric series, we have

$$0.222. . . = 0.2 + 0.02 + 0.002 + \cdots$$
$$= 0.2 + 0.2(0.1) + 0.2(0.1)^2 + \cdots .$$

Thus, $a_1 = 0.2$ and $r = 0.1$, and by the formula for the sum of an infinite geometric series, we have

$$0.2\overline{2} = \frac{a_1}{1 - r} = \frac{0.2}{1 - 0.1} = \frac{0.2}{0.9} = \frac{2}{9}.$$

(b) Since the first two digits are not part of the repeating pattern, we rewrite the number as follows.

$$2.1425425425. . . = 2.1 + 0.0425425425. . .$$

For the repeating portion, we write

$$0.0\overline{425} = 0.0425 + 0.0000425 + 0.0000000425 + \cdots$$
$$= 0.0425 + 0.0425(0.001) + 0.0425(0.001)^2 + \cdots$$
$$= \frac{a_1}{1 - r} = \frac{0.0425}{1 - 0.001} = \frac{0.0425}{0.999} = \frac{425}{9990}.$$

Finally, by adding 2.1 to this result, we have

$$2.1\overline{425} = \frac{21}{10} + \frac{425}{9990} = \frac{20979}{9990} + \frac{425}{9990}$$

$$= \frac{21404}{9990} = \frac{10702}{4995}.$$

WARM UP

Evaluate the expressions.

1. $\left(\frac{4}{5}\right)^3$

2. $\left(\frac{3}{4}\right)^2$

3. 2^{-4}

4. $\dfrac{5}{3^4}$

Simplify the expressions.

5. $(2n)(3n^2)$

6. $n(3n)^3$

7. $\dfrac{4n^5}{n^2}$

8. $\dfrac{(2n)^3}{8n}$

9. $[2(3)^{-4}]^n$

10. $3(4^2)^{-n}$

EXERCISES 12.3

In Exercises 1–10, determine whether the sequence is geometric. If it is, find its common ratio.

1. 5, 15, 45, 135, . . .

2. 3, 12, 48, 192, . . .

3. 3, 12, 21, 30, . . .

4. 1, −2, 4, −8, . . .

5. 1, $-\frac{1}{2}$, $\frac{1}{4}$, $-\frac{1}{8}$, . . .

6. 5, 1, 0.2, 0.04, . . .

7. $\frac{1}{2}$, $\frac{2}{3}$, $\frac{3}{4}$, $\frac{4}{5}$, . . .

8. 9, −6, 4, $-\frac{8}{3}$, . . .

9. 1, $\frac{1}{2}$, $\frac{1}{3}$, $\frac{1}{4}$, . . .

10. $\frac{1}{5}$, $\frac{2}{3}$, $\frac{3}{9}$, $\frac{4}{11}$, . . .

In Exercises 11–18, write the first five terms of the geometric sequence.

11. $a_1 = 2$, $r = 3$

12. $a_1 = 6$, $r = 2$

13. $a_1 = 1$, $r = \frac{1}{2}$

14. $a_1 = 1$, $r = \frac{1}{3}$

15. $a_1 = 5$, $r = -\frac{1}{10}$

16. $a_1 = 1$, $r = -x$

17. $a_1 = 1$, $r = \dfrac{x}{2}$

18. $a_1 = 2$, $r = \sqrt{3}$

In Exercises 19–30, find the nth term of the geometric sequence.

19. $a_1 = 4$, $r = \frac{1}{2}$, $n = 10$

20. $a_1 = 5$, $r = \frac{3}{2}$, $n = 8$

21. $a_1 = 6$, $r = -\frac{1}{3}$, $n = 12$

22. $a_1 = 1$, $r = -\dfrac{x}{3}$, $n = 7$

23. $a_1 = 100$, $r = e^x$, $n = 9$

24. $a_1 = 8$, $r = \sqrt{5}$, $n = 9$

25. $a_1 = 500$, $r = 1.02$, $n = 40$

26. $a_1 = 1000$, $r = 1.005$, $n = 60$

27. $a_1 = 16$, $a_4 = \frac{27}{4}$, $n = 3$

28. $a_2 = 3$, $a_5 = \frac{3}{64}$, $n = 1$

29. $a_2 = -18$, $a_5 = \frac{2}{3}$, $n = 6$

30. $a_3 = \frac{16}{3}$, $a_5 = \frac{64}{27}$, $n = 7$

31. A sum of \$1000 is invested at 10% interest. Find the amount after 10 years if the interest is compounded
   (a) annually   (b) semiannually
   (c) quarterly   (d) monthly
   (e) daily.

32. A total of \$2500 is invested at 12% interest. Find the amount after 20 years if the interest is compounded
   (a) annually   (b) semiannually
   (c) quarterly   (d) monthly
   (e) daily.

33. A company buys a machine for \$135,000 that depreciates at the rate of 30% per year. (In other words, at the end of each year the depreciated value is 70% of what it was at the beginning of the year.) Find the depreciated value of the machine after five full years.

34. A city of 250,000 people is growing at the rate of 1.3% per year. Estimate the population of the city 30 years from now.

In Exercises 35–40, find the given sum.

35. $\sum_{n=0}^{20} 3(\frac{3}{2})^n$

36. $\sum_{n=0}^{15} 2(\frac{4}{3})^n$

37. $\sum_{i=1}^{10} 8(\frac{-1}{4})^{i-1}$

38. $\sum_{i=1}^{10} 5(\frac{-1}{3})^{i-1}$

39. $\sum_{n=0}^{8} 2^n$

40. $\sum_{n=0}^{8} (-2)^n$

41. A deposit of $100 is made at the beginning of each month for five years in an account that pays 10%, compounded monthly. What is the balance A in the account at the end of five years?

$$A = 100\left(1 + \frac{0.10}{12}\right)^1 + \cdots + 100\left(1 + \frac{0.10}{12}\right)^{60}$$

42. A deposit of $50 is made at the beginning of each month for five years in an account that pays 12%, compounded monthly. What is the balance A in the account at the end of five years?

$$A = 50\left(1 + \frac{0.12}{12}\right)^1 + \cdots + 50\left(1 + \frac{0.12}{12}\right)^{60}$$

43. A deposit of P dollars is made at the beginning of each month for T years in an account that pays R percent interest, compounded monthly. Let $N = 12T$ be the total number of deposits. The balance A after T years is

$$A = P\left(1 + \frac{R}{12}\right) + P\left(1 + \frac{R}{12}\right)^2 + \cdots + P\left(1 + \frac{R}{12}\right)^N.$$

Show that the balance is given by

$$A = P\left[\left(1 + \frac{R}{12}\right)^N - 1\right]\left(1 + \frac{12}{R}\right).$$

44. Use the formula in Exercise 43 to find the amount in an account earning 9% compounded monthly, after monthly deposits of $50 have been made for 40 years.

45. Use the formula in Exercise 43 to find the amount in an account earning 12% compounded monthly, after monthly deposits of $50 have been made for 40 years.

46. Suppose that you went to work at a company that pays $0.01 for the first day, $0.02 for the second day, $0.04 for the third day, and so on. If the daily wage keeps doubling, what would your total income be for working 29 days? 30 days?

In Exercises 47–56, find the sum of the infinite geometric series.

47. $\sum_{n=0}^{\infty} (\frac{1}{2})^n = 1 + \frac{1}{2} + \frac{1}{4} + \frac{1}{8} + \cdots$

48. $\sum_{n=0}^{\infty} 2(\frac{2}{3})^n = 2 + \frac{4}{3} + \frac{8}{9} + \frac{16}{27} + \cdots$

49. $\sum_{n=0}^{\infty} (-\frac{1}{2})^n = 1 - \frac{1}{2} + \frac{1}{4} - \frac{1}{8} + \cdots$

50. $\sum_{n=0}^{\infty} 2(-\frac{2}{3})^n = 2 - \frac{4}{3} + \frac{8}{9} - \frac{16}{27} + \cdots$

51. $\sum_{n=0}^{\infty} 4(\frac{1}{4})^n = 4 + 1 + \frac{1}{4} + \frac{1}{16} + \cdots$

52. $\sum_{n=0}^{\infty} (\frac{1}{10})^n = 1 + 0.1 + 0.01 + 0.001 + \cdots$

53. $8 + 6 + \frac{9}{2} + \frac{27}{8} + \cdots$

54. $3 - 1 + \frac{1}{3} - \frac{1}{9} + \cdots$

55. $4 - 2 + 1 - \frac{1}{2} + \cdots$

56. $2 + \sqrt{2} + 1 + \dfrac{1}{\sqrt{2}} + \cdots$

In Exercises 57 and 58, find the sum of the infinite series.

57. $\sum_{n=0}^{\infty} \left(\frac{1}{2^n} - \frac{1}{3^n}\right)$

58. $\sum_{n=0}^{\infty} [(0.7)^n + (0.9)^n]$

59. A ball is dropped from a height of 16 feet. Each time it drops h feet, it rebounds $0.81h$ feet. Find the total distance traveled by the ball.

60. The ball in Exercise 59 takes the following times for each fall.

$s_1 = -16t^2 + 16,$	$s_1 = 0$ if $t = 1$
$s_2 = -16t^2 + 16(0.81),$	$s_2 = 0$ if $t = 0.9$
$s_3 = -16t^2 + 16(0.81)^2,$	$s_3 = 0$ if $t = (0.9)^2$
$s_4 = -16t^2 + 16(0.81)^3,$	$s_4 = 0$ if $t = (0.9)^3$
$\vdots$	$\vdots$
$s_n = -16t^2 + 16(0.81)^{n-1},$	$s_n = 0$ if $t = (0.9)^{n-1}$

Beginning with s_2, the ball takes the same amount of time to bounce up as it does to fall, and thus the total time elapsed before it comes to rest is

$$t = 1 + 2\sum_{n=1}^{\infty} (0.9)^n.$$

Find this total.

Sequences, Series, and Probability

In Exercises 61–68, write the repeating decimal as the ratio of two integers by considering it to be the sum of an infinite geometric series.

61. 0.1111. . .

62. 0.4444. . .

63. 0.363636. . .

64. 0.212121. . .

65. 0.432432432. . .

66. 0.46666. . .

67. 1.363636. . .

68. 1.185185185. . .

12.4 *Mathematical Induction*

In this section we look at a form of mathematical proof, called the principle of **mathematical induction.** It is important that you clearly see the logical need for it, so let's take a closer look at a problem we discussed earlier.

In Section 12.2 we looked at the following pattern for the sum of the first n odd integers.

$$S_1 = 1 = 1^2$$
$$S_2 = 1 + 3 = 2^2$$
$$S_3 = 1 + 3 + 5 = 3^2$$
$$S_4 = 1 + 3 + 5 + 7 = 4^2$$
$$S_5 = 1 + 3 + 5 + 7 + 9 = 5^2$$

Judging from the pattern formed by these first five partial sums, we decided that the nth partial sum was of the form

$$S_n = 1 + 3 + 5 + 7 + 9 + \cdots + 2n - 1 = n^2.$$

Recognizing a pattern and then simply *jumping to the conclusion* that the pattern must be true for all values of n is *not* a logically valid method of proof. There are many examples in which a pattern appears to be developing for small values of n and then at some point the pattern fails. One of the most famous cases of this was the conjecture by the French mathematician Pierre de Fermat (1601–1655), who speculated that all numbers of the form

$$F_n = 2^{2^n} + 1, \qquad n = 0, 1, 2, \ldots$$

are prime. For $n = 0, 1, 2, 3,$ and 4, the conjecture is true.

$$F_0 = 3, \ F_1 = 5, \ F_2 = 17, \ F_3 = 257, \ F_4 = 65{,}537$$

The size of the next Fermat number ($F_5 = 4{,}294{,}967{,}297$) is so great that it was difficult for Fermat to determine whether it was prime or not. However, another well-known mathematician, Leonhard Euler (1707–1783), later found a factorization

$$F_5 = 4{,}294{,}967{,}297 = 641(6{,}700{,}417)$$

which proved that F_5 is not prime, and therefore Fermat's conjecture was false.

Just because a rule, pattern, or formula seems to work for several values of n, we cannot simply decide that it is valid for all values of n without going through a *legitimate proof*. Let's see how we can prove such statements by the principle of **mathematical induction.**

The Principle of Mathematical Induction

Let P_n be a statement involving the positive integer n. If

1. P_1 is true, and
2. the truth of P_k implies the truth of P_{k+1}, for every positive integer k,

then P_n must be true for all positive integers n.

Remark: It is important to recognize that both parts of the Principle of Mathematical Induction are necessary.

To apply the Principle of Mathematical Induction we need to be able to find P_{k+1} for a given P_k. This skill is demonstrated in Example 1.

EXAMPLE 1   A Preliminary Example

Find S_{k+1} for the following.

(a) $S_k = \dfrac{k^2(k + 1)^2}{4}$

(b) $S_k = 1 + 5 + 9 + \cdots + [4(k - 1) - 3] + (4k - 3)$

(c) S_k: $3^k \geq 2k + 1$

SOLUTION

(a) Substituting $k + 1$ for k, we have

$$S_{k+1} = \frac{(k + 1)^2(k + 1 + 1)^2}{4} \qquad \textit{Replace k by k + 1}$$

$$= \frac{(k + 1)^2(k + 2)^2}{4}. \qquad \textit{Simplify}$$

(b) In this case we have

$$S_{k+1} = 1 + 5 + 9 + \cdots + (4[(k + 1) - 1] - 3) + [4(k + 1) - 3]$$
$$= 1 + 5 + 9 + \cdots + (4k - 3) + (4k + 1).$$

(c) Replacing k by $k + 1$ in the statement $3^k \geq 2k + 1$, we have

$$S_{k+1}: 3^{k+1} \geq 2(k + 1) + 1$$
$$3^{k+1} \geq 2k + 3.$$

Sequences, Series, and Probability

EXAMPLE 2 *Using Mathematical Induction*

Use mathematical induction to prove the following formula.

$$S_n = 1 + 3 + 5 + 7 + \cdots + (2n - 1) = n^2$$

SOLUTION

Mathematical induction consists of two distinct parts. First, we must show that the formula is true when $n = 1$.

1. When $n = 1$, the formula is valid, since

$$S_1 = 1 = 1^2.$$

The second part of mathematical induction has two steps. The first step is to assume that the formula is valid for *some* integer k. The second step is to use this assumption to prove that the formula is valid for the next integer, $k + 1$.

2. Assuming that the formula

$$S_k = 1 + 3 + 5 + 7 + \cdots + (2k - 1) = k^2$$

is true, we must show that the formula $S_{k+1} = (k + 1)^2$ is true.

$$\begin{aligned}
S_{k+1} &= 1 + 3 + 5 + 7 + \cdots + (2k - 1) + [2(k + 1) - 1] \\
&= [1 + 3 + 5 + 7 + \cdots + (2k - 1)] + (2k + 2 - 1) \\
&= S_k + (2k + 1) \\
&= k^2 + 2k + 1 \\
&= (k + 1)^2
\end{aligned}$$

Combining the results of parts (1) and (2), we conclude by mathematical induction that the formula is valid for *all* positive integer values of n.

Remark: When using mathematical induction to prove a *summation* formula (like the one in Example 2), it is helpful to think of S_{k+1} as $S_{k+1} = S_k + a_{k+1}$, where a_{k+1} is the $(k + 1)$ term of the original sum.

A well-known illustration used to explain why the principle of mathematical induction works is the unending line of dominoes shown in Figure 12.1. If the line actually contains infinitely many dominoes, then it is clear that we could not knock the entire line down by knocking down only *one domino* at a time. However, suppose it were true that each domino would knock down the next one as it fell. Then we could knock them all down simply by pushing the first one and starting a chain reaction. Mathematical

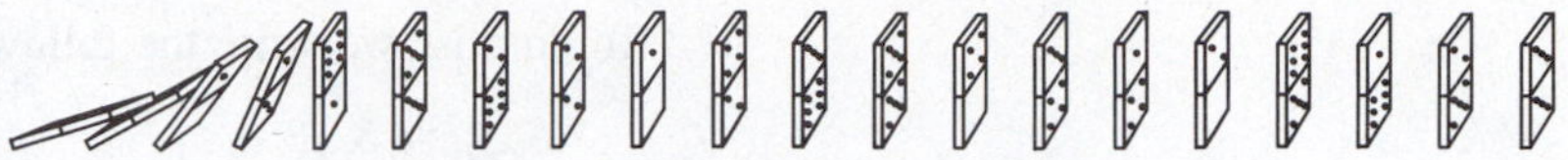

FIGURE 12.1 The first domino knocks over the second which knocks over the third which knocks over the fourth and so on.

induction works in the same way. If the truth of P_k implies the truth of P_{k+1} and if P_1 is true, then the chain reaction proceeds as follows:

P_1 implies P_2
P_2 implies P_3
P_3 implies P_4

and so on.

It occasionally happens that a statement involving natural numbers is not true for the first $k - 1$ positive integers but is true for all values of $n \geq k$. In these instances, we use a slight variation of the principle of mathematical induction in which we verify P_k rather than P_1. This variation is called the **extended principle of mathematical induction.** To see the validity of this, note from Figure 12.1 that all but the first $k - 1$ dominoes can be knocked down by knocking over the kth domino. This suggests that we can prove a statement P_n to be true for $n \geq k$ by showing that P_k is true and that P_k implies P_{k+1}. In Exercises 31–34 of this section you are asked to apply this extension of mathematical induction.

EXAMPLE 3 *Using Mathematical Induction*

Use mathematical induction to prove the following formula.

$$S_n = 1^2 + 2^2 + 3^2 + 4^2 + \cdots + n^2 = \frac{n(n + 1)(2n + 1)}{6}$$

SOLUTION

1. When $n = 1$, the formula is valid, because

$$S_1 = 1^2 = \frac{1(2)(3)}{6}.$$

2. Assuming that

$$S_k = 1^2 + 2^2 + 3^2 + 4^2 + \cdots + k^2 = \frac{k(k + 1)(2k + 1)}{6}$$

we must show that

$$S_{k+1} = \frac{(k + 1)(k + 2)(2k + 3)}{6}.$$

Sequences, Series, and Probability

To do this, we write the following.

$$S_{k+1} = S_k + a_{k+1}$$

$$= (1^2 + 2^2 + 3^2 + 4^2 + \cdots + k^2) + (k + 1)^2$$

$$= \frac{k(k + 1)(2k + 1)}{6} + (k + 1)^2$$

$$= \frac{k(k + 1)(2k + 1) + 6(k + 1)^2}{6}$$

$$= \frac{(k + 1)[k(2k + 1) + 6(k + 1)]}{6}$$

$$= \frac{(k + 1)[2k^2 + 7k + 6]}{6}$$

$$= \frac{(k + 1)(k + 2)(2k + 3)}{6}$$

Combining the results of parts (1) and (2), we conclude by mathematical induction that the formula is valid for *all n* $\geq$ 1.

The formula in Example 3 is one of a collection of useful summation formulas. We summarize this and other formulas dealing with the sum of various powers of the first *n* positive integers as follows.

Sums of Powers of Integers

1. $1 + 2 + 3 + 4 + \cdots + n = \dfrac{n(n + 1)}{2}$

2. $1^2 + 2^2 + 3^2 + 4^2 + \cdots + n^2 = \dfrac{n(n + 1)(2n + 1)}{6}$

3. $1^3 + 2^3 + 3^3 + 4^3 + \cdots + n^3 = \dfrac{n^2(n + 1)^2}{4}$

4. $1^4 + 2^4 + 3^4 + 4^4 + \cdots + n^4 = \dfrac{n(n + 1)(2n + 1)(3n^2 + 3n - 1)}{30}$

5. $1^5 + 2^5 + 3^5 + 4^5 + \cdots + n^5 = \dfrac{n^2(n + 1)^2(2n^2 + 2n - 1)}{12}$

Remark: Each of these formulas for sums can be proved by mathematical induction. (See Exercises 17–21.)

Mathematical Induction

Although choosing a formula on the basis of a few observations does *not* guarantee the validity of the formula, pattern recognition *is* important. Once you have a pattern or formula that you think works, you can try using mathematical induction to prove your formula. We outline these steps as follows.

Finding a Formula for the nth Term of a Sequence

1. Calculate the first several terms of the sequence. (It is often a good idea to write the terms in both simplified and factored form.)
2. Try to find a recognizable pattern from these terms, and write down a formula for the *n*th term of the sequence. (This is your hypothesis. You might try computing one or two more terms in the sequence to test your hypothesis.)
3. Use mathematical induction to attempt to prove your hypothesis.

EXAMPLE 4   Finding a Formula for the nth Partial Sum of a Sequence

Find a formula for the *n*th *partial sum* of the sequence

$$\frac{1}{1 \cdot 2}, \frac{1}{2 \cdot 3}, \frac{1}{3 \cdot 4}, \frac{1}{4 \cdot 5}, \cdots, \frac{1}{n(n+1)}, \cdots$$

SOLUTION

We begin by writing out a few of the partial sums.

$$S_1 = \frac{1}{1 \cdot 2} = \frac{1}{2}$$

$$S_2 = \frac{1}{1 \cdot 2} + \frac{1}{2 \cdot 3} = \frac{4}{6} = \frac{2}{3}$$

$$S_3 = \frac{1}{1 \cdot 2} + \frac{1}{2 \cdot 3} + \frac{1}{3 \cdot 4} = \frac{9}{12} = \frac{3}{4}$$

$$S_4 = \frac{1}{1 \cdot 2} + \frac{1}{2 \cdot 3} + \frac{1}{3 \cdot 4} + \frac{1}{4 \cdot 5} = \frac{48}{60} = \frac{4}{5}$$

Now, from these first four partial sums, it appears that the formula for the *k*th partial sum is

$$S_k = \frac{1}{1 \cdot 2} + \frac{1}{2 \cdot 3} + \frac{1}{3 \cdot 4} + \frac{1}{4 \cdot 5} + \cdots + \frac{1}{k(k+1)} = \frac{k}{k+1}.$$

To prove the validity of this formula, we use mathematical induction, as follows. Note that we have already verified the formula for $n = 1$, so we begin by assuming that the formula is valid for $n = k$ and try to show that it is valid for $n = k + 1$.

$$S_{k+1} = \left[\frac{1}{1 \cdot 2} + \frac{1}{2 \cdot 3} + \frac{1}{3 \cdot 4} + \cdots + \frac{1}{k(k+1)} \right] + \frac{1}{(k+1)(k+2)}$$

$$= \frac{k}{k+1} + \frac{1}{(k+1)(k+2)}$$

$$= \frac{k(k+2) + 1}{(k+1)(k+2)}$$

$$= \frac{k^2 + 2k + 1}{(k+1)(k+2)}$$

$$= \frac{(k+1)^2}{(k+1)(k+2)}$$

$$= \frac{k+1}{k+2}.$$

Thus, the formula is valid.

EXAMPLE 5 *Proving an Inequality by Mathematical Induction*

Prove that $n < 2^n$ for all positive integers n.

SOLUTION

1. For $n = 1$, the formula is true, since

   $1 < 2^1$.

2. Assuming that

   $k < 2^k$

   we need to show that $k + 1 < 2^{k+1}$. For $n = k + 1$, we have

   $2^{k+1} = 2(2^k) > 2(k) = 2k.$ *By assumption*

   Since $2k = k + k > k + 1$ for all $k > 1$, it follows that

   $2^{k+1} > 2k > k + 1$

   or

   $k + 1 < 2^{k+1}.$

   Hence, $n < 2^n$ for all integers $n \geq 1$.

WARM UP

Find the sums.

1. $\displaystyle\sum_{k=3}^{6} (2k - 3)$

2. $\displaystyle\sum_{j=1}^{5} (j^2 - j)$

3. $\displaystyle\sum_{k=2}^{5} \frac{1}{k}$

4. $\displaystyle\sum_{i=1}^{2} \left(1 + \frac{1}{i}\right)$

Simplify the expressions.

5. $\dfrac{2(k + 1) + 3}{5}$

6. $\dfrac{3(k + 1) - 2}{6}$

7. $2 \cdot 2^{2(k+1)}$

8. $\dfrac{3^{2k}}{3^{2(k+1)}}$

9. $\dfrac{k + 1}{k^2 + k}$

10. $\dfrac{\sqrt{32}}{\sqrt{50}}$

EXERCISES 12.4

In Exercises 1–6, find the indicated sum using the formulas for the sums of powers of integers.

1. $\displaystyle\sum_{n=1}^{6} n^2$

2. $\displaystyle\sum_{n=1}^{10} n^2$

3. $\displaystyle\sum_{n=1}^{5} n^3$

4. $\displaystyle\sum_{n=1}^{8} n^3$

5. $\displaystyle\sum_{n=1}^{6} n^4$

6. $\displaystyle\sum_{n=1}^{4} n^5$

In Exercises 7–10, find S_{k+1} for the given S_k.

7. $S_k = \dfrac{5}{k(k + 1)}$

8. $S_k = \dfrac{1}{(k + 1)(k + 3)}$

9. $S_k = \dfrac{k^2(k + 1)^2}{4}$

10. $S_k = \dfrac{k}{2}(3k - 1)$

In Exercises 11–24, use mathematical induction to prove the given formula for every positive integer n.

11. $2 + 4 + 6 + 8 + \cdots + 2n = n(n + 1)$

12. $3 + 7 + 11 + 15 + \cdots + (4n - 1) = n(2n + 1)$

13. $2 + 7 + 12 + 17 + \cdots + (5n - 3) = \dfrac{n}{2}(5n - 1)$

14. $1 + 4 + 7 + 10 + \cdots + (3n - 2) = \dfrac{n}{2}(3n - 1)$

15. $1 + 2 + 2^2 + 2^3 + \cdots + 2^{n-1} = 2^n - 1$

16. $2(1 + 3 + 3^2 + 3^3 + \cdots + 3^{n-1}) = 3^n - 1$

17. $1 + 2 + 3 + 4 + \cdots + n = \dfrac{n(n + 1)}{2}$

18. $1^2 + 2^2 + 3^2 + 4^2 + \cdots + n^2 = \dfrac{n(n + 1)(2n + 1)}{6}$

19. $1^3 + 2^3 + 3^3 + 4^3 + \cdots + n^3 = \dfrac{n^2(n + 1)^2}{4}$

20. $\displaystyle\sum_{i=1}^{n} i^4 = \dfrac{n(n + 1)(2n + 1)(3n^2 + 3n - 1)}{30}$

21. $\displaystyle\sum_{i=1}^{n} i^5 = \dfrac{n^2(n + 1)^2(2n^2 + 2n - 1)}{12}$

22. $\left(1 + \dfrac{1}{1}\right)\left(1 + \dfrac{1}{2}\right)\left(1 + \dfrac{1}{3}\right) \cdots \left(1 + \dfrac{1}{n}\right) = n + 1$

23. $\displaystyle\sum_{i=1}^{n} i(i + 1) = \dfrac{n(n + 1)(n + 2)}{3}$

24. $\displaystyle\sum_{i=1}^{n} \dfrac{1}{(2i - 1)(2i + 1)} = \dfrac{n}{2n + 1}$

In Exercises 25–30, find a formula for the *n*th partial sum of the sequence.

25. 3, 7, 11, 15, . . .

26. 25, 22, 19, 16, . . .

27. 1, $\frac{9}{10}$, $\frac{81}{100}$, $\frac{729}{1000}$, . . .

28. 3, $-\frac{9}{2}$, $\frac{27}{4}$, $-\frac{81}{8}$, . . .

29. $\frac{1}{4}$, $\frac{1}{12}$, $\frac{1}{24}$, $\frac{1}{40}$, . . . , $\frac{1}{2n(n+1)}$, . . .

30. $\frac{1}{2 \cdot 3}$, $\frac{1}{3 \cdot 4}$, $\frac{1}{4 \cdot 5}$, $\frac{1}{5 \cdot 6}$, . . . , $\frac{1}{(n+1)(n+2)}$, . . .

In Exercises 31–34, use mathematical induction to prove the given inequality for the indicated integer values of *n*.

31. $\left(\frac{4}{3}\right)^n > n$, $n \geq 7$

32. $\frac{1}{\sqrt{1}} + \frac{1}{\sqrt{2}} + \frac{1}{\sqrt{3}} + \cdots + \frac{1}{\sqrt{n}} > \sqrt{n}$, $n \geq 2$

33. $n! > 2^n$, $n \geq 4$

34. If $0 < x < y$, then $\left(\frac{x}{y}\right)^{n+1} < \left(\frac{x}{y}\right)^n$, $n \geq 1$.

In Exercises 35–46, use mathematical induction to prove the given property for all positive integers *n*.

35. $(ab)^n = a^n b^n$

36. $\left(\frac{a}{b}\right)^n = \frac{a^n}{b^n}$

37. If $x_1 \neq 0$, $x_2 \neq 0$, $x_3 \neq 0$, . . . , $x_n \neq 0$, then
$$(x_1 x_2 x_3 \cdots x_n)^{-1} = x_1^{-1} x_2^{-1} x_3^{-1} \cdots x_n^{-1}.$$

38. If $x_1 > 0$, $x_2 > 0$, $x_3 > 0$, . . . , $x_n > 0$, then
$$\ln(x_1 x_2 x_3 \cdots x_n) = \ln x_1 + \ln x_2 + \ln x_3 + \cdots + \ln x_n.$$

39. Generalized Distributive Law:
$$x(y_1 + y_2 + \cdots + y_n) = xy_1 + xy_2 + \cdots + xy_n$$

40. $x^n - y^n = (x - y)(x^{n-1} + x^{n-2}y + \cdots + xy^{n-2} + y^{n-1})$
[*Hint:* $x^{n+1} - y^{n+1} = x^n(x - y) + y(x^n - y^n)$.]

41. $\sin(\theta + n\pi) = (-1)^n \sin \theta$

42. $\cos(\theta + n\pi) = (-1)^n \cos \theta$

43. DeMoivre's Theorem:
$$[r(\cos \theta + i \sin \theta)]^n = r^n[\cos n\theta + i \sin n\theta]$$

44. $(a + bi)^n$ and $(a - bi)^n$ are complex conjugates for all $n \geq 1$.

45. 3 is a factor of $(n^3 + 3n^2 + 2n)$ for all $n \geq 1$.

46. 5 is a factor of $(2^{2n-1} + 3^{2n-1})$ for all $n \geq 1$.

12.5 *The Binomial Theorem*

Recall that a **binomial** is an expression that has two terms. In this section we will look at a formula that gives us a quick method of raising a binomial to a power.

To begin, let's look at the expansion of $(x + y)^n$ for a few values of *n*.

$$(x + y)^0 = 1$$
$$(x + y)^1 = x + y$$
$$(x + y)^2 = x^2 + 2xy + y^2$$
$$(x + y)^3 = x^3 + 3x^2y + 3xy^2 + y^3$$
$$(x + y)^4 = x^4 + 4x^3y + 6x^2y^2 + 4xy^3 + y^4$$
$$(x + y)^5 = x^5 + 5x^4y + 10x^3y^2 + 10x^2y^3 + 5xy^4 + y^5$$

There are several observations we can make about these expansions of $(x + y)^n$.

1. In each expansion, there are $n + 1$ terms.
2. In each expansion, *x* and *y* have symmetrical roles. The powers of *x* decrease by 1 in successive terms, whereas the powers of *y* increase by 1.

3. The sum of the powers of each term in a binomial expansion is n. For example,

$$4 + 1 = 5 \qquad 3 + 2 = 5$$

$$(x + y)^5 = x^5 + 5x^4y^1 + 10x^3y^2 + 10x^2y^3 + 5xy^4 + y^5.$$

4. The first term is x^n. The last term is y^n. Each has a coefficient of 1 because $x^n = 1(x^n)$ and $y^n = 1(y^n)$.

The most difficult part of a binomial expansion is finding the coefficients of the interior terms. To find these **binomial coefficients** we use a well-known theorem called the **Binomial Theorem.** It tells us that the coefficient of $x^{n-m}y^m$ is

$$_nC_m = \frac{n!}{m!(n - m)!}$$

$$= \frac{n(n - 1)(n - 2) \cdots (n - m + 1)}{m!}$$

where $m = 0, 1, 2, \ldots, n$.

Remark: The symbol $\binom{n}{m}$ is often used in place of $_nC_m$ to denote binomial coefficients.

The Binomial Theorem

$$(x + y)^n = x^n + nx^{n-1}y + \cdots + {_nC_m}x^{n-m}y^m + \cdots + nxy^{n-1} + y^n$$

where the coefficient of $x^{n-m}y^m$ is given by

$$_nC_m = \frac{n!}{(n - m)!m!} = \frac{n(n - 1)(n - 2) \cdots (n - m + 1)}{m!}.$$

PROOF

The Binomial Theorem can be proved quite nicely using mathematical induction. The steps are straightforward but look a little messy, so we will only present an outline of the proof.

1. If $n = 1$, then we have

$$(x + y)^1 = x^1 + y^1 = {_1C_0}x + {_1C_1}y$$

and the formula is valid.

2. Assuming the formula is true for $n = k$, then the coefficient of $x^{k-m}y^m$ is given by

$$_kC_m = \frac{k!}{(k-m)!m!}$$

$$= \frac{k(k-1)(k-2)\cdots(k-m+1)}{m!}.$$

To show that the formula is true for $n = k + 1$, we look at the coefficient of $x^{k+1-m}y^m$ in the expansion of

$$(x + y)^{k+1} = (x + y)^k(x + y).$$

From the right-hand side, we can determine that the term involving $x^{k+1-m}y^m$ is the sum of two products.

$$(_kC_m x^{k-m}y^m)(x) + (_kC_{m-1}x^{k+1-m}y^{m-1})(y)$$

$$= \left[\frac{k!}{(k-m)!m!} + \frac{k!}{(k-m+1)!(m-1)!}\right] x^{k+1-m}y^m$$

$$= \left[\frac{(k+1-m)k!}{(k+1-m)!m!} + \frac{k!m}{(k+1-m)!m!}\right] x^{k+1-m}y^m$$

$$= \left[\frac{k!(k+1-m+m)}{(k+1-m)!m!}\right] x^{k+1-m}y^m$$

$$= \left[\frac{(k+1)!}{(k+1-m)!m!}\right] x^{k+1-m}y^m$$

$$= {}_{k+1}C_m x^{k+1-m}y^m$$

Thus, by mathematical induction, the Binomial Theorem is valid for all positive integers n.

Be sure you see that the expansion of $(x + y)^n$ has $n + 1$ coefficients.

$_nC_0,\ _nC_1,\qquad _nC_2,\qquad \cdots,\qquad\qquad\qquad _nC_m,\qquad\qquad\qquad \cdots,\ _nC_{n-1},\ _nC_n$

$\downarrow\quad\downarrow\qquad\quad\downarrow\qquad\qquad\qquad\qquad\qquad\qquad\downarrow\qquad\qquad\qquad\qquad\qquad\downarrow\quad\downarrow$

$$1,\ n,\ \frac{n(n-1)}{2},\ \ldots,\ \frac{n(n-1)(n-2)\cdots(n-m+1)}{m!},\ \ldots,\ n,\ 1$$

EXAMPLE 1 *Finding Binomial Coefficients*

Find the following binomial coefficients.

(a) $_7C_3$ (b) $_{12}C_5$ (c) $_{12}C_7$ (d) $_8C_0$

The Binomial Theorem

SOLUTION

(a) $_7C_3 = \dfrac{7 \cdot 6 \cdot 5}{3 \cdot 2 \cdot 1} = 7(5) = 35$

Note that the denominator of the binomial coefficient $_nC_m$ is $m!$ and the numerator has m factors (provided $m \neq 0$).

(b) $_{12}C_5 = \dfrac{12 \cdot 11 \cdot 10 \cdot 9 \cdot 8}{5 \cdot 4 \cdot 3 \cdot 2 \cdot 1} = 11(9)(8) = 792$

(c) $_{12}C_7 = \dfrac{12 \cdot 11 \cdot 10 \cdot 9 \cdot 8 \cdot 7 \cdot 6}{7 \cdot 6 \cdot 5 \cdot 4 \cdot 3 \cdot 2 \cdot 1} = 11(9)(8) = 792$

(d) $_8C_0 = \dfrac{8!}{8!(0!)} = 1$

Remark: In Example 1, note that the coefficients in parts (b) and (c) are the same. This is not a coincidence, since it is true in general that

$$_nC_m = {_nC_{n-m}}, \qquad m = 0, 1, 2, \ldots, n.$$

This is in keeping with our observation that the coefficients of a binomial expansion occur in a symmetrical pattern.

EXAMPLE 2 *Using the Binomial Theorem*

Use the Binomial Theorem to expand $(x + 2)^5$.

SOLUTION

$$\begin{aligned}
(x + 2)^5 &= {_5C_0}x^5 + {_5C_1}x^4 2 + {_5C_2}x^3 2^2 + {_5C_3}x^2 2^3 + {_5C_4}x 2^4 + {_5C_5}2^5 \\
&= x^5 + 5(2)x^4 + 10(4)x^3 + 10(8)x^2 + 5(16)x + 32 \\
&= x^5 + 10x^4 + 40x^3 + 80x^2 + 80x + 32
\end{aligned}$$

EXAMPLE 3 *Using the Binomial Theorem*

Use the Binomial Theorem to expand $(1 - a^2)^4$.

SOLUTION

Since this binomial is a difference, rather than a sum, we write

$$(1 - a^2)^4 = [1 + (-a^2)]^4.$$

From this we see that the signs of the terms will alternate, as follows.

$(1 - a^2)^4$

$= {_4C_0}(1^4) + {_4C_1}(1^3)(-a^2) + {_4C_2}(1^2)(-a^2)^2 + {_4C_3}(1)(-a^2)^3 + {_4C_4}(-a^2)^4$

$= 1 - 4a^2 + 6a^4 - 4a^6 + a^8$

Sequences, Series, and Probability

Pascal's Triangle

There is a convenient way to remember the patterns for binomial coefficients. By arranging the coefficients in a triangular pattern, we have the following array, which is called **Pascal's Triangle.**

$$
\begin{array}{ccccccccccccc}
 & & & & & & 1 & & & & & & \\
 & & & & & 1 & & 1 & & & & & \\
 & & & & 1 & & 2 & & 1 & & & & \\
 & & & 1 & & 3 & & 3 & & 1 & & & \\
 & & 1 & & 4 & & 6 & & 4 & & 1 & & \\
 & 1 & & 5 & & 10 & & 10 & & 5 & & 1 & \\
1 & & 6 & & 15 & & 20 & & 15 & & 6 & & 1 \\
\end{array}
$$
$$
\begin{array}{ccccccccccccccc}
1 & & 7 & & 21 & & 35 & & 35 & & 21 & & 7 & & 1
\end{array}
$$

The first and last number in each row is 1, and every other number in the triangle is formed by adding the two numbers immediately above that number. For example, the two numbers above 35 are 15 and 20.

$$
\begin{matrix}
15 & & 20 \\
 & \searrow \swarrow & \\
 & 35 &
\end{matrix}
$$

$15 + 20 = 35$

Since the top row in Pascal's Triangle corresponds to the binomial expansion $(x + y)^0 = 1$, we call it the **zero row.** Similarly, the next row corresponds to the binomial expansion $(x + y)^1 = 1(x) + 1(y)$, and we call it the **first row.** In general, the **nth row** in Pascal's Triangle gives us the coefficients of $(x + y)^n$.

EXAMPLE 4 *Using Pascal's Triangle in a Binomial Expansion*

Use Pascal's Triangle to expand

$$\left(x - \frac{1}{2}\right)^6.$$

SOLUTION

Because this binomial is a difference, we write

$$\left(x - \frac{1}{2}\right)^6 = \left[x + \left(-\frac{1}{2}\right)\right]^6.$$

Using the sixth row of Pascal's triangle

$$
\begin{array}{ccccccc}
1 & 6 & 15 & 20 & 15 & 6 & 1
\end{array}
$$

we have

$$\left(x - \frac{1}{2}\right)^6 = x^6 + 6x^5\left(-\frac{1}{2}\right) + 15x^4\left(-\frac{1}{2}\right)^2 + 20x^3\left(-\frac{1}{2}\right)^3$$

$$+ 15x^2\left(-\frac{1}{2}\right)^4 + 6x\left(-\frac{1}{2}\right)^5 + \left(-\frac{1}{2}\right)^6$$

$$= x^6 - \frac{6}{2}x^5 + \frac{15}{4}x^4 - \frac{20}{8}x^3 + \frac{15}{16}x^2 - \frac{6}{32}x + \frac{1}{64}$$

$$= x^6 - 3x^5 + \frac{15}{4}x^4 - \frac{5}{2}x^3 + \frac{15}{16}x^2 - \frac{3}{16}x + \frac{1}{64}.$$

EXAMPLE 5 Finding a Specified Term in a Binomial Expansion

Find the sixth term in the expansion of $(3a + 2b)^{12}$.

SOLUTION

Using the Binomial Theorem, we let $x = 3a$ and $y = 2b$ and note that in the *sixth* term the exponent of y is $m = 5$ and the exponent of x is $n - m = 12 - 5 = 7$. Consequently, the sixth term of the expansion is

$$_{12}C_5 x^7 y^5 = \frac{12 \cdot 11 \cdot 10 \cdot 9 \cdot 8}{5!}(3a)^7(2b)^5.$$

WARM UP

Perform the indicated operations and/or simplify.

1. $5x^2(x^3 + 3)$

2. $(x + 5)(x^2 - 3)$

3. $(x + 4)^2$

4. $(2x - 3)^2$

5. $x^2 y(3xy^{-2})$

6. $(-2z)^5$

Evaluate the expressions.

7. $5!$

8. $\dfrac{8!}{5!}$

9. $\dfrac{10!}{7!}$

10. $\dfrac{6!}{3! \, 3!}$

Sequences, Series, and Probability

EXERCISES 12.5

In Exercises 1–10, evaluate $_nC_m$.

1. $_5C_3$

2. $_8C_6$

3. $_{12}C_0$

4. $_{20}C_{20}$

5. $_{20}C_{15}$

6. $_{12}C_5$

7. $_{100}C_{98}$

8. $_{10}C_4$

9. $_{100}C_2$

10. $_{10}C_6$

In Exercises 11–24, use the Binomial Theorem to expand the given binomial. Simplify your answer.

11. $(x + y)^5$

12. $(x + y)^6$

13. $(a + 2)^4$

14. $(s + 3)^5$

15. $(r + 3s)^6$

16. $(x + 2y)^4$

17. $(x - y)^5$

18. $(2x - y)^5$

19. $(1 - 2x)^3$

20. $\left(\dfrac{x}{2} - 3y\right)^3$

21. $(x^2 + 5)^4$

22. $(x^2 + y^2)^6$

23. $\left(\dfrac{1}{x} + y\right)^5$

24. $\left(\dfrac{1}{x} + 2y\right)^6$

In Exercises 25–30, use the Binomial Theorem to expand the given complex number. Simplify your answer by using the fact that $i^2 = -1$.

25. $(1 + i)^4$

26. $(2 - i)^5$

27. $(2 - 3i)^6$

28. $(5 + \sqrt{-9})^3$

29. $\left(\dfrac{-1}{2} + \dfrac{\sqrt{3}}{2}i\right)^3$

30. $(5 - \sqrt{3}\,i)^4$

In Exercises 31–34, expand the binomial using Pascal's Triangle to determine the coefficients.

31. $(2t - s)^5$

32. $\left(\dfrac{x}{2} + 2y\right)^5$

33. $(3 - 2z)^4$

34. $(3y + 2)^5$

In Exercises 35–42, find the required term in the expansion of the binomial.

	Binomial	Term
35.	$(x + 3)^{12}$	x^5
36.	$(x^2 + 3)^{12}$	x^8
37.	$(x - 2y)^{10}$	x^8y^2
38.	$(4x - y)^{10}$	x^2y^8
39.	$(3x - 2y)^{15}$	x^4y^{11}
40.	$(x^2 - 5)^8$	Middle
41.	$(\sqrt{x} + \sqrt{y})^{12}$	Middle
42.	$\left(\dfrac{2}{x} - \dfrac{3}{y}\right)^{10}$	Middle

In Exercises 43–48, use the Binomial Theorem to expand the given expression. In the study of probability, it is sometimes necessary to use the expansion of $(p + q)^n$, where $p + q = 1$.

43. $\left(\dfrac{1}{2} + \dfrac{1}{2}\right)^7$

44. $\left(\dfrac{1}{4} + \dfrac{3}{4}\right)^{10}$

45. $\left(\dfrac{1}{3} + \dfrac{2}{3}\right)^8$

46. $(0.3 + 0.7)^{12}$

47. $(0.6 + 0.4)^5$

48. $(0.35 + 0.65)^6$

In Exercises 49–52, prove the given property for all integers m and n, $0 \le m \le n$.

49. $_nC_m = {_nC_{n-m}}$

50. $_nC_0 - {_nC_1} + {_nC_2} - \cdots \pm {_nC_n} = 0$

51. $_{n+1}C_m = {_nC_m} + {_nC_{m-1}}$

52. $_{2n}C_n = ({_nC_0})^2 + ({_nC_1})^2 + ({_nC_2})^2 + \cdots + ({_nC_n})^2$

53. Prove that the sum of the numbers in the nth row of Pascal's Triangle is 2^n. [*Hint:* Consider $2^n = (1 + 1)^n$.]

12.6 Counting Principles, Permutations, and Combinations

In the last two sections of this chapter, we give a brief introduction to some fundamental counting principles and their application to probability. Much of probability has to do with counting the number of ways an event can occur. The following example describes some simple cases.

EXAMPLE 1 *Some Simple Counting Problems*

A random number generator (on a computer) selects an integer from 1 to 40. Find the number of ways the following events can occur.

(a) An even integer is selected.
(b) A number less than 10 is selected.
(c) A square number is selected.
(d) A prime number is selected.

SOLUTION

(a) Since half the numbers between 1 and 40 (inclusive) are even, this event can occur in 20 ways.
(b) The numbers that are less than 10 are

$$\{1, 2, 3, 4, 5, 6, 7, 8, 9\}.$$

Therefore, there are nine ways this event can happen.
(c) The square numbers from 1 to 40 are

$$\{1, 4, 9, 16, 25, 36\}.$$

Therefore, there are six ways this event can happen.
(d) The prime numbers from 1 to 40 are

$$\{2, 3, 5, 7, 11, 13, 17, 19, 23, 29, 31, 37\}.$$

Therefore, there are twelve ways this event can happen.

Each of the parts in Example 1 consists of a single event. The situation becomes somewhat more complicated when we try to count the number of ways two or more events can occur in succession (or in order). To do this, we make use of the Fundamental Counting Principle.

Fundamental Counting Principle

If $E_1, E_2, E_3, \ldots, E_n$ is a sequence of events such that E_1 can occur in m_1 ways, and after E_1 has occurred E_2 can occur in m_2 ways, and after E_2 has occurred E_3 can occur in m_3 ways, and so forth, then the number of ways the sequence can occur is given by

$$m_1 \cdot m_2 \cdot m_3 \cdots m_n.$$

Be sure you see that the Fundamental Counting Principle applies to a *sequence* of events. This means that the order of the events is important. It also means that when you are counting the number of ways the second event can occur, you must take into consideration the fact that the first event has already occurred.

The following example describes two similar sequences of events that differ in a subtle, yet important, way.

EXAMPLE 2 *Applying the Fundamental Counting Principle*

(a) A certain auto license number is made using three digits. How many different numbers are possible? (Leading zeros such as 001 or 027 are legitimate.)

(b) The digits from 0 to 9 are written on slips of paper and placed in a box. Three of the slips of paper are drawn and placed in the order in which they were drawn. How many different outcomes are possible?

SOLUTION

(a) By considering the selection of each digit as a separate event, we have the following.

> Event 1 = Choice of first digit
> Event 2 = Choice of second digit
> Event 3 = Choice of third digit

Because there are ten choices for each position, we can apply the Fundamental Counting Principle to conclude that there are

$$10 \cdot 10 \cdot 10 = 1000 \text{ different license numbers.}$$

(b) This problem is quite like the first, except for one very important distinction. Once the first slip of paper has been drawn, there are only nine slips left for the second draw. Moreover, once the second slip has been drawn, there are only eight slips left for the third draw. Hence, the Fundamental Counting Principle tells us that there are

$$10 \cdot 9 \cdot 8 = 720 \text{ different outcomes.}$$

Remark: The distinction between the two parts of Example 2 is that in part (a) we considered different **arrangements with repetition** and in part (b) we considered different **arrangements without repetition.**

EXAMPLE 3   Applying the Fundamental Counting Principle

(a) Some versions of the BASIC programming language allow variable names that can be one or two characters long. The first character can be any letter, and the second character can be any letter *or* digit. How many different variable names are possible?

(b) Most versions of the FORTRAN programming language allow variable names that can be up to six characters long. The first character can be any letter, and the other characters can be any alphanumeric character (the letters, a dollar sign, or a digit). How many different variable names are possible?

SOLUTION

(a) We count the two cases separately.

One-character length: Because the first character must be a letter of the alphabet, there are

26 different one-character names.

Two-character length: There are 26 choices for the first character, and 36 (any letter or digit) for the second, which gives a total of

$26 \cdot 36 = 936$ different two-character names.

Thus, the total number of variable names is

$26 + 936 = 962$ different names.

(b) The solution is similar to that in part (a), except that the dollar sign ($) gives us 37 choices for the second through sixth characters. The total number of variable names is as follows.

1-character length:	26	=	26
2-character length:	$26 \cdot 37$	=	962
3-character length:	$26 \cdot 37 \cdot 37$	=	35,594
4-character length:	$26 \cdot 37 \cdot 37 \cdot 37$	=	1,316,978
5-character length:	$26 \cdot 37 \cdot 37 \cdot 37 \cdot 37$	=	48,728,186
6-character length:	$26 \cdot 37 \cdot 37 \cdot 37 \cdot 37 \cdot 37$	=	1,802,942,882
Total:			1,853,024,628

Sequences, Series, and Probability

One important application of the Fundamental Counting Principle is in determining the number of ways that n elements can be arranged (in order). We call an ordering of n elements a **permutation** of elements.

Definition of Permutation

A **permutation** of n distinct elements is an ordering of the elements such that one element is first, one is second, and so on.

$$a_1, a_2, a_3, a_4, \ldots, a_n$$

EXAMPLE 4 *Finding the Number of Permutations of n Elements*

A horse race has five entries. In how many different orders can the horses finish? (Assume that there are no ties.)

SOLUTION

For the five horses, we consider the following five possibilities.

1st place:	Any of the *five* horses
2nd place:	Any of the remaining *four* horses
3rd place:	Any of the remaining *three* horses
4th place:	Any of the remaining *two* horses
5th place:	The *one* remaining horse

Multiplying these five numbers together, we find the total number of orders for the horses to be

$$5 \cdot 4 \cdot 3 \cdot 2 \cdot 1 = 5! = 120 \text{ orders.}$$

The result obtained in Example 4 can be generalized to conclude that the number of permutations (orderings) of n distinct elements is $n!$.

Number of Permutations of n Elements

There are $n!$ different permutations of n distinct elements.

PROOF

We can use the Fundamental Counting Principle as follows.

1st position:	Any of the n elements
2nd position:	Any of the remaining $n - 1$ elements
3rd position:	Any of the remaining $n - 2$ elements
4th position:	Any of the remaining $n - 3$ elements
$\vdots$	$\vdots$
$(n - 1)$th position:	Any of the remaining two elements
nth position:	The one remaining element

By multiplying these n numbers together, we find the total number of permutations to be

$$n(n - 1)(n - 2)(n - 3) \cdots 3 \cdot 2 \cdot 1 = n!.$$

Occasionally, we are interested in ordering a subset of a collection of elements rather than the entire collection. For example, we might want to choose (and order) m elements out of a collection of n elements. We call such an ordering a **permutation of n elements taken m at a time.** The following example illustrates such a case.

EXAMPLE 5   *Permutations of n Elements Taken m at a Time*

Eight horses are running in a race. In how many different ways can these horses come in first, second, and third? (Assume that there are no ties.)

SOLUTION

We have the following possibilities.

Win	(1st position):	*eight* choices
Place	(2nd position):	*seven* choices
Show	(3rd position):	*six* choices

Using the Fundamental Counting Principle, we multiply these three numbers together to obtain

$$8 \cdot 7 \cdot 6 = 336 \text{ orders.}$$

We can generalize the result of Example 5 as follows. The proof of this result closely parallels that given for the number of permutations of n elements.

Sequences, Series, and Probability

Number of Permutations of n Elements Taken m at a Time

The number of permutations of n elements taken m at a time is

$$_nP_m = \frac{n!}{(n-m)!} = n(n-1)(n-2)\cdots(n-m+1).$$

To help visualize the Fundamental Counting Principle, we can make a **tree diagram** which actually lists the various permutations of three elements. In the diagram shown in Figure 12.2, we can see in the right-hand branches of the tree the six possible permutations of the letters A, B, and C.

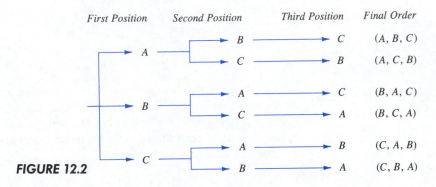

FIGURE 12.2

EXAMPLE 6 Listing Permutations

List the 24 different permutations of the letters A, B, C, and D.

SOLUTION

The various permutations are as follows.

A in 1st Place	B in 1st Place	C in 1st Place	D in 1st Place
(A, B, C, D)	(B, A, C, D)	(C, A, B, D)	(D, A, B, C)
(A, B, D, C)	(B, A, D, C)	(C, A, D, B)	(D, A, C, B)
(A, C, B, D)	(B, C, A, D)	(C, B, A, D)	(D, B, A, C)
(A, C, D, B)	(B, C, D, A)	(C, B, D, A)	(D, B, C, A)
(A, D, B, C)	(B, D, A, C)	(C, D, A, B)	(D, C, A, B)
(A, D, C, B)	(B, D, C, A)	(C, D, B, A)	(D, C, B, A)

Remember that for permutations order is important. Thus, if we are looking at the possible permutations of the letters $A, B, C,$ and D taken three at a time, the permutations

$$(A, B, D)$$

and

$$(B, A, D)$$

are different, since the order of the elements is different.

In Example 6, each letter is distinguishable from the other. But suppose there were four A's and one each of $B, C,$ and D.

$$A, A, A, A, B, C, D$$

The *total* number of permutations of the seven letters would be

$$_7P_7 = 7!.$$

However, not all of these arrangements would be *distinguishable,* since the four A's are alike. In other words, any rearrangement of the A's, keeping the $B, C,$ and D in the same locations, would not be a distinguishable permutation. Thus, the total number of *distinguishable* permutations of the seven letters would be less than 7!. To be more precise, let P be the number of distinguishable permutations of the seven letters. Then, since the four A's have $4! = 24$ permutations, the total number of permutations would be $24P$. Consequently, we have

$$24P = 7!$$

$$P = \frac{7!}{24}$$

$$= \frac{7!}{4!}.$$

This argument can be generalized to obtain the following result.

Distinguishable Permutations

Suppose a set of n objects has n_1 of one kind of object, n_2 of a second kind, n_3 of a third kind, and so on, with

$$n = n_1 + n_2 + n_3 + \cdots + n_k.$$

Then the number of **distinguishable permutations** of the n objects is

$$\frac{n!}{n_1! \cdot n_2! \cdot n_3! \cdots n_k!}.$$

Sequences, Series, and Probability

EXAMPLE 7 *Distinguishable Permutations*

In how many distinguishable ways can the product x^2y^3z be written without using exponents?

SOLUTION

Without exponents, this product can be written

$$x \cdot x \cdot y \cdot y \cdot y \cdot z.$$

Since there are two x's, three y's, and only one z, the number of distinguishable permutations is

$$\frac{6!}{2!(3!)(1!)} = \frac{6 \cdot 5 \cdot 4}{2}$$

$$= 6 \cdot 5 \cdot 2$$

$$= 60.$$

Combinations

As a final topic in this section, we look at a method of selecting subsets of a larger set in which order is *not* counted. We call such subsets **combinations of n elements taken m at a time.** We denote combinations with braces (rather than parentheses). Thus, the combinations

$$\{A, B, D\}$$

and

$$\{B, A, D\}$$

are equivalent, since both sets contain the same elements and order is not important. A common example of how a combination occurs is a card game in which the player is free to reorder the cards after they have been dealt.

Counting Principles, Permutations, and Combinations

Number of Combinations of n Elements Taken m at a Time

The number of combinations of n elements taken m at a time is

$$_nC_m = \frac{n!}{(n-m)!m!}$$

$$= \frac{n(n-1)(n-2)\cdots(n-m+1)}{m!}.$$

Remark: Note that the formula for $_nC_m$ is the same one given for binomial coefficients. This means that we can use Pascal's Triangle as an alternative way of computing the number of combinations of n elements taken m at a time.

EXAMPLE 8 *Combinations of n Elements Taken m at a Time*

A standard poker hand consists of five cards dealt from a deck of 52. How many different poker hands are possible? (After the cards are dealt, the player may reorder them, and therefore order is not important.)

SOLUTION

We use the formula for the number of combinations of 52 elements taken five at a time to obtain

$$\frac{52!}{47!(5!)} = \frac{52 \cdot 51 \cdot 50 \cdot 49 \cdot 48}{5 \cdot 4 \cdot 3 \cdot 2 \cdot 1}$$

$$= 2{,}598{,}960 \text{ different hands.}$$

Even though the formula for $_nC_m$ is given in fractional form, we know that the fraction must always reduce to an integer. (Remember that Pascal's Triangle consists entirely of integers.) For large values of $_nC_m$ we suggest that you lessen your chance of calculator overflow by reducing the fraction before multiplying the factors of the numerator. For example, the calculation in Example 8 could be simplified as follows.

$$\frac{52 \cdot 51 \cdot 50 \cdot 49 \cdot 48}{5 \cdot 4 \cdot 3 \cdot 2 \cdot 1} = \left(\frac{52}{4}\right)\left(\frac{51}{3}\right)\left(\frac{50}{5 \cdot 2}\right)(49 \cdot 48) = 13(17)(5)(49)(48)$$

To decide whether a problem calls for the number of permutations or the number of combinations, you may find the following guidelines helpful.

Guidelines for Use of Permutations and Combinations

1. Use *permutations* if a problem calls for the number of arrangements of objects and different orders are counted.
2. Use *combinations* if a problem calls for the number of ways of selecting objects and the order of selection is not counted.

EXAMPLE 9 **Using Combinations with the Fundamental Counting Principle**

The traveling squad for a college basketball team consists of two centers, five forwards, and four guards. In how many ways can the coach select a starting team of one center, two forwards, and two guards?

SOLUTION

The number of ways to select one center is

$$_2C_1 = \frac{2!}{1!(1!)} = 2.$$

The number of ways to select two forwards from among five is

$$_5C_2 = \frac{5!}{3!(2!)} = 10.$$

The number of ways to select two guards from among four is

$$_4C_2 = \frac{4!}{2!(2!)} = 6.$$

Therefore, the total number of ways to select a starting team is

$$_2C_1 \cdot {}_5C_2 \cdot {}_4C_2 = 2 \cdot 10 \cdot 6 = 120.$$

WARM UP

Evaluate the expressions.

1. $13 \cdot 8^2 \cdot 2^3$

2. $10^2 \cdot 9^3 \cdot 4$

3. $\dfrac{12!}{2!(7!)(3!)}$

4. $\dfrac{25!}{22!}$

Find the binomial coefficients.

5. $_{12}C_7$

6. $_{25}C_{22}$

Simplify the expressions.

7. $\dfrac{n!}{(n-4)!}$

8. $\dfrac{(2n)!}{4(2n-3)!}$

9. $\dfrac{2 \cdot 4 \cdot 6 \cdot 8 \cdots (2n)}{2^n}$

10. $\dfrac{3 \cdot 6 \cdot 9 \cdot 12 \cdots (3n)}{3^n}$

EXERCISES 12.6

1. A small college needs two additional faculty members, a chemist, and a statistician. In how many ways can these positions be filled if there are three applicants for the chemistry position and four for the position in statistics?

2. A customer in a computer store can choose one of three monitors, one of two keyboards, and one of four computers. If all the choices are compatible, how many different systems could be chosen?

3. Four people are lining up for a ride on a toboggan, but only two of the four are willing to take the first position. With that constraint, in how many ways can the four people be seated on the toboggan?

4. A college student is preparing her course schedule for the next semester. She may select one of two mathematics courses, one of three science courses, and one of five courses from the social sciences and humanities. In how many ways can she select her schedule?

5. In a certain state the automobile license plates consist of two letters followed by a four-digit number. How many distinct license plate numbers can be formed?

6. In how many ways can a six-question true-false exam be answered? (Assume that no questions are omitted.)

7. How many three-digit numbers (leading digits cannot be zero) can be formed under the following conditions?
(a) There are no restrictions.
(b) No repetition of digits is allowed.
(c) The number must be a multiple of 5.

8. A combination lock will open when the right choice of three numbers (from 1 to 40, inclusive) is selected. How many different lock combinations are possible?

9. Three couples have reserved seats in a given row for a concert. In how many different ways can they be seated, given the following conditions?
(a) There are no seating restrictions.
(b) The two members of each couple wish to sit together.

10. In how many orders can three girls and two boys walk through a doorway single-file, given the following conditions?
(a) There are no restrictions.
(b) The boys go before the girls.
(c) The girls go before the boys.

In Exercises 11–20, evaluate $_nP_m$. (See Section 12.5, Exercises 1–10, for problems on evaluating $_nC_m$.)

11. $_4P_4$ **12.** $_5P_5$

13. $_8P_3$ **14.** $_{20}P_2$

15. $_{20}P_5$ **16.** $_{100}P_1$

17. $_{100}P_2$ **18.** $_{10}P_2$

19. $_5P_4$ **20.** $_7P_4$

In Exercises 21–24, use permutations to solve the given counting problem.

21. In how many ways can five children line up in one row to have their picture taken?

22. In how many ways can six people sit in a six-passenger car?

23. From a pool of 12 candidates, the offices of president, vice-president, secretary, and treasurer will be filled. In how many different ways can the offices be filled, if each of the 12 candidates can hold any office?

24. There are four processes involved in assembling a certain product, and these can be performed in any order. The management wants to test each order to determine which is the least time-consuming. How many different orders will have to be tested?

In Exercises 25–28, use combinations to solve the given counting problem.

25. In order to conduct a certain experiment, four students are randomly selected from a class of 20. How many different groups of four students are possible?

26. A student may answer any ten questions from a total of twelve questions on an exam. How many different ways can the student select the questions?

27. There are 40 numbers in a particular state lottery. In how many ways can a player select six of the numbers? (The order of selection is not important.)

28. How many subsets of four elements can be formed from a set of 100 elements?

In Exercises 29–32, determine the number of three-digit numbers that can be formed from the ten digits 0, 1, 2, 3, 4, 5, 6, 7, 8, 9 under the given conditions. (The leading digit cannot be zero.)

29. No restrictions

30. The number is a multiple of 2.

31. The number is at least 400.

32. The number is less than 600.

33. A committee composed of three graduate students and two undergraduate students is to be selected from a group of eight graduates and five undergraduates. How many different committees can be formed?

34. A shipment of twelve microwave ovens contains three defective units. In how many ways can a vending company purchase four of these units and receive (a) all good units, (b) two good units, and (c) at least two good units?

35. An employer interviews eight people for four openings in the company. Three of the eight people are from a minority group. If all eight are qualified, in how many ways could the employer fill the four positions if (a) the selection is random and (b) exactly two are selected from the minority group?

36. Five cards are selected from an ordinary deck of 52 playing cards. In how many ways can you get a full house? (A full house consists of three of one kind and two of another. For example, A-A-A-5-5 and K-K-K-10-10 are full houses.)

37. Four people are to be selected at random from a group of four couples. In how many ways can this be done, given the following conditions?
(a) There are no restrictions.
(b) There is to be at least one couple in the group of four.
(c) The selection must include one member from each couple.

38. Repeat Exercise 37 if there are five couples from which to make the selection. (Assume that there are still only four to be selected.)

In Exercises 39–42, find the number of diagonals of the given polygon. (A line segment connecting any two non-adjacent vertices is called a *diagonal* of the polygon.)

39. Pentagon   **40.** Hexagon

41. Octagon   **42.** Decagon (ten sides)

In Exercises 43–48, find the number of distinguishable permutations of the given groups of letters.

43. *A, A, G, E, E, E, M*

44. *B, B, B, T, T, T, T, T*

45. *A, A, Y, Y, Y, Y, X, X, X*

46. *K, K, M, M, M, L, L, N, N*

47. *A, L, G, E, B, R, A*

48. *M, I, S, S, I, S, S, I, P, P, I*

In Exercises 49 and 50, solve for n.

49. $14 \cdot {}_nP_3 = {}_{n+2}P_4$ **50.** ${}_nP_5 = 18 \cdot {}_{n-2}P_4$

In Exercises 51–55, prove the identity.

51. ${}_nP_{n-1} = {}_nP_n$ **52.** ${}_nP_1 = {}_nC_1$

53. ${}_nC_{n-1} = {}_nC_1$ **54.** ${}_nC_n = {}_nC_0$

55. ${}_nC_m = \dfrac{{}_nP_m}{m!}$

12.7 Probability

As a member of a complex society, you are used to living with varying amounts of uncertainty. For example, you may be questioning the likelihood of getting a good job after graduation, of winning a state lottery, of having an accident on your next trip home, or of any of several other possibilities.

In assigning measurements to uncertainties in everyday life, we often use ambiguous terminology, such as "fairly certain," "probable," or "highly unlikely." In mathematics, we attempt to remove this ambiguity by assigning a number to the likelihood of the occurrence of an event. We call this measurement the **probability** that the event will occur. For example, if we toss a fair coin, we say that the probability that it will land heads up is one-half, or 50%.

In the study of probability, we call any happening whose result is uncertain an **experiment.** The various possible results of the experiment are called **outcomes,** and the collection of all possible outcomes of an experiment is called the **sample space** of the experiment. Finally, any subcollection of a sample space is called an **event.** In this section we will deal only with sample spaces in which each outcome is equally likely, such as flipping a fair coin or tossing a fair die.

EXAMPLE 1   Finding the Sample Space

An experiment consists of tossing a six-sided die.

(a) What is the sample space?
(b) Describe the event corresponding to a number greater than 2 turning up.

SOLUTION

(a) The sample space consists of six outcomes, which we represent by the numbers 1 through 6. That is,

$$S = \{1, 2, 3, 4, 5, 6\}.$$

Note that each of the outcomes in the sample space is equally likely (assuming that the die is balanced).

(b) The *event* corresponding to a number greater than 2 turning up is the following subset of *S*.

$$A = \{3, 4, 5, 6\}$$

To describe sample spaces in such a way that each outcome is equally likely, we must sometimes distinguish between various outcomes in ways that appear artificial. The next example illustrates such a situation.

EXAMPLE 2 *Finding the Sample Space*

Find the sample spaces for the following.

(a) One coin is tossed.
(b) Two coins are tossed.
(c) Three coins are tossed.

SOLUTION

(a) Since the coin will land either heads up (denoted by *H*) or tails up (denoted by *T*), the sample space is

$$S = \{H, T\}.$$

(b) Since either coin can land heads up or tails up, the possible outcomes are as follows.

HH = heads up on both coins
HT = heads up on first coin and tails up on second coin
TH = tails up on first coin and heads up on second coin
TT = tails up on both coins

Thus, the sample space is

$$S = \{HH, HT, TH, TT\}.$$

Note that we must distinguish between the two cases *HT* and *TH*, even though these two outcomes appear to be similar.

(c) Following the notation of part (b), the sample space is

$$S = \{HHH, HHT, HTH, HTT, THH, THT, TTH, TTT\}.$$

To calculate the probability of an event, we must count the number of outcomes in the event and in the sample space. We denote the *number of outcomes* in Event *A* by $n(A)$ and the number of outcomes in the sample space

S by $n(S)$. Moreover, we will assume that every sample space has a finite number of outcomes.

The Probability of an Event

If an event *A* has $n(A)$ equally likely outcomes and its sample space has $n(S)$ equally likely outcomes, then the **probability** of event *A* is

$$P(A) = \frac{n(A)}{n(S)}.$$

Because the number of outcomes in an event must be less than or equal to the number of outcomes in the sample space, we can see that the probability of an event must be a number between 0 and 1. That is, for any event *A*, it must be true that

$$0 \le P(A) \le 1.$$

The statement $P(A) = 0$ means that event *A* *cannot occur*, whereas the statement $P(A) = 1$ means that event *A* *must occur*.

Historically, a great deal of interest in probability stemmed from its applications to gambling.

EXAMPLE 3 *Finding the Probability of an Event*

Find the probability of the following events.

(a) Two coins are tossed. What is the probability that both land heads up?
(b) A card is drawn from a standard deck of playing cards. What is the probability that it is an ace?

SOLUTION

(a) Following the procedure in Example 2(b), we let

$$A = \{HH\} \quad \text{and} \quad S = \{HH, HT, TH, TT\}.$$

The probability of getting two heads is

$$P(A) = \frac{n(A)}{n(S)} = \frac{1}{4}.$$

(b) Since there are 52 cards in a standard deck of playing cards and there are four aces (one in each suit), the probability of drawing an ace is

$$P(A) = \frac{n(A)}{n(S)} = \frac{4}{52} = \frac{1}{13}.$$

EXAMPLE 4 *Finding the Probability of an Event*

Two six-sided dice are tossed. What is the probability that the total of the two dice is 7?

SOLUTION

Since there are six possible outcomes on each die, we use the Fundamental Counting Principle to conclude that there are

$6 \cdot 6 = 36$ different outcomes

when two dice are tossed. To find the probability of rolling a total of 7, we must first count the number of ways this can occur.

	Total of 7					
First die	1	2	3	4	5	6
Second die	6	5	4	3	2	1

Thus, a total of 7 can be rolled in six ways, which means that the probability of rolling a 7 is

$$P(A) = \frac{n(A)}{n(S)} = \frac{6}{36} = \frac{1}{6}.$$

We could have written out each sample space in Examples 3 and 4 and simply counted the outcomes in the desired events. For larger sample spaces, however, we must make more use of the counting principles discussed in the previous section.

FIGURE 12.3

EXAMPLE 5

Twelve-sided dice can be constructed (in the shape of regular dodecahedrons) so that each of the numbers from 1 to 6 appears twice on each die, as shown in Figure 12.3. Prove that these dice can be used in any game requiring ordinary six-sided dice without changing the probabilities of different outcomes.

SOLUTION

For an ordinary six-sided die, each of the numbers 1, 2, 3, 4, 5, and 6 occurs only once, so the probability of any particular number coming up is

$$P(A) = \frac{n(A)}{n(S)} = \frac{1}{6}.$$

For one of the twelve-sided dice, each number occurs twice, so the probability of any particular number coming up is

$$P(A) = \frac{n(A)}{n(S)} = \frac{2}{12} = \frac{1}{6}.$$

Thus, the twelve-sided dice can be used in place of the six-sided dice without changing the probabilities.

EXAMPLE 6 *Finding the Probability of an Event*

A state lottery is set up so that each player chooses six numbers from 1 to 40. If these six numbers match the six numbers drawn by the lottery commission, the player wins (or shares) the top prize. What is the probability of winning the top prize in this game?

SOLUTION

Since the order of the numbers is not important, we use our formula for the number of combinations of 40 elements taken six at a time to determine the size of the sample space.

$$n(S) = {}_{40}C_6 = \frac{40 \cdot 39 \cdot 38 \cdot 37 \cdot 36 \cdot 35}{6 \cdot 5 \cdot 4 \cdot 3 \cdot 2 \cdot 1} = 3,838,380$$

If a person buys only one ticket, the probability of winning is

$$P(A) = \frac{n(A)}{n(S)} = \frac{1}{3,838,380}.$$

Two events A and B (from the same sample space) are called **mutually exclusive** if A and B have no outcomes in common. For instance, if two dice are tossed, the event A of rolling a total of 6 and the event B of rolling a total of 9 are mutually exclusive because the following sets have no elements in common.

$A = \{(1, 5), (2, 4), (3, 3), (4, 2), (5, 1)\}$
$B = \{(3, 6), (4, 5), (5, 4), (6, 3)\}$

Sequences, Series, and Probability

(The ordered pairs represent the numbers on the dice.) To find the probability that one or the other of two mutually exclusive events will occur, we *add* their individual probabilities.

Probability of Mutually Exclusive Events

If A and B are mutually exclusive events, then for a given experiment the probability that A *or* B will occur is

$$P(A \text{ or } B) = P(A) + P(B).$$

EXAMPLE 7 *Probability of Mutually Exclusive Events*

A card is drawn from a standard deck of 52 cards. What is the probability that it is a face card (event A) or a 4 (event B)?

SOLUTION

The two events are mutually exclusive because no one card can be both a face card and a 4. Therefore, since

$$P(A) = \frac{n(A)}{n(S)} = \frac{12}{52} = \frac{3}{13}$$

and

$$P(B) = \frac{n(B)}{n(S)} = \frac{4}{52} = \frac{1}{13}$$

it follows that

$$P(A \text{ or } B) = P(A) + P(B) = \frac{3}{13} + \frac{1}{13} = \frac{4}{13}.$$

We say that two events are **independent** if the occurrence of one has no effect on the occurrence of the other. To find the probability that two independent events will both occur, we *multiply* the probabilities of each.

Probability of Independent Events

If A and B are independent events, then the probability that both A and B will occur is

$$P(A \text{ and } B) = P(A) \cdot P(B).$$

EXAMPLE 8 *Probability of Independent Events*

A random number generator on a computer selects three integers from 1 to 20. What is the probability that all three numbers are less than (or equal to) 5?

SOLUTION

If the random number generator is truly random, then we can conclude that the selection of any given number will not affect the selection of the next number. This means that the three choices represent independent events. Furthermore, since the probability of selecting a number from 1 to 5 is

$$P(A) = \frac{5}{20} = \frac{1}{4}$$

we can conclude that the probability of selecting all three numbers less than or equal to 5 is

$$P(A) \cdot P(A) \cdot P(A) = \left(\frac{1}{4}\right)\left(\frac{1}{4}\right)\left(\frac{1}{4}\right) = \frac{1}{64}.$$

The **complement of an event** A is the collection of all outcomes in the sample space that are not in A. We denote the complement of event A by A'. Since $P(A \text{ or } A') = 1$ and since A and A' are mutually exclusive, we have $P(A) + P(A') = 1$. Therefore, the probability of A' is given by

$$P(A') = 1 - P(A).$$

For instance, if the probability of winning a certain game is

$$P(A) = \frac{1}{4}$$

then the probability of losing the game is

$$P(A') = 1 - \frac{1}{4} = \frac{3}{4}.$$

EXAMPLE 9 *Finding the Probability of the Complement of an Event*

A manufacturer has determined that a certain machine averages one faulty unit for every 1000 it produces. What is the probability that an order of 200 units will have one or more faulty units?

Sequences, Series, and Probability

SOLUTION

To solve this problem as stated, we would need to find the probability of having exactly one faulty unit, exactly two faulty units, exactly three faulty units, and so on. However, using complements, we can simply find the probability that all units are perfect and then subtract this value from 1. Since the probability that any given unit is perfect is 999/1000, the probability that all 200 units are perfect is

$$P(A) = \left(\frac{999}{1000} \right)^{200} \approx 0.8186.$$

Therefore, the probability that at least one unit is faulty is

$$P(A') = 1 - P(A) \approx 0.1814.$$

WARM UP

Evaluate the expressions.

1. $\dfrac{1}{4} + \dfrac{5}{8} - \dfrac{5}{16}$

2. $\dfrac{4}{15} + \dfrac{3}{5} - \dfrac{1}{3}$

3. $\dfrac{5 \cdot 4}{5!}$

4. $\dfrac{5!22!}{27!}$

5. $\dfrac{4!8!}{12!}$

6. $\dfrac{9 \cdot 8 \cdot 7 \cdot 6 \cdot 5}{9!}$

7. $\dfrac{{}_5C_3}{{}_{10}C_3}$

8. $\dfrac{{}_{10}C_2 \cdot {}_{10}C_2}{{}_{20}C_4}$

Evaluate the expressions. (Round your answers to three decimal places.)

9. $\left(\dfrac{99}{100} \right)^{100}$

10. $1 - \left(\dfrac{89}{100} \right)^{50}$

EXERCISES 12.7

In Exercises 1–4, find the indicated probabilities in the experiment of tossing a coin three times. Use the sample space S = {HHH, HHT, HTH, HTT, THH, THT, TTH, TTT}.

1. The probability of getting exactly one tail.
2. The probability of getting a head on the first toss.
3. The probability of getting at least one head.
4. The probability of getting at least two heads.

In Exercises 5–8, find the indicated probabilities in the experiment of selecting one card from a standard deck of 52 playing cards.

5. The probability of getting a face card.
6. The probability of not getting a face card.
7. The probability of getting a black card that is not a face card.
8. The probability that the card will be a 6 or less.

In Exercises 9–14, find the indicated probabilities in the experiment of tossing a six-sided die twice.

9. The probability that the sum is 4.

10. The probability that the sum is less than 11.

11. The probability that the sum is at least 7.

12. The probability that the total is 2, 3, or 12.

13. The probability that the sum is odd and no more than 7.

14. The probability that the sum is odd or a prime.

In Exercises 15–18, find the indicated probabilities in the experiment of drawing two marbles (the first is *not* replaced before the second is drawn) from a bag containing one green, two yellow, and three red marbles.

15. The probability of drawing two red marbles.

16. The probability of drawing two yellow marbles.

17. The probability of drawing neither yellow marble.

18. The probability of drawing marbles of different colors.

19. Repeat Exercise 18 assuming that the marble drawn first is replaced prior to drawing the second.

20. Three people have been nominated for president of a college class. From a small poll, it is estimated that the probability of Jane's winning the election is 0.37, and the probability of Larry's winning the election is 0.44. What is the probability of the third candidate's winning the election?

21. Taylor, Moore, and Jenkins are candidates for public office. It is estimated that Moore and Jenkins have about the same probability of winning, and Taylor is believed to be twice as likely to win as either of the others. Find the probability of each candidate's winning the election.

22. In a high school graduating class of 72 students, 28 are on the honor roll. Of these, 18 are going on to college, and of the other 44 students, 12 are going on to college. If a student is selected at random from the class, what is the probability that the person chosen is (a) going to college, (b) not going to college, and (c) on the honor roll, but not going to college?

23. Two integers (between 1 and 30 inclusive) are chosen by a random number generator on a computer. What is the probability that (a) the numbers are both even, (b) one number is even and one is odd, (c) both numbers are less than 10, and (d) the same number is chosen twice?

24. An instructor gives her class a list of eight study problems, from which she will select five to be answered on an exam. If a given student knows how to solve six of the problems,

find the probability that the student will be able to answer (a) all five questions on the exam, (b) exactly four questions on the exam, and (c) at least four questions on the exam.

25. Two cards are selected at random from an ordinary deck of 52 playing cards. Find the probability that two aces are selected, given the following conditions.
 (a) The cards are drawn in sequence, with the first card being replaced and the deck reshuffled prior to the second drawing.
 (b) The two cards are drawn consecutively, without replacement.

26. On a game show you are given four digits to arrange in the proper order to give the price of a car. If you are correct, you win the car. What is the probability of winning, given the following conditions?
 (a) You guess the position of each digit.
 (b) You know the first digit, but must guess the remaining three.

27. Four letters and envelopes are addressed to four different people. If the letters are randomly inserted into the envelopes, what is the probability that (a) exactly one will be inserted in the correct envelope and (b) at least one will be inserted in the correct envelope?

28. Five cards are drawn from an ordinary deck of 52 playing cards. What is the probability of getting a full house? (See Exercise 36 in Section 12.6.)

29. A shipment of twelve microwave ovens contains three defective units. A vending company has ordered four of these units, and since each is identically packaged, the selection will be random. (See Exercise 34 in Section 12.6.) What is the probability that (a) all four units are good, (b) exactly two units are good, and (c) at least two units are good?

30. A fire company keeps two rescue vehicles to serve the community. Because of the demand on the company's time and the chance of mechanical failure, the probability that a specific vehicle is available when needed is 90%. If the availability of one vehicle is *independent* of the other, find the probability that (a) both vehicles are available at a given time, (b) neither vehicle is available at a given time, and (c) at least one is available at a given time.

31. A space vehicle has an independent back-up system for one of its communication networks. The probability that either system will function satisfactorily for the duration of a flight is 0.985. What is the probability that during a given flight (a) both systems function satisfactorily, (b) at least one system functions satisfactorily, and (c) both systems fail?

32. A door-to-door sales representative makes a sale at approximately one-third of the homes she calls on. If, on a given day, she goes to four homes, what is the probability that she will make a sale at (a) all four homes, (b) none of the homes, and (c) at least one home?

33. Assume that the probability of the birth of a child of a particular sex is 50%. In a family with six children, what is the probability that (a) all the children are girls, (b) all the children are the same sex, and (c) there is at least one girl?

34. The probability of getting at least one defective unit in a shipment of ten units is p. What is the probability that all ten are good?

[*Optional*] In Exercises 35–38, use the following information about binomial experiments. A **binomial experiment** is distinguished by the following four conditions.

(1) There are n identical trials.
(2) There are two possible outcomes to each trial, one of which is called success and the other failure.
(3) The probability of a success is the same for every trial and is denoted by p; the probability of failure is denoted by $q = 1 - p$.
(4) The n trials are independent.

Each term in the expansion of $(p + q)^n$ gives the probability of a certain number of successes. For example, if there are ten trials, then the probability of seven successes is given by $_{10}C_7 p^7 q^3$.

35. A particular binomial experiment has ten trials. The probability of success is 10%.
(a) Find the probability that all ten trials fail.
(b) Find the probability that all ten trials succeed.
(c) Find the probability that at least one trial succeeds.
(d) Find the probability that exactly one trial succeeds.

36. Use the information in Exercise 32 ($p = 1/3$, $q = 2/3$, $n = 4$) to expand $(p + q)^n$ and find the probability of the following.
(a) four sales   (b) three sales
(c) two sales   (d) one sale
(e) no sale

37. Use the information in Exercise 33 ($p = q = 1/2$, $n = 6$) to expand $(p + q)^n$ and find the probability of each of the following.

Number of Girls	0	1	2	3	4	5	6
Probability							

38. A basketball player makes 65% of her shots from the floor. Use the expansion of $(p + q)^n$ to find the probability that she makes x of her next five shots.

x	0	1	2	3	4	5
Probability						

CHAPTER 12 REVIEW EXERCISES

In Exercises 1–4, use sigma notation to write the given sum.

1. $\dfrac{1}{2(1)} + \dfrac{1}{2(2)} + \dfrac{1}{2(3)} + \cdots + \dfrac{1}{2(20)}$

2. $2(1^2) + 2(2^2) + 2(3^2) + \cdots + 2(9^2)$

3. $\dfrac{1}{2} + \dfrac{2}{3} + \dfrac{3}{4} + \cdots + \dfrac{9}{10}$

4. $1 - \dfrac{1}{3} + \dfrac{1}{9} - \dfrac{1}{27} + \cdots$

In Exercises 5–18, find the sum.

5. $\displaystyle\sum_{i=1}^{6} 5$

6. $\displaystyle\sum_{k=2}^{5} 4k$

7. $\displaystyle\sum_{j=3}^{10} (2j - 3)$

8. $\displaystyle\sum_{j=1}^{8} (20 - 3j)$

9. $\displaystyle\sum_{i=0}^{6} 2^i$

10. $\displaystyle\sum_{i=0}^{4} 3^i$

11. $\displaystyle\sum_{i=0}^{\infty} \left(\tfrac{7}{8}\right)^i$

12. $\displaystyle\sum_{i=0}^{\infty} \left(\tfrac{1}{3}\right)^i$

13. $\displaystyle\sum_{k=0}^{\infty} 4\left(\tfrac{2}{3}\right)^k$

14. $\displaystyle\sum_{k=0}^{\infty} 1.3\left(\tfrac{1}{10}\right)^k$

15. $\displaystyle\sum_{k=1}^{11} \left(\tfrac{2}{3}k + 4\right)$

16. $\displaystyle\sum_{k=1}^{25} \left(\dfrac{3k + 1}{4}\right)$

17. $\displaystyle\sum_{n=0}^{10} (n^2 + 3)$

18. $\displaystyle\sum_{n=1}^{100} \left(\dfrac{1}{n} - \dfrac{1}{n + 1}\right)$

In Exercises 19 and 20, write the first five terms of the arithmetic sequence.

19. $a_1 = 3$, $d = 4$

20. $a_4 = 10$, $a_{10} = 28$

In Exercises 21 and 22, write the first five terms of the geometric sequence.

21. $a_1 = 4$, $r = -\frac{1}{4}$ **22.** $a_1 = 9$, $a_3 = 4$

23. Find the sum of the first 100 positive multiples of 5.

24. Find the sum of the integers from 20 to 80 (inclusive).

In Exercises 25–30, write the repeating decimal as the ratio of two integers by considering each to be the sum of an infinite geometric series.

25. 0.454545. . .   **26.** 0.151515. . .
27. 1.0666. . .   **28.** 0.0222. . .
29. 0.01333. . .   **30.** 2.7333. . .

In Exercises 31–34, use mathematical induction to prove the given formula for every positive integer n.

31. $1 + 4 + \cdots + (3n - 2) = \frac{n}{2}(3n - 1)$

32. $1 + \frac{3}{2} + 2 + \frac{5}{2} + \cdots + \frac{1}{2}(n + 1) = \frac{n}{4}(n + 3)$

33. $\sum_{i=0}^{n-1} ar^i = \frac{a(1 - r^n)}{1 - r}$

34. $\sum_{k=0}^{n-1} [a + kd] = \frac{n}{2}[2a + (n - 1)d]$

In Exercises 35–40, use the Binomial Theorem to expand the binomial. Simplify your answer. [Remember that $i = \sqrt{-1}$.]

35. $\left(\frac{x}{2} + y\right)^4$ **36.** $(a - 3b)^5$

37. $\left(\frac{2}{x} - 3x\right)^6$ **38.** $(3x + y^2)^7$

39. $(5 + 2i)^4$ **40.** $(4 - 5i)^3$

41. As a family increases in number, the number of different interpersonal relationships that exist increases at an even faster rate. Find the number of different interpersonal relationships that exist (between two people) if the number of members in the family is (a) 2, (b) 4, and (c) 6.

42. In Morse Code, all characters are transmitted using a sequence of dits or dahs. How many different characters can be formed by using a sequence of three dits and dahs? (These can be repeated. For example, dit-dit-dit represents the letter s.)

43. A Novice Amateur Radio license consists of two letters, one digit, and then three more letters. How many different licenses can be issued if no restrictions are placed on the letters or digits?

44. How many different straight-line segments are determined by (a) five points and (b) ten points? (Assume that no three of the given points are collinear and that two points determine a line.)

45. A man has five pairs of socks (no two pairs are the same color). If he randomly selects two socks from the drawer, what is the probability that he gets a matched pair?

46. A child carries a five-volume set of books to a bookshelf. She is not able to read, and hence cannot distinguish one volume from another. What is the probability that she puts the books on the shelf in the correct order?

47. Are the chances of rolling a 3 with one die the same as that of rolling a total of 6 with two dice? If not, which has the higher probability?

48. A die is rolled six times. What is the probability that each side of the die will appear exactly once?

49. Five cards are drawn from an ordinary deck of 52 playing cards. Find the probability of getting two pairs. (For example, the hand could be A-A-5-5-Q or 4-4-7-7-K.)

50. (a) What is the probability that, in a group of ten people, at least two people have the same birthday? (Assume there are 365 different birthdays in a year.)

(b) How large must the group be before the probability that at least two people have the same birthday is at least 50%?

Appendix A

Using Logarithmic and Trigonometric Tables

Although it is more efficient to use calculators than tables in computations with logarithms or trigonometric functions, it is instructive to see how to work with tables. We begin with a discussion of the use of tables of logarithms.

Using Logarithmic Tables

Using base 10, we note first that every positive real number can be written as a product $c \times 10^k$, where $1 \leq c < 10$ and k is an integer. For example, $1989 = 1.989 \times 10^3$, $5.37 = 5.37 \times 10^0$, and $0.0439 = 4.39 \times 10^{-2}$. Suppose we apply the properties of logarithms to the number 1989.

$$1989 = 1.989 \times 10^3$$
$$\log_{10} 1989 = \log_{10} (1.989 \times 10^3)$$
$$= \log_{10} (1.989) + \log_{10} (10^3)$$
$$= \log_{10} (1.989) + 3 \log_{10} (10)$$
$$= \log_{10} (1.989) + 3$$

In general, for any positive real number x (expressible as $x = c \times 10^k$), its common logarithm has the **standard form**

$$\log_{10} x = \log_{10} c + \log_{10} (10^k) = \log_{10} c + k$$

where $1 \leq c < 10$. We call $\log_{10} c$ the **mantissa** and k the **characteristic** of

$\log_{10} x$. Since the function $f(x) = \log_{10} x$ increases as x increases and since $1 \le c < 10$, it follows that

$$\log_{10} 1 \le \log_{10} c < \log_{10} 10$$
$$0 \le \log_{10} c < 1$$

which means that the *mantissa* of $\log_{10} x$ lies between 0 and 1.

The common logarithm table in Appendix D gives four-decimal-place approximations of the *mantissa* for the logarithm of every three-digit number between 1.00 and 9.99. The next example shows how to use the table in Appendix D to approximate common logarithms.

EXAMPLE 1 **Approximating Common Logarithms with Tables**

Use the tables in Appendix D to approximate the following.

(a) $\log_{10} 85.6$ (b) $\log_{10} 0.000329$

SOLUTION

(a) Since $85.6 = 8.56 \times 10^1$, the characteristic is 1. Using the common logarithm table, we see that the mantissa is $\log_{10} 8.56 \approx 0.9325$. Therefore,

$$\begin{aligned} \log_{10} 85.6 &= \text{(mantissa)} + \text{(characteristic)} \\ &= \log_{10} 8.56 + 1 \\ &\approx 0.9325 + 1 \\ &= 1.9325. \end{aligned}$$

(b) Since $0.000329 = 3.29 \times 10^{-4}$, the characteristic is -4. From the common logarithm table for the mantissa 3.29, we obtain

$$\begin{aligned} \log_{10} 0.000329 &= \log_{10} 3.29 + (-4) \\ &\approx 0.3598 - 4 \\ &= -3.6402. \end{aligned}$$

The next example shows how to combine the use of properties of logarithms with tables to evaluate logarithms.

EXAMPLE 2 **Combining Properties of Logarithms with Tables**

Use the tables in Appendix D to approximate $\log_{10} \sqrt[3]{38.6}$.

SOLUTION

$$\log_{10} \sqrt[3]{38.6} = \frac{1}{3} \log_{10} 38.6$$

$$= \frac{1}{3}(\log_{10} 3.86 + 1)$$

$$\approx \frac{1}{3}(0.5866 + 1)$$

$$\approx 0.5289$$

The table for common logarithms can be used in the *reverse* manner to find the number (called an **antilogarithm**) that has a given logarithm. We demonstrate this procedure in Example 3.

EXAMPLE 3   *Finding the Antilogarithm of a Number*

Use the tables in Appendix D to approximate the value of x in each of the following.

(a) $\log_{10} x = 2.6571$ 　　　　　　　　(b) $x = 10^{-3.6364}$

SOLUTION

(a) We know that $\log_{10} x = 2.6571 = 0.6571 + 2$. Thus, the mantissa is 0.6571 and the characteristic is 2. From the table, we find that the mantissa 0.6571 corresponds approximately to $\log_{10} 4.54$. Since the characteristic is 2, it follows that x is given by

$$x \approx 4.54 \times 10^2 = 454.$$

(b) In logarithmic form, this exponential equation can be written as $\log_{10} x = -3.6364$. To obtain the standard form, we add and subtract 4 to obtain

$$\log_{10} x = (4 - 3.6364) - 4 = 0.3636 - 4.$$

Thus, the mantissa is 0.3636 and the characteristic is -4. From the table, we find that

$$x \approx 2.31 \times 10^{-4} = 0.000231.$$

For numbers with more than three nonzero digits, we can still use the common logarithm tables by applying a procedure called **linear interpolation.** This procedure is based on the fact that changes in $\log_{10} x$ are approximately proportional to the corresponding changes in x. We demonstrate the procedure in Example 4.

EXAMPLE 4 *Linear Interpolation*

Use linear interpolation and the tables in Appendix D to approximate the value of $\log_{10} 5.382$.

SOLUTION

The three-digit x-values in the table that are closest to 5.382 are $x = 5.38$ and $x = 5.39$. We use the logarithms of these two values in the following arrangement.

$$0.01 \left\{ 0.002 \left\{ \begin{array}{l} \log_{10} 5.38 \approx 0.7308 \\ \log_{10} 5.382 \approx \; ? \\ \log_{10} 5.39 \approx 0.7316 \end{array} \right\} d \right\} 0.0008$$

Note that the differences between the x-values are beside the left braces and the differences between the corresponding logarithms are beside the right braces. From this arrangement, we can write the following proportion.

$$\frac{d}{0.0008} = \frac{0.002}{0.01}$$

$$d = (0.0008)\left(\frac{0.002}{0.01}\right) = 0.00016 \approx 0.0002$$

Therefore,

$$\log_{10} 5.382 \approx \log_{10} 5.38 + d \approx 0.7308 + 0.002 = 0.7310.$$

The next example shows how to perform numerical computations with logarithms.

EXAMPLE 5 *Using Logarithms to Perform Numerical Computations*

Use the tables in Appendix D to approximate the value of

$$x = \frac{(1.9)^3}{\sqrt{82.7}}.$$

SOLUTION

Using the properties of logarithms, we can write

$$\log_{10} x = 3 \log_{10} 1.9 - \frac{1}{2} \log_{10} 82.7$$

$$\approx 3(0.2788) - \frac{1}{2}(1.9175) \approx -0.12235.$$

By adding and subtracting 1, we obtain the standard form

$$\log_{10} x \approx (1 - 0.12235) - 1 = 0.87765 - 1.$$

Finally, since the antilogarithm of 0.87765 is approximately 7.54, we find that

$$x \approx 7.54 \times 10^{-1} = 0.754.$$

Using Trigonometric Tables

Linear interpolation can also be used with tables of trigonometric functions. We demonstrate this procedure in Example 6.

EXAMPLE 6 Approximating Values of Trigonometric Functions

Use the trigonometric tables in Appendix E to approximate the following.

(a) $\cos 132° \, 14'$ 　　　　　　　　(b) $\tan(-20° \, 23')$

SOLUTION

(a) The given angle lies in Quadrant II with a reference angle of $\theta = 179° \, 60' - 132° \, 14' = 47° \, 46'$, so that

$$\cos 132° \, 14' = -\cos 47° \, 46'.$$

Using the table in Appendix E, we obtain the following arrangement.

$$10\left\{ 6\left\{ \begin{matrix} \cos 47° \, 40' \approx 0.6734 \\ \cos 47° \, 46' \approx \, ? \end{matrix} \right\} d \atop \cos 47° \, 50' \approx 0.6713 \right\} 0.0021$$

Thus, $d = 0.0021(6/10) \approx 0.0013$, and we obtain

$$\cos 132° \, 14' = -\cos 47° \, 46'$$
$$\approx -(\cos 47° \, 40' - 0.0013)$$
$$\approx -(0.6734 - 0.0013)$$
$$= -0.6721.$$

(b) We know that $\tan(-20° \, 23') = -\tan 20° \, 23'$, and we interpolate as follows.

$$10\left\{ 3\left\{\begin{array}{l} \tan 20° \, 20' \approx 0.3706 \\ \tan 20° \, 23' \approx \; ? \\ \tan 20° \, 30' \approx 0.3739 \end{array}\right\} d \right\} 0.0033$$

Thus, $d = 0.0033(3/10) \approx 0.0010$, and we obtain

$$\begin{aligned} \tan(-20° \, 23') &= -\tan 20° \, 23' \\ &\approx -(\tan 20° \, 20' + 0.0010) \\ &\approx -(0.3706 + 0.0010) \\ &= -0.3716. \end{aligned}$$

As with antilogarithms, the trigonometric tables can be used in the reverse manner, as demonstrated in Example 7.

EXAMPLE 7   Approximating Angles Using Trigonometric Tables

Approximate the value of t in the interval $0 \leq t \leq \pi/2$ such that $\sin t = 0.8619$.

SOLUTION

From the column corresponding to the sine in the tables in Appendix E, we choose the two values closest to 0.8619 and interpolate as follows. (Remember to use radians.)

$$0.0029\left\{ d\left\{\begin{array}{l} \sin 1.0385 \approx 0.8616 \\ \sin \, (?) \approx 0.8619 \\ \sin 1.0414 \approx 0.8631 \end{array}\right\} 3 \right\} 15$$

Thus, $d = 0.0029(3/15) \approx 0.0006$, and we obtain

$$t \approx 1.0385 + 0.0006 = 1.0391 \text{ radians.}$$

EXERCISES FOR APPENDIX A

In Exercises 1–4, approximate the common logarithm of the given number by using the table in Appendix D.

1. (a) 417   (b) 0.0417

2. (a) 985   (b) 9.85

3. (a) 6300   (b) 1000

4. (a) 0.0001   (b) 41.3

In Exercises 5–8, approximate the common logarithm of the given quantity by using the table in Appendix D.

5. (a) $\dfrac{5.30}{21.5}$ (b) $(30500)(0.258)$

6. (a) $\sqrt[3]{5.33}$ (b) $(1.02)^{36}$

7. (a) $\sqrt[5]{7200}$ (b) $(3.4)^{8}$

8. (a) $\dfrac{(3.6)^{6}}{500}$ (b) $(0.245)^{4}(8.7)^{3}$

In Exercises 9–12, find N (antilogarithm) by using the table in Appendix D.

9. (a) $N = 10^{4.3979}$ (b) $N = 10^{-1.6021}$

10. (a) $\log_{10} N = 3.6702$ (b) $\log_{10} N = -2.3298$

11. (a) $\log_{10} N = 6.1335$ (b) $\log_{10} N = 8.1335 - 10$

12. (a) $\log_{10} N = 4.8420$ (b) $\log_{10} N = 7.8420 - 10$

In Exercises 13 and 14, use linear interpolation to approximate the common logarithm of the given number by using the table in Appendix D.

13. (a) 4385   (b) 0.6058

14. (a) 125.2   (b) 0.08675

In Exercises 15 and 16, use linear interpolation to approximate the antilogarithm N by using the table in Appendix D.

15. (a) $\log_{10} N = 5.6175$ (b) $\log_{10} N = -2.1503$

16. (a) $N = 10^{0.5743}$ (b) $N = 10^{9.9317}$

In Exercises 17–22, approximate the given quantity by using common logarithms.

17. $\dfrac{(86.4)(8.09)}{38.6}$ **18.** $\dfrac{1243}{(42.8)(67.9)}$

19. $\sqrt[3]{86.5}$ **20.** $\sqrt[4]{(4.705)(18.86)}$

21. $500(1.03)^{20}$ **22.** $(0.2313)^{6}$

In Exercises 23 and 24, approximate the natural logarithm of the given number by using the table in Appendix C.

23. (a) 6.24   (b) 9.55

24. (a) 2.605   (b) 3.005

In Exercises 25 and 26, approximate the antilogarithm N by using the table in Appendix C.

25. (a) $\ln N = 2.0096$ (b) $\ln N = 1.4422$

26. (a) $\ln N = 1.1233$ (b) $\ln N = 0.2271$

In Exercises 27 and 28, approximate the exponential by using the table in Appendix B.

27. (a) $e^{3.5}$ (b) $e^{-3.5}$

28. (a) $e^{6.2}$ (b) $e^{-6.2}$

In Exercises 29–32, approximate the trigonometric function by using the table in Appendix E.

29. (a) $\cos 34°$ (b) $\tan 56°$

30. (a) $\sin 12° 30'$ (b) $\cot 77° 30'$

31. (a) $\sin 1.0472$ (b) $\cot 0.5236$

32. (a) $\tan 233°$ (b) $\cos(-56°)$

In Exercises 33–37, approximate two values of θ in degrees $(0 \le \theta < 360°)$ corresponding to the given functions by using the table in Appendix E.

33. (a) $\sin \theta = 0.4226$ (b) $\cot \theta = 2.145$

34. (a) $\cos \theta = 0.7585$ (b) $\tan \theta = 1.921$

35. (a) $\cos \theta = 0.4230$ (b) $\cot \theta = 4$

36. (a) $\sin \theta = 0.5140$ (b) $\tan \theta = 0.5$

Appendix B

EXPONENTIAL TABLES

x	e^x	e^{-x}	x	e^x	e^{-x}	x	e^x	e^{-x}
0.0	1.0000	1.0000	3.5	33.115	0.0302	7.0	1096.63	0.0009
0.1	1.1052	0.9048	3.6	36.598	0.0273	7.1	1211.97	0.0008
0.2	1.2214	0.8187	3.7	40.447	0.0247	7.2	1339.43	0.0007
0.3	1.3499	0.7408	3.8	44.701	0.0224	7.3	1480.30	0.0007
0.4	1.4918	0.6703	3.9	49.402	0.0202	7.4	1635.98	0.0006
0.5	1.6487	0.6065	4.0	54.598	0.0183	7.5	1808.04	0.0006
0.6	1.8221	0.5488	4.1	60.340	0.0166	7.6	1998.20	0.0005
0.7	2.0138	0.4966	4.2	66.686	0.0150	7.7	2208.35	0.0005
0.8	2.2255	0.4493	4.3	73.700	0.0136	7.8	2440.60	0.0004
0.9	2.4596	0.4066	4.4	81.451	0.0123	7.9	2697.28	0.0004
1.0	2.7183	0.3679	4.5	90.017	0.0111	8.0	2980.96	0.0003
1.1	3.0042	0.3329	4.6	99.484	0.0101	8.1	3294.47	0.0003
1.2	3.3201	0.3012	4.7	109.95	0.0091	8.2	3640.95	0.0003
1.3	3.6693	0.2725	4.8	121.51	0.0082	8.3	4023.87	0.0002
1.4	4.0552	0.2466	4.9	134.29	0.0074	8.4	4447.07	0.0002
1.5	4.4817	0.2231	5.0	148.41	0.0067	8.5	4914.77	0.0002
1.6	4.9530	0.2019	5.1	164.02	0.0061	8.6	5431.66	0.0002
1.7	5.4739	0.1827	5.2	181.27	0.0055	8.7	6002.91	0.0002
1.8	6.0496	0.1653	5.3	200.34	0.0050	8.8	6634.24	0.0002
1.9	6.6859	0.1496	5.4	221.41	0.0045	8.9	7331.97	0.0001
2.0	7.3891	0.1353	5.5	244.69	0.0041	9.0	8103.08	0.0001
2.1	8.1662	0.1225	5.6	270.43	0.0037	9.1	8955.29	0.0001
2.2	9.0250	0.1108	5.7	298.87	0.0033	9.2	9897.13	0.0001
2.3	9.9742	0.1003	5.8	330.30	0.0030	9.3	10938.02	0.0001
2.4	11.023	0.0907	5.9	365.04	0.0027	9.4	12088.38	0.0001
2.5	12.182	0.0821	6.0	403.43	0.0025	9.5	13359.73	0.0001
2.6	13.464	0.0743	6.1	445.86	0.0022	9.6	14764.78	0.0001
2.7	14.880	0.0672	6.2	492.75	0.0020	9.7	16317.61	0.0001
2.8	16.445	0.0608	6.3	544.57	0.0018	9.8	18033.74	0.0001
2.9	18.174	0.0550	6.4	601.85	0.0017	9.9	19930.37	0.0001
3.0	20.086	0.0498	6.5	665.14	0.0015	10.0	22026.47	0.0000
3.1	22.198	0.0450	6.6	735.10	0.0014			
3.2	24.533	0.0408	6.7	812.41	0.0012			
3.3	27.113	0.0369	6.8	897.85	0.0011			
3.4	29.964	0.0334	6.9	992.27	0.0010			

Appendix C

NATURAL LOGARITHMIC TABLES

	0.00	0.01	0.02	0.03	0.04	0.05	0.06	0.07	0.08	0.09
1.0	0.0000	0.0100	0.0198	0.0296	0.0392	0.0488	0.0583	0.0677	0.0770	0.0862
1.1	0.0953	0.1044	0.1133	0.1222	0.1310	0.1398	0.1484	0.1570	0.1655	0.1740
1.2	0.1823	0.1906	0.1989	0.2070	0.2151	0.2231	0.2311	0.2390	0.2469	0.2546
1.3	0.2624	0.2700	0.2776	0.2852	0.2927	0.3001	0.3075	0.3148	0.3221	0.3293
1.4	0.3365	0.3436	0.3507	0.3577	0.3646	0.3716	0.3784	0.3853	0.3920	0.3988
1.5	0.4055	0.4121	0.4187	0.4253	0.4318	0.4383	0.4447	0.4511	0.4574	0.4637
1.6	0.4700	0.4762	0.4824	0.4886	0.4947	0.5008	0.5068	0.5128	0.5188	0.5247
1.7	0.5306	0.5365	0.5423	0.5481	0.5539	0.5596	0.5653	0.5710	0.5766	0.5822
1.8	0.5878	0.5933	0.5988	0.6043	0.6098	0.6152	0.6206	0.6259	0.6313	0.6366
1.9	0.6419	0.6471	0.6523	0.6575	0.6627	0.6678	0.6729	0.6780	0.6831	0.6881
2.0	0.6931	0.6981	0.7031	0.7080	0.7129	0.7178	0.7227	0.7275	0.7324	0.7372
2.1	0.7419	0.7467	0.7514	0.7561	0.7608	0.7655	0.7701	0.7747	0.7793	0.7839
2.2	0.7885	0.7930	0.7975	0.8020	0.8065	0.8109	0.8154	0.8198	0.8242	0.8286
2.3	0.8329	0.8372	0.8416	0.8459	0.8502	0.8544	0.8587	0.8629	0.8671	0.8713
2.4	0.8755	0.8796	0.8838	0.8879	0.8920	0.8961	0.9002	0.9042	0.9083	0.9123
2.5	0.9163	0.9203	0.9243	0.9282	0.9322	0.9361	0.9400	0.9439	0.9478	0.9517
2.6	0.9555	0.9594	0.9632	0.9670	0.9708	0.9746	0.9783	0.9821	0.9858	0.9895
2.7	0.9933	0.9969	1.0006	1.0043	1.0080	1.0116	1.0152	1.0188	1.0225	1.0260
2.8	1.0296	1.0332	1.0367	1.0403	1.0438	1.0473	1.0508	1.0543	1.0578	1.0613
2.9	1.0647	1.0682	1.0716	1.0750	1.0784	1.0818	1.0852	1.0886	1.0919	1.0953
3.0	1.0986	1.1019	1.1053	1.1086	1.1119	1.1151	1.1184	1.1217	1.1249	1.1282
3.1	1.1314	1.1346	1.1378	1.1410	1.1442	1.1474	1.1506	1.1537	1.1569	1.1600
3.2	1.1632	1.1663	1.1694	1.1725	1.1756	1.1787	1.1817	1.1848	1.1878	1.1909
3.3	1.1939	1.1969	1.2000	1.2030	1.2060	1.2090	1.2119	1.2149	1.2179	1.2208
3.4	1.2238	1.2267	1.2296	1.2326	1.2355	1.2384	1.2413	1.2442	1.2470	1.2499
3.5	1.2528	1.2556	1.2585	1.2613	1.2641	1.2669	1.2698	1.2726	1.2754	1.2782
3.6	1.2809	1.2837	1.2865	1.2892	1.2920	1.2947	1.2975	1.3002	1.3029	1.3056
3.7	1.3083	1.3110	1.3137	1.3164	1.3191	1.3218	1.3244	1.3271	1.3297	1.3324
3.8	1.3350	1.3376	1.3403	1.3429	1.3455	1.3481	1.3507	1.3533	1.3558	1.3584
3.9	1.3610	1.3635	1.3661	1.3686	1.3712	1.3737	1.3762	1.3788	1.3813	1.3838
4.0	1.3863	1.3888	1.3913	1.3938	1.3962	1.3987	1.4012	1.4036	1.4061	1.4085
4.1	1.4110	1.4134	1.4159	1.4183	1.4207	1.4231	1.4255	1.4279	1.4303	1.4327
4.2	1.4351	1.4375	1.4398	1.4422	1.4446	1.4469	1.4493	1.4516	1.4540	1.4563
4.3	1.4586	1.4609	1.4633	1.4656	1.4679	1.4702	1.4725	1.4748	1.4770	1.4793
4.4	1.4816	1.4839	1.4861	1.4884	1.4907	1.4929	1.4951	1.4974	1.4996	1.5019
4.5	1.5041	1.5063	1.5085	1.5107	1.5129	1.5151	1.5173	1.5195	1.5217	1.5239
4.6	1.5261	1.5282	1.5304	1.5326	1.5347	1.5369	1.5390	1.5412	1.5433	1.5454
4.7	1.5476	1.5497	1.5518	1.5539	1.5560	1.5581	1.5602	1.5623	1.5644	1.5665
4.8	1.5686	1.5707	1.5728	1.5748	1.5769	1.5790	1.5810	1.5831	1.5851	1.5872
4.9	1.5892	1.5913	1.5933	1.5953	1.5974	1.5994	1.6014	1.6034	1.6054	1.6074
5.0	1.6094	1.6114	1.6134	1.6154	1.6174	1.6194	1.6214	1.6233	1.6253	1.6273
5.1	1.6292	1.6312	1.6332	1.6351	1.6371	1.6390	1.6409	1.6429	1.6448	1.6467
5.2	1.6487	1.6506	1.6525	1.6544	1.6563	1.6582	1.6601	1.6620	1.6639	1.6658
5.3	1.6677	1.6696	1.6715	1.6734	1.6752	1.6771	1.6790	1.6808	1.6827	1.6845
5.4	1.6864	1.6882	1.6901	1.6919	1.6938	1.6956	1.6974	1.6993	1.7011	1.7029

NATURAL LOGARITHMIC TABLES (Continued)

	0.00	0.01	0.02	0.03	0.04	0.05	0.06	0.07	0.08	0.09
5.5	1.7047	1.7066	1.7084	1.7102	1.7120	1.7138	1.7156	1.7174	1.7192	1.7210
5.6	1.7228	1.7246	1.7263	1.7281	1.7299	1.7317	1.7334	1.7352	1.7370	1.7387
5.7	1.7405	1.7422	1.7440	1.7457	1.7475	1.7492	1.7509	1.7527	1.7544	1.7561
5.8	1.7579	1.7596	1.7613	1.7630	1.7647	1.7664	1.7681	1.7699	1.7716	1.7733
5.9	1.7750	1.7766	1.7783	1.7800	1.7817	1.7834	1.7851	1.7867	1.7884	1.7901
6.0	1.7918	1.7934	1.7951	1.7967	1.7984	1.8001	1.8017	1.8034	1.8050	1.8066
6.1	1.8083	1.8099	1.8116	1.8132	1.8148	1.8165	1.8181	1.8197	1.8213	1.8229
6.2	1.8245	1.8262	1.8278	1.8294	1.8310	1.8326	1.8342	1.8358	1.8374	1.8390
6.3	1.8405	1.8421	1.8437	1.8453	1.8469	1.8485	1.8500	1.8516	1.8532	1.8547
6.4	1.8563	1.8579	1.8594	1.8610	1.8625	1.8641	1.8656	1.8672	1.8687	1.8703
6.5	1.8718	1.8733	1.8749	1.8764	1.8779	1.8795	1.8810	1.8825	1.8840	1.8856
6.6	1.8871	1.8886	1.8901	1.8916	1.8931	1.8946	1.8961	1.8976	1.8991	1.9006
6.7	1.9021	1.9036	1.9051	1.9066	1.9081	1.9095	1.9110	1.9125	1.9140	1.9155
6.8	1.9169	1.9184	1.9199	1.9213	1.9228	1.9242	1.9257	1.9272	1.9286	1.9301
6.9	1.9315	1.9330	1.9344	1.9359	1.9373	1.9387	1.9402	1.9416	1.9430	1.9445
7.0	1.9459	1.9473	1.9488	1.9502	1.9516	1.9530	1.9544	1.9559	1.9573	1.9587
7.1	1.9601	1.9615	1.9629	1.9643	1.9657	1.9671	1.9685	1.9699	1.9713	1.9727
7.2	1.9741	1.9755	1.9769	1.9782	1.9796	1.9810	1.9824	1.9838	1.9851	1.9865
7.3	1.9879	1.9892	1.9906	1.9920	1.9933	1.9947	1.9961	1.9974	1.9988	2.0001
7.4	2.0015	2.0028	2.0042	2.0055	2.0069	2.0082	2.0096	2.0109	2.0122	2.0136
7.5	2.0149	2.0162	2.0176	2.0189	2.0202	2.0215	2.0229	2.0242	2.0255	2.0268
7.6	2.0281	2.0295	2.0308	2.0321	2.0334	2.0347	2.0360	2.0373	2.0386	2.0399
7.7	2.0412	2.0425	2.0438	2.0451	2.0464	2.0477	2.0490	2.0503	2.0516	2.0528
7.8	2.0541	2.0554	2.0567	2.0580	2.0592	2.0605	2.0618	2.0631	2.0643	2.0656
7.9	2.0669	2.0681	2.0694	2.0707	2.0719	2.0732	2.0744	2.0757	2.0769	2.0782
8.0	2.0794	2.0807	2.0819	2.0832	2.0844	2.0857	2.0869	2.0882	2.0894	2.0906
8.1	2.0919	2.0931	2.0943	2.0956	2.0968	2.0980	2.0992	2.1005	2.1017	2.1029
8.2	2.1041	2.1054	2.1066	2.1078	2.1090	2.1102	2.1114	2.1126	2.1138	2.1150
8.3	2.1163	2.1175	2.1187	2.1199	2.1211	2.1223	2.1235	2.1247	2.1258	2.1270
8.4	2.1282	2.1294	2.1306	2.1318	2.1330	2.1342	2.1353	2.1365	2.1377	2.1389
8.5	2.1401	2.1412	2.1424	2.1436	2.1448	2.1459	2.1471	2.1483	2.1494	2.1506
8.6	2.1518	2.1529	2.1541	2.1552	2.1564	2.1576	2.1587	2.1599	2.1610	2.1622
8.7	2.1633	2.1645	2.1656	2.1668	2.1679	2.1691	2.1702	2.1713	2.1725	2.1736
8.8	2.1748	2.1759	2.1770	2.1782	2.1793	2.1804	2.1815	2.1827	2.1838	2.1849
8.9	2.1861	2.1872	2.1883	2.1894	2.1905	2.1917	2.1928	2.1939	2.1950	2.1961
9.0	2.1972	2.1983	2.1994	2.2006	2.2017	2.2028	2.2039	2.2050	2.2061	2.2072
9.1	2.2083	2.2094	2.2105	2.2116	2.2127	2.2138	2.2148	2.2159	2.2170	2.2181
9.2	2.2192	2.2203	2.2214	2.2225	2.2235	2.2246	2.2257	2.2268	2.2279	2.2289
9.3	2.2300	2.2311	2.2322	2.2332	2.2343	2.2354	2.2364	2.2375	2.2386	2.2396
9.4	2.2407	2.2418	2.2428	2.2439	2.2450	2.2460	2.2471	2.2481	2.2492	2.2502
9.5	2.2513	2.2523	2.2534	2.2544	2.2555	2.2565	2.2576	2.2586	2.2597	2.2607
9.6	2.2618	2.2628	2.2638	2.2649	2.2659	2.2670	2.2680	2.2690	2.2701	2.2711
9.7	2.2721	2.2732	2.2742	2.2752	2.2762	2.2773	2.2783	2.2793	2.2803	2.2814
9.8	2.2824	2.2834	2.2844	2.2854	2.2865	2.2875	2.2885	2.2895	2.2905	2.2915
9.9	2.2925	2.2935	2.2946	2.2956	2.2966	2.2976	2.2986	2.2996	2.3006	2.3016

Appendix D

COMMON LOGARITHMIC TABLES

	0.00	0.01	0.02	0.03	0.04	0.05	0.06	0.07	0.08	0.09
1.0	0.0000	0.0043	0.0086	0.0128	0.0170	0.0212	0.0253	0.0294	0.0334	0.0374
1.1	0.0414	0.0453	0.0492	0.0531	0.0569	0.0607	0.0645	0.0682	0.0719	0.0755
1.2	0.0792	0.0828	0.0864	0.0899	0.0934	0.0969	0.1004	0.1038	0.1072	0.1106
1.3	0.1139	0.1173	0.1206	0.1239	0.1271	0.1303	0.1335	0.1367	0.1399	0.1430
1.4	0.1461	0.1492	0.1523	0.1553	0.1584	0.1614	0.1644	0.1673	0.1703	0.1732
1.5	0.1761	0.1790	0.1818	0.1847	0.1875	0.1903	0.1931	0.1959	0.1987	0.2014
1.6	0.2041	0.2068	0.2095	0.2122	0.2148	0.2175	0.2201	0.2227	0.2253	0.2279
1.7	0.2304	0.2330	0.2355	0.2380	0.2405	0.2430	0.2455	0.2480	0.2504	0.2529
1.8	0.2553	0.2577	0.2601	0.2625	0.2648	0.2672	0.2695	0.2718	0.2742	0.2765
1.9	0.2788	0.2810	0.2833	0.2856	0.2878	0.2900	0.2923	0.2945	0.2967	0.2989
2.0	0.3010	0.3032	0.3054	0.3075	0.3096	0.3118	0.3139	0.3160	0.3181	0.3201
2.1	0.3222	0.3243	0.3263	0.3284	0.3304	0.3324	0.3345	0.3365	0.3385	0.3404
2.2	0.3424	0.3444	0.3464	0.3483	0.3502	0.3522	0.3541	0.3560	0.3579	0.3598
2.3	0.3617	0.3636	0.3655	0.3674	0.3692	0.3711	0.3729	0.3747	0.3766	0.3784
2.4	0.3802	0.3820	0.3838	0.3856	0.3874	0.3892	0.3909	0.3927	0.3945	0.3962
2.5	0.3979	0.3997	0.4014	0.4031	0.4048	0.4065	0.4082	0.4099	0.4116	0.4133
2.6	0.4150	0.4166	0.4183	0.4200	0.4216	0.4232	0.4249	0.4265	0.4281	0.4298
2.7	0.4314	0.4330	0.4346	0.4362	0.4378	0.4393	0.4409	0.4425	0.4440	0.4456
2.8	0.4472	0.4487	0.4502	0.4518	0.4533	0.4548	0.4564	0.4579	0.4594	0.4609
2.9	0.4624	0.4639	0.4654	0.4669	0.4683	0.4698	0.4713	0.4728	0.4742	0.4757
3.0	0.4771	0.4786	0.4800	0.4814	0.4829	0.4843	0.4857	0.4871	0.4886	0.4900
3.1	0.4914	0.4928	0.4942	0.4955	0.4969	0.4983	0.4997	0.5011	0.5024	0.5038
3.2	0.5052	0.5065	0.5079	0.5092	0.5105	0.5119	0.5132	0.5145	0.5159	0.5172
3.3	0.5185	0.5198	0.5211	0.5224	0.5237	0.5250	0.5263	0.5276	0.5289	0.5302
3.4	0.5315	0.5328	0.5340	0.5353	0.5366	0.5378	0.5391	0.5403	0.5416	0.5428
3.5	0.5441	0.5453	0.5465	0.5478	0.5490	0.5502	0.5514	0.5527	0.5539	0.5551
3.6	0.5563	0.5575	0.5587	0.5599	0.5611	0.5623	0.5635	0.5647	0.5658	0.5670
3.7	0.5682	0.5694	0.5705	0.5717	0.5729	0.5740	0.5752	0.5763	0.5775	0.5786
3.8	0.5798	0.5809	0.5821	0.5832	0.5843	0.5855	0.5866	0.5877	0.5888	0.5899
3.9	0.5911	0.5922	0.5933	0.5944	0.5955	0.5966	0.5977	0.5988	0.5999	0.6010
4.0	0.6021	0.6031	0.6042	0.6053	0.6064	0.6075	0.6085	0.6096	0.6107	0.6117
4.1	0.6128	0.6138	0.6149	0.6160	0.6170	0.6180	0.6191	0.6201	0.6212	0.6222
4.2	0.6232	0.6243	0.6253	0.6263	0.6274	0.6284	0.6294	0.6304	0.6314	0.6325
4.3	0.6335	0.6345	0.6355	0.6365	0.6375	0.6385	0.6395	0.6405	0.6415	0.6425
4.4	0.6435	0.6444	0.6454	0.6464	0.6474	0.6484	0.6493	0.6503	0.6513	0.6522
4.5	0.6532	0.6542	0.6551	0.6561	0.6571	0.6580	0.6590	0.6599	0.6609	0.6618
4.6	0.6628	0.6637	0.6646	0.6656	0.6665	0.6675	0.6684	0.6693	0.6702	0.6712
4.7	0.6721	0.6730	0.6739	0.6749	0.6758	0.6767	0.6776	0.6785	0.6794	0.6803
4.8	0.6812	0.6821	0.6830	0.6839	0.6848	0.6857	0.6866	0.6875	0.6884	0.6893
4.9	0.6902	0.6911	0.6920	0.6928	0.6937	0.6946	0.6955	0.6964	0.6972	0.6981
5.0	0.6990	0.6998	0.7007	0.7016	0.7024	0.7033	0.7042	0.7050	0.7059	0.7067
5.1	0.7076	0.7084	0.7093	0.7101	0.7110	0.7118	0.7126	0.7135	0.7143	0.7152
5.2	0.7160	0.7168	0.7177	0.7185	0.7193	0.7202	0.7210	0.7218	0.7226	0.7235
5.3	0.7243	0.7251	0.7259	0.7267	0.7275	0.7284	0.7292	0.7300	0.7308	0.7316
5.4	0.7324	0.7332	0.7340	0.7348	0.7356	0.7364	0.7372	0.7380	0.7388	0.7396

COMMON LOGARITHMIC TABLES (Continued)

	0.00	0.01	0.02	0.03	0.04	0.05	0.06	0.07	0.08	0.09
5.5	0.7404	0.7412	0.7419	0.7427	0.7435	0.7443	0.7451	0.7459	0.7466	0.7474
5.6	0.7482	0.7490	0.7497	0.7505	0.7513	0.7520	0.7528	0.7536	0.7543	0.7551
5.7	0.7559	0.7566	0.7574	0.7582	0.7589	0.7597	0.7604	0.7612	0.7619	0.7627
5.8	0.7634	0.7642	0.7649	0.7657	0.7664	0.7672	0.7679	0.7686	0.7694	0.7701
5.9	0.7709	0.7716	0.7723	0.7731	0.7738	0.7745	0.7752	0.7760	0.7767	0.7774
6.0	0.7782	0.7789	0.7796	0.7803	0.7810	0.7818	0.7825	0.7832	0.7839	0.7846
6.1	0.7853	0.7860	0.7868	0.7875	0.7882	0.7889	0.7896	0.7903	0.7910	0.7917
6.2	0.7924	0.7931	0.7938	0.7945	0.7952	0.7959	0.7966	0.7973	0.7980	0.7987
6.3	0.7993	0.8000	0.8007	0.8014	0.8021	0.8028	0.8035	0.8041	0.8048	0.8055
6.4	0.8062	0.8069	0.8075	0.8082	0.8089	0.8096	0.8102	0.8109	0.8116	0.8122
6.5	0.8129	0.8136	0.8142	0.8149	0.8156	0.8162	0.8169	0.8176	0.8182	0.8189
6.6	0.8195	0.8202	0.8209	0.8215	0.8222	0.8228	0.8235	0.8241	0.8248	0.8254
6.7	0.8261	0.8267	0.8274	0.8280	0.8287	0.8293	0.8299	0.8306	0.8312	0.8319
6.8	0.8325	0.8331	0.8338	0.8344	0.8351	0.8357	0.8363	0.8370	0.8376	0.8382
6.9	0.8388	0.8395	0.8401	0.8407	0.8414	0.8420	0.8426	0.8432	0.8439	0.8445
7.0	0.8451	0.8457	0.8463	0.8470	0.8476	0.8482	0.8488	0.8494	0.8500	0.8506
7.1	0.8513	0.8519	0.8525	0.8531	0.8537	0.8543	0.8549	0.8555	0.8561	0.8567
7.2	0.8573	0.8579	0.8585	0.8591	0.8597	0.8603	0.8609	0.8615	0.8621	0.8627
7.3	0.8633	0.8639	0.8645	0.8651	0.8657	0.8663	0.8669	0.8675	0.8681	0.8686
7.4	0.8692	0.8698	0.8704	0.8710	0.8716	0.8722	0.8727	0.8733	0.8739	0.8745
7.5	0.8751	0.8756	0.8762	0.8768	0.8774	0.8779	0.8785	0.8791	0.8797	0.8802
7.6	0.8808	0.8814	0.8820	0.8825	0.8831	0.8837	0.8842	0.8848	0.8854	0.8859
7.7	0.8865	0.8871	0.8876	0.8882	0.8887	0.8893	0.8899	0.8904	0.8910	0.8915
7.8	0.8921	0.8927	0.8932	0.8938	0.8943	0.8949	0.8954	0.8960	0.8965	0.8971
7.9	0.8976	0.8982	0.8987	0.8993	0.8998	0.9004	0.9009	0.9015	0.9020	0.9025
8.0	0.9031	0.9036	0.9042	0.9047	0.9053	0.9058	0.9063	0.9069	0.9074	0.9079
8.1	0.9085	0.9090	0.9096	0.9101	0.9106	0.9112	0.9117	0.9122	0.9128	0.9133
8.2	0.9138	0.9143	0.9149	0.9154	0.9159	0.9165	0.9170	0.9175	0.9180	0.9186
8.3	0.9191	0.9196	0.9201	0.9206	0.9212	0.9217	0.9222	0.9227	0.9232	0.9238
8.4	0.9243	0.9248	0.9253	0.9258	0.9263	0.9269	0.9274	0.9279	0.9284	0.9289
8.5	0.9294	0.9299	0.9304	0.9309	0.9315	0.9320	0.9325	0.9330	0.9335	0.9340
8.6	0.9345	0.9350	0.9355	0.9360	0.9365	0.9370	0.9375	0.9380	0.9385	0.9390
8.7	0.9395	0.9400	0.9405	0.9410	0.9415	0.9420	0.9425	0.9430	0.9435	0.9440
8.8	0.9445	0.9450	0.9455	0.9460	0.9465	0.9469	0.9474	0.9479	0.9484	0.9489
8.9	0.9494	0.9499	0.9504	0.9509	0.9513	0.9518	0.9523	0.9528	0.9533	0.9538
9.0	0.9542	0.9547	0.9552	0.9557	0.9562	0.9566	0.9571	0.9576	0.9581	0.9586
9.1	0.9590	0.9595	0.9600	0.9605	0.9609	0.9614	0.9619	0.9624	0.9628	0.9633
9.2	0.9638	0.9643	0.9647	0.9652	0.9657	0.9661	0.9666	0.9671	0.9675	0.9680
9.3	0.9685	0.9689	0.9694	0.9699	0.9703	0.9708	0.9713	0.9717	0.9722	0.9727
9.4	0.9731	0.9736	0.9741	0.9745	0.9750	0.9754	0.9759	0.9764	0.9768	0.9773
9.5	0.9777	0.9782	0.9786	0.9791	0.9795	0.9800	0.9805	0.9809	0.9814	0.9818
9.6	0.9823	0.9827	0.9832	0.9836	0.9841	0.9845	0.9850	0.9854	0.9859	0.9863
9.7	0.9868	0.9872	0.9877	0.9881	0.9886	0.9890	0.9894	0.9899	0.9903	0.9908
9.8	0.9912	0.9917	0.9921	0.9926	0.9930	0.9934	0.9939	0.9943	0.9948	0.9952
9.9	0.9956	0.9961	0.9965	0.9969	0.9974	0.9978	0.9983	0.9987	0.9991	0.9996

Appendix E

TRIGONOMETRIC TABLES

1 degree ≈ 0.01745 radians
1 radian ≈ 57.29578 degrees

For $0 \le \theta \le 45$, read from upper left.
For $45 \le \theta \le 90$, read from lower right.
For $90 \le \theta \le 360$, use the identities:

θ	Quadrant II	Quadrant III	Quadrant IV
sin θ	$\sin(180-\theta)$	$-\sin(\theta-180)$	$-\sin(360-\theta)$
cos θ	$-\cos(180-\theta)$	$-\cos(\theta-180)$	$\cos(360-\theta)$
tan θ	$-\tan(180-\theta)$	$\tan(\theta-180)$	$-\tan(360-\theta)$
cot θ	$-\cot(180-\theta)$	$\cot(\theta-180)$	$-\cot(360-\theta)$

Degrees	Radians	sin	cos	tan	cot		
0°00'	.0000	.0000	1.0000	.0000	—	1.5708	90°00'
10	.0029	.0029	1.0000	.0029	343.774	1.5679	50
20	.0058	.0058	1.0000	.0058	171.885	1.5650	40
30	.0087	.0087	1.0000	.0087	114.589	1.5621	30
40	.0116	.0116	.9999	.0116	85.940	1.5592	20
50	.0145	.0145	.9999	.0145	68.750	1.5563	10
1°00'	.0175	.0175	.9998	.0175	57.290	1.5533	89°00'
10	.0204	.0204	.9998	.0204	49.104	1.5504	50
20	.0233	.0233	.9997	.0233	42.964	1.5475	40
30	.0262	.0262	.9997	.0262	38.188	1.5446	30
40	.0291	.0291	.9996	.0291	34.368	1.5417	20
50	.0320	.0320	.9995	.0320	31.242	1.5388	10
2°00'	.0349	.0349	.9994	.0349	28.636	1.5359	88°00'
10	.0378	.0378	.9993	.0378	26.432	1.5330	50
20	.0407	.0407	.9992	.0407	24.542	1.5301	40
30	.0436	.0436	.9990	.0437	22.904	1.5272	30
40	.0465	.0465	.9989	.0466	21.470	1.5243	20
50	.0495	.0494	.9988	.0495	20.206	1.5213	10
3°00'	.0524	.0523	.9986	.0524	19.081	1.5184	87°00'
10	.0553	.0553	.9985	.0553	18.075	1.5155	50
20	.0582	.0581	.9983	.0582	17.169	1.5126	40
30	.0611	.0610	.9981	.0612	16.350	1.5097	30
40	.0640	.0640	.9980	.0641	15.605	1.5068	20
50	.0669	.0669	.9978	.0670	14.924	1.5039	10
	cos	sin	cot	tan	Radians	Degrees	

Degrees	Radians	sin	cos	tan	cot		
4°00'	.0698	.0698	.9976	.0699	14.301	1.5010	86°00'
10	.0727	.0727	.9974	.0729	13.727	1.4981	50
20	.0756	.0756	.9971	.0758	13.197	1.4952	40
30	.0785	.0785	.9969	.0787	12.706	1.4923	30
40	.0814	.0814	.9967	.0816	12.251	1.4893	20
50	.0844	.0843	.9964	.0846	11.826	1.4864	10
5°00'	.0873	.0872	.9962	.0875	11.430	1.4835	85°00'
10	.0902	.0901	.9959	.0904	11.059	1.4806	50
20	.0931	.0929	.9957	.0934	10.712	1.4777	40
30	.0960	.0958	.9954	.0963	10.385	1.4748	30
40	.0989	.0987	.9951	.0992	10.078	1.4719	20
50	.1018	.1016	.9948	.1022	9.788	1.4690	10
6°00'	.1047	.1045	.9945	.1051	9.514	1.4661	84°00'
10	.1076	.1074	.9942	.1080	9.255	1.4632	50
20	.1105	.1103	.9939	.1110	9.010	1.4603	40
30	.1134	.1132	.9936	.1139	8.777	1.4573	30
40	.1164	.1161	.9932	.1169	8.556	1.4544	20
50	.1193	.1190	.9929	.1198	8.345	1.4515	10
7°00'	.1222	.1219	.9925	.1228	8.144	1.4486	83°00'
10	.1251	.1248	.9922	.1257	7.953	1.4457	50
20	.1280	.1276	.9918	.1287	7.770	1.4428	40
30	.1309	.1305	.9914	.1317	7.596	1.4399	30
40	.1338	.1334	.9911	.1346	7.429	1.4370	20
50	.1367	.1363	.9907	.1376	7.269	1.4341	10
	cos	sin	cot	tan	Radians	Degrees	

TRIGONOMETRIC TABLES (Continued)

Degrees	Radians	sin	cos	tan	cot			Degrees	Radians	sin	cos	tan	cot		
8°00′	.1396	.1392	.9903	.1405	7.115	1.4312	82°00′	18°00′	.3142	.3090	.9511	.3249	3.078	1.2566	72°00′
10	.1425	.1421	.9899	.1435	6.968	1.4283	50	10	.3171	.3118	.9502	.3281	3.047	1.2537	50
20	.1454	.1449	.9894	.1465	6.827	1.4254	40	20	.3200	.3145	.9492	.3314	3.018	1.2508	40
30	.1484	.1478	.9890	.1495	6.691	1.4224	30	30	.3229	.3173	.9483	.3346	2.989	1.2479	30
40	.1513	.1507	.9886	.1524	6.561	1.4195	20	40	.3258	.3201	.9474	.3378	2.960	1.2450	20
50	.1542	.1536	.9881	.1554	6.435	1.4166	10	50	.3287	.3228	.9465	.3411	2.932	1.2421	10
9°00′	.1571	.1564	.9877	.1584	6.314	1.4137	81°00′	19°00′	.3316	.3256	.9455	.3443	2.904	1.2392	71°00′
10	.1600	.1593	.9872	.1614	6.197	1.4108	50	10	.3345	.3283	.9446	.3476	2.877	1.2363	50
20	.1629	.1622	.9868	.1644	6.084	1.4079	40	20	.3374	.3311	.9436	.3508	2.850	1.2334	40
30	.1658	.1650	.9863	.1673	5.976	1.4050	30	30	.3403	.3338	.9426	.3541	2.824	1.2305	30
40	.1687	.1679	.9858	.1703	5.871	1.4021	20	40	.3432	.3365	.9417	.3574	2.798	1.2275	20
50	.1716	.1708	.9853	.1733	5.769	1.3992	10	50	.3462	.3393	.9407	.3607	2.773	1.2246	10
10°00′	.1745	.1736	.9848	.1763	5.671	1.3963	80°00′	20°00′	.3491	.3420	.9397	.3640	2.747	1.2217	70°00′
10	.1774	.1765	.9843	.1793	5.576	1.3934	50	10	.3520	.3448	.9387	.3673	2.723	1.2188	50
20	.1804	.1794	.9838	.1823	5.485	1.3904	40	20	.3549	.3475	.9377	.3706	2.699	1.2159	40
30	.1833	.1822	.9833	.1853	5.396	1.3875	30	30	.3578	.3502	.9367	.3739	2.675	1.2130	30
40	.1862	.1851	.9827	.1883	5.309	1.3846	20	40	.3607	.3529	.9356	.3772	2.651	1.2101	20
50	.1891	.1880	.9822	.1914	5.226	1.3817	10	50	.3636	.3557	.9346	.3805	2.628	1.2072	10
11°00′	.1920	.1908	.9816	.1944	5.145	1.3788	79°00′	21°00′	.3665	.3584	.9336	.3839	2.605	1.2043	69°00′
10	.1949	.1937	.9811	.1974	5.066	1.3759	50	10	.3694	.3611	.9325	.3872	2.583	1.2014	50
20	.1978	.1965	.9805	.2004	4.989	1.3730	40	20	.3723	.3638	.9315	.3906	2.560	1.1985	40
30	.2007	.1994	.9799	.2035	4.915	1.3701	30	30	.3752	.3665	.9304	.3939	2.539	1.1956	30
40	.2036	.2022	.9793	.2065	4.843	1.3672	20	40	.3782	.3692	.9293	.3973	2.517	1.1926	20
50	.2065	.2051	.9787	.2095	4.773	1.3643	10	50	.3811	.3719	.9283	.4006	2.496	1.1897	10
12°00′	.2094	.2079	.9781	.2126	4.705	1.3614	78°00′	22°00′	.3840	.3746	.9272	.4040	2.475	1.1868	68°00′
10	.2123	.2108	.9775	.2156	4.638	1.3584	50	10	.3869	.3773	.9261	.4074	2.455	1.1839	50
20	.2153	.2136	.9769	.2186	4.574	1.3555	40	20	.3898	.3800	.9250	.4108	2.434	1.1810	40
30	.2182	.2164	.9763	.2217	4.511	1.3526	30	30	.3927	.3827	.9239	.4142	2.414	1.1781	30
40	.2211	.2193	.9757	.2247	4.449	1.3497	20	40	.3956	.3854	.9228	.4176	2.394	1.1752	20
50	.2240	.2221	.9750	.2278	4.390	1.3468	10	50	.3985	.3881	.9216	.4210	2.375	1.1723	10
13°00′	.2269	.2250	.9744	.2309	4.331	1.3439	77°00′	23°00′	.4014	.3907	.9205	.4245	2.356	1.1694	67°00′
10	.2298	.2278	.9737	.2339	4.275	1.3410	50	10	.4043	.3934	.9194	.4279	2.337	1.1665	50
20	.2327	.2306	.9730	.2370	4.219	1.3381	40	20	.4072	.3961	.9182	.4314	2.318	1.1636	40
30	.2356	.2334	.9724	.2401	4.165	1.3352	30	30	.4102	.3987	.9171	.4348	2.300	1.1606	30
40	.2385	.2363	.9717	.2432	4.113	1.3323	20	40	.4131	.4014	.9159	.4383	2.282	1.1577	20
50	.2414	.2391	.9710	.2462	4.061	1.3294	10	50	.4160	.4041	.9147	.4417	2.264	1.1548	10
14°00′	.2443	.2419	.9703	.2493	4.011	1.3265	76°00′	24°00′	.4189	.4067	.9135	.4452	2.246	1.1519	66°00′
10	.2473	.2447	.9696	.2524	3.962	1.3235	50	10	.4218	.4094	.9124	.4487	2.229	1.1490	50
20	.2502	.2476	.9689	.2555	3.914	1.3206	40	20	.4247	.4120	.9112	.4522	2.211	1.1461	40
30	.2531	.2504	.9681	.2586	3.867	1.3177	30	30	.4276	.4147	.9100	.4557	2.194	1.1432	30
40	.2560	.2532	.9674	.2617	3.821	1.3148	20	40	.4305	.4173	.9088	.4592	2.177	1.1403	20
50	.2589	.2560	.9667	.2648	3.776	1.3119	10	50	.4334	.4200	.9075	.4628	2.161	1.1374	10
15°00′	.2618	.2588	.9659	.2679	3.732	1.3090	75°00′	25°00′	.4363	.4226	.9063	.4663	2.145	1.1345	65°00′
10	.2647	.2616	.9652	.2711	3.689	1.3061	50	10	.4392	.4253	.9051	.4699	2.128	1.1316	50
20	.2676	.2644	.9644	.2742	3.647	1.3032	40	20	.4422	.4279	.9038	.4734	2.112	1.1286	40
30	.2705	.2672	.9636	.2773	3.606	1.3003	30	30	.4451	.4305	.9026	.4770	2.097	1.1257	30
40	.2734	.2700	.9628	.2805	3.566	1.2974	20	40	.4480	.4331	.9013	.4806	2.081	1.1228	20
50	.2763	.2728	.9621	.2836	3.526	1.2945	10	50	.4509	.4358	.9001	.4841	2.066	1.1199	10
16°00′	.2793	.2756	.9613	.2867	3.487	1.2915	74°00′	26°00′	.4538	.4384	.8988	.4877	2.050	1.1170	64°00′
10	.2822	.2784	.9605	.2899	3.450	1.2886	50	10	.4567	.4410	.8975	.4913	2.035	1.1141	50
20	.2851	.2812	.9596	.2931	3.412	1.2857	40	20	.4596	.4436	.8962	.4950	2.020	1.1112	40
30	.2880	.2840	.9588	.2962	3.376	1.2828	30	30	.4625	.4462	.8949	.4986	2.006	1.1083	30
40	.2909	.2868	.9580	.2994	3.340	1.2799	20	40	.4654	.4488	.8936	.5022	1.991	1.1054	20
50	.2938	.2896	.9572	.3026	3.305	1.2770	10	50	.4683	.4514	.8923	.5059	1.977	1.1025	10
17°00′	.2967	.2924	.9563	.3057	3.271	1.2741	73°00′	27°00′	.4712	.4540	.8910	.5095	1.963	1.0996	63°00′
10	.2996	.2952	.9555	.3089	3.237	1.2712	50	10	.4741	.4566	.8897	.5132	1.949	1.0966	50
20	.3025	.2979	.9546	.3121	3.204	1.2683	40	20	.4771	.4592	.8884	.5169	1.935	1.0937	40
30	.3054	.3007	.9537	.3153	3.172	1.2654	30	30	.4800	.4617	.8870	.5206	1.921	1.0908	30
40	.3083	.3035	.9528	.3185	3.140	1.2625	20	40	.4829	.4643	.8857	.5243	1.907	1.0879	20
50	.3113	.3062	.9520	.3217	3.108	1.2595	10	50	.4858	.4669	.8843	.5280	1.894	1.0850	10
		cos	sin	cot	tan	Radians	Degrees			cos	sin	cot	tan	Radians	Degrees

TRIGONOMETRIC TABLES (Continued)

Degrees	Radians	sin	cos	tan	cot		
28°00'	.4887	.4695	.8829	.5317	1.881	1.0821	62°00'
10	.4916	.4720	.8816	.5354	1.868	1.0792	50
20	.4945	.4746	.8802	.5392	1.855	1.0763	40
30	.4974	.4772	.8788	.5430	1.842	1.0734	30
40	.5003	.4797	.8774	.5467	1.829	1.0705	20
50	.5032	.4823	.8760	.5505	1.816	1.0676	10
29°00'	.5061	.4848	.8746	.5543	1.804	1.0647	61°00'
10	.5091	.4874	.8732	.5581	1.792	1.0617	50
20	.5120	.4899	.8718	.5619	1.780	1.0588	40
30	.5149	.4924	.8704	.5658	1.767	1.0559	30
40	.5178	.4950	.8689	.5696	1.756	1.0530	20
50	.5207	.4975	.8675	.5735	1.744	1.0501	10
30°00'	.5236	.5000	.8660	.5774	1.732	1.0472	60°00'
10	.5265	.5025	.8646	.5812	1.720	1.0443	50
20	.5294	.5050	.8631	.5851	1.709	1.0414	40
30	.5323	.5075	.8616	.5890	1.698	1.0385	30
40	.5325	.5100	.8601	.5930	1.686	1.0356	20
50	.5381	.5125	.8587	.5969	1.675	1.0327	10
31°00'	.5411	.5150	.8572	.6009	1.664	1.0297	59°00'
10	.5440	.5175	.8557	.6048	1.653	1.0268	50
20	.5469	.5200	.8542	.6088	1.643	1.0239	40
30	.5498	.5225	.8526	.6128	1.632	1.0210	30
40	.5527	.5250	.8511	.6168	1.621	1.0181	20
50	.5556	.5275	.8496	.6208	1.611	1.0152	10
32°00'	.5585	.5299	.8480	.6249	1.600	1.0123	58°00'
10	.5614	.5324	.8465	.6289	1.590	1.0094	50
20	.5643	.5348	.8450	.6330	1.580	1.0065	40
30	.5672	.5373	.8434	.6371	1.570	1.0036	30
40	.5701	.5398	.8418	.6412	1.560	1.0007	20
50	.5730	.5422	.8403	.6453	1.550	.9977	10
33°00'	.5760	.5446	.8387	.6494	1.540	.9948	57°00'
10	.5789	.5471	.8371	.6536	1.530	.9919	50
20	.5818	.5495	.8355	.6577	1.520	.9890	40
30	.5847	.5519	.8339	.6619	1.511	.9861	30
40	.5876	.5544	.8323	.6661	1.501	.9832	20
50	.5905	.5568	.8307	.6703	1.492	.9803	10
34°00'	.5934	.5592	.8290	.6745	1.483	.9774	56°00'
10	.5963	.5616	.8274	.6787	1.473	.9745	50
20	.5992	.5640	.8258	.6830	1.464	.9716	40
30	.6021	.5664	.8241	.6873	1.455	.9687	30
40	.6050	.5688	.8225	.6916	1.446	.9657	20
50	.6080	.5712	.8208	.6959	1.437	.9628	10
35°00'	.6109	.5736	.8192	.7002	1.428	.9599	55°00'
10	.6138	.5760	.8175	.7046	1.419	.9570	50
20	.6167	.5783	.8158	.7089	1.411	.9541	40
30	.6196	.5807	.8141	.7133	1.402	.9512	30
40	.6225	.5831	.8124	.7177	1.393	.9483	20
50	.6254	.5854	.8107	.7221	1.385	.9454	10
36°00'	.6283	.5878	.8090	.7265	1.376	.9425	54°00'
10	.6312	.5901	.8073	.7310	1.368	.9396	50
20	.6341	.5925	.8056	.7355	1.360	.9367	40
30	.6370	.5948	.8039	.7400	1.351	.9338	30
40	.6400	.5972	.8021	.7445	1.343	.9308	20
50	.6429	.5995	.8004	.7490	1.335	.9279	10
	cos	sin	cot	tan	Radians	Degrees	

Degrees	Radians	sin	cos	tan	cot		
37°00'	.6458	.6018	.7986	.7536	1.327	.9250	53°00'
10	.6487	.6041	.7969	.7581	1.319	.9221	50
20	.6516	.6065	.7951	.7627	1.311	.9192	40
30	.6545	.6088	.7934	.7673	1.303	.9163	30
40	.6574	.6111	.7916	.7720	1.295	.9134	20
50	.6603	.6134	.7898	.7766	1.288	.9105	10
38°00'	.6632	.6157	.7880	.7813	1.280	.9076	52°00'
10	.6661	.6180	.7862	.7860	1.272	.9047	50
20	.6690	.6202	.7844	.7907	1.265	.9018	40
30	.6720	.6225	.7826	.7954	1.257	.8988	30
40	.6749	.6248	.7808	.8002	1.250	.8959	20
50	.6778	.6271	.7790	.8050	1.242	.8930	10
39°00'	.6807	.6293	.7771	.8098	1.235	.8901	51°00'
10	.6836	.6316	.7753	.8146	1.228	.8872	50
20	.6865	.6338	.7735	.8195	1.220	.8843	40
30	.6894	.6361	.7716	.8243	1.213	.8814	30
40	.6923	.6383	.7698	.8292	1.206	.8785	20
50	.6952	.6406	.7679	.8342	1.199	.8756	10
40°00'	.6981	.6428	.7660	.8391	1.192	.8727	50°00'
10	.7010	.6450	.7642	.8441	1.185	.8698	50
20	.7039	.6472	.7623	.8491	1.178	.8668	40
30	.7069	.6494	.7604	.8541	1.171	.8639	30
40	.7098	.6517	.7585	.8591	1.164	.8610	20
50	.7127	.6539	.7566	.8642	1.157	.8581	10
41°00'	.7156	.6561	.7547	.8693	1.150	.8552	49°00'
10	.7185	.6583	.7528	.8744	1.144	.8523	50
20	.7214	.6604	.7509	.8796	1.137	.8494	40
30	.7243	.6626	.7490	.8847	1.130	.8465	30
40	.7272	.6648	.7470	.8899	1.124	.8436	20
50	.7301	.6670	.7451	.8952	1.117	.8407	10
42°00'	.7330	.6691	.7431	.9004	1.111	.8378	48°00'
10	.7359	.6713	.7412	.9057	1.104	.8348	50
20	.7389	.6734	.7392	.9110	1.098	.8319	40
30	.7418	.6756	.7373	.9163	1.091	.8290	30
40	.7447	.6777	.7353	.9217	1.085	.8261	20
50	.7476	.6799	.7333	.9271	1.079	.8232	10
43°00'	.7505	.6820	.7314	.9325	1.072	.8203	47°00'
10	.7534	.6841	.7294	.9380	1.066	.8174	50
20	.7563	.6862	.7274	.9435	1.060	.8145	40
30	.7592	.6884	.7254	.9490	1.054	.8116	30
40	.7621	.6905	.7234	.9545	1.048	.8087	20
50	.7650	.6926	.7214	.9601	1.042	.8058	10
44°00'	.7679	.6947	.7193	.9657	1.036	.8029	46°00'
10	.7709	.6967	.7173	.9713	1.030	.7999	50
20	.7738	.6988	.7153	.9770	1.024	.7970	40
30	.7767	.7009	.7133	.9827	1.018	.7941	30
40	.7796	.7030	.7112	.9884	1.012	.7912	20
50	.7825	.7050	.7092	.9942	1.006	.7883	10
45°00'	.7854	.7071	.7071	1.0000	1.000	.7854	45°00'
	cos	sin	cot	tan	Radians	Degrees	

Appendix F

Graphs of Inverse Trigonometric Functions

Domain: $[-1, 1]$
Range: $[-\pi/2, \pi/2]$

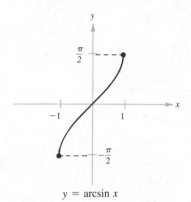

$y = \arcsin x$

Domain: $(-\infty, -1]$ and $[1, \infty)$
Range: $[-\pi/2, 0)$ and $(0, \pi/2]$

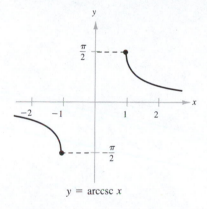

$y = \operatorname{arccsc} x$

Domain: $(-\infty, \infty)$
Range: $(-\pi/2, \pi/2)$

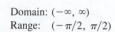

$y = \arctan x$

Domain: $[-1, 1]$
Range: $[0, \pi]$

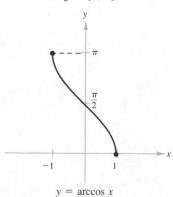

$y = \arccos x$

Domain: $(-\infty, -1]$ and $[1, \infty)$
Range: $[0, \pi/2)$ and $(\pi/2, \pi]$

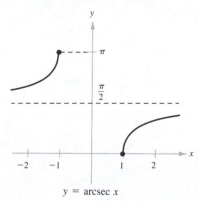

$y = \operatorname{arcsec} x$

Domain: $(-\infty, \infty)$
Range: $(0, \pi)$

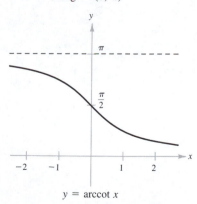

$y = \operatorname{arccot} x$

Graphs of Inverse Trigonometric Functions

Definition of the Inverse Trigonometric Functions

Function	*Domain*	*Range*
$y = \arcsin x$ iff $\sin y = x$	$-1 \leq x \leq 1$	$-\pi/2 \leq y \leq \pi/2$
$y = \arccos x$ iff $\cos y = x$	$-1 \leq x \leq 1$	$0 \leq y \leq \pi$
$y = \arctan x$ iff $\tan y = x$	$-\infty < x < \infty$	$-\pi/2 < y < \pi/2$
$y = \text{arccot } x$ iff $\cot y = x$	$-\infty < x < \infty$	$0 < y < \pi$
$y = \text{arcsec } x$ iff $\sec y = x$	$|x| \geq 1$	$0 \leq y \leq \pi, \quad y \neq \pi/2$
$y = \text{arccsc } x$ iff $\csc y = x$	$|x| \geq 1$	$-\pi/2 \leq y \leq \pi/2, \quad y \neq 0$

Answers to Warm Ups
and Odd-Numbered Exercises

CHAPTER 1

Section 1.1 *(page 12)*

1. Commutative (addition) **3.** Inverse (addition)
5. Inverse (multiplication) **7.** Distributive property
9. Identity (addition) **11.** Identity (multiplication)
13. Associative (addition)
15. Associative and Commutative (multiplication) **17.** 0
19. Division by zero is undefined. **21.** -6 **23.** 2
25. 6 **27.** -14 **29.** 96 **31.** $\frac{2}{7}$ **33.** $\frac{7}{20}$
35. $\frac{3}{10}$ **37.** $\frac{1}{12}$ **39.** 48

41. $\frac{3}{2} < 7$

43. $-4 > -8$

45. $\frac{5}{6} > \frac{2}{3}$

47. $x < 0$ **49.** $A \geq 30$ **51.** $3.5\% \leq R \leq 6\%$
53. (a) 10 (b) $\pi - 3$ **55.** (a) -1 (b) -6
57. 4 **59.** $\frac{5}{2}$ **61.** 51 **63.** 14.99
65. $|x - 5| \leq 3$ **67.** $|z - \frac{3}{2}| > 1$ **69.** $|y| \geq 6$
73. $\frac{24}{35}$ **75.** (a) 0.625 (b) $0.3\overline{3}$ (c) $0.123\overline{123}$
 (d) $0.54\overline{54}$ (e) $0.113\overline{3}$
77. (a) $\frac{127}{90}, \frac{584}{413}, \frac{7071}{5000}, \sqrt{2}, \frac{47}{33}$ (b) $\frac{7071}{5000}$

Section 1.2 *(page 18)*

WARM UP				
1. 1	**2.** 5	**3.** 4	**4.** $\frac{1}{4}$	**5.** 0
6. 1	**7.** $\frac{3}{7}$	**8.** 4	**9.** $-\frac{1}{8}$	**10.** 1

1. 64 **3.** 125 **5.** 729 **7.** 5184 **9.** $-\frac{3}{5}$
11. 36 **13.** $\frac{16}{3}$ **15.** 1 **17.** -24 **19.** $\frac{1}{2}$
21. 5 **23.** $9x^2$ **25.** $5x^6$ **27.** $10x^4$ **29.** $24y^{10}$

31. $3x^2$ **33.** $\frac{7}{x}$ **35.** $\frac{4}{3}(x + y)^2$ **37.** 1 **39.** $\frac{1}{4x^4}$
41. $-2x^3$ **43.** $\frac{10}{x}$ **45.** $\frac{x^2}{9z^4}$ **47.** $\frac{a^6}{64b^9}$ **49.** 1
51. $\frac{b^5}{a^5}$ **53.** 9.3×10^7 **55.** 4.35×10^{-6}
57. 4.392×10^{-3} **59.** 1.637×10^9 **61.** 1,910,000
63. 6.21 **65.** 0.00852 **67.** 0.00000003798
69. (a) 7697.125 (b) 954.448
71. (a) 2854.573 (b) 93.370
73. (a) 3.0981×10^6 (b) 3.077×10^{10}
75. $8\frac{1}{3}$ min

77.

t	5	10	20
A	\$910.97	\$1659.73	\$5509.41

t	30	40	50
A	\$18,288.29	\$60,707.30	\$201,515.58

79. (a) \$3499.20 (b) \$3514.98 (c) \$3518.66
 (d) \$3520.47

Section 1.3 *(page 29)*

WARM UP			
1. $\frac{4}{27}$	**2.** 48	**3.** $-8x^3$	**4.** $6x^7$
5. $28x^6$ ·	**6.** $\frac{1}{5}x^2$	**7.** $3z^4$	**8.** $\frac{25}{4x^2}$
9. 1	**10.** $(x + 2)^{10}$		

1. $9^{1/2} = 3$ **3.** $\sqrt[5]{32} = 2$ **5.** $\sqrt{196} = 14$
7. $(-216)^{1/3} = -6$ **9.** $\sqrt[3]{27^2} = 9$ **11.** $81^{3/4} = 27$
13. 6 **15.** 3 **17.** $\frac{1}{2}$ **19.** 64 **21.** 562 **23.** $\frac{2}{3}$

A19

25. $2\sqrt{2}$ **27.** $2x\sqrt[3]{2x^2}$ **29.** $\dfrac{5|x|\sqrt{3}}{y^2}$ **31.** $\dfrac{\sqrt{3}}{3}$

33. $4\sqrt[3]{4}$ **35.** $\dfrac{x(5 + \sqrt{3})}{11}$ **37.** $3(\sqrt{6} - \sqrt{5})$

39. 2 **41.** $\dfrac{2}{3(\sqrt{5} - \sqrt{3})}$ **43.** $-\dfrac{1}{2(\sqrt{7} + 3)}$

45. $2\sqrt{x}$ **47.** $34\sqrt{2}$ **49.** $8\sqrt{y}$

51. $|x|y\sqrt{15}$ **53.** $5\sqrt[6]{2^5}$ **55.** $\sqrt[6]{x^5}$

57. $2\sqrt[4]{2}$ **59.** $\sqrt[8]{2x}$ **61.** $3^{1/2} = \sqrt{3}$

63. $(x + 1)^{2/3} = \sqrt[3]{(x + 1)^2}$ **65.** 7.5498

67. 2.2361 **69.** 9.1370

71. $\sqrt{5} + \sqrt{3} > \sqrt{5 + 3}$ **73.** $5 > \sqrt{3^2 + 2^2}$

75. $5 = \sqrt{3^2 + 4^2}$ **77.** **(a)** 1.72 **(b)** 1.08

Section 1.4   *(page 36)*

WARM UP

1. $42x^3$ **2.** $-20z^2$ **3.** $-27x^6$ **4.** $-3x^6$

5. $\dfrac{9}{4}z^3$ **6.** $4\sqrt{3}$ **7.** $\dfrac{9}{4x^2}$ **8.** 8

9. $\sqrt{2}$ **10.** $-3x$

	Degree	Leading Coefficient
1.	2	2
3.	5	1
5.	5	4

7. $-2x - 10$ **9.** $3x^3 - 2x + 2$

11. $8x^3 + 29x^2 + 11$ **13.** $12z + 8$

15. $3x^3 - 6x^2 + 3x$ **17.** $-15z^2 + 5z$

19. $30x^3 + 12x^2$ **21.** $x^2 + 7x + 12$

23. $6x^2 - 7x - 5$ **25.** $x^2 + 12x + 36$

27. $4x^2 - 20xy + 25y^2$

29. $x^2 + 2xy + y^2 - 6x - 6y + 9$ **31.** $x^2 - 100$

33. $x^2 - 4y^2$ **35.** $m^2 - n^2 - 6m + 9$ **37.** $4r^4 - 25$

39. $x^3 + 3x^2 + 3x + 1$ **41.** $8x^3 - 12x^2y + 6xy^2 - y^3$

43. $x - y$ **45.** $16r^6 - 24r^3s^2 + 9s^4$

47. $x^4 - 5x^3 - 2x^2 + 11x - 5$

49. $x^4 - x^3 + 5x^2 - 9x - 36$ **51.** $x^3 + 27$

53. $x^4 - 1$ **55.** $2x^4 + 3x^3 + 7x^2 - 7x - 5$

57. $x^3 + x^2y + xy^2 + y^3$ **59.** $17z - xz$

61. $x^3 + 4x^2 - 5x - 20$ **63.** $10p^5 - 20p^4 + 10p^3$

65. $\frac{1}{4}mr^2 + \frac{1}{12}ml^2$

Section 1.5   *(page 43)*

WARM UP

1. $15x^2 - 6x$ **2.** $-2y^2 - 2y$

3. $4x^2 + 12x + 9$ **4.** $9x^2 - 48x + 64$

5. $2x^2 + 13x - 24$ **6.** $-5z^2 - z + 4$

7. $4y^2 - 1$ **8.** $x^2 - a^2$

9. $x^3 + 12x^2 + 48x + 64$

10. $8x^3 - 36x^2 + 54x - 27$

1. $3(x + 2)$ **3.** $2x(x^2 - 3)$ **5.** $(x - 1)(x + 5)$

7. $(x + 6)(x - 6)$ **9.** $(4y + 3)(4y - 3)$

11. $(x + 1)(x - 3)$ **13.** $(x - 2)^2$ **15.** $(2t + 1)^2$

17. $(5y - 1)^2$ **19.** $(x + 2)(x - 1)$

21. $(s - 3)(s - 2)$ **23.** $(y + 5)(y - 4)$

25. $(x - 20)(x - 10)$ **27.** $(3x - 2)(x - 1)$

29. $(3z + 1)(3z - 2)$ **31.** $(5x + 1)(x + 5)$

33. $(x - 2)(x^2 + 2x + 4)$ **35.** $(y + 4)(y^2 - 4y + 16)$

37. $(2t - 1)(4t^2 + 2t + 1)$ **39.** $(x - 1)(x^2 + 2)$

41. $(2x - 1)(x^2 - 3)$ **43.** $(3 + x)(2 - x^3)$

45. $x^2(x - 4)$ **47.** $(x - 1)^2$ **49.** $(1 - 2x)^2$

51. $2x(x + 1)(x - 2)$ **53.** $(9x + 1)(x + 1)$

55. $(2x - 1)(6x - 1)$ **57.** $-(x + 1)(x - 3)(x + 9)$

59. $(3x + 1)(x^2 + 5)$ **61.** $x(x - 4)(x^2 + 1)$

63. $-z(z + 10)$ **65.** $(x + 1)^2(x - 1)^2$

67. $2(t - 2)(t^2 + 2t + 4)$

69. $(x + 2)(x - 2)(x^2 + 2x + 4)(x^2 - 2x + 4)$

71. $(x + 2)^3$ **73.** $(x - 2)^5$

Section 1.6   *(page 53)*

WARM UP

1. $5x^2(1 - 3x)$ **2.** $(4x + 3)(4x - 3)$

3. $(3x - 1)^2$ **4.** $(2y + 3)^2$

5. $(z + 3)(z + 1)$ **6.** $(x - 5)(x - 10)$

7. $(3 - x)(1 + 3x)$ **8.** $(3x - 1)(x - 15)$

9. $(s + 1)(s + 2)(s - 2)$

10. $(y + 4)(y^2 - 4y + 16)$

1. $15x$ **3.** $(x + 1)(x - 2), x \neq 2$

5. $3x(x + 2), x \neq -2$ **7.** $\dfrac{3x}{2}$ **9.** $\dfrac{3y}{y + 1}$ **11.** $-\dfrac{1}{2}$

13. $\dfrac{x(x+3)}{x-2}$ **15.** $\dfrac{y-4}{y+6}$ **17.** $-(x^2+1)$

19. $z-2$ **21.** $\dfrac{1}{5(x-2)}$ **23.** $-\dfrac{x(x+7)}{x+1}$

25. $\dfrac{r+1}{r}$ **27.** $\dfrac{t-3}{(t+3)(t-2)}$ **29.** $\dfrac{x-y}{x(x+y)^2}$

31. $\dfrac{3}{2}$ **33.** $xy(x+y)$ **35.** $\dfrac{x+5}{x-1}$ **37.** $\dfrac{6x+13}{x+3}$

39. $-\dfrac{2}{x-2}$ **41.** $\dfrac{x-4}{(x+2)(x-2)(x-1)}$

43. $-\dfrac{x^2+3}{(x+1)(x-2)(x-3)}$ **45.** $-\dfrac{(x-1)^2}{x(x^2+1)}$

47. $\dfrac{1}{y}$ **49.** $\dfrac{(x+3)^3}{2x(x-3)}$ **51.** $\dfrac{1}{2y+1}$

53. $-\dfrac{2x+h}{x^2(x+h)^2}$ **55.** $\dfrac{3\sqrt{x+1}}{x+1}$ **57.** $\dfrac{2x\sqrt{x^2-1}}{x^2-1}$

59. $\dfrac{1}{\sqrt{x+2}+\sqrt{x}}$ **61.** $\dfrac{2x-1}{2x}$ **63.** $-\dfrac{1}{t^2\sqrt{t^2+1}}$

65. $-\dfrac{1}{x^2(x+1)^{3/4}}$ **67.** $\dfrac{x^2-2}{x^3(1-x^2)^{1/2}}$

69. $\dfrac{R_1R_2}{R_2+R_1}$

Section 1.7   (*page* 60)

1. $2x-(3y+4)=2x-3y-4$
3. $5z+3(x-2)=5z+3x-6$
5. $-\dfrac{x-3}{x-1}=\dfrac{3-x}{x-1}$ **7.** $a\left(\dfrac{x}{y}\right)=\dfrac{ax}{y}$
9. $(4x)^2=16x^2$ **11.** $\sqrt{x+9}$ cannot be simplified.
13. $\dfrac{6x+y}{6x-y}$ cannot be simplified.   **15.** $\dfrac{1}{x+y^{-1}}=\dfrac{y}{xy+1}$
17. $x(2x-1)^2=x(4x^2-4x+1)$
19. $\sqrt[3]{x^3+7x^2}$ cannot be simplified.
21. $\dfrac{3}{x}+\dfrac{4}{y}=\dfrac{3y+4x}{xy}$ **23.** $\dfrac{1}{2y}=\dfrac{1}{2}\cdot\dfrac{1}{y}$ **25.** $3x+2$
27. $2x^2+x+15$ **29.** $\frac{1}{3}$ **31.** $\frac{1}{2}$ **33.** $-\frac{1}{4}$
35. 2   **37.** $\frac{1}{2}$ **39.** $\dfrac{1}{2x^2}$ **41.** $\frac{25}{9}, \frac{49}{16}$ **43.** 1, 2
45. $1+x$ **47.** $10x+3$ **49.** -1 **51.** $3x-1$
53. $\dfrac{16}{x}-5-x$ **55.** $4x^{8/3}-7x^{5/3}+\dfrac{1}{x^{1/3}}$

57. $\dfrac{3}{\sqrt{x}}-5x^{3/2}-x^{7/2}$

59. $\dfrac{x^2}{x^4+1}+\dfrac{4x}{x^4+1}+\dfrac{8}{x^4+1}$

61. $\dfrac{4x}{x+3}-\dfrac{11}{x+3}$

Chapter 1 Review Exercises   (*page* 61)

1. $\frac{7}{16}+\frac{3}{16}=\frac{10}{16}$ **3.** $10(4\cdot7)=10(28)=280$
5. $4\left(\frac{3}{7}\right)=\frac{12}{7}$ **7.** $\frac{15}{16}\div\frac{2}{3}=\frac{45}{32}$ **9.** $\dfrac{x-1}{1-x}=-1$
11. $2[5-(3-2)]=2(5-3+2)$ **13.** $(2x)^4=16x^4$
15. $(5+8)^2=13^2$ **17.** $(3^4)^4=3^{16}$
19. $\sqrt{3^2+4^2}=\sqrt{25}=5$ **21.** -20 **23.** -11
25. -11 **27.** 25   **29.** -144 **31.** $\frac{15,625}{729}$
33. 50   **35.** 9×10^8 **37.** $|x-7|\ge4$
39. $|y+30|<5$ **41.** $2\sqrt{2}$ **43.** $3\sqrt[3]{x}+4x$
45. $2\sqrt{2}$ **47.** $x^5-2x^4+x^3-x^2+2x-1$
49. $x^4+x^2-x-\dfrac{1}{x}$ **51.** $y^6+y^4-y^3-y$
53. $\dfrac{1}{x^2(x^2+1)(x+1)}$ **55.** $\dfrac{1}{x^2}$ **57.** $\dfrac{3}{(x-1)(x+2)}$
59. $\dfrac{x^3-x+3}{(x+2)(x-1)}$
61. $\dfrac{x^3+x^2+x+1}{(x-1)^2}$ **63.** $\dfrac{x+1}{x(x^2+1)}$
65. $\dfrac{2x^2-3x+2}{(x-2)^2(x+2)}$ **67.** $\dfrac{x(x+1)}{x^2+x+1}$ **69.** $\dfrac{x(5x-6)}{5}$
71. $-\dfrac{1}{xy(x+y)}$ **73.** $\dfrac{3ax^2}{(a^2-x)(a-x)}$
75. $\dfrac{1}{x(x-4)(x-5)}$ **77.** $(x+1)(x-1)$
79. $(2x+1)$ **81.** (x^2+x+1) **83.** $(x-1)$
85. $9x^2-10x+48$ **87.** (x^2+2) **89.** -1
91. 50,625   **93.** 0
95.

n	1	10	10^2	10^4	10^6	10^{10}
$\dfrac{5}{\sqrt{n}}$	5	1.5811	0.5	0.05	0.005	0.00005

$5/\sqrt{n}$ approaches 0 as n increases without bound.

CHAPTER 2

Section 2.1 (page 71)

WARM UP

1. $-3x - 10$ 2. $5x - 12$ 3. x 4. $x + 26$

5. $\dfrac{8x}{15}$ 6. $\dfrac{3x}{4}$ 7. $-\dfrac{1}{x(x + 1)}$ 8. $\dfrac{5}{x}$

9. $\dfrac{7x - 8}{x(x - 2)}$ 10. $-\dfrac{2}{x^2 - 1}$

---

1. (a) No (b) No (c) Yes (d) No
3. (a) Yes (b) Yes (c) No (d) No
5. (a) Yes (b) No (c) No (d) No
7. (a) No (b) No (c) Yes (d) Yes
9. 5 11. -4 13. 3 15. 9 17. -26
19. -4 21. $-\frac{6}{5}$ 23. 9 25. No solution
27. 10 29. 4 31. 3 33. 5
35. No solution 37. $\frac{11}{6}$ 39. 3 41. 0
43. All real numbers 45. No solution
47. $\dfrac{1 + 4b}{2 + a}, a \neq -2$ 49. Conditional
51. Identity 53. Conditional 55. Conditional
57. $x \approx 138.889$ 59. $x \approx 62.372$ 61. $x \approx -2.386$
63. (a) 6.46 (b) 6.41 65. (a) 1.06 (b) 1.06
67. (a) 1.00 (b) 1.01

Section 2.2 (page 83)

WARM UP

1. 14 2. 4 3. -3 4. 4 5. -2
6. 1 7. $\frac{2}{5}$ 8. $\frac{10}{3}$ 9. 6 10. $-\frac{11}{5}$

---

1. $\dfrac{2A}{b}$ 3. $\dfrac{V}{WH}$ 5. $\dfrac{S}{1 + R}$ 7. $\dfrac{A - P}{PT}$

9. $\dfrac{2A - ah}{h}$ 11. $\dfrac{3V + \pi h^3}{3\pi h^2}$ 13. $\dfrac{L - L_0}{L_0 \Delta t}$

15. $\dfrac{Fr^2}{\alpha m_1}$ 17. $\dfrac{(n - 1)fR_2}{R_2 + (n - 1)f}$ 19. $\dfrac{L - a + d}{d}$

21. $\dfrac{S - a}{S - L}$ 23. $\dfrac{CC_2}{C_2 - C}$ 25. $2n + 1$

27. $d = 50t$ 29. $6x$ 31. 262, 263

33. 37, 185 35. $-5, -4$ 37. (a) 15 (b) 35
39. 15 in × 22.5 in 41. 97 43. 13.5
45. 1192.5 47. 135% 49. 175 51. $22,316.98
53. $\frac{1}{3}$ hr 55. (a) 3.8 hr, 3.2 hr (b) 1.1 hr
  (c) 25.6 mi 57. $66\frac{2}{3}$ mph 59. 1.29 sec
61. 27 quarters, 8 half dollars
63. $4500 at 10.5%, $7500 at 13% 65. 11.43%
67. 8824 units per month 69. 14 ft
71.

	Solution 1	Solution 2
(a)	25 gal	75 gal
(b)	4 L	1 L
(c)	5 qt	5 qt
(d)	18.75 gal	6.25 gal

73. 0.48 gal

Section 2.3 (page 92)

WARM UP

1. $\dfrac{\sqrt{14}}{10}$ 2. $4\sqrt{2}$ 3. 14 4. $\dfrac{\sqrt{10}}{4}$

5. $x(3x + 7)$ 6. $(2x - 5)(2x + 5)$
7. $-(x - 7)(x - 15)$ 8. $(x - 2)(x + 9)$
9. $(5x - 1)(2x + 3)$ 10. $(6x - 1)(x - 12)$

---

1. $2x^2 + 5x - 3 = 0, a = 2, b = 5, c = -3$
3. $x^2 - 25x = 0, a = 1, b = -25, c = 0$
5. $x^2 - 6x + 7 = 0, a = 1, b = -6, c = 7$
7. $2x^2 - 2x + 1 = 0, a = 2, b = -2, c = 1$
9. $3x^2 - 60x - 10 = 0, a = 3, b = -60, c = -10$
11. $0, -\frac{1}{2}$ 13. $4, -2$ 15. -5 17. $3, -\frac{1}{2}$
19. $-a$ 21. ± 4 23. $\pm\sqrt{7}$ 25. $\pm 2\sqrt{3}$
27. $12 \pm 3\sqrt{2}$ 29. 2 31. $0, 2$ 33. $4, -8$
35. $-3 \pm \sqrt{7}$ 37. $1 \pm \dfrac{\sqrt{6}}{3}$ 39. $2 \pm 2\sqrt{3}$
41. ± 8 43. $1 \pm \sqrt{2}$ 45. $\pm\frac{3}{4}$ 47. $\frac{3}{2}$
49. $6, -12$ 51. $\frac{3}{2}, -\frac{1}{2}$ 53. $5, -\frac{10}{3}$ 55. $9, 3$
57. $\frac{1}{2} \pm \sqrt{3}$ 59. $\frac{3}{5} \pm \dfrac{\sqrt{2}}{2}$ 61. $-1, -5$
63. $-\frac{1}{2}$ 65. $\dfrac{1}{(x + 1)^2 + 4}$ 67. $\dfrac{1}{(x - 2)^2 - 16}$
69. $\dfrac{1}{\sqrt{9 - (x - 3)^2}}$

Section 2.4 *(page 102)*

```
┌─────────────────────────────────────────────┐
│ WARM UP                                       │
│                                               │
│ 1. $3\sqrt{17}$   2. $2\sqrt{3}$   3. $4\sqrt{6}$   4. $3\sqrt{73}$ │
│ 5. $2, -1$   6. $\frac{3}{2}, -3$   7. $5, -1$ │
│ 8. $\frac{1}{2}, -7$   9. $3, 2$   10. $4, -1$ │
└─────────────────────────────────────────────┘
```

1. One real solution   3. Two real solutions
5. No real solutions   7. Two real solutions
9. $\frac{1}{2}, -1$ 11. $\frac{1}{4}, -\frac{3}{4}$ 13. $1 \pm \sqrt{3}$
15. $-7 \pm \sqrt{5}$ 17. $-4 \pm 2\sqrt{5}$ 19. $\frac{2}{3} \pm \frac{\sqrt{7}}{3}$
21. $-\frac{1}{3} \pm \frac{\sqrt{11}}{6}$ 23. $-\frac{1}{2} \pm \sqrt{2}$ 25. $\frac{2}{7}$
27. $-\frac{8}{5} \pm \frac{\sqrt{3}}{5}$ 29. $6 \pm \sqrt{11}$ 31. $-\frac{1}{2} \pm \frac{\sqrt{21}}{6}$
33. $4, -5$ 35. $x \approx 0.976, -0.643$
37. $x \approx 0.561, 0.126$ 39. $x \approx 1.687, -0.488$
41. $50, 50$ 43. $7, 8$ 45. 200 units
47. 2000 units   49. 10 sec   51. 2.56 sec
53. 35 ft × 20 ft or 15 ft × $\frac{140}{3}$ ft
55. 19.1 ft, 9.5 times   57. 14 in × 14 in
59. 6 in × 6 in   61. 550 mph, 600 mph
63. $30\sqrt{6}$ ft ≈ 73.5 ft   65. \$6.00   67. 9.5%

Section 2.5 *(page 110)*

```
┌─────────────────────────────────────────────┐
│ WARM UP                                       │
│                                               │
│ 1. $2\sqrt{3}$   2. $10\sqrt{5}$   3. $\sqrt{5}$   4. $-6\sqrt{3}$ │
│ 5. 12   6. 48   7. $\frac{\sqrt{3}}{3}$   8. $\sqrt{2}$ │
│ 9. $-\frac{1}{2} \pm \frac{\sqrt{5}}{2}$   10. $-1 \pm \sqrt{2}$ │
└─────────────────────────────────────────────┘
```

1. $i, -1, -i, 1, i, -1, -i, 1,$
   $i, -1, -i, 1, i, -1, -i, 1$
3. $a = -10, b = 6$ 5. $a = 6, b = 5$
7. $4 + 3i, 4 - 3i$ 9. $2 - 3\sqrt{3}i, 2 + 3\sqrt{3}i$
11. $5\sqrt{3}i, -5\sqrt{3}i$ 13. $-1 - 6i, -1 + 6i$
15. $-5i, 5i$ 17. $8, 8$ 19. $11 - i$
21. 4   23. $3 - 3\sqrt{2}i$ 25. $\frac{1}{6} + \frac{7}{6}i$
27. $-2\sqrt{3}$ 29. -10 31. $5 + i$ 33. 41
35. $12 + 30i$ 37. $-40 + 16i$ 39. 24

41. $-9 + 40i$ 43. $\frac{16}{41} + \frac{20}{41}i$ 45. $\frac{3}{5} + \frac{4}{5}i$
47. $-7 - 6i$ 49. $\frac{1}{8}i$ 51. $-\frac{5}{4} - \frac{5}{4}i$
53. $\frac{35}{29} + \frac{595}{29}i$ 55. -10 57. $1 \pm i$ 59. $-2 \pm \frac{1}{2}i$
61. $-\frac{3}{2}, -\frac{5}{2}$ 63. $\frac{1}{8} \pm \frac{\sqrt{11}}{8}i$

Section 2.6 *(page 117)*

```
┌─────────────────────────────────────────────┐
│ WARM UP                                       │
│                                               │
│ 1. 11   2. $20, -3$   3. $5, -45$   4. $0, -\frac{1}{5}$ │
│ 5. $\frac{2}{3}, -2$   6. $\frac{11}{6}, -\frac{5}{2}$   7. $1, -5$   8. $\frac{3}{2}, -\frac{5}{2}$ │
│ 9. $\frac{3}{4}$   10. $\frac{5}{2}$ │
└─────────────────────────────────────────────┘
```

1. $0, \pm\frac{3\sqrt{2}}{2}$ 3. $3, -1, 0$ 5. $\pm 3, \pm 3i$
7. $-3, 0$ 9. $3, 1, -1$ 11. $\pm 1, \frac{1}{2} \pm \frac{\sqrt{3}}{2}i$
13. $\pm 3, \pm 1$ 15. $\pm 2, \pm 3i$ 17. $\pm\frac{1}{2}, \pm 4$
19. $1, -2, 1 \pm \sqrt{3}i, -\frac{1}{2} \pm \frac{\sqrt{3}}{2}i$ 21. $-\frac{1}{5}, -\frac{1}{3}$
23. $\frac{1}{4}$ 25. $1, -\frac{125}{8}$ 27. 50   29. 26
31. -16 33. $6, 5$ 35. $2, -5$ 37. 0
39. 36   41. $\frac{101}{4}$ 43. $-59, 69$ 45. 1
47. $\pm\sqrt{69}, \pm\sqrt{59}i$ 49. 1
51. $\pm 1, \pm\frac{\sqrt{33}}{11}$ 53. $2, -\frac{3}{2}$ 55. -1
57. $1, -3$ 59. $1, -3$ 61. $3, -2$
63. $\sqrt{3}, -3$ 65. $10, -1$ 67. $x \approx \pm 1.038$
69. $x \approx 16.756$ 71. 26,250
73. 0.26 mi or 1 mi   75. 8 in × 6 in
77. $h = \frac{1}{\pi r}\sqrt{S^2 - \pi^2 r^4}$

Section 2.7 *(page 127)*

```
┌─────────────────────────────────────────────┐
│ WARM UP                                       │
│                                               │
│ 1. $-\frac{1}{2}$   2. $-\frac{1}{6}$   3. $-3$   4. $-6$   5. $x \geq 0$ │
│ 6. $-3 < z < 10$   7. $P \leq 2$   8. $W \geq 200$ │
│ 9. $2, 7$   10. $0, 1$ │
└─────────────────────────────────────────────┘
```

1. (a) Yes   (b) No   (c) Yes   (d) No
3. (a) Yes   (b) No   (c) No   (d) Yes

5. c **6.** h **7.** f **8.** e
9. g **10.** a **11.** b **12.** d

13. $x < 3$

15. $x > -4$

17. $x \geq 12$

19. $x < -\frac{1}{2}$

21. $x \geq \frac{1}{2}$

23. $x > \frac{1}{2}$

25. $-1 < x < 3$

27. $-\frac{9}{2} < x < \frac{15}{2}$

29. $-\frac{3}{4} < x < -\frac{1}{4}$

31. $-5 < x < 5$

33. $x < -6, x > 6$

35. $16 \leq x \leq 24$

37. $x \leq 16, x \geq 24$

39. $x \leq -7, x \geq 13$

41. $4 < x < 5$

43. $x \leq -\frac{29}{2}, x \geq -\frac{11}{2}$

45. No solution **47.** $[5, \infty)$ **49.** $[-3, \infty)$
51. $(-\infty, \frac{7}{2}]$ **53.** $|x| \leq 2$ **55.** $|x - 9| \geq 3$
57. $|x - 12| \leq 10$ **59.** $|x + 3| > 5$ **61.** $r > 0.125$
63. $x \geq 36$ units **65.** $[65.8, 71.2]$

Section 2.8 *(page 136)*

WARM UP

1. $y < -6$ **2.** $z > -\frac{9}{2}$ **3.** $-3 \leq x < 1$
4. $x \leq -5$ **5.** $-3 < x$ **6.** $5 < x < 7$
7. $-\frac{7}{2} \leq x \leq \frac{7}{2}$ **8.** $x < 2, x > 4$
9. $x < -6, x > -2$ **10.** $-2 \leq x \leq 6$

1. $-3 \leq x \leq 3$

3. $x < -2, x > 2$

5. $-7 < x < 3$

7. $x \leq -5, x \geq 1$

9. $-3 < x < 2$

11. $x < -1, x > 1$

13. $-3 < x < 1$

15. $x < 0, 0 < x < \frac{3}{2}$

17. $-2 \leq x \leq 0, x \geq 2$

19. $x < -1, 0 < x < 1$

21. $x < -1, x > 4$

23. $5 < x < 15$

25. $-5 < x < -\frac{3}{2}, x > -1$

27. $-\frac{3}{4} < x < 3, x \geq 6$

29. $[-2, 2]$ **31.** $(-\infty, 3], [4, \infty)$ **33.** $[-4, 3]$
35. All real numbers
37. (a) 10 sec   (b) 4 sec $< t <$ 6 sec
39. Between 13.8 meters and 36.2 meters
41. $R_1 \geq 2$

Chapter 2 Review Exercises *(page 137)*

1. 20   **3.** $0, \frac{12}{5}$ **5.** $0, 2$ **7.** $\frac{1}{5}$
9. $\pm\dfrac{\sqrt{2}}{2}$ **11.** $2, -1 \pm \sqrt{3}i$ **13.** $0, 1, 2$
15. $1, \frac{5}{3}$ **17.** $0, 2$ **19.** $2, 6$ **21.** $-5, 15$
23. $1, 3$ **25.** 5   **27.** $-124, 126$
29. $-4, -2 \pm \dfrac{\sqrt{95}}{5}$ **31.** $2, \frac{2}{5}$
33. No solution   **35.** $x > -\frac{5}{3}$ **37.** $-2 \leq x \leq 2$
39. $x < 3, x > 5$ **41.** $1 < x < 3$
43. $x \leq 0, x \geq 3$ **45.** $x < -1, 0 < x < 2$
47. $r = \sqrt{\dfrac{3V}{\pi h}}$ **49.** $t = \dfrac{V_0 \pm \sqrt{V_0^2 - 64S}}{32}$
51. $p = \dfrac{k}{3\pi r^2 L}$ **53.** $3 + 7i$ **55.** $-\sqrt{2}i$
57. $40 + 65i$ **59.** $-4 - 46i$ **61.** $1 - 6i$
63. $\frac{4}{3}i$ **65.** $x \approx 123.1$ mi
67. 4 farmers   **69.** $L \geq \dfrac{32}{\pi^2}$

CHAPTER 3
Section 3.1 *(page 146)*

WARM UP
1. 5   **2.** $3\sqrt{2}$ **3.** 1   **4.** -2
5. $3(\sqrt{2} + \sqrt{5})$ **6.** $2(\sqrt{3} + \sqrt{11})$
7. $-3, 11$ **8.** 9, 1   **9.** 11   **10.** 4

1.
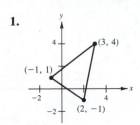

3.
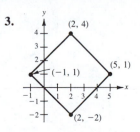

5. 8   **7.** 5   **9.** $a = 4, b = 3, c = 5$
11. $a = 10, b = 3, c = \sqrt{109}$
13. (a)
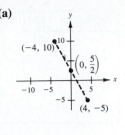

15. (a)
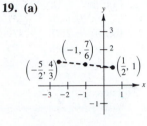

(b) 10   (c) $(5, 4)$ (b) 17   (c) $(0, \frac{5}{2})$
17. (a)

19. (a)

(b) $2\sqrt{10}$ (c) $(2, 3)$ (b) $\dfrac{\sqrt{82}}{3}$ (c) $(-1, \frac{7}{6})$
21. (a)

23. (a)
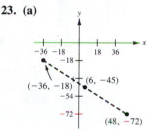

(b) $\sqrt{110.97}$ (b) $6\sqrt{277}$
(c) $(1.25, 3.6)$ (c) $(6, -45)$
25. $(\sqrt{5})^2 + (\sqrt{45})^2 = (\sqrt{50})^2$
27. All sides have a length of $\sqrt{5}$.   **29.** $x = 6, -4$
31. $y = \pm 15$ **33.** $3x - 2y - 1 = 0$
35. Quadrant IV   **37.** Quadrant I
39. Quadrant II   **41.** Quadrant III or IV
43. $(2x_m - x_1, 2y_m - y_1)$
45. $\left(\dfrac{3x_1 + x_2}{4}, \dfrac{3y_1 + y_2}{4}\right), \left(\dfrac{x_1 + x_2}{2}, \dfrac{y_1 + y_2}{2}\right),$
$\left(\dfrac{x_1 + 3x_2}{4}, \dfrac{y_1 + 3y_2}{4}\right)$

47. \$630,000
49. (a) \$200   (b) \$180   (c) \$173   (d) \$162
51. (a) \$12.6 billion   (b) \$14.5 billion

Section 3.2 *(page* 157)

WARM UP

1. $y = \dfrac{3x - 2}{5}$ 2. $y = -\dfrac{(x - 5)(x + 1)}{2}$
3. 2   4. 1, -5 5. 0, ±3 6. ±2
7. $y = x^3 + 4x$ 8. $x^2 + y^2 = 4$
9. $y = 4x^2 + 8$ 10. $y^2 = 3x^2 + 4$

1. **(a)** Yes   **(b)** Yes   3. **(a)** No   **(b)** Yes
5. **(a)** Yes   **(b)** Yes   7. 2   9. 4
11. $(5, 0), (0, -5)$ 13. $(-2, 0), (1, 0), (0, -2)$
15. $(0, 0), (-2, 0)$ 17. $(1, 0), \left(0, \frac{1}{2}\right)$
19. y-axis symmetry   21. x-axis symmetry
23. Origin symmetry   25. Origin symmetry
27. c   28. f   29. d   30. a   31. e   32. b
33. Intercepts: 35. Intercepts:
$\left(\frac{2}{3}, 0\right), (0, 2)$ $(-1, 0), (1, 0), (0, 1)$
Symmetry: y-axis

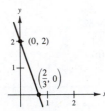

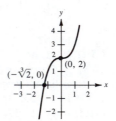

37. Intercepts: 39. Intercepts:
$(3, 0), (1, 0), (0, 3)$ $(-\sqrt[3]{2}, 0), (0, 2)$

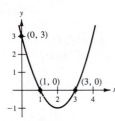

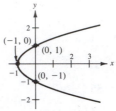

41. Intercepts: $(0, 0)$,   43. Intercept: $(3, 0)$
$(2, 0)$

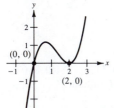

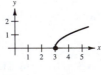

45. Intercept: $(0, 0)$ 47. Intercepts: $(2, 0), (0, 2)$
Symmetry: Origin

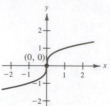

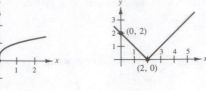

49. Intercepts: $(-1, 0)$,   51. Intercepts: $(-2, 0)$,
$(0, 1), (0, -1)$ $(2, 0), (0, 2), (0, -2)$
Symmetry: x-axis   Symmetry: x-axis,
y-axis, origin

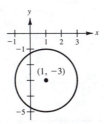

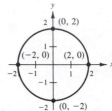

53. $x^2 + y^2 = 9$
55. $(x - 2)^2 + (y + 1)^2 = 16$
57. $(x + 1)^2 + (y - 2)^2 = 5$
59. $(x - 3)^2 + (y - 4)^2 = 25$
61. $(x - 1)^2 + (y + 3)^2 = 4$

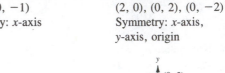

63. $(x - 1)^2 + (y + 3)^2 = 0$

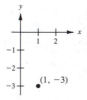

65. $\left(x - \frac{1}{2}\right)^2 + \left(y - \frac{1}{2}\right)^2 = 2$

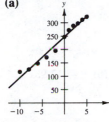

67. $\left(x + \frac{1}{2}\right)^2 + \left(y + \frac{5}{4}\right)^2 = \frac{9}{4}$

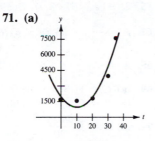

69. (a)

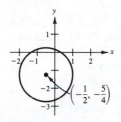

(b) $y = 399.55$

71. (a)

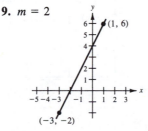

(b) $y = \$9,706.40$

Section 3.3 (*page* 169)

WARM UP

1. $-\frac{9}{2}$ **2.** $-\frac{13}{3}$ **3.** $-\frac{5}{4}$ **4.** $\frac{1}{2}$
5. $y = \frac{2}{3}x - \frac{5}{3}$ **6.** $y = -2x$
7. $y = 3x - 1$ **8.** $y = \frac{2}{3}x + 5$
9. $y = -2x + 7$ **10.** $y = x + 3$

1. 1   **3.** 0   **5.** -3
7.

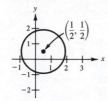

9. $m = 2$

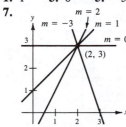

11. m is undefined

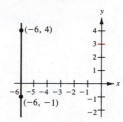

13. $m = \frac{4}{3}$

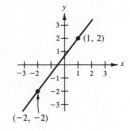

15. Perpendicular   **17.** Parallel
19. $(0, 1), (3, 1), (-1, 1)$
21. $(6, -5), (7, -4), (8, -3)$
23. $(-8, 0), (-8, 2), (-8, 3)$ **25.** $m = 5, (0, 3)$
27. m is undefined, no y-intercept
29. $m = -\frac{7}{6}, (0, 5)$
31. $3x + 5y - 10 = 0$ **33.** $x + 2y - 3 = 0$

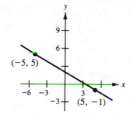

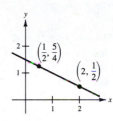

35. $x + 8 = 0$ **37.** $2x - 5y + 1 = 0$

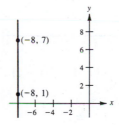

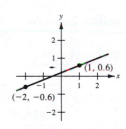

39. $3x - y - 2 = 0$ **41.** $2x + y = 0$

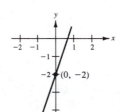

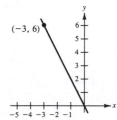

43. $x + 3y - 4 = 0$ **45.** $x - 6 = 0$

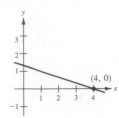

47. $8x - 6y - 17 = 0$ **51.** $3x + 2y - 6 = 0$

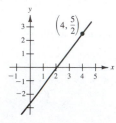

53. $12x + 3y + 2 = 0$ **55.** $x + y - 3 = 0$
57. (a) $2x - y - 3 = 0$ **(b)** $x + 2y - 4 = 0$
59. (a) $3x + 4y + 2 = 0$ **(b)** $4x - 3y + 36 = 0$
61. (a) $y = 0$ **(b)** $x + 1 = 0$
63. $F = \frac{9}{5}C + 32$ **65.** $I = 95t$
67. $S = 0.85L$ **69.** $W = 0.07S + 2500$
71. (a) $C = 16.75t + 36,500$ **(b)** $R = 27t$
     (c) $P = 10.25t - 36,500$ **(d)** $t \approx 3561$ hr

Section 3.4 *(page 180)*

WARM UP

1. -73 **2.** 13   **3.** $2(x + 2)$ **4.** $-8(x - 2)$
5. $y = \frac{7}{5} - \frac{2}{5}x$ **6.** $y = \pm x$
7. $x \le -2, x \ge 2$ **8.** $-3 \le x \le 3$
9. All real numbers   **10.** $x \le 1, x \ge 2$

1. (a) $6 - 4(3)$ **(b)** $6 - 4(-7)$
   (c) $6 - 4t$ **(d)** $6 - 4(c + 1)$
3. (a) $\dfrac{1}{4 + 1}$ **(b)** $\dfrac{1}{0 + 1}$
   (c) $\dfrac{1}{4x + 1}$ **(d)** $\dfrac{1}{(x + h) + 1}$
5. (a) -1 **(b)** -9 **(c)** $2x - 5$ **(d)** $-\frac{5}{2}$
7. (a) 0 **(b)** 3 **(c)** $x^2 + 2x$ **(d)** -0.75

9. (a) 1   **(b)** -7 **(c)** $3 - 2|x|$ **(d)** 2.5
11. (a) $\frac{1}{7}$ **(b)** $-\frac{1}{9}$ **(c)** Undefined   **(d)** $\dfrac{1}{y^2 + 6y}$
13. (a) 1   **(b)** -1 **(c)** 1   **(d)** $\dfrac{|x - 1|}{x - 1}$
15. (a) -1 **(b)** 2   **(c)** 4   **(d)** 6
17. $3 + h$ **19.** $3x\Delta x + 3x^2 + (\Delta x)^2$
21. 3   **23.** 5   **25.** ± 3 **27.** $\frac{10}{7}$
29. All real numbers
31. All real numbers except $t = 0$
33. $y \ge 10$ **35.** $-1 \le x \le 1$
37. All real numbers except $x = 0, -2$
39. Not a function   **41.** Function   **43.** Function
45. Not a function   **47.** Function
49. (a) Function   **(b)** Not a function
   (c) Function   **(d)** Not a function
51. $(-2, 4), (-1, 1), (0, 0), (1, 1), (2, 4)$
53. $(-2, 0), (-1, 1), (0, \sqrt{2}), (1, \sqrt{3}), (2, 2)$
55. $2, -1$ **57.** $3, 0$ **59.** $V = e^3$
61. $A = \dfrac{C^2}{4\pi}$ **63.** $V = 4x(6 - x)^2$
65. $A = \dfrac{x^2}{x - 1}, x > 1$ **67.** $h = \sqrt{d^2 - 2000^2}$
69. (a) $C = 12.30x + 98,000$ **(b)** $R = 17.98x$
   (c) $P = 5.68x - 98,000$
71. (a) $R = \dfrac{240n - n^2}{20}$
   (b)

n	90	100	110	120
$R(n)$	\$675	\$700	\$715	\$720

n	130	140	150
$R(n)$	\$715	\$700	\$675

73.

y	5	10	20
$F(y)$	26,474	149,760	847,170

y	30	40
$F(y)$	2,334,527	4,792,320

Section 3.5 (*page* 194)

WARM UP

1. 2   **2.** 0   **3.** $-\dfrac{3}{x}$ **4.** $x^2 + 3$ **5.** 0, ± 4

6. $\frac{1}{2}$, 1   **7.** All real numbers except $x = 4$

8. All real numbers except $x = 4, 5$

9. $t \le \frac{5}{3}$ **10.** All real numbers

1. Domain: $[1, \infty)$
  Range: $[0, \infty)$

3. Domain: $(-\infty, -2]$, $[2, \infty)$
  Range: $[0, \infty)$

5. Domain: $[-5, 5]$
  Range: $[0, 5]$

7. Function   **9.** Not a function   **11.** Function

13. (a) Increasing on $(-\infty, \infty)$ **(b)** Odd function

15. (a) Increasing on $(-\infty, 0)$, $(2, \infty)$, decreasing on $(0, 2)$
  (b) Neither even nor odd

17. (a) Increasing on $(-1, 0)$, $(1, \infty)$, decreasing on
  $(-\infty, -1)$, $(0, 1)$ **(b)** Even function

19. (a) Increasing on $(-2, \infty)$, decreasing on $(-3, -2)$
  (b) Neither even nor odd

21. Even   **23.** Odd   **25.** Neither even nor odd

27. Even   **29.** Neither even nor odd

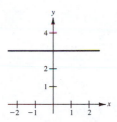

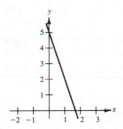

31. Odd   **33.** Neither even nor odd

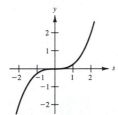

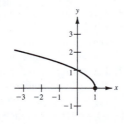

35. Neither even nor odd   **37.** Neither even nor odd

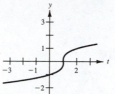

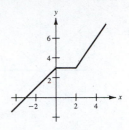

39. $(-\infty, 4]$ **41.** $(-\infty, -3]$, $[3, \infty)$

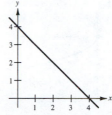

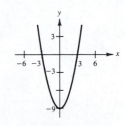

43. $[-1, 1]$ **45.** $(-\infty, \infty)$

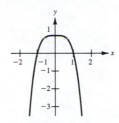

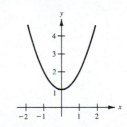

47. $f(x) < 0$ for all x

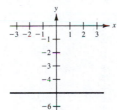

49. (a)   **(b)**

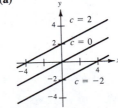

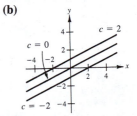

(c)

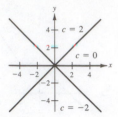

51. (a)

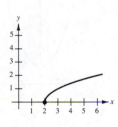

(b)

(c)

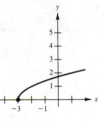

(d)

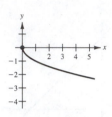

(e)

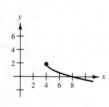

(f)

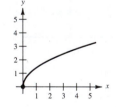

53. (a) $g(x) = (x - 1)^2 + 1$ **(b)** $g(x) = -(x + 1)^2$
55. $h = (4x - x^2) - 3$ **57.** $h = 4x - 2x^2$
59. $L = (4 - y^2) - (y + 2)$

Section 3.6   (*page* 203)

WARM UP

1. $\dfrac{1}{x(1 - x)}$ **2.** $-\dfrac{12}{(x + 3)(x - 3)}$

3. $\dfrac{3x - 2}{x(x - 2)}$ **4.** $\dfrac{4x - 5}{3(x - 5)}$

5. $\sqrt{\dfrac{x - 1}{x + 1}}$ **6.** $\dfrac{x + 1}{x(x + 2)}$ **7.** $5(x - 2)$

8. $\dfrac{x + 1}{(x - 2)(x + 3)}$ **9.** $\dfrac{1 + 5x}{3x - 1}$ **10.** $\dfrac{x + 4}{4x}$

1. (a) $2x$ **(b)** 2 **(c)** $x^2 - 1$ **(d)** $\dfrac{x + 1}{x - 1}, x \neq 1$

3. (a) $x^2 - x + 1$ **(b)** $x^2 + x - 1$

　　(c) $x^2 - x^3$ **(d)** $\dfrac{x^2}{1 - x}, x \neq 1$

5. (a) $x^2 + 5 + \sqrt{1 - x}$ **(b)** $x^2 + 5 - \sqrt{1 - x}$

　　(c) $(x^2 + 5)\sqrt{1 - x}$ **(d)** $\dfrac{x^2 + 5}{\sqrt{1 - x}}, x < 1$

7. (a) $\dfrac{x + 1}{x^2}$ **(b)** $\dfrac{x - 1}{x^2}$

　　(c) $\dfrac{1}{x^3}$ **(d)** $x, x \neq 0$

9. 9 **11.** $4t^2 - 2t + 5$ **13.** 0
15. 26 **17.** 5 **19.** $\frac{3}{5}$
21. (a) $(x - 1)^2$ **(b)** $x^2 - 1$ **(c)** x^4
23. (a) $20 - 3x$ **(b)** $-3x$ **(c)** $9x + 20$
25. (a) $\sqrt{x^2 + 4}$ **(b)** $x + 4$
27. (a) $x - \frac{8}{3}$ **(b)** $x - 8$
29. (a) $\sqrt[4]{x}$ **(b)** $\sqrt[4]{x}$
31. (a) $|x + 6|$ **(b)** $|x| + 6$
33. (a) 3 **(b)** 0 **35. (a)** 0 **(b)** 4
37. $f(x) = x^2, g(x) = 2x + 1$
39. $f(x) = \sqrt[3]{x}, g(x) = x^2 - 4$
41. $f(x) = \dfrac{1}{x}, g(x) = x + 2$
43. $f(x) = x^2 + 2x, g(x) = x + 4$
45. (a) $x \geq 0$
　　(b) All real numbers
　　(c) All real numbers
47. (a) All real numbers except $x = \pm 1$
　　(b) All real numbers
　　(c) All real numbers except $x = -2, 0$
49. $(A \circ r)(t) = \pi(0.6t)^2$

51. $s = \sqrt{(200 - 450t)^2 + (150 - 450t)^2}$

57. (a) $\underbrace{(x^2 + 1)}_{\text{Even}} + \underbrace{(-2x)}_{\text{Odd}}$ **(b)** $\underbrace{-\dfrac{1}{x^2 - 1}}_{\text{Even}} + \underbrace{\dfrac{x}{x^2 - 1}}_{\text{Odd}}$

Section 3.7   *(page 212)*

WARM UP

1. All real numbers   **2.** $[-1, \infty)$
3. All real numbers except $x = 0, 2$
4. All real numbers except $x = -\frac{5}{3}$
5. x **6.** x **7.** x **8.** x
9. $x = \frac{3}{2}y + 3$ **10.** $x = \dfrac{y^3}{2} + 2$

1.
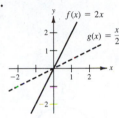

3.
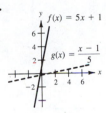

5.
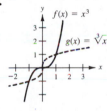

7.
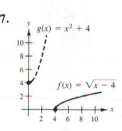

9.
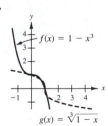

11. One-to-one

13. Not one-to-one   **15.** One-to-one
17. Not one-to-one   **19.** Not one-to-one

21. $f^{-1}(x) = \dfrac{x + 3}{2}$

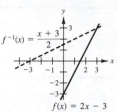

23. $f^{-1}(x) = \sqrt[5]{x}$

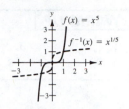

25. $f^{-1}(x) = x^2, x \geq 0$

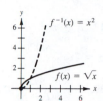

27. $f^{-1}(x) = \sqrt{4 - x^2},$
$0 \leq x \leq 2$

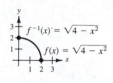

29. $f^{-1}(x) = x^3 + 1$

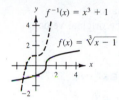

31. Not one-to-one

33. $g^{-1}(x) = 8x$ **35.** Not one-to-one
37. $f^{-1}(x) = \sqrt{x} - 3, x \geq 0$ **39.** $h^{-1}(x) = \dfrac{1}{x}$
41. $f^{-1}(x) = \dfrac{x^2 - 3}{2}, x \geq 0$ **43.** Not one-to-one
45. $f^{-1}(x) = -\sqrt{25 - x}$

47.

x	0	1	2	3	4
$f^{-1}(x)$	-2	0	1	2	4

49. 32   **51.** 600
53. $(g^{-1} \circ f^{-1})(x) = \dfrac{x + 1}{2}$

55. $(f \circ g)^{-1}(x) = \dfrac{x + 1}{2}$

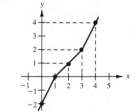

Section 3.8 *(page 219)*

WARM UP

1. $\frac{1}{3}$ **2.** $\frac{9}{16}$ **3.** $\frac{128}{3}$ **4.** 75 **5.** $\frac{275}{27}$
6. $\frac{105}{128}$ **7.** 27 **8.** $\frac{2}{9}$ **9.** $\frac{28}{13}$ **10.** 157.5

1. $A = kr^2$ **3.** $y = \dfrac{k}{x^2}$ **5.** $z = k\sqrt[3]{u}$

7. $z = kuv$ **9.** $F = \dfrac{kg}{r^2}$ **11.** $P = \dfrac{k}{V}$

13. $F = \dfrac{km_1m_2}{r^2}$ **15.** $y = \frac{5}{2}x$ **17.** $A = \pi r^2$

19. $y = \dfrac{75}{x}$ **21.** $h = \dfrac{12}{t^3}$ **23.** $z = 2xy$

25. $F = 14rs^3$ **27.** $z = \dfrac{2x^2}{3y}$ **29.** $S = \dfrac{4L}{3(L-S)}$

31. (a) 2 in **(b)** 15 lb **33.** 39.47 lb
35. 0.61 mi/hr **37.** 400 ft
39. One-fourth the original illumination
41. 0.054 in **43.** Four-thirds the original velocity

Chapter 3 Review Exercises *(page 221)*

1. (a) 10 **(b)** (0, 5) **(c)** $x = 0$
  (d) $x^2 + (y - 5)^2 = 25$
3. (a) 13 **(b)** $(8, \frac{7}{2})$ **(c)** $5x - 12y + 2 = 0$
  (d) $(x - 8)^2 + (y - \frac{7}{2})^2 = \frac{169}{4}$
5. (a) $\sqrt{53}$ **(b)** $(\frac{5}{2}, 1)$ **(c)** $2x - 7y + 2 = 0$
  (d) $(x - \frac{5}{2})^2 + (y - 1)^2 = \frac{53}{4}$
7. $t = \frac{7}{3}$ **9.** $t = 3$
15. Intercept: (0, 0) **17.** Intercepts: (0, 0),
  Symmetry: x-axis $(2\sqrt{2}, 0), (-2\sqrt{2}, 0)$
          Symmetry: y-axis
19. Intercepts: (0, 0), **21.** Intercepts: (0, 0),
  $(-2, 0), (2, 0)$ (3, 0)
  Symmetry: Origin
23. Intercept: (0, 0) **25.** (6, 4), $r = 3$
  Symmetry: x-axis

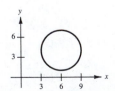

27. $(\frac{1}{2}, 5), r = \frac{3}{2}$ **29.**

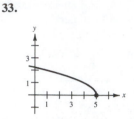

31. **33.**

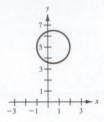
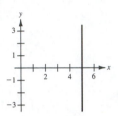

35. **37.**

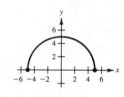
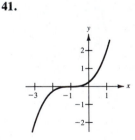

39. **41.**

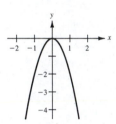
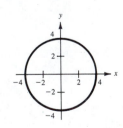

43. (a) 5 **(b)** 17 **(c)** $t^4 + 1$ **(d)** $-x^2 - 1$
45. (a) -14 **(b)** $-5x^2 - 30x - 39$
  (c) -30 **(d)** $-10x - 5\Delta x$
47. $[-5, 5]$ **49.** All real numbers except $s = 3$
51. All real numbers except $x = -2, 3$

53. (a) $f^{-1}(x) = 2x + 6$
(b)

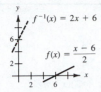

$f^{-1}(x) = 2x + 6$

$f(x) = \dfrac{x - 6}{2}$

55. (a) $f^{-1}(x) = x^2 - 1, x \geq 0$
(b)

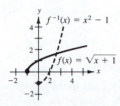

$f^{-1}(x) = x^2 - 1$

$f(x) = \sqrt{x + 1}$

57. (a) $f^{-1}(x) = \sqrt{x + 5}, x \geq -5$
(b)

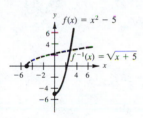

$f(x) = x^2 - 5$

$f^{-1}(x) = \sqrt{x + 5}$

59. $x \geq 4, f^{-1}(x) = \sqrt{\dfrac{x}{2}} + 4$

61. $x \geq 2, f^{-1}(x) = \sqrt{x^2 + 4}$

63. -7 **65.** 5   **67.** 23   **69.** 9

71. $F = \frac{1}{3}x\sqrt{y}$ **73.** $z = \dfrac{32x^2}{25y}$

75. (a) 16 ft/sec   **(b)** 1.5 sec   **(c)** -16 ft/sec

77. $A = x(12 - x), (0, 6]$

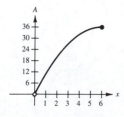

CHAPTER 4
Section 4.1   (*page* 231)

WARM UP

1. $\frac{1}{2}, -6$ **2.** $-\frac{3}{5}, 3$ **3.** $\frac{3}{2}, -1$ **4.** -10

5. $3 \pm \sqrt{5}$ **6.** $-2 \pm \sqrt{3}$ **7.** $4 \pm \dfrac{\sqrt{14}}{2}$

8. $-5 \pm \dfrac{\sqrt{3}}{3}$ **9.** $-\dfrac{3}{2} \pm \dfrac{\sqrt{3}}{2}i$ **10.** $-\dfrac{3}{2} \pm \dfrac{\sqrt{21}}{2}$

1. f   **2.** d   **3.** c   **4.** e   **5.** b   **6.** a

7. $f(x) = (x - 2)^2$ **9.** $f(x) = -(x + 2)^2 + 4$

11. $f(x) = -2(x + 3)^2 + 3$

13. Intercepts: $(\pm\sqrt{5}, 0),$ **15.** Intercepts: $(\pm4, 0),$
(0, −5)   (0, 16)
Vertex: (0, −5)   Vertex: (0, 16)

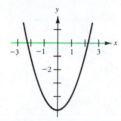
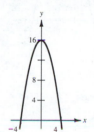

17. Intercepts:
$(-5 \pm \sqrt{6}, 0), (0, 19)$
Vertex: $(-5, -6)$

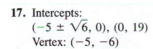
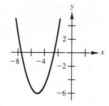

19. Intercepts: (4, 0),
(0, 16)
Vertex: (4, 0)

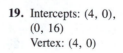
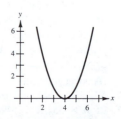

21. Intercepts: $(1, 0)$,
$(-3, 0)$, $(0, 3)$
Vertex: $(-1, 4)$

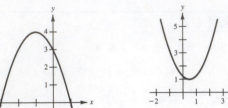

23. Intercept: $(0, \frac{5}{4})$
Vertex: $(\frac{1}{2}, 1)$

25. Intercepts:
$(1 \pm \sqrt{6}, 0)$, $(0, 5)$
Vertex: $(1, 6)$

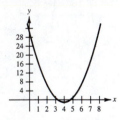

27. Intercept: $(0, 21)$
Vertex: $(\frac{1}{2}, 20)$

29. Intercepts: $\left(4 \pm \dfrac{\sqrt{2}}{2}, 0\right)$, $(0, 31)$

Vertex: $(4, -1)$

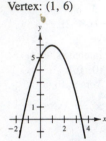

31. $f(x) = -\frac{1}{2}(x - 3)^2 + 4$
33. $f(x) = \frac{3}{4}(x - 5)^2 + 12$
35. $f(x) = x^2 - 2x - 3$ **37.** $f(x) = x^2 - 10x$
  $g(x) = -x^2 + 2x + 3$ $g(x) = -x^2 + 10x$
  (Answer not unique)   (Answer not unique)
39. $f(x) = 2x^2 + 7x + 3$
  $g(x) = -2x^2 - 7x - 3$
  (Answer not unique)
41. 55, 55   **43.** 12, 6   **45.** 25 ft × 25 ft
47. 25 ft × $33\frac{1}{3}$ ft   **49.** 4500 units   **51.** $2000

Section 4.2   *(page 242)*

WARM UP

1. $(3x - 2)(4x + 5)$ **2.** $x(5x - 6)^2$
3. $z^2(12z + 5)(z + 1)$
4. $(y + 5)(y^2 - 5y + 25)$
5. $(x + 3)(x + 2)(x - 2)$ **6.** $(x + 2)(x^2 + 3)$
7. No real solution   **8.** $3 \pm \sqrt{5}$
9. $-\frac{1}{2} \pm \sqrt{3}$ **10.** ± 3

1. e   **2.** c   **3.** b   **4.** f
5. a   **6.** g   **7.** d   **8.** h
9. Rises to the left.   **11.** Falls to the left.
    Rises to the right.   Falls to the right.
13. Falls to the left.   **15.** Rises to the left.
    Rises to the right.   Falls to the right.
17. Falls to the left.   **19.** ± 5 **21.** 3
    Falls to the right.
23. 1, -2 **25.** $2 \pm \sqrt{3}$ **27.** 2, 0   **29.** ± 1
31. $\pm\sqrt{5}$ **33.** No real zeros
35. $f(x) = x^2 - 10x$
37. $f(x) = x^2 + 4x - 12$
39. $f(x) = x^3 + 5x^2 + 6x$
41. $f(x) = x^4 - 4x^3 - 9x^2 + 36x$
43. $f(x) = x^2 - 2x - 2$
45. **(a)**   **(b)**

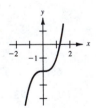

(c)   **(d)**

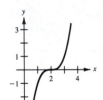

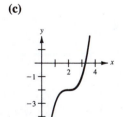

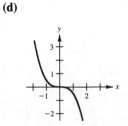

47.

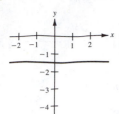

49.

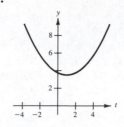

51.

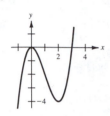

53.

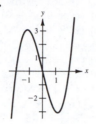

55.

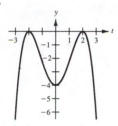

57.

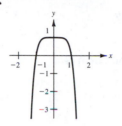

59. 0.7   **61.** 3.3

11. $x^2 + 2x + 4 + \dfrac{2x - 11}{x^2 - 2x + 3}$

13. $2x - \dfrac{17x - 5}{x^2 - 2x + 1}$ **15.** $3x^2 - 2x + 5$

17. $4x^2 - 9$ **19.** $-x^2 + 10x - 25$

21. $5x^2 + 14x + 56 + \dfrac{232}{x - 4}$

23. $10x^3 + 10x^2 + 60x + 360 + \dfrac{1360}{x - 6}$

25. $x^2 - 8x + 64$

27. $-3x^3 - 6x^2 - 12x - 24 - \dfrac{48}{x - 2}$

29. $-x^2 + 3x - 6 + \dfrac{11}{x + 1}$ **31.** $4x^2 + 14x - 30$

33. $(x - 2)(x + 3)(x - 1)$ **35.** $(2x - 1)(x - 5)(x - 2)$

37. $(x - 1)(x - 1 - \sqrt{3})(x - 1 + \sqrt{3})$

39. $(x - 1)(x + 2 - \sqrt{3})(x + 2 + \sqrt{3})$

41. $(x - 4)(x^2 + 3x - 2) + 3, f(4) = 3$

43. $(x - \sqrt{2})[x^2 + (3 + \sqrt{2})x + 3\sqrt{2}] - 8, f(\sqrt{2}) = -8$

45. (a) 1   (b) 4   (c) 4   (d) 1954

47. (a) 97   (b) $-\frac{5}{3}$ (c) 17   (d) -199

49. (a) 72   (b) 0   (c) 37.648   (d) 30

51. (a) 27   (b) -274.625

53. (a) 11.705   (b) -5646.972

Section 4.3   (*page* 253)

WARM UP

1. $x^3 - x^2 + 2x + 3$ **2.** $2x^3 + 4x^2 - 6x - 4$
3. $x^4 - 2x^3 + 4x^2 - 2x - 7$
4. $2x^4 + 12x^3 - 3x^2 - 18x - 5$
5. $(x - 3)(x - 1)$ **6.** $2x(2x - 3)(x - 1)$
7. $x^3 - 7x^2 + 12x$ **8.** $x^2 + 5x - 6$
9. $x^3 + x^2 - 7x - 3$
10. $x^4 - 3x^3 - 5x^2 + 9x - 2$

1. $2x + 4$ **3.** $x^2 - 3x + 1$ **5.** $x^3 + 3x^2 - 1$

7. $7 - \dfrac{11}{x + 2}$ **9.** $3x + 5 - \dfrac{2x - 3}{2x^2 + 1}$

Section 4.4   (*page* 262)

WARM UP

1. $f(x) = 3x^3 - 8x^2 - 5x + 6$
2. $f(x) = 4x^4 - 3x^3 - 16x^2 + 12x$

3. $x^4 - 3x^3 + 5 + \dfrac{3}{x + 3}$

4. $3x^3 + 15x^2 - 9 - \dfrac{2}{x + \frac{2}{3}}$

5. $\frac{1}{2}, -3 \pm \sqrt{5}$ **6.** $10, -\frac{2}{3}, -\frac{3}{2}$
7. $-\frac{3}{4}, 2 \pm \sqrt{2}$ **8.** $\frac{2}{5}, -\frac{7}{2}, -2$
9. $\pm\sqrt{2}, \pm 1$ **10.** $\pm 2, \pm\sqrt{3}$

1. One negative zero   **3.** No real zeros
5. One positive zero   **7.** Three or one positive zeros
9. Two or no positive zeros   **11.** $\pm 1, \pm 2, \pm 4$
13. $\pm 1, \pm 3, \pm\frac{1}{2}, \pm\frac{3}{2}, \pm\frac{1}{4}, \pm\frac{3}{4}$
15. $\pm 1, \pm 2, \pm 4, \pm 8, \pm\frac{1}{2}$

17. (a) Upper bound
(b) Lower bound
(c) Neither

19. (a) Neither
(b) Lower bound
(c) Upper bound

21. 1, 2, 3   **23.** 1, −1, 4   **25.** −1, −10

27. 1, 2   **29.** $\frac{1}{2}$, −1   **31.** 1, −$\frac{1}{2}$ **33.** −$\frac{3}{4}$

35. ±1, ±$\sqrt{2}$ **37.** −1, 2   **39.** 0, −1, −3, 4

41. −2, 4, −$\frac{1}{2}$ **43.** 0, 3, 4, ±$\sqrt{2}$

45. (a) ±1, ±3, ±$\frac{1}{2}$,
$±\frac{3}{2}, ±\frac{1}{4}, ±\frac{3}{4}$,
$±\frac{1}{8}, ±\frac{3}{8}, ±\frac{1}{16}$,
$±\frac{3}{16}, ±\frac{1}{32}, ±\frac{3}{32}$
(c) 1, $\frac{3}{4}$, −$\frac{1}{8}$

(b)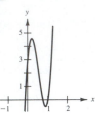

47. (a) ±1, ±2, ±3, ±6,
$±9, ±\frac{1}{2}, ±\frac{3}{2}$,
$±\frac{9}{2}, ±\frac{1}{4}, ±\frac{3}{4}, ±\frac{9}{4}$,
±18
(c) −2, $\frac{1}{8} ± \frac{\sqrt{145}}{8}$

(b)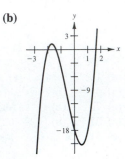

49. ±2, ±$\frac{3}{2}$ **51.** ±1, $\frac{1}{4}$

53. 3.77 in × 7.77 in × 0.614 in

13. −10, −7 ± 5i; $(x + 10)(x + 7 − 5i)(x + 7 + 5i)$

15. −$\frac{3}{4}$, 1 ± $\frac{1}{2}i$; $(4x + 3)(2x − 2 + i)(2x − 2 − i)$

17. −2, 1 ± $\sqrt{2}i$; $(x + 2)(x − 1 + \sqrt{2}i)(x − 1 − \sqrt{2}i)$

19. −$\frac{1}{5}$, 1 ± $\sqrt{5}i$; $(5x + 1)(x − 1 + \sqrt{5}i)(x − 1 − \sqrt{5}i)$

21. 2, ±2i; $(x − 2)^2(x + 2i)(x − 2i)$

23. ±i, ±3i; $(x + i)(x − i)(x + 3i)(x − 3i)$

25. −2, −$\frac{1}{2}$, ±i; $(x + 2)(2x + 1)(x + i)(x − i)$

27. $x^3 − x^2 + 25x − 25$ **29.** $x^3 − 10x^2 + 33x − 34$

31. $x^4 + 37x^2 + 36$

33. $x^4 + 8x^3 + 9x^2 − 10x + 100$

35. $16x^4 + 36x^3 + 16x^2 + x − 30$

37. (a) $(x^2 + 9)(x^2 − 3)$
(b) $(x^2 + 9)(x + \sqrt{3})(x − \sqrt{3})$
(c) $(x + 3i)(x − 3i)(x + \sqrt{3})(x − \sqrt{3})$

39. (a) $(x^2 − 2x − 2)(x^2 − 2x + 3)$
(b) $(x − 1 + \sqrt{3})(x − 1 − \sqrt{3})(x^2 − 2x + 3)$
(c) $(x − 1 + \sqrt{3})(x − 1 − \sqrt{3})$
$(x − 1 + \sqrt{2}i)(x − 1 − \sqrt{2}i)$

41. ±5i, −$\frac{3}{2}$ **43.** ±2i, 1, −$\frac{1}{2}$

45. −3 ± i, $\frac{1}{4}$ **47.** 1, 2, −3 ± $\sqrt{2}i$

49. $\frac{3}{4}$, $\frac{1}{2} ± \frac{\sqrt{5}}{2}i$ **51.** $x^2 + b$

Section 4.6   *(page 276)*

1. 0.68   **3.** 0.21   **5.** 2.77   **7.** −1.16, 1.45

9. 0.53   **11.** 1.56   **13.** 4.5 hr

Section 4.5   *(page 271)*

WARM UP

1. $4 − \sqrt{29}i$, $4 + \sqrt{29}i$
2. $−5 − 12i$, $−5 + 12i$
3. $−1 + 4\sqrt{2}i$, $−1 − 4\sqrt{2}i$ **4.** $6 + \frac{1}{2}i$, $6 − \frac{1}{2}i$
5. $−13 + 9i$ **6.** $12 + 16i$ **7.** $26 + 22i$
8. 29   **9.** i **10.** $−9 + 46i$

1. ±5i; $(x + 5i)(x − 5i)$
3. 2 ± $\sqrt{3}$; $(x − 2 − \sqrt{3})(x − 2 + \sqrt{3})$
5. ±3, ±3i; $(x + 3)(x − 3)(x + 3i)(x − 3i)$
7. 1 ± i; $(z − 1 + i)(z − 1 − i)$
9. 2, 2 ± i; $(x − 2)(x − 2 + i)(x − 2 − i)$
11. −5, 4 ± 3i; $(t + 5)(t − 4 + 3i)(t − 4 − 3i)$

Chapter 4 Review Exercises   *(page 277)*

1.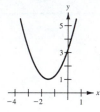

3.

5.

7.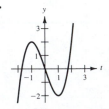

9.

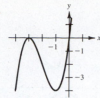

11. Minimum: -1 **13.** Maximum: 9
15. Minimum: $-\frac{41}{4}$ **17.** Maximum: 3
19. Minimum: $-\frac{41}{4}$ **21.** $8x + 5 + \dfrac{2}{3x - 2}$

23. $x^2 - x + 1$ **25.** $x^2 - 2$

27. $x^2 - 3x + 2 - \dfrac{1}{x^2 + 2}$

29. $0.25x^3 - 3.5x^2 - 7x - 14 - \dfrac{28}{x - 2}$

31. $6x^3 - 27x$ **33.** $2x^2 - (3 - 4i)x + (1 - 2i)$
35. **(a)** No   **(b)** Yes   **(c)** Yes   **(d)** No
37. **(a)** No   **(b)** Yes   **(c)** Yes   **(d)** No
39. **(a)** 580   **(b)** 0   **41.** **(a)** -421 **(b)** 96
43. $6x^4 + 13x^3 + 7x^2 - x - 1$ **45.** $1, \frac{3}{4}$
47. $\frac{5}{6}, \pm 2i$ **49.** $-1, \frac{3}{2}, 3, \frac{2}{3}$ **51.** 0.48   **53.** 3.26
55. 4500 units   **57.** \$680   **59.** $(3, \frac{3}{2})$

CHAPTER 5
Section 5.1   *(page* 289)

WARM UP

1. $(x - 5)(x + 2)$ **2.** $(x - 5)(x - 2)$
3. $x(x + 1)(x + 3)$ **4.** $(x^2 - 2)(x - 4)$
5.

6.

7.

8.

9. $x + 9 + \dfrac{42}{x - 4}$ **10.** $x + 1 + \dfrac{2}{x + 4}$

1. f   **2.** g   **3.** a   **4.** b   **5.** c   **6.** d
7. h   **8.** e
9. Domain: all $x \neq 0$
   Vertical asymptote:
   $x = 0$
   Horizontal asymptote:
   $y = 0$
11. Domain: all $x \neq 2$
   Vertical asymptote:
   $x = 2$
   Horizontal asymptote:
   $y = -1$
13. Domain: all $x \neq \pm 1$
   Vertical asymptote:
   $x = \pm 1$
   Slant asymptote: $y = x$
15. Domain: all reals
   Horizontal asymptote:
   $y = 3$
17. Domain: all reals

19. **(a)**

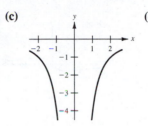

(b)

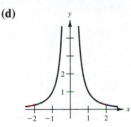

(c)

(d)

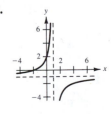

21.

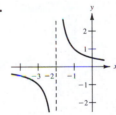

23.

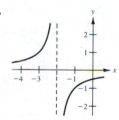

25.

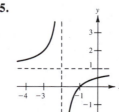

27.

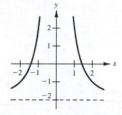

29.

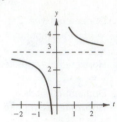

31.

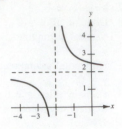

49.

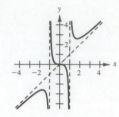

51.

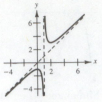

33.

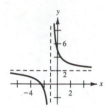

35.

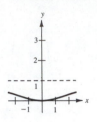

53.

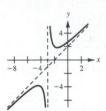

37.

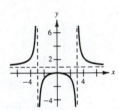

39.

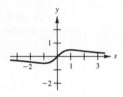

55. (a) 167, 250, 400
(b) 750

41.

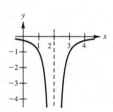

43.

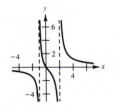

57. (a)

n	1	2	3	4	5
P	0.50	0.74	0.82	0.86	0.89

n	6	7	8	9	10
P	0.91	0.92	0.93	0.94	0.95

(b) 100%

45.

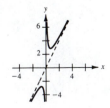

47.

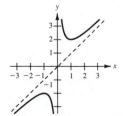

59. (c) $a = 4$

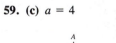
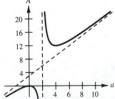

Section 5.2 (*page* 297)

WARM UP

1. $\dfrac{5x + 2}{x(x + 1)}$ 2. $\dfrac{2(4x + 3)}{x(x + 2)}$

3. $\dfrac{11x - 1}{(x - 2)(2x - 1)}$ 4. $-\dfrac{3x + 1}{(x + 5)(x + 12)}$

5. $\dfrac{x^2 - 3x - 5}{(x - 3)^3}$ 6. $-\dfrac{5x + 6}{(x + 2)^2}$

7. $-\dfrac{x + 9}{x(x^2 + 3)}$ 8. $\dfrac{4x^2 + 5x + 31}{(x + 1)(x^2 + 5)}$

9. $\dfrac{x(3x + 1)}{(x^2 + 1)^2}$ 10. $\dfrac{x^3 + x^2 + 1}{(x^2 + x + 1)^2}$

1. $\dfrac{1}{2}\left(\dfrac{1}{x - 1} - \dfrac{1}{x + 1}\right)$ 3. $\dfrac{1}{x} - \dfrac{1}{x + 1}$

5. $\dfrac{1}{x} - \dfrac{2}{2x + 1}$ 7. $\dfrac{1}{x - 1} - \dfrac{1}{x + 2}$

9. $\dfrac{3}{2x - 1} - \dfrac{2}{x + 1}$ 11. $-\dfrac{3}{x} - \dfrac{1}{x + 2} + \dfrac{5}{x - 2}$

13. $\dfrac{3}{x} - \dfrac{1}{x^2} + \dfrac{1}{x + 1}$ 15. $\dfrac{2}{x} - \dfrac{1}{x^2} - \dfrac{2}{x + 1}$

17. $\dfrac{3}{x - 3} + \dfrac{9}{(x - 3)^2}$ 19. $-\dfrac{1}{x} + \dfrac{2x}{x^2 + 1}$

21. $\dfrac{1}{3(x^2 + 2)} - \dfrac{1}{6(x + 2)} + \dfrac{1}{6(x - 2)}$

23. $\dfrac{1}{8(2x + 1)} + \dfrac{1}{8(2x - 1)} - \dfrac{x}{2(4x^2 + 1)}$

25. $\dfrac{1}{x^2 + 2} + \dfrac{x}{(x^2 + 2)^2}$ 27. $\dfrac{1}{x + 1} + \dfrac{2}{x^2 - 2x + 3}$

29. $2x + \dfrac{1}{2}\left(\dfrac{3}{x - 4} - \dfrac{1}{x + 2}\right)$

31. $x + 3 + \dfrac{6}{x - 1} + \dfrac{4}{(x - 1)^2} + \dfrac{1}{(x - 1)^3}$

33. $\dfrac{1}{2a}\left(\dfrac{1}{a + x} + \dfrac{1}{a - x}\right)$ 35. $\dfrac{1}{L}\left(\dfrac{1}{y} + \dfrac{1}{L - y}\right)$

Section 5.3 (*page* 308)

WARM UP

1. $9x^2 + 16y^2 = 144$ 2. $x^2 + 4y^2 = 32$

3. $16x^2 - y^2 = 4$ 4. $243x^2 + 4y^2 = 9$

5. $c = 2\sqrt{2}$ 6. $c = \sqrt{13}$ 7. $c = 2\sqrt{3}$

8. $c = \sqrt{5}$ 9. $d = 4$ 10. $d = 2$

1. a 2. c 3. d 4. h 5. g 6. f
7. e 8. b

9. Vertex: $(0, 0)$ 11. Vertex: $(0, 0)$
   Focus: $(0, \frac{1}{16})$ Focus: $(-\frac{3}{2}, 0)$

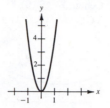

13. Vertex: $(0, 0)$ 15. Vertex: $(0, 0)$
    Focus: $(0, -2)$ Focus: $(2, 0)$

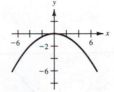
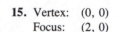

17. $x^2 = -6y$ 19. $y^2 = -8x$ 21. $x^2 = 4y$
23. $x^2 = -8y$ 25. $y^2 = 9x$
27. Center: $(0, 0)$ 29. Center: $(0, 0)$
    Vertices: $(\pm 5, 0)$ Vertices: $(0, \pm 5)$

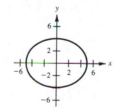
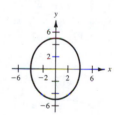

31. Center: $(0, 0)$ 33. Center: $(0, 0)$
    Vertices: $(\pm 3, 0)$ Vertices: $(0, \pm\sqrt{5})$

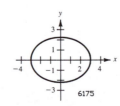
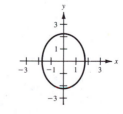

6175

35. $\dfrac{x^2}{1} + \dfrac{y^2}{4} = 1$ **37.** $\dfrac{x^2}{25} + \dfrac{y^2}{21} = 1$

39. $\dfrac{x^2}{36} + \dfrac{y^2}{11} = 1$ **41.** $\dfrac{21x^2}{400} + \dfrac{y^2}{25} = 1$

43. Center: (0, 0)   **45.** Center: (0, 0)
    Vertices: $(\pm 1, 0)$ Vertices: $(0, \pm 1)$

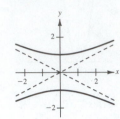

47. Center: (0, 0)   **49.** Center: (0, 0)
    Vertices: $(0, \pm 5)$ Vertices: $(\pm\sqrt{3}, 0)$

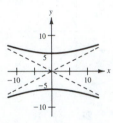
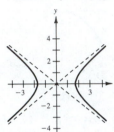

51. $\dfrac{y^2}{4} - \dfrac{x^2}{12} = 1$ **53.** $\dfrac{x^2}{1} - \dfrac{y^2}{9} = 1$

55. $\dfrac{17y^2}{1024} - \dfrac{17x^2}{64} = 1$ **57.** $\dfrac{y^2}{9} - \dfrac{4x^2}{9} = 1$

59. $x^2 = 12y$ **61.** $\left(\pm\frac{3}{2}, 0\right)$, 5 ft

63. (a)   **(b)**

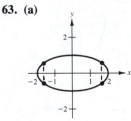
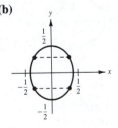

(c)

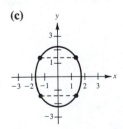

Section 5.4   (*page* 314)

WARM UP

1. Hyperbola	**2.** Ellipse	**3.** Parabola
4. Hyperbola	**5.** Ellipse	**6.** Circle
7. Hyperbola	**8.** Parabola	**9.** Parabola
10. Ellipse		

1. Vertex: (1, −2)   **3.** Vertex: $\left(5, -\frac{1}{2}\right)$
    Focus: (1, −4)   Focus: $\left(\frac{11}{2}, -\frac{1}{2}\right)$
    Directrix: $y = 0$ Directrix: $x = \frac{9}{2}$

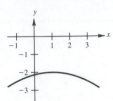
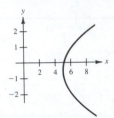

5. Vertex: (1, 1)   **7.** Vertex: (8, −1)
    Focus: (1, 2)   Focus: (9, −1)
    Directrix: $y = 0$ Directrix: $x = 7$

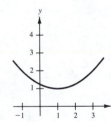
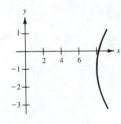

9. Vertex: (−2, −3)   **11.** Vertex: (−1, 2)
    Focus: (−4, −3)   Focus: (0, 2)
    Directrix: $x = 0$ Directrix: $x = -2$

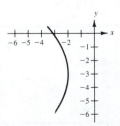
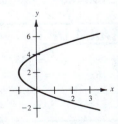

13. $(y - 2)^2 = -8(x - 3)$ **15.** $x^2 = 8(y - 4)$

17. $(y - 2)^2 = 8x$ **19.** $x^2 = -(y - 4)$

21. Center: $(1, 5)$
Vertices: $(1, 5 \pm 5)$
Foci: $(1, 5 \pm 4)$

23. Center: $(-2, 3)$
Vertices: $(-2, 3 \pm 3)$
Foci: $(-2, 3 \pm \sqrt{5})$

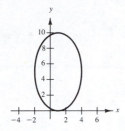

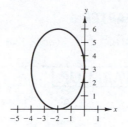

25. Center: $(1, -1)$
Vertices: $\left(1 \pm \frac{5}{4}, -1\right)$
Foci: $\left(1 \pm \frac{3}{4}, -1\right)$

27. Center: $\left(\frac{1}{2}, -1\right)$
Vertices: $\left(\frac{1}{2} \pm \sqrt{5}, -1\right)$
Foci: $\left(\frac{1}{2} \pm \sqrt{2}, -1\right)$

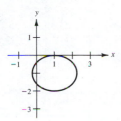

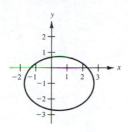

29. $\dfrac{(x - 2)^2}{4} + \dfrac{(y - 2)^2}{1} = 1$

31. $\dfrac{x^2}{48} + \dfrac{(y - 4)^2}{64} = 1$

33. $\dfrac{(x - 3)^2}{9} + \dfrac{(y - 5)^2}{16} = 1$ **35.** $\dfrac{x^2}{16} + \dfrac{(y - 4)^2}{12} = 1$

37. Center: $(1, -2)$
Vertices: $(1 \pm 2, -2)$
Foci: $(1 \pm \sqrt{5}, -2)$

39. Center: $(2, -6)$
Vertices: $(2, -6 \pm 1)$
Foci: $(2, -6 \pm \sqrt{2})$

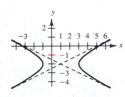

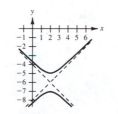

41. Center: $(2, -3)$
Vertices: $(2 \pm 1, -3)$
Foci: $(2 \pm \sqrt{10}, -3)$

43. Center: $(1, -3)$
Vertices: $(1, -3 \pm \sqrt{2})$
Foci: $(1, -3 \pm 2\sqrt{5})$

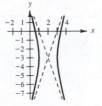

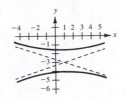

45. Degenerate hyperbola
Lines: $y + 3 = \pm\frac{1}{3}(x + 1)$

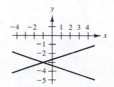

47. $\dfrac{(x - 4)^2}{4} - \dfrac{y^2}{12} = 1$ **49.** $\dfrac{(y - 5)^2}{16} - \dfrac{(x - 4)^2}{9} = 1$

51. $\dfrac{y^2}{9} - \dfrac{4(x - 2)^2}{9} = 1$ **53.** $\dfrac{(x - 3)^2}{9} - \dfrac{(y - 2)^2}{4} = 1$

55. Circle   **57.** Hyperbola   **59.** Ellipse

61. Parabola

63. (a) $17{,}500\sqrt{2}$ mi/hr   (b) $x^2 = -16{,}400(y - 4100)$

65. $\dfrac{x^2}{25} + \dfrac{y^2}{16} = 1$ **67.** 91,419,000 mi; 94,581,000 mi

Chapter 5 Review Exercises   *(page* 317)

1.

3.

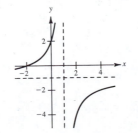

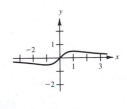

5.

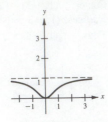

7.

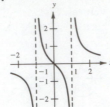

9.

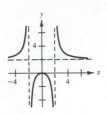

11.

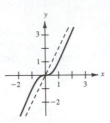

13.

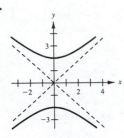

15.

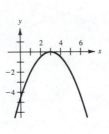

17.

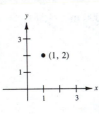

19.

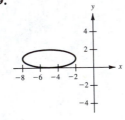

21. $\dfrac{3}{x+2} - \dfrac{4}{x+4}$ **23.** $1 - \dfrac{25}{8(x+5)} + \dfrac{9}{8(x-3)}$

25. $\dfrac{1}{2}\left(\dfrac{3}{x-1} - \dfrac{x-3}{x^2+1}\right)$ **27.** $\dfrac{3x}{x^2+1} + \dfrac{x}{(x^2+1)^2}$

29. $(x-4)^2 = -8(y-2)$ **31.** $(y-2)^2 = -4x$

33. $\dfrac{(x-2)^2}{25} + \dfrac{y^2}{21} = 1$ **35.** $\dfrac{2x^2}{9} + \dfrac{y^2}{36} = 1$

37. $\dfrac{y^2}{1} - \dfrac{x^2}{8} = 1$ **39.** $\dfrac{5(x-4)^2}{16} - \dfrac{5y^2}{64} = 1$

41. (a) \$176 million   (b) \$528 million
(c) \$1584 million   (d) C approaches infinity

43. (0, 50)

CHAPTER 6
Section 6.1   (*page* 328)

WARM UP

1. 5^x **2.** 3^{2x} **3.** 4^{3x} **4.** 10^x **5.** 4^{2x}
6. 4^{10x} **7.** $\left(\dfrac{3}{2}\right)^x$ **8.** 4^{3x} **9.** 2^{-x} **10.** $16^{x/4}$

1. 946.852   **3.** 747.258   **5.** 5.256
7. 472,369.379   **9.** 673.639   **11.** 7.389
13. 0.472   **15.** g   **16.** e   **17.** b   **18.** h
19. d   **20.** a   **21.** f   **22.** c

23.

25.

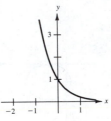

27.

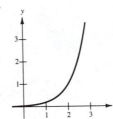

29.

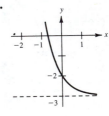

31.

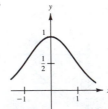

33.

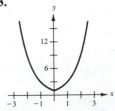

35.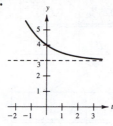

37.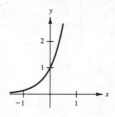

39.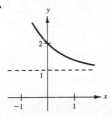

41.

n	1	2	4
A	\$7,764.62	\$8,017.84	\$8,155.09

n	12	365	Continuous compounding
A	\$8,250.97	\$8,298.66	\$8,300.29

43.

n	1	2	4
A	\$24,115.73	\$25,714.29	\$26,602.23

n	12	365	Continuous compounding
A	\$27,231.38	\$27,547.07	\$27,557.94

45.

t	1	10	20
P	\$91,393.12	\$40,656.97	\$16,529.89

t	30	40	50
P	\$6,720.55	\$2,732.37	\$1,110.90

47.

t	1	10	20
P	\$90,521.24	\$36,940.70	\$13,646.15

t	30	40	50
P	\$5,040.98	\$1,862.17	\$687.90

49. (a) \$472.70 **(b)** \$298.29
51. (a) 100 **(b)** 300 **(c)** 900

Section 6.2 (*page* 339)

WARM UP

1. 3 **2.** 0 **3.** -1 **4.** 1 **5.** 7.389
6. 0.368 **7.** Graph is shifted 2 units to the left.
8. Graph is reflected about the x-axis.
9. Graph is shifted down 1 unit.
10. Graph is reflected about the y-axis.

1. 4 **3.** -2 **5.** $\frac{1}{2}$ **7.** 0 **9.** -2 **11.** 3
13. -2 **15.** 2 **17.** $\log_5 125 = 3$ **19.** $\log_{81} 3 = \frac{1}{4}$
21. $\log_6 \frac{1}{36} = -2$ **23.** $\ln 20.0855 \approx 3$ **25.** $\ln 4 = x$
27. 2.538 **29.** 2.913 **31.** 1.005
33. 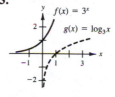 **35.**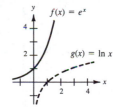

37. d **38.** e **39.** a **40.** c **41.** f **42.** b
43. Domain: $(0, \infty)$
Vertical asymptote: $x = 0$
Intercept: $(1, 0)$

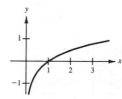

45. Domain: $(3, \infty)$
Vertical asymptote: $x = 3$
Intercept: $(4, 0)$

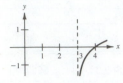

47. Domain: $(2, \infty)$
Vertical asymptote: $x = 2$
Intercept: $(3, 0)$

49. Domain: $(-\infty, 0)$
Vertical asymptote: $x = 0$
Intercept: $(-1, 0)$

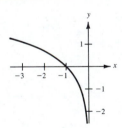

51. $\dfrac{\log_{10} 5}{\log_{10} 3}$ **53.** $\dfrac{\log_{10} x}{\log_{10} 2}$ **55.** $\dfrac{\ln 5}{\ln 3}$ **57.** $\dfrac{\ln x}{\ln 2}$

59. 1.771   **61.** -2.000 **63.** -0.417 **65.** 2.633

67. (a) 80.0   **(b)** 68.1   **(c)** 62.3   **69.** 23.68 yr

71.

r	0.005	0.010	0.015
t	138.6 yr	69.3 yr	46.2 yr

r	0.020	0.025	0.030
t	34.7 yr	27.7 yr	23.1 yr

73. (a)

x	1	5	10
$f(x)$	0	0.322	0.230

x	10^2	10^4	10^6
$f(x)$	0.046	0.00092	0.0000138

(b) 0

Section 6.3   *(page 346)*

WARM UP

1. 2   **2.** -5 **3.** -2 **4.** -3 **5.** e^5

6. $\dfrac{1}{e}$ **7.** e^6 **8.** 1   **9.** x^{-2} **10.** $x^{1/2}$

1. $\log_2 5 + \log_2 x$ **3.** $\log_3 5 - \log_3 x$ **5.** $4 \log_8 x$

7. $\frac{1}{2} \ln z$ **9.** $\log_2 x + \log_2 y + \log_2 z$

11. $\frac{1}{2} \ln(a - 1)$ **13.** $\ln z + 2 \ln(z - 1)$

15. $2 \log_b x - 2 \log_b y - 3 \log_b z$ **17.** $\frac{1}{3} \ln x - \frac{1}{3} \ln y$

19. $4 \log_9 x + \frac{1}{2} \log_9 y - 5 \log_9 z$ **21.** $\ln 2x$

23. $\log_4 \dfrac{z}{y}$ **25.** $\log_2(x + 4)^2$

27. $\ln \dfrac{x}{(x + 1)^3}$ **29.** $\log_3 \sqrt[3]{5x}$ **31.** $\log_3 \dfrac{x - 2}{x + 2}$

33. $\ln \dfrac{x}{(x^2 - 4)^2}$ **35.** $\ln \sqrt[3]{\dfrac{x(x + 3)^2}{x^2 - 1}}$

37. $\ln \dfrac{\sqrt[3]{y(y + 4)^2}}{y - 1}$ **39.** $\ln \dfrac{9}{\sqrt{x^2 + 1}}$ **41.** 0.9208

43. 0.2084   **45.** 1.6542   **47.** 0.1781   **49.** 1.8957

51. -0.7124 **53.** 0.9136   **55.** 2.0367   **57.** 2

59. 2.4   **61.** 4.5   **63.** $\frac{3}{2}$ **65.** $\frac{1}{2} + \frac{1}{2} \log_7 10$

67. $-3 - \log_5 2$ **69.** $6 + \ln 5$

73.

x	y	$\dfrac{\ln x}{\ln y}$	$\ln \dfrac{x}{y}$	$\ln x - \ln y$
1	2	0	-0.6931	-0.6931
3	4	0.7925	-0.2877	-0.2877
10	5	1.4307	0.6931	0.6931
4	0.5	-2.000	2.0794	2.0794

Section 6.4 (*page* 354)

WARM UP

1. $\dfrac{\ln 3}{\ln 2}$ **2.** $1 + \dfrac{2}{\ln 4}$ **3.** $\dfrac{e}{2}$ **4.** $2e$
5. $2 \pm i$ **6.** $\frac{1}{2}, 1$ **7.** $2x$ **8.** $3x$ **9.** $2x$
10. $-x^2$

1. 2   **3.** -2 **5.** 3   **7.** 64   **9.** $\frac{1}{10}$ **11.** x^2
13. $5x + 2$ **15.** x^2 **17.** $\log_{10} 42$ **19.** $\frac{1}{2} \log_{10} 36$
21. $1 + \log_{10} \frac{2}{3}$ **23.** $\ln 10$ **25.** $\ln \frac{5}{3}$ **27.** $\frac{2}{3} \ln \frac{40}{3}$
29. $-\frac{1}{2} + \frac{1}{2} \ln \frac{962}{25}$ **31.** $\dfrac{\ln 3}{4 \ln\left(1 + \frac{0.09}{4}\right)}$ **33.** $\dfrac{\ln 3}{0.09}$
35. $\dfrac{\ln 2}{12 \ln\left(1 + \frac{0.10}{12}\right)}$ **37.** $-5 \ln \frac{4}{19}$ **39.** $-\dfrac{\ln 0.2247}{\ln 1.0775}$
41. $\dfrac{7 - \log_{10} 5}{1 + \log_{10} 5}$ **43.** $\frac{1}{2} \ln \frac{1}{3}$ **45.** 0   **47.** $\frac{1}{2} \ln 3$
49. e^5 **51.** $e^{7/2}$ **53.** $\frac{1}{4}$ **55.** 103
57. $1 \pm \sqrt{1 + e}$ **59.** $-\dfrac{1}{2} \pm \dfrac{\sqrt{17}}{2}$
61. $-\dfrac{3}{2} \pm \dfrac{\sqrt{9 + 4e}}{2}$ **63.** $1, e^2$ **65.** 8.155 yr
67. (a) 1426 units
(b) 1498 units
69. (a) 29.33 yr
(b) 39.79 yr

Section 6.5 (*page* 363)

1.

Initial investment	Annual % rate	Effective yield	Time to double	Amount after 10 years
$1000	12%	12.75%	5.78 yr	$3,320.12

3.

Initial investment	Annual % rate	Effective yield	Time to double	Amount after 10 years
$750	8.94%	9.35%	7.75 yr	$1,833.67

5.

Initial investment	Annual % rate	Effective yield	Time to double	Amount after 10 years
$500	9.5%	9.97%	7.30 yr	$1,292.85

7.

Initial investment	Annual % rate	Effective yield	Time to double	Amount after 10 years
$6,392.79	11%	11.63%	6.30 yr	$19,205.00

9.

Initial investment	Annual % rate	Effective yield	Time to double	Amount after 10 years
$5000	8%	8.33%	8.66 yr	$11,127.70

11. $112,087.09   **13.** (a) 6.642 yr   (b) 6.330 yr
(c) 6.302 yr   (d) 6.301 yr

15.

r	2%	4%	6%
t	54.93 yr	27.47 yr	18.31 yr

r	8%	10%	12%
t	13.73 yr	10.99 yr	9.16 yr

17. $26,111.12   **19.** $640,501.62

21. 2009   **23.** $\dfrac{\ln 2}{r}, \dfrac{\ln 3}{r}$

25.

Half-life (in years)	Initial quantity	Amount after 1000 years	Amount after 10,000 years
1620	10 g	6.52 g	0.14 g

27.

Half-life (in years)	Initial quantity	Amount after 1000 years	Amount after 10,000 years
5730	6.71 g	5.95 g	2 g

29.

Half-life (in years)	Initial quantity	Amount after 1000 years	Amount after 10,000 years
24,360	2.16 g	2.1 g	1.62 g

31. 95.8%   **33.** $\frac{1}{4} \ln 10 \approx 0.5756$

35. $\frac{1}{4} \ln \frac{1}{4} \approx -0.3466$

37. (a) $S(t) = 100(1 - e^{-0.1625t})$ (b) 55,625

39. (a) 2137   (b) 4.6 months

41. (a) 20   (b) 70   (c) 95   (d) 120

43. (a) 7.91   (b) 7.68

45. 4.64   **47.** 1.58×10^{-6} moles per liter

49. (a) $T(t) = 70 + 280e^{-0.02784t}$ (b) 122.7°

(c) 119.7 min

Chapter 6 Review Exercises   (*page* 365)

1. 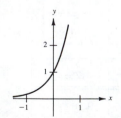   **3.**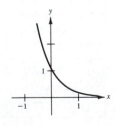

5.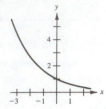

7.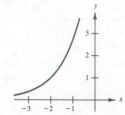

9. 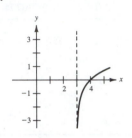   **11.**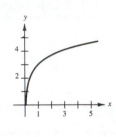

13. 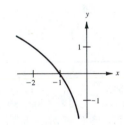   **15.**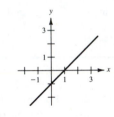

17. $1 + 2 \log_5 x$ **19.** $\log_{10} 5 + \frac{1}{2} \log_{10} y - 2 \log_{10} x$

21. $\ln(x^2 + 1) - \ln |x|$ **23.** $\ln(x^2 + 1) + \ln(x - 1)$

25. $\log_{10} \dfrac{5}{(x + 4)^2}$ **27.** $\ln \dfrac{\sqrt{|2x - 1|}}{(x + 1)^2}$ **29.** $\ln x^{7/3}$

31. $\ln \dfrac{3\sqrt[3]{4 - x^2}}{x}$ **33.** False   **35.** False   **37.** True

39. True   **41.** 229.2 units per milliliter

43. (a) 15.3%   (b) 48.7%   (c) 81.1%

45.

Speed	50	55	60	65	70
Miles per gallon	28	26.4	24.8	23.4	22.0

47. $y = 2e^{0.1014t}$ **49.** $y = 4e^{-0.4159t}$

51. **(a)** 1151 units   **(b)** 1325 units

53. **(a)** 8.94%   **(b)** \$1,834.37   **(c)** 9.36%

55.

x	2	5
e^{-x}	1.3534×10^{-1}	6.7379×10^{-3}

x	10	15
e^{-x}	4.5399×10^{-5}	3.0590×10^{-7}

57. **(a)** Sum: 2.7083, Calculator: 2.7183
   (b) Sum: 1.6484, Calculator: 1.6487
   (c) Sum: 0.6068, Calculator: 0.6065

CHAPTER 7

Section 7.1   (*page* 378)

(*page* 378)

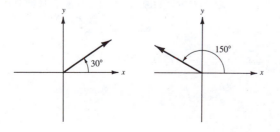

WARM UP

1. 45   **2.** 70   **3.** $\frac{\pi}{6}$ **4.** $\frac{\pi}{3}$ **5.** $\frac{\pi}{4}$

6. $\frac{4\pi}{3}$ **7.** $\frac{\pi}{9}$ **8.** $\frac{11\pi}{6}$ **9.** 45   **10.** 45

1. **(a)** Quadrant II   **(b)** Quadrant IV
3. **(a)** Quadrant III   **(b)** Quadrant I
5. **(a)** Quadrant I   **(b)** Quadrant III
7. **(a)** Quadrant IV   **(b)** Quadrant II
9. **(a)** Quadrant III   **(b)** Quadrant II
11. **(a)**   **(b)**

13. **(a)** 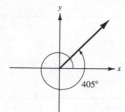   **(b)**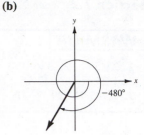

15. **(a)** 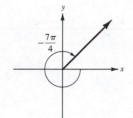   **(b)**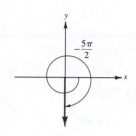

17. **(a)** 396°, −324°   **(b)** 315°, −405°
19. **(a)** 660°, −60°   **(b)** 20°, −340°
21. **(a)** $\frac{19\pi}{9}$, $-\frac{17\pi}{9}$ **(b)** $\frac{10\pi}{3}$, $-\frac{2\pi}{3}$
23. **(a)** $\frac{7\pi}{4}$, $-\frac{\pi}{4}$ **(b)** $\frac{28\pi}{15}$, $-\frac{32\pi}{15}$
25. **(a)** $\frac{\pi}{6}$ **(b)** $\frac{5\pi}{6}$ **27.** **(a)** $-\frac{\pi}{9}$ **(b)** $-\frac{4\pi}{3}$
29. **(a)** 270°   **(b)** 210°   **31.** **(a)** 420°   **(b)** −66°
33. 2.007   **35.** −3.776   **37.** 9.285   **39.** −0.014
41. 25.714°   **43.** 337.5°   **45.** −756°
47. −114.592°   **49.** **(a)** 54.75°   **(b)** −128.5°
51. **(a)** 85.308°   **(b)** 330.007°
53. **(a)** 240° 36′   **(b)** −145° 48′
55. **(a)** 143° 14′ 22″   **(b)** −205° 7′ 8″
57. $\frac{2}{5}$ rad   **59.** 1.724 rad
61. 15π in ≈ 47.12 in   **63.** 12 m
65. 591.72 mi   **67.** 1141.02 mi   **69.** 4.655°
71. 171.89°
73. **(a)** 560.2 rev/min   **(b)** 3520 rad/min
75. **(a)** 2 in: 3400π rad/min   **(b)** 850 rev/min
         4 in: 1700π rad/min

Section 7.2 *(page 389)*

WARM UP

1. $2\sqrt{5}$ **2.** $3\sqrt{10}$ **3.** 10   **4.** $3\sqrt{2}$
5. 1.24   **6.** 317.55   **7.** 63.13   **8.** 133.57
9. 2,785,714.29   **10.** 28.80

1. $\sin \theta = \frac{1}{2}$ $\csc \theta = 2$

   $\cos \theta = \dfrac{\sqrt{3}}{2}$ $\sec \theta = \dfrac{2\sqrt{3}}{3}$

   $\tan \theta = \dfrac{\sqrt{3}}{3}$ $\cot \theta = \sqrt{3}$

3. $\sin \theta = \frac{3}{5}$ $\csc \theta = \frac{5}{3}$

   $\cos \theta = \frac{4}{5}$ $\sec \theta = \frac{5}{4}$

   $\tan \theta = \frac{3}{4}$ $\cot \theta = \frac{4}{3}$

5. $\sin \theta = \dfrac{\sqrt{161}}{15}$ $\csc = \dfrac{15\sqrt{161}}{161}$

   $\cos \theta = \frac{8}{15}$ $\sec \theta = \frac{15}{8}$

   $\tan \theta = \dfrac{\sqrt{161}}{8}$ $\cot \theta = \dfrac{8\sqrt{161}}{161}$

7. $\sin \theta = \dfrac{8\sqrt{73}}{73}$ $\csc \theta = \dfrac{\sqrt{73}}{8}$

   $\cos \theta = \dfrac{3\sqrt{73}}{73}$ $\sec \theta = \dfrac{\sqrt{73}}{3}$

   $\tan \theta = \frac{8}{3}$ $\cot \theta = \frac{3}{8}$

9. $\cos \theta = \dfrac{\sqrt{5}}{3}$

   $\tan \theta = \dfrac{2\sqrt{5}}{5}$

   $\csc \theta = \frac{3}{2}$

   $\sec \theta = \dfrac{3\sqrt{5}}{5}$

   $\cot \theta = \dfrac{\sqrt{5}}{2}$

11. $\sin \theta = \dfrac{\sqrt{3}}{2}$

   $\cos \theta = \frac{1}{2}$

   $\tan \theta = \sqrt{3}$

   $\csc \theta = \dfrac{2\sqrt{3}}{3}$

   $\cot \theta = \dfrac{\sqrt{3}}{3}$

13. $\sin \theta = \dfrac{3\sqrt{10}}{10}$

   $\cos \theta = \dfrac{\sqrt{10}}{10}$

   $\csc \theta = \dfrac{\sqrt{10}}{3}$

   $\sec \theta = \sqrt{10}$

   $\cot \theta = \frac{1}{3}$

15. $\sin \theta = \dfrac{2\sqrt{13}}{13}$

   $\cos = \dfrac{3\sqrt{13}}{13}$

   $\tan \theta = \frac{2}{3}$

   $\csc \theta = \dfrac{\sqrt{13}}{2}$

   $\sec \theta = \dfrac{\sqrt{13}}{3}$

17. **(a)** $\sqrt{3}$ **(b)** $\frac{1}{2}$ **(c)** $\dfrac{\sqrt{3}}{2}$ **(d)** $\dfrac{\sqrt{3}}{3}$

19. **(a)** $\frac{1}{3}$ **(b)** $\dfrac{2\sqrt{2}}{3}$ **(c)** $\dfrac{\sqrt{2}}{4}$ **(d)** 3

21. **(a)** $\frac{1}{2}$ **(b)** $\dfrac{\sqrt{3}}{3}$ **23.** **(a)** 1   **(b)** $\dfrac{\sqrt{2}}{2}$

25. **(a)** 0.1736   **(b)** 0.1736
27. **(a)** 1.3499   **(b)** 1.3432
29. **(a)** 0.2815   **(b)** 3.5523   **31.** **(a)** 5.0273
   (b) 0.1989   **33.** **(a)** 1.1884   **(b)** 1.1884

35. **(a)** $30° = \dfrac{\pi}{6}$ **(b)** $30° = \dfrac{\pi}{6}$

37. **(a)** $60° = \dfrac{\pi}{3}$ **(b)** $45° = \dfrac{\pi}{4}$

39. **(a)** $60° = \dfrac{\pi}{3}$ **(b)** $45° = \dfrac{\pi}{4}$

41. **(a)** $55° \approx 0.96$ **(b)** $89° \approx 1.55$
43. **(a)** $50° \approx 0.87$ **(b)** $25° \approx 0.44$
45. 57.74   **47.** $\dfrac{25\sqrt{3}}{3}$ **49.** 15.56
51. 9.19   **53.** 15 ft   **55.** 19.32 ft   **57.** 2145 ft
59. True, $\csc x = \dfrac{1}{\sin x}$ **61.** False, $\dfrac{\sqrt{2}}{2} + \dfrac{\sqrt{2}}{2} \neq 1$
63. False, $\dfrac{\sin 60°}{\sin 30°} = 1.7321 \neq \sin 2°$

Section 7.3 *(page 401)*

WARM UP

1. $\sin 30° = \frac{1}{2}$ **2.** $\tan 45° = 1$ **3.** $\cos \frac{\pi}{4} = \frac{\sqrt{2}}{2}$

4. $\cot \frac{\pi}{3} = \frac{\sqrt{3}}{3}$ **5.** $\sec \frac{\pi}{6} = \frac{2\sqrt{3}}{3}$

6. $\csc \frac{\pi}{4} = \sqrt{2}$

7. $\sin \theta = \frac{3\sqrt{13}}{13}$ **8.** $\sin \theta = \frac{\sqrt{5}}{3}$

$\cos \theta = \frac{2\sqrt{13}}{13}$ $\tan \theta = \frac{\sqrt{5}}{2}$

$\csc \theta = \frac{\sqrt{13}}{3}$ $\csc \theta = \frac{3\sqrt{5}}{5}$

$\sec \theta = \frac{\sqrt{13}}{2}$ $\sec \theta = \frac{3}{2}$

$\cot \theta = \frac{2}{3}$ $\cot \theta = \frac{2\sqrt{5}}{5}$

9. $\cos \theta = \frac{2\sqrt{6}}{5}$ **10.** $\sin \theta = \frac{2\sqrt{2}}{3}$

$\tan \theta = \frac{\sqrt{6}}{12}$ $\cos \theta = \frac{1}{3}$

$\csc \theta = 5$ $\tan \theta = 2\sqrt{2}$

$\sec \theta = \frac{5\sqrt{6}}{12}$ $\csc \theta = \frac{3\sqrt{2}}{4}$

$\cot \theta = 2\sqrt{6}$ $\cot \theta = \frac{\sqrt{2}}{4}$

3. (a) $\sin \theta = \frac{1}{2}$ **(b)** $\sin \theta = -\frac{\sqrt{2}}{2}$

$\cos \theta = -\frac{\sqrt{3}}{2}$ $\cos \theta = -\frac{\sqrt{2}}{2}$

$\tan \theta = -\frac{\sqrt{3}}{3}$ $\tan \theta = 1$

$\csc \theta = 2$ $\csc \theta = -\sqrt{2}$

$\sec \theta = -\frac{2\sqrt{3}}{3}$ $\sec \theta = -\sqrt{2}$

$\cot \theta = -\sqrt{3}$ $\cot \theta = 1$

5. (a) $\sin \theta = \frac{24}{25}$ **(b)** $\sin \theta = -\frac{24}{25}$

$\cos \theta = \frac{7}{25}$ $\cos \theta = \frac{7}{25}$

$\tan \theta = \frac{24}{7}$ $\tan \theta = -\frac{24}{7}$

$\csc \theta = \frac{25}{24}$ $\csc \theta = -\frac{25}{24}$

$\sec \theta = \frac{25}{7}$ $\sec \theta = \frac{25}{7}$

$\cot \theta = \frac{7}{24}$ $\cot \theta = -\frac{7}{24}$

7. (a) $\sin \theta = \frac{5\sqrt{29}}{29}$ **(b)** $\sin \theta = -\frac{5\sqrt{34}}{34}$

$\cos \theta = -\frac{2\sqrt{29}}{29}$ $\cos \theta = \frac{3\sqrt{34}}{34}$

$\tan \theta = -\frac{5}{2}$ $\tan \theta = -\frac{5}{3}$

$\csc \theta = \frac{\sqrt{29}}{5}$ $\csc \theta = -\frac{\sqrt{34}}{5}$

$\sec \theta = -\frac{\sqrt{29}}{2}$ $\sec \theta = \frac{\sqrt{34}}{3}$

$\cot \theta = -\frac{2}{5}$ $\cot \theta = -\frac{3}{5}$

9. (a) $c_1 = 5, b_2 = 12, c_2 = 15$

(b) $\sin \alpha_1 = \sin \alpha_2 = \frac{3}{5}$

$\cos \alpha_1 = \cos \alpha_2 = \frac{4}{5}$

$\tan \alpha_1 = \tan \alpha_2 = \frac{3}{4}$

$\csc \alpha_1 = \csc \alpha_2 = \frac{5}{3}$

$\sec \alpha_1 = \sec \alpha_2 = \frac{5}{4}$

$\cot \alpha_1 = \cot \alpha_2 = \frac{4}{3}$

1. (a) $\sin \theta = \frac{4}{5}$ **(b)** $\sin \theta = -\frac{15}{17}$

$\cos \theta = \frac{3}{5}$ $\cos \theta = \frac{8}{17}$

$\tan \theta = \frac{4}{3}$ $\tan \theta = -\frac{15}{8}$

$\csc \theta = \frac{5}{4}$ $\csc \theta = -\frac{17}{15}$

$\sec \theta = \frac{5}{3}$ $\sec \theta = \frac{17}{8}$

$\cot \theta = \frac{3}{4}$ $\cot \theta = -\frac{8}{15}$

11. (a) $b_1 = \sqrt{3}$, $a_2 = \dfrac{5\sqrt{3}}{3}$, $c_2 = \dfrac{10\sqrt{3}}{3}$

   (b) $\sin \alpha_1 = \sin \alpha_2 = \frac{1}{2}$

      $\cos \alpha_1 = \cos \alpha_2 = \dfrac{\sqrt{3}}{2}$

      $\tan \alpha_1 = \tan \alpha_2 = \dfrac{\sqrt{3}}{3}$

      $\csc \alpha_1 = \csc \alpha_2 = 2$

      $\sec \alpha_1 = \sec \alpha_2 = \dfrac{2\sqrt{3}}{3}$

      $\cot \alpha_1 = \cot \alpha_2 = \sqrt{3}$

13. (a) Quadrant III   (b) Quadrant II

15. (a) Quadrant II   (b) Quadrant IV

17. $\sin \theta = \frac{3}{5}$ $\csc \theta = \frac{5}{3}$

    $\cos \theta = -\frac{4}{5}$ $\sec \theta = -\frac{5}{4}$

    $\tan \theta = -\frac{3}{4}$ $\cot \theta = -\frac{4}{3}$

19. $\sin \theta = -\frac{15}{17}$ $\csc \theta = -\frac{17}{15}$

    $\cos \theta = \frac{8}{17}$ $\sec \theta = \frac{17}{8}$

    $\tan \theta = -\frac{15}{8}$ $\cot \theta = -\frac{8}{15}$

21. $\sin \theta = \dfrac{\sqrt{3}}{2}$ $\csc \theta = \dfrac{2\sqrt{3}}{3}$

    $\cos \theta = -\frac{1}{2}$ $\sec \theta = -2$

    $\tan \theta = -\sqrt{3}$ $\cot \theta = -\dfrac{\sqrt{3}}{3}$

23. $\sin \theta = 0$ $\csc \theta$ is undefined.

    $\cos \theta = -1$ $\sec \theta = -1$

    $\tan \theta = 0$ $\cot \theta$ is undefined.

25. $\sin \theta = -\dfrac{2\sqrt{5}}{5}$ $\csc \theta = \dfrac{-\sqrt{5}}{2}$

    $\cos \theta = -\dfrac{\sqrt{5}}{5}$ $\sec \theta = -\sqrt{5}$

    $\tan \theta = 2$ $\cot \theta = \frac{1}{2}$

27. (a) $\theta' = 23°$ (b) $\theta' = 53°$

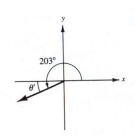
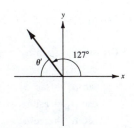

29. (a) $\theta' = 65°$ (b) $\theta' = 72°$

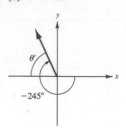
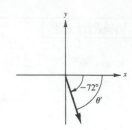

31. (a) $\theta' = \dfrac{\pi}{3}$ (b) $\theta' = \dfrac{\pi}{6}$

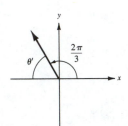
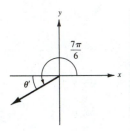

33. (a) $\theta' = 3.5 - \pi$ (b) $\theta' = 2\pi - 5.8$

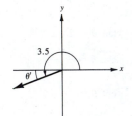
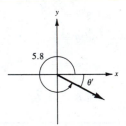

35. (a) $\sin \theta = -\dfrac{\sqrt{2}}{2}$, $\cos \theta = -\dfrac{\sqrt{2}}{2}$, $\tan \theta = 1$

   (b) $\sin \theta = \dfrac{\sqrt{2}}{2}$, $\cos \theta = -\dfrac{\sqrt{2}}{2}$, $\tan \theta = -1$

37. (a) $\sin \theta = \dfrac{1}{2}$, $\cos \theta = \dfrac{\sqrt{3}}{2}$, $\tan \theta = \dfrac{\sqrt{3}}{3}$

   (b) $\sin \theta = \dfrac{1}{2}$, $\cos \theta = -\dfrac{\sqrt{3}}{2}$, $\tan \theta = -\dfrac{\sqrt{3}}{3}$

39. (a) $\sin \theta = -\dfrac{\sqrt{3}}{2}$, $\cos \theta = -\dfrac{1}{2}$, $\tan \theta = \sqrt{3}$

   (b) $\sin \theta = \dfrac{\sqrt{3}}{2}$, $\cos \theta = -\dfrac{1}{2}$, $\tan \theta = -\sqrt{3}$

41. (a) $\sin \theta = -\dfrac{1}{2}$, $\cos \theta = \dfrac{\sqrt{3}}{2}$, $\tan \theta = -\dfrac{\sqrt{3}}{3}$

(b) $\sin \theta = \dfrac{1}{2}$, $\cos \theta = -\dfrac{\sqrt{3}}{2}$, $\tan \theta = -\dfrac{\sqrt{3}}{3}$

43. (a) $\sin \theta = \dfrac{\sqrt{2}}{2}$, $\cos \theta = -\dfrac{\sqrt{2}}{2}$, $\tan \theta = -1$

(b) $\sin \theta = -\dfrac{1}{2}$, $\cos \theta = \dfrac{\sqrt{3}}{2}$, $\tan \theta = -\dfrac{\sqrt{3}}{3}$

45. (a) 0.1736   (b) 5.7588
47. (a) 0.3640   (b) 0.3640
49. (a) -0.3420 (b) -0.3420
51. (a) 1.7321   (b) 1.7321
53. (a) $30° = \dfrac{\pi}{6}$, $150° = \dfrac{5\pi}{6}$

(b) $210° = \dfrac{7\pi}{6}$, $330° = \dfrac{11\pi}{6}$

55. (a) $60° = \dfrac{\pi}{3}$, $120° = \dfrac{2\pi}{3}$

(b) $135° = \dfrac{3\pi}{4}$, $315° = \dfrac{7\pi}{4}$

57. (a) $45° = \dfrac{\pi}{4}$, $225° = \dfrac{5\pi}{4}$

(b) $150° = \dfrac{5\pi}{6}$, $330° = \dfrac{11\pi}{6}$

59. (a) 54.99°, 125.01°   (b) 195°, 345°
61. (a) 0.175, 6.109   (b) 2.201, 4.083
63. (a) 0.873, 4.014   (b) 1.693, 4.835
65. 1   **67.** 0
69. (a) 25.2°F   (b) 65.1°F   (c) 50.8°F
71. (a) 0.25 ft   (b) 0.0177 ft   (c) -0.2475 ft

Section 7.4   *(page 414)*

WARM UP

1. 6π **2.** $\frac{1}{2}$ **3.** $\dfrac{\pi}{6}$ **4.** $\dfrac{7\pi}{6}$ **5.** -2

6. $-\frac{4}{3}$ **7.** 1   **8.** 0   **9.** 1   **10.** 0

1. Period: π
    Amplitude: 2

3. Period: 4π
    Amplitude: $\frac{3}{2}$

5. Period: 2
    Amplitude: $\frac{1}{2}$

7. Period: 2π
    Amplitude: 2

9. Period: $\dfrac{\pi}{5}$
    Amplitude: 2

11. Period: 3π
    Amplitude: $\frac{1}{2}$

13. Period: $\frac{1}{2}$
    Amplitude: 3

15. *Shift* graph of f π units to the right to obtain the graph of g.
17. *Reflect* the graph of f about the x-axis to obtain the graph of g.
19. The *period* of f is twice the period of g.
21. *Shift* the graph of f two units up to obtain the graph of g.

23.

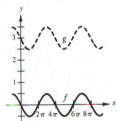

25.

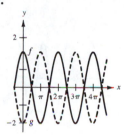

27.

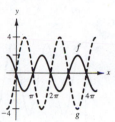

29.

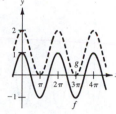

31.

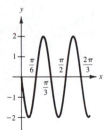

33.

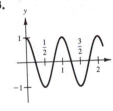

35.

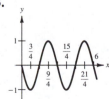

37.

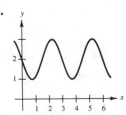

39.

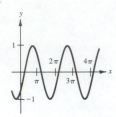

41.

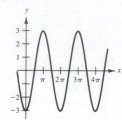

43.

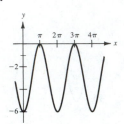

45.

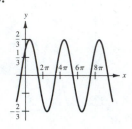

47.

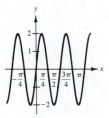

49.

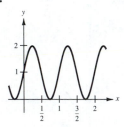

51.

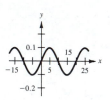

53.

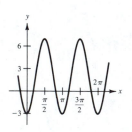

55.

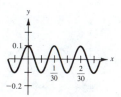

57. $y = 2 \sin 4x$

59. $y = \cos\left(2x + \dfrac{\pi}{2}\right)$

61.

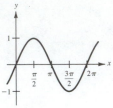

63.

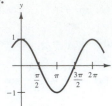

65. (a) $t = 6$ sec

(b) 10

(c)

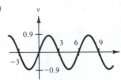

67. (a) $p = \dfrac{1}{440}$

(b) $f = 440$

(c)

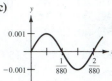

Section 7.5 *(page 425)*

WARM UP

1. 0   **2.** $\dfrac{\sqrt{2}}{2}$ **3.** 1   **4.** 0   **5.** 0   **6.** 0

7.

8.

9.

10.

1. c, π **2.** g, $\dfrac{\pi}{3}$ **3.** e, 2π **4.** a, 4π **5.** d, 1

6. h, 2   **7.** b, 2π **8.** f, 1

9.

11.

13.

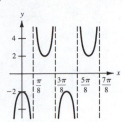

15.

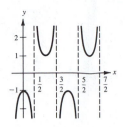

33.

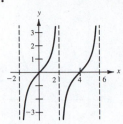

35.

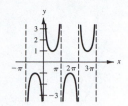

37. $d = 6 \cot x$

17.

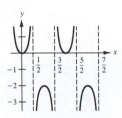

19.

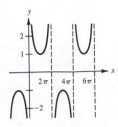

Section 7.6   (*page* 432)

WARM UP

1. $f(x) = -1: \dfrac{3\pi}{2}$

$f(x) = 0: 0, \pi, 2\pi$

$f(x) = 1: \dfrac{\pi}{2}$

2. $f(x) = -1: \pi$

$f(x) = 0: \dfrac{\pi}{2}, \dfrac{3\pi}{2}$

$f(x) = 1: 0, 2\pi$

3. $f(x) = -1: \dfrac{3\pi}{4}, \dfrac{7\pi}{4}$

$f(x) = 0: 0, \dfrac{\pi}{2}, \pi, \dfrac{3\pi}{2}, 2\pi$

$f(x) = 1: \dfrac{\pi}{4}, \dfrac{5\pi}{4}$

4. $f(x) = -1: 2\pi$

$f(x) = 0: \pi$

$f(x) = 1: 0$

21.

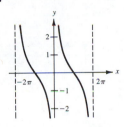

23.

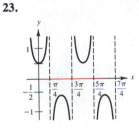

5.

6.

25.

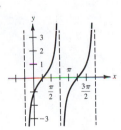

27.

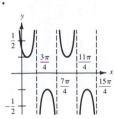

7.

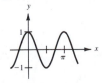

8.

29.

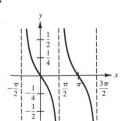

31.

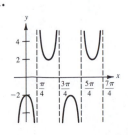

9. $0, \dfrac{\sqrt{3}\pi}{12}, \dfrac{\sqrt{2}\pi}{8}, \dfrac{\pi}{6}, 0$

10. $0, \dfrac{3 + \pi}{6}, \dfrac{2\sqrt{2} + \pi}{4}, \dfrac{3\sqrt{3} + 2\pi}{6}, \dfrac{\pi + 2}{2}$

1.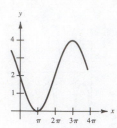

3.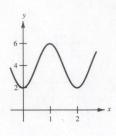

21.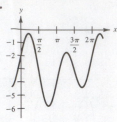

23.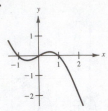

5.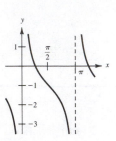

7.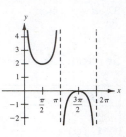

25.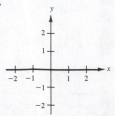

27.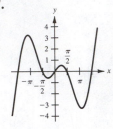

9.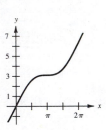

11.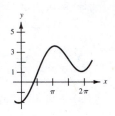

29.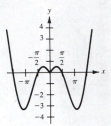

31.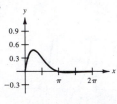

13.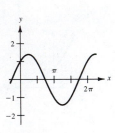

15.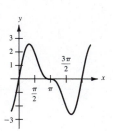

33.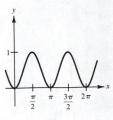

17.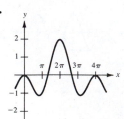

19.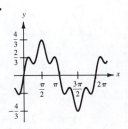

35.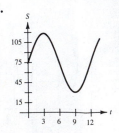

37.

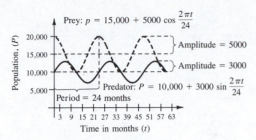

We can explain the cycles of this predator-prey population by noting the following cause and effect pattern:

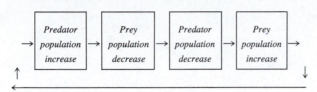

39.

x	-0.5	-0.4	-0.3	-0.2	-0.1
$\dfrac{1 - \cos x}{x}$	-0.245	-0.197	-0.149	-0.100	-0.050

x	0.1	0.2	0.3	0.4	0.5
$\dfrac{1 - \cos x}{x}$	0.050	0.100	0.149	0.197	0.245

$f(x)$ approaches 0 as x approaches 0.

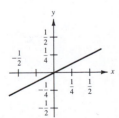

Section 7.7 (*page* 442)

WARM UP

1. -1 **2.** -1 **3.** -1 **4.** $\dfrac{\sqrt{2}}{2}$ **5.** 0

6. $\dfrac{\pi}{6}$ **7.** π **8.** $\dfrac{\pi}{4}$ **9.** 0 **10.** $\dfrac{\pi}{4}$

1. $\dfrac{\pi}{6}$ **3.** $\dfrac{\pi}{3}$ **5.** $\dfrac{\pi}{6}$ **7.** $\dfrac{5\pi}{6}$ **9.** $-\dfrac{\pi}{3}$ **11.** $\dfrac{2\pi}{3}$

13. $\dfrac{\pi}{3}$ **15.** 0 **17.** 1.29 **19.** -0.85

21. -1.11 **23.** 0.32 **25.** 1.99 **27.** 0.74

29. 0.3 **31.** -0.1 **33.** 0 **35.** $\dfrac{3}{5}$ **37.** $\dfrac{\sqrt{5}}{5}$

39. $\dfrac{12}{13}$ **41.** $\dfrac{\sqrt{34}}{5}$ **43.** $\dfrac{1}{x}$ **45.** $\sqrt{1 - 4x^2}$

47. $\sqrt{1 - x^2}$ **49.** $\dfrac{\sqrt{9 - x^2}}{x}$ **51.** $\dfrac{\sqrt{x^2 + 2}}{x}$

53. $\dfrac{9}{\sqrt{x^2 + 81}}$ **55.** $\dfrac{|x - 1|}{\sqrt{x^2 - 2x + 10}}$

57. 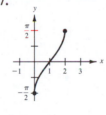 **59.**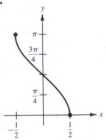

61. (a) $\beta \approx 40.6°$ **(b)** $\beta \approx 30.3°$

65. (a) $\dfrac{\pi}{4}$ **(b)** 0

Section 7.8 (*page* 452)

WARM UP

1. 8.45 **2.** 78.99 **3.** 1.06 **4.** 1.24
5. 4.88 **6.** 34.14 **7.** $4; \pi$ **8.** $\dfrac{1}{2}; 2$
9. $3; \dfrac{2}{3}$ **10.** $0.2; 8\pi$

1. $a \approx 3.64$
$c \approx 10.64$
$B = 70°$

3. $a \approx 8.26$
$c \approx 25.38$
$A = 19°$

5. $a \approx 91.34$
$b \approx 420.70$
$B = 77°45'$

7. $c = 2\sqrt{34}$
$A \approx 30.96°$
$B \approx 59.04°$

9. $a = 12\sqrt{17}$
$A \approx 72.08°$
$B \approx 17.92°$

11. 15.4 ft   **13.** 2.56 in

15. $56.3°$ **17.** $12.68°$ **19.** 19.9 ft
21. 508 miles north; 650 miles east   **23.** N 56.3° W
25. (a) N 58° E   (b) 68.8 yd   **27.** 1657.13 ft
29. 17,054 ft $\approx$ 3.23 mi   **31.** 29.389 in
33. $y = \sqrt{3}r$ **35.** $a \approx 7$, $c \approx 12.2$
37. (a) 4   (b) 4   (c) $t = \frac{1}{16}$
39. (a) $\frac{1}{16}$ (b) 60   (c) $t = \frac{1}{120}$
41. $\omega = 528\pi$

27. $\sin \theta = -\dfrac{3\sqrt{13}}{13}$ $\csc \theta = -\dfrac{\sqrt{13}}{3}$

$\cos \theta = -\dfrac{2\sqrt{13}}{13}$ $\sec \theta = -\dfrac{\sqrt{13}}{2}$

$\tan \theta = \frac{3}{2}$ $\cot \theta = \frac{2}{3}$

29. $\sin \theta = -\dfrac{\sqrt{11}}{6}$ $\cos \theta = \frac{5}{6}$

$\tan \theta = -\dfrac{\sqrt{11}}{5}$ $\cot \theta = -\dfrac{5\sqrt{11}}{11}$

$\csc \theta = -\dfrac{6\sqrt{11}}{11}$

31. $\cos \theta = -\dfrac{\sqrt{55}}{8}$ $\sec \theta = -\dfrac{8\sqrt{55}}{55}$

$\tan \theta = -\dfrac{3\sqrt{55}}{55}$ $\cot \theta = -\dfrac{\sqrt{55}}{3}$

$\csc \theta = \frac{8}{3}$

33. $-\dfrac{\sqrt{3}}{2}$ **35.** $-\dfrac{\sqrt{2}}{2}$ **37.** 0.65   **39.** 3.24

41. $135° = \dfrac{3\pi}{4}$, $225° = \dfrac{5\pi}{4}$

43. $210° = \dfrac{7\pi}{6}$, $330° = \dfrac{11\pi}{6}$

45. $57.0° \approx 0.9949$, $123.0° \approx 2.1470$
47. $165.0° \approx 2.8798$, $195.0° \approx 3.4034$

49. $\dfrac{\sqrt{-x^2 + 2x}}{-x^2 + 2x}$ **51.** $\dfrac{2\sqrt{4 - 2x^2}}{4 - x^2}$

53.

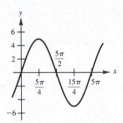

55.

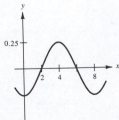

Chapter 7 Review Exercises   (*page* 456)

1. $\dfrac{3\pi}{4}$, $-\dfrac{5\pi}{4}$

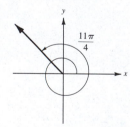

3. $250°$, $-470°$

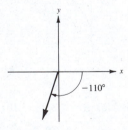

5. $135.28°$ **7.** $5.38°$ **9.** $135° \ 16' \ 12''$
11. $-85° \ 9'$ **13.** $128.57°$ **15.** $-200.54°$

17. 8.3776   **19.** -0.5890 **21.** $\dfrac{\pi}{5}$ **23.** $72°$

25. $\sin \theta = \dfrac{2\sqrt{53}}{53}$ $\csc \theta = \dfrac{\sqrt{53}}{2}$

$\cos \theta = -\dfrac{7\sqrt{53}}{53}$ $\sec \theta = -\dfrac{\sqrt{53}}{7}$

$\tan \theta = -\frac{2}{7}$ $\cot \theta = -\frac{7}{2}$

57.

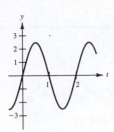

59.

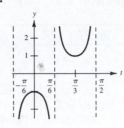

61.

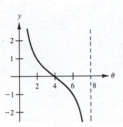

63.

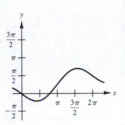

65.

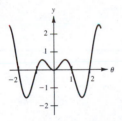

67.

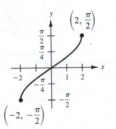

69.

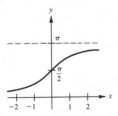

71. 1.33 mi

73. 268.8 ft

CHAPTER 8
Section 8.1 (*page* 465)

WARM UP

1. $\sin \theta = \dfrac{3\sqrt{13}}{13}$ $\qquad$ $\sec \theta = \dfrac{\sqrt{13}}{2}$

$\cos \theta = \dfrac{2\sqrt{13}}{13}$ $\qquad$ $\cot \theta = \frac{2}{3}$

$\csc \theta = \dfrac{\sqrt{13}}{3}$

2. $\sin \theta = \dfrac{2\sqrt{2}}{3}$ $\qquad$ $\csc \theta = \dfrac{3\sqrt{2}}{4}$

$\cos \theta = \frac{1}{3}$ $\qquad$ $\cot \theta = \dfrac{\sqrt{2}}{4}$

$\tan \theta = 2\sqrt{2}$

3. $\sin \theta = -\dfrac{3\sqrt{58}}{58}$ $\qquad$ $\csc \theta = -\dfrac{\sqrt{58}}{3}$

$\cos \theta = \dfrac{7\sqrt{58}}{58}$ $\qquad$ $\sec \theta = \dfrac{\sqrt{58}}{7}$

$\tan \theta = -\frac{3}{7}$ $\qquad$ $\cot \theta = -\frac{7}{3}$

4. $\sin \theta = \dfrac{\sqrt{5}}{5}$ $\qquad$ $\csc \theta = \sqrt{5}$

$\cos \theta = -\dfrac{2\sqrt{5}}{5}$ $\qquad$ $\sec \theta = -\dfrac{\sqrt{5}}{2}$

$\tan \theta = -\frac{1}{2}$ $\qquad$ $\cot \theta = -2$

5. $\frac{1}{2}$ $\quad$ **6.** $\frac{5}{4}$ $\quad$ **7.** $\dfrac{\sqrt{73}}{8}$ $\quad$ **8.** $\frac{2}{3}$

9. $\dfrac{x^2 + x + 16}{4(x + 1)}$ $\qquad$ **10.** $\dfrac{8x - 2}{1 - x^2}$

1. $\sin x = \frac{1}{2}$ $\qquad$ $\cot x = \sqrt{3}$

$\cos x = \dfrac{\sqrt{3}}{2}$ $\qquad$ $\sec x = \dfrac{2\sqrt{3}}{3}$

$\tan x = \dfrac{\sqrt{3}}{3}$ $\qquad$ $\csc x = 2$

3. $\sec \theta = \sqrt{2}$ $\qquad$ $\tan \theta = -1$

$\sin \theta = -\dfrac{\sqrt{2}}{2}$ $\qquad$ $\cot \theta = -1$

$\cos \theta = \dfrac{\sqrt{2}}{2}$ $\qquad$ $\csc \theta = -\sqrt{2}$

5. $\tan x = \frac{5}{12}$ $\sin x = -\frac{5}{13}$

$\sec x = -\frac{13}{12}$ $\cot x = \frac{12}{5}$

$\cos x = -\frac{12}{13}$ $\csc x = -\frac{13}{5}$

7. $\sec \phi = -1$ $\tan \phi = 0$

$\sin \phi = 0$ $\cot \phi$ is undefined.

$\cos \phi = -1$ $\csc \phi$ is undefined.

9. $\sin x = \frac{2}{3}$ $\cot x = -\frac{\sqrt{5}}{2}$

$\tan x = -\frac{2\sqrt{5}}{5}$ $\sec x = -\frac{3\sqrt{5}}{5}$

$\cos x = -\frac{\sqrt{5}}{3}$ $\csc x = \frac{3}{2}$

11. $\tan \theta = 2$ $\sin \theta = -\frac{2\sqrt{5}}{5}$

$\sec \theta = -\sqrt{5}$ $\csc \theta = -\frac{\sqrt{5}}{2}$

$\cos \theta = -\frac{\sqrt{5}}{5}$ $\cot \theta = \frac{1}{2}$

13. $\sin \theta = -1$ $\csc \theta = -1$

$\cot \theta = 0$ $\tan \theta$ is undefined.

$\cos \theta = 0$ $\sec \theta$ is undefined.

15. d   **16.** e   **17.** a   **18.** f   **19.** b   **20.** c
21. b   **22.** c   **23.** e   **24.** a   **25.** f   **26.** d
27. $\sec \phi$ **29.** $\sin \beta$ **31.** $\cos x$ **33.** 1
35. $-\tan x$ **37.** $\tan x$ **39.** $1 + \sin y$ **41.** $\sin^2 x$
43. $\sin^2 x \tan^2 x$ **45.** $\sec^4 x$ **47.** $\sin^2 x - \cos^2 x$
49. $1 + 2 \sin x \cos x$ **51.** $\tan^2 x$ **53.** $2 \csc^2 x$
55. $2 \sec x$ **57.** $1 + \cos y$ **59.** $3(\sec x + \tan x)$
61. $5 \cos \theta$ **63.** $3 \tan \theta$ **65.** $5 \sec \theta$ **67.** $\cos \theta$

69. $27 \sec^3 \theta$ **71.** $0 \le \theta < \frac{\pi}{2}, \frac{3\pi}{2} < \theta < 2\pi$

73. $\ln |\cot \theta|$ **75.** False, $\dfrac{\sin k\theta}{\cos k\theta} = \tan k\theta$

77. True, provided $\sin \theta \ne 0$

79. (a) $\csc^2 132° - \cot^2 132° \approx 1.8107 - 0.8107 = 1$
  (b) $1.6360 - 0.6360 = 1$

81. (a) $\cos (90° - 80°) = \sin 80° \approx 0.9848$

  (b) $\cos\left(\frac{\pi}{2} - 0.8\right) = \sin 0.8 \approx 0.7174$

83. $\cos \theta = \pm\sqrt{1 - \sin^2 \theta}$ $\sec \theta = \pm\dfrac{1}{\sqrt{1 - \sin^2 \theta}}$

$\tan \theta = \pm\dfrac{\sin \theta}{\sqrt{1 - \sin^2 \theta}}$ $\csc \theta = \dfrac{1}{\sin \theta}$

$\cot \theta = \pm\dfrac{\sqrt{1 - \sin^2 \theta}}{\sin \theta}$

Section 8.2   *(page 473)*

WARM UP

1. (a) $x^2(1 - y^2)$ (b) $\sin^4 x$
2. (a) $x^2(1 + y^2)$ (b) 1
3. (a) $(x^2 + 1)(x^2 - 1)$ (b) $\sec^2 x(\tan^2 x - 1)$
4. (a) $(z + 1)(z^2 - z + 1)$
  (b) $(\tan x + 1)(\tan^2 x - \tan x + 1)$
5. (a) $(x - 1)(x^2 + 1)$ (b) $(\cot x - 1) \csc^2 x$
6. (a) $(x^2 - 1)^2$ (b) $\cos^4 x$
7. (a) $\dfrac{y^2 - x^2}{x}$ (b) $\tan x$
8. (a) $\dfrac{x^2 - 1}{x^2}$ (b) $\sin^2 x$
9. (a) $\dfrac{y^2 + (1 + z)^2}{y(1 + z)}$ (b) $2 \csc x$
10. (a) $\dfrac{y(1 + y) - z^2}{z(1 + y)}$ (b) $\dfrac{\tan x - 1}{\sec x(1 + \tan x)}$

61. $\theta = -\dfrac{\pi}{3}; -0.8660 \ne 0.8660$ **63.** $x = -\dfrac{\pi}{4}; 1 \ne -1$

Section 8.3   *(page 483)*

WARM UP

1. $\dfrac{2\pi}{3}, \dfrac{4\pi}{3}$ **2.** $\dfrac{\pi}{3}, \dfrac{2\pi}{3}$ **3.** $\dfrac{\pi}{4}, \dfrac{7\pi}{4}$

4. $\dfrac{7\pi}{4}, \dfrac{5\pi}{4}$ **5.** $\dfrac{\pi}{3}$ **6.** $\dfrac{3\pi}{4}$ **7.** $\dfrac{15}{8}$ **8.** $-3, \dfrac{5}{2}$

9. $\dfrac{2 \pm \sqrt{14}}{2}$ **10.** $-1, 3$

7. $\dfrac{2\pi}{3} + 2n\pi, \dfrac{4\pi}{3} + 2n\pi$ **9.** $\dfrac{\pi}{3} + 2n\pi, \dfrac{2\pi}{3} + 2n\pi$

11. $\dfrac{\pi}{4} + \dfrac{n\pi}{2}$ **13.** $\dfrac{\pi}{6} + n\pi, \dfrac{5\pi}{6} + n\pi$

15. $n\pi, \dfrac{\pi}{4} + n\pi$ **17.** $n\pi, \dfrac{3\pi}{2} + 2n\pi$

19. $\dfrac{\pi}{2} + n\pi, \dfrac{2\pi}{3} + 2n\pi, \dfrac{4\pi}{3} + 2n\pi$ **21.** $\dfrac{\pi}{3}, \dfrac{5\pi}{3}$

23. $\dfrac{7\pi}{6}, \dfrac{3\pi}{2}, \dfrac{11\pi}{6}$ **25.** $0, \dfrac{\pi}{2}, \pi, \dfrac{3\pi}{2}$

27. $\dfrac{\pi}{2}, \dfrac{2\pi}{3}, \dfrac{3\pi}{2}, \dfrac{4\pi}{3}$ **29.** $\dfrac{\pi}{6}, \dfrac{5\pi}{6}, \dfrac{7\pi}{6}, \dfrac{11\pi}{6}$

31. No solution **33.** $\dfrac{2\pi}{3}, \dfrac{5\pi}{6}, \dfrac{5\pi}{3}, \dfrac{11\pi}{6}$ **35.** $\dfrac{\pi}{2}$

37. π **39.** $\dfrac{\pi}{3}, \dfrac{5\pi}{3}$

41. 0.9828, 1.7682, 4.1244, 4.9098

43. 0.3398, 0.8481, 2.2935, 2.8018 **45.** 0.8411, 5.4421

47. 1.1555, 1.3981, 4.2971, 4.5397 **49.** 0.4271, 2.7145

51. $\dfrac{\pi}{4}, \dfrac{5\pi}{4}$ **53.** 0.04, 0.43, 0.83

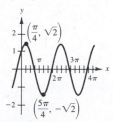

55. 37°

Section 8.4 *(page 492)*

WARM UP

1. $\dfrac{\sqrt{10}}{10}$ **2.** $\dfrac{-5\sqrt{34}}{34}$ **3.** $-\dfrac{\sqrt{7}}{4}$ **4.** $\dfrac{2\sqrt{2}}{3}$

5. $\dfrac{\pi}{4} + 2n\pi, \dfrac{3\pi}{4} + 2n\pi$ **6.** $\dfrac{\pi}{2} + n\pi$

7. $\tan^3 x$ **8.** $\cot^2 x$ **9.** $\sec x$

10. $(2\cos^2 x - 1)(\cos^2 x - 1)$

1. $\sin 75° = \dfrac{\sqrt{2}}{4}(1 + \sqrt{3})$

$\cos 75° = \dfrac{\sqrt{2}}{4}(\sqrt{3} - 1)$

$\tan 75° = \sqrt{3} + 2$

3. $\sin 105° = \dfrac{\sqrt{2}}{4}(\sqrt{3} + 1)$

$\cos 105° = \dfrac{\sqrt{2}}{4}(1 - \sqrt{3})$

$\tan 105° = -2 - \sqrt{3}$

5. $\sin 195° = \dfrac{\sqrt{2}}{4}(1 - \sqrt{3})$

$\cos 195° = -\dfrac{\sqrt{2}}{4}(\sqrt{3} + 1)$

$\tan 195° = 2 - \sqrt{3}$

7. $\sin \dfrac{11\pi}{12} = \dfrac{\sqrt{2}}{4}(\sqrt{3} - 1)$

$\cos \dfrac{11\pi}{12} = -\dfrac{\sqrt{2}}{4}(\sqrt{3} + 1)$

$\tan \dfrac{11\pi}{12} = -2 + \sqrt{3}$

9. $\sin \dfrac{17\pi}{12} = -\dfrac{\sqrt{2}}{4}(\sqrt{3} + 1)$ **11.** $\cos 40°$

$\cos \dfrac{17\pi}{12} = \dfrac{\sqrt{2}}{4}(1 - \sqrt{3})$

$\tan \dfrac{17\pi}{12} = 2 + \sqrt{3}$

13. $\sin 200°$ **15.** $\tan 239°$ **17.** $\sin 1.8$

19. $\tan 3x$ **21.** $\dfrac{33}{65}$ **23.** $-\dfrac{56}{65}$ **25.** $-\dfrac{3}{5}$ **27.** $\dfrac{44}{125}$

49. (a) $\sqrt{2} \sin\left(\theta + \dfrac{\pi}{4}\right)$ (b) $\sqrt{2} \cos\left(\theta - \dfrac{\pi}{4}\right)$

51. (a) $13 \sin(3\theta + 0.3948)$ (b) $13 \cos(3\theta - 1.1760)$

53. $\sqrt{2} \sin \theta + \sqrt{2} \cos \theta$ **55.** 1 **57.** $\dfrac{\pi}{2}$

59. $\dfrac{\pi}{4}, \dfrac{7\pi}{4}$ **61.** $0, \dfrac{\pi}{3}, \pi, \dfrac{5\pi}{3}$

Section 8.5 *(page 502)*

WARM UP

1. $\sin x(2 + \cos x)$ **2.** $(\cos x - 2)(\cos x + 1)$

3. $0, \dfrac{\pi}{2}, \pi, \dfrac{3\pi}{2}$ **4.** $\dfrac{\pi}{4}, \dfrac{3\pi}{4}, \dfrac{5\pi}{4}, \dfrac{7\pi}{4}$ **5.** π

6. 0 **7.** $\dfrac{2 - \sqrt{2}}{4}$ **8.** $\dfrac{3}{4}$ **9.** $\tan 3x$

10. $\cos x(1 - 4\sin^2 x)$

1. $\sin 90° = 1$ **3.** $\sin 60° = \dfrac{\sqrt{3}}{2}$

$\cos 90° = 0$ $\cos 60° = \dfrac{1}{2}$

$\tan 90°$ is undefined. $\tan 60° = \sqrt{3}$

5. $\sin \dfrac{2\pi}{3} = \dfrac{\sqrt{3}}{2}$ **7.** $\sin 2u = \dfrac{24}{25}$

$\cos \dfrac{2\pi}{3} = -\dfrac{1}{2}$ $\cos 2u = \dfrac{7}{25}$

$\tan \dfrac{2\pi}{3} = -\sqrt{3}$ $\tan 2u = \dfrac{24}{7}$

9. $\sin 2u = \dfrac{4}{5}$

$\cos 2u = \dfrac{3}{5}$

$\tan 2u = \dfrac{4}{3}$

11. $\sin 2u = -\dfrac{4\sqrt{21}}{25}$

$\cos 2u = -\dfrac{17}{25}$

$\tan 2u = \dfrac{4\sqrt{21}}{17}$

13. $\sin 105° = \dfrac{1}{2}\sqrt{2 + \sqrt{3}}$

$\cos 105° = -\dfrac{1}{2}\sqrt{2 - \sqrt{3}}$

$\tan 105° = -2 - \sqrt{3}$

15. $\sin 112° \, 30' = \dfrac{1}{2}\sqrt{2 + \sqrt{2}}$

$\cos 112° \, 30' = -\dfrac{1}{2}\sqrt{2 - \sqrt{2}}$

$\tan 112° \, 30' = -1 - \sqrt{2}$

17. $\sin \dfrac{\pi}{8} = \dfrac{1}{2}\sqrt{2 - \sqrt{2}}$

$\cos \dfrac{\pi}{8} = \dfrac{1}{2}\sqrt{2 + \sqrt{2}}$

$\tan \dfrac{\pi}{8} = \sqrt{2} - 1$

19. $\sin \dfrac{u}{2} = \dfrac{5\sqrt{26}}{26}$

$\cos \dfrac{u}{2} = \dfrac{\sqrt{26}}{26}$

$\tan \dfrac{u}{2} = 5$

21. $\sin \dfrac{u}{2} = \sqrt{\dfrac{\sqrt{89} - 8}{2\sqrt{89}}}$

$\cos \dfrac{u}{2} = -\sqrt{\dfrac{\sqrt{89} + 8}{2\sqrt{89}}}$

$\tan \dfrac{u}{2} = \dfrac{8 - \sqrt{89}}{5}$

23. $\sin \dfrac{u}{2} = \dfrac{3\sqrt{10}}{10}$

$\cos \dfrac{u}{2} = -\dfrac{\sqrt{10}}{10}$

$\tan \dfrac{u}{2} = -3$

25. $\sin 3x$ **27.** $-\tan 4x$

29. (a) $\dfrac{1}{8}(3 + 4\cos 2x + \cos 4x)$

(b) $\dfrac{1}{32}(2 + \cos 2x - 2\cos 4x - \cos 6x)$

31. $3(\sin \dfrac{\pi}{2} + \sin 0)$ **33.** $\dfrac{1}{2}(\sin 8\theta + \sin 2\theta)$

35. $\dfrac{5}{2}(\cos 8\beta + \cos 2\beta)$

37. $\dfrac{1}{2}(\cos 2y - \cos 2x)$

39. $\dfrac{1}{2}[\sin 2\theta + \sin 2\pi]$ **41.** $2\sin 45° \cos 15°$

43. $-2\sin \dfrac{\pi}{2} \sin \dfrac{\pi}{4}$ **45.** $2\cos 4x \cos 2x$

47. $2\cos \alpha \sin \beta$ **49.** $2\cos(\phi + \pi)\cos \pi$

73. $\dfrac{\pi}{12}, \dfrac{5\pi}{12}, \dfrac{13\pi}{12}, \dfrac{17\pi}{12}$ **75.** $0, \dfrac{2\pi}{3}, \dfrac{4\pi}{3}$

77. $0, \dfrac{\pi}{2}, \pi, \dfrac{3\pi}{2}$ **79.** $0, \dfrac{\pi}{4}, \dfrac{\pi}{2}, \dfrac{3\pi}{4}, \pi, \dfrac{5\pi}{4}, \dfrac{3\pi}{2}, \dfrac{7\pi}{4}$

81. $0, \dfrac{\pi}{4}, \dfrac{\pi}{2}, \dfrac{3\pi}{4}, \pi, \dfrac{5\pi}{4}, \dfrac{3\pi}{2}, \dfrac{7\pi}{4}$ **83.** $\dfrac{\pi}{6}, \dfrac{5\pi}{6}$

85.

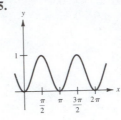

87. $2x\sqrt{1 - x^2}$

Chapter 8 Review Exercises   (*page* 504)

1. $\sin^2 x$ **3.** $1 + \cot \alpha$ **5.** 1   **7.** 1

9. $\cos^2 2x$ **41.** $\dfrac{\sqrt{2}}{4}(\sqrt{3} + 1)$ **43.** $-\dfrac{1}{2}\sqrt{2 + \sqrt{2}}$

45. $-\dfrac{3}{52}(5 + 4\sqrt{7})$ **47.** $\dfrac{1}{52}(36 + 5\sqrt{7})$

49. $\dfrac{1}{4}\sqrt{2}(4 - \sqrt{7})$ **51.** $0, \pi$ **53.** $\dfrac{\pi}{3}, \dfrac{5\pi}{3}$

55. $0, \dfrac{3\pi}{4}, \pi, \dfrac{5\pi}{4}$ **57.** $0, \dfrac{\pi}{2}, \pi$ **59.** $\dfrac{\pi}{4}, \dfrac{5\pi}{4}$

61. $2\cos \dfrac{5\theta}{2}\cos \dfrac{\theta}{2}$ **63.** $\dfrac{1}{2}(\cos \alpha - \cos 5\alpha)$

65.

x	-2.029	0	2.029
$\tan x + x$	-0.001	0	0.001
$x \sin x$	1.820	0	1.820

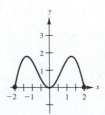

CHAPTER 9

Section 9.1   (*page* 513)

WARM UP

1. $b = 3\sqrt{3}$, $A = 30°$, $B = 60°$
2. $c = 5\sqrt{2}$, $A = 45°$, $B = 45°$
3. $a = 8$, $A \approx 28.07°$, $B \approx 61.93°$
4. $b \approx 8.33$, $c \approx 11.21$, $B = 48°$
5. $a \approx 22.69$, $c \approx 23.04$, $A = 80°$
6. $a \approx 45.73$, $b \approx 142.86$, $A = 17° \, 45'$
7. 8.48   8. 12.94   9. 2.25   10. 91.06

1. $C = 105°$, $b \approx 14.1$, $c \approx 19.3$
3. $C = 110°$, $b \approx 22.4$, $c \approx 24.4$
5. $B \approx 21.6°$, $C \approx 122.4°$, $c \approx 11.5$
7. $B = 10°$, $b \approx 69.5$, $c \approx 136.8$
9. $B = 42° \, 4'$, $a \approx 22.1$, $b \approx 14.9$
11. $A \approx 10° \, 11'$, $C \approx 154° \, 19'$, $c \approx 11.0$
13. $A \approx 25.6°$, $B \approx 9.4°$, $a \approx 10.5$
15. $B \approx 18° \, 13'$, $C \approx 51° \, 32'$, $c \approx 40.1$
17. No solution
19. Two solutions $B \approx 70.4°$, $C \approx 51.6°$, $c \approx 4.16$;
    $B \approx 109.6°$, $C \approx 12.4°$, $c \approx 1.14$
21. No solution
23. (a) $b \le 5$, $b = \dfrac{5}{\sin 36°}$ (b) $5 < b < \dfrac{5}{\sin 36°}$

    (c) $b > \dfrac{5}{\sin 36°}$ 25. 10.4   27. 1675.2
29. 474.9   31. 6.0 units   33. 77 yd   35. 5 mi
37. 26.1 mi, 15.9 mi   39. 4.55 mi   41. No

Section 9.2   (*page* 521)

WARM UP

1. $2\sqrt{13}$ 2. $3\sqrt{5}$ 3. $4\sqrt{10}$ 4. $3\sqrt{13}$
5. 20   6. 48   7. $a \approx 4.62$, $c \approx 26.20$, $B = 70°$
8. $a \approx 34.20$, $b \approx 93.97$, $B = 70°$ 9. No solution
10. $a \approx 15.09$, $B \approx 18.97°$, $C \approx 131.03°$

1. $A \approx 27.7°$, $B \approx 40.5°$, $C \approx 111.8°$
3. $B \approx 23.8°$, $C \approx 126.2°$, $a \approx 12.4$
5. $A \approx 36.9°$, $B \approx 53.1°$, $C = 90°$
7. $A \approx 92.9°$, $B \approx 43.55°$, $C \approx 43.55°$

9. $a \approx 11.79$, $B \approx 12.73°$, $C \approx 47.27°$
11. $A \approx 158°36'$, $C \approx 12°38'$, $b \approx 10.4$
13. $A = 27°10'$, $B = 27°10'$, $c \approx 56.9$ 15. 16.25
17. 54   19. 96.82   21. S 52°37' E, S 25°20' W
23. 43.3 mi   25. 116.35 ft, 133.09 ft   27. 114.95°
29. $\overline{PQ} \approx 9.4$ ft, $\overline{QS} \approx 5.0$ ft, $\overline{RS} \approx 12.8$ ft
31. (a) 63.7 ft   (b) 47.6 ft
33. (a) N 58.3° W   (b) S 81.6° W
35. (a) 570.60   (b) 5910.08   (c) 177.09

Section 9.3   (*page* 534)

WARM UP

1. $7\sqrt{10}$ 2. $\sqrt{58}$ 3. $3x + 5y - 14 = 0$
4. $4x - 3y - 1 = 0$ 5. 111.8°   6. 323.1°
7. $\dfrac{1}{2}, \dfrac{\sqrt{3}}{2}$ 8. $\dfrac{\sqrt{3}}{2}, -\dfrac{1}{2}$ 9. $-\dfrac{\sqrt{3}}{2}, \dfrac{1}{2}$

10. $-\dfrac{1}{2}, -\dfrac{\sqrt{3}}{2}$

1.

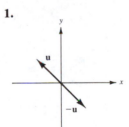

3.

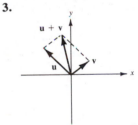

5.

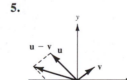

7. $\mathbf{v} = \langle 3, 4 \rangle$, $\|\mathbf{v}\| = 5$
9. $\mathbf{v} = \langle -3, 2 \rangle$, $\|\mathbf{v}\| = \sqrt{13}$
11. $\mathbf{v} = \langle 0, 5 \rangle$, $\|\mathbf{v}\| = 5$

13. $\mathbf{v} = \langle 16, -3 \rangle$, $\|\mathbf{v}\| = \sqrt{265}$
15. $\mathbf{v} = \langle 8, 4 \rangle$, $\|\mathbf{v}\| = 4\sqrt{5}$
17. (a) $\langle 4, 3 \rangle$ (b) $\langle -2, 1 \rangle$ (c) $\langle -7, 1 \rangle$
19. (a) $\langle -4, 4 \rangle$ (b) $\langle 0, 2 \rangle$ (c) $\langle 2, 3 \rangle$
21. (a) $\langle 4, -2 \rangle$ (b) $\langle 4, -2 \rangle$ (c) $\langle 8, -4 \rangle$
23. (a) $3\mathbf{i} - 2\mathbf{j}$ (b) $-\mathbf{i} + 4\mathbf{j}$ (c) $-4\mathbf{i} + 11\mathbf{j}$
25. (a) $2\mathbf{i} + \mathbf{j}$ (b) $2\mathbf{i} - \mathbf{j}$ (c) $4\mathbf{i} - 3\mathbf{j}$

27. $\mathbf{v} = \langle 3, 0 \rangle$

29. $\mathbf{v} = \left\langle -\dfrac{\sqrt{3}}{2}, \dfrac{1}{2} \right\rangle$

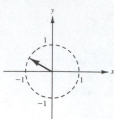

31. $\mathbf{v} = \left\langle -\dfrac{3\sqrt{6}}{2}, \dfrac{3\sqrt{2}}{2} \right\rangle$

33. $\mathbf{v} = \left\langle \dfrac{\sqrt{10}}{5}, \dfrac{3\sqrt{10}}{5} \right\rangle$

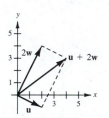

35. $\mathbf{v} = \left\langle 3, -\dfrac{3}{2} \right\rangle$

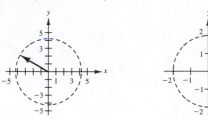

37. $\mathbf{v} = \langle 4, 3 \rangle$

39. $\mathbf{v} = \left\langle \dfrac{7}{2}, -\dfrac{1}{2} \right\rangle$

41. $\langle 5, 5 \rangle$

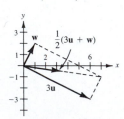

43. $\langle (10\sqrt{2} - 50), 10\sqrt{2} \rangle$ **45.** $\frac{4}{5}\mathbf{i} - \frac{3}{5}\mathbf{j}$ **47.** $\mathbf{j}$
49. $90°$ **51.** $63.4°$ **53.** 82.2 lb
55. $71.3°$, 228.5 lb **57.** 425 ft-lb
59. N $52.1°$ W, 569.5 mph **61.** N $25.2°$ E, 82.8 mph
63. $62.7°$

Section 9.4 (*page* 541)

WARM UP

1. $-5 - 10i$ **2.** $7 + 3\sqrt{6}i$ **3.** $-1 - 4i$
4. $-3i$ **5.** $6 - 14i$ **6.** $6 + 4\sqrt{2}i$
7. $-22 + 16i$ **8.** 13 **9.** $-\frac{3}{2} + \frac{5}{2}i$
10. $-\frac{5}{2} - \frac{3}{2}i$

1. $4\left(\cos\dfrac{\pi}{2} + i \sin\dfrac{\pi}{2}\right)$ **3.** $3\sqrt{2}\left(\cos\dfrac{5\pi}{4} + i \sin\dfrac{5\pi}{4}\right)$

5. $3\sqrt{2}\left(\cos\dfrac{7\pi}{4} + i \sin\dfrac{7\pi}{4}\right)$ **7.** $2\left(\cos\dfrac{\pi}{6} + i \sin\dfrac{\pi}{6}\right)$

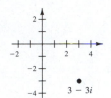

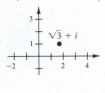

9. $4\left(\cos\dfrac{4\pi}{3} + i \sin\dfrac{4\pi}{3}\right)$ **11.** $6\left(\cos\dfrac{\pi}{2} + i \sin\dfrac{\pi}{2}\right)$

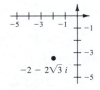

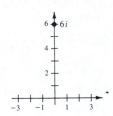

13. $\sqrt{65}(\cos 2.62 + i \sin 2.62)$

15. $7(\cos 0 + i \sin 0)$

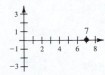

17. $\sqrt{37}(\cos 1.41 + i \sin 1.41)$

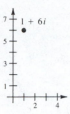

19. $\sqrt{10}(\cos 3.46 + i \sin 3.46)$

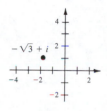

21. $-\sqrt{3} + i$

23. $\dfrac{3}{4} - \dfrac{3\sqrt{3}}{4}i$

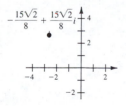

25. $\dfrac{-15\sqrt{2}}{8} + \dfrac{15\sqrt{2}}{8}i$

27. $-4i$

29. $2.8408 + 0.9643i$

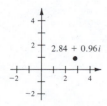

31. $12(\cos 90° + i \sin 90°)$
33. $\dfrac{10}{9}(\cos 200° + i \sin 200°)$
35. $0.27(\cos 150° + i \sin 150°)$

39. $\cos \dfrac{2\pi}{3} + i \sin \dfrac{2\pi}{3}$
41. $4[\cos(-58°) + i \sin(-58°)]$
43. $4(\cos 0° + i \sin 0°) = 4$
45. $2\sqrt{2}\left(\cos \dfrac{7\pi}{4} + i \sin \dfrac{7\pi}{4}\right) = 2 - 2i$
47. $\dfrac{5\sqrt{13}}{13}[\cos(-56.3°) + i \sin(-56.3°)] \approx \dfrac{5}{13}(2 - 3i)$
51. (a) r^2 (b) $\cos 2\theta + i \sin 2\theta$
53.

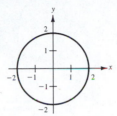

Section 9.5   *(page 548)*

WARM UP

1. $3\sqrt[3]{2}$ **2.** $2\sqrt{2}$
3. $5\sqrt{2}(\cos 135° + i \sin 135°)$
4. $3(\cos 270° + i \sin 270°)$
5. $12(\cos 180° + i \sin 180°)$
6. $12(\cos 0° + i \sin 0°)$
7. $\cos \dfrac{3\pi}{4} + i \sin \dfrac{3\pi}{4}$ **8.** $\cos \dfrac{11\pi}{12} + i \sin \dfrac{11\pi}{12}$
9. $2\left(\cos \dfrac{\pi}{2} + i \sin \dfrac{\pi}{2}\right)$ **10.** $\dfrac{2}{3}(\cos 45° + i \sin 45°)$

1. $-4 - 4i$ **3.** $-32i$ **5.** $-128\sqrt{3} - 128i$
7. $\dfrac{125}{2} + \dfrac{125\sqrt{3}}{2}i$ **9.** i
11. $608.02 + 144.69i$

13. $3\left(\cos\dfrac{\pi}{3} + i\sin\dfrac{\pi}{3}\right) = \dfrac{3}{2} + \dfrac{3\sqrt{3}}{2}i$

$3\left(\cos\dfrac{4\pi}{3} + i\sin\dfrac{4\pi}{3}\right) = -\dfrac{3}{2} - \dfrac{3\sqrt{3}}{2}i$

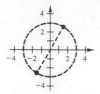

15. $2\left(\cos\dfrac{\pi}{3} + i\sin\dfrac{\pi}{3}\right) = 1 + \sqrt{3}i$

$2\left(\cos\dfrac{5\pi}{6} + i\sin\dfrac{5\pi}{6}\right) = -\sqrt{3} + i$

$2\left(\cos\dfrac{4\pi}{3} + i\sin\dfrac{4\pi}{3}\right) = -1 - \sqrt{3}i$

$2\left(\cos\dfrac{11\pi}{6} + i\sin\dfrac{11\pi}{6}\right) = \sqrt{3} - i$

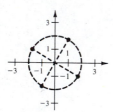

17. $5\left(\cos\dfrac{3\pi}{4} + i\sin\dfrac{3\pi}{4}\right) = -\dfrac{5\sqrt{2}}{2} + \dfrac{5\sqrt{2}}{2}i$

$5\left(\cos\dfrac{7\pi}{4} + i\sin\dfrac{7\pi}{4}\right) = \dfrac{5\sqrt{2}}{2} - \dfrac{5\sqrt{2}}{2}i$

19. $5\left(\cos\dfrac{4\pi}{9} + i\sin\dfrac{4\pi}{9}\right) = 0.8682 + 4.9240i$

$5\left(\cos\dfrac{10\pi}{9} + i\sin\dfrac{10\pi}{9}\right) = -4.6985 - 1.7101i$

$5\left(\cos\dfrac{16\pi}{9} + i\sin\dfrac{16\pi}{9}\right) = 3.8302 - 3.2139i$

21. $2(\cos 0 + i\sin 0) = 2$

$2\left(\cos\dfrac{2\pi}{3} + i\sin\dfrac{2\pi}{3}\right) = -1 + \sqrt{3}i$

$2\left(\cos\dfrac{4\pi}{3} + i\sin\dfrac{4\pi}{3}\right) = -1 - \sqrt{3}i$

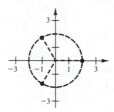

23. $\cos 0 + i\sin 0 = 1$

$\cos\dfrac{2\pi}{5} + i\sin\dfrac{2\pi}{5} = 0.3090 + 0.9510i$

$\cos\dfrac{4\pi}{5} + i\sin\dfrac{4\pi}{5} = -0.8090 + 0.5878i$

$\cos\dfrac{6\pi}{5} + i\sin\dfrac{6\pi}{5} = -0.8090 - 0.5878i$

$\cos\dfrac{8\pi}{5} + i\sin\dfrac{8\pi}{5} = 0.3090 - 0.9510i$

25. $\cos \dfrac{\pi}{8} + i \sin \dfrac{\pi}{8}$

$\cos \dfrac{5\pi}{8} + i \sin \dfrac{5\pi}{8}$

$\cos \dfrac{9\pi}{8} + i \sin \dfrac{9\pi}{8}$

$\cos \dfrac{13\pi}{8} + i \sin \dfrac{13\pi}{8}$

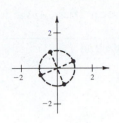

27. $3\left(\cos \dfrac{\pi}{5} + i \sin \dfrac{\pi}{5}\right)$

$3\left(\cos \dfrac{3\pi}{5} + i \sin \dfrac{3\pi}{5}\right)$

$3(\cos \pi + i \sin \pi)$

$3\left(\cos \dfrac{7\pi}{5} + i \sin \dfrac{7\pi}{5}\right)$

$3\left(\cos \dfrac{9\pi}{5} + i \sin \dfrac{9\pi}{5}\right)$

29. $4\left(\cos \dfrac{\pi}{2} + i \sin \dfrac{\pi}{2}\right)$

$4\left(\cos \dfrac{7\pi}{6} + i \sin \dfrac{7\pi}{6}\right)$

$4\left(\cos \dfrac{11\pi}{6} + i \sin \dfrac{11\pi}{6}\right)$

31. $\sqrt[6]{2}(\cos 105° + i \sin 105°)$

$\sqrt[6]{2}(\cos 225° + i \sin 225°)$

$\sqrt[6]{2}(\cos 345° + i \sin 345°)$

Chapter 9 Review Exercises (*page 549*)

1. $A \approx 29.7°$, $B \approx 52.4°$, $C \approx 97.9°$

3. $C = 110°$, $b \approx 20.4$, $c \approx 22.6$

5. $A = 35°$, $C = 35°$, $b \approx 6.6$ **7.** No solution

9. $A \approx 25.9°$, $C \approx 39.1°$, $c \approx 10.1$

11. $B \approx 31.2°$, $C \approx 133.8°$, $c \approx 13.9$;

$B \approx 148.8°$, $C \approx 16.2°$, $c \approx 5.38$

13. $A \approx 9.9°$, $C \approx 20.1°$, $b \approx 29.1$

15. $A \approx 40.9°$, $C \approx 114.1°$, $c \approx 8.6$;

$A \approx 139.1°$, $C \approx 15.9°$, $c \approx 2.6$

17. $4\sqrt{6}$ **19.** 9.08

21. 31 ft   **23.** 31.1 m   **25.** 1135 mi

27. $\langle 7, -7 \rangle$ **29.** $\langle -4, 4\sqrt{3} \rangle$

31. $\left\langle \dfrac{6}{\sqrt{61}}, -\dfrac{5}{\sqrt{61}} \right\rangle$ **33.** $\langle -26, -35 \rangle$

35. 133.92 lb   **37.** 92.3°, 117.0

39. 460.3 mph, N 32.2° E

41. $5\sqrt{2}(\cos 315° + i \sin 315°)$

43. $13(\cos 67.38° + i \sin 67.38°)$

45. $-50 - 50\sqrt{3}i$

47. (a) $z_1 = 5(\cos \pi + i \sin \pi)$, $z_2 = 5\left(\cos \dfrac{\pi}{2} + i \sin \dfrac{\pi}{2}\right)$

(b) $z_1 z_2 = 25\left(\cos \dfrac{3\pi}{2} + i \sin \dfrac{3\pi}{2}\right)$,

$\dfrac{z_1}{z_2} = \cos \dfrac{\pi}{2} + i \sin \dfrac{\pi}{2}$

49. (a) $z_1 = 3\sqrt{2}\left(\cos \dfrac{5\pi}{4} + i \sin \dfrac{5\pi}{4}\right)$

$z_2 = 4\left(\cos \dfrac{\pi}{6} + i \sin \dfrac{\pi}{6}\right)$

(b) $z_1 z_2 = 12\sqrt{2}\left(\cos \dfrac{17\pi}{12} + i \sin \dfrac{17\pi}{12}\right)$

$\dfrac{z_1}{z_2} = \dfrac{3\sqrt{2}}{4}\left(\cos \dfrac{13\pi}{12} + i \sin \dfrac{13\pi}{12}\right)$

51. $\dfrac{5^4}{2} + \dfrac{5^4 \sqrt{3}}{2}i$ **53.** $2035 - 828i$

55. $3\left(\cos \dfrac{\pi}{4} + i \sin \dfrac{\pi}{4}\right)$

$3\left(\cos \dfrac{7\pi}{12} + i \sin \dfrac{7\pi}{12}\right)$

$3\left(\cos \dfrac{11\pi}{12} + i \sin \dfrac{11\pi}{12}\right)$

$3\left(\cos \dfrac{5\pi}{4} + i \sin \dfrac{5\pi}{4}\right)$

$3\left(\cos \dfrac{19\pi}{12} + i \sin \dfrac{19\pi}{12}\right)$

$3\left(\cos \dfrac{23\pi}{12} + i \sin \dfrac{23\pi}{12}\right)$

57. $\cos \dfrac{\pi}{3} + i \sin \dfrac{\pi}{3} = \dfrac{1}{2} + \dfrac{\sqrt{3}}{2}i$

$\cos \pi + i \sin \pi = -1$

$\cos \dfrac{5\pi}{3} + i \sin \dfrac{5\pi}{3} = \dfrac{1}{2} - \dfrac{\sqrt{3}}{2}i$

CHAPTER 10

Section 10.1 *(page 559)*

WARM UP

1.

2.

3.

4.

5. x **6.** $-37v$ **7.** $2x^2 + 9$ **8.** -1
9. $x = 6$ **10.** $y = 1$

1. $(1, 2)$ **3.** $(-1, 2), (2, 5)$ **5.** $(0, 5), (3, 4)$
7. $(0, 0), (2, 4)$ **9.** $(-1, 1), (8, 4)$ **11.** $(5, 5)$
13. $\left(\dfrac{1}{2}, 3\right)$ **15.** $(1.5, 0.3)$ **17.** $\left(\dfrac{20}{3}, \dfrac{40}{3}\right)$ **19.** $(0, 0)$
21. $(1, 2)$ **23.** $\left(\dfrac{29}{10}, \dfrac{21}{10}\right), (-2, 0)$
25. $(-1, -2), (2, 1)$ **27.** $(-1, 0), (0, 1), (1, 0)$
29. $\left(\dfrac{1}{2}, 2\right), \left(-4, -\dfrac{1}{4}\right)$ **31.** $(2, 2), (4, 0)$
33. No points of intersection **35.** $(0, 1)$
37. $\left(\dfrac{\pi}{2}, 0\right), \left(\dfrac{3\pi}{2}, 0\right), \left(\dfrac{\pi}{6}, \dfrac{\sqrt{3}}{2}\right), \left(\dfrac{5\pi}{6}, -\dfrac{\sqrt{3}}{2}\right)$
39. $x = 5, y = 5, \lambda = -5$
41. $x = 0, \pm\dfrac{\sqrt{2}}{2}; y = 0, \dfrac{1}{2}; \lambda = 0, 1$ **43.** 193 units
45. 6400 units **47.** 8, 12 **49.** 8 mi $\times$ 12 mi
51. $48\sqrt{2}$ ft/sec, 3 **53.** 68.71 ft/sec, 68.7°

Section 10.2 *(page 566)*

WARM UP

1.

2.

3. $x - y + 4 = 0$ **4.** $5x + 3y - 28 = 0$
5. $-\dfrac{1}{2}$ **6.** $\dfrac{7}{4}$ **7.** Perpendicular **8.** Parallel
9. Neither parallel nor perpendicular
10. Perpendicular

1. $(2, 0)$ **3.** $(-1, -1)$ **5.** Inconsistent
7. $(2a, 3a - 3)$ **9.** $\left(-\dfrac{1}{3}, -\dfrac{2}{3}\right)$ **11.** $\left(\dfrac{5}{2}, \dfrac{3}{4}\right)$
13. $(3, 4)$ **15.** $(4, -1)$ **17.** $(40, 40)$
19. Inconsistent **21.** $\left(\dfrac{18}{5}, \dfrac{3}{5}\right)$ **23.** $(5, -2)$
25. $\left(a, \dfrac{5}{6}a - \dfrac{1}{2}\right)$ **27.** $\left(\dfrac{90}{31}, -\dfrac{67}{31}\right)$ **29.** $\left(-\dfrac{6}{35}, \dfrac{43}{35}\right)$
31. $(79{,}400, 398)$ **33.** 11, 9 **35.** 550 mph, 50 mph
37. $\dfrac{20}{3}$ gal of 20% solution, $\dfrac{10}{3}$ gal of 50% solution
39. 375 adults, 125 children **41.** 8 ft $\times$ 12 ft
43. (a) $y = \dfrac{3}{4}x + \dfrac{4}{3}$
(b)

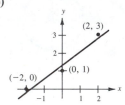

45. (a) $y = -2x + 4$
(b)

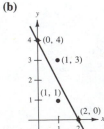

Section 10.3 *(page 578)*

WARM UP
1. (15, 10) **2.** $\left(-2, -\frac{8}{3}\right)$ **3.** (28, 4)
4. (4, 3) **5.** Not a solution **6.** Not a solution
7. Solution **8.** Solution **9.** $5a + 2$
10. $a + 13$

1. (1, 2, 3) **3.** (2, −3, −2) **5.** (5, −2, 0)
7. Inconsistent **9.** $\left(1, -\frac{3}{2}, \frac{1}{2}\right)$
11. $(-3a + 10, 5a − 7, a)$ **13.** $\left(13 − 4a, \frac{45}{2} − \frac{15}{2}a, a\right)$
15. $(-a, 2a − 1, a)$ **17.** $\left(\frac{1}{2} − \frac{3}{2}a, 1 − \frac{2}{3}a, a\right)$
19. (1, 1, 1, 1) **21.** Inconsistent **23.** (0, 0, 0)
25. $\left(-\frac{3}{5}a, \frac{4}{5}a, a\right)$ **27.** $y = 2x^2 + 3x − 4$
29. $y = x^2 − 4x + 3$ **31.** $x^2 + y^2 − 4x = 0$
33. $x^2 + y^2 − 6x − 8y = 0$
35. $a = −32, v_0 = 0, s_0 = 144$
37. $a = −32, v_0 = −32, s_0 = 500$
39. $\frac{1}{2}\left(-\frac{2}{x} + \frac{1}{x−1} + \frac{1}{x+1}\right)$
41. $\frac{1}{2}\left(\frac{1}{x} − \frac{1}{x−2} + \frac{2}{x+3}\right)$
43. 4 medium *or* 2 large, 1 medium, 2 small
45. $y = −\frac{5}{24}x^2 − \frac{3}{10}x + \frac{41}{6}$ **47.** $y = x^2 − x$

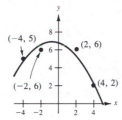

Section 10.4 *(page 586)*

WARM UP
1. Line **2.** Line **3.** Parabola **4.** Parabola
5. Circle **6.** Ellipse **7.** (1, 1) **8.** (2, 0)
9. (2, 1), $\left(-\frac{5}{2}, -\frac{5}{4}\right)$ **10.** (2, 3), (3, 2)

1. f **2.** h **3.** e **4.** c **5.** a **6.** g **7.** b
8. d

9. 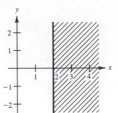 **11.**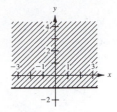

13. 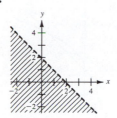 **15.**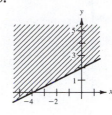

17. 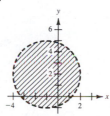 **19.**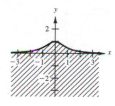

21. 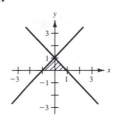 **23.**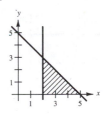

25. 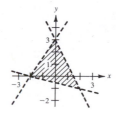 **27.** No solution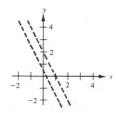

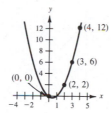

29.

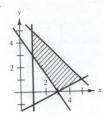

31.

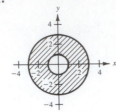

33.

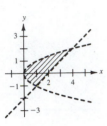

35.

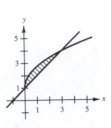

37.

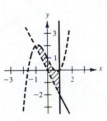

39.

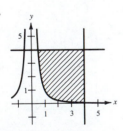

41. $2 \le x \le 5$, $1 \le y \le 7$

43. $y \le \frac{3}{2}x$, $y \le -x + 5$, $y \ge 0$

45. $x^2 + y^2 \le 16$, $x \ge 0$, $y \ge 0$

47. $x + \frac{3}{2}y \le 12$, $\frac{4}{3}x + \frac{3}{2}y \le 15$, $x \ge 0$, $y \ge 0$

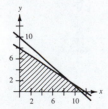

49. $x + y \le 20{,}000$, $x \ge 5{,}000$, $y \ge 5{,}000$, $y \ge 2x$

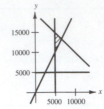

51. $xy \ge 500$, $2x + \pi y \ge 125$, $x \ge 0$, $y \ge 0$

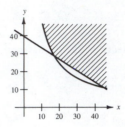

Section 10.5 *(page 593)*

WARM UP

1.

2.

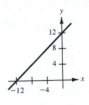

3.

4.

5. $(0, 4)$ **6.** $(12, 0)$ **7.** $(3, 1)$ **8.** $(2, 5)$

9.

10.

1. Minimum at (0, 0): 0
   Maximum at (3, 4): 17
3. Minimum at (0, 0): 0
   Maximum at (4, 0): 20
5. Minimum at (0, 0): 0
   Maximum at (60, 20): 740
7. Minimum at (0, 0): 0
   Maximum at any point on line segment connecting
   (60, 20) and (30, 45): 2100
9. Minimum at (5, 3): 35
   No maximum
11. Minimum at (10, 0): 20
    No maximum
13. Minimum at (24, 8): 104
    Maximum at (40, 0): 160
15. Minimum at (36, 0): 36
    Maximum at (24, 8): 56
17. 200 units (at \$250),   50 units (at \$400)
19. 3 bags Brand X, 6 bags Brand Y, Minimum cost: \$195
21. C is maximum at any point on the line $5x + 2y = 10$,
    $\frac{20}{19} \le x \le 2$

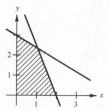

23. The constraint $x \le 10$ is extraneous. Maximum 14 at
    (0, 7)

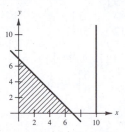

25. (a) No value of t
    (b) $t < 0$
    (c) $t > 0$
    (d) No value of t

Chapter 10 Review Exercises   (*page* 595)

1. (1, 1)   3. (5, 4)   5. (0, 0), (2, 8), (−2, 8)
7. (0, 0), (−3, 3)   9. $\left(\frac{5}{2}, 3\right)$ 11. (−0.5, 0.8)
13. (0, 0)   15. (4.8, 4.4, −1.6)
17. (3a + 4, 2a + 5, a)   19. (−3a + 2, 5a + 6, a)
21. 16, 24, 48   23. (X, Y, Z) = (10, 5, 12)

25.   27.

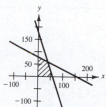

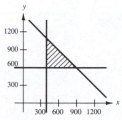

29.

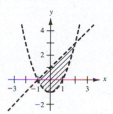

31.

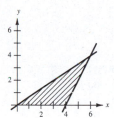

33. $x + y \le 1500,\ x \ge 400,\ y \ge 600$

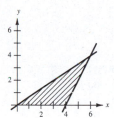

35. Maximum at (5, 8): 47
37. Minimum at (15, 0): 26.25
39. 5 units of A and 2 units of B

CHAPTER 11

Section 11.1   (*page* 607)

WARM UP

1. -3 **2.** 30   **3.** 6   **4.** $-\frac{1}{9}$ **5.** Solution
6. Not a solution   **7.** (5, 2)   **8.** $\left(\frac{12}{5}, -3\right)$
9. (40, 14, 2)   **10.** $\left(\frac{15}{2}, 4, 1\right)$

1. 3×2 **3.** 5×1 **5.** 2×2
7. Reduced row-echelon form
9. Not in row-echelon form

11. (a) $\begin{bmatrix} 1 & 2 & 3 \\ 0 & -5 & -10 \\ 3 & 1 & -1 \end{bmatrix}$ (b) $\begin{bmatrix} 1 & 2 & 3 \\ 0 & -5 & -10 \\ 0 & -5 & -10 \end{bmatrix}$

(c) $\begin{bmatrix} 1 & 2 & 3 \\ 0 & -5 & -10 \\ 0 & 0 & 0 \end{bmatrix}$ (d) $\begin{bmatrix} 1 & 2 & 3 \\ 0 & 1 & 2 \\ 0 & 0 & 0 \end{bmatrix}$

(e) $\begin{bmatrix} 1 & 0 & -1 \\ 0 & 1 & 2 \\ 0 & 0 & 0 \end{bmatrix}$

13. $\begin{bmatrix} 1 & 1 & 0 & 5 \\ 0 & 1 & 2 & 0 \\ 0 & 0 & 1 & -1 \end{bmatrix}$

15. $\begin{bmatrix} 1 & -1 & -1 & 1 \\ 0 & 1 & 6 & 3 \\ 0 & 0 & 0 & 0 \end{bmatrix}$

17. $\begin{bmatrix} 1 & 0 & 0 \\ 0 & 1 & 0 \\ 0 & 0 & 1 \end{bmatrix}$ **19.** $\begin{bmatrix} 1 & 2 & 0 & 0 \\ 0 & 0 & 1 & 0 \\ 0 & 0 & 0 & 1 \\ 0 & 0 & 0 & 0 \end{bmatrix}$

21. $4x + 3y = 8, x - 2y = 3$
23. $x + 2z = -10, 3y - z = 5, 4x + 2y = 3$

25. $\begin{bmatrix} 4 & -5 & \vdots & -2 \\ -1 & 8 & \vdots & 10 \end{bmatrix}$ **27.** $\begin{bmatrix} 1 & 10 & -3 & \vdots & 2 \\ 5 & -3 & 4 & \vdots & 0 \\ 2 & 4 & 0 & \vdots & 6 \end{bmatrix}$

29. (3, 2)   **31.** (4, −2)   **33.** $\left(\frac{1}{2}, -\frac{3}{4}\right)$
35. Inconsistent   **37.** (4, −3, 2)
39. $(2a + 1, 3a + 2, a)$ **41.** $(5a + 4, -3a + 2, a)$
43. $(0, 2 - 4a, a)$ **45.** $(1, 0, 4, -2)$ **47.** (0, 0)
49. $(-2a, a, a)$
51. \$800,000 at 8%, \$500,000 at 9%, \$200,000 at 12%
53. $a = \frac{1}{2}, b = -3, c = \frac{7}{2}$
55. $D = -5, E = -3, F = 6$

Section 11.2   (*page* 620)

WARM UP

1. -5 **2.** -7
3. Not in reduced row-echelon form
4. Not in reduced row-echelon form
5. $\begin{bmatrix} -5 & 10 & \vdots & 12 \\ 7 & -3 & \vdots & 0 \\ -1 & 7 & \vdots & 25 \end{bmatrix}$
6. $\begin{bmatrix} 10 & 15 & -9 & \vdots & 42 \\ 6 & -5 & 0 & \vdots & 0 \end{bmatrix}$
7. (0, 2)   **8.** $(2 + a, 3 - a, a)$
9. $(1 - 2a, a, -1)$ **10.** $(2, -1, -1)$

1. $x = -4, y = 22$ **3.** $x = 2, y = 3$

5. (a) $\begin{bmatrix} 3 & -2 \\ 1 & 7 \end{bmatrix}$ (b) $\begin{bmatrix} -1 & 0 \\ 3 & -9 \end{bmatrix}$

(c) $\begin{bmatrix} 3 & -3 \\ 6 & -3 \end{bmatrix}$ (d) $\begin{bmatrix} -1 & -1 \\ 8 & -19 \end{bmatrix}$

7. (a) $\begin{bmatrix} 7 & 3 \\ 1 & 9 \\ -2 & 15 \end{bmatrix}$ (b) $\begin{bmatrix} 5 & -5 \\ 3 & -1 \\ -4 & -5 \end{bmatrix}$

(c) $\begin{bmatrix} 18 & -3 \\ 6 & 12 \\ -9 & 15 \end{bmatrix}$ (d) $\begin{bmatrix} 16 & -11 \\ 8 & 2 \\ -11 & -5 \end{bmatrix}$

9. (a) $\begin{bmatrix} 3 & 3 & -2 & 1 & 1 \\ -2 & 5 & 7 & -6 & -8 \end{bmatrix}$
(b) $\begin{bmatrix} 1 & 1 & 0 & -1 & 1 \\ 4 & -3 & -11 & 6 & 6 \end{bmatrix}$
(c) $\begin{bmatrix} 6 & 6 & -3 & 0 & 3 \\ 3 & 3 & -6 & 0 & -3 \end{bmatrix}$
(d) $\begin{bmatrix} 4 & 4 & -1 & -2 & 3 \\ 9 & -5 & -24 & 12 & 11 \end{bmatrix}$

11. (a) $\begin{bmatrix} 0 & 15 \\ 6 & 12 \end{bmatrix}$ (b) $\begin{bmatrix} -2 & 2 \\ 31 & 14 \end{bmatrix}$ (c) $\begin{bmatrix} 9 & 6 \\ 12 & 12 \end{bmatrix}$

13. (a) $\begin{bmatrix} 0 & -10 \\ 10 & 0 \end{bmatrix}$ (b) $\begin{bmatrix} 0 & -10 \\ 10 & 0 \end{bmatrix}$ (c) $\begin{bmatrix} 8 & -6 \\ 6 & 8 \end{bmatrix}$

15. (a) $\begin{bmatrix} 6 & -21 & 15 \\ 8 & -23 & 19 \\ 4 & 7 & 5 \end{bmatrix}$

(b) $\begin{bmatrix} 9 & 0 & 13 \\ 7 & -2 & 21 \\ 1 & 4 & -19 \end{bmatrix}$

(c) $\begin{bmatrix} 20 & 7 & -8 \\ 24 & 7 & -2 \\ 2 & -5 & 30 \end{bmatrix}$

17. Not possible

19. $\begin{bmatrix} -1 & 19 \\ 4 & -27 \\ 0 & 14 \end{bmatrix}$

21. $\begin{bmatrix} 1 & 0 & 0 \\ 0 & 1 & 0 \\ 0 & 0 & \frac{7}{2} \end{bmatrix}$ **23.** $\begin{bmatrix} 60 & 72 \\ -20 & -24 \\ 10 & 12 \\ 60 & 72 \end{bmatrix}$

25. $\begin{bmatrix} -6 & -9 \\ -1 & 0 \\ 17 & -10 \end{bmatrix}$ **27.** $\begin{bmatrix} 3 & 3 \\ -\frac{1}{2} & 0 \\ -\frac{13}{2} & \frac{11}{2} \end{bmatrix}$

29. $A = \begin{bmatrix} -1 & 1 \\ -2 & 1 \end{bmatrix}, X = \begin{bmatrix} x \\ y \end{bmatrix}, B = \begin{bmatrix} 4 \\ 0 \end{bmatrix}, x = 4, y = 8$

31. $A = \begin{bmatrix} 1 & -2 & 3 \\ -1 & 3 & -1 \\ 2 & -5 & 5 \end{bmatrix}, X = \begin{bmatrix} x \\ y \\ z \end{bmatrix}, B = \begin{bmatrix} 9 \\ -6 \\ 17 \end{bmatrix},$
$x = 1, y = -1, z = 2$

33. $AC = BC = \begin{bmatrix} 12 & -6 & 9 \\ 16 & -8 & 12 \\ 4 & -2 & 3 \end{bmatrix}$

35. (a) No, $AB \ne BA$ (b) No, $AB \ne BA$

37. $BA = [\$1250 \quad \$1331.25 \quad \$981.25]$

39. $\begin{bmatrix} -4 & 0 \\ 8 & 2 \end{bmatrix}$ **41.** $\begin{bmatrix} 0 & 0 & 0 \\ 0 & 0 & 0 \\ 0 & 0 & 0 \end{bmatrix}$

Section 11.3 *(page 630)*

WARM UP

1. $\begin{bmatrix} 4 & 24 \\ 0 & -16 \\ 48 & 8 \end{bmatrix}$ **2.** $\begin{bmatrix} \frac{11}{2} & 5 & 24 \\ \frac{1}{2} & 0 & 8 \\ 0 & 1 & 4 \end{bmatrix}$

3. $\begin{bmatrix} -5 & -2 & -13 \\ 4 & -13 & -2 \end{bmatrix}$ **4.** $\begin{bmatrix} -13 & 11 \\ -19 & 21 \end{bmatrix}$

5. $\begin{bmatrix} 1 & 0 \\ 0 & 1 \end{bmatrix}$ **6.** $\begin{bmatrix} 6 & 5 \\ 3 & -2 \end{bmatrix}$ **7.** $\begin{bmatrix} 1 & 0 & 0 \\ 0 & 1 & 0 \\ 0 & 0 & 1 \end{bmatrix}$

8. $\begin{bmatrix} 1 & 0 & 0 \\ 0 & 1 & 0 \\ 0 & 0 & 1 \end{bmatrix}$ **9.** $\begin{bmatrix} 1 & 0 & \vdots & 3 & -2 \\ 0 & 1 & \vdots & 4 & -3 \end{bmatrix}$

10. $\begin{bmatrix} 1 & 0 & 0 & \vdots & -6 & -4 & 3 \\ 0 & 1 & 0 & \vdots & 11 & 6 & -5 \\ 0 & 0 & 1 & \vdots & -2 & -1 & 1 \end{bmatrix}$

5. $\begin{bmatrix} \frac{1}{2} & 0 \\ 0 & \frac{1}{3} \end{bmatrix}$ **7.** $\begin{bmatrix} -3 & 2 \\ -2 & 1 \end{bmatrix}$ **9.** $\begin{bmatrix} 1 & -1 \\ 2 & -1 \end{bmatrix}$

11. Does not exist **13.** Does not exist

15. $\begin{bmatrix} 1 & 1 & -1 \\ -3 & 2 & -1 \\ 3 & -3 & 2 \end{bmatrix}$ **17.** $\begin{bmatrix} -175 & 37 & -13 \\ 95 & -20 & 7 \\ 14 & -3 & 1 \end{bmatrix}$

19. $\begin{bmatrix} -24 & 7 & 1 & -2 \\ -10 & 3 & 0 & -1 \\ -29 & 7 & 3 & -2 \\ 12 & -3 & -1 & 1 \end{bmatrix}$ **21.** $\frac{1}{2}\begin{bmatrix} -3 & 3 & 2 \\ 9 & -7 & -6 \\ -2 & 2 & 2 \end{bmatrix}$

23. $\frac{5}{11}\begin{bmatrix} 0 & -4 & 2 \\ -22 & 11 & 11 \\ 22 & -6 & -8 \end{bmatrix}$ **25.** $\begin{bmatrix} -\frac{1}{8} & 0 & 0 & 0 \\ 0 & 1 & 0 & 0 \\ 0 & 0 & \frac{1}{4} & 0 \\ 0 & 0 & 0 & -\frac{1}{5} \end{bmatrix}$

27. $\begin{bmatrix} 1 & 0 & 0 \\ -0.75 & 0.25 & 0 \\ 0.35 & -0.25 & 0.2 \end{bmatrix}$ **29.** Does not exist

31. (a) $(4, 8)$ (b) $(-8, -11)$ (c) $(10, 30)$
(d) $(-7, -7)$

33. (a) $(2, 1, 0, 0)$ (b) $(-32, -13, -37, 15)$

Section 11.4 *(page 639)*

WARM UP

1. $\begin{bmatrix} 3 & 5 \\ 4 & 0 \end{bmatrix}$ **2.** $\begin{bmatrix} -2 & 8 \\ 2 & -4 \end{bmatrix}$

3. $\begin{bmatrix} 9 & -12 & 6 \\ 3 & 0 & -3 \\ 0 & 3 & -6 \end{bmatrix}$ **4.** $\begin{bmatrix} 0 & 8 & 12 \\ -4 & 8 & 12 \\ -8 & 4 & -8 \end{bmatrix}$

5. -22 **6.** 35 **7.** -15 **8.** $-\frac{1}{8}$ **9.** -45
10. -16

1. 5 **3.** 5 **5.** 27 **7.** -24 **9.** 6 **11.** 0
13. -2 **15.** 0 **17.** 0 **19.** $-7x + 3y - 8$
21. (a) $M_{11} = -5, M_{12} = 2, M_{21} = 4, M_{22} = 3$
(b) $C_{11} = -5, C_{12} = -2, C_{21} = -4, C_{22} = 3$
23. (a) $M_{11} = 30, M_{12} = 12, M_{13} = 11, M_{21} = -36,$
$M_{22} = 26, M_{23} = 7, M_{31} = -4, M_{32} = -42,$
$M_{33} = 12$
(b) $C_{11} = 30, C_{12} = -12, C_{13} = 11, C_{21} = 36,$
$C_{22} = 26, C_{23} = -7, C_{31} = -4, C_{32} = 42,$
$C_{33} = 12$
25. -75 **27.** 96 **29.** 170 **31.** -58 **33.** -30
35. -108 **37.** 0 **39.** 412 **41.** $x = -1, x = 4$
43. $8uv - 1$ **45.** 1 **47.** e^{5x} **49.** $1 - \ln x$

Section 11.5   (*page* 648)

WARM UP

1. $\begin{bmatrix} 1 & -3 \\ 0 & 1 \end{bmatrix}$ 2. $\begin{bmatrix} 1 & -3 \\ 0 & 1 \end{bmatrix}$ 3. $\begin{bmatrix} 1 & 3 & 4 \\ 0 & 1 & 1 \\ 0 & 0 & 0 \end{bmatrix}$

4. $\begin{bmatrix} 1 & 2 & 4 \\ 0 & 1 & \frac{10}{7} \\ 0 & 0 & 0 \end{bmatrix}$ 5. -2 6. 0   7. -8 8. x^2

9. 8   10. 60

---

1. Column 2 is a multiple of Column 1.
3. Row 2 has only zero entries.
5. The interchange of Columns 2 and 3 results in a change of sign of the determinant.
7. Multiplying any row by a constant multiplies the value of the determinant by that constant.
9. Multiplying the entries of all three rows by 5 produces a determinant that is 5^3 times the determinant of the second matrix.
11. Adding -4 times the entries of Row 1 to the elements of Row 2 leaves the determinant unchanged.
13. Adding multiples of Column 2 to Columns 1 and 3 leaves the determinant unchanged.
15. 1   17. -6 19. -26 21. -126 23. 236
25. -3740 27. 7441   29. 410   31. Not invertible
33. Invertible   35. Invertible   39. Equal

Section 11.6   (*page* 659)

WARM UP

1. $(1, 1)$ 2. $(1, 2)$ 3. $(3, 0, -4)$
4. $(-2, 1, 1)$ 5. 8   6. -49 7. -3
8. 20   9. 9   10. 35

---

1. $(1, 2)$ 3. $(2, -2)$ 5. $\left(\frac{3}{4}, -\frac{1}{2}\right)$
7. Cramer's rule does not apply.   9. $\left(\frac{2}{3}, \frac{1}{2}\right)$
11. $(-1, 3, 2)$ 13. $\left(1, \frac{1}{2}, \frac{3}{2}\right)$ 15. $\left(0, -\frac{1}{2}, \frac{1}{2}\right)$
17. Cramer's rule does not apply.   19. $(5, 0, -2, 3)$
21. 7   23. 14   25. $\frac{33}{8}$ 27. Collinear
29. Not collinear   31. Collinear   33. $3x - 5y = 0$
35. $x + 3y - 5 = 0$ 37. $2x + 3y - 8 = 0$
39. 1 -25 -65 17 15 -9 -12 -62 -119 27 51 48 43 67 48 57 111 117

41. -5 -41 -87 91 207 257 11 -5 -41 40 80 84 76 177 227
43. SEND PLANES

Chapter 11 Review Exercises   (*page* 661)

1. $(10, -12)$ 3. $(0.6, 0.5)$ 5. $(2, -3, 3)$
7. $\left(\frac{1}{2}, -\frac{1}{3}, 1\right)$ 9. $\left(-2a + \frac{3}{2}, 2a + 1, a\right)$
11. Inconsistent

13. $\begin{bmatrix} -13 & -8 & 18 \\ 0 & 11 & -19 \end{bmatrix}$ 15. $\begin{bmatrix} 14 & -2 & 8 \\ 14 & -10 & 40 \\ 36 & -12 & 48 \end{bmatrix}$

17. $\begin{bmatrix} 44 & 4 \\ 20 & 8 \end{bmatrix}$ 19. $\begin{bmatrix} 4 & 6 & 3 \\ 0 & 6 & -10 \\ 0 & 0 & 6 \end{bmatrix}$

21. $\begin{bmatrix} -14 & -4 \\ 7 & -17 \\ -17 & -2 \end{bmatrix}$ 23. $\frac{1}{3}\begin{bmatrix} 9 & 2 \\ -4 & 11 \\ 10 & 0 \end{bmatrix}$ 25. 550

27. 279

29. $\begin{bmatrix} \frac{1}{5} & \frac{1}{5} \\ \frac{1}{10} & -\frac{1}{15} \end{bmatrix}$ 31. $\begin{bmatrix} \frac{1}{2} & -1 & -\frac{1}{2} \\ \frac{1}{2} & -\frac{2}{3} & -\frac{5}{6} \\ 0 & \frac{2}{3} & \frac{1}{3} \end{bmatrix}$

33. $5x + 4y = 2, -x + y = -22$
35. $(-3, 1)$ 37. $(1, 1, -2)$ 39. $(2, -4, 6)$
41. 16   43. 7   45. $x - 2y + 4 = 0$
47. $2x + 6y - 13 = 0$
49. 128; Each of the three rows is multiplied by 4.

CHAPTER 12
Section 12.1   (*page* 671)

WARM UP

1. $\frac{4}{5}$ 2. $\frac{1}{3}$ 3. $(2n + 1)(2n - 1)$
4. $(2n - 1)(2n - 3)$ 5. $(n - 1)(n - 2)$
6. $(n + 1)(n + 2)$ 7. $\frac{1}{3}$ 8. 24   9. $\frac{13}{24}$
10. $\frac{3}{4}$

---

1. 2, 4, 8, 16, 32   3. $-\frac{1}{2}, \frac{1}{4}, -\frac{1}{8}, \frac{1}{16}, -\frac{1}{32}$
5. 3, $\frac{9}{2}, \frac{9}{2}, \frac{27}{8}, \frac{81}{40}$ 7. $-1, \frac{1}{4}, -\frac{1}{9}, \frac{1}{16}, -\frac{1}{25}$
9. 2, $\frac{3}{2}, \frac{4}{3}, \frac{5}{4}, \frac{6}{5}$ 11. 0, 1, 2, 3, 4   13. 0, 1, 0, $\frac{1}{2}$, 0
15. $\frac{5}{2}, \frac{11}{4}, \frac{23}{8}, \frac{47}{16}, \frac{95}{32}$ 17. $-\frac{1}{2}, \frac{2}{3}, -\frac{3}{4}, \frac{4}{5}, -\frac{5}{6}$
19. 3, 4, 6, 10, 18   21. 90   23. $n + 1$

25. $\dfrac{1}{2n(2n+1)}$ **27.** $a_n = 3n - 2$ **29.** $a_n = n^2 - 1$

31. $a_n = \dfrac{n+1}{n+2}$ **33.** $a_n = \dfrac{(-1)^{n+1}}{2^n}$ **35.** $a_n = 1 + \dfrac{1}{n}$

37. $a_n = \dfrac{1}{n!}$ **39.** $a_n = (-1)^{n+1}$ **41.** 35   **43.** $\frac{9}{5}$

45. 40   **47.** 238   **49.** $4x^2 + 20$ **51.** $\displaystyle\sum_{i=1}^{9} \dfrac{1}{3i}$

53. $\displaystyle\sum_{i=1}^{8} \left[2\left(\dfrac{i}{8}\right) + 3 \right]$ **55.** $\displaystyle\sum_{i=1}^{6} (-1)^{i+1} 3^i$

57. $\displaystyle\sum_{i=1}^{20} \dfrac{(-1)^{i+1}}{i^2}$ **59.** $\displaystyle\sum_{i=1}^{5} \dfrac{2^i - 1}{2^{i+1}}$

63. (a) $A_1 = \$5100.00$, $A_2 = \$5202.00$, $A_3 = \$5306.04$,
  $A_4 = \$5412.16$, $A_5 = \$5520.40$, $A_6 = \$5630.81$,
  $A_7 = \$5743.43$, $A_8 = \$5858.30$

  (b) \$11,040.20

Section 12.2   (page 680)

WARM UP

1. 36   **2.** 240   **3.** $\frac{11}{2}$ **4.** $\frac{10}{3}$ **5.** 18
6. 4   **7.** 143   **8.** 160   **9.** 430   **10.** 256

1. Arithmetic sequence, $d = 3$
3. Not an arithmetic sequence
5. Arithmetic sequence, $d = -\frac{1}{4}$
7. Not an arithmetic sequence
9. Arithmetic sequence, $d = 0.4$
11. 8, 11, 14, 17, 20; arithmetic sequence, $d = 3$
13. $\frac{1}{2}, \frac{1}{3}, \frac{1}{4}, \frac{1}{5}, \frac{1}{6}$; not an arithmetic sequence
15. 97, 94, 91, 88, 85; arithmetic sequence, $d = -3$
17. 1, 1, 2, 3, 5; not an arithmetic sequence
19. 28   **21.** 44   **23.** $99x$ **25.** $-\frac{37}{2}$
27. 5, 11, 17, 23, 29, . . .
29. $-2.6, -3.0, -3.4, -3.8, -4.2, . . .$
31. $\frac{3}{2}, \frac{5}{4}, 1, \frac{3}{4}, \frac{1}{2}, . . .$
33. 2, 6, 10, 14, 18, . . .
35. $-2, 2, 6, 10, 14, . . .$
37. 620   **39.** 4600   **41.** 265   **43.** 4000
45. 1275   **47.** 25,250   **49.** 126,750   **51.** 520
53. 44,625   **55.** 10,000
57. (a) \$35,000   (b) \$187,500   **59.** 2340
61. 9, 13   **63.** $\frac{15}{4}, \frac{9}{2}, \frac{21}{4}$

Section 12.3   (page 690)

WARM UP

1. $\frac{64}{125}$ **2.** $\frac{9}{16}$ **3.** $\frac{1}{16}$ **4.** $\frac{5}{81}$ **5.** $6n^3$
6. $27n^4$ **7.** $4n^3$ **8.** n^2 **9.** $\dfrac{2^n}{81^n}$ **10.** $\dfrac{3}{16^n}$

1. Geometric sequence, $r = 3$
3. Not a geometric sequence
5. Geometric sequence, $r = -\frac{1}{2}$
7. Not a geometric sequence
9. Not a geometric sequence
11. 2, 6, 18, 54, 162, . . .
13. $1, \frac{1}{2}, \frac{1}{4}, \frac{1}{8}, \frac{1}{16}, . . .$
15. $5, -\frac{1}{2}, \frac{1}{20}, -\frac{1}{200}, \frac{1}{2000}, . . .$
17. $1, \dfrac{x}{2}, \dfrac{x^2}{4}, \dfrac{x^3}{8}, \dfrac{x^4}{16}, . . .$
19. $\left(\dfrac{1}{2}\right)^7$ **21.** $-\dfrac{2}{3^{10}}$ **23.** $100e^{8x}$ **25.** $500(1.02)^{39}$
27. 9   **29.** $-\frac{2}{9}$
31. (a) \$2593.74   (b) \$2653.30   (c) \$2685.06
  (d) \$2707.04   (e) \$2717.91
33. \$22,689.45   **35.** 29,921.31   **37.** 6.4   **39.** 511
41. \$7808.24   **45.** \$594,121.01   **47.** 2   **49.** $\frac{2}{3}$
51. $\frac{16}{3}$ **53.** 32   **55.** $\frac{8}{3}$ **57.** $\frac{1}{2}$ **59.** 152.42 ft
61. $\frac{1}{9}$ **63.** $\frac{4}{11}$ **65.** $\frac{16}{37}$ **67.** $\frac{15}{11}$

Section 12.4   (page 699)

WARM UP

1. 24   **2.** 40   **3.** $\frac{77}{60}$ **4.** $\frac{7}{2}$ **5.** $\dfrac{2k+5}{5}$
6. $\dfrac{3k+1}{6}$ **7.** $8 \cdot 2^{2k} = 2^{2k+3}$ **8.** $\frac{1}{9}$ **9.** $\dfrac{1}{k}$
10. $\frac{4}{5}$

1. 91   **3.** 225   **5.** 2275   **7.** $\dfrac{5}{(k+1)(k+2)}$

9. $\dfrac{(k+1)^2(k+2)^2}{4}$ **25.** $n(2n+1)$

27. $10[1 - (0.9)^n]$ **29.** $\dfrac{n}{2(n+1)}$

Section 12.5   *(page 705)*

> **WARM UP**
>
> **1.** $5x^5 + 15x^2$ **2.** $x^3 + 5x^2 - 3x - 15$
>
> **3.** $x^2 + 8x + 16$ **4.** $4x^2 - 12x + 9$ **5.** $\dfrac{3x^3}{y}$
>
> **6.** $-32z^5$ **7.** 120 **8.** 336 **9.** 720 **10.** 20

1. 10   **3.** 1   **5.** 15,504   **7.** 4950   **9.** 4950

11. $x^5 + 5x^4y + 10x^3y^2 + 10x^2y^3 + 5xy^4 + y^5$

13. $a^4 + 8a^3 + 24a^2 + 32a + 16$

15. $r^6 + 18r^5s + 135r^4s^2 + 540r^3s^3 + 1215r^2s^4 + 1458rs^5$ $+ 729s^6$

17. $x^5 - 5x^4y + 10x^3y^2 - 10x^2y^3 + 5xy^4 - y^5$

19. $1 - 6x + 12x^2 - 8x^3$

21. $x^8 + 20x^6 + 150x^4 + 500x^2 + 625$

23. $\dfrac{1}{x^5} + \dfrac{5y}{x^4} + \dfrac{10y^2}{x^3} + \dfrac{10y^3}{x^2} + \dfrac{5y^4}{x} + y^5$

25. -4 **27.** $2035 + 828i$ **29.** 1

31. $32t^5 - 80t^4s + 80t^3s^2 - 40t^2s^3 + 10ts^4 - s^5$

33. $81 - 216z + 216z^2 - 96z^3 + 16z^4$

35. $1{,}732{,}104x^5$ **37.** $180x^8y^2$

39. $-226{,}437{,}120x^4y^{11}$ **41.** $924x^3y^3$

43. $\dfrac{1}{128} + \dfrac{7}{128} + \dfrac{21}{128} + \dfrac{35}{128} + \dfrac{35}{128} + \dfrac{21}{128} + \dfrac{7}{128} + \dfrac{1}{128}$

45. $\dfrac{1}{6561} + \dfrac{16}{6561} + \dfrac{112}{6561} + \dfrac{448}{6561} + \dfrac{1120}{6561} + \dfrac{1792}{6561} + \dfrac{1792}{6561} +$ $\dfrac{1024}{6561} + \dfrac{256}{6561}$

47. $0.07776 + 0.25920 + 0.34560 + 0.23040 + 0.07680$ $+ 0.01024$

Section 12.6   *(page 717)*

> **WARM UP**
>
> **1.** 6656   **2.** 291,600   **3.** 7920   **4.** 13,800
>
> **5.** 792   **6.** 2300   **7.** $n(n-1)(n-2)(n-3)$
>
> **8.** $n(n-1)(2n-1)$ **9.** $n!$ **10.** $n!$

1. 12   **3.** 12   **5.** 6,760,000

7. (a) 900   (b) 648   (c) 180

9. (a) 720   (b) 48   **11.** 24   **13.** 336

15. 1,860,480   **17.** 9900   **19.** 120   **21.** 120

23. 11,880   **25.** 4845   **27.** 3,838,380   **29.** 900

31. 600   **33.** 560   **35.** (a) 70   (b) 30

37. (a) 70   (b) 54   (c) 16   **39.** 5   **41.** 20

43. 420   **45.** 1260   **47.** 2520   **49.** $n = 5$ or $n = 6$

Section 12.7   *(page 726)*

> **WARM UP**
>
> **1.** $\dfrac{9}{16}$ **2.** $\dfrac{8}{15}$ **3.** $\dfrac{1}{6}$ **4.** $\dfrac{1}{80{,}730}$ **5.** $\dfrac{1}{495}$
>
> **6.** $\dfrac{1}{24}$ **7.** $\dfrac{1}{12}$ **8.** $\dfrac{135}{323}$ **9.** 0.366   **10.** 0.997

1. $\dfrac{3}{8}$ **3.** $\dfrac{7}{8}$ **5.** $\dfrac{3}{13}$ **7.** $\dfrac{5}{13}$ **9.** $\dfrac{1}{12}$ **11.** $\dfrac{7}{12}$

13. $\dfrac{1}{3}$ **15.** $\dfrac{1}{5}$ **17.** $\dfrac{2}{5}$ **19.** $\dfrac{11}{18}$

21. $P(\{\text{Taylor wins}\}) = \dfrac{1}{2}$

     $P(\{\text{Moore wins}\}) = P(\{\text{Jenkins wins}\}) = \dfrac{1}{4}$

23. (a) $\dfrac{1}{4}$ (b) $\dfrac{1}{2}$ (c) $\dfrac{9}{100}$ (d) $\dfrac{1}{30}$

25. (a) $\dfrac{1}{169}$ (b) $\dfrac{1}{221}$ **27.** (a) $\dfrac{1}{3}$ (b) $\dfrac{5}{8}$

29. (a) $\dfrac{14}{55}$ (b) $\dfrac{12}{55}$ (c) $\dfrac{54}{55}$

31. (a) 0.9702   (b) 0.9998   (c) 0.0002

33. (a) $\dfrac{1}{64}$ (b) $\dfrac{1}{32}$ (c) $\dfrac{63}{64}$

35. (a) 0.9^{10} (b) 0.1^{10} (c) $1 - 0.9^{10}$ (d) 0.9^9

37.

Number of girls	0	1	2	3	4	5	6
Probability	$\dfrac{1}{64}$	$\dfrac{6}{64}$	$\dfrac{15}{64}$	$\dfrac{20}{64}$	$\dfrac{15}{64}$	$\dfrac{6}{64}$	$\dfrac{1}{64}$

Chapter 12 Review Exercises   *(page 728)*

1. $\displaystyle\sum_{k=1}^{20} \dfrac{1}{2k}$ **3.** $\displaystyle\sum_{k=1}^{9} \dfrac{k}{k+1}$ **5.** 30   **7.** 80

9. 127   **11.** 8   **13.** 12   **15.** 88   **17.** 418

19. $3, 7, 11, 15, 19, \ldots$ **21.** $4, -1, \dfrac{1}{4}, -\dfrac{1}{16}, \dfrac{1}{64}, \ldots$

23. 25,250   **25.** $\dfrac{5}{11}$ **27.** $\dfrac{16}{15}$ **29.** $\dfrac{1}{75}$

35. $\dfrac{x^4}{16} + \dfrac{x^3y}{2} + \dfrac{3x^2y^2}{2} + 2xy^3 + y^4$

37. $\dfrac{64}{x^6} - \dfrac{576}{x^4} + \dfrac{2160}{x^2} - 4320 + 4860x^2 - 2916x^4 +$ $729x^6$

39. $41 + 840i$ **41.** (a) 1   (b) 6   (c) 15

43. 118,813,760   **45.** $\dfrac{1}{9}$

47. $P(\{3\}) = \dfrac{1}{6}$

     $P(\{(1, 5), (5, 1), (2, 4), (4, 2), (3, 3)\}) = \dfrac{5}{36}$

49. 0.0475

APPENDIX A

1. (a) 2.6201　　(b) −1.3799
3. (a) 3.7993　　(b) 3
5. (a) −0.6081　　(b) 3.8959
7. (a) 0.7715　　(b) 4.2520
9. (a) 25,000　　(b) 0.025
11. (a) 1,360,000　　(b) 0.0136
13. (a) 3.6420　　(b) −0.2176
15. (a) 414,500　　(b) 0.007075　　**17.** 18.10
19. 4.42　　**21.** 901.5
23. (a) 1.8310　　(b) 2.2565
25. (a) 7.46　　(b) 4.23
27. (a) 33.115　　(b) 0.0302
29. (a) 0.8290　　(b) 1.483
31. (a) 0.8660　　(b) 1.732
33. (a) 25°, 155°　　(b) 25°, 205°
35. (a) 64° 58′, 294° 2′　　(b) 14° 2′, 194° 2′

Index

FORMULAS FROM GEOMETRY

Triangle:

$h = a \sin\theta$

$\text{Area} = \dfrac{1}{2}bh$

(Law of Cosines)

$c^2 = a^2 + b^2 - 2ab \cos\theta$

Right Triangle:

(Pythagorean Theorem)

$c^2 = a^2 + b^2$

Equilateral Triangle:

$h = \dfrac{\sqrt{3}\,s}{2}$

$\text{Area} = \dfrac{\sqrt{3}\,s^2}{4}$

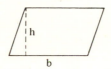

Parallelogram:

$\text{Area} = bh$

Trapezoid:

$\text{Area} = \dfrac{h}{2}(a+b)$

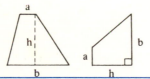

Circle:

$\text{Area} = \pi r^2$

$\text{Circumference} = 2\pi r$

Sector of Circle:

(θ in radians)

$\text{Area} = \dfrac{\theta r^2}{2}$

$s = r\theta$

Circular Ring:

(p = average radius,
 w = width of ring)

$\text{Area} = \pi(R^2 - r^2)$

$\qquad = 2\pi pw$

Sector of Circular Ring:

(p = average radius,
 w = width of ring,
 θ in radians)

$\text{Area} = \theta pw$

Ellipse:

$\text{Area} = \pi ab$

$\text{Circumference} \approx 2\pi \sqrt{\dfrac{a^2+b^2}{2}}$

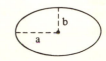

Cone:

(A = area of base)

$\text{Volume} = \dfrac{Ah}{3}$

Right Circular Cone:

$\text{Volume} = \dfrac{\pi r^2 h}{3}$

$\text{Lateral Surface Area} = \pi r \sqrt{r^2 + h^2}$

Frustum of Right Circular Cone:

$\text{Volume} = \dfrac{\pi(r^2 + rR + R^2)h}{3}$

$\text{Lateral Surface Area} = \pi s(R+r)$

Right Circular Cylinder:

$\text{Volume} = \pi r^2 h$

$\text{Lateral Surface Area} = 2\pi rh$

Sphere:

$\text{Volume} = \dfrac{4}{3}\pi r^3$

$\text{Surface Area} = 4\pi r^2$

Wedge:

(A = area of upper face,
 B = area of base)

$A = B \sec\theta$

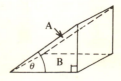